The

Divine Savage
Revealing the Miracle of Being

Anthony Hernandez

The Divine Savage
Revealing the Miracle of Being

Anthony Hernandez

All rights reserved.
Copyright © 2012

Cover art by Don Ellis Aguillo (www.donaguillo.com)

Dawnstar Books
San Francisco, CA
www.DawnstarBooks.com

Hernandez, Anthony 1968-
The Divine Savage/Anthony Hernandez
    ISBN: 978-0-9855793-1-9 (paperback)
    ISBN: 978-0-9855793-2-6 (Amazon® Kindle®)

Advance Praise for *The Divine Savage*

An ambitious—dare I say epic—work encompassing deep examinations of all our social constructs, starting with the world's religions and moving on from there to such areas as neuroscience and quantum physics. If you enjoyed Phillip Pullman's *His Dark Materials* trilogy, this well-researched and beautifully written opus provides a non-fiction view of Pullman's fictional universe.

Shel Horowitz
Award-winning author of eight books, including *Guerrilla Marketing Goes Green*
www.GreenAndProfitable.com

The Divine Savage awakens our souls and minds to limitless possibilities of the human spirit. I believe that those who shed this mortal coil go to a place that is part of our Universe that is peaceful and all loving. I believe that persons who loved us here love us there as well. They never leave us. *The Divine Savage* provides the scientific explanation for how this is possible. It debunks myths and open minds and hearts.

Kathi Papaleo

The Divine Savage is a beautifully researched and scholarly book bringing together world religions and scientific research to demonstrate their synergy in support of the author's belief in a spiritual universe.

Yvonne McClung
Former Principal
School of The Arts, San Francisco, CA

Table of Contents

Advance Praise for *The Divine Savage*	i
Dedication	xxi
Acknowledgements	xxiii
Introduction	xxv

PART ONE
Getting Started

1: Dying — 3
- The Dying Process — 6
  - Accepting the End — 6
  - Physical Death — 7
  - The Body after Death — 11
  - What is Death? — 13
- The Meaning of Death — 17
  - The Materialist View — 18
  - Beyond Materialism — 20

2: Questions and Assumptions — 23
- Conflation — 23
  - Evolution — 24
  - Creation — 25
  - Intelligent Design — 26
  - Religion — 27
  - God — 28
  - Soul — 28
  - Afterlife — 29
  - Supernatural — 29
  - Paranormality — 30
  - Mysticism — 31
- The Truth of Evolution — 31
- Any Room for God? — 33
- Mapping Our Journey — 34
- A Note for People of Faith — 37
- On the Shoulders of Giants — 38

PART TWO
Religion

3:	**In the Beginning**	**41**
	Myth as Science	41
	Modern Genesis	44
	Genesis Around the World	50
	Egypt	51
	Sumer	51
	Iran/Middle East	51
	Greece	52
	Rome	52
	India	53
	China	55
	Japan	56
	Pacific Islands	56
	Central America	57
	North America	58
	African Tribes	59
	Madagascar	61
	Universal Symbols	61
	Universal Knowledge	62
4:	**From Myth to Religion**	**63**
	The Perennial Philosophy	64
	The Evolution of Myth	70
	Why Myth?	72
	Primitive Religion	74
	Shamanism	76
	Religion	78
	The Challenges of Religion	81
	Universal Truths	83
5:	**The World's Religions**	**85**
	Judaism	86
	Monotheism	87
	The Vengeful God	90
	The Exodus	91
	God's Chosen People	92
	David & Solomon	93

Modern Israel	95
Life after Death	96
Christianity	96
Paul	98
Constantine	99
Orthodox Christianity	100
The Gnostics	104
East/West Split	108
The Protestants	109
Women	110
Revelations	110
Islam	111
A Brief History	112
Koran and Hadith	115
Jihad	117
The Afterlife	117
Jewish Roots	118
Hinduism	119
Brahman and Atman	120
Jainism	122
The Bhagavad Gita	122
Plato	123
Buddhism	124
Nothingness and Everything	125
Four Noble Truths	128
The Eightfold Path	129
Ten Bad Actions	130
Nonviolence and Compassion	130
Other Religions	132
Confucianism	132
Daoism	133
Zoroastrianism	134
Sikhism	134
Taoism	135
6: Jesus Christ	**137**
Foretelling Christ	138
Lineage and Breeding	140
Birth	141

Life and Training 143
    John the Baptist 143
    Plans within Plans 144
    Essenes and Therapeutae 144
    Egyptian Spirituality 146
    Initiations 147
    Nearer and Farther Still 149
Return and Betrayal 150
The Crucifixion 152
Mary Magdalene 155
Exile to France 156
    The Cathars 157
The Jesus Family Tomb 158
Other Messiahs 159
The Legacy of Jesus Christ 159

7: The One God — 165

Progress? 165
    Civilization 165
    Gender 166
    Sexuality 166
    Devotion 167
    Culture Wars 167
    Thought Control 168
    Intolerance 169
    Latecomers 170
Pitfalls and Promises 171
Conversion 172
The History of Monotheism 173
    Babylon 174
    Egypt 174
    Judaism 176
    Christianity 181
    Islam 188
Verdict 189

8: In God's Name — 191

Science 193
Altering the Bible 194

	Heresy & Inquisition	209
	Irenaeus	212
	Timeline	214
	Cruel Legacies	220
9:	**The Biology of Religion**	**223**
	Legitimacy	224
	Genetics	226
	Religion in the Brain	229
	Beliefs	235
	Logic	239
	Mysticism	246
	Marketing Religion	249
	Thought Control	250
	Who Believes?	251
	Is God in Our Brains?	253
10:	**Religion, Interrupted**	**255**
	Eastern vs. Western Religions	256
	The Perils of Tolerance	258
	Religion Behaving Badly	261
	Deliberate Ignorance	265
	Indoctrinating Children	269
	God's Will	272
	Forsaking Reason	275
	Dubious Heroes	280
	Corrupted Sexuality	282
	Genital Mutilation	284
	Rape and Sexual Torture	285
	Criminalizing Sexuality	287
	Sexually Transmitted Diseases	289
	Fundamentalism and Extremism	290
	False Moderation	292
	Religious Persecution	294
	Civilization on the Brink	302
	Breaking the Spell	306

11: Religion Fights Back — 311
- Claims Against Materialism — 313
- The Origin of Knowledge — 315
- The Moral Law — 317
- Evidence of Divine Design — 321
  - Thousands or Billions? — 322
  - Guided Evolution — 322
  - The Anthropic Principle — 326
  - The Orderly Universe — 334
  - The Mathematical Universe — 337
  - The Contingent Universe — 342
- Why Christianity is Correct — 343
- Against Atheism — 345
- At Loggerheads — 346

12: Appraising Religion — 349
- Mystical Wisdom — 350
  - RSMEs — 353
- Is Faith Valid? — 356
- Separate Magisteria? — 359
- Probability — 362
- The Probability of God — 364
- Religious Behavior — 366
- Which Religion is Correct? — 368
- The Legacy of Religion — 371

PART THREE
God

13: Can God Exist? — 379
- A Few Arguments Against God — 380
- Can We Prove God? — 381
  - Toward a Divine Hypothesis — 384
- Do We Need God? — 386
  - Differing Opinions — 387
  - Evolution — 389
  - Intelligent Design — 393
  - Quantum Mechanics — 398
  - Cosmology — 402

	Multiverse	404
	Consciousness	405
	Finding God	412
	Free Will	414
	The Case for Idealism	415
	What of the Soul?	418
	The Possibility of God	420
14:	**On the Nature of God**	**423**
	What Kind of God?	424
	Reductionism	427
	The Limits of Science	429
	The Illusion of Matter	430
	Perfection?	432
	Universal Grandeur	432
	A Hypothesis of God	433

PART FOUR
Paranormality

15:	**Through the Looking Glass**	**437**
	Paranormal Research	438
	Quantum Mechanics	442
	The Copenhagen Interpretation	442
	Many Worlds Interpretation	444
	Hidden Variables Interpretations	445
	Beyond the Second Law	446
	Quantum Consciousness	446
	Out of Time	447
	Paranormality Defined	451
	Clairvoyance	451
	Extra Sensory Perception	452
	Ghosts	453
	Mediums	454
	Near Death Experience	455
	Out of Body Experiences	457
	Precognition	458
	Psychokinesis	458
	Telepathy	458

	Super Psi	458
	Paranormal Encounters	459
	Subtle Bodies	464
	Egyptian	464
	Theosophy	465
	Tibetan	465

16: Paranormal Evidence — 467

- Psi Research — 467
- NDE Research — 472
- Mediumship Experiments — 481
  - Reports from the Afterlife — 482
- Lives Before Life — 484
- In Between Lives — 486

17: Weighing the Evidence — 489

- Common Objections — 489
- Rebutting the Skeptics — 494
- Who Are You? — 497
- Revisiting Death — 502
  - The Meaning of Death — 503
  - Implications for Life — 506
- More Study Needed — 506
- Assessing Psi — 509

PART FIVE
Evolution

18: Defining Life — 515

- What is Life? — 516
  - The Monk and the Bean Sprouts — 517
  - Organized Complexity — 518
- Cruel Indifference? — 519
- Ghosts in the Machine — 521

19: Our Evolutionary History — 527

- Order and Structure — 528
  - Common Descent — 530
  - Extraterrestrial Origins? — 533

History of a Theory	535
Jean-Baptiste Lamarck	535
Charles Darwin	536
Natural Selection	538
Junk in our Genes?	544
Eye See You	545
How Vision Works	547
Primitive Eyes	548
Directional Vision	548
Pinhole Cameras	549
Lenses	549
Irises	550
Other Features	551
Forward or Backward?	552

20: Accident or Design? — 553

Arguments Against Design	555
Probability	557
Origins of Life	558
Thermodynamics	561
Examples of Bad Design	562
Genetics	564
Mutation and Selection	567
Competition	571
Transitional Species	573
Arguments For Design	575
Materialism	576
Probability	577
Irreducible Complexity	579
Beyond Competition	580
Transitional Species	581
An Apology	581
The Idealist View	583
Shared Reality	587
Quantum Evolution	588
Chakras	590
Life Goes On	590
Well?	592

PART SIX
The Physics

21:	**Ancient Physics**	**597**
	Aristotle	598
	Aristarchus	599
	Hipparchus	599
	Claudius Ptolemy	600
	Nicolaus Copernicus	602
	Tycho Brahe	603
	Johannes Kepler	604
	Galileo Galilei	604
22:	**Sir Isaac Newton**	**607**
	Newton's Three Laws of Motion	609
	The First Law	609
	The Second Law	609
	The Third Law	609
	Reference Frames	609
	Limitations	610
	Newtonian Nonlocality	611
	Other Contributions	612
	Mathematics	612
	Optics	612
23:	**Quantum Mechanics**	**615**
	The Quantum Revolution	616
	Black Body Radiation	616
	Light Propagation	618
	The End of Determinism	618
	Grainy Reality	621
	Layers of Science	622
	Particles	623
	A Slight Surplus	624
	The "God" Particle	626
	Exclusion and Stability	626
	Entanglement	627
	Nonlocality	628
	Delayed Choice	630

	Dual Slit	631
	Weirdness	635
	Heisenberg	636
	Waveform	637
	Shrödinger's Cat	641
	The Zero Point Field	643
	Implications	644
24:	**It's All Relative**	**647**
	Special Relativity	649
	All Together Now	649
	Unification of Space and Time	650
	Time Dilation	650
	Length Contraction	651
	Mass–Energy Equivalence	652
	Ultimate Speed Limit	652
	General Relativity	653
	Curves in Space	654
	The Expanding Universe	654
	Linking Space and Gravity	655
	Implications of General Relativity	655
25:	**Strings, Branes, and Loops**	**657**
	Superstrings	658
	Grand Unification?	659
	Higher Dimensions	660
	Limitations of String Theory	662
	Insane in the M-Brane	663
	Thrown for a Loop	664
	The Infinity Problem	665
	Unifying Space, Time, and Matter	667
	Infinity Avoided	668
	Universal Evolution	670
	What About Time?	671
	Qumans	672

26:	**Universal Origins**	**673**
	Timeline	674
	Our Expanding Universe	677
	The Big Bang	678
	Black Holes	680
	Cosmic Background Radiation	681
	Inflation	682
	The Higgs Field	685
	A Zero Sum Game	686
	Dark Energy and Matter	687
	Multiverse?	688
	Revisiting the Anthropic Principle	688
	Self-Service Evolution	690
	God in the Machine	690
	The Dying Universe	691
	De Temporum Fine Comedia	694
27:	**Holography**	**695**
	Making a Hologram	696
	Grainy Holographic Reality	697
	Black Holes and Information	697
	Structuring the Universal Hologram	698
	Finding the Evidence	699
	Implications	700
28:	**About Time**	**703**
	What is Time?	704
	Moving Through Time	705
	Time's Arrow	705
	Relative Time	707
	Bidirectional Time	708
	Rewriting History	709
	Beyond Time	710
	Endless Nows	711
	Sensing the Nonexistent Flow	713
	Choosing Nows	714
	Young Earth Creationism	715

PART SEVEN
Mind and Matter

29: **Interpretations** — **719**
- Copenhagen — 721
  - Observer-created Reality — 722
  - The Role of Consciousness — 723
  - Biocentrism Revisited — 725
  - Problems with Copenhagen — 729
- Hidden Variables — 729
  - David Bohm — 730
  - Bell's Theorem — 731
  - Problems with Hidden Variables — 733
- Many Worlds — 733
  - Sidestepping the Observer — 734
  - Problems with Many Worlds — 735

30: **The Universe in our Heads** — **737**
- The Human Brain — 737
  - Anatomy — 738
  - Neurons — 741
  - Quantum Processing — 743
  - The Holographic Brain — 745
  - Memory — 747
  - To Sleep, Perchance — 748
  - A Delicate Electrochemical Balance — 750
  - Are We In Our Heads? — 752
- Mind — 755
  - Where Is the Mind? — 755
  - A World of Ideas — 757
  - The Power of Suggestion — 759
  - The Power of Belief — 761
- Consciousness — 763
  - The Illusion of Self — 764
  - What is Consciousness? — 765
- Quantum Consciousness — 774

A Theory of Quantum Consciousness 779
   The Conscious Electron 779
   A Quantum Theory of Sleep and Identity 782
   The Consciousness Data Stream 786
   Sharing a Consensus Reality 788

31: Models of Reality — 795

Materialism 797
   Denying the Anthropic Principle 798
   Against Quantum Spirituality 799
   No God Need Apply 800
   Not the Final Word 802
Dualism 806
   The Attraction of Dualism 807
   Problems with Dualism 808
Idealism 808
   We Are All One 810
   Surfing the Cosmic Quantum Waves 813
   The Quantum Godhead 817
   Beyond Birth and Death 820
   Refuting Materialism 823
   Fractured Unity 826
   Nothing and Everything 826

32: Philosophy — 829

Metaphysics 830
   Nothing and Everything 831
   Exposing the Illusion 832
A Model of Reincarnation 834
   Pondering the Soul 835
   Quantum Processing 837
   Revisiting the Quantum Monad 839
   Do We Need Subtle Bodies? 841
   Karma 842
   The Case for Reincarnation 844
Knowledge Limits 846
   The Drive to Know 846
   Inherent Imprecision 848
   Gödel's Incompleteness Theorem 849
   No Proof of Meaning 849
   Intellectual Limits 850

Perceptual Limits	852
The Theory of Everything	852
Accounting for Consciousness	853
Requirements for a Final Theory	853
Can We Find a Final Theory?	854
What If We Succeed?	856

PART EIGHT
The Thesis

33: Evidence from Religion — 859
Religious Shortcomings	860
False Atheism	861
No Sacred Texts	861
Limitations	862
Universal Truths	863
The Afterlife	863
Young Earth	864
All Have Sinned	865
In Conclusion	867

34: Evidence from Evolution — 869
The Blind Watchmaker	870
Mind and Brain	871
Electromagnetism?	872
The Elusive Mind	873
Idealism in our Heads	875
Waste Not	879
Quantum DNA	880
What Evolution Tells Us	881

35: God and Intelligent Design — 883
Do We Need God?	883
Unintelligent Design	885
What God Is	886
Constrained Omnipotence	887
Free Will	888
Suffering and Evil	889
How God Acts	889
Praising God	890

      Finding God   890
          Debunking Religion   891
          On Immortality   892
      Is There a God?   893

36: Evidence from Physics   895
      Where We Came From   896
      The Lawful Universe   896
      The Nature of Reality   897
          The Discontinuous Universe   897
          Out of Time   898
          Out of Space   899
          Out of Matter   899
          Out of Energy   900
      Very Much in Mind   900
          On the Nature of Consciousness   903
          No Place for Snake Oil   904

37: Evidence from Paranormality   905
      Beyond Normal Explanation   905
      Revisiting NDEs   906
          Universal Experiences   906
          Beyond Normality?   907
      Beyond Skepticism   907
      Facing Death   908
      What Paranormality Tells Us   909

38: The Multivariate Monoverse   911
      Methodology   911
      The Inescapable Conclusion   913

39: Zero and Infinity   915
      Nothing to See Here   915
      Everything to See Here   917

40: The Divine Savage   919
      Everything Stands   920
      Who Are You?   920
      Why Are We Here?   921

What Happens When We Die?	922
Other Opinions	924
A Note on Suicide	924
The Joy of Death	925
Where Did We Come From?	926
Where Are We Going?	926
The Divine Savage	926

Appendix A: Words to Live By — 929

Excerpts from the Gospel of Thomas	929
The Upanishads	931
Tibetan Book of the Dead	932
Socrates	933

Appendix B: The Real Secret: Beyond the "Law of Attraction" — 935

What is the "Law of Attraction?"	937
A Brief History	937
Inside The Secret	938
The "Law of Attraction" Explained	939
Idealist Roots	940
The Power of Positivism	940
Placing Your Cosmic Pizza Order	941
Miraculous Possibilities	942
The Power of Gratitude	944
Always Look on the Bright Side	944
The Universal Mind	945
The Flaws in the "Law"	945
Creating our Reality	946
Heisenberg Would Not Approve	948
The Universe is Not Stupid	948
Shared Reality	949
Disentanglement	950
The Quantum Tsunami	950
Paranormality is No Excuse	952
Debunking the "Law of Attraction"	952

    The Law of Collapse   953
        Skewing the Waveform   954
        Beliefs Don't Matter (Much)   957
        Accept no Limitations   958
    Living the Real Secret   958

Appendix C: Additional Resources   961
    Books   961
    Web Sites   967

Dedication

This book is dedicated to my son Logan and my partner Jennifer.

xxii | *The Divine Savage*
Revealing the Miracle of Being

Acknowledgements

This book traces its beginnings to November 12th, 1999, the date of the experience described in this introduction. Many people since then have contributed their support and encouragement, including:

- My wonderful and beloved partner Jennifer, who has supported my endeavors by pointing out relevant books and other sources, tolerating my writing at all hours, and putting up with my many musings and ramblings on esoteric subjects.

- My son Logan, whose love and enthusiasm inspire me to keep going and whose insightful and probing questions keep me on my mental toes.

- Gretchen Holmes, a longtime friend who has taught me much about religion, friendship, and love.

- Larry Loebig, who encouraged me to continue my studies and who started me on the path to writing the entire *Savage* series.

- John Bilorusky of the Western Institute for Social Research, who mentored me and helped me through some tough spots during my research and thesis.

- Andrew Davis, whose staunch support and genuine interest in my work has been a huge inspiration.

- Yvonne McClung, whose positive feedback helped me get through writing this book when it seemed like the light at the end of the tunnel was an oncoming train.

- Jim Saltzman, with whom I can (and do!) talk about most anything and everything and who consistently challenges my research and forces me to have my facts straight.

- Dan and Erin Kellem, who educated me about Christianity and referred me to several of the sources used in writing this book. Dan was the first person to read the essay that sparked this book back in 1999.

- Downtown 65 Toastmasters, who has patiently listened to and critiqued my many speeches on the various topics contained within this book.

- Fred Roach, for crunching the big numbers and helping me make sense of some of the probabilities and odds contained in this book.

- Rose Caraet, for introducing me to Fred and for always being ready with encouragement and/or comic relief as needed.

- Alexa Pagonas, whose intelligence and humor are sources of great joy and inspiration in my life and who introduced me to Rose.

- Rose Wahlin, without whom Dawnstar Books would not be a reality.

- Jim Britt, whose life experiences and insight have helped me immeasurably and whose counsel has helped me more than he will ever know.

- Sarah Nelms, who is like a second mother to me and who has proven beyond any doubt that this world is not all there is to existence.

Thank you all. I really could not have done this without you!

Introduction

"Go read Genesis 1."

I opened my eyes with a start and looked around. Where had that voice come from? The TV was still playing the inane comedy that had lulled me to sleep in the recliner that Friday evening in 1999. My then wife was asleep upstairs and there was no one within a hundred yards of our country home; the dogs provided ample warning of all visitors. So where had that voice come from? I looked around for a few moments before dismissing it as a fragment of a dream and returning my attention to the TV.

"Go read Genesis 1."

That was no dream. I stood up and walked through the silent house, even opening the front door to peer out into the gloom. I was alone. Nobody could have uttered those words and yet I heard them inside my head just as clearly as I hear Jennifer chewing gum and the wind howling past the windows as I recount this story more than 10 years later. I am not exactly the religious type; on the contrary, I considered myself a staunch *materialist* at the time. My worldview was simple: People are ambulatory bags of meat who are born, live a little while, and die without a trace beyond their decomposing remains. Sure, I'd read a few science books that said otherwise and had heard several New Age "gurus" talking about the miraculous implications of quantum physics, but had never given the issue much thought. The few paranormal experiences I had were easily explainable for the most part, and the few I could not explain were the exceptions that proved the rule. Religious people were deluded but mostly harmless fools in my never-humble opinion. Genesis 1 was therefore one of the last things I would ever want to read. Besides, I already knew the story. So why on Earth would I be hearing voices telling me to...

"Go read Genesis 1."

Grumbling, I picked up the Bible I kept for reference purposes, wiped off the thick layer of dust obscuring its cover, and opened it up.

"In the beginning, God created the heavens and the Earth. And the Earth was formless and empty; darkness was over the surface of the deep, and blah, blah, blah…"

I read down to Genesis 2:2 where God rests on the seventh day and shrugged my shoulders. This episode had wasted five minutes of my life and I wanted nothing more than to return to my TV-induced stupor.

"Read it again."

What the…? I reopened the Bible and read the story again. God created the world in seven days. Yeah, I get it. So?

"You don't understand. Read it again."

Now I was angry. I read Genesis 1 again. Then I read it again. And again. Each time, my frustration grew. Each time, the voice in my head told me to read it again in the same even tone. I soon had the entire passage just about memorized and was starting to question my sanity.

Then it hit me: Replacing the Bible's metaphoric language with scientific terms revealed the scientifically accepted history of the universe from the Big Bang to the ascent of man perfectly! The moment I realized this, I had the sudden urge to write it down as quickly as possible. I fired up my computer and had a four-page essay done within minutes. A refined version of those hastily typed words appears in Chapter 3.

Serious questions soon overshadowed my initial thrill of discovery. If my hunch was right and the Bible really did describe the history of the universe perfectly, then this could mean any one of the following:

- The Bible is correct and some version of Judeo-Christianity is indeed the "true" religion. If this is the case, I would expect other creation myths to be far off the mark. I would also expect there to be significant archaeological and other evidence to support some of the key events in the Bible. After all, one cannot reasonably expect the book of the one "true" religion to slack off after getting its single most important story right.

- Western science languished in the shadow of religion for many hundreds of years before the Renaissance and Enlightenment ushered in the Age of Reason that broke the Church's stranglehold on knowledge. Einstein, Bohr, Planck, and most of the world's other great modern scientists are Westerners who grew up surrounded by a very long and deep religious tradition. Perhaps the scientists unconsciously gerrymandered their observations and/or conclusions to fit the religiously accepted truth. It wouldn't be the first time such a thing has happened. If this is the case, I would expect to see fundamentally different theories of the universe from various sources. After all, the West is only one part of a much larger world.

- Science is correct and the Bible has no monopoly on the truth. If this is the case, I would expect to see creation myths from different parts of the world that tell all or part of the scientifically accepted history of the universe. This would imply that primitive people around the world had grasped the truth long before modern science showed up to prove them right. This was an intriguing possibility, but I gave it long odds against being correct because primitive people and modern science seem quite at odds with each other.

This episode was not the only thing challenging my materialist beliefs. I had met a woman named Sarah on a computer chat room soon after moving into my home in La Honda, California in 1997 and we had struck up a virtual acquaintance. One of our first conversations involved a chain saw blade that I had purchased to remove some trees from my property. The blade had gone missing. Sarah claimed to be able to "see" things remotely so I decided to humor her by asking her to find it for me. At that time, Sarah had not yet visited La Honda nor had she ever received any description about the house or property.

"You have three storage areas in your carport," she said. "The blade is in the middle area on the second shelf."

I went cold. The carport did indeed have three doors with shelves behind them. I went downstairs, opened the middle door... and there was my chain saw blade underneath a plastic bag!

Then there was the night Sarah and I were chatting. We used a program called Freetel that required people to type a message and then press [ENTER] to send it to the other person. No data was sent unless and until someone pressed [ENTER], a fact I verified by asking the program's creator who happened to be a friend of mine. For almost two hours, I typed messages without pressing [ENTER] once and Sarah replied, not in generalities, but to my exact words. I would type a sentence and she would answer. I would delete the sentence, type another, and she would answer again. The screen slowly filled with her replies to dozens of messages I had never sent.

These are just two of the many feats of precognition, remote viewing, etc. that Sarah performed and continues to perform to this day. I long ago made it a habit to call her before boarding an airplane. If she ever tells me not to get on, I'm not going to get on... and I won't begrudge her at all if the plane ends up landing safely at its destination.

My list of paranormal experiences does not stop there. I've lost count of the number of times I've had a completely mundane dream only to find myself living that dream some weeks or months later. These are not examples of simple déjà vu where one has a feeling of having been here and done this before. These are dreams that I distinctly remember having and later circumstances that follow those scripts to the letter. Still, none of these decidedly odd events had spurred me to move beyond my comfortable beliefs. They certainly did not inspire me to read Genesis 1.

The Bible incident was the final straw separating me from what has since become a quest to find "The Truth," whatever that Truth may be. This quest has led me through the full spectrum of sciences from astronomy to zoology and through paranormal, spiritual, and religious texts from around the world. My initial goal was to prove the veracity of materialism to my own satisfaction. With this in mind, I researched human evolution. I also cracked open some physics books I'd abandoned years before. These studies yielded two important facts:

- Evolution is not just a theory. In cosmic terms, humans are just another animal species inhabiting a tiny blue planet in a backwater solar system in a backwater corner of a backwater galaxy in a backwater corner of the universe that makes the hillbillies from *Deliverance* look positively urbane by comparison. Looking at humans as just another species of chimpanzee yields some profound truths about how we operate and reveals the extent to which most of us are unsuccessfully trying to go against millions of years of evolutionary programming. Understanding and working within that programming has the potential to radically transform our definitions of mental health and ideas about human potential and show us how to lead truly rewarding and fulfilling lives.

- A good percentage of the New Age "gurus" who claim that quantum physics validates their ideas about wealth and happiness are snake oil peddlers. Still, the standard Copenhagen interpretation of quantum physics really does say some astonishing things about how the universe works and about our place in it.

The Enlightened Savage was born of that early research, and I am pleased to say that the science therein has been validated by my extensive subsequent research. My next book, *The Natural Savage*, explains the science behind *The Enlightened Savage* in detail by breaking life down into its six core components: predator avoidance, group status, food, shelter, reproduction, and death. I have since continued my research with the aim of writing six more books, one for each of the six core life components. Until recently, my self-imposed task was simple: Provide a materialistic, evolutionarily correct view of life and the human condition. We're born, we live, we die; it's that simple. My plan was to write *The Social Savage* first and finish up with this book, which I initially envisioned as having a much less uplifting title like *The Finite Savage* for obvious reasons.

I began working on the *The Social Savage* but soon found myself being drawn inexorably back to books on astronomy, cosmology, quantum physics, superstrings, neurology, religion, mysticism, the paranormal, etc. The more I studied, the more I realized that my materialist beliefs may have been just plain wrong. This does not invalidate any of the science or personal development model contained in *The Enlightened Savage* by any definition. Rather, I came to realize that the model was analogous to Newtonian physics: an approximation that yields excellent results under most circumstances but that starts to break down when pushed beyond certain limits. Just as quantum physics contains Newtonian physics within it, I needed a new theory that would pick up where evolution left off by dealing with those exceptional circumstances that evolution cannot touch. The model contained in *The Enlightened Savage* is absolutely viable for normal daily life but breaks down when one starts talking about death, the single most profound thing all of us will experience in this life.

But what is the Truth? On one side, scientists like Richard Dawkins, Daniel Dennett, Steven Weinberg, and others preach materialism and the lack of any design or purpose to the universe or life. Others like Michael Behe speak of "intelligent design" and "irreducible complexity." Some, like physicists Amit Goswami and Evan Walker Harris (1935-2006), go even further, postulating that mind creates matter instead of the other way around. Religious texts proclaim the superiority of their particular sect *über alles* and issue stern warnings about false gods and false prophets backed by visions of hellfire and other divine punishments. Some scientists explain the paranormal as just another fact of ordinary life; others attempt to explain it away altogether. Some claim to offer evidence for God, others to deny God's existence. Some claim there is one universe, some say there are many, and some even say that the universe splits into copies whenever a quantum event occurs (the Everett many-worlds theory). More than one author misquotes his sources to make a particular point.

My research progressed until I eventually found myself reading the same basic arguments over and over again. I kept going in hopes of finding something new, but eventually realized that the law of diminishing returns had reared its ugly head. By then, I was fairly certain that materialism is just not sufficient to have a serious shot at being The Truth but I just couldn't decide which alternative to accept. I did, however, figure out that many people make the mistake of conflating concepts that are not necessarily related. For example, disproving the existence of the Judeo-Christian-Islamic God does not disprove creationism. Proving evolution correct does not rule out creationism. The purported reality of a physical body does not rule out the possibility of a soul. The possibility of *idealism* (the opposite of materialism) does not mean we live beyond this lifetime or that we have ever lived before. The potential reality of ghosts, clairvoyance, telepathy, etc. does not necessarily rule out the possibility of alternative explanations. In short, A does not imply B no matter how badly some authors want us to believe otherwise. Religion, science, body, soul, creation, evolution, etc. are all separate concepts that can exist (or not) in virtually any imaginable combination.

My methodology primarily consisted of reading a book or article, circling the parts I thought relevant, making notes in the margins, and then proceeding to the next book. Online resources got the same treatment after I printed them out. I found my beliefs swinging back and forth like a pendulum from one extreme of monistic materialism to monistic idealism as I went through this exercise. Idealism is like materialism, only backward: where materialism posits that mind and consciousness are by-products of a physical brain and a physical universe, idealism posits that the physical brain and universe are by-products of mind that is itself a by-product of consciousness. I quickly rejected dualism, which posits the reality of both mind and brain, because I agree with the authors who see no way for mind and brain to exchange information without violating the First Law of Thermodynamics (the conservation of mass and energy).

Once I decided to stop reading, my next task was to go back through every one of my many resources and summarize the circled portions of text while also capturing my many notes. This endeavor required several more months of diligent work, during which I found myself seesawing back and forth again, only this time I was disagreeing with people I had previously agreed with while agreeing with others I had previously disagreed with. It was challenging on many levels, not the least of which was being this far along in the project with little to no idea how it was going to turn out. All I knew is that some people were right and others were wrong. But who? Everyone I read had solid reasons for their beliefs backed by at least moderately good evidence. To make matters more interesting, everyone seemed to be finding what they were looking for. Creationists found God. Materialists found matter. Everyone seemed to be on the right path, or at least on a right path toward completely different destinations.

It eventually dawned on me that looking for evidence to back a preselected conclusion is anything but scientific. The only way to find the truth was to toss all my chips in the air and see where they landed. What if everyone is wrong? If so, then all of our intellectual and technological endeavors over millions of years have been for nothing. On the other extreme, what if everyone is correct? That question sounds ludicrous, but I had to ask it. What if the cosmologists are right when they say that the universe is 13.7 billion years old and the Earth is 4.5 billion years old? What if groups like the Seventh Day Adventists are just as correct when they claim the Earth is only 6,000 years old? What if those who say that the universe began from random quantum fluctuations are just as correct as those who claim it began from the Word of God? What if intelligent design is just as correct as evolution? What if *Jesus Christ* (literally, "Anointed Savior") was both human and divine? To put this another way, what if the only way to solve the puzzle was to remove a few pieces from many different boxes to form a coherent image?

I reread my notes with growing excitement. The pieces fit! Strip away the dogma of both hard-core science and religion (which are astonishingly similar in tone and hubris), and the many seemingly contradictory pieces of evidence fit together into a puzzle that is self explanatory, logically coherent, scientifically and spiritually valid, and which reveals a reality that is far richer and more miraculous than I had ever imagined.

My original quest to prove the universe a cold, pointless, mundane place is an utter failure, and my original reasons for writing this book have been turned on their head. Dear reader, I hope and trust that you will forgive me for this. I also hope that you will not see my admission as a reason to stop reading this book. On the contrary, I hope you read it the entire book with an open mind no matter which subset of beliefs currently forms your personal reality. Whether you are religious or atheist, spiritual or not, materialist or dualist or anything in between, Eastern or Western, male or female, young or old, I hope that you will find your beliefs affirmed and enriched. If I succeed, however minimally, in this new quest then it will be an abject success.

Welcome to the Multivariate Monoverse!

Anthony Hernandez

PART ONE

Getting Started

2 | *The Divine Savage*
Revealing the Miracle of Being

Chapter 1

Dying

I shall die, but that is all I shall do for death.
Edna St. Vincent Millay

When we look back in time and study old cultures and people, we are impressed that death has always been distasteful to man and will probably always be.
Elizabeth Kübler-Ross

You are going to die.

You may not die today, tomorrow, or the next day. You may even go on living long after I've passed on, but die you must and die you will. Chances are good that you won't be alive 50 years from now, and you will almost certainly be dead sometime within the next 100 years. Contemplate this simple truth for a few minutes as you try to imagine utter nothingness: not black, not silence, but utter annihilation. In other words, imagine ceasing to exist forever.

You may find this exercise disquieting or even terrifying. You may even carry this fear around with you like a silent invisible weight. If so, you're not alone. Humanity seems to be one of the few species blessed and cursed by our awareness of our own mortality. Knowing we will eventually die can motivate us to live our lives to the fullest so that death will come as a fulfillment instead of an intruder. This knowledge can also contribute to anxiety, depression, and a host of other ills. The Greek philosopher Epicurus believed that the fear of death is the root cause of all human misery. He may well be right: Millions of people every year experience the so-called *mid-life crisis* where they come to realize that they are the strongest and healthiest they will ever be and that life is all downhill from there in terms of health, vitality, and living itself. From then on, life continues in the growing shadow of our own approaching demise.

But what exactly do we fear? Is it the prospect of nonexistence or the idea that we may pass on before we have had our fill of life? Interestingly, studies have shown a strong correlation between a person's fear of death and the sense that her or his life will end before it's truly over, before s/he has finished living. In other words, the fear of death is really the fear that one's life has not been fully lived.

Humans are very social animals, to the point that ostracism is the single most traumatic thing that can befall anyone. Getting fired from a job, divorce, expulsion from a group—All of these social rejections traumatize us on a very deep level, because loneliness equals increased risk of predation. Loneliness is one of the worst things that can happen to a person, and dying is the loneliest thing we will ever experience. No matter how many people are by our sides when we draw our final breaths, each of us is doomed to die alone and thus to experience the ultimate separation and isolation from the world, and even from ourselves.

According to Epicurus, death itself is nothing because our souls are mortal and nothing survives death. Why then should we fear something we will never perceive? This is fine, except that some enterprising folks dreamed up a tale of eternal damnation and punishment waiting for all those who did not live by certain rules and pray to certain deities in the prescribed manner. Co-opting the fear of death with both the threat and promise of eternal existence based on following (or not following) a certain set of rules governing all aspects of life is one of the central tenets of most religions. This ultimate carrot and stick proposition enraged Epicurus, who pointed out that human self-esteem is so low that we credit God for things we do ourselves—a direct insult to God, for reasons we will examine throughout this book.

We have just seen both extremes, those who think this life is all we get and those who think existence is eternal. This begs the question: Is there more to life than our short time on this planet, or is this all we get? Countless authors have tackled this question from all angles from the religious to the scientific and anywhere in between, and have produced endless reams of

---

Today is a good day to die for all the things of my life are present.
Crazy Horse

You want to know of death? Well, I shall save my breath. When you know of Life, why then, we'll talk of death again.
Confucius

> *To science, a human being is nothing more than a complicated machine.*
> David Darling

speculation on this subject. As for me, if you've read *The Enlightened Savage*, *The Natural Savage*, and/or any of the other *Savage* books, you know that I am a staunch believer in evolution and so-called *materialism* (defined as, "the philosophical theory that regards matter and its motions as constituting the universe, and all phenomena, including those of mind, as due to material agencies"). You would therefore be justified in thinking that my goal in this book is to drive the proverbial nail into the coffin of any notion that anything exists for us beyond this lifetime. You would also be wrong.

As I said in the introduction, classical materialist physics and evolution provide excellent models of human life and behavior under ordinary conditions, just as Euclidean geometry provides an excellent model for constructing skyscrapers. The problem is that these models break down when pushed to extremes at either end. Classical (Newtonian) physics breaks down at very small and very large scales. Evolution fails to explain our origins and also leaves open some nagging questions. Euclidean geometry fails to account for the curvature of space-time. Likewise, the model I present in the other *Savage* books does an admirable job of explaining normal life and how to get the absolute most out of it but reaches its limits at birth and death, the extreme ends of familiar human existence. The model contained in this book does not invalidate anything I say in the other *Savage* books; on the contrary, this model contains and expands upon my previous model within itself, just as quantum mechanics contains and expands upon classical Newtonian physics.

Whether or not any part of us survives death, we do know that our bodies will eventually expire and that we must face the ultimate leap into the void that lies so near yet so far beyond our perception. Let us therefore begin our journey by familiarizing ourselves with the process of dying and what will happen to our bodies after they cease to function.

> *Some people want to achieve immortality through their works or their descendants. I want to achieve it through not dying.*
> Woody Allen

If you're feeling sad and/or frightened right now, cheer up! If there was conclusive evidence that this life is all we get then I would not have wasted my time writing this book, nor would I have asked you to waste your time reading it. I cannot promise pearly gates, angels, or any of the trappings of any religious afterlife. What I can promise is that the best available evidence points to the most sublime reality one can imagine.

The Dying Process

Dying is the single most profound experience any of us will have. Some deaths can be so rapid that the victim may literally never know what hit them. Even when death is somewhat slower, our minds can be preoccupied with trying to survive and thus may never truly grasp the fact that life is ending. Still, most of us will have plenty of time to let the concept of our own mortality sink in and to make preparations to leave this world. Here's what we have to look forward to:

The mere act of dying is not in itself a passport to eternity.
Aldous Huxley

Accepting the End

Dr. Elisabeth Kübler-Ross (1926-2004) spent many years comforting and studying dying people. She identified what has become known as the Grief Cycle, which consists of five stages. It is interesting to note that this cycle applies to anyone who is experiencing a severe loss or setback, such as losing a job or loved one, leaving a relationship, etc. Let's look at this cycle from the point of view of someone who has been diagnosed with a terminal illness.

You can be a king or a street sweeper, but everyone dances with the grim reaper.
Robert Alton Harris

1. **Denial:** The person denies the diagnosis. S/he may or may not feel sick. The initial reaction is to think that the diagnosis is a mistake or that the situation cannot possibly be happening to them. As the news sinks in, the person generally becomes very aware of what is happening and starts worrying about the people s/he will leave behind.
2. **Anger:** When the person realizes that denial is useless, s/he will often become very angry and may even fly into a rage, feeling that the situation is not fair, that s/he does not deserve to die, etc. It is very difficult to be around someone in this stage because s/he is prone to resent and envy anything that represents life and health.
3. **Bargaining:** As the anger fades, the dying person tries to buy more time, to see their children grow up, do the things they never got a chance to do, or simply savor life for just a few more years. This is a clear sign that the fear of death comes from the feeling of not having fully lived. The negotiation often involves a higher power with the promise of a changed life in exchange for more time. By this point, the dying person knows s/he is dying and is simply trying to postpone the inevitable. Henry Thoreau summed

Death hath ten thousand several doors for men to take their exits.
John Webster

up the feeling of a life not fully lived when he said, "Oh God, to have reached the point of death only to find that you have never lived at all."

4. **Depression:** Bargaining has failed and the end is relentlessly coming closer. Death is no longer a matter of if but when. As the full weight of this reality crashes through the last barriers of mental resistance, the dying person can lose all hope and succumb to sadness as s/he grieves for the life that will soon be over. The dying person may become silent and listless, may not want to see anyone, and may spend a good deal of time crying and mourning. As awful as this sounds, this part of the dying process is essential because it allows the dying person to remove her or his attachments to this world. Caregivers should allow this process to run its course and not derail it by trying to cheer the dying person.

5. **Acceptance:** After all the denial, anger, hope, and grief have run their course, the dying person has expended her or his emotional energy. S/he finally understands and has come to terms with her or his impending death and has let go of her or his attachments to this world. Part of this letting go may include no longer feeling physical pain and the desire to be left alone. This stage marks the end of the struggle against one's own death.

I must emphasize that these stages are different for everyone. A dying person may not experience all of the stages, may experience them in a different order, or may even cycle back and forth through various stages before finally washing up on the shores of acceptance and preparation. I believe that the degree to which one accepts the inherent neutrality of all things in life, owns the power to consciously decide what to label as good or bad in life, and remains free of attachments to anyone or anything is the extent to which the dying process can be eased significantly. Of course, there is only one way to find out and I confess to being in no great hurry. Please read *The Enlightened Savage* for more about learning to let go of attachments and consciously observe and guide your life.

The uniqueness of each of us extends even to the way we die.
Sherwin B. Nuland

Physical Death

If there is one constant in life, it is change. Our bodies change as we age and eventually begin the inevitable decline toward physical death. This decline can be spread over many decades or can be accelerated by disease or injury. It is a slippery slope that begins gradually and keeps on accelerating until the end. As we discussed above, the end stages of life involve many changes to our bodies, intellects and emotions. A dying person may even reevaluate her or his spirituality.

Now death has come; what does it mean? What is it that is dying? This body dies.
Ramana Maharshi

It is very important to understand that each person is unique and will experience dying in her or his own way over a time period that may span anywhere from years to seconds. We can only discuss generalities and must adapt to every dying person we come across, including ourselves.

Physical weakness is normal. The dying person must expend more effort for everyday tasks and thus tires more easily. This can lead to frustration, depression, and embarrassment over needing to depend on others for things s/he used to be able to do for her or himself. On the up side, some people use this time to focus on life's bigger questions and deeper realities of intellect, emotions, and spirit as they try to figure out the meaning of life and their role in the grand scheme of things. The best way to help is by helping the person with tasks and errands while preserving her or his dignity and comfort. Controlling symptoms and preventing injury are key priorities. Caregivers must be patient, compassionate, and non-judgmental. They must also realize that the dying process can be a time of significant personal growth and development.

Letting go of worldly attachments is an important part of preparing for death because we can't take anyone or anything with us when we die. This can be a time-consuming process that requires privacy. Visitors mean well but can be overwhelming because the dying person may feel socially obligated to be a good host despite her or his needs for solitude. The energy expended on entertainment can lead to exhaustion and increased sleep. The person may also withdraw and become uncommunicative, a clear sign that s/he needs more alone time. Visitors or caregivers should focus on simply being there without feeling the need to say or do anything unless asked. A quiet loving presence can be a huge help. Also, be aware of what you say: Just because someone seems to be sleeping or comatose does not mean that they cannot hear you. In fact, hearing is one of the last senses to shut down. Of course, some parts of letting go may require interaction and discus-

As soon as I became conscious of life, I began the long process of watching someone gradually die of old age.
Sherwin B. Nuland

sion. The dying person may want to tie up loose ends, such as accounts or relationships, provide final service and other instructions, create or amend a will, seek spiritual guidance, etc.

Remember that 95% of communication is nonverbal. The words you say convey only 5% of the message. Roughly 35% consists of tone, volume, intonation, etc. and the remaining 60% is purely physical. Breathing, expression, posture, eyes, etc. all betray our thoughts and emotions and provide the context for our communications. This is why it is so important to check your ideas, judgments, opinions, etc. at the door when visiting or caring for someone in the final stages of life. They are experiencing the most frightening and exhilarating thing one can ever experience in this lifetime. The last thing they need is you interfering with that, no matter how well intentioned you may be. Don't ever make the mistake of assuming that someone you are caring for can't be or isn't very aware of the messages you are sending out.

Man has given a false importance to death. Any animal, plant, or man who dies adds to nature's compost heap, becomes the manure without which nothing could grow, nothing could be created. Death is simply part of the process.
Peter Weiss

Don't be surprised if the person's appetite decreases. Digestion consumes an ever greater percentage of a weakening body's available energy and the overall lessening of physical activity means that less food is required. The sense of taste can be the first to shut down, and medications can also interfere with the taste and/or desire for food. This is a very sensitive subject for friends and family, because we are conditioned to eat certain amounts of food and will naturally become concerned when someone doesn't seem able or willing to eat their share. Hungry people will eat if they are at all able to. I've learned this lesson watching my son grow up through phases where he refuses to eat followed by phases where he refuses to stop eating. Also, food may not be that big a priority for a person who knows that s/he is about to die. This does not mean that s/he does not want or need emotional and spiritual nourishment. The person could also have difficulty swallowing and may need her or his food cut up, blended, etc.

It is not uncommon for dying people to experience confusion, which can be caused by diseases such as Alzheimer's and/or by the person's becoming increasingly unattached from the world. They may hallucinate, not recognize loved ones, forget the day and time, where they are, etc. The best thing one can do is remain accepting and non-judgmental. For example, a hallucination may be every bit as real to the person experiencing it as your reality is to you or my reality to me. Besides, who can say for sure that what we call hallucinations are not actual otherworldly communications?

Restlessness can be a sign of distress that is as simple as needing to go to the bathroom or as complex as the beginning of the actual death itself. One must do one's best to determine and correct the source of the discomfort while realizing there may be nothing at all that can be done. The dying person may feel unable or unwilling to finish letting go of this life, and it is important to tell her or him that it's OK to let go, that there is nothing left to do unless she or he wants to do it. Just as the dying person must let go of this life, so must friends and family let go knowing that there is nothing good or bad about death; it just is.

A summons has come and I am ready for my journey.
Rabindranath Tagore

Urinary and/or fecal incontinence may be a problem as the body becomes increasingly weak and unable to regulate itself. It should go without saying that this can be extremely embarrassing and that any such incidents should be handled with the utmost dignity and respect. Constipation can also be a problem, and it's therefore important to track bowel movements.

Gradual kidney shutdown means that more elimination happens through the skin. This can cause itching and can combine with the body's loss of ability to regulate temperature to cause intense discomfort. The person may sweat despite being cold or the skin may be dry despite being very hot. Limbs may become cool and nails may turn blue as the heart becomes increasingly unable to maintain circulation. The skin itself may become gray or ashen in color.

Breathing may also become increasingly labored and difficult and the person may need supplemental oxygen. Exhalations that are longer than inhalations are a clear sign that the final dying process has begun. From there, breathing will become increasingly irregular, eventually involve the entire rib cage, and speed up to 30-50 breaths per minute with possible 10-15 second *apnic* (non-breathing) pauses. Once breathing reaches this point, there is little chance of recovery and death is anywhere from a few minutes to a few days away. Keeping the lips and mouth moist and positioning the person with her or his head elevated can help with breathing.

Our need for consolation is impossible to satisfy.
Stig Dagerman

When the end is imminent, saliva stuck deep in the throat may cause a rattling noise, the so-called death rattle. This saliva is often too deep for suction to remove it but does not seem to bother the dying person. Turning the head and/or body to one side may help keep the throat clear.

A person can become energetic, alert, talkative, hungry, and social in the days or hours before death. This is the final opportunity to spend time with loved ones and finish preparations for death. Many people become comatose and unresponsive just before death.

Some of the signs of imminent death include a gray or blue-gray skin color (especially on the lips and extremities), cold hands and feet, eyes that are glassy, fixed, and dilated, and rapid mouth breathing with apnic pauses. The dying person may no longer respond to voice or pain. Clinical death occurs when breathing and heartbeat cease and the eyes become fixed and dilated. The bladder and/or bowel may release.

Loud voices, arguments, crying, etc. are not appropriate; everyone deserves to die in peace. If you are present at this time, it is critical to remain quiet and calm while doing your best to follow her or his last wishes or instructions. I see it this way: If this life is all we get and death is the descent into oblivion, then I want my final moments of existence to be as calm and peaceful as possible. The last thing I or anyone needs at the point of death is any more trauma and drama than we are already experiencing. If death is simply a doorway to a new plane of existence, then what better way to pass through that door than at peace with all the love and support of everyone around you? Buddhists believe that one's state of mind at the moment of death is critically important and that one must die in a calm, peaceful, positive state. To them, death is the separation of body and mind. As we will see throughout this book, there is every reason to suspect that the Buddhists are absolutely correct.

The Body after Death

Birth and death are purely biological processes, a fact we spend our entire lives trying to ignore. Thanks to the Second Law of Thermodynamics, life is a struggle against decay and decomposition—a struggle that our bodies will eventually

The obituary pages tell us of the news that we are dying away while the birth announcements in finer print, off at the side of the page, inform us of our replacements, but we get no grasp from this of the enormity of the scale.
Lewis Thomas

Oh lord, give each of us his own death, the dying that issue forth out of the life in which he had love, meaning, and despair.
Rilke

lose. Decomposition begins at the moment of death and will continue until there is nothing left of our bodies and our atoms have been recycled to create new life. Embalming can delay decomposition and special circumstances such as cremation and deliberate or accidental mummification can avert it, but sooner or later the material that makes up our body is destined to be put to other uses. In that sense, we are immortal.

Bodies that are not cremated or embalmed may be put to scientific use in anything from basic anatomy lessons to practicing surgery, crash tests, etc. Parts of a body that are in usable condition may be transplanted into other people to help save or enhance their lives. Some transplant recipients experience significant personality changes, and some even have memories that seem to have come from the donor. Our organs may carry some remnants of our personality and identity with them, which could account for at least some of the changes. This is not nearly as farfetched as it sounds: On a physical level, our emotions are caused by chemical "cocktails" that circulate in our bloodstream and attach themselves to opioid receptors located throughout the body. Transplanted organs carrying this cocktail could conceivably cause the reported changes. (See *The Enlightened Savage* for more information about how this works.)

The time line presented here is based on numerous assumptions. Decomposition happens in stages and can last anywhere from days to years. For example, all else being equal, a body in a tropical environment will decompose much more quickly than a body in a cold or dry environment. Insects play a vital role in breaking the body down at each stage of the process.

Initially, there are no visible signs of decomposition, but the body's regulatory mechanisms have ceased to function. The body cools to the temperature of its environment. Cells and organs break down in a process called *autolysis*, which is the digestion of tissues by their own enzymes. Autolysis eventually stops as oxygen in the body is used up, at which point the bacteria normally present in the body begin breaking down carbohydrates, proteins, and lipids. This creates various by-products including *volatile organic compounds* (VOCs), which are essentially vaporized organic molecules. These by-products attract insects that begin laying eggs on the body. If the body is on or lying in dirt, soil-dwelling insects will also move in to feast on

I died as a mineral and became a plant. I died as a plant and rose to animal. I died as an animal and became a man. Why should I fear? When was I less by dying?
Jalal-uddin Rumi

For, as I draw closer and closer to the end, I travel in a circle nearer and near to the beginning.
Charles Dickens

the remains. A condition called *rigor mortis* (stiffening of the muscles after death) may begin approximately 3 hours after death, peak after about 12 hours, and then dissipate until about 72 hours after death.

We are already in the kingdom. Eternity is now.
Andre Compte-Sponville

All of this chemical and insect activity eventually leads to *putrefaction*, or outright rotting. The abdomen turns green as bacteria convert hemoglobin to sulfhemoglobin and swells as organic gases form under the skin. These gases force any remaining urine and feces out of the body. By about five or six days after death, the abdomen is swollen and blisters are forming all over the body. Bacteria in the veins cause red streaks to form along the veins. These streaks soon turn green. By 14 days after death, the abdomen is very tight and swollen. After a few more days, the skin becomes very fragile and can slip easily, making moving the body difficult. Body hair and fingernails fall off. Eventually, the green color turns to brown or even black and tissues begin to liquefy. The body splits open and the trapped gases escape. Tissues liquefy, and the body is no longer recognizable. Insects swarm in to complete eating and breaking down the soft tissues. The bones become visible.

As the putrefaction winds down, the body begins to *mummify* or dry out. The stench of death caused by the escaping VOCs disappears and *adipocere* (grave wax) forms as bacteria break down fats in a process called *hydrolysis*, in which water molecules (H_2O) are split into hydrogen (H^+, or protons) and hydroxide (OH^-). Adipocere has a cheesy appearance.

Finally, when all of the soft tissue has been consumed, *skeletonization* (the breakdown of skeletal remains) begins. Bones consist of protein-mineral bonds, making skeletonization the slowest part of the decomposition process. The protein slowly leaches away, leaving the minerals behind. The minerals themselves then leach away at a rate that depends on many factors such as soil type, pH, moisture, temperature, etc. Bones disintegrate as the minerals disperse. Eventually, no sign remains of the body; its components have been fully recycled into the great circle of life.

There is an extraordinary relation between consciousness and experience.
Jerry Wheatley

What is Death?

Imagining one's own death is difficult—so difficult, in fact, that most of us find it impossible to imagine total blackness that goes beyond void or vacuum into utter nothingness and nonexistence. Try it yourself; the blackest black you can imagine is probably nowhere near total nonexistence.

We like to believe that we will remain clearheaded and equanimous in the face of our own deaths and that dying is simply slipping into painless unconsciousness after we have successfully concluded our business and said goodbye to the people we love. We understand on a certain level that death is not unnatural, but part of life itself. But what exactly is death?

What I am about to say may sound like New Age mumbo jumbo, and you may be tempted to dismiss it as utter nonsense. You may also interpret my words as coming from or endorsing a particular religious view. This is not the case. On the contrary, my beliefs about the nature of life and death are based on solid scientific principles. Keep reading, and you will eventually discover what astronomer Robert Jastrow (1925-2008) meant when he said, "For a scientist who has lived by his faith in the power of reason, the story ends like a bad dream. He has scaled the mountains of ignorance; he is about to conquer the highest peak; as he pulls himself over the final rock, he is greeted by a band of theologians who have been sitting there for centuries."

We are extremely accustomed to identifying ourselves with our bodies and identify the combination of body and mind as "I." Of course, our bodies and minds are always in motion, both literally and figuratively. We think we own this seething mass of tissue, but the truth is that we're just tenants living on borrowed time. Also, as described in *The Enlightened Savage* and *The Natural Savage*, we never experience objective reality, whatever that may be. What we think of as "reality" is nothing more than a by-product of our limited perceptual abilities filtered through our own beliefs. In short, everything we think of as life and reality is nothing more than a reflection of our deepest levels of mental programming. But, as we'll discover later, our true nature may be pure existence with no movement, no name, no reputation, nothing. It is important to understand that by *nothing*, I mean "no thing," not "nothingness" or "nonexistent" as normally defined in the dictionary.

Getting in touch with our own inner stillness requires nonattachment to everything and everyone in our lives. Don't con-

---

Death punctures a hole in the tight fabric of the ego, which allows us to slip through in a moment outside of time to experience ourselves as infinite perfection.
Kenneth Ring

The most formidable weapon against errors of every kind is Reason. I have never used any other, and trust I never shall.
Thomas Paine

The brain is merely a transmitter and receiver of information, but not the main place for storage and processing of information (i.e. memories).
Simon Berkovich

> *Thine own consciousness, shining, void, and inseparable from the Great Body of Radiance, hath no birth, no death, and is the immutable light, Buddha Amitabha.*
> Tibetan Book of the Dead

> *The universe is immaterial—mental and spiritual. Live, and enjoy.*
> Richard Conn Henry

> *Awareness cannot be different than the object. Awareness is the essential nature of the object.*
> Utpaladeva

fuse non-attachment with a total lack of desire or with listlessness and apathy. Think of it as giving yourself room to allow your thoughts and feelings to flow freely, to actively witness your life and receive whatever comes. As things happen to us and we experience thoughts and feelings, it is important to understand that these are merely things happening to and in ourselves; they are not part of us unless and until we internalize them. Developing the art of conscious observation allows us to choose what we internalize and how we label the things we choose to let in. Being both unattached and conscious reveals that we are pure awareness. In other words, each of us comes to understand that I AM. We also begin to sense that awareness is timeless, that it does not come or go. It just is. Likewise, this lifetime is neither good nor bad; it too just is.

Young children tend to be very comfortable about the idea of death, because they believe that death itself is just another moment in life. I've talked about death with my son Logan as he has grown up, and he has simply accepted it without any fear or question. When I ask him what happens after death, he talks about ghosts and somehow continuing on after this life. When I ask him why he believes this, he replies, "How can something just stop being?" Sadly, this connection fades as we grow up and identify ourselves more and more with our bodies. Children are not so intimately connected to their bodies and thus have much less of a problem with the idea of death.

Non-attachment must extend to letting go of the will to live. I'm not saying you need to harbor a death wish, simply that the will to live comes from identification of self with the physical body. This of course begs the question: Is our awareness, our I AM, a product of physical processes occurring in the body and brain? Are mind (soul) and matter (body) distinct entities that interact as separate but equal partners? Or is matter a product of mind? The fact that everything we see, hear, feel, smell, and taste only exists because we perceive it to exist provides a powerful clue. This is where we will begin our search for additional clues that will point us to the ultimate answers about life, death, the universe, and our place in it.

What is death? For reasons that will take me the rest of this book to explain, death may not exist. Birth and death may be illusions caused by our attachments to our own bodies that we also project onto others. These attachments cause us to believe that birth and death are real events. Seeing people seemingly coming into and out of existence completes a feedback loop that strengthens the attachment and makes the illusion seem all the more real.

If death is just an illusion, then what "survives" death? Does our intellect and emotion dissolve along with the rest of the illusion, or are they real and therefore *transcendent* (beyond the limits of experience, the universe, and time)? Materialists point to loss of consciousness during sleep as proof that our brains are the source of our consciousness and identity. What they fail to point out is that both return immediately upon waking and sometimes even during our sleep as dreams. But what about time? Surely our consciousness is extinguished for hours at a time every night or whenever it is we sleep. To this, I ask, what if time itself is an illusion? As you will discover later in Chapter 28, there are very solid scientific reasons to think that is indeed the case. Also, *ego* (the subjective experience we all have of a separate and distinct "I"), thought, and emotion may shut down during sleep but awareness remains. This is why we can wake up when we sense danger or when something or someone disturbs us.

They believe that death is something different form and opposed to life and not perceive or embrace their own death daily so that they might know the freedom of it.
Marge Koenig

Is birth also an illusion? One must be born in order to die. If you believe that birth is real, then you will fear death; however, if death itself is an illusion, then it stands to reason that birth is just as much of an illusion... and if you were never born, then it also stands to reason that you will never die.

Let go of your attachment to your body and your ego. Let your mind become still. In this state, you will begin to discover your true, timeless self. You will begin to realize that there is no such thing as birth or death. You will begin to understand your true, eternal nature. This nature extends far beyond your current identity. For example, Anthony Hernandez will eventually cease to exist on the mundane level we are all accustomed to. And that is perfectly OK with me because Anthony Hernandez may well be only a temporary manifestation of my true self.

It is because you believe you are born that you fear death. Who is it that was born? Who is it that dies? Who you are, in reality, was never born and never dies. Let go of who you think you are and become who you have always been.
Stephen Levine

The Meaning of Death

I was once as you are; you will soon be as I am.
Anonymous epitaph

Imagine looking into the eyes of your own child as she or he looks up at you from a hospital or hospice bed in the final moments of her or his struggle with a terminal illness. What will you tell her when s/he asks you what will happen to her when s/he dies? How can anyone answer such a question? The question of what, if anything, happens when we die is the ultimate question, because it holds the key to unraveling many of the mysteries of life. We may not be the only species to ponder such things; elephants bury their dead with fruit, flowers, and other mementos, which may be evidence of belief in an afterlife. At the very least, it proves that we are not the only species to mourn our dead.

The simple truth is that there is no way to prove what happens when we die until each of us dies. We can speculate, hypothesize, theorize, and examine all available evidence from all imaginable viewpoints all we like, but in the end there is only one way to find out. *NDEs* (Near Death Experiences) may provide powerful evidence of survival after death, and many people who have those experiences are clinically dead when they occur... but they all return to tell the tale. We cannot know for sure whether these "almost" situations are close enough to the real thing to give reliable answers; for all we know, the point of no return many people report reaching deep into the experience could well be the last step before oblivion. (See Chapter 16.)

Humankind is born with two incurable diseases, life from which it inevitably dies, and hope which hints that death may not be the end.
Adam Smith

The medical profession expends enormous sums of money and resources prolonging the final three weeks of patients' lives in an all-out attempt to delay the end just as long as possible. A complicated labyrinth of medical, legal, and ethical issues surrounds this practice, which is largely based on the materialist idea that death is the end of existence. This same materialist viewpoint staunchly refuses to ask just what or who an individual person is or to examine the question of any possibility of some existence after death while dismissing any such evidence as "supernatural." This is regrettable, because a serious investigation into the question of existence after death could shed light on individual identities and revolutionize our evasive psychology, spirituality, medicine, and more. If the "I" in each of us is merely a phenomenon caused by our physical bodies, then death must be end. But if the reverse is true—if our physical bodies are a by-product of consciousness—then death becomes meaningless; consciousness can manifest itself through sequential identities but cannot die because there is nothing to do the dying.

Belief and speculation are not enough. Religious and materialist people alike mourn at funerals despite their beliefs about what happens at death. Grieving loved ones don't want platitudes or reassurances; they want knowledge. In fact, the uncertainty may be the worst part of all. This book will present my answer to this question. It will also present a logical, self-explanatory framework that uses all of the best scientific and spiritual evidence available. It is my hope that the model I present in these pages will prove to be, if not totally accurate, then at least close enough for the major points to work as described. After all, I will someday have to put them to the ultimate test.

In the meantime, it is important to understand that each of us is all we have. We do not own other people, nor do we share their lives directly. All we have are mental and emotional impressions of them caused by our beliefs filtering sensory data. In a very literal sense, the only reason we have other people around us is because we believe we do. (See *The Enlightened Savage*.) This extends to objects. None of us really has anything at all beyond conscious representations of things that we call "property." Death is the ultimate reminder that we don't truly own any of it or anything beyond our own selves. In the end, all we own is information, and this is our only source of potential security. If there is an afterlife and if we do maintain some personal identity after death, then that information will be all we have. Think about that the next time you feel tempted to buy something fancy. Pamper yourself; just don't base your identity on anything or anyone outside yourself.

Sorrow enters my heart. I am afraid of death.
Gilgamesh

The Materialist View

According to materialists, life consists of higher-level systems that consist of various arrangements of lower-level systems in various tiers that include genes, cells, organs, and ultimately entire bodies. A malfunction in any lower-level system can cause the whole system to stop working, just like the failure of

And almost everyone, when age, disease, or sorrows strike him, inclines to think there is a god or something very like him.
Arthur H. Clough

Chapter 1
Dying   19

a small component inside a car engine can quickly render the entire vehicle irreparable. Any question of what might happen after death is dualism, which contradicts materialist science; death is end of life and existence, and that's all there is to it. Any notion we may have of a person's continued existence is the afterglow we carry in our minds that fades over time and subsequent generations until it eventually winks out altogether and nothing more is left.

But what if the materialists are wrong? There is a well-known story of a man walking down a street at night who encounters a friend searching for something under a street lamp.

"What are you looking for?" The man asks, "Let me help you."

"I lost my keys in the house," the friend replies.

"In the house! But then why are you looking out here?"

"Because it's too dark to look in the house."

The material world is intimately familiar to anyone who has been alive for any length of time. We have explained almost all there is to explain about how this world works and are very much at home here. This is the streetlight of knowledge under which scientists gather to look for new discoveries. But what if they are looking in the wrong place? What if the answers are inside the house? What undiscovered countries lie inside those walls? It is telling that materialism does a great job of explaining life from conception to death, just like classical physics explains objects and movements of objects on familiar scales; however, classical physics breaks down at the very small scales of atoms and subatomic particles and must be replaced with quantum mechanics. On the opposite end, relativity must take over when we are discussing very large and/or very fast-moving objects. The even deeper levels of superstring and brane theories unite both extremes. Similarly, materialism cannot explain life's ultimate origins, nor can it tell us for sure what happens when we die. Going beyond these limitations requires abandoning materialism in favor of dualism or idealism. Materialism's ultimate failure is its inability to explain why the universe has gone to the bother of evolving intelligent life only to limit it to short bursts of existence.

Death is everything and it is nothing. The worms crawl in, the worms crawl out.
Unknown

If you ask, 'why is death happening to me (or to anyone)?' the answer is; because the universe is happening to you; you are an event of the universe; you are a child of the stars, as well as of your parents, and you could not be a child in any other way. Even while you live, and certainly when you die, the atoms and molecules which are at present locked into your shape and appearance are being unlocked and scattered into other shapes and forms of construction.
John Bowker

Is this wishful thinking, or do we have good reason to suspect that materialism is not the end all, be all of human knowledge? Consider that clinical death is defined as unreadable vital signs combined with a total lack of electrical activity in the brain cortex. This occurs within 10-20 seconds after cutting off oxygen to the brain. A clinically dead person has no heartbeat, no breathing, no corneal reflex, fixed and dilated pupils, no gag reflex, and is completely unresponsive to stimuli. More than one person in this state has accurately recounted their surgeries in detail down to describing instruments and procedures and conversations taking place among the surgeons. Some have even described objects and events taking place outside the operating room. This is only one type of evidence that suggests that *RSMEs* (Religious, Spiritual, or Mystical Experiences) can occur when the brain is not functioning—and that is only one of many challenges facing materialism.

Why assign perception to a hovering soul if we can find it in the biological machinery?
Judith Hooper and Dick Teresi

Beyond Materialism

Poet Rabindranth Tagore (1861-1941) said, "On the day when death will knock at my door, what will thou offer him? I will set before my guest the full vessel of my life, I will never let him go with empty hands." These are not the words of a despondent person, nor of one who feels that life is pointless. On the contrary, they seem to be the words of a person who knows something most of the rest of us don't know: that death is not the end by any definition. The quest to answer the question of life after death has been going strong since the dawn of human history. Ancient Egyptian schools used hypnotic trances to learn about immortality and offered mystical training programs that taught death does not exist. Some graduates of these programs went on to become teachers and priests, including Jesus Christ, as we will learn in Chapter 6. Psychiatrist Carl Jung (1875-1961) saw death as the ultimate creative force.

There is no end. There is no beginning. There is only the passion of life.
Frederico Fellini

Interestingly, many of the spiritual *traditions* (continuing patterns of cultural beliefs or practices) about life after death do not appear in the Bible; according to the Old Testament, death meant the end of all existence. In fact, as we will see throughout this book, much of the evidence for the existence of a soul does not lie where we might normally expect it to lie. Immortality, if it exists, may come in the form of "long bodies" or

> *The perspective of hell is less disturbing than that of nothingness.*
> Andre Compte-Sponville

quantum monads, which philosopher Gottfried Leibniz (1646-1716) defined as, "an unextended, indivisible, and indestructible entity that is the basic or ultimate constituent of the universe and a microcosm of it." A monad could survive physical death and persist across multiple selves in the same way that the same actor persists across multiple roles. Each role seems to "die" when the production wraps up, but the actor behind that role lives on. Such a view is very compatible with both dualist religions and idealist philosophy, as we will see throughout this book; it is also very compatible with scientific evidence.

Western religions postulate a one-way trip where the monad reunited with God after this life is over. The Eastern religions postulate a long series of rebirths into new bodies, until the monad has reached the point of being able to reunite with God. The Eastern God is infinitely patient and will give each monad as many chances as needed to get it right, whereas the Western God has less tolerance for those who don't walk the straight and narrow. This approach has profound consequences, as we will discover in Chapter 10.

Many hospice workers report that patients experience joyful visions shortly before dying. One theory holds that this happens because the dying person is identifying less and less with her or his physical body and more with eternal consciousness. Death begins with losing one's attachment to, and identification with, the physical body, and rediscovering one's monad. According to the Eastern religions, this process culminates in the unlimited bliss of rediscovering *Brahman* (God). The monad's journey is thus a jagged path that leads toward freedom. We see a glimpse of this freedom when we die before again descending into the bondage of another body and life. If this is correct, then death is simply a rite of passage.

> *In my own life there have been few moments more frightening, more life-affirming, and more enlightening than holding the hand of people who are about to find out what comes next.*
> Irwin Kula

Believe it or not, there is evidence that this philosophy may be correct. In this book, we will look at the evidence offered by religion, evolution, paranormality, neurology, psychology, physics, mysticism, philosophy, and more as we construct a vision of reality that is both logical and self-explanatory. I call this the Multivariate Monoverse.

Materialism has benefitted each of us immensely. We owe our understanding of everything from medicine to the cosmos to materialist scientists. Even this book is only possible thanks to materialism; however, materialism is only one side of a very large coin. Fully understanding the universe and our place in it requires us to flip that coin and look at the other side. If we are correct, then this inward journey will eventually answer all of our questions and free us from the cycle of birth, death, and rebirth.

What if we're wrong? What if this life is the only one we get and death brings total annihilation? In that case, why worry? We will not be around to know that we are dead. In the meantime, we will have learned something about ourselves and the universe in which we live. No matter the outcome, that can only be a valuable pursuit.

In America, we've decided that death is optional.
Woody Allen

Chapter 2

Questions and Assumptions

> *If we choose only to expose ourselves to opinions and viewpoints that are in line to our own, we become more polarized, more set in our own ways. It will only reinforce and deepen the political divides in our country. But if we choose to actively seek out information that challenges our assumptions and beliefs, perhaps we can begin to understand where the people who disagree with us are coming from.*
> Barack Obama

Having set the stage for the central theme of this book by describing the dying process, I must now openly share the core assumptions and methods that went into my research and that underline the thesis that forms my answer to the question, "What is the ultimate nature of reality?" and its corollary, "What happens when we die?"

My three core assumptions have to do with *conflation* (the process or result of fusing separate items into one entity), the truth of *evolution* (changes in the genetic composition of a population during successive generations as a result of natural selection), and whether there is any room for any sort of creative process or power that we might call *God* (a supreme being or consciousness).

Conflation

As I said above, conflation is the act of fusing separate concepts. For example, saying that A being true requires B to also be true is an example of conflation, as is saying that B must be false if A is false or any possible combination of statements that create a relationship between A and B. Virtually every source I researched for this book relied on at least some degree of conflation to try to prove a particular point. This is both appropriate and necessary in many cases; however, conflation is also very easy to abuse when one assumes that connections can and must exist.

Saying that A implies B, B requires A, and so forth is an attempt to create causal relationships where no such relationship may exist or where any such relationships that do exist are merely coincidental. In general, the more extreme an author's viewpoint, the more s/he will attempt to conflate concepts to try and prove her or his point (but this is not always the case). Many researchers have reached mutually contradictory conclusions from the same evidence, and the culprit usually turns out to be conflation.

Here are some of the many concepts I have seen conflated in every imaginable way. We will discuss each of these concepts and more in great detail throughout this book; for now, what you need to understand is that any of the following (and more) can theoretically exist in any imaginable combination without violating the laws of nature as we understand them. I will do my utmost throughout the remainder of this book to examine each of the following on its own merits to see where the evidence leads and whether the collected pieces of evidence can be used to paint a logical picture of the nature of reality.

> *All parts should go together without forcing. You must remember that the parts you are reassembling were disassembled by you. Therefore, if you can't get them together again, there must be a reason. By all means, do not use a hammer.*
> IBM maintenance manual, 1925

Evolution

Evolution is the theory of change in living creatures over time and the emergence and extinction of species as environmental pressures change over time. This theory is controversial because some religious people believe that it contradicts their faith and is therefore wrong. Conversely, many (if not most) scientists believe that evolution is true, and that creation and any idea of a god must therefore be false. The co-called "intelligent design" movement attempts to bridge this gap by reinterpreting the theory of evolution in creation-friendly terms.

Intelligent design has been roundly discredited for reasons we shall explore later. The assumptions behind this idea may be flawed, but the central premise is that evolution and creation are not mutually incompatible concepts. In fact, as we will see in Chapter 3, creation myths from around the world are rife with descriptions of evolution. These myths may or may not explain evolution in scientifically valid terms, but they all

> *Life appears simply as a disease of matter.*
> Gerald Feinberg

acknowledge that all was not always as it is today... and they are all far more accurate than not, as we will see in Chapter 3.

The theory of evolution neither requires nor precludes a creator. Still, it cannot yet explain how life began in the first place. Some people try to avoid the problem by saying that life on Earth originated elsewhere. This may be, but that still does not explain how life itself began. Similarly, evolution does not of itself preclude the possibility that humans have souls, or that death is not the ultimate end of some form of personal existence. All the theory of evolution can tell us that the earliest forms of life began very simply, were fruitful and multiplied, and that this process gave rise to the wondrous diversity of life we see today, life that includes an ape capable of building a machine on which to record his thoughts for posterity using a standardized form of indirect communication. That's about it.

Creation

A large percentage of people believe that the universe was created by God. Different religions describe God in different ways, and some religions have more than one god. These differences aside, the God of religion is described as *omnipotent* (all-powerful), *omniscient* (all-knowing), *omnipresent* (everywhere at once), and *eternal* (beyond the normal flow of time). Some religious cults even believe that the Earth is less than 10,000 years old (as opposed to 4.5 billion), that the universe is also less than 10,000 years old (as opposed to about 13.7 billion), and that dinosaurs coexisted with humans. There is even a museum in Kentucky that includes a sculpture of a *Triceratops* with a saddle on it. Needless to say, such "Young Earth" views are very much at odds with the best available evidence.

All of this aside, the fact remains that the universe could be the result of a creation event orchestrated by an intelligent power that we call God. In theory, God could have done anything s/he wanted, including creating the universe 10,000 years ago and making it look much older. For reasons we will explore later, it is even possible that creation occurred both 10,000 and 13,700,000,000 years ago—an idea that seems ludicrous until one examines the nature of time.

Creationists tend to conflate the idea of God with such concepts as the existence of an immortal soul, an afterlife of end-

---

We do not live in a Pleistocene environment, but our minds were built there and often function as if we do.
Michael Shermer

Science has proof without any certainty. Creationists have certainty without any proof.
Ashley Montague

If we are going to teach creation science as an alternative to evolution, then we should also teach the stork theory as an alternative to biological reproduction.
Judith Hayes

less bliss for those who obey God's rules and one of eternal torment for those who defy God, and so on. This is a very tempting ideology for religious people, but the fact remains that the presence of God does no more guarantee the existence of survival after death, heavenly and hellish dimensions, etc. than the absence of God renders such things nonexistent. Also, just because God exists does not mean that s/he created the universe any more than I created my houseplants.

> *I shall die, but that is all I shall do for death.*
> Edna St. Vincent Millay

Intelligent Design

Intelligent design is a concept created by religiously minded scientists such as biochemist Michael Behe (1952-) that attempts to bridge the perceived gap between religion and evolution (a gap created by conflation). The basic idea is that evolution occurs under God's watchful gaze. From time to time, God steps in to "design" a biological mechanism (such as an eye or a flagellum) of extraordinary beauty and complexity. Adherents of intelligent design claim that removing components from these mechanisms would render them inoperable. The claimed Catch-22 is that blind evolution cannot build an eye by stages, nor can it create a fully functional eye from scratch in one step.

> *The problem with intelligent design theory is not that it is false but that it is not falsifiable: Not being susceptible to contradicting evidence, it is not a testable hypothesis. Hence it is not a scientific but a creedal tenet—a matter of faith, unsuited to a public school's curriculum.*
> George F. Will

The first problem with the intelligent design concept is the false assumption that the theories of evolution and creation are mutually exclusive, that there is a gap that needs to be filled. As we discussed above, the validity of one theory does not invalidate the other in and of itself.

The second problem with intelligent design is the false assumption that a partially functional biological mechanism (such as a partially evolved eye) is less useful than none at all. Natural history abounds with partially formed mechanisms in different species. For example, eyes have evolved in completely separate evolutionary lines and run the gamut from light-sensitive cells that can only distinguish light from dark to so-called "pinhole" eyes in animals such as the nautilus that offer relatively clear vision, to human eyes and beyond.

> *Evolution is Darwin's great gift to theology.*
> John Haught

Even if intelligent design is correct, all that does is validate the idea that God exists in some form. It does not and cannot say anything whatsoever about the existence of a soul, survival after death, what the afterlife is like (if any), etc. There is, however, a way in which intelligent design adherents could be proven at least partially correct, and that is to turn traditional materialist models of reality on their head by postulating a *biocentric* (life-centered) universe. As we will see later, this is not nearly so farfetched an idea as it may seem.

Religion

I define *religion* as a set of beliefs about the origin and purpose of the universe as a creation of God and how God is to be worshipped in devotional and ritual observances. These beliefs often contain a set of purportedly divine laws about how humans must behave and the rewards and punishments that await believers and *infidels* (anyone who does not believe the religion's precepts) alike. The "Big 3" religions (Judaism, Christianity, and Islam) claim approximately 3.7 billion followers, or well over half of the world's total population. Almost 1.8 billion more subscribe to Eastern religions (Hinduism, Buddhism, and traditional Chinese practices). This represents almost 80% of the world's population of 6.9 billion people. Approximately 16% of the world's population is *secular* (not connected to religion; this includes atheists, agnostics, and spiritual people who do not belong to any particular religion). There are plenty of secular examples that have all the trappings of religion without believing in God. I will mention these in passing, but my focus is on God-centered religion.

It would appear that religion has a lot going for it, but that's not necessarily the case. The mere fact that 80% of the world's population is religious says nothing about the validity of religion. Appeals to belief, tradition, and consequences of belief are logical fallacies that do nothing to prove or disprove religion. Likewise, the existence of religious beliefs about where the universe came from, how it operates, why it exists, where people fit in, the nature of God, and God's laws does not say anything at all about whether any of this is true. If you're having a hard time with this, consider that a child's belief in Santa Claus, the Easter Bunny, or monsters under the bed doesn't mean that any of those things exist. To be fair, the existence of any of these things does not mean that people will believe in them. For example, few people believe in the Loch Ness monster, which does not by itself mean that Nessie is not real.

If a biblical figure has a tête-a-tête with God on the road to Damascus, well, okay, that's in the Bible, but if the conversation occurs today on the highway to Peoria, then cerebrospinal fluid samples are ordered.
Charles Tart

Faith backed by knowledge is much stronger than faith based on an emotionally driven gossamer hope, whether that faith be secular or religious.
Gerald L. Shroeder

The presence or lack of religion therefore does not imply anything whatsoever about the existence of God or what happens when we die. If that's not enough proof for you, consider that different religions offer different—and often contradictory—explanations. Any judgment about which religion is the correct one must therefore be purely subjective. Religious readers may have a hard time accepting this idea, because their beliefs seem as natural to them as breathing. To them, I say that your beliefs don't reflect any kind of objective reality for reasons I discussed in *The Enlightened Savage* and will discuss further here. Imagine being born in a tribal area where people have never heard of your current religion. It is entirely reasonable to predict that you would grow up with the same level of belief in that tribal religion that you carry for your current religion. Which one is correct? Who can say for sure? Of course, the reverse argument also holds, in that one cannot invalidate any particular religion just because it is one among many.

Mainstream science has itself become dangerously dogmatic and dismissive of evidence that does not accord with its philosophical beliefs.
Bernard Haisch

God

Many people conflate God with religion and evolution with the nonexistence of God. Those who believe in both God and evolution generally try to explain one in terms of the other, such as explaining evolution in a way that does not conflict with their beliefs about God—in other words, to explain evolution within their religious contexts. This is yet another example of conflation.

The fool says in his heart, 'there is no God.'
Psalms 14:1

God may exist but that does not mean that s/he created the universe, nor does it mean that God inspired or endorsed any particular religious beliefs. The existence of God does not mean that humans have a soul, that heaven or hell exist, or that there is any form of personal existence after death. Just because God can do something does not mean that she should or that she did, and it is hubris to think otherwise. Of course, the inverse is also true: The presence of any combination of the other conflated terms in this chapter does not do anything to prove God's existence or nonexistence.

Soul

A *soul* is generally considered to be an immaterial part of the human body that has the capacity to survive after death. Most

Good men spiritualize their bodies; bad men incarnate their souls.
Benjamin Whichcote

beliefs about the human soul involve *dualism* (the belief that both mind and matter are irreducible). This is generally considered a religious concept. Christians and Muslims believe that our beliefs, thoughts, and actions during this lifetime merit either eternal punishment or reward according to their particular rituals and laws. Eastern religions such as Hinduism and Buddhism believe that the soul keeps *reincarnating* (coming back to inhabit a succession of new bodies) until it reunites with God. As you can see, the typical view of the human soul conflates at least three separate concepts.

> *Those whose souls are filled with desires, intent only on heaven as their goal, their way offers only rebirth as the result of their actions. They offer only various rituals for obtaining their goals of pleasure and power.*
> Krsna

Another view sees the soul as an emergent phenomenon, a product of our brains. It is possible for the soul to exist as an irreducible and immaterial by-product of the human brain just as heat can exist as an emergent immaterial by-product of fire. If true, this would mean that the inverse is also true: just like heat ceases to radiate when a fire is extinguished, the soul would cease to exist when the brain ceases to function at death. Thus, nothing about the existence of a human soul implies anything about the existence of God, *immortality* (freedom from death), reincarnation, the validity of religious beliefs, or much of anything else.

Afterlife

> *He hoped and prayed that there wasn't an afterlife. Then he realized there was a contradiction involved here and merely hoped that there wasn't an afterlife.*
> Douglas Adams

Many people want to believe that some part of our individual identities survives the death of our physical selves, that death is simply a portal to a new plane of existence that could range from an eternity spent in some other world to coming back to Earth for another go. These beliefs usually come bundled with ideas about God, religion, and the idea that a person's post-death destination depends on her or his observance of certain laws and rites.

This conflation is entirely possible, in theory. It is also possible for an afterlife to exist without a God, and possibly even without something most of us would recognize as a soul.

Supernatural

> *One man's magic is another man's engineering. "supernatural" is a null word.*
> Robert Heinlein

My dictionary defines *supernatural* as, "of, pertaining to, or being above or beyond what is natural; unexplainable by natural law or phenomena; abnormal." The word "supernatural" is therefore a catchall term used to describe anything outside or beyond the laws of nature. But what are those laws? Talk to a materialist and s/he will tell you that the "supernatural" encompasses everything that does not have an explanation rooted in *material* (of or concerned with the physical as distinct from the intellectual or spiritual) terms, where anything intellectual or spiritual is caused by, and depends on, physical structures and processes. Such things as ESP, ghosts, angels, soul, and God are considered "supernatural" by materialists. Dualists believe in the reality of both physical and spiritual worlds, and thus have a much broader idea of what is natural, because they tend to be more open to believing in some or all of the above. Idealists posit that consciousness creates matter, that mind causes brain and not vice-versa, and therefore tend to have an even more open idea about what is, or could be, natural. Each worldview entails its own set of beliefs about what is natural and part of the order of things, and what is "supernatural" and thus the stuff of fairy tales and movies.

Clearly, the definition of just what is and is not "supernatural" is more than a little subjective because it relies on beliefs about how the universe operates that in turn rely on conflation more often than not. Of course, none of this says that there is no such thing as the "supernatural" (the faster-than-light spaceships of science fiction fame come to mind) or that many things classified as "supernatural" are not in fact part of the natural order, and thus not "supernatural" by definition.

At first cock-crow the ghosts must go back to their quiet graves below.
Theodosia Garrison

Paranormality

Paranormality is another very broad term that loosely describes purported human abilities beyond our normal five senses. Things such as *ESP* (Extra Sensory Perception, which includes such examples as telepathy, clairvoyance, and mediumship), NDEs, *OBEs* (Out of Body Experiences), are all examples of so-called paranormality. It should, of course, be immediately obvious that this is another subjective term that is often defined by one's beliefs in what constitutes normality versus the "supernatural". It should also go without saying that paranormality may exist in conjunction with none, some, or all of the things we are discussing, and that the existence of some or all of these things does not mean that the paranormal is real. Calling something paranormal is thus just as subjective as calling it "supernatural".

Far from being irrational, the paranormal is postulated by today's physics.
Brian Josephson

Mysticism

> *Man, if thy spirit rise above Time and Space, each moment canst thou be in eternity.*
> Angelus Silesius

Many thousands of people have had mystical experiences, myself included. *Mysticism* is defined as, "a belief in the existence of realities beyond perceptual or intellectual apprehension that are central to being and directly accessible by subjective experience." I know for a fact that mystical experiences exist because of my own experiences. I also know that it's almost impossible to say for sure whether these deeper levels of reality are in fact real or merely the by-products of a brain in a very unusual state. Of course, the key word in that last sentence is "almost." Quantum mechanics and cosmological ideas such as superstring and brane theory work because they postulate dimensions beyond the four we are familiar with. We also know that everything in the universe was in contact with everything else at the instant of the Big Bang. One could even say that the universe was only one thing before the explosion began the evolution of our familiar surroundings.

Monks, yogis, and others devote their lives to meditative and other practices designed to deepen their connections to the ultimate reality, which we can call God for now. Those who manage to achieve a complete union with this one reality or God are said to be "enlightened," and there appears to be some controversy over whether *enlightenment* (awakening to the ultimate truth) is a transitory or persistent state.

Mysticism can certainly exist without religion; however, all religions are ultimately based on mystical experiences that are codified into dogmas, doctrines, and institutions where those roots are lost more often than not. Still, there is nothing about mysticism that proves that God exists or that humans have souls, or that anything that makes any of us who we are survives physical death.

The Truth of Evolution

> *Darwin's discovery caused him great, personal grief and serves as an exemplar of a scientist following the evidence wherever it leads and whatever the consequences.*
> Victor Stenger

Accepting the theory of evolution as fact is my second core assumption. Everything changes, whether we want to admit it or not. You are literally not the same person you were yesterday; you have evolved. The universe is expanding, a clear sign that it was once much smaller—too small to support life. Even if we accept the idea of a god who created the universe to appear older than it really is, we must acknowledge that all creation myths speak of change over time.—of evolution.

Saying that evolution as a law or process of nature is a fact is not the same as saying that the theory of evolution is correct. Theories of evolution such as Darwinism and Lamarckism attempt to explain the process of evolution; they do not directly address whether evolution occurs.

Evolution is smarter than you are.
Leslie Orgel

Evolutionary theory is also silent on the idea of God. Evolutionary processes appear to be able to work without the need for any divine intervention. Evolution and natural selection are responses to external forces such as predation, climate, and pollution, and are not nearly as random as they may seem. Think of a balloon flying around a room propelled by escaping air. Its motion seems random, but we could predict its flight path with remarkable precision if we calculated all of the internal and external forces acting on the balloon. This oversimplified example helps explain how complex, highly organized life forms (such as yourself) can evolve from the most primitive organisms by demonstrating that evolution seems random until we peek under the hood. All of this can work—and seems to work—without any need for divine intervention; however, just because something does not seem necessary does not mean that it does not exist.

Classical evolutionary theory has one critical flaw: It fails to explain its own origins. We can construct some very plausible scenarios for how the very first life form survived and began evolving, and we can even know some of the possible mechanisms by which this evolution could have occurred. Primitive species stretching back almost to the dawn of life itself are still with us today, and we all bear the legacy of our earliest single ancestor cell. Where did that cell come from? Well... we don't know.

Penetrating so many secrets, we cease to believe in the unknowable. But there it sits, nevertheless, calmly licking its chops.
H. L. Mencken

Plenty of theories attempt to explain this yawning gap in our knowledge, but none has ever been shown to be a likely candidate. As I mentioned earlier, some scientists attempt to sidestep the problem by saying that life on Earth arrived from outer space. Interstellar space is indeed full of organic molecules, including amino acids, but all such speculation does is sweep the problem under the rug without really cleaning it up. Life on Earth may have come from somewhere else, but where did that life come from? Creationists pounce on this as pointing to the necessity of God but this is only speculation.

The dilemma seems clear: Evolution is a fact, but classical theories cannot account for how the process got started. So what if there are was a theory of evolution that could not only explain its own existence but also address the God question? Even better, what if this theory could also help shed light on the question of survival after death? We will see later in this book that there is just such a theory.

Any Room for God?

Modern science has done a lot to dispel our ancient ideas of God and nature. What primitive man saw as the raging of an angry sky god we now understand as electrical discharges that cause sonic booms and often start fires. Lightning storms have been demystified, as have virtually all natural phenomena. The effects of an earthquake or volcanic eruption may be no less devastating to us today than they've always been, but we now accept them as part of the price we pay for living on a geologically active planet, and have developed methods of coping with all but the worst that it can throw at us. The recent magnitude 8.8-8.9 earthquakes in Chile and Japan caused awesome destruction. What's even more awesome is how much destruction did not occur thanks to us being better prepared to anticipate and respond to these disasters than we have ever been. Science continues to push our understanding of everything from subatomic to cosmic scales to new heights, and seems to be crowding out God in the process.

In response, some creationists point to the inevitable gaps in our scientific knowledge and say, "See? We have found God!" The only problem is that the gaps in our knowledge are much smaller than they were even a few years ago, not to mention a few decades or centuries ago. If God resides in the gaps, then s/he is continually becoming ever more weak and impotent; hardly inspiring for the supposed source of all that is.

Other creationists take a more direct approach by going so far as to try to discredit the entire theory of evolution and any other aspect of science that they deem to be in conflict with their sacred texts. The Texas Board of Education is notorious

---

For thousands of years the greatest minds of every generation have worked diligently to prove the existence of god, and for thousands of years equally great minds have produced valid refutations of those proofs.
Michael Shermer

There is a very good saying that if triangles invented a god, they would make him three-sided.
Baron de Montesquieu

for its repeated attempts to water down scientific advances as "mere theories" while also presenting creationism as an equally valid alternative. There are two problems with this approach: One, the people who attempt to make such arguments usually display an almost breathtaking ignorance about even the most basic scientific principles. It honestly makes me wonder whether they are not trying to outright deceive people. Two, presenting the argument as one of either/or conflates two concepts that have no business being conflated.

On the other extreme, the undeniable success of science in explaining so much of the universe we live in is also a source of conflation, to the point that some see a universe entirely without purpose or beneficence. "The more we understand about the universe," says physicist Steven Weinberg (1933-), "the more it also seems pointless." I see it differently: I see the progression of human knowledge from the shadows of myths born in ignorance to the ever-growing light of knowledge. Each new discovery reveals a universe that is more complex, more intricate, more wondrous, and more marvelous than we previously imagined. Philosopher William Paley (1743-1805) coined the famous *watchmaker's argument* when he said that anyone finding a watch on the moor would assume that the watch had a designer and builder. The universe is infinitely more complex than a watch, therefore it too must have a designer and builder. Paley's argument has been roundly discredited for reasons we will explore later; still, the more closely we look at this watch and the more we tease apart its innermost workings, the more fantastic those workings become. The watch evolves before our eyes.

Avoiding needless conflation is my first core assumption in this book. Accepting evolution as fact is the second core assumption. As for the question of whether modern science (which I ardently subscribe to) leaves room for God and what that means for us in this life and anything that comes next, keep reading.

> *One of the greatest favors bestowed on the soul transiently in this life is to enable it to see so distinctly and to see so distinctly and to feel so profoundly that it cannot comprehend god at all.*
> John of the Cross

Mapping Our Journey

It is impossible to construct a comprehensive theory of reality without examining a wide assortment of evidence gathered by many different scientific, philosophical, spiritual and religious

> *Even with the best of maps and instruments, we can never fully chart our journeys.*
> Gail Pool

The road goes ever on and on, down from the door where it began.
J.R.R. Tolkien

Every day you may make progress. Every step may be fruitful. Yet there will stretch out before you an ever-lengthening, ever-ascending, ever-improving path. You know you will never get to the end of the journey. But this, so far from discouraging, only adds to the joy and glory of the climb.
Winston Churchill

sources. Presenting the results of this examination to a reader is no easy task. First, there is a lot of evidence, as the sheer length of this book should tell you. Second, it is extremely difficult to discuss any given concept in isolation. For example, one cannot talk about God without talking about evolution and physics. It is likewise impossible to talk about evolution without talking about intelligent design, the idea that evolution has been guided by God. I have done my best to organize the material by subject and minimize (but not eliminate) both crossover (discussing one topic in another section) and repetition while still presenting credible discussions on each of the many topics in this book, which I have structured into 40 chapters in eight parts as follows:

- **Part One - Getting Started (Chapters 1 and 2):** This book is all about finding out what happens when we die. It is therefore fitting to begin our exploration by discussing the dying process in depth and then framing the discussions to come by defining and separating concepts and exposing the core assumption behind this book.

- **Part Two - Religion (Chapters 3-12):** My quest for answers came when I heard a voice telling me to read Genesis 1. This prompted me to explore creation myths from all around the world and delve into the history of religion from its earliest beginnings. Next, we will learn some of the fundamentals about the world's major religions and discover what Jesus Christ's life may really have been like once fact has been separated from myth. The glaring differences between Christ's actual life (whether it followed the story in this myth or not) and *dogma* (authoritative principles, beliefs, or statement of ideas or opinion, especially those considered to be absolutely true) are a good foundation for looking at the history of monotheism and learning that belief in a single god is not our natural state. We will then examine the impact of religion on society and the arguments for and against religion.

- **Part Three - God (Chapters 13 and 14):** Can the God of religion exist? If not, can any form of deity exist in a manner that is consistent with what we know of the laws of nature? What would such a being look like? Answering these questions is an important milestone on our journey as we then start looking for evidence to support a possible hypothesis of what God may be and how we as individuals fit into the larger scheme of things.

- **Part Four - Paranormality (Chapters 15-17):** Campfire stories and horror movies aside, there does seem to be a significant amount of credible evidence to support the idea that paranormal events can and do occur, and that the human mind is not nearly as confined to the body as traditional science might want to believe. I personally know for a fact that paranormality is real based on the experiences I shared about my friend Sarah in the Introduction; however, I have no empirical evidence to back up my assertion and must therefore look to other documented cases and studies. This section defines various types of paranormality, shares a very few samples of paranormal events, and examines the evidence for and against the idea that there is more to reality than we normally see.

- **Part Five - Evolution (Chapters 18-20):** The great debate of our time is that of evolution versus creation, whether we are simply accidents or the result of some divine plan. This section attempts to provide a workable definition of what life is versus nonlife and then reveals both some interesting details about how life and evolution might have gotten started and some arguments for and against the idea of "intelligent design."

- **Part Six - The Physics (Chapters 21-28):** These chapters recount the history of physics from ancient Greek philosophers and scientists to Sir Isaac Newton, quantum mechanics, Einstein's relativity, and modern theories such as superstrings and holography. We conclude this section with a look at a novel theory of time.

- **Part Seven - Mind and Matter (Chapters 29-33):** The problem of consciousness is one of the most persistent challenges facing scientists of many disciplines. Here we learn that each of the major interpretations of quantum mechanics involves consciousness at some level and then discuss the brain, the mind, and some ideas about how quantum consciousness might work and its implications for all of our lives and eventual deaths.

The real voyage of discovery consists not in seeking new landscapes but in having new eyes.
Marcel Proust

One glance at a book and you hear the voice of another person, perhaps someone dead for 1,000 years.
Carl Sagan

- **Part Eight: - The Thesis (Chapters 34-40):** The final chapters of this book summarize the evidence collected during our journey and presents a new model of reality that encompasses and embraces divergent viewpoints in a logical, self-contained, self-explanatory model and demonstrates that each of us is both far less than we seem and far more than we ever imagined possible.

One's destination is never a place but rather a new way of looking at things.
Henry Miller

For each topic in this book, I have done my best to provide both a history and opposing viewpoints to help you understand how and why I am building my thesis in the way that I am, and to help spur your own thinking and explorations. This makes for a much longer book than most; however, you have purchased this book and read this far because you have some deep questions about the universe we live in and where you fit in. You deserve nothing less than the most complete answer I know how to give you, which I hope will encourage you to continue your own explorations.

A Note for People of Faith

Faith is a cop-out. If the only way you can accept an assertion is by faith, then you are conceding that it can't be taken on its own merits.
Dan Barker

There is a very good chance that you consider yourself Jewish, Christian, Islamic, Hindu, Buddhist, or otherwise affiliated with some organized religion. Certain parts of this book will be extremely difficult for you to read, such as the history of your own beliefs, a more historically plausible biography of Jesus, and the tremendous impact your faith has had on the entire world. I must assume that you picked up this book and have read this far because you have questions and/or needs that faith alone is not able to provide you. You may be seeking to validate your faith through this book. Either way, you will find your faith challenged and yourself personally impugned for what has been done in the name of your religion. This will be very tough reading, and you will probably be tempted to put this book down. I submit to you that whatever urge you may feel to stop reading at that point will be exactly the reason why you should keep reading, because you will discover that your faith is a model of reality that is perfectly compatible with the model of reality that I am building in this book. To the extent that you can see your faith and religious figures as metaphors and interpretations, you will find your faith validated on levels you may not have imagined possible.

Alternatively, you may consider yourself a staunch atheist. I have listened to many devout religious people and many equally devout atheists. These two groups sound amazingly alike both in their stridency and in the logical fallacies they employ. Denying evidence that something exists beyond the material world we are all familiar with because it does not fit your worldview is just as much an act of blind faith as accepting religious teachings without question. Dismissing certain phenomena as "superstition" and evidence as "fake" or "nonexistent" does nothing to support your position and everything to reveal you as a zealot, albeit a Godless zealot.

> *Treat the other man's faith gently; it is all he has to believe with. His mind was created for his own thoughts, not yours or mine.*
> Henry S. Haskins

Only someone with a truly open mind can examine all of the evidence to decide for her or himself where that evidence leads. Only someone who looks beyond the concept of right and wrong to seek a framework within which all of these disparate ideas can coexist in relative equality can claim to have opened the doors to true knowledge. I am not suggesting that specializing in a single discipline is wrong, or that all viewpoints can coexist no matter what; I am saying that prejudice has no place in our search for the truth.

On the Shoulders of Giants

This book represents my study of a vast body of research conducted by countless people over thousands of years. None of the concepts I use to create my theory are new. I have drawn my evidence directly from the work of many people who have devoted their lives to research and at times paraphrased them for lack of a better way to present their findings. It is my hope and belief that the final thesis of this book has room for all of the work done both by the many sources on which I have relied and by the many other people involved in the search for truth whose work is not included. It is also my hope and belief that the way in which I combine all of this work forms a logical and self-explanatory framework of reality that can further our quest for the biggest answers to the biggest questions.

> *You learn more quickly under the guidance of experienced teachers. You waste a lot of time going down blind alleys if you have no one to lead you.*
> W. Somerset Maugham

This book is not intended as a scientific treatise and thus does not contain footnotes; however, the Additional Resources section at the end lists my sources, who rightfully retain full credit for their work.

PART TWO

Religion

The Divine Savage
Revealing the Miracle of Being

Chapter 3

In the Beginning

> *All great deeds and all great thoughts have a ridiculous beginning.*
> Albert Camus

Let's go back to the beginning of our understanding about the beginning of the universe to see both how far we have come since those primal *myths* (fictions or half-truths, especially those forming part of an ideology), and whether any of these ancient tales of our debut have any modern scientific merit to them. If not, then we can dismiss these myths as either early attempts at understanding that have become obsolete or nothing more than fairy tales told for moral and/or religious purposes. On the other hand, discovering that creation myths have scientific validity would force us to reconsider both the state of human knowledge and where that knowledge could have come from. Such an exploration would ultimately force us to rethink what it means to be human and even our place in the universe.

> *A creature that thinks knows next to nothing of the substrate allowing its thinking to happen, but nonetheless it knows all about its symbolic interpretation of the world, and knows very intimately something it calls 'I.'*
> Douglas Hofstadter

Myth as Science

It is tempting to think of myth as simply old wives tales of dubious value. Some specific myths may fall into this category, but myth is where human attempts to understand the universe began. Myth is nothing less than the earliest form of scientific inquiry. Mythic tales may not represent reality as we currently understand it, but that is no reason to dismiss or disrespect our ancestors' attempts to explain what was then otherwise unexplainable in the context of their own cultures and levels of understanding. Modern astronomy long ago rendered Ptolemaic astronomy—with its epicycles, equants, and deferents—hopelessly obsolete, but that does not lessen the validity of that ancient system as an honest attempt to explain the workings of our solar system. Myth also has the power to enshrine values and bond societies in a common cause. All nations, even contemporary ones, are founded on myth. Patriotism, systems of government, mores, laws, religion, and more are at least partially based on myth. In many ways, myth is the glue that holds society together.

As I said, myth is the original human science, a universal language that attempts to describe the many levels of reality that exist beyond our normal five senses. All human societies past and present include myth because all humans have pondered the deepest questions of life. The answers provided by myths form a pattern of belief that provide meaning to life in the form of a foundation for shared experiences that contains codes of conduct.

Most myths describe events that occurred before written history existed; however, plenty of myths exist to describe more recent events, albeit in the context of more ancient times. Consider the mythology of the founding of the United States for example. The Founding Fathers are revered both for their actions and for how those actions fit into mythological interpretations of events said to have happened a very long time ago. Even our individual lives are largely based on myths; our inner beliefs that color our perceptions (such as my belief that fish doesn't taste good) say nothing at all about the world "out there" and everything about how we see, experience, and interact with the world. Myth therefore becomes our own way of explaining and relating to the world, our own personal science.

It is one thing to say that thunderstorms are caused by an angry god and to extend that concept to mundane things like the rhythms of night and day, seasons, rains, etc. But answering "what" or "who" only tackles part of the problem; the ever-curious human mind naturally wants or even needs to know "how" and "why." In other words, myth must not only explain the existence of a thunder god but also how and why that god makes the skies come alive with bolts of electricity.

Myth embodies the nearest approach to absolute truth that can be expressed in words.
Ananda Coomaraswamy

Myth is a symbolic story which demonstrates the inner meaning of the universe and of human life.
Alan Watts

We need to know what makes this god tick, where she or he came from, how s/he interacts with and relates to the other gods, and—ultimately—where all these gods come from, and how and why we came to be a part of it all.

Any great endeavor we undertake must literally start wherever were we happen to be at the time. Primitive people, lacking a modern scientific understanding of how the world and universe work, had to start with (presumably) a detailed knowledge of how their surroundings worked (which plants and animals were edible, how to hunt, where and how to take shelter, etc.) and observations gleaned from observing and participating in human society. In short, they had to *anthropomorphize*, or project human qualities onto what they were trying to explain. Thus did human features, traits, and personalities become projected onto the divine.

Only by stepping outside of logic, so the theory goes, can one make the leap to enlightenment.
Douglas Hofstadter

We made God (or rather our packaged, comprehensible idea of God) in our image. Our gods look and act like us and demand all of the tributes due to the highest ranking people among us, and then some—the birthplace of *rites* (formal or ceremonial acts or procedures), some of which would eventually become dogma. Like any alpha personality, such gods must possess human-like strengths and weaknesses, the latter often manifesting themselves in behavior that no modern person would find acceptable. Were the ancient Greek, Egyptian, and other gods human, many of them would be jailed for life without parole, including the Jewish, Christian, and Islamic gods. Secular regimes that claim to disavow religion are no different; these leaders simply stand in for God and demand all of the same loyalties. Hitler, Stalin, Pol Pot, Allende, Mugabe, Kim Il Sung, Kim Jong Il, and others are all alike in one crucial respect: Their regimes are all theocracies with earthly gods.

The human quest for meaning cannot be satisfied without concurrently pursuing that for intelligibility. Our intellectual, moral, and spiritual capacities are not nearly separable into discrete compartments.
Arthur Peacocke

Why gods? Why religion? Why did the earliest explanations for natural phenomena involve so-called "supernatural" beings? We will return to this discussion in Chapter 4. Meanwhile, the important points to remember are:

- Myth is the earliest form of human scientific inquiry.
- Ancient myths ascribed natural phenomena to one or more gods and then anthropomorphized those gods to explain how and why they acted the ways they did.

- These myths gave rise to a sense of tribal and natural identity and a corresponding set of morals and taboos, the violation of which could be punishable by death or by something far worse: banishment.

Technology and methods of inquiry continue to evolve. Each step gives us deeper insights into how the universe works and where we fit into the picture. These tremendous advances give us the illusion that we have abandoned the shadows of myth for the shining light of knowledge, but that too is a myth. Some people bemoan what they perceive as a decline in ethics and blame it on a growing disrespect for myth. The truth is that myth is just as alive and well as it has always been, and my prognosis is that state will continue for as long we have beliefs standing in the way of our perceptions—in short, for as long as we remain human.

Scientific theory has some fascinating parallels with the myths, including the Genesis account.
J.F. Bierlin

Let's see what myths from around the world have to say about the origins of the universe and humankind. This exploration will prove crucial throughout the rest of this book.

Modern Genesis

In the introduction, I recounted the tale of the voice in my head that told me to read Genesis 1 repeatedly until I realized that it was telling the scientifically accepted history of the universe perfectly. This made me wonder whether Genesis 1 is unique in its verisimilitude. If so, that would either mean that Judaism, Christianity, and Islam had more going for them than most creation myths, or that scientists over the centuries have unconsciously gerrymandered their investigations to make them fit the myths. If not, then all bets are off, for that could only mean that ancient humans possessed either a much more advanced state of knowledge than we normally credit them with, or that there may well be a deeper level of reality that our ancestors have been able to connect to throughout the ages.

The account of the beginning (Genesis 1) is natural science but so profound that it is cloaked in parables.
Maimonides

If we could demonstrate that such a deeper level of reality does in fact exist, then that could go a long way toward opening up possible avenues by which we might survive physical death. It could also go a long way toward bridging the gap most people perceive between creationism and evolutionism. We might begin to see that the difference between creation myths and modern cosmology may have far more to do with semantics than with the actual facts being described. I am not for a moment suggesting that science has erred in its unraveling of the history of the universe or of life on Earth. There is simply far too much fossil evidence and far too many reliable dating methods for anyone to reasonably conclude that the universe is only 10,000 years old. The argument that God created the universe to look *as if* it were much older simply doesn't hold water... or does it? We will explore this question further in Chapter 38. Meanwhile, Genesis 1 is the most widely known creation myth in the world. How well does it stand up to modern scientific knowledge of cosmology and evolution?

> *In the beginning there was nothing. God said, 'Let there be light!' And there was light. There was still nothing, but you could see it a whole lot better.*
> Ellen DeGeneres

As promised, here is my (much refined) Genesis essay:

Genesis 1:1-2: In the beginning, God created the heavens and the Earth. Now the Earth was formless and empty, darkness was over the surface of the deep, and the spirit of God was hovering over the waters. And the earth was without form, and void; and darkness was upon the face of the deep. And the Spirit of God moved upon the face of the waters.

Theoretical and observational evidence corroborate the idea that the universe began as a zero-dimensional *singularity* (a point in space-time at which gravitational forces cause matter to have extreme density and volume, and space-time to become highly distorted), of incredible mass and energy. Black holes distort space and time to the point where it's meaningless to speak of form or time once one enters the hole, and the universe before the Big Bang was the blackest hole imaginable. The Earth was indeed formless, the universe one unimaginable void. As we will see in Chapter 24, space and time are simply different aspects of each other; without space, there is no time. It is therefore meaningless to ask what happened before the Big Bang in the context of this universe. As for moving across water, no particle is ever entirely at rest; the point of lowest possible energy is not zero. This means that the singularity was subject to quantum fluctuations, and that one of those fluctuations was strong enough to get things going.

> *The world is round and the place which may seem like the end may also be only the beginning.*
> Ivy Baker Priest

Genesis 1:3 And God said, Let there be light: and there was light.

It is safe to assume that the most brilliant light imaginable was an immediate by-product of the Big Bang, along with many other forms of radiation across the entire electromagnetic spectrum. This radiation has been detected and mapped.

Genesis 1:4-5: And God saw the light, that it was good. And God divided the light from the darkness. And God called the light Day, and the darkness he called Night. And the evening and the morning were the first day.

The universe expanded and cooled to the point where matter precipitated out of the initial mass of energy. The first stars were formed as matter clumped together under its own mass, separating the universe into inky blackness punctuated with points of brilliant light—a literal division of light from darkness. Calling this process the "first day" is appropriate because space was an opaque glowing ball for approximately the first 377,000 years until the universe cooled down enough for space itself to become dark and transparent.

Genesis 1:6-8: And God said, Let there be a firmament in the midst of the waters, and let it divide the waters from the waters. And God made the firmament, and divided the waters which were under the firmament from the waters which were above the firmament: and it was so. And God called the firmament heaven. And the evening and the morning were the second day.

Today's superstring and brane theories (see Chapter 25) postulate 10 or 11 dimensions, the four we are familiar with (length, width, height, time) and six more that are cut off from direct perception. It is appropriate to assume that the term "water" could refer to these higher dimensions for reasons we will cover later in this chapter. This passage is therefore telling us that God created a division in space-time. We cannot directly observe or measure the fifth or higher dimensions in the same way that Edwin Abbott's (1838-1926) two-dimensional creatures living in Flatland cannot directly observe or measure anything beyond length and width. The Flatlander's "firmament" thus contains three dimensions (length, width, time). Many religions describe heaven as lying beyond space and time, an idea that is perfectly compatible with contemporary science.

Genesis 1:9-10: And God said, Let the waters under the heaven be gathered together unto one place, and let the dry land appear: and it was so. And God called the dry land Earth; and the gathering together of the waters he called Seas: and God saw that it was good.

I will love the light for it shows me the way, yet I will endure the darkness for it shows me the stars.
Og Mandino

If there is, in fact, a Heaven and a Hell, all we know for sure is that Hell will be a viciously overcrowded version of Phoenix.
Hunter S. Thompson

> *Once you've learned to study in a bathing suit on the grass with muscled men throwing frisbees over your head, you can accomplish almost anything.*
> Susan Rice

Loose drifting matter clumped together to form stars, planets, etc. As the matter that would eventually form Earth coalesced, our home planet became increasingly differentiated from the vast sea of space. This passage becomes even more relevant when we understand the mythical meaning of water, which we will look at later in this chapter.

Genesis 1:11-13: And God said, Let the earth bring forth grass, the herb yielding seed, and the fruit tree yielding fruit after his kind, whose seed is in itself, upon the earth: and it was so. And the earth brought forth grass, and herb yielding seed after his kind, and the tree yielding fruit, whose seed was in itself, after his kind: and God saw that it was good. And the evening and the morning were the third day.

Most evolutionists agree that plants were the first species to evolve on Earth. The initial parent species eventually split into multiple species that to this day continue to attempt to reproduce new members of the same species... to reproduce themselves in kind. This can also be seen as a metaphor for different universal structures and generations of stars that created the elements necessary for life.

Genesis 1:14-15: And God said, Let there be lights in the firmament of the heaven to divide the day from the night; and let them be for signs, and for seasons, and for days, and years. And let them be for lights in the firmament of the heaven to give light upon the earth and it was so.

Stars were created, and always are being created, and their cycles across our sky have always heralded our seasons and years. This is possible because the Earth moves in several regular and predictable ways: a nearly perfect circular orbit around the Sun, regular revolutions on its axis to form day and night, a wobble that alternately aligns the northern and southern hemispheres with the Sun's rays to create our seasons, and a slight amount of *precession* (the slow, conical motion of the earth's axis of rotation) that repeats itself every 26,000 years—a fact that will become extremely important in Chapter 24.

> *We may go to the Moon, but that's not very far. The greatest distance we have to cover still lies within us.*
> Charles de Gaulle

Genesis 1:16-19: And God made two great lights; the greater light to rule the day, and the lesser light to rule the night; he made the stars also. And God set them in the firmament of the heaven to give light upon the earth. And to rule over the day and over the night, and to divide the light from the darkness: and God saw that it was good. And the evening and the morning were the fourth day.

The Sun is above us in daylight, the Moon at night, our greater and lesser lights. Stars existed before the Earth formed; the heavy elements forged in their nuclear furnaces were essential for forming the planet and allowing life to evolve. Previous verses already alluded to this. Also, stars are continually being born, and they do cast light upon the Earth. It is also important to note that our "lesser light" did not exist when the Earth formed; a huge meteor colliding with the Earth removed a chunk of material that eventually became the Moon, another celestial body whose nearly perfect circular orbit gives us regular tides and provides an almost exact way to mark passing seasons and years with almost 97% accuracy. This can also be seen as a metaphor for the formation of our Sun, which is a third-generation star.

Genesis 1:20-21: And God said, Let the waters bring forth abundantly the moving creature that hath life, and fowl that may fly above the earth in the open firmament of heaven. And God created great whales, and every living creature that moveth, which the waters brought forth abundantly, after their kind, and every winged fowl after his kind: and God saw that it was good.

Life evolved in the seas and eventually moved onto land. It is safe to speculate that the conquest of the skies soon began, possibly long before dinosaurs grew feathers and wings. Mammals evolved from reptiles, and some of them eventually returned to the as whales. As always, each species reproduces after its own kind; pigeons give rise to pigeons, sperm whales to sperm whales, and so forth.

Genesis 1:22-25: And God blessed them, saying, Be fruitful, and multiply, and fill the waters in the seas, and let fowl multiply in the earth. And the evening and the morning were the fifth day. And God said, Let the earth bring forth the living creature after his kind, cattle, and creeping thing, and beast of the earth after his kind: and it was so. And God made the beast of the earth after his kind, and cattle after their kind, and every thing that creepeth upon the earth after his kind: and God saw that it was good.

Species grew and multiplied, and every corner of Earth became host to life. New research has even discovered fish living miles beneath the surface of the seas and microorganisms living miles beneath the surface of the Earth. Life on Earth is indeed fruitful and multiplying, which is life's top priority, as

---

Last year I went fishing with Salvador Dali. He was using a dotted line. He caught every other fish.
Steven Wright

It's a lot like nature. You only have as many animals as the ecosystem can support and you only have as many friends as you can tolerate the bitching of.
Randy K. Milholland

we saw in *The Natural Savage*. Cattle, modern species that are both edible and domesticable, evolved fairly recently, as did horses and other species we commonly associate with our own history.

Genesis 1:26-28: And God said, Let us make man in our image, after our likeness: and let them have dominion over the fish of the sea, and over the fowl of the air, and over the cattle, and over all the earth, and over every creeping thing that creepeth upon the earth. So God created man in his own image, in the image of God he created him; male and female he created them. And God blessed them, and God said unto them, Be fruitful, and multiply, and replenish the earth, and subdue it: and have dominion over the fish of the sea, and over the fowl of the air, and over every living thing that moveth upon the earth.

All of the current families of plants and animals existed before humans appeared on the scene, having evolved from other mammals. Here again, Genesis gets the story right as far as evolutionary theory is concerned. As for dominion, no other species has spread across the globe and enjoyed such unparalleled success: Humans have penetrated the depths of the oceans, the Earth's crust, subatomic structures, and the furthest reaches of the universe. We have forever altered huge swaths of the environment, causing mass extinctions, deforestation, pollution, etc. The massive oil spill in the Gulf of Mexico is just one sensational example of how human actions have dominated the natural environment, much to the latter's detriment.

Rewrite Genesis in contemporary language, and you would have a thoroughly understandable and demonstrably correct account of the history of the universe, Earth, and evolution. It is true that Genesis uses metaphors and archaic language to tell its story, but this does not lessen its value or accuracy.

So why does the Bible use words like "deep," "water," and "day" to describe events taking billions of years to unfold? The answer is painfully obvious: Primitive humans had no knowledge of evolution, astronomy, relativity, biology, ten-dimensional space-time, etc. and would therefore have been unable to grasp such words, just as a two year old is unable to grasp the workings of a computer on a firmware level. Thus, the Bible story had to be told in terms that its audience could understand. Unfortunately, we lost sight of this simple expla-

---

I know that there are people who do not love their fellow man, and I hate people like that!
Tom Lehrer

I've often thought the Bible should have a disclaimer in the front saying this is fiction.
Ian McKellen

nation, even as we read simple bedtime stories to our children. Also, words such as "water" have special meaning in the context of myth.

On one hand, Genesis tells a decidedly materialist version of events. On the other hand, it portrays creation as being caused by God's word or *Logos*. The Old Testament is not alone in this: John of Patmos wrote that, "In the beginning was the word and the word was God." (John 1:1.) This seems to be the foundation of the dualism that lies at the core of the Judeo-Christian-Islamic faiths. One would expect the same story told from a materialist or idealist viewpoint to have a decidedly different slant. It is important to understand that the slant neither validates nor invalidates the story's intrinsic validity.

Life appears simply as a disease of matter.
Gerald Feinberg

Genesis Around the World

Genesis tells the scientifically accepted history of the universe not closely but perfectly, opening the door to the Judeo-Christian-Islamic family of religions, whose values shaped Western societies and both religious and secular values. Do any other creation myths contain the same level of accuracy at any level? That is the million-dollar question. It is not necessary for a myth to tell the whole story perfectly to be considered scientifically valid, provided that its central claims or assumptions meet the test of according with modern knowledge.

A joyful life is an individual creation that cannot be copied from a recipe.
Mihaly Csikszentmihalyi

It may seem that I am cutting other creation myths too much slack by not insisting that they do a Genesis-like job of telling the entire story perfectly. This is not my intent; on the contrary, my goal is to level the playing field. Genesis was born of earlier myths that originated in Egypt, Sumer, and other advanced societies that eventually helped form the foundation of modern Western society, whose views and methods have come to dominate science. Few of the world's other cultures enjoyed the same level of technological advancement—a fact that would cause their downfall more often than not. It is therefore impractical to insist that their myths contain the same level of detail in order to deem them valid. This is especially true because a growing body of evidence suggests that Egypt's astronomical knowledge was far more advanced far earlier than most Egyptologists are willing to admit.

Many of the myths you are about to read go beyond the standard materialist account of the history of the universe and are more compatible with idealist and/or biocentric theories. For reasons we will continue exploring throughout this book, there is every reason to think that these more expansive theories are much closer to the truth than mere materialism. Just as superstring and brane theories include and contain quantum physics that in turn includes and contains classical Newtonian physics, so do idealist and biocentric theories include and contain the classic materialism story about the Big Bang and ensuing history of the universe.

With all this in mind, how do the world's creation myths stack up? Hold on to your hat...

Egypt

> *Would I had phrases that are not known, utterances that are strange, in new language that has not been used, free from repetition, not an utterance which has grown stale, which men of old have spoken.*
> Ancient Egyptian inscription

The very short version of this myth says that Ra emerged from the watery abyss and that all things came into existence from the words of his mouth. This corroborates the John 1:1 account of, "In the beginning was the Word." Both myths have the universe being willed into existence—a very idealist (and possibly even biocentric) viewpoint that absolutely meets the requirement of scientific accuracy.

Sumer

This myth has the god Anu creating heaven, which created the Earth, which created the rivers, and so forth. "Heaven" in the form of the higher dimensions and the universe itself, did come into being and did in turn create the Earth, on which rivers eventually began to flow. This too is scientifically accurate.

Iran/Middle East

> *The Earth made us no promises.*
> Alain

The Zoroastrian myth speaks of eternal, limitless time as the ultimate creator. In the beginning, there was no one to call time "creator," for nothing had been created. Time created fire and water from which emerged the good god Ohrmazd, at which point time became the creator; everything Ohrmazd does with regard to creation is done with the aid of time. The sky was made first, then Earth, and then water, plants, and animals. Ohrmazd's physical form is our father, his spiritual form our mother. The evil god Ahriman declared war on creation, and people began saying that Ahriman created them. This was the first sin, which was punished by making people work and reproduce on their own—a story very analogous to (and pre-dating) the Eden myth of Genesis, both of which have strong parallels to the Buddhist ideas of the illusion of *maya*, an illusion that deludes us into thinking that we are separate, finite individuals. This myth absolutely meets the accuracy test.

The myth of Mani says that two natures, good and evil, existed before creation. Matter is evil—a clear predecessor to the Gnostic beliefs we will encounter in Chapter 5. The importance of saying that matter is evil and knowledge good ties into idealism, and the idea of matter and separation from the Godhead as an illusion, an idea that has strong scientific merit.

Greece

In the beginning was chaos and darkness, a great sea where all mixed without form. Eurynome rose from this sea and danced on its surface. Ophion saw the dance and assumed the form of a bird that laid the egg from which came the universe. A snake coiled its tail around the egg until it cracked, giving birth to all species. Eurynome and Ophion could be thought of as the quantum fluctuations that gave rise to the singularity that in turn became the universe and everything in it.

The Gaia and Uranus myth speaks of Gaia emerging from the chaos and having her son Uranus, who ascended to the heavens and rained down on the Earth. This fertilized the Earth and brought dormant seeds to life. Interstellar space contains amino acids and other organic molecules, and it is not far-fetched to think that life on Earth may have begun elsewhere—a theory put forth by biochemist Christian de Duve (1917-) in his book *Vital Dust*.

> *The Scripture vouches Solomon for the wisest of men; and his proverbs prove him so, The seven wise men of Greece, so famous for their wisdom all the world over, acquired all that fame each of them by a single sentence, consisting of two or three words.*
> Unknown

Rome

The Roman sage Ovid spoke of a shapeless inert bulk where atoms jostled until God or nature separated heaven from earth, water from land, etc. and molded Earth into a globe with plants and animals. Man was born out of the need for a linear being, a clear reference to the human tendency to think

> *The Romans would never have found time to conquer the world if they had been obliged first to learn Latin..*
> Heinrich Heine

India

> *Keep five yards from a carriage, ten yards from a horse, and a hundred yards from an elephant; but the distance one should keep from a wicked man cannot be measured.*
> Indian proverb

In Indian mythology, Brahman is the creator, Vishnu the preserver, and Shiva the destroyer; but all are part of Brahman. In the beginning, there was neither existence nor nonexistence. (It is, after all, meaningless to ask what came before the Big Bang.) Creation came from an egg (singularity) that was created by churning waters (quantum forces). Heat was generated from these churning waters, and the universe came into existence (in the Big Bang). Initially there was only Brahman, the Great Self. Brahman eventually yearned for (desired) company. This desire was the first seed of mind (a very idealist/Buddhist concept). Brahman temporarily split into male and female halves who recognized each other and made love, creating man. The woman then played coy and changed into all other species. Thus is Brahman found in every little thing, for all comes from Brahman. Brahman has created and destroyed the world many times; only s/he knows how many times. The cycle consists of goodness, energy, a mix of the two, and eventual darkness, over and over again.

The scientific accuracy of the account of creation is beyond dispute. There is some ambiguity around the idea that humans came first; however, there is also a scientific idea that the history of the universe is only the way it is because we believe it is—a tangled hierarchy that lies at the heart of biocentric theories. The idea in this myth that all things are connected and part of the One is consistent with both Big Bang theory (since everything in the universe was originally connected to everything else) and idealism (in the sense that we are each slivers of one consciousness, the Godhead). The *Rig-Veda X* corroborates this idea by saying that man is the whole universe, what was and will be, the lord of immortality.

> *And do you think that unto such as you, a maggot-minded, starved, fanatic crew, God gave a secret, and denied it me?*
> The Rubiyat of Omar Khayyam

Rig-Veda C explores these concepts in detail, saying, "When neither being nor not being was, what did it encompass, where, and under whose protection? Neither death nor immortality was there then; One breathed windless by its own energy. Nothing else existed, all was unmanifested water. Whatever was, the One coming into being was generated by

of time as a unidirectional river. The scientific accuracy of this theory is self-evident.

heat. In the beginning this One evolved and became desire, the first seed of mind. Who knows? Who can declare it?"

The *Chandogya Upanishad* says that the world was merely being in the beginning and turned into an egg. This egg was one being. Some say it was non-being that came into being, but how can this be so? This is a clear reference to the theory of idealism, that the Ground of being is pure awareness.

The *Kena Upanishad* asks, "Who sends mind to wander far away, drives life to the start of its journey, and is the spirit behind eye and ear? It is the ear of the ear, the eye of the eye, the word of the words, the mind of mind, the life of life. Those who follow wisdom pass beyond and become immortal. There eyes and ears don't go, nor words nor mind. We don't know and can't understand that the immortal is above known and unknown, but this is what ancient sages have been telling us. What can't be said with words but that whereby words are spoken, know that alone to be Brahman, the spirit, and not what people here adore." This is another reference to idealism.

The Jinasena (Jain) myth states that, "Some fools say a creator made the world. If God created the world, where was s/he before creation? No single being could make the world, for how can the immaterial make the material? How could God make the world without material? If we say that A made B, then we fall into the trap of infinite regression. It is another fallacy to simply postulate that the raw material was already there. Also, how can anyone believe the silliness that God created the world by will alone? If God is perfect and complete, how could will arise? On the other hand, an imperfect/incomplete God could not create the universe any more than could a potter. The Earth (universe) simply is uncreated as time itself, eternal, based on principles of life and the rest, and endures under the compulsion of its own nature, divided into hell, earth, and heaven."

At face value, the Jain myth seems to contradict the other myths, but this is not actually the case. Forget the idea that the physical universe or anything/anyone in it is real and start to wrap your mind around the idea of reality being, as physicist Albert Einstein (1879-1955) put it, an illusion, albeit a very persistent one. Thinking in this way reconciles the differences

---

As a man, when in the embrace of a beloved wife, knows nothing within or without, so this person, when in the embrace of the intelligent soul, knows nothing within or without.
Upanishads

Things derive their being and nature by mutual dependence and are nothing in themselves.
Nagarjuna

and allows one to see that all of the Indian myths described above are perfectly compatible with idealist explanations of the universe. In other words, this myth meets the criterion of scientific accuracy.

China

The "Creation from Chaos" myth speaks of pure light coming from chaos and building the sky. Heavy dim matter moved and formed the Earth, and sky and Earth brought forth creations (species). Likewise, the Huai-Nan Tzu myth says that all was amorphous and vague before heaven and Earth formed. The Great Beginning produced emptiness, which produced the universe. The bright clear light became heaven, the dark and turbid became Earth. The clear light was lighter and heaven therefore formed before the Earth, which assumed its shape only later. All things come from an initial Oneness. When something moves, it is called living; when it grows exhausted, it is called dead. Man is born out of non-being to assume form in being; he who can return to his original state is called a true man.

Just as it is meaningless to ask what came before the Big Bang, it is also meaningless to ask where the Big Bang occurred. Both space and time formed at the instant of creation; the Big Bang therefore existed in a state of anywhere/nowhere and anytime/no time. This myth is also correct that lighter elements formed the heavens first and that heavier dark (non-glowing) elements later formed the Earth. The claim that all comes from an initial Oneness is synonymous with the primordial singularity. As for both sky and Earth bringing forth creations, all life that evolves on Earth depends on the Sun either directly (plants) or indirectly (animals). Solar radiation and other radiation from deep space are known to cause genetic *mutations* (random genetic copying errors), which means that our evolution depends on both terrestrial and extraterrestrial drivers. This myth is therefore also scientifically accurate.

The Kuo Hsiang myth explains that the music of nature does not exist independent of things. Nonexistence is non-themselves. This does not mean that there is an "I" to produce because the "I" cannot produce things, nor can things produce the "I." It is impossible for non-being to become being

---

There are a billion people in China. It's not easy to be an individual in a crowd of more than a billion people. Think of it. More than a BILLION people. That means even if you're a one-in-a-million type of guy, there are still a thousand guys exactly like you.
A. Whitney Brown

He who knows does not speak; he who speaks does not know.
Lao Tzu

and for being to become non-being. It is important to understand that things can be spontaneous, to avoid infinite regression. This myth speaks to both idealism and the illusion of mundane existence, concepts you will become intimately familiar with as you keep reading.

There is a Mongolian myth that tells of a lama coming from heaven and stirring the water. The influence of the wind and fire that was thus created caused the water to thicken into material things (matter). In other words, the quantum potential underwent a fiery transformation from which matter evolved—a perfectly accurate description of what happened from both a materialist and dualist perspective.

Japan

The Japanese have a myth about a vast oily sea of chaos that contained a mixture of all elements. A spear was thrust into chaos and some drops congealed, which became the world. This refers to the primordial quantum state that was perturbed and became the universe.

The reverse side also has a reverse side.
Japanese proverb

Pacific Islands

The many islands strewn throughout the Pacific Ocean are home to an assortment of myths, including:

- The Marshall Islands creation myth says that the only god made a command followed by a magical sound that created the islands (and presumably everything else).

- According to the Maori, "From the conception the increase, from the increase the thought, from the thought the remembrance, from the remembrance the consciousness, from the consciousness the desire."

- According to the Tahitians, Taaroa existing alone called out and became the universe.

- In New Zealand, Tane-mahuta stood and pushed to separate Earth and sky. The parents of the gods did not like being pushed apart, but space was enlarged, and day and night were separated. Now there was room for both gods and humans to grow.

All of us have in our veins the exact same percentage of salt in our blood that exists in the ocean, and, therefore, we have salt in our blood, in our sweat, in our tears. We are tied to the ocean. And when we go back to the sea—whether it is to sail or to watch it—we are going back from whence we came.
John F. Kennedy

- In Hawaii, the entire universe is dual. Ao is light and male and Po is darkness, female, and nurturing (not evil). In the beginning, there was one great watery chaos until chants separated Ao from Po. The first person was made from clay.

All of these myths are scientifically accurate. The Maori myth in particular is compatible with Buddhist and other Eastern thought, which in turn is perfectly compatible with our contemporary understanding of physics. The New Zealand myth is correct that space expanded, that "night" (space) and "day" (stars) separated and that the space created by the expanding universe provides the room we need to exist.

Central America

The Aztec myth has five eras: In the first, the world was in darkness, and humans lived by animal instinct alone without reason. (Humans did indeed evolve from other animals and did not possess any superior reason or any other trait that we would identify as uniquely human until very recently in our evolutionary history.) The second era was one of spirits and transparent beings; humans did not understand how to be redeemed, and the gods therefore changed them into monkeys. (Superstition and accompanying myths are the earliest forms of scientific inquiry and our attempts to be "redeemed" in the light of understanding; and, of course, we did evolve from—and remain—monkeys/apes.) In the third era, people were ignorant of the gods and all creatures were burned. (Scientific and technological "progress" has a done a fantastic job of eliminating God as a necessary force in many people's minds, and our actions are indeed threatening to burn or otherwise destroy all life on Earth.) In the fourth era, all was destroyed by a flood paving the way for the fifth era—the present epoch. (The latter two eras don't seem to be accurate at face value but may have some predictive merit in that they could be pointing the way to a replacement of destructive actions by a new way of thinking—the so-called "Age of Aquarius."

The Quiche Maya believed that in the beginning there was only immobility and silence in the darkness. There was only the creator in the water surrounded by light... and then came the Logos.

To be a mystic is simply to participate here and now in that real and eternal life; in the fullest, deepest sense which is possible... as a free and conscious agent.
Evelyn Underhill

North America

Many of the North American tribal myths talk about successive development of culture, consciousness, and evolution among other things. Here is a sampling:

- **Abanaki:** Humans came from union of sea and land. This accords with what we know of evolution.

- **Apache:** In the beginning, there was nothing where the world now is, only *hactcin* or personifications of grand forces.

- **Arikara:** The great sky spirit Nesaru (sometimes called the Great Mystery) was the master of all creation. The sky was an endless body of water where two ducks swam. Nesaru made twin brothers, who told the ducks to swim down and bring up earth. The brothers found two spiders and told them how to reproduce and create new life forms. This myth also features a flood that wiped out evil that is very similar to the Biblical story of Noah.

- **Blackfoot:** An old man floated on a log on the water. He found four animals and sent them down to see what they could find. Three of the animals vanished, but the turtle came up with mud. The man took the mud and rolled it into a ball to make the Earth, which fell into the waters and grew to its present size. The story of Poia parallels the Biblical story of Eden and the fall of Eve.

- **Cherokee:** Humans and animals used to communicate, and then humans started killing them for food and fur. This angered the animals. The birds and insects met and named diseases they could spread among humans. Plants offered remedies.

- **Iroquois:** The first humans lived in the sky as there was no Earth.

- **Mandan:** A tree climbed from underground to the surface (evolution), but a fat woman broke the root. (Humans lost our connections to our past)

- **Pima:** At first there was only darkness and water, which congealed in some places to make the Creator. The Creator wandered, eventually becoming fully conscious of

The Puritans gave thanks for being preserved from the Indians, and we give thanks for being preserved from the Puritans.
Finley Peter Dunne

It is the winners who write history-their way.
Elaine Pagels

who he was and what he had to do. He took his walking stick and used the resin that gathered on the tip to create the Earth. He then took a great rock and threw it into the sky where it became the stars.

- **Salishan-Sahaptin:** The chief above made the Earth. The Earth was small at first, and the Creator let it increase in size. He covered it with white dust, which became soil. He then created animals, with humans coming last.

- **Sioux:** The great spirit used to kill and eat buffalo and had the form of a bird. A snake tried to eat the bird's eggs, and one hatched in a clap of thunder. The great spirit molded a stone into a man, and the man grew very old before the snake ate the roots and the man and woman wandered off together.

- **Zuni:** At first there was only water, which became clouds. Awonawilona, maker and container of all, conceived in himself and thought outward in space whereupon the mists increased in size into a sea. Awonawilona fertilized the sea, and algae grew on it and produced Earth and sky. The marriage of Earth, sky, and the action of the Sun produced all living things.

The Indian humbles himself before all of creation because all visible things were created before him and, being older than he, deserve respect.
Joseph E. Brown

At first glance, many of the above myths seem somewhat off the mark, however, remember that these tribes were less technologically advanced than even their Central American contemporaries, let alone the Middle and Far East. Read between the lines, and their accuracy will become much more apparent, both in the context of traditional science and emerging ideas about the primacy of consciousness (idealism) and the entire universe as we know it being nothing but an illusion. I am repeating these ideas on purpose without necessarily expecting you to believe me. Extraordinary claims require extraordinary evidence, as the late astronomer Carl Sagan (1934-1996) pointed out. As you will gradually learn, the extraordinary nature of the evidence makes the claims based on that evidence seem more than a little pedestrian by comparison.

African Tribes

If you refuse to be made straight when you are green, you will not be made straight when you are dry.
African proverb

If so many North American tribes were able to get the story right in their myths, it stands to reason that the African tribes would be at least as successful. Here is an assortment of African myths:

- **Barotse:** The god Nysmbi made all things and lived on Earth, but left because of man's behavior.

- **Bushman:** Cagn was the first being. He gave orders and all things appeared: The Sun, Moon, stars, wind, mountains, animals, and his wife Coti.

- **Bushongo:** In the dark there was nothing but water; the god Bumba was alone. He vomited up the Sun and dried up the water until the dry edges of the world began to show, but there were no living things. Bumba vomited up the Moon and stars to give the night its light. He then created all animals, and the last was man.

- **Dogon:** Stars came from pellets thrown into space by Amma, the one god. He created the Sun and Moon by a more complicated pottery process.

- **Fon:** The world was created by a god that is both male and female.

- **Swahili:** At the beginning of time was an eternal god. If this god wishes something, it happens. The god created light and was so pleased that he made that light his prophet. (Remember this when you read about Lord Kelvin's "two small clouds" on the horizon of physics in Chapter 23!) This god also created a well-preserved table with a record of all that can ever possibly happen anywhere in both the past and future. (Remember this when you read about the nature of time in Chapter 28!)

- **Yao:** The god Mulungu went fishing and caught humans in trap. He set them on the Earth and they grew.

- **Yoruba:** The beginning was a watery formless chaos like a marsh. The supreme god Olorun called Orish Nla and gave him magic earth, a pigeon, and a 5-toed hen to make the Earth. Orish Nla threw the dirt into a patch where the pigeon and hen scratched until they separated land and sea.

Here again, the accuracy of these myths is extraordinary.

There are three things which if one does not know, one cannot live long in the world: what is too much for one, what is too little for one, and what is just right for one.
Swahili proverb

Madagascar

The Creator watched his daughter, Mother Earth, making dolls out of clay, and breathed life into them. The resulting humans worshiped the Creator, but then forgot and worshipped the Earth. The Creator takes people's souls back, usually when the people grow old, because he is patient, and the Earth gets back the bodies she made. This myth is very much in tune with some of the concepts contained in quantum mechanics.

Universal Symbols

The very idea of a bird is a symbol and a suggestion to the poet. A bird seems to be at the top of the scale, so vehement and intense his life. The beautiful vagabonds, endowed with every grace, masters of all climes, and knowing no bounds—how many human aspirations are realized in their free, holiday-lives—and how many suggestions to the poet in their flight and song!
John Burroughs

It is important not to read the above myths literally. Of course no animal could literally swim down under water and bring up enough mud to build dry land. The animals in the myths are meant as symbols. Rodents, lizards, snakes, fish, ducks, etc. are intermediaries that combine underwater ability with flight and/or that mediate between heaven and Earth. Water too is a symbol. Ancient people knew that water is essential for life. When their myths speak of the universe emerging from water, they are talking about life emerging from a prerequisite structure. Thus, the water becomes a symbol of the quantum fluctuations that scientists believe sparked the Big Bang.

Trees are another potent symbol. Their roots go deep into the Earth and branches reach up toward heaven, making trees another form of mediation between heaven and Earth. Trees are firmly rooted yet can touch the sky. They also live far longer than people. Buddha sat under a tree, Eve sinned under a tree, Jesus was crucified on a tree, etc.

Many myths have humans being made out of clay. Interestingly enough, there are emerging theories of life's origins that speculate that life could have originated in clay formations.

Ashes to ashes, and dust to dust: Some see this as referring to the transient nature of human life, where we are an assemblage of "dust" that dissipates upon our death and personal oblivion. Given that human skin cells and other tissues make up a significant percentage of the dust in any building, this analogy makes perfect sense, especially since anyone witnessing a body decaying will eventually be unable to distinguish it from the soil in which it lies. There is, however, another meaning to "dust," and that is "source." We come from source and return to source. Interpret the word "dust" in this way and the saying "dust to dust" takes on a far less apocalyptic meaning.

Many myths from around the world also feature floods and arks. Some of these myths speak of ravens being dispatched to find land. It is widely assumed that these societies were not all in contact with one another; we must therefore suppose that they thought of these similarities on their own without any outside interference.

Universal Knowledge

If flood stories were the only similarity among so many ancient myths, then it would be an astonishing coincidence; however, as we have seen, the similarities go much further. Not only do creation myths from around the world contain many of the same elements and features, they also accord extremely well with what we know from today's sciences.

The beginning of knowledge is the discovery of something we do not understand.
Frank Herbert

The conclusion is inescapable: Genesis has no monopoly on the truth. Ancient people all over the world knew far more about the history of the universe and human evolution than most anyone alive today would want to admit. We might expect this in places like Egypt that boasted extremely advanced astronomy, but we should not necessarily expect it to exist among nomadic tribes in North America, Africa, or clustered on strings of tiny islands far out in the Pacific Ocean. And yet the parallels are there. All of the creation myths we just looked at are scientifically accurate after their own fashion. Each of these myths also connects to a rich web of additional myths that continue telling the story of how reality works and explaining the eternal, transcendent divinity that lies at the core of existence. How did ancient people around the world know this? How could they know? What does this imply about the universe and of reality itself? While we're at it, what does any of this have to do with whether or not we survive the death of our bodies? Dear reader, we stand on the edge of Alice's rabbit hole.

Chapter 4

From Myth to Religion

There is nothing truer than myth: history, in its attempt to realize myth, distorts it, stops halfway; when history claims to have succeeded, this is nothing but humbug and mystification. Everything we dream is realizable. Reality does not have to be: it is simply what it is.
Eugene Ionesco

As we saw in Chapter 3, myths deal with—and attempt to explain—the deepest levels of reality and the interdependence of being and not being. Both myths and the religions built on those myths probe beneath all of the temporal change in the universe to find that which is eternal and changeless. Myth and mystical experiences were humanity's earliest attempts to find the ultimate "why," despite lacking the scientific tools to do so. This led to anthropomorphic explanations, which necessitated gods made in our images, which evolved—or devolved—into religions that claimed to know all of the answers down to what God thinks we should wear and what we should eat. The results have often been far less than perfect, but the religions have tried their best to explain ultimate reality and the Ground or Godhead in ways we could understand, while also trying to help us establish (or reestablish) our own connections to the divine through *avatars* (human incarnations of God) that purport to show us the way back to our eternal roots.

Myth and religion say that everything that exists partakes of the divine and attempt to show how the eternal Ground connects to daily life. We humans tend to characterize ourselves relatively (to other people, animals, religion, etc.), but these labels have no absolute reality. People who define themselves relatively end up only having relative worth. The only way out of this trap is to replace, "I am _____" with simply, "I am."

The source of reality is absolute. A glass of salty water tastes salty whether you drink from the top, middle, or bottom... but let the water evaporate and the salt remains. In this example, the water is transient but the salt remains.

The problem is that the original message has been corrupted far too often. To prove this, we need look no further than religions, which, as I said, claim to know the ultimate reality of the Godhead and its desires, preferences, and intentions down to what we should wear, and eat.

Myth is neither a lie nor a confession: it is an inflexion.
Roland Barthes

The Perennial Philosophy

Imagine owning a large company. One day, you set off on a trip of unknown duration leaving your employees with a long list of detailed instructions on what they should do in your absence. You return to find your company in shambles, your workers bickering and sometimes outright fighting over your instructions, what they meant, and who is following them or not. Thousands of variations of your instructions have been created and disseminated throughout your organization, some of which bear little resemblance to your original instructions or what you intended to convey. Even worse, imagine that this same problem has spread to other firms, causing bitter infighting and competition within and among the companies, to the point that research and development has been all but abandoned, and more and more of the general public are openly questioning what's going on. Meanwhile, all but a few people have forgotten that there is a simple, universal set of guidelines by which a business, any business, should be run.

Wisdom ceases to be wisdom when it becomes too proud to weep, too grave to laugh, and too selfish to seek other than itself.
Kahlil Gibran

Welcome to the state of modern religion: an assortment of sects rife with often-violent internal and external strife over who is following the correct set of instructions and who has "the answer" to life's questions—including what happens when we die. This sorry state of affairs seems even worse when one remembers that the philosophical principles underlying each of these religions and their variants are virtually universal across all cultures from all corners of the globe throughout all of known human history. These principles are referred to as the "perennial philosophy" (a term coined by Leibniz and popularized by author Aldous Huxley (1894-1963) in his book of the same name) because even primitive

I believe it because it is absurd.
Tertulliant

myths contain at least some aspects of this philosophy. (Some examples include the idea of people being made of different divine elements and the existence of a soul that reunites with the Ground of being when the body dies.) The perennial philosophy is about an eternal spiritual reality, and is the foundation of both most myth and all religions that I have examined. In general, mystical experiences lead seekers to the perennial philosophy, which becomes embodied in myth, which can become codified and institutionalized into religion.

It is important to note that just because the perennial philosophy is *perennial* (lasting for an indefinitely long time) does not necessarily mean that it is uniform, logical, or even consistent. It tends to have paradoxes and contradictions, because hints and peeks at the Ground are the best we have been able to do until the advent of modern science. Another key challenge is that people tend to find what they are looking for; a religious person sees signs of God everywhere, while a materialist sees only mechanical natural processes at work. Each side can back up its argument with reams of evidence that the other will not admit as valid, often because of varying degrees of conflation.

Modern science has the tools to validate the perennial philosophy. Many scientists have done just that; however, many other scientists have assailed it, claiming that there is no eternal now and that time and change, as evidenced by the increase of *entropy* (randomness/disorder) are fundamental. This assumes that time flows like a river, but there is ample reason to believe that time does not flow at all, but simply exists as a series of frames that we experience as a movie-like flow because we can only be in one frame at a time. We will return to this in Chapter 28.

In the perennial philosophy and its associated mythology, the human psyche is rooted in simple awareness. God gave things their becoming and illusion of separation with the goal of them returning to the eternal Ground or Godhead. God is everywhere and nowhere; everywhen and nowhen; aware of all possibilities. The human mind with its reasoning ability can anticipate which frame(s) in the eternal movie will come next, and can remember what frame(s) it has seen before, albeit without any direct awareness of the whole. People who assert that the human mind can know the Godhead are mistaken, because this would require us to be both "in time" (the experi-

---

Sit down before fact like a little child, and be prepared to give up every preconceived notion, follow humbly wherever and to whatever abyss nature leads, or you shall know nothing.
Thomas Huxley

You can ponder and analyze till the cows come home, but the real question is whether all your ponderings and analyses will convince you that life is worth living. That's what it all comes down to. Everything else is detail.
Brian Greene

---

ence of flowing time) and "out of time" in the eternal now. It is the human spirit that transits these two worlds, coming into time when identified with a body, and stepping back out of time when the association ends. We cannot have it both ways. As an aside, our eternal nature resolves the problem of good and evil. Evil is only a problem if birth and death exist. Thus, those aspects of ourselves attached to a body may experience evil, but our eternal nature is not evil at all.

Many primitive cultures emphasize group identity and interconnectedness. In the modern world, children grow out of their awareness of the Ground because they are taught to think analytically instead of spiritually. Analysis is not generally compatible with spirituality. (Analysis can assess spirituality as I am attempting to do in this book, but can never substitute for the actual experience.) Thus, we accept as given the idea that we are a separate form that is at once part of and yet distinct from the rest of the universe. I believe this happens when we develop the sense of "I" as young children.

All are lunatics, but he who can analyze his delusion is called a philosopher.
Ambrose Bierce

I will never forget my son Logan as he neared his first birthday. He could point to things when asked, "Where is _____?" and come when he was called, but simply looked confused when I asked, "Where is Logan?" Then one day he pointed to himself; he had lost his connection with the Ground in favor of an identification with a temporary ego called the "I." On a purely mechanical level, this is essential to survival for reasons I've discussed in *The Natural Savage*. On a spiritual level, it's like being cast adrift to find one's own way home. The extent of one's attachment to "I" is the extent of one's separation from—and ignorance of—the Ground. Asserting selfhood can be harmful to both the individual and collective organism. Consider for example the recent outpouring of so-called patriotism in the United States, where millions of people are championing ideas that are blatantly against everyone's individual and collective best interests.

Dear! Dear! How strange everything is today. I wonder if I've changed in the night. Let me think: Was I the same when i got up this morning?
Lewis Carroll

If mind as we understand it cannot experience the timeless Ground directly, and if spirit is the vehicle for transiting these two worlds, then it follows that immortality consists of a spiritual participation in the eternal now of the Ground, which entails a total deliverance from the concept of time. This is different from mere survival where those who have not learned to reconnect to the Ground during life are relegated,

> *Death. It doesn't have to be boring.*
> Mary Roach

not to the heaven that comes from reconnecting, but to the hell of ongoing separation and remaining in time. Jews, Christians, and Muslims believe that this life is our only chance to achieve this deliverance, and that heaven and hell represent a one-way trip. Buddhists and others believe that those who have not learned keep being reborn in an ongoing cycle, until at last they are liberated from the wheel of birth and death and achieve *nirvana*, the state of total reunion with the Ground. Don't mistake survival for immortality: Enduring the cycles of rebirth means dying many, many times. Still, in Eastern religions, the Ground displays infinite patience and mercy, features that are distinctly lacking in Christianity despite their claims to the contrary.

What survives death? Which parts of you or me will carry on when our bodies cease to function? In Western religions, the concept seems to be that a distinct soul survives, carrying the "I" with it. In other words, after I die, I will remain Anthony Hernandez. The Eastern concept of reincarnation says that personality does not survive. What we do in this life creates a new instance in a new body. This is not to say that souls are interchangeable and lack identity. There is a unique consciousness in each soul from which an "I" arises at each birth. Personality could thus be said to equal soul plus body, which means that soul itself cannot carry one's personality. In other words, Anthony Hernandez will indeed cease to exist when my body dies, but my unique soul will continue. We will take this topic up again in Chapter 5.

> *All that we see or dream is but a dream within a dream.*
> Edgar Allen Poe

We already know that mind can affect body. This can happen in four ways: subconsciously (physiological), consciously (will), subconsciously by reaction to emotional states, and/or as a result of so-called "supernatural" phenomena. Minds can interact with minds, either directly through various psychic abilities, or indirectly via speech, writing, etc. Minds can also affect physical objects, such as rolling dice or random number generators. (See Chapter 16.) This is where the concept of *miracles* might come from (divine or otherwise "supernatural" intercessions into the material world). For example, faith healing cannot be entirely discounted; there are studies showing that prayer aids healing. (In fairness, there are other studies that claim to show no such effect, but still...) If mind can influence body and world, then it follows that an *immanent* (within the universe, time, and the limits of experience; also sometimes interpreted as, "ever-present," as in, "God is with us at all times.") or transcendent form (God) should be able to influence that mind. It is therefore entirely possible that God is ultimately responsible for some or all human behavior.

All of these ideas make perfect sense to the spiritually inclined, but the problem of actually learning how to reconnect to the Ground remains. Physical death does not do this; it simply consigns us to an eternal otherworldly heaven or hell (Judaism, Christianity, Islam) or to a cycle of rebirth (Buddhism, et al). To help us, religions provide one or more avatars that guide us by showing us both that knowing the Ground is possible and how to attain such knowledge. The underlying assumption is that the Ground has a personal aspect, such as the Hindu gods or the Christian trinity. This aspect incarnates as a god with human limitations (Jesus Christ, for example). In this way, the transcendent is made immanent.

Thinking of God as absolute and transcendent risks losing the perennial philosophy beneath layers of rituals, sacrifices, and laws. As we will see in Chapters 8 and 10, this is precisely what has happened in far too many cases. It is often better to regard God as an immanent, loving mother or father, which allows one to discard external practices and the trappings of organized religion in order to focus on one's own inner development. The danger of this path is that it can lead to laziness or quiescence that inhibits or even prevents the character change required for spiritual growth.

Religious rites, sacraments, and ceremonies can be useful if they remind us of the truth of the Ground and if the actual acts don't matter much at all. Sadly, many people become attached to the need for rites and sacraments even though this distances themselves from the teachings of the relevant avatars. For example, modern Christianity bears very little resemblance to the actual teachings of Christ, a fact we will explore in Chapter 6.

It is all but impossible for a single person to construct a complete system of belief, myth, and religion. Rather, as we will see in the coming chapters, this is a collective effort that involves many people with many viewpoints, some of which may be in open conflict with each other. It is natural that these

> *I remember when I lost my mind. There was something so pleasant about that space. Even your emotions have an echo in so much space.*
> Gnarls Barkley

> *Any serious attempt to try to do something worthwhile is ritualistic.*
> Derek Walcott

> *Were triangles to invent a god, they would give him three sides.*
> Montesquieu

viewpoints will reflect a wide range of approaches from moderate to extreme. Of these, the extreme positions are clearer, more easily recognized, and more readily understood than moderate positions that do not necessarily countermand the extreme positions. The danger increases as the perennial philosophy becomes expressed through myths that in turn evolve into religions that then fragment into denominations and sects.

No religion is perfect, but this has not stopped people from taking them far too seriously throughout the ages, often with disastrous results. History is littered with religions that have lost their way and become self-absorbed, seeking power and gain for their own sake (including violent or coerced conversions). These religions have forgotten all about the Ground of being, the Godhead. In India, the caste system subordinates secular rule to the religious. Christianity used to emphasize contemplation, but now emphasizes action and thinking—the best ways to lose the connection to the Ground.

This is not to say that there is only one correct way to think of the Ground. Theological speculation is perfectly acceptable—even necessary—but one must always remember that theology is not reality; it is at best a map. Maps are useful but are not the destination. As far as the map is concerned, the route one takes shouldn't matter. Theological imperialism is one of the most dire threats to world peace and the ongoing survival of the human species, and indeed of all life on Earth. (See Chapter 10.) Religious violence will only end when our planet is a smoking lifeless hulk or when people return to the perennial philosophy and reject religion as we know it.

> *There is something absurd about Biblical and Koranic literalists flying around the world on jets instead of magic carpets, communicating via cell phones and the Internet instead of smoke signals, and publishing anti-science tracts on laser printers rather than carved rocks.*
> Victor Stenger

The Western religions have tended to be more violent, domineering, and proselytizing than their Eastern counterparts. The Christian attitude toward nature often seems like outright contempt because of the remark in Genesis 1:26-28 about God giving humans dominion over the Earth. The Western humanitarian movement has been primarily secular, while that in the East has been mostly religious. The difference may exist because Christianity has only one avatar, Jesus Christ. It is therefore an all-or-nothing proposition, unlike the Eastern religions that have many possible avatars. The East has seen religious wars, but they have been few and far between compared to their Western counterparts.

The Evolution of Myth

I provided a brief example of how myths originated and evolved in Chapter 3. This concept is so important that it bears reexamination in more depth. As I said, myth represents the earliest form of human science that tried to answer questions about how the world works, our place in the universe, where we come from, and where we are going. The fact that scientific discoveries have rendered many of these myths obsolete does not invalidate those myths any more than quantum mechanics invalidates Newtonian physics. Myth codifies standards of behavior and real or imagined histories, thus forming a framework that helps define a society and strengthen the bonds between members of that society. Myth is also ubiquitous; everything we are aware of perceiving, thinking, or feeling is the product of our own inner myths; it is therefore almost our only way of knowing about the world.

Knowledge is not a series of self-consistent theories that converges toward an ideal view; it is rather an ever increasing ocean of mutually incompatible (and perhaps even incommensurable) alternatives, each single theory, each fairy tale, each myth that is part of the collection forcing the others into greater articulation and all of them contributing, via this process of competition, to the development of our consciousness.
Frank Herbert

The earliest social myths probably involved natural phenomena such as the diurnal cycle (day and night), rain, storms, fire, etc. that were of interest to the entire group. Answering these questions seems simple at first blush. For example, it is easy enough to say that rain comes from clouds. But where do the clouds come from and why? Glib answers only go so far; any attempt to fully explain a phenomenon must entail at least some of the following questions:

- What is happening?
- What or who is causing it?
- Why is this thing or person causing this event?
- Where does this thing or person comes from?
- When does this event happen?
- How does this thing or person relate to other events and their causes?
- How can this event be encouraged or discouraged?

It is natural to give a clear view of the world after accepting the idea that it must be clear.
Albert Camust

This partial list should be enough to demonstrate that explanations are not necessarily as simple as they seem. A modern adult with even a modest education understands enough about nature to know that rain is formed from evaporated seawater that forms clouds, which then condense and release

The fact that I can plant a seed and it becomes a flower, share a bit of knowledge and it becomes another's, smile at someone and receive a smile in return, are to me continual spiritual exercises.
Leo Buscaglia

their moisture. If s/he remains curious, s/he can then research how the Sun's heat and Earth's rotation heat the water and cause wind, how the Sun and Earth formed, and so on all the way back to the beginning of time. Primitive humans had no such knowledge or educational resources but had no less curiosity. In fact, as I explained in *The Natural Savage*, there is an inverse relationship between fear of predation and curiosity. (Remember that humans are prey animals.) The fear keeps us from being killed and eaten, while the curiosity ensures that we don't go through life afraid of our own shadows. Combining these elements with our own growing intellectual and imaginative capabilities gives us the foundation of our almost insatiable need to know. Children who incessantly ask, "Why?" until the parent can't or won't answer anymore provide both material for comedians and a window into our own insatiable need to know our world.

Natural phenomena were beyond our control at the dawn of myth, and many remain so today. Humans are inherently social and hierarchical animals. It is relatively rare for a person to countermand real or perceived authority, even when s/he has reason to believe that the order is improper. (The Milgram shock experiments are a well-known example.) Anything or anyone capable of controlling things beyond human control must be incredibly powerful, more so than even the most powerful human chieftain or king. As for whether the agency is personal or impersonal, ancient humans had no modern scientific knowledge and thus no way to know about blind forces that operate without any intelligent input. They did know that both everything they and other animals did required intelligence, and the intelligence they were most familiar with was their own. The agency causing any given phenomenon therefore had to be intelligent. Since ancient humans knew human intelligence best, it followed that the causative agency had to have human-like traits. In short, the controlling entity had to be both *personal* (of the nature of an individual rational being) and *superhuman* (having a higher nature or greater powers than humans have). For example, the Klamath tribe believed that the west wind was caused by a gassy 30" tall woman wearing a buckskin dress.

The general root of superstition is that men observe when things hit, and not when they miss; and commit to memory the one, and forget and pass over the other.
Francis Bacon

The ancient human world consisted of (and the modern world still consists of) doing things, including tasks taken at the chief's behest. The chief gave orders and those orders were carried out, often without question. In fact, questioning or otherwise resisting the chief could carry potentially severe consequences—a clear predecessor to, "God said it, and it was done." A superhuman intellect capable of controlling things that humans still have not figured out how to harness can only be described as a "super chief" or "über-alpha." All humans have relationships and stories behind their existence, so it stands to reason that our gods would have the same attributes; thus, as I said in Chapter 3, we created our gods in our own images, complete with strengths, flaws, families, and histories. If one believes that gods control various natural phenomena like wind and rain, then it is only a small leap to assume that the beginning of the universe was a divine creative act.

One does not become a guru by accident.
James Fentont

Tribal chiefs enact and enforce laws and customs. Why should the chief(s) in the sky be any different? There is just one problem: Human chiefs can make their wishes known to their followers directly and in no uncertain terms. But how can an entity who cannot be seen, heard, or otherwise directly detected make her or his wishes known? We will return to this question presently. Also, some people speculate that there is a biological predisposition to religious beliefs; we will examine this in Chapter 9.

Why Myth?

All people seek a sense of unity with both their peers and the world. We all need some sense of security in this inherently insecure life. A system of stories that provide (or purport to offer) enduring, timeless truths can be a huge source of comfort and guidance, because they give us workable answers to mysteries. This is where myth and religion began; the moral component (in the forms of "divine" laws and penalties) came later. It is important to note that creation myths are about confining and channeling chaos, not getting rid of it. As rabbi Irwin Kula (1957-) says, they recognize that the infinite beauty of creation is inseparable from its destructiveness. There is a *yin* (female) and *yang* (male) component to everything in the universe. The mythical triumph of the gods over chaos may not have been a literal attempt to describe the Big Bang, but we saw in Chapter 3 that one cannot deny the scientific validity of these myths.

Storytelling reveals meaning without committing the error of defining it.
Hannah Arendt

Chapter 4
From Myth to Religion

We have seen how myths look at who we are and where we've been but past and present are only part of the story. For as long as humans have been aware of our mortality, we have asked, "What comes next? What happens when we die?" It is both easy and simplistic to say that afterlife myths and religion rose from our fear of death and deep need to assuage that fear by inventing stories about the afterlife. Life is often difficult and involves much toil, but those who please the chief earn favors. It is but a small step to apply these facts to a mythical god to concoct a tale of an otherworldly paradise. Some 100,000 years ago, Neanderthals had complex death rituals that included burying tools and food with the deceased. Food was hard to come by in the era before agriculture, and toolmaking consumed valuable resources that could be better spent using those tools to find or process food. It makes no sense to bury these valuables with a corpse, unless one believes that at least some part of said corpse survives physical death and thus requires those resources. Did these beliefs come from fears soothed by creative storytelling? Or is there a deeper source?

It so happens that there is such a source: mysticism, or, more precisely, a mystical state of consciousness. People in these states report feelings of bliss and deep connection with the Ground of the universe, to the point where their egos and desires can almost entirely vanish. The details vary, but the main thrust of the story is universal: The Ground of being, the Godhead, God, whatever you choose to call the ultimate source of reality, is composed of pure, eternal consciousness. Everything in the universe stems from—and is part of—that consciousness. Anyone who understands this understands that birth, death, and the appearance of separation between any two things/beings in the universe is nothing but the illusion of maya. This Ground defies description. Poetry, metaphor, and allegory are the best anyone can do to describe these mystical experiences and the nature of the Godhead. In other words, myth is as close as we can come to the Ground without having a mystical experience of our own. Mysticism is a far more likely candidate for the origin of afterlife beliefs than the simple fear of death. We will look at this more in Chapter 12.

Mystical experiences combined with the realities of human nature and life are the origins of myth.

We boast our emancipation from many superstitions; but if we have broken any idols, it is through a transfer of idolatry.
Ralph Waldo Emerson

I know there's a ghost in the machine. What I don't know is, is there a machine in the ghost?.
Daniel N. Robinson

Primitive Religion

Naturalist Charles Darwin (1809-1882) said that the separation between humans and animals is one of degree and not of kind. Thus, one could argue that a bear gazing at a sunset or a wolf howling at the Moon is just as religious to them as any ritual is to us. It is entirely possible that animals view and worship natural forces with just as much awe as our ancient ancestors did. Animals also have fewer mental filters between raw sensory input and conscious perception. Are animals more in touch with the Ground by default? This digression is far from frivolous, because the underlying question concerns where and why we draw the line between human and animal, how we view our relationships with other species, and by extension how we view the entire world.

> *It is from the study of the true theology that all our knowledge of science is derived, and it is from that knowledge that all the arts have originated.*
> Thomas Paine

Since religion began with myths about how the world works, it follows that the earliest religions were *polytheistic* (believing in multiple gods). To date, anthropologists have not found a single natively *monotheistic* (believing in a single god) hunter-gatherer society. Many tribal religions may have a chief god leading a divine *pantheon* (collection of gods within a single mythology) that often resembles the tribal leadership structure. Some gods involve themselves in daily life, and some don't. Some gods can be explained in human terms, others can't.

Despite these differences, hunter-gatherer religions do have a lot going for them. These "primitive" beliefs often do a better job of tackling thorny issues than their modern counterparts. For example, primitive religions can easily explain good and evil, because they have no single all-powerful being, and none of their beings are expected to be morally perfect. Meanwhile, monotheistic religions that postulate a single perfect God have been struggling with the questions of evil and suffering for thousands of years. Primitive gods are not expected to solve moral problems that would exist with or without them. Primitive tribes often have excellent moral standards that stand on their own based on tradition and public opinion, without the need for religious bolstering. The fact that each member of the tribe has to coexist with the same people for her or his entire life provides pretty powerful incentive to treat each other well.

> *Why should my spirit be subjected to a faith I do not share, a religion in which I do not believe, laws that were formulated hundreds or even thousands of years ago by a tribal chief or warrior?*
> Andre Compte-Sponville

Some other general traits of primitive religions include:

- Primitive tribes tend to revere women, a pattern that faded and reversed itself as civilizations spread, often violently. For example, ancient mother goddess cults praised the feminine fertility of agriculture.

- Tribal myths are not intended to be taken literally. They are attempts to describe a reality that is too complex and elusive to explain in any other way. This is how a tale of a turtle diving underwater and bringing up dry land can be scientifically valid; the truth emerges as one "reverse engineers" the myth by applying modern terminology and seeing if the pieces fit.

- In some cultures, diagnosing an illness by figuring out which god is responsible and how to induce that god to reverse the disease is just as valid for them as searching for pathogens and antidotes is for "advanced" cultures.

- The Aborigines measured the world in time (seasons, generations, etc.) against the backdrop of "everywhen" stability.

- Western religions look forward, and Eastern religions see cycles. Primal religions tend to look back, but they don't consider time linear; there is only "now." Anything in the past is closer to the Ground for them, a concept that has plenty of merit, as we will see in Chapter 28. Totems blur the lines between forms of life into a continuum with no separation (a view that modern evolutionary science embraces).

- The Navajo Indians have no word for God, because God in an unknown power (which is different than being unknowable). They worship through creation, for God is everything in creation. In fact, many primitive religions tend to attach divinity not to a supreme being, but to all of creation. This is known as *pantheism* (believing that one or more gods are identical to the entire natural world, and/or that everything is a god).

This very brief glimpse should be enough to tell you that so-called primitive religions may not be all that primitive. In fact, they may be far more connected with—and relevant to—daily life than many of the large religions we take for granted.

It is the customary fate of new truths to begin as heresies and to end as superstitions.
Thomas Huxley

All I'm trying to do is not join my ancestral spirits just yet.
Joshua Nkomo

Shamanism

The fleeting nature of life, the certainty of death, and the larger transcendent reality can be both terrible and fascinating at once. There is nothing rational about this, nor can it adequately expressed by most people. This leaves the door open for people who have—or who at least claim to have—a direct connection to the divine/supernatural. Untold thousands of people have profited from having the reputation that they can contact God or other entities beyond this world since before the dawn of agriculture.

Shamanism is a critical step on the path from myth to organized religion. The earliest religions had flexible beliefs about flexible spirits, as opposed to the later bodies of doctrinal beliefs and practice enforced by authoritarian institutions. A *shaman* (intermediary between humans and gods) is the predecessor to an archbishop or an ayatollah. The shaman need not have a perfect track record of making predictions or healing people. The idea that power waxes and wanes like the Sun, Moon, and seasons explains away failures and weaknesses without calling the very idea of shamanistic power into question. This same logic applies to modern experts. For example, a stockbroker who makes a series of bad guesses has simply "lost his touch," which does not impugn the overall expertise of stockbrokers in general.

Getting rid of failing experts is key to sustaining the faith in expertise. A Chippewa religious leader who did not maintain a demonstrable connection with the spirit world was replaced. Individual leaders were expendable in order to preserve faith in the idea of leadership. In this context, the Catholic belief that popes are *infallible* (beyond all error) seems more like devolution than progress. Either way, religious leaders from tribal shamans and medicine men to popes and televangelists wield real power that sometimes extends around the world. Whether this power comes from a divine source or is merely the result of human acquiescence is practically meaningless. What is important is the context of that power. Primitive polytheism equates natural phenomena with gods, thus connecting humans directly to nature. By contrast, the "Big 3" religions separate God from humans and humans from nature. (See Chapters 5 and 8.)

The depth of your mythology is the extent of your effectiveness.
John Maxwell

It is superstitious to put one's hopes in formalities, but arrogant to refuse to submit to them.
Blaise Pascal

> *Superstition is only the fear of belief, while religion is the confidence.*
> Marguerite Blessington

Why would anyone in a primitive tribe believe anything a shaman has to say? Why would stories about gods controlling the weather, disease being caused by evil spirits, or spiritual contact with long-departed ancestors be anything but a source of great amusement? I have already given two explanations: that myth represents our earliest scientific inquiries into the big questions of life, and that mystical experiences are real (albeit neither ubiquitous nor proof of any greater reality beyond ourselves). Let me also remind you that humans are prey animals. According to anthropologist Stewart Guthrie (1948-1990), humans are biased toward making false sightings because the potential cost of not seeing a real predator is far higher than the cost of a false alarm. Thus, our brains are predisposed to false positives. Did someone claiming to see something extraordinary actually see that thing? Sort of.

Coerced credulity (Stockholm syndrome is an example) is another factor. To put it bluntly, humans seem designed to be easily brainwashed—a helpful feature for a hierarchical social animal to have, since it allows followers to more easily accept and act on their leaders' directives. This may seem shocking to some, but the fact remains that isolation in small groups is the natural human condition. Primitive tribes usually numbered some 40-60 people who depended on group cohesion for their survival. Alienating one's peers by disparaging their beliefs lowers one's chances of finding a willing mate, and can even risk expulsion or death.

> *Religion has certain ideas at the heart of it which we call sacred or holy or whatever. What it means is, '"Here is an idea or notion that you're not allowed to say anything bad about; you're just not. Why not? Because you're not!"*
> Douglas Adams

Brainwashing is as easy as listening to someone in a position of authority. In this context, one has no choice but to accept what those in authority are saying; actual coercion is hardly ever necessary, and is usually counterproductive. Any good interrogator will tell you that the best way to get prisoners to divulge factual information is through building rapport. Runaway children given shelter and some hot meals are amazingly receptive to cult beliefs that they would probably reject under other circumstances. People are wired to recognize and obey authority. I used to be a volunteer firefighter and EMT and have stopped at many accidents to assist. The knot of people I always found at the scene obeyed my every command without hesitation; not a single person ever asked me who I was or what my qualifications were.

Xenophobia (unreasonable fear or hatred of strangers or outsiders) also seems to be part of the human condition. Each tribe evolves to think of itself as somehow special, unique, and superior to surrounding tribes. Any outsider could see that the differences among the tribes in a given area are superficial at best. In fact, there are far more similarities than differences across all human cultures. Marriage, death, and/or religious rituals may be carried out differently across cultures, but all cultures have these rituals, which serve the same purposes in all cultures. Still, every tribe and culture thinks that it is superior to all others. Wariness of outsiders helped us survive throughout our evolutionary history. The implications for this discussion are that each culture will regard its gods and religious leaders as superior to those of other cultures.

Religion

Agriculture changed everything. As I explained in *The Natural Savage*, I believe that agriculture represents both the single largest disturbance in our evolutionary history and a prime example of negative stability, because its effects continue to ripple across our planet. Like everything else in life, agriculture changed the reasons and uses for religion. Even the gods we worshipped underwent profound changes. Agriculture led to the concept of private property and enabled the formation of ever-larger permanent settlements. Power and wealth could be consolidated and passed down across generations. The growing complexity of society required people to specialize into professions. Even tribal shamans have to hunt and provide food for the tribe. Agriculture means that a relatively small percentage of the overall population can feed everyone else, who must occupy themselves with other professions, including the priesthood. For the first time in human history, societies could afford to maintain a permanent priestly class. Writing and record-keeping was initially invented to tally and track resources, and could be readily adapted to codify laws and rituals, including a society's myths, gods, and how those gods wanted their people to behave.

Civilization brought new challenges. Wars grew increasingly protracted and violent. Slavery was rampant. The masses were taxed to provide luxury for those in charge. Disease caused by lack of sanitation was a problem that did not get resolved until

---

Myth is an attempt to narrate a whole human experience, of which the purpose is too deep, going too deep in the blood and soul, for mental explanation or description.
David Herbert Lawrence

Men rarely (if ever) manage to dream up a god superior to themselves. Most gods have the manners and morals of a spoiled child.
Robert Heinlein

Our species is far too clever to survive without wisdom.
E. F. Schumacher

> *A population weakened and exhausted by battling against so many obstacles, whose needs are never satisfied and desires never fulfilled, is vulnerable to manipulation and regimentation.*
> Ryszard Kapuscinski

the 20th century AD. For the first time, humans were divided into classes, the haves and the have-nots. No longer could hunting or foraging guarantee a successful life. In agricultural societies, success was as much a matter of which family/caste you were born into and whether one's city could repel invaders as it was on individual skill and cunning. In short, the insecurities of hunter-gatherer life were replaced by new insecurities that remain with us today, and that are less controllable than those we faced in our ancestral tribes. Religion reflected these changes as the gods' roles evolved from simply controlling natural phenomena to actively participating in people's daily lives and taking an active interest in their behaviors. Gods became ersatz parents. Religions that favored the poor, meek, hungry and thirsty, etc. did very well for themselves. Christianity (which essentially consists of existing spiritual ideas that were rehashed and recombined) spread rapidly because it spoke to and exalted the powerless masses.

The separation of church and state that many of us take for granted today is a relatively new invention. The two institutions have been joined and sometimes virtually indistinguishable for most of our history. Kings and emperors were worshipped as gods, either posthumously or in life. State religions gave divine authority to earthly rulers. Godly pantheons bore striking resemblances to kingly courts on Earth and the politicking, bickering, and outright fighting among the gods mirrored that on Earth. Church and state did clash from time to time, (such as when Pharaoh Amenhotep IV, died ca. 1334BC) tried to replace the Egyptian pantheon with the one god Amon, but the overall pattern of politicians and clergy controlling access to divine knowledge, power, and favors persists to this day. The United States is no exception; politicians today can be made or broken by their faith or lack thereof.

> *Christianity is the most perverted system that ever shone on man.*
> Thomas Jefferson

Warfare may have been common among city-states and early nations, but religious war was almost unheard of. Individual cities were not originally part of a regional government; they contacted each other through trade and war. Interestingly, they handled each other's religious claims pretty well overall, because they did not see different religions as competing with or infringing on their own religions in the slightest. Polytheism has no fixed upper limit to the number of gods it can accommodate, so there is no *a priori* (preexisting) reason to contest anyone's gods. On the contrary, conquerors were far more likely to worship the conquered gods than to desecrate or destroy them. Two cities with trade or other alliances between them could benefit by affirming both sets of gods and merging both sets into a single pantheon. Gods of different cities thus became blood relatives and the cities came to agree on the details of the family tree. Creating such common cause served everyone's interests and left everyone free to worship as they choose. Contrast that with later monotheistic religions that fought for the exclusive rights to control territory. Jerusalem remains at the center of a long-simmering dispute that may yet trigger the next world war, with catastrophic consequences for all life on Earth. Polytheism was good for empires and good for business when relations were positive. The transition to monotheism was a protracted, bloody affair that had no guarantee of succeeding. We will look at this in more detail in Chapter 7. Meanwhile, it bears mentioning that negative relations between cities could spawn theologies of hate and intolerance. Religion thus mirrored the facts on the ground; theology responded and adapted to current market conditions.

A myth is a religion in which no one any longer believes.
James Feibleman

As an aside, belief in an afterlife makes death less harrowing. It also makes dying in a holy war more attractive. This one fact becomes increasingly important as religions grow in size, power, and ability to shape the market to their own needs. Think about the largest corporations you know. The world's largest hamburger brand began life as a single stand in southern California. The world's largest chain of coffee shops began as a single café near the Pike Place Market in Seattle, Washington. Their initial expansion was fueled by quality and customer demand; however, they eventually grew large enough to manufacture demand and market share through the sheer force of marketing budgets that made each company a household name. (I am not making any comment about the quality of either company's products, merely using them as examples.) Religions initially reflected the facts on the ground; as they expanded, they became increasingly able to create facts to their own ends using the power of belief in the afterlife to recruit hordes of willing soldiers to enforce their goals.

I have a problem about being nearly sixty: I keep waking up in the morning and thinking I'm thirty-one.
Elizabeth Janeway

I freely confess that it sounds cynical to talk of religion in commercial terms, but the truth is that many religions resemble network marketing companies more than they would care to admit. For example, the Mormon Church helps facilitate commercial contracts. Churches grow by being nice to outsiders and inviting them in with promises of fellowship and other benefits in this world and the next. Once the new person joins, s/he is expected to give as well as get—a business model that has a lot in common with drug pushers. Paul's Christian church was easy to get into, and easy to be expelled from as well if someone did not contribute their share. The edict to "love thy neighbor" sounds wonderful, until one realizes that the entire concept hinges on the definition of just who one's neighbor is. Attend any Christian service, and you will quickly learn that the term "neighbor" has nothing to do with kinship or place of residence, and everything to do with belonging to the Christian religion.

Religion is just mind control.
George Carlin

Thinking up a welcoming message of inclusion in a special group of people and rewards in the afterlife is only part of the marketing challenge for any religion. Members must be attracted and then induced to behave in ways that will sustain and grow the religion. Defining *sin* as anything that does not promote church cohesion is a powerful expedient: Anyone who does not live a "righteous" life that keeps the church robust fails to earn a ticket to God's eternal kingdom, and may even be condemned to eternal punishment. This is not a universal message by any means; the Eastern religions don't consign anyone to eternal punishment. Still, all religions have codes of conduct that preserve cohesion among their members when followed.

The Challenges of Religion

We are all instruments endowed with feeling and memory. Our senses are so many strings that are struck by surrounding objects and that also frequently strike themselves.
Denis Diderot

We will be looking at the promises and pitfalls of religion in much more detail in Chapters 8 through 12, but a brief glimpse at a few selected topics is merited here.

The fundamental promise of all religion is that knowing the truth can set us free. The challenge is that discovering truth is an ongoing process. The moment we stop seeking truth, we begin to stagnate. Ancient people knew this; for example, removing the first "e" from the Hebrew word *emet* (truth) leaves *met* (death)—a clear reminder that truth is never absolute and must be sought with humility. Early religions saw creativity and discovery as divine, and encouraged exploration and discovery. These attitudes are clearly reflected in the tendency for different cultures to embrace and incorporate outside gods into their own religions.

Much of the mysticism and wisdom of religion has been buried under thousands of years of dogma that has become more about touting the religion's supposed superiority over others than about encouraging spiritual and scientific exploration. Ideas meant to illuminate humanity have been distorted and stifled, and religion has become one of the largest obstacles to human growth. Thus, the original inspirations, images, and promise of religion have been all but lost. History proves that lusting after spiritual gains is just as bad as any lust for material possessions, and often worse.

> *We have been taught to think in limited, shame-producing ways since we were children. This is the work of religion, not spirituality. Religion is comprised of concrete dogmas, while spirituality is comprised of direct realization of the Real.*
> Michael Beckwith

Changing attitudes toward sex are another example. Many early religions celebrated fertility, femininity, and sexuality. Many early Indian temples had prostitutes. The goddess Ishtar never tired of sex, Egyptian gods are rife with sexuality, and even Clement of Alexandria (one of the early fathers of the Catholic Church) says that the real eunuchs are those who are able but unwilling to have sex. This early acceptance of the human sex drive (an instinct so powerful it trumps the drive to eat and drink) gave way to a growing disconnect between happiness and goodness.

Today's *synagogues, mosques,* and *churches* (houses of worship) preach that what feels good can't be good. Sigmund Freud (1856-1939) referred to this as exchanging happiness for security. Nowhere is this more evident than in the United States, where a strong religious presence duels against an equally strong voyeuristic tendency. Far too many Americans don't enjoy sex, which may be one reason why the divorce rate is so high. (There are other reasons that I will discuss in The Romantic Savage.) Combining the churchly obsession with the human sex drive with the carrot/stick of eternal bliss/damnation gives religion a tremendous amount of control over its followers, control that extends to and includes acts of war undertaken with the promise of a debauchery-filled afterlife.

> *For most people today, the term 'spirituality' means going to church and swearing allegiance to a set of doctrines.*
> R. Craig Hogan

We accept religious intolerance as a given these days, often failing to realize that it's a fairly novel invention. Monotheism is by nature intolerant (see Chapter 7) in a way that paganism cannot match. It's hard for one god to threaten another god when there is no limit to how many gods can exist. Underlying ideals also play a huge role. Hinduism and Buddhism encourage believers to go beyond the gods and find the Ground/Godhead that lies beyond. By contrast, the Western religions forbid this, and demand total faith and separation from other systems of belief. Toning down the claims about how special they are could bridge many of the gaps between the Western religions and between West and East. The divide between religious people and nonreligious people is far larger than the divide between any two religious people.

But who prays for Satan? Who, in eighteen centuries, has had the common humanity to pray for the one 'sinner' that needed it most?
Mark Twain

Darwin investigated the creative force behind plants and animals, and generated a comprehensive theory of evolution. Religious believers are fully entitled to ask about the mechanism of natural selection, an idea so powerful that it demands explanation. With that said, they should remember that their religions—all religions—began as science. The divide between what we know as religion and science today is an artificial one. Religious people pooh-pooh evolution. On the other side, serious speculating about God's existence is a great way to throw a wrench into an otherwise promising scientific career. What started as a single pursuit has evolved—or devolved—into mutual dismissal and ridicule. Religious people may need to abandon the quest to know God fully, especially in light of recent scientific discoveries; however, as I said in Chapter 1, they can take heart in knowing that the inability to know God does not for a moment mean that God does not exist.

In short, the central challenge facing religion is that it must mature more if the world is to survive.

Universal Truths

Man is harder than iron, stronger than stone and more fragile than a rose.
proverb

Despite the serious challenges I outlined above (plus those that I will describe more in Chapter 12), religion at its heart is an attempt to connect to and come to terms with some of the universal truths that have confronted us since the beginning. As philosopher Franz Rosenweig said (1886-1929), life is a series of leaps into pathlessness, never a final decision or choice. The future is always unknown, and we are moving from known to unknown all the time.

Psychologist William James (1842-1910) pointed out that normal waking consciousness is only one type of consciousness, while other forms lie barely hidden beneath the surface. On the mundane level we are all accustomed to, Genesis happens every morning when we wake up. The idea that we consist of a single unified self is an illusion, just like so many other things in life. We think we need an identity to get through the day, but what we call "self" is only a momentary thing. These fleeting moments string together into a pattern that spans a few decades of the universe's 13.7-billion-year history before dissolving. Somewhere deep in their psyches, our ancestors understood that this temporary coming together and eventual dissolution is part of a grand illusion, and that there is no "core self" to lose when we die, because that eternal part of us lies beyond our time-bound ideas about who we are in this lifetime.

Having demonstrated the validity of creation myths from around the world and briefly described how they evolved into religions, our next step is to look at those religions.

To the enlightened man whose consciousness embraces the universe, to him the universe becomes his 'body,' while his physical body becomes a manifestation of the universal mind, his inner vision an expression of the highest reality, and his speech a reflection of eternal truth and mantric power.
Lama Govinda

Chapter 5

The World's Religions

Let the hero born of women crush the serpent with his heel since god is marching on.
Battle Hymn of the Republic

The philosopher Karl Jaspers (1883-1969) coined the term "Axial Age" to describe the era between roughly 800BC and 200BC when a wave of new religious and philosophical ideas swept the globe. People seemed to be aware that gods did not intervene in daily affairs, a view that has increasingly given way in the West to the idea of a paternal God who watches over us and steps in when needed. These glaring differences between religions belie the extraordinary similarities that appear when one peels back layers of dogma to reveal the underlying mystical traditions.

These traditions seem different because religions define the context of mystical visions and experiences. A Buddhist won't see the Virgin Mary, nor will a Christian see the Hindu god Krishna. Still, all mystics will agree that God is both very real and defies description. Mysticism tends to be viewed with some suspicion, because it relies on subjective reports of individual experiences that cannot be empirically verified, leaving the door open for massive fraud; however, *electroencephalograms* (brain wave recordings) of meditating people show profound changes compared to normal waking consciousness. Mystics universally report moving beyond their egos to see a deep unity pervading the entire universe.

The Eastern religions therefore see the self as an illusion, making birth and death equally illusory and thus nothing to fear. Eastern deities do not require conversion and rejection of prior gods, nor do they help people attain salvation. A person has as many chances as they need to go through the cycle of birth and death to achieve enlightenment and rejoin the deity. By contrast, the Greeks thought that a self-identified soul continues on after physical death, an idea that lies at the core of the *JCI* (Judeo-Christian-Islamic) religions. Adam and Eve fell from grace by becoming self-conscious, and were therefore separated from a God who is not automatically on His people's side and who only grants people one chance to get it right or face eternal damnation and torture. Read between the lines, and you will see plenty of similarities between Adam and Eve's fall from grace and the Buddhist ideas of suffering, impermanence, and the illusion of self—the same story being told in a radically different way that begets radically differing philosophies and behaviors.

This chapter provides brief glimpses at some of the world's major religions, a fundamental understanding of which provides a crucial piece of the puzzle we are assembling. Later chapters will continue exploring religion, focusing primarily on the JCI religions.

Take the bee gathering honey from different flowers, the wise man accepts the influence of different scriptures and sees only the good in all religions.
Srimad Bhagavatan

Judaism

Judaism is both a religion unto itself and the foundation for both Christianity and Islam. Roughly half of the world's population belongs to one of the JCI religions. Ancient maps show Jerusalem at the center of the world, and it remains important to all three religions today. In Chapter 3, we saw how the Genesis story tells the scientifically accepted history of the Universe perfectly. We also know that the mythical Garden of Eden was thought to exist in an area known today as the cradle of modern civilization. One could be forgiven for thinking that Judaism and its offshoots are the "true" religions.

One might also think that the Hebrews were a great civilization, but just the opposite is true: They were a smattering of tiny Bronze Age nomadic tribes, mere country yokels compared to the technologically superior Philistines (Palestinians) who also occupied the tiny strip of land between large empires

The first war in Iraq was between Cain and Abel.
Bruce Feiler

> *How odd of God to choose the Jews.*
> Anonymous

such as the Akkadians, Babylonians, and Egyptians. This strip of land (Israel) served as the world's corridor for travel between these empires, which explains why the Jewish religion seems to have been cobbled together from so many different sources. It also explains why Israel's history seems to be one of permanent crisis. The only way to survive in such a situation is to band together by creating a strong shared identity under a strong leader. It is no wonder that the first king to do right in God's eyes was Josiah, who may well deserve credit as the true founder of the Jewish religion. It is also no wonder that the Bible enjoins the Jews against mingling with foreigners and their gods. (See Joshua 24, for example.) According to the Dead Sea scrolls, the children of Israel are really the children of El, the most high god, who created different ethnic groups and gave each to a different god, with Yahweh receiving the people of Israel. So far, Yahweh is one god of many.

Early in the Old Testament story, God becomes sorry that he made the Earth and wipes everyone and everything out but for a boatload of the faithful and mating pairs of all animal species to repopulate the world. From then on, God makes a covenant with His people: God would grant peace and prosperity so long as the people followed God's law. The people obey for a time but wander, only to be struck by calamity after calamity. God sends a *messiah* (savior) to bail them out, and the cycle begins again. The people's success depends on the strength of their unity, which is measured in their belief in the one God and obeisance to God's laws. The Old Testament therefore focuses far more on the waxing and waning fortunes of the Jewish people than on miracles. It also helps establish the idea that some people receive direct inspiration or revelations from God to write down His word for humans to use—an idea that transformed the world.

Monotheism

> *With such commonality in core beliefs between and within the major monotheistic faiths, it is to many of us absurd that differing theologies can be the source of tension, debate, and much worse.*
> Stephen D. Unwin

As we saw in Chapter 4, humans tend to be naturally polytheistic. As we will see in Chapter 7, the conversion to monotheism was a long, drawn out, and often painful affair driven by politics and economics. The basic idea was that worshipping only one God would give Israel the national identity and unity needed to dominate the world. (Polytheism implies split loyalties.) To this day, much of the world is preoccupied with Israel, a scrap of real estate smaller than the state of California. There is a very real possibility that the next world war will be fought because of Israel. Not world domination, perhaps, but still.

Josiah was literally the godfather of modern monotheism. He transferred the people's allegiance from their traditional pantheon to Yahweh, which also consolidated his own power. He centralized Yahweh worship at the temple in Jerusalem (where he "found" the book of Deuteronomy) to avoid different temples spawning different interpretations. It is no coincidence that the Bible sings Josiah's praises. These efforts were effective but not totally successful; differences persisted between the official religion and popular beliefs. For example, the Pharisees believed that God could be everywhere walking among them in every moment of daily life. Needless to say, the Bible excoriates these progressive beliefs.

> *Israel, the world's greatest hallway.*
> Bruce Feiler

Why erase past myths? Consider the huge theological and political stakes: Myths deal with plenty of formidable gods who can sometimes thwart your own ambitions. Having one all-powerful God solves this problem, especially when accompanied by a king blessed by this God who acts according to God's will. Any and all setbacks can be attributed to a breakdown in faith or obedience. As ironic as this sounds, the worse things get, the more one can believe in the same system that caused the problem in the first place. This is why Isaiah 43:10 quotes Yahweh as saying that, "Before me there was no god formed, neither shall there be after me." This directly contradicts Exodus 20:3, in which Yahweh's very first of the Ten Commandments is, "You shall have no other gods before me." The latter is a clear admission that Yahweh is not the only game in town.

In fact, none of the JCI religions are truly monotheistic. Jews, Christians, and Muslims believe in the evil Satan, a being powerful enough to both negotiate and battle with God. The story of Job shows the "one true God" torturing one of his most faithful followers just to prove a point to Satan—a clear show of deference to Satan's power. To this, Christians add Jesus Christ and the "Holy Spirit" while trying to preserve the label of monotheism by describing them as aspects of a *triune* (three in one) God, or Trinity. Many a church sage has pondered the "mystery" of the Trinity and written it off as beyond understanding because they refuse to acknowledge the polytheistic

> *The continued existence of Jewry down the centuries is rationally inexplicable.*
> Nicolas Berdyaev

nature of their beliefs. Christianity and Islam also have troupes of angels, saints, and jinn. One could go so far as to say that these religions are monotheistic in name only. They certainly retain traces of the polytheistic past.

If a man extols his own faith and disparages another because of devotion to his own and because he wants to glorify it, he seriously injures his own faith.
Akosha

The Old Testament is rife with references to other gods. Jews are warned not to anger Yahweh by following these other gods. Why include these warnings if there are no other gods? Why should Yahweh be jealous of other gods if they don't exist? Even if they do exist, why must Yahweh act like a playground bully? The ancient Jews clearly practiced *monolatry*, acknowledging the presence of other gods while prohibiting their worship. For example, if the Moabites wanted to worship Chemosh, that was their business—even if passages like Jeremiah 48:46 warn them, "Woe be unto thee, O Moab! the people of Chemosh perisheth: for thy sons are taken captives, and thy daughters captives." Keep in mind that the Bible was written and edited over centuries by people who decided what to keep and what to discard; thus, the many hints of polytheism that remain are probably only the tip of a very large iceberg.

Baal is Yahweh's chief rival for Jewish hearts and minds. Yahweh absorbed Baal's personality but had to renounce it in order to become the one God. The prophet Elijah tells of a showdown between Yahweh and Baal, in which the faithful prepared a bull sacrifice and invited each god to ignite it from heaven. One would think that Baal could simply throw down a lightning bolt and seal the deal with his 450 cheering supporters, but Yahweh is the one who roasts the bull even after it had been soaked in water. (How dousing the bull in water would challenge the God who made the Sun and other stars is not explained.) The triumphant Yahweh then makes rain (Baal's specialty), and the Jews are convinced. From then on, Yahweh softens his tone, knowing that he is Top God.

Politics and economics gave us the one true god of the Abrahamic faiths.
Robert Wright

Defeating Baal did not mean letting Baal's stories and exploits go to waste. According to H.L. Ginsberg, Psalm 29 may have originally been written to Baal. Another scholar changed all Yahweh references to Baal and found that the amount of alliteration in the Bible increased dramatically—yet another example of how Judaism does nothing more than amalgamate preexisting beliefs from sources such as Egypt, Sumer, Akkadia, and Babylon. The clear implication is that the knowledge contained in Genesis that we saw in Chapter 3 existed long before Genesis was written, an idea we will return to later.

In Second Isaiah, Yahweh is loud and proud, having won the people's hearts and minds. In some ways, monotheism is the ultimate revenge: The Jewish God is both universal and national; He created the universe but places Israel above all others because He has a covenant with His people. No other group can possibly ever have the same special status or relationship with God. All other people are second-class citizens at best. Anyone setting out to create propaganda that legitimized and even deified Josiah's agenda would have come up with something very similar to the Old Testament we know today. Josiah didn't write the entire story, but he wrote at least some of the crucial parts, as we will see in Chapter 7.

If the historical faith of Israel is not founded in history, such faith is erroneous, and therefore, our faith is also.
Roland de Vaux

The conversion to monotheism was neither easy nor quick. The Exodus story tells of the Jews needing 40 years of hardship for Hebrews to embrace one God and finally move on to the Promised Land, a distance that ordinarily requires about three weeks on foot. Even this was not enough to stamp out polytheism once and for all, because Satan is still thought to exist. Even if one assumes that Satan exists at God's whim, the best we can say is that Judaism is a monolatric religion, not entirely monotheistic.

One might think that the one true God would be a pretty peaceful and self-assured fellow once His place on the universal throne was secure. The truth, however, is a little different.

The Vengeful God

Yahweh is a warrior god. We cannot be sure what combination of invention, co-opting, or amalgamation went into this deity, but the fact remains that Yahweh starts off as a savage, warlike, hypocritical god. Whereas the Greeks, Romans, Syrians, and most others considered their gods amoral and indifferent to humans, Yahweh was an in-your-face god who used life and death to control his subjects—a clear precedent for the establishment of the legal death penalty. Of course, God would rather not kill the sinner (Ezekiel 20:21), but will not hesitate to strike down the unrepentant or one who strays into sin (Ezekiel 20:22). Still, God is so determined to have the

While the Judeo Christian Islamic god described in scripture is hardly benevolent, the faithful of those religions are far more likely to ignore unpleasant scriptural passages than abandon belief in a benevolent god.
Victor Stenger

Jews worship him that he sends the Babylonians to destroy them because they aren't good enough.

Faced with the choice of obeying God or following Eve, Adam responded like billions of men by doing the evolutionarily correct thing and siding with his wife, thus "choosing sin." If God designed and created humans, then God is responsible for making reproduction our #1 urge above even food (see *The Natural Savage*); to condemn Adam for behaving as designed is tantamount to buying a car only to fault it for not being an airplane. Still, both Judaism and Christianity took this absurd concept and ran with it. For example, Augustine (354-430AD) used this story to justify imperial rule, saying that humans need tyrants because they had sinned so badly. The Catholic Church ran with this concept; Augustine marks the beginning of the Church's obsession with sex and alliances with governments.

I've often thought the Bible should have a disclaimer in the front saying this is fiction.
Ian McKellen

Similarly, the story of the Tower of Babel, where God scatters the people and makes them speak mutually unintelligible languages, could well mark the beginning of God's opposition to science and reason, because God is worried that they will learn too much. Overall, the main Old Testament narrative is the Deuteronomic history that explains all misfortunes as occurring because the people strayed from God's will. Human curiosity and sexuality are the strongest drives our species possesses. The Old Testament would have you believe that God created both of these drives only to prohibit us from using them in favor of blind obedience and faith—just the kind of traits a young king needs to consolidate his power by fomenting a xenophobic nationalism and a model for most, if not all, secular and religious excesses that have occurred since then. Yahweh thus continues to be vengeful today.

The Exodus

Thou shalt not bear false witness.
Exodus 20:16

The Exodus was intended to establish a national myth for the Jews, who did not distinguish between natural laws and divine intervention. The story of the flight from Egypt is the foundation for the idea of one true God and the notion that the Jews are God's "chosen" people. Jews pray facing Jerusalem and pray to be able to have Passover (the feast of the Exodus) there—a Jewish version of the Muslim *hajj*, or pilgrimage.

According to Exodus, the Egyptians forcibly removed the Canaanites to Egypt to perform slave labor. Yahweh intervened, and the people escaped back to Canaan to become the Jews. A compelling story. There's just one hitch: There is no physical evidence to indicate that anything about the Exodus is based in fact. Decades of painstaking research have discovered no traces of the Old Testament battles, nor of a sudden influx of desert wanderers or Jewish displacement of the Canaanites. In fact, we can safely conclude that the Jews *were* Canaanites; the lack of fortifications at settlement sites indicates a mostly peaceful existence. There is simply no sign of any mass migration from Egypt nor proof that Moses existed.

God's Chosen People

Many—if not most—tribes consider themselves special for reasons that make perfect evolutionary sense. In this light, the Jewish idea they are somehow special is nothing exceptional. What is exceptional is the concept that a lone God who created all of the tribes on Earth would choose a small backward group sandwiched between large empires as His special children. The basic idea is that Adam and Eve's sin upset the cosmic order of things and God chose the Jews to help Him restore order. The Jews therefore have a special mission that God is depending on them to complete.

Interfaith problems are rooted in Abraham.
Bruce Feiler

Getting and maintaining religious faith is therefore crucial. The Bible sees marriage as the way to have and raise children. (The idea of reuniting man with his missing rib may have been intended as humor.) Intermarriage risks diluting the faith and exposing children to different faiths—not the way to preserve the kind of cohesive group needed to perform God's work. The Bible therefore contains numerous restrictions on intermarriage. Jews are therefore defined more by blood (having a Jewish mother) than by faith. In fact, people of Jewish descent have genetic markers that can be detected using DNA testing.

When compared to the regime of Moses, the regime of the Taliban comes off looking like the ACLU.
Michael Earl

DNA testing was not available in the ancient world, but there were several ways to help Jews identify each other and avoid any temptation to breed outside the faith. Circumcision was an easy way to spot a Jewish male and also demonstrated people's commitment to God because of the difficult (and painful) sacrifice of genital mutilation. Untold millions of babies, Jewish and others alike, have been circumcised at countless hospitals and during countless ceremonies around the world. Ethnic cleansing is another way to keep the tribe pure, which is why the Bible repeatedly instructs the Jews to destroy other people living in their land; polytheistic impulses were strong enough without the added weight of living followers. There could be no peace between Israel and her enemies, a concept that appears to be alive and well today. According to the Jews, much of their historical misfortune has come because they strayed too far from God's law. The question of whether Jews should assimilate into their host nations or remain separate remains a thorny issue. Orthodox Jews insist on following the Old Testament literally; Reform Jews think they must evolve to meet changing circumstances; Conservative Jews are somewhere in the middle.

The messiah is not coming, and he's not even going to call.
popular Israeli song

When the Babylonian exile ended in 536BC, the Jews attempted to restore the line of David. Their failure led to the belief that God would send a future messiah to establish His perfect kingdom. Later, the Seleucids tried to Hellenize the Jews, but a successful Maccabean revolt led to the rededication of the Temple in 146BC, an event still commemorated as Hanukkah. The Romans took over in 63BC and no amount of rebellion could dislodge them. Why was God abandoning His chosen people? Perhaps those dying for God gained special status. This led to the idea of an afterlife, the precursor to Christian and Islamic ideas of heaven and hell.

As we will see, the myth of a divinely chosen people created in a narrow, xenophobic theology has persisted from the time of Deuteronomy to modern fundamentalism.

David & Solomon

Archaeological data have now definitively confirmed that the empire of David and Solomon never existed.
Niels Peter

Jesus of Nazareth, Maimonides (a Jewish philosopher, 1135-1204), and others are claimed as descendants of David, the greatest king of ancient Israel. Christ's Davidic lineage is particularly important, as we will see in Chapter 6. This lineage began and continued in ways that would shock most modern people and result in the arrest and conviction of those responsible.

Incest was common in pre-Babylonian Jerusalem. For example, a widow had to marry her late husband's brother and was not allowed to marry outside the family. This requirement stemmed from the view of women as chattel commodities, making marriage a business arrangement. Men were not required to be virgins but women were. A woman who did not bleed on her wedding night was to be stoned to death on her father's doorstep; a false accusation could cost a man a whipping and a monetary fine. Families retained the bloody sheets as evidence of their women's virtues. Of course, women were expected to obey their husbands without question. My theory is that accepting incestuous relationships increased the pool of potential mates while also preserving racial purity.

Abraham, the reputed father of the JCI religions, lived in an incestuous relationship with his half-sister Sarah. Sarah was childless and gave her slave Hagar to Abraham to have a baby. Hagar gave birth to Ishmael and became contemptuous of Sarah, who complained to Abraham. He responded by returning Hagar to Sarah to do with as she willed. Sarah's mistreatment drives Hagar into the desert with Ishmael, but an angel tells her to return with the promise that Ishmael might avenge her. Sarah gives birth to Isaac 13 years later and kicks Hagar out over Abraham's doubts. Hagar and Ishmael head into the desert with bread and a jug of water until Hagar becomes exhausted and leaves her son crying under a bush. Again the angel appears and tells Hagar to keep going, because she will soon find a spring. Ishmael goes on to have 12 sons who settle all over the land. Meanwhile, God commands Abraham to kill Isaac for a sacrificial offering. Abraham complies, but God stops him at the last moment. I can only marvel that one Deanna Laney was sentenced to life in prison for killing her children on God's orders in conservative Texas, while Abraham is still revered as a holy man and the foundation beneath the world's three largest religions... including in Texas.

Is what you hear at church religion? Is that which can bend, turn, and descend and ascend, to fit every crooked phrase of selfish, worldly society religion? Is that religion which is scrupulous, less generous, less just, less considerate for man, than even my own ungodly, worldly, blinded nature? No! When I look for religion, I must look for something above me, and not something beneath.
Harriet Beecher Stowe

After the destruction of Sodom and Gomorrah, Lot's daughters fear they will not bear heirs. They solve this problem by getting their father drunk on two consecutive nights and having sex with him. David's mother is a descendant of this relationship. In another story, Tamar disguises herself as a prostitute and seduces her father-in-law Judah to obtain an heir. She has twins, Perez and Zarah, and Perez is one of David's ancestors.

As for David, he spies the married Bathsheba taking a ritual bath one morning to purify herself after menstruation. David

> *There are two sorts of hypocrites; ones that are deceived with their outward morality and external religion; and the other are those that are deceived with false discoveries and elevation; and men's own righteousness, and talk much of free grace; but at the same time make righteousness of their discoveries, and of their humiliation, and exalt themselves to heaven with them.*
> Jonathan Edwards

seduces her, and she becomes pregnant. To cover himself, David orders her husband to return the favor by sleeping with his own wife, but the husband refuses because that would make him ritually impure for battle. David tries bribery and intoxication, but nothing works. Finally, David sends him to the front to be killed and takes Bathsheba for his own. Later, one of David's sons rapes his sister and is killed by David's other son, who then stages a coup against his father. Tens of thousands are killed in the ensuing battles. When David becomes impotent, Bathsheba convinces him to name Solomon as his heir. Despite all of this, God promises an endless dynasty if they follow the law—just one of many examples of the religious elite being seemingly exempt from the laws the laity are supposed to live by (the Ten Commandments).

David wanted to create a trinity of one God, one king, and one city. This is the origin of the later Trinity of Father, Son, and Holy Spirit. As part of this effort, Solomon built the Jewish temple on a Canaanite holy site. The later building of the Dome of the Rock on the same spot by Muslims is therefore just another link in the chain of usurping supposedly sacred spots for one's own purposes. The Temple Mount in Jerusalem has borne witness to many historical and mythical events, from Christ's Passover pilgrimage to Mohammed's ascent to heaven. The Temple itself was impressive but hardly unique. All religions had houses for their gods; few passerby would think that Yahweh was anything special just by seeing the Temple. Solomon even built other temples to other gods, an ecumenical gesture that earned him Yahweh's wrath and stripped him of his kingdom. (More evidence that the transition to monotheism was not fast or easy; see Chapter 7.)

Modern Israel

> *Let me tell you something that we Israelis have against Moses. He took us 40 years through the desert in order to bring us to the one spot in the Middle East that has no oil!*
> Golda Meir

In the late 1890s, Zionist Theodore Herzl (1860-1904) decided that anti-Semitism was a permanent part of non-Jewish society, and called for the formation of a Jewish homeland in what was then Palestine. He asked Sultan Abdul Hamid II (1842-1918) to cede the land, but the Ottoman Sultan refused, saying, "If one day the Islamic State falls apart then you can have Palestine for free, but as long as I am alive I would rather have my flesh be cut up than cut out Palestine from the Muslim land." Herzl received a more sympathetic response from the British in the form of the Balfour Declaration of 1917 that supported his dream of a Jewish state. This dream took a huge step forward when Britain took over ruling Palestine from 1920 to 1948 after World War I ended the Ottoman Empire.

The Holocaust of World War II led many Jews to wonder if they had again strayed from God's law and were being punished. Others thought that Hitler may have been an agent of God spurring the Jews to return to their ancestral homeland. Still, the tragic events of the 1930s and early 1940s meant that the Jews could wait no longer. They took matters into their own hands and began migrating en masse to Palestine to begin rebuilding Israel. The modern nation of Israel was formally born on May 14th, 1948, when the last British troops left Haifa and David Ben-Gurion announced the new state. Israel and the Arabs have been in conflict ever since—a modern-day reenactment of the old wars with the Philistines with much higher stakes, thanks to international alliances, the rise of Islam, and the proliferation of weapons of mass destruction.

Our generation is realistic, for we have come to know man as he really is. After all, man is that being who invented the gas chambers of Auschwitz; however, he is also that being who entered those gas chambers upright, with the Lord's Prayer or the Shema Yisrael on his lips.
Victor Frankl

Life after Death

The ancient Jews initially had no belief in a soul, resurrection, or rebirth. Passages such as Genesis 2:7, Genesis 3:19, and Psalms 115:17 say that death is the end. Genesis 3:19 is particularly blunt: "By the sweat of your brow you will eat your food until you return to the ground, since from it you were taken; for dust you are and to dust you will return." Dust is also used in another context with a far different meaning, as we saw in Chapter 3.

I knew a man once who said, Death smiles at us all. All a man can do is smile back.
Russell Crowe as Maximus

Psalms 146:4 reverses this idea by saying, "When their spirit departs, they return to the ground; on that very day their plans come to nothing." This passage explicitly describes a dualist belief. Other indications of this emerging dualism include the dead Samuel appearing to Saul in physical form to give advice about a looming battle with the Philistines. In this example, Samuel clearly had foreknowledge despite being already dead.

Christianity

Christianity is a consequence of the Jewish belief that a *christos* (anointed one) or messiah would be sent by God to defeat the

God forbid that we should believe in this for Christ died once for us for our sins and rising again dies no more.
Augustine

> *Take this nation for Jesus. Whoever stands with the messiah will rule with him.*
> Bill McCartney

Romans. Jesus Christ was born of the Virgin Mary as God's son and the incarnation of God on Earth. From the Jewish perspective, Christ's mission to defeat the Romans was an abject failure since he was thought to have been crucified by them. To the Christians, Christ died to atone for humanity's sins, making salvation available to all. We will look at Christ's life in Chapter 6; meanwhile, it is important to note that Christianity did not exist while Christ was alive. Christ himself was a Jew, and Christians therefore see their history as extending back to the moment of creation described in Genesis 1:1.

The belief that Christ rose from the grave after three days is part of the bedrock of the Christian religion. Most other religions are based on cycles of life and death (such as the Buddhist belief in reincarnation), but Christianity broke this cycle. They believe that humans get one lifetime to either become good Christians and enjoy a blissful eternity or stray from the path and endure eternal torment. God's love is a one-shot deal, unlike the infinite love of Brahman. Christians unite with Christ through the rituals of *baptism* (ritual purification in water) and the Eucharist, in which crackers and wine are thought to *transubstantiate* (change) into the actual flesh and blood of Christ himself. Cannibalism, anyone?

The Romans allowed people to worship any god(s) they chose, provided they paid token homage to the official state gods. The Christians refused to do this, and both refused and challenged the legitimacy of all other gods. They also proselytized heavily to gain converts. Unlike the other gods in the Roman Empire who were mostly aloof, the Christian god was a ferocious primitive who stuck his nose into human affairs; to the Romans, a Jew who died a dishonorable death in some far corner of the land was hardly the role model for a god. If that was not enough, the Christian community was more than a little insular; becoming a Christian meant breaking with one's families and old beliefs, values, and practices—a process called being "born again" that is virtually identical to modern cults. Once a Christian, a person refused to deal with outsiders except to convert them; the commandment, "Love thy neighbor" really means, "Love thy fellow Christian."

> *Well, you think the Christian is capable of every crime—an enemy of the gods, of the laws, of good morals, of all nature.*
> Tertullian

Tertullian (c. 160-220AD), credited as "the father of Latin Christianity" summed up the growing Roman antipathy to Christianity as follows: "The husband casts the wife out of the house; the father disinherits the son; the master commands the slave to depart from his presence; it is a huge offense for anyone to be reformed by this hated name [Christians]."

Paul

Paul of Tarsus is one of the key architects of Christianity despite never seeing Christ and being either unknown to or reviled by the apostles. He wrote letters to Corinth and other places claiming to be an apostle of Christ and saying that he had been taken up into a paradise that he could not describe because no mortal is allowed to speak of it. (If you think this sounds awfully similar to Joseph Smith's story about magical stones that only he could interpret or just about any other cult leader, you're not alone.) Other people spreading Christ's message specified that qualifying for His salvation required following Jewish law, including diet restrictions and circumcision (a strong disincentive in the days before anesthesia). Seeing these barriers to entry, Paul said that neither Judaism nor circumcision was required; anyone could become a Christian and receive Christ's salvation. In fact, he railed against people who would castrate themselves, a rebuke that may have been directed at the followers of Cybele, and one that certainly went against the practice of circumcision practiced by the Jews.

If nothing else, Paul was a consummate marketer. Whereas pagans distribute their faith, Christians worship in one congregation and develop close bonds with their peers. Paul emphasized this theme, preaching the concept of brotherly love to anyone who would listen. Who does not want to join an instant family where one is respected and loved? Who would not want to find friendly, welcoming faces wherever s/he travels? By building communities of Christians who loved each other based on their faith, Paul was creating one of the earliest known franchises, despite the fact that his views of Christianity were probably far from the original views of Christ. For example, the Ebionites believed that Christ was a thoroughly human messiah born to normal parents but adopted by God because of his exemplary conduct—a belief that is more faithful to the Bible than modern Christianity! Unlike Paul, the Ebionites made it hard to join their movement; they died out and Paul's vision carried the day. By the end of the 1st century AD, Christianity was no longer seen as an offshoot of Juda-

What I am saying is true and reasonable.
Paul

I often found myself preferring the company of people outside my congregation, men and women who did not follow Jesus. Or worse, preferring the company of my sovereign self. But soon I found that my preferences were honored by neither Scripture nor Jesus. I didn't come to the conviction easily, but finally there was no getting around it: there can be no maturity in the spiritual life, no obedience in following Jesus, no wholeness in the Christian life apart from immersion and embrace of community. I am not myself by myself. Community, not the highly vaunted individualism of our culture, is the setting in which Christ is at play.
Eugene Peterson

ism, and the first signs of anti-Semitism became apparent in the Church.

Christian initiates to Paul's church were stripped naked, baptized, given new names, fed milk and honey (baby food), told the name of Jesus Christ, given new clothes, and greeted with kisses by her or his new family—a process of "rebirth" that has analogies in many cults around the world. Paul also said that Christ had commanded the Eucharist ritual, something that early Christians would have recognized as typical of cult worship of the time. In fact, Justin Martyr fretted that pagans might think the Christians copycats, with priests of "devils" (other gods) recognizing baptism and communion in "sacred" meals as knockoffs. Justin's answer? He claimed Christians invented these rituals first! Paul's beliefs about the afterlife became the earliest Christian view: Christ returned to heaven, but ordinary people must wait for Him to come back and resurrect the dead, who will then live on an improved Earth. (No mention of where all these people were supposed to fit.)

Should anyone, then, employ a teacher? For how could anyone be helped, if there is no truth even in them?
Justin Martyr

What if Paul had failed? It is safe to assume that another monotheistic religion would have taken Christianity's place. For example, the Marcionites believed that one Appolonius of Tyana was the son of God with the gift of prophesy who traveled around performing miracles. Appolonius told his followers to focus on their souls above material items and preached an ethic of sharing. Like Christ, he was persecuted by Rome and ascended to heaven. The parallels between Marcionism and Christianity are probably not coincidental, and it is fascinating to speculate about what might have happened had Paul not triumphed.

Constantine

Empires built on force will always be destroyed. Those built on trust in Christ will remain.
Joseph R. Sizoo

Emperor Constantine may well have been Christianity's most influential convert. He is largely responsible for transforming Christianity from a spiritual following to a formal organization. This history-changing development owes itself to Constantine's victory over Maxentius at the Battle of Milvian Bridge in 312AD. Having seen a vision of the cross the night before the battle, Constantine believed that the Christian God had brought him victory and converted to Christianity. He ended the Roman persecution of Christians and elevated the Church to a powerful position that it still enjoys today (albeit somewhat tempered). His law of 333 ordered imperial officials to enforce bishops' decisions and to accept bishops as better witnesses than others. He also donated the Lateran property to the Bishop of Rome, which is where the Church of Saint Peter was built and the sovereign Vatican nation established. The Christian prejudice against Jews became the law of the land.

Constantine sought to end the many theological disputes among the bishops and summoned them to the Council of Nicaea in 325 to hammer out their differences. Among other things, this meeting made Christ divine by fiat and adopted the belief in creation *ex nihilo* (from nothing). The Nicene Creed became official Church law. Today's Catholics must still profess their belief in this creed. More importantly, the Catholic Church became the only church recognized by the emperor. To quote Constantine, "The privileges that have been granted in consideration of religion must benefit only the adherents of the Catholic faith or law. It is our will, furthermore, that heretics and schismatics shall not only be alien from these privileges, but also shall be bound and subjected to various compulsory public services." This was all the bishops needed to set what Irenaeus (the Bishop of Lugdunum in Gaul, ca. 202) had begun putting into motion almost a century earlier.

Every country gets the circus it deserves. Spain gets bullfights. Italy gets the Catholic Church. America gets Hollywood.
Erica Jong

Orthodox Christianity

"Pay attention all families of nations and observe. An extraordinary murder has occurred in the center of Jerusalem in the city devoted to God's law. In the city of the Hebrews, in the city of the prophets, the city thought by us as just. Who has been murdered? Who is the murderer? I am ashamed to answer, but I must: The one who hung Earth in space was himself hung, the one who fixed the heavens in place is Himself impaled, the one who firmly fixed all things is Himself fixed to the tree. The Lord is insulted, God has been murdered, the king of Israel has been destroyed by the right hand of Israel!" This is an excerpt of a sermon given by Melito, the Bishop of Sardis, c. 170AD.

Leaving aside both the obvious question of how mere mortals could possibly kill the creator of the entire Universe, and the fact that Christ's death was required to both forgive the Christians and found their religion, the real irony is that Christ him-

Where orthodoxy is optional, orthodoxy will sooner or later be proscribed.
John Neuhaus

self was a Jew who prayed to the Jewish God, followed Jewish customs, knew the Jewish law, and had Jewish disciples who took him to be a Jewish messiah. All of this aside, it only took a few decades for Christians to form a religion that strongly opposed the Jews. This point bears repeating: There was no such thing as Christianity prior to Christ; all that happened occurred within the context of Jewish prophesies from the Old Testament for practical reasons that will become clear in Chapter 6; a new religion was the last thing on anyone's mind.

> *Universal orthodoxy is enriched by every new discovery of truth: what at first appeared universal, by wishing to stand still, sooner or later becomes a sect.*
> Edgar Quinet

Christ's followers thought of him as the man who would defeat Israel's enemies and establish a new kingdom in Israel to be ruled by God Himself, albeit through human agents. Christians ran into problems trying to sell this story to others, because it was common knowledge that Christ had been a powerless itinerant preacher who got on the wrong side of the law and died the most humiliating death Rome could dream up, making any claim of his being a messiah ludicrous. Christians therefore claimed that his death was the will of God, carried out in order to save the world. Both versions could not be right, nor could the Christians admit that they were wrong. The obvious solution was to make the Jews wrong by claiming that they had rejected the salvation offered by their own God. The arrogance of this move is simply breathtaking.

It was but a short step from making the Jews wrong to drubbing them for rejecting Christ's salvation. Justin Martyr claimed that God wanted the Jews circumcised to mark them as deserving of persecution. Tertullian and Origen (c. 185-254AD) claimed that the Romans destroyed Jerusalem in 70AD to punish the Jews who killed Christ. As we saw above, some Christians even accused the Jews of *deicide* (killing God). The path led away from a Palestinian Jew until, by the 2nd century AD, most Christians were converted non-Jewish pagans who believed that they belonged to an anti-Jewish religion that nevertheless worshipped the Jewish god. Wow.

> *How can we tell the difference between the word of god and human words?*
> Irenaeus

The Gospel of John is the foundation for a unified Church, which stands in stark opposition to the Gnostic Gospel of Thomas (see below), which emphasizes an individual search for God. The New Testament says that Christ appeared to many people, but that only certain encounters conferred authority. Belief in the bodily resurrection of Christ supports the legitimacy of men claiming to be the successors of Christ or Peter, which validates the apostolic succession of bishops and is still the basis of the pope's authority.

Christ told Peter to lead the apostles before rising to heaven and has not been seen or heard from since, aside from indirect communications. The idea that authority flows from apostolic experiences that are closed to the rest of us helps restrict Church leadership to a small group of chosen people, who alone can establish a chain of command and name their successors without any possibility of challenge. This belief has guided the Catholic Church for over 2,000 years. Today's popes trace their line all the way back to Peter. Pope Clement 1 (ca. 90-100AD) said that the God of Israel rules all things and delegates his rule to men (the Church leaders); anyone who disobeys the Church disobeys God Himself and deserves the death penalty. Christianity was thus divided into clergy and laity, with a strong pecking order among the former. (Of course, it was the bishops themselves who were responsible for concocting the entire system.) Ignatius of Antioch (1st century AD) went one step further: The hierarchy on Earth reflects the hierarchy in heaven; there is only one God so there can be only one Church leader. All had to obey this leader as though he was God. Tertullian went so far as to ridicule those who were seekers instead of believers, a direct violation of Christ's own words in Matthew 7:7 when he admonishes us to, "Seek and ye shall find." Oh, the hubris!

Tertullian also created the orthodox idea that all believers would be physically resurrected because Christ rose from the grave. He believed that stories of Christ's ordeal and resurrection should be told precisely because they are absurd, and that anyone who denies bodily resurrection is a heretic. Different passages in the Bible imply different things; interestingly, in 1 Corinthians 15:05, Paul himself denies bodily resurrection, saying that the perishable does not inherit the imperishable.

The Gospel of Mark (c. 70-80AD) portrays Christ being betrayed by Judas, arrested, tried, convicted of sedition in front of Pilate, crucified, and crying out before dying. Later Gospels paint Christ in a more heroic light. All of the Gospels try to show Christ's innocence. Orthodox Christians must believe that Christ's crucifixion as a human being was an actual event, despite many hints to the contrary in the *Koran* (the book chronicling Mohammed's revelations) and else-

When I was a young boy, my father taught me that to be a good Catholic, I had to confess at church if I ever had impure thoughts about a girl. That very evening I had to rush to confess my sin. And the next night, and the next. After a week, I decided religion wasn't for me.
Fidel Castro

The Gateway to Christianity is not through an intricate labyrinth of dogma, but by a simple belief in the person of Christ.
William Lyon Phelps

where. This was important, because it related to how Christians should react to persecution. To the Romans, Christians followed a wandering magician who denounced Roman gods, refuted the emperor's divinity, performed ritual cannibalism in the form of the Eucharist, and was executed for treason. Anyone who confessed to be a Christian was killed; someone who recanted and cursed Christ was pardoned. The Christians admired their martyrs, which was fine for the Romans, because gladiators were expensive and the games had to go on. Despite Rome's best efforts, persecution actually made the Church stronger. "We grow as much as you cut us down," proclaimed Tertullian. "Blood is the seed of Christians."

> *The church must be reminded that it is not the master or the servant of the state, but rather the conscience of the state.*
> Martin Luther King, Jr.

Bishops assumed political roles as part of the effort to protect and expand the Church. The pope was originally a judgeship that became a spiritual leader. Not everyone accepted papal power, which caused rifts among Christians. For example, Arius of Alexandria (ca. 250-336AD) believed that people should follow God directly, while Athanasius of Alexandria (ca. 293-373AD) argued on behalf of the Church; after all, a *catholic* (universal) church can tell converts that it possesses both truth and God's official blessing. Conveniently, the Gospel of John tells people that the only way to truth is by believing in Christ. John also praises people who believe without seeing and calls Thomas's faith into question because he tried to find truth from his own experiences. The Catholic assault on science and reason continues to this day, possibly emboldened by theologians who believe that the Holy Spirit guides the Church; any tradition that survives today must therefore be God's will.

Athanasius made a list of Old and New Testament books that he deemed divine and others that he felt were polluted. This is one of the first attempts at creating a Christian *canon* (body of authoritative writings) of Holy Scripture. The New Testament that we know today was agreed upon at the Councils of Carthage in 393AD and Hippo in 397AD. The Council of Oxford in 1407 prohibited translating the Bible, but it was too late: John Wycliffe translated the Bible into *vulgate* (common language) and used the power of the newly invented printing press to spread copies far and wide. This at last broke the priestly stranglehold on all aspects of daily life, because for the first time people could see the differences between the Bible

> *A heresy can spring only from a system that is in full vigor.*
> Eric Hoffer

and the Church for themselves. Still, the Catholic Church soldiers on. As an excerpt from a letter to Diognetius written circa 150AD says, "Christians are in the world what the soul is in the body. The soul is dispersed throughout the body, Christians throughout the world. Christians under daily punishment flourish all the more."

The Gnostics

Gnostic Christianity picks up where orthodoxy ends. Gnosticism is based on self-observation and prayer (both mystical states), and relies on self-knowledge to know God. The idea that one finds God only through Christ implies that one needs a church and its attendant institutions. *Gnosis* (knowledge) is the path to God; ignorance causes suffering. One sees the vision through the mind, which lies between the spirit and the soul. (Compare this to Buddhism, which we will see later in this chapter.)

> *For one of those Gnostics the visible universe was an illusion or (more precisely) a sophism.*
> Jorge Luis Borges

Gnostic Christians came together as equals. While their orthodox counterparts built institutions of command and control, the Gnostics made few distinctions between clerics and laity. All people of both genders were equal and could be selected to serve as a priest, bishop, or prophet by casting lots—a system of chance they perceived as expressing the will of God.

The Gnostic God was transcendent, above the imperfection of material creation. God created agents until He found one strong enough to create the world. Knowing this shows one how to find God's pure spirit, which is the root of everything, the ineffable One who is alone in silence because no one came before him. This God is the perfect, preexistent, uncontainable, and invisible Godhead (Brahman). The Gospel of Truth says that whatever we see in God is what we need to see at that time; God is ineffable but presents Himself in ways we can all understand. According to the Secret Gospel, Eve symbolizes spiritual awareness; her awareness of being naked was her awareness of lacking spiritual knowledge. The Gospel of Thomas implies that only a few ever truly know God (find enlightenment). The philosopher Plotinus said that all existence owes itself to the primal One, which is all of existence; lesser beings only exist as they partake of the One. The universe consists of mind dividing itself into form and substance;

> *I shall give you what no hand has seen and what no ear has heard and what no hand has touched and what has never occurred to human mind.*
> Nag Hammadi text

animal movements that seem mechanical are actually psychical descending into the material (idealism).

The Gnostics criticized those who saw Christ as external and above his disciples. The Gospel of Philip makes distinctions between Christians who have been merely baptized, and those who have experienced a spiritual awakening and transformation. One who achieves *gnosis* (knowledge or enlightenment) is not a Christian, but a Christ. Christ himself was born normally to Mary and Joseph, but then also born spiritually; his resurrection was both literal and spiritual. The virgin birth is therefore not to be taken literally

In the Gospel of Thomas, Christ says that God cannot be perceived or understood in human terms and rebukes those who tried to find God outside themselves—including by following Christ. According to verse 70, "If you bring forth what is within you, what you bring forth will save you. If you do not bring forth what is within you then what you do not bring forth will destroy you." (Compare this to the Buddhist idea of karma.)

The Gnostic theologian Valentinus (c. 100-160 AD) said that the anticipation of death begins the search for gnosis. One Nag Hammadi text says that resurrection is real, yet the world is an illusion. Human existence is spiritual death, and resurrection occurs at the moment of enlightenment. Thus, anyone can be resurrected at any time; those who think they have to die first are mistaken. Christ invited James and Peter to accompany him to heaven in life and to surpass him in gnosis. In the Gospel of Mary Magdalene, Christ says there are seven powers of wrath that challenge one's soul and how to overcome them. (Compare this with Tibetan *bardos* and near death experiences.)

The first people to investigate the Gnostics were their orthodox contemporaries. The Gnostics believed that the orthodox God of Israel was only an image of God, and that one can know this through initiation into the secrets of gnosis. The practical implication of this belief was discrediting the orthodox bishops, who the Gnostics accused of teaching only elementary doctrines—a clear challenge to Catholic authority. Some Gnostics even denounced the Catholics as heretics who

> *Each person recognizes the lord in his own way, not all alike.*
> Theodotus

> *They really have no gospel which is not full of blasphemy.*
> Irenaeus

"do not understand the mystery of truth while claiming it belongs to them alone."

"All of them are arrogant, all of them offer you gnosis!" lamented Irenaeus. And yet, orthodox Christianity does contain some ideas that seem even stranger than gnosis, such as a perfect God who created an imperfect world, that Christ was born through *parthenogenesis* (from a virgin), that he died and rose from the grave, etc. According to Valentinus, only personal experience contains the truth, all people are spiritually alive, and authority must therefore be flexible and open. Contrast that with Tertullian, who claimed that no one judging from personal experience would ever believe that Christ rose from the grave; the people therefore had to accept what the priests told them.

The Gnostics recognized the political implications of believing that whoever sees God can claim equal or greater authority than the disciples and their successors. Of course, those who are wiser than the apostles are wiser than the priests. Gnostics made no secret of their disdain for orthodox priests, accusing them of making up stories in order to grab power

Orthodox apologists like Irenaeus accused the Gnostics of fraud, claiming that the canonical Gospels were written by Christ's own disciples and others with firsthand knowledge of the described events. (Few modern scholars think this is true.) How could someone write the Secret Gospel of John (for example) over 100 years later? Someone could ponder questions, get answers, and attribute this to God. Irenaeus accused the Gnostics of making it all up as best they could, because no one could be initiated unless s/he came up with some real whoppers; to Irenaeus, the only truth came from the apostles and was handed down by the Catholic Church.

Irenaeus took the structure of heavenly and earthly authority seriously. If there is only one God, then there can only be one Church and only one representative (pope); all orthodox Christians therefore had to believe in one God and obey the priests in the churches who carry the apostolic succession and receive the real truth along with their rank. As he said, "All (such as the Gnostics) who harm the Church's teachings have fallen from the Church and are spreading Satan's poison." Even orthodox followers who did not repent and fall into line

---

Others outside our number call themselves bishops and also deacons, as if they had received their authority from god. These people are waterless canals.
The Apocalypse of Peter

It is necessary that every church should agree with this church, on account of its preeminent authority.
Irenaeus

had much to fear when it was their turn to be judged. It should go without saying that none of the Gnostic gospels was admitted into the New Testament.

> *Where questions of religion are concerned, people are guilty of every possible sort of dishonesty and intellectual misdemeanor.*
> Sigmund Freud

Tertullian accused Gnostics of making up stories to avoid persecution and death. He called their arguments against *martyrdom* (dying for one's religion) heretical, saying that good Christians would go to their deaths while the heretics went about their business as usual. In Peter's Apocalypse, he sees Catholics coming for him and is frightened until he sees a vision of God, who tells him that all who follow the Church will become prisoners because of their ignorance. Catholics send each other to die in the delusion that clinging to the name of a dead man (Christ) will make them pure. Peter was horrified at the violence being done to children and believers, saying that the Catholics would grind many believers to pieces, a clear vision of the Inquisition to come. He eventually learns to face suffering, because the intelligent spirit is released to join with God when the body dies. The Gospel of Truth speaks of Christ stripping himself of his perishable body and taking on an imperishable spirit form.

Irenaeus conceded that the Gnostics were trying to elevate theological understanding, but said that it was impossible for them to make up for the harm they were causing; any argument against martyrdom undermined all of Christianity. Orthodox beliefs affirm the physical body as central to life; the Church loves God so much that it regularly sends Him martyrs. This idea has enjoyed remarkable staying power: Christians who have suffered over the last 2,000 years have taken comfort in the idea of a human Christ.

> *Reason is the devil's harlot, who can do nought but slander and harm whatever God does.*
> Martin Luther

Most Christian persecution may have happened at Rome's hands, but Christians also persecuted other Christians. Gnostic texts speak of orthodox persecutions and blast them for spreading fear and slavery. The Gnostics tried to use the Church's own qualities to demonstrate the falseness of orthodoxy, but failed. By the 2nd century AD, the Catholics began defining criteria for Church membership that included baptism, professing belief in the Nicene Creed, prayer, and obeying Catholic clergy. Bishops enacted rules and procedures to unify the Catholic Churches. According to Irenaeus, true gnosis lies in accepting the Catholic Church lock, stock, and barrel, because Church doctrine is the sole source of salvation—an attitude that Tertullian agreed with wholeheartedly.

In the 3rd century AD, the Persian Gnostic prophet Mani taught that avoiding the evils of the material required one to give up work, marriage, and fighting. These ideas spread to the Roman Empire, where the Manicheans took root in Armenia in the 5th century. They eventually grew strong enough to provoke a Byzantine attack that saw many of them deported to Greece, where their ideas were adopted by the followers of a Slav priest named Bogomil. Bogomil founded a church in the Balkans that rejected the Old Testament, baptism, the Eucharist, the cross, and the sacraments and structure of the Catholic Church. Having children was seen as collaborating with the Devil, and they practiced anal intercourse to prevent the problem—hence the derisive word *bugger* (anal sex, often homosexual), a term coined by Bogomil's orthodox detractors.

We sacrifice the intellect to god.
Ignatius Loyola

In 567 AD, Athanasius, the orthodox Bishop of Alexandria and an admirer of Irenaeus, ordered all non-canonical or accepted writings destroyed in an effort to stamp out *heresy* (dissent or disagreement with orthodox teachings or methods). Gnostic Christianity was denounced as heresy. The Gnostic writings were thought to have been all but lost until the Nag Hammadi texts were discovered in Egypt in 1945. These texts contain additional Gospels not included in the orthodox New Testament.

In the end, as author Elaine Pagels (1943-) says, the real value of the Gnostic Gospels discovered at Nag Hammadi might lie in showing people how poorly we understand and follow Christ's teachings. I couldn't agree more.

[The Roman Catholic Church is] one Universal Church outside of which there is absolutely no salvation.
Fourth Lateran Council

East/West Split

The Roman Empire divided into East and West, and the Church reflected that division. Those in the West increasingly saw the pope in Rome as the head of the Church, but the Eastern Church did not recognize papal authority. The Eastern Church was organized around patriarchs with a senior in Constantinople (Istanbul), as per the Council of Chalcedon in 451AD. The Byzantine Empire extended the Church's reach to Russia, where Vladimir of Kiev pronounced it the state religion in 988. Tensions between East and West over how to define the Trinity grew until both churches excommunicated each other in the Great Schism of 1054. These decrees were finally withdrawn in 1965, but the split remains.

While the Eastern Church preached peace, joy, and a return to God, the West took a much harder line. Western Christianity is unique for having alternating periods of repression and permissiveness. Repression tends to coincide with revivals, such as with the recent resurgence of the Evangelical Christian movement in the United States. We will look at religious repression in Chapters 8 and 10.

The Protestants

Where the Renaissance of the 14th through 17th centuries tried to reconcile Heaven and Earth, the Protestant and Catholic Reformations tore them apart. Priest Martin Luther (1483-1546) believed that the popes had strayed from Christ's message by making faith overly transactional, such as through the selling of *indulgences* (purchased full or partial forgiveness of sins). He nailed his *95 Theses* to the door of Castle Church in Wittenberg, Germany. To him, the Bible was the only source of divine knowledge and only God's grace and faith in Christ could save people.

Pastor John Calvin (1509-1654) joined the fray by denouncing Catholicism as tyranny. He believed that one can only know God by studying the Bible. Sin originated with Adam and spread to all humanity. The urge to sin is so powerful that people require redemption through faith as the solid knowledge of God in Christ. This act of repentance forgives past sins, but the redeemed person cannot attain complete perfection.

Both Luther and Calvin believed nature to be as passive as Christians who could only accept salvation from God without doing anything for themselves. Calvin was for the sciences; after all, God adapted Himself to His audience's changing capabilities over time in the Bible. Thus, one could think of the Genesis story of creation as baby talk that broke complex science into easy concepts that simpletons could understand, and should not be taken literally. A key test for any religion is how well it integrates into daily life. Calvin inspired the Puri-

---

When I told the people of Northern Ireland that I was an atheist, a woman in the audience stood up and said, yes, but is it the God of the Catholics or the God of the Protestants in whom you don't believe?
Quentin Crisp

One of my favorite fantasies is that next Sunday not one single woman, in any country of the world, will go to church. If women simply stop giving our time and energy to the institutions that oppress, they could cease to be.
Sonia Johnson

Women

Christianity teaches that men and women are fundamentally flawed because of Eve's "original sin" of eating the apple in the Garden of Eden, which gave humans the gift of reason and thus the ability to question God and His many lieutenants. These teachings have all too often alienated people from themselves and each other, especially when it comes to sexuality and misogyny. Irenaeus, Tertullian, Jerome, and many other Church elders openly hated women. Augustine was baffled, saying that a man needing a companion would do better with another man than a woman.

There is a strong correlation between religious theory and social practice. While the Gnostics revered women as equals to men and considered God a dyad of both genders, women played no role in orthodox Christianity after the year 200—an amazing development since Christ himself spoke with women, included them among his companions, and was probably married to Mary Magdalene. Contrast that with the Church's portrayal of her as a prostitute and Bible passages such as 2 Timothy 14:35-36 and 1 Corinthians 2:11-12. Irenaeus complained that women are especially attracted to heretical groups and that many women had been taken, even from his own congregation. Tertullian was similarly outraged, saying that heretical women have no modesty and are bold enough to teach, argue, exorcise, cure, and even baptize. The Catholic Church forbade women to so much as speak in church. This institutionalized misogyny is still alive and well; in 1977, Pope Paul VI ruled that women cannot be priests because God is male. Even more recently, Pope John Paul II ruled that Mary's hymen remained intact before, during, and after the birth of Christ—an anatomical impossibility.

Anyone who believes that his god came out of a woman's privates is quite mad; he should not be spoken to, and has neither intelligence nor faith.
Bahr al-Favas'id

When the end comes, Armageddon outta here!
J.F. Bierlin

Revelations

Revelations, the final book of the New Testament, provides an apocalyptic vision taken by modern Christians to mean the end of the world and the final defeat of Satan, with the faithful being lifted by Christ to heaven—an idea of the "Rapture" that comes from a bad interpretation of the book of Thessalonians. The truth is that Revelations may actually have been predicting the liberation of Christians from Roman persecution. The number 666 is a reference to Nero, who persecuted the Christians horribly. All persecuted people dream of vengeance, and this particular dream became part of the Christian theology. What they fail to understand is that Rome was indeed defeated, that Christianity was saved by Constantine, and that it went on to become an oppressor of historic proportions. This could also refer to the Roman sack of Jerusalem in the year 70. Either way, Revelations was fulfilled a long time ago!

To you your religion, and to me my religion.
Mohammed

Islam

According to Islamic tradition, the prophet Mohammed had his first revelation from Allah (God) in the year 610AD during the month of Ramadan, and for 21 years thereafter. He kept quiet about these revelations for two years before consulting his Christian wife and cousin, who felt that God was actually talking to him. (This may go a long way toward explaining Mohammed's monotheism and acceptance of Abrahamic traditions.) The message was simple: Mohammed was not ushering in a new doctrine about God but simply bringing God's message to the Arabs. Thus, he never received godlike powers like Christ's, nor did he rule a mature state like Josiah. Whether his revelations were real or not is beside the point; plenty of people have believed that they are under a divine guidance that just happens to be very wise to the current political facts on the ground. Humans are, after all, political animals.

I was a hidden treasure and desired to be known; therefore I created the creation in order to be known.
Sufi creation myth

Mohammed attracted the young, disillusioned, and marginalized. All was well with the new religion for the first three years, until Mohammed forbade the worship of pagan gods. Most of his followers deserted him, and Islam became despised and persecuted. Mohammed did accept other monotheistic religions and mentioned other prophets (Jesus Christ being one of them), but saw the Christian trinity as a prime example of *zanna* (confusion) and self-indulgent guesswork about the unknowable. He felt that Muslims should see through this illusion to transcendent reality (a very Buddhist idea). Thus, the Koran stresses the need to actively decipher

God's signs; a seeker of truth should, "Shun no creed, scorn no book, nor cling fanatically to a single creed." This openness inspired Muslims to build the natural sciences and math that some Christians still see as dangerous to this day.

Mohammed broke from the idolatry of Mecca and the divisions between Jews and Christians to start a direct faith and trust in Allah, the God above all other deities. He began a political movement based on his revelations, making treaties with those who recognized him and war with those who did not. He eventually conquered Mecca itself, avenging an earlier defeat that drove him 300 miles away to Medina during the *hijrah* (migration or flight) of 622. Muslim years are dated AH, for After Hijrah. Today, Islam is the world's second-largest religion with over 1 billion followers.

Mohammed did not set out to create a new religion or override the older religions. His messages were the same as the prophets who came before him. Had he known of the Eastern religions, he would have incorporated them as well. The whole idea was simply to create a single unified package by which the Arabs could come to know God.

Some Muslim thought bears a strong resemblance to the Gnostics and Buddhism. For example, Abu Bakr Al-Baqillani said that said there is no God or reality but Allah; the universe is reduced to atoms with discontinuous time and space. (Compare this to relativity, which we will see in Chapter 24.) Jalāl ad-Dīn Mohammed Rūmī said that, "All theologies are straws that God's sun burns to death. Knowledge takes you to the threshold, but not through the door." Ibn al-Arabi seemed to be warning against orthodoxy when he said, "Don't attach to any creed exclusively to disbelieve all the rest, or you will lose much good and fail to realize the full truth of the matter. God is not for one creed; wherever you turn, there is the face of Allah." Islamic mystics saw God as the all-enveloping reality and ultimate existence that cannot be perceived by normal means. They sought to lose their self-consciousness to unite with God and achieve enlightenment.

A Brief History

In the year 610AD, the Quraysh nomads in Arabia transitioned away from their harsh struggle for survival and made

---

Traveling in the desert brought me closer to god but further away from organized religion.
Bruce Feiler

To fight in the defense of religion and belief is a collective duty. There is no duty after belief than fighting, the enemy who is corrupting the life and the religion.
Ibn Taymiyya

Chapter 5
The World's Religions

Mecca the most important Arabic city. Greedy capitalism replaced their tribal life, leaving the common people feeling lost and left out. Mohammed knew they needed a new ideology to fit their new situations and took it upon himself to deliver. Religion at the time was a loose pagan polytheism that did not play a major role in life. Beliefs about an afterlife were dismissed as fairy tales. To the polytheistic Meccan elite, denouncing the rich and espousing monotheism were bad ideas. They were not exactly enamored of Mohammed, and it did not take them long to mount serious opposition. In 622, Mohammed escaped Mecca with his followers and began the hijrah. War ensued, with the Battle of Badr giving Mohammed a clear victory and the power he needed to begin implementing his vision and revealing his revelations. By the time Mohammed died in 632, the Arabic tribes were united under Allah, the same God worshipped by Jews and Christians.

Islam has bloody borders.
Samuel Huntington

Mohammed died with no obvious successor, which caused a rift among Muslims. Sunnis saw Abu Bakr as the rightful heir who was divinely ordained by Mohammed in accordance with God's will—a viewpoint similar to that used to justify papal power and succession among Catholics. Shias believe that all *imams* (priests) have spiritual qualities similar to Mohammed and are free from sin. This schism was similar to that between orthodox and Gnostic Christians and persists to this day.

The Arabs went on to build an empire that extended from Spain to Nepal in the century following Mohammed's death. Where Christians saw God in defeat, Muslims saw God in victory, which validated the Koran's message that society can only prosper by having the correct relationship with God. As *ahl al-kitab* (people of the Book of Abraham, the Old Testament), Jews and Christians were allowed to continue practicing their religions, unlike pagans who had to either convert or face death. Jews and Christians, did, however, have to pay the *jizyah* or religious tax. Many converted to Islam to avoid paying. The Arabs eventually got wise to this and started banning conversions around 700 to keep the money flowing. Apparently, God needed the money more than the souls.

No one ever curled up on a rainy weekend to read the Koran.
Huston Smith

The Dome of the Rock was built on the ground formerly occupied by the Jewish Temple (and the Canaanite Temple before that) in 691, an act that still remains a source of tension between Jews and Muslims. Like it or not, Islam was here to stay. Muslim *caliphs* (rulers) gradually acquired all of the pomp and circumstance of any traditional ruler. Caliphs were seen as shadows of God on Earth and functioned much like popes, with additional secular roles and executioners to enforce their laws. The advent of *sharia* (laws based on the Koran, hadith, etc.) around the year 870 made religion part of all facets of life.

Disputes among *emirs* (regional rulers) left them unprepared for the First Crusade of 1096-1099 and the Christian attack on Jerusalem that signified Rome's emergence from the Dark Ages and desire to avenge Christ's death. Popes promised crusaders forgiveness for past sins and a place in Heaven should they fall in battle. While the Crusades had profound impacts on the Arabs living in the war zone, they were merely border skirmishes to the Arabic empire as a whole. The Crusaders did manage to win some victories—and even regain control of Jerusalem for a time—but the Crusades ultimately failed. Islam has been seen negatively in the West ever since and vice-versa. Christians were not the only invaders. Mongols attacked in 1200s and conquered large tracts of land. Many converted to Islam and rebuilt conquered cities on a grand scale.

Islam was the greatest power bloc in the entire world by the end of the 15th century. From Malaysia to Africa and the Arabic Peninsula, the entire world seemed to be becoming Muslim; other religions were increasingly seen as obsolete and irrelevant. Non-Muslims were prohibited from visiting Muslim holy cities. Insulting Mohammed became a capital offense. The Mongols had brought art and science with them, but the West caught up with and surpassed the Muslims in the 15th and 16th centuries, probably thanks to the Chinese. (*1421* and *1434* by Gavin Menzies should be required reading for any history buff.) The new European progress was fueled by industrialism, which is a far more effective driver of progress and efficiency than an economy based on agriculture and surplus labor. New weapons gave religion more coercive power.

The rise of the Ottoman Empire in the 1700s and the conquest of Muslim lands by industrialized European empires led to artificial national boundaries being drawn that are still a source of conflict among the inhabitants. For example, India and Pakistan were created in 1947 by partitioning British India into Hindu and Islamic nations; these two countries remain at each other's throats today, and both have acquired nuclear

The idea of a rigged crucifixion has been around a long time; even the Koran mentions it.
Michael Baigent

I simply laugh when I read the Koran with its endless prohibitions on sex and its corrupt promise of infinite debauchery in the life to come.
Christopher Hitchens

weapons. Europe was the dominant power by the end of the 18th century, and the rest of the world was unable to compete with the combination of industrialism and secularism. (More on this in Chapter 10.) Muslims had gone from world leaders to dependency on the West in just a few hundred years. I can only wonder how much of this was due to conservative religion slowing scientific progress for the Muslims as it did in Europe until a huge Chinese ship docked in Venice in 1434 and jolted the West out of its stupor.

> *Muhammad professed to derive from Heaven, and he has inserted in the Koran, not only a body of religious doctrines, but political maxims, civil and criminal laws, and theories of science.*
> Alexis de Tocqueville

Industrialism had other effects. Among other things, it required an educated and literate workforce, who in turn demanded more for their efforts. The middle class was born. Religion was increasingly pushed to the sidelines with science becoming increasingly prominent.

The Koran had promised that a society revealed by God could not fail. Muslims had turned to religion time and time again through history, but religion had failed them. Like the Jews, some Muslims felt that they had not been true to their religion. Reformers tried to get Islam back on the right track. Failures were seen not as indications of being on the wrong path, but as indications that the reforms had not gone far enough, and thus Muslim fundamentalism was born. Like all who try to recapture some mythical bygone golden age, fundamentalists are the throwbacks who cannot adapt to change and who see God's glory in progress and knowledge.

As an aside, Muslims believe that God gives legitimacy to governments, not the people. This places politics at the heart of Islam and makes it difficult to create a democratic Islamic state that sidelines religion. American attempts to foster democracy in the Middle East are therefore viewed as absurd.

Koran and Hadith

> *If Woody Allen were a Muslim, he'd be dead by now.*
> Salman Rushdie

To Muslims, the Koran existed in God's mind since the beginning; creating the book allowed them to hear God directly. This foundation cannot be altered or translated, only interpreted. In fairness, the Koran has a better claim to be an accurate record of Mohammed's sayings than the Bible has of being an accurate record of Christ's teachings (but less than the Gnostic gospels). Parts may have been written during Mohammed's life, perhaps even under his direct supervision.

The entire Koran was mostly complete within 20 years of his death. Still, the Koran is not chronological; one must read it in the order of composition to get the full picture, which explains why the Koran seems to swing back and forth when read cover to cover.

The Koran is not the only revelation. It acknowledges the earlier Book of Moses (Bible). Prior prophets such as Moses and Christ also got the word of God, but their words were corrupted as the Bible was translated and modified from its original form. Jewish and Christian theology was given to Jews and Christians, and then directly to Mohammed from God. Thus, the Koran was received and interpreted sans corruption; Mohammed is the final prophet; it is the Logos, the literal Word of God. Preserving Mohammed's direct connection to God is crucial; merely being tutored by a Christian relative is not enough to gain a following. Islam therefore sees itself as being of the same lineage as Judaism and Christianity, but not descended from those religions. Muslims see the Bible as historical, but not divine. To them, the Koran is the direct word of Allah; it embodies Christ's teachings in law. (As an aside, Sura 4.256 also says that Christ did not die on the cross.)

Little new theological ground is broken in the Koran over its predecessors. Humanity, once united, became divided, and all must live as God intends. The Koran contains plenty of threats against those who do not believe. As with Jews and Christians, there is no separation between religion and the rest of daily life. Despite the admonition that we are created to understand God's works that led to the preservation of math and science during the Christian Dark Ages, some Muslims preached *fideism* (the acceptance of religious authority), because they saw reason as an attempt to make God inferior when reason could never stand up to divine revelation. Conservative theologians said that God creates all possibilities, from which humans choose like items in a supermarket (an idea very similar to that of *quantum superposition* (where an object is in all possible states at once until observed) and *waveforms* (sets of possibilities), as we will learn in Chapter 23). Any seeming contradictions were God tailoring the Koran to the varying circumstances Mohammed faced. Of course, the other view is that Mohammed himself did the tailoring, however

I have condemned Khomeini's fatwa to kill Salman Rushdie as a breach of international relations and as an assault on Islam as we know it in the era of apostasy.
Naguib Mahfouz

Myth is neither a lie nor a confession: it is an inflexion.
Roland Barthes

unconsciously and earnestly. Which is easier to believe? I leave that up to you.

The *hadith* are anecdotes based on Mohammed's words and actions. Muslims see the hadith as important tools for understanding the Koran and Islamic law. Many hundreds of thousands of these stories were gathered and evaluated for authenticity, mostly during the reign of Umar bin Abdul Aziz in the 8th and 9th centuries. Those deemed legitimate were placed into one or more collections that are still referred to in matters of law and history today. A *muhaddith* is someone who has memorized at least 400,000 narrations and their individual chains of narrators. Which hadith are divine? A science of hadith exists to evaluate narrations, but the fact remains that the hadith are based on humans evaluating hearsay evidence.

> *Myth is neither a lie nor a confession: it is an inflexion.*
> Roland Barthes

Jihad

If the Koran is God's word and does not contain the doctrine of *jihad* (holy struggle or war), then where did this idea come from? The first step in figuring this out was to resolve conflicting information in the Koran itself. The process was simple: The more recent an utterance, the more likely it represented God's will. Leaving aside the obvious contradiction of assigning relative values to different parts of a book that supposedly existed in God's mind from the beginning of time, Mohammed did grow more aggressive over time as his patience and tolerance waned. From there, the architects of jihad could pick and choose hadith narrations to suit their cause.

> *You may even see that the ideas you're willing to fight to the death for are the very ones you're most unsure about; the fierceness of your answer a mask for uncertainty.*
> Irwin Kula

Jihad has greater and lesser forms. Living as God commands is the greater form, and war is the lesser form. One who dies in God's service is a martyr, but so is one who lives according to the greater jihad. The West sees jihad as something carried out by crazed fanatics in the name of Islam, but Muslims point out that Christianity also promises Heaven to its martyrs and says that the acts of lesser *jihadis* should not be a mark against Islam itself. We will revisit this in Chapters 10 and 11.

The Afterlife

Plato (428BC-347BC) thought that people need laws they believed to be divine in order to preserve social order.

Mohammed simply told people to follow God's law or go to hell—a far more persuasive argument than any amount of logic, and yet another example of the long relationship between religion and politics. Keeping the people faithful and in line during hard times presented a serious challenge, which religion responded to by creating images of a lush afterlife. Irenaeus spoke of a hereafter brimming with grain, delicacies, and highly fertile women. Both Christianity and Islam have a habit of denigrating living women while making their services freely available in Heaven. According to Irenaeus, "Vines shall grow, each with 10,000 branches, in each branch 10,000 twigs, in each twig 10,000 shoots, in each shoot 10,000 clusters, in every cluster 10,000 grapes, and in every grape 25 metretes of wine." Mohammed took a similar tack, promising his followers a very cushy afterlife with plenty of virgins.

Isn't it enough to see that a garden is beautiful without having to believe that there are fairies at the bottom of it too?
Douglas Adams

Jewish Roots

Like Judaism and Christianity before it, Islam is far more than belief in a set of creeds; it is a way of life that extends to the most intimate details.

Mohammed is said to come from the line of Ishmael, making him a direct descendant of Jews. It is thanks to this lineage that he became the last of the authentic prophets. Depicting Ishmael as the beloved son of Abraham allowed Mohammed to expand on this theme and claim that Abraham traveled to Mecca to build the *Kaaba* (the most holy building in all of Islam) so that Muslims could worship Abraham's god. This sweeping gesture brought Muslims into the covenant between God and Abraham and made Judaism and Christianity part of Islam.

Mohammed knew that monotheism was the best way to squash competing religious ideas. Allah means *"The God."* The word *Allah* comes from the Syriac word *allaha*, which leads back to kinship with *Elohim*, which is one of the Hebrew names for God, who remains one and the same despite the name changes. The methods Mohammed used to make Allah the one true God are similar to those employed by the Jews and Christians. The Koran says that the old polytheistic religions in the Arab lands were not working and calls the new religion *Islam* for "surrender" to the struggle to live as God intended. He destroyed the idols in Mecca and attached

It would seem to border on sheer irrationality to insist that all of Adam's experiences in genesis 2:15-22 could have been crowded into the last hour or two of a literal twenty-four-hour day.
Gleason L. Archer

Islamic significance to pagan rituals by linking them to Abraham. Formal prayer occurs five times a day, and all Muslims are expected to visit Mecca at least once in their lifetime for the hajj pilgrimage.

In Medina, Mohammed decreed that Muslims should fast for 24 hours every year and even called it *yom kippur*. Muslims were banned from eating pork, and the *halal* ritual of purifying meat is identical to the Jewish *kosher* laws. He even insisted that people face Jerusalem when praying, which was later changed to Mecca. So thoroughly did Mohammed embrace Jewish customs and rituals that some scholars have called him an Arab rabbi who taught the Old Testament to the Arabs. He was even willing to accept the virgin birth of Christ and Christ's word as law. In return, he wanted Jews and Christians to accept Islam and understand that their books were but preludes to the Koran, their prophets mere forbears of himself, and their entire religions as steps on the path to his culminating vision. In short, Mohammed wanted to merge the JCI religions, but not as a coming together of equals—a position that sounds awfully similar to Pope Benedict XVI's invitation in 2009 to fold parts of the Anglican Church into Catholicism under his leadership.

> *I owe what is best in my own development to the impression made by Kant's works, the sacred writings of the Hindus, and Plato.*
> Arthur Schopenhauer

Hinduism

Hinduism is a blanket term for many religions that share many commonalities and belong to the same family that includes Buddhism, Jainism, Sikhism, and others. Its roots begin around 2500BC in the Indus Valley, preceding the Aryan settlement that began around 2000BC. Hindu religions spread across much of India between 1200 and 600BC. Their beliefs are based on the *Vedas* (a body of sacred writings that includes the *Rig-Veda*, *Sama-Veda*, *Atharva-Veda*, and *Yajur-Veda*) and the religions are therefore called Vedic religions. Unlike the Bible and other texts, the *Vedas* ask more questions than they answer and do not seek to explain how the universe works.

> *Do you spend time with your family? Good. Because a man that doesn't spend time with his family can never be a real man.*
> Marlon Brando as Don Corleone

Religion is at least partly responsible for holding India together in the face of Muslim—and later British—interference. To Hindus, each individual religion may vary, but all religions are valid paths to God. After all, there are many roads up a mountain but only one summit. Claiming that one religion is better or true is like claiming that God is in one particular place. Mahatma Gandhi tried to include Muslims in post-British India and was assassinated for his efforts.

The family is the foundation of Indian life and belief. Women have limited public roles, but huge family roles. Stories are important; the village storyteller remains respected to this day. *Dharma* (the essential quality or character of the cosmos and all within it) defines how one should behave. The *artha* dharma refers to material goods, the *kama* dharma refers to sensual enjoyment including sex (hence *kama sutra*), and *moksha* refers to release from rebirth.

In the root divine wisdom is all-Brahman; in the stem she is all-illusion; in the flower she is all-world; and in the fruit, all liberation.
Tantra Tattva

The universe is governed by order and balance that is reliable despite occasional disorder and being beyond human control. Hindus revere cows. The five gifts (milk, curds, clarified butter called *ghee*, urine, dung) come from cows. Killing a cow is a crime equal to killing a *Brahmin* (member of the highest caste). Cows remind us that humans depend both on non-humans and on order.

Brahman and Atman

Brahman is the source of all appearances, the unproduced producer of all there is. Brahman is beyond description or characteristics and is not a part of the universe; we therefore cannot say what Brahman is, but through the created world we can know what Brahman is not. The universe is a consequence of Brahman. Very loosely, one can say that Brahman is the eternal consciousness at the Ground of reality, the Godhead. Hinduism thus postulates an idealist universe where matter is a product of mind (an idea that has much scientific validity, as we will see in Chapter 20). Hindus are sometimes thought to be polytheistic, but this not actually the case. Brahman creates appearances in deities that can be worshipped—a close parallel to the Catholic belief in saints.

That which is the finest essence, this whole world has that as its soul. That is reality. That is Atman. That thou art.
Upanishads

A person's real Self (as opposed to the self we identify with in this or any lifetime) is called the *atman*, which is both eternal and identical to Brahman. Realizing this frees us from the notion that this world and body are important; they come in and out of being, but our atman cannot die. Realizing that we are nothing but Brahman is to realize that death cannot touch us. Our everyday self is the *jiva*. The jiva and atman are like two birds on the same branch. Brahman and atman are eternal and important, the jiva transitory. Realizing one's true identity as atman/Brahman leads to moksha, or freedom from the cycle of birth and death—a goal that may require many millions of incarnations to achieve.

> *It's really pretty simple: if we don't have great sex, we can't have a great spiritual life.*
> Irwin Kula

The *Upanishads* argue that Brahman as creator projects itself on the world and reveals itself in creation but also remains hidden behind creation. Anyone who mistakes the maya illusion of creation for reality suffers from ignorance. This belief fits with the holographic universe theory, which we will examine in Chapter 27. Water flowing from ocean to ocean in the rain circle seems like separate droplets, rivers, etc. but all come from the same source, which is atman.

The human spirit does not depend on the body any more than the body depends on clothes. When we outgrow or wear out our clothes, we get new ones. The spirit's path is determined by its choices; actions during a lifetime create good or bad karma that influences *samsara*, the birth of a new incarnation. We can return either higher or lower depending on our karma, a fact that is often seen as punishment and reward, but which is actually a completely neutral moral law. At death, a person's Self (called a "subtle body" or quantum monad, see Chapter 15) goes to a new body. We will discuss the difference between Self and self when we look at Buddhism later in this chapter.

All castes and genders of people can seek *bhakti* (selfless devotion to Brahman). According to the *Bhagavad Gita*, "whoever takes refuge in me whether they have been born in sin or are women or traders or laborers, they too shall reach the highest goal." This passage directly comparable with John 3:15-16. Each person's body is a temple and no other is needed, a concept also seen in passages such as 1 Corinthians 3:16 and John 2:18-21.

> *It's the pleasure haters who become unjust.*
> W.H. Auden

Both *linga* (male) and *yoni* (female) exist in creation and are the origin of life. The power of sex to unite carries the fundamental realization that uniting opposites is required to bring anything into being. Sex can unite both people and gods, hence *tantra*, which is found in almost all Hindu beliefs. The human body is a miniature version of the cosmos, and the human Self is absolute, unchanging consciousness (*purusha* or atman) that is aware of the history of the universe but not involved in it. History occurs because of *prakriti* (the unconscious cause of all that appears and happens). The relationship between purusha/atman and prakriti is like a blind man with strong legs carrying a sighted man with bad legs.

There are several schools of Vedantic thought based on the part of the *Vedas* known as the *Upanishads*:

- **Advaita:** There is only Brahman. All comes from Brahman like a dream that is called maya. We believe that maya is true because we are ignorant.

- **Vishishtavaita:** Brahman is real but all is not an illusion. For example, individual selves are real but depend on Brahman for life and function. This is not a dualist belief; we can think of it as a qualified form of idealism.

- **Dvaita:** Brahman is utterly transcendent and distinct. Souls can unite with Brahman and yet remain distinct in their final state where their relation to Brahman is that of lover and loved.

Truth, like divinity, is never to be known directly.
Goethe

Jainism

The Jains thought that Indian religion failed to bring one to enlightenment. They rejected the structure and immorality of society and withdrew to practice asceticism. According to the Jains, we should live at peace with the whole world and practice nonviolence. Jiva is a spiritual reality that descended from bliss and understanding under the weight of karma; they therefore teach how to escape karma. As part of their asceticism, Jains have five great vows (nonviolence in mind, speaking and being in truth, relying 100% on alms, renouncing sex, and detaching from possessions and objects of the senses). Those who are not yet ready or capable of taking these vows can take the five lesser vows of nonviolence, truthfulness, honesty, control of sex, and a moderate lifestyle.

When a man lacks discrimination, his will wanders in all directions, after innumerable aims.
Bhagavad Gita

The Bhagavad Gita

The *Bhagavad Gita* is taken from the world's longest epic poem, the *Mahabharata*, which tells the story of ancient Indians and is similar to the Old Testament for Jews or the *Iliad* and *Odyssey* for the Greeks. Like the *Upanishads*, the *Bhagavad*

Gita states that everyone eventually gets her or his due; all actions have consequences. For example, if you help someone in this life, you will get good karma in your next life.

Here are just a few of the many passages that struck me as I read the *Bhagavad Gita*. As you read, you may be reminded of nearly identical sayings from the Bible.

- He with no attachments to anything and who, having gotten good or evil, neither delights nor hates but has a steadfast mind.

- Churning senses can carry off the mind of even the most ardent striver for perfection.

- When man dwells on objects of the senses, attachment to those objects is born. Desire is born from attachment, and unfulfilled desire is the source of hatred. From hatred comes obsession, and from obsession comes the destruction of reason itself. When reason is destroyed, man himself is destroyed. Compare this to the *Gospel of Thomas*.

- Krishna's devotees never perish. Those who take refuge in me, even though born in the womb of sin, go to the highest goal.

- This is a fleeting and sorrowful world. While you are in it, devote yourself lovingly to Krishna.

- Fix your mind on me (Krishna), be dedicated to me, lay yourself devoutly before me, discipline yourself and with me as your supreme goal, to me you will surely come.

Plato

Plato's idea of liberation is very similar to Hinduism. He believed that each person has three soul natures: rational, spirited, and appetitive. A person's soul nature can be discovered empirically; someone living according to her or his soul nature is happy. Unhappy states (nations) exist because of unhappy people, who in turn exist because of unhappy souls. A proper education helps people find their soul nature and discover their best vocation.

In its broadest terms, religion says that there is an unseen order, and that our supreme good lies in right full relations to it.
William James

When not enlightened, Buddhas are no other than ordinary beings; when there is enlightenment, ordinary beings turn at once into Buddhas.
Hui Neng

Buddhism

"Ztt! I entered. I lost the boundary of my physical body. I had my skin, of course, but I felt I was standing in the center of the cosmos. I saw people coming toward me, but all were the same man. All were myself. I had never known this world before. I had believed that I was created, but now I must change my opinion. I was never created. I was the cosmos. No individual existed." This is how one anonymous Zen master described his experience of being enlightened—an account that sounds surprisingly similar to quantum entanglement. (See Chapter 23).

Buddhism is part of the Hindu family of religions. Contrary to popular belief, it is not atheistic, because Buddha did not deny the gods, but rather believed that reality is beyond them. Adherents believe that the samsara cycle of rebirth will continue until one gains release. A person can be reborn as human or animal, and can experience heaven or hell, according to the law of karma. This sounds a lot like Hinduism, except that Buddha did not believe that there was any soul or self to be reborn—at least not in the JCI sense we are familiar with.

Buddha was born Prince Siddartha Gautama circa 480 BC Looking around, he saw his rich life as full of decay. He felt no joy when his child was born, because he knew that all things would eventually decay and die. As he saw it, human life was a circle of suffering caused by attachment to things and people that are all impermanent. Many people, especially atheists, get stuck here. Attachment and desire are incompatible with high spirituality; however, Buddha was convinced that enlightenment is possible in this lifetime and decided to set out to prove it. After all, if normal life is so fragile and full of suffering and death, there must be another mode of existence.

Buddha left his family to go into the world and solve the riddle of existence, an apparently successful quest. (I don't know how his family reacted to his departure.) Unlike many spiritual leaders, Buddha did not want anyone to take his teachings on faith; he admonished people to see for themselves because anyone is capable of doing so.

And the light shone in the darkness and against the Word the unstilled world still whirled about the center of the silent word.
T.S. Eliot

We need the courage as well as the inclination to consult, and profit from, the wisdom traditions of mankind.
E. F. Schumacher

In 250BC, the movement Buddha founded split into two major "schools" called *Theravada* (the "lower" literal school) and *Mahayana* (the less rigid "higher" school). In the former, selflessness extends to persons and personal identities but not to things and events in general. In the latter, selflessness encompasses all phenomena as a universal principle. Tibetan Buddhism follows Mahayana. It is important to note that the Buddhist split, unlike the JCI splits, involved substance but not control. This split was also remarkably peaceful compared to the JCI splits.

Each human being is the dwelling place of infinite power, the root of the universe.
Mani

Buddhism spread from India to China. It arrived in Korea in the 4th century AD and in Japan in about 538AD, when Prince Shotoky adapted it as a means to unite his people in loyalty to the imperial house. Buddhist and Taoist thought merged into Zen Buddhism during this period. Today, there are over 400 million Buddhists on Earth.

Nothingness and Everything

The central tenet of Buddhism is that nirvana, the state of reunion with Brahman, or Ground, is nothingness with no inherent existence or characteristics.

This is not the same as saying that there is no existence, because "nothing" and "no thing" are different. For example, a soul that has achieved liberation from the cycles of rebirth and reached the state of union with the Ground does not cease to exist. Brahman transcends subject/object relationships and is non-dual. Duality is the illusion of maya. We achieve liberation from the cycle of birth and death when we lose our fear and desire in the knowledge of nonduality—when we realize, as philosopher Arthur Schopenhauer (1788-1860) said, that "We are all one and the same being."

My religion is kindness.
The Dalai Lama

Accepting that the first moment of existence simply exists without a cause and that a moment of consciousness requires a preceding moment of consciousness leads to the conclusion that consciousness is eternal. On the normal physical/mental plane, it is impossible to achieve ultimate fulfillment on Earth because of the inevitability of disease and death. The best we can do is live a good life and behave well to ensure a good rebirth. The mental continuum is the basis of self-identity and the level on which we move through cycles of rebirth. Pursuing a spiritual path based on the continuity of eternal consciousness allows us to make mental improvements that move us closer to liberation from the cycle and union with the Ground. By "us" and "we," I don't mean Anthony Hernandez or whoever you happen to be in this lifetime; it is important not to confuse Self (our eternal nature) with self (our present identities). Thus, as I said in Chapter 1, Anthony Hernandez will cease to exist at death; but the Self that contains and experiences being Anthony Hernandez continues. Just what this Self consists of is open to debate. Still, the self that we all think we are familiar with is a delusion, and the path to liberation therefore lies in living as if the ego did not exist. We will return to this in Chapter 30.

If everything we experience in this life is a product of our own minds, then how can we have shared experiences? For example, why would two people standing on a street agree about virtually all of the details of their surroundings? External phenomena do exist separately from individual minds, but they lack objective reality because we label them the moment we see them. This discrepancy and subjectivity is what leads to phenomena being called illusions—a point of view that is consistent with the *Heisenberg Uncertainty Principle* (there is a limit to how precisely we can measure something), the *observer effect* (the concept that an observer can affect the outcome of an experiment), quantum superposition, and *nonlocality* (one thing instantaneously affecting another regardless of distance) as we will see in Chapter 23.

Pain and suffering exist because of our attachment to our bodies. Each person must endure a long number of lives to experience all that can be experienced. Once we reunite with our eternal, absolute state, pain and suffering will no longer touch us. Nirvana is not self-extinction but the elimination of desire, hate, and attachments. Meanwhile, Nirvana is both natural and within each of us, whether we are aware of it or not.

Buddha did not believe that consciousness is permanent. The analogy would be like a flame being passed from candle to candle. Upon liberation, the candle is removed from all drafts and burning steadily. Unlike the JCI heaven, Nirvana is not a location but a state. Saying that consciousness is not extinguished is therefore not the same as saying that a soul arrives

---

Nirvana or lasting enlightenment or true spiritual growth can be achieved only through persistent exercise of real love.
M. Scott Peck

The age-old theological view of the universe is all existence is the manifestation of a transcendent wisdom, with a universal consciousness being its manifestation.
Gerald L. Shroeder

at a specific location such as the "Pearly Gates," because there is no soul and no place to go.

Every emotional state is a complex web of emotional and cognitive experiences. We feel like an existing, self-sufficient person, an "I" who is totally separate from our surroundings. When we can see our own absence, we begin to let go of friends, possessions, and even subjective identity, the first stirrings of liberation. We must also learn to see objects as empty of meanings and labels to begin to free ourselves of delusion and to perceive the nature of an eternal fundamental mind of clear light that is empty of those delusions. I believe author Lynne McTaggart would call this clear-light mind "the Field." We will look at this more in Chapter 23.

Am I a man dreaming I'm a butterfly, or a butterfly dreaming I'm a man?
Chuang-Tzu

Consider a dream: Both the "you" in your dream and everything in your dream world are manifestations created by your brain. As far as the "you" in the dream is concerned, the dream world is every bit as real as the normal waking world. In fact, the "you" in the dream world probably thinks s/he is awake and having the experiences taking place in the dream. Of course, neither "you" nor the dream world can be said to have any real existence. This "you" may be on a cross-country road trip, but there is no real you nor is there a country to cross or a destination to arrive at. Thus, the "you" in the dream is deluded into thinking s/he is a separate and distinct entity from the other entities and phenomena in that world, when no such separation actually exists. In this example, your mind is the primal consciousness that is manifesting both the dream "you" and the entire dream universe. Meanwhile, your mind is not aware of all of this; it simply is. From the point of view of the dream universe, your mind has always existed and always is; it is meaningless for someone in the dream world to ask what came before the dream. Apply this concept to the waking world we are all familiar with, and you will start to understand how Buddhists look at the world. Put simply, individuation is only appearance.

Had I been absorbed by the universe, or had the universe penetrated me?
Marius Favre

At death, we take another turn on the merry-go-round. We go through an intermediate state and then are reborn. At death, the grossest levels of mind and energy dissolve into the subtlest levels, where the person experiences the clear light. From there, one takes a new subtle body for the duration of the intermediate state, and then returns to a gross material level when one takes a new physical body. It is said that a practiced meditator can experience this clear light state just as we all will when we die. The actual transference of consciousness is like traveling from town to town as the individual consciousness (Self) settles into a new body and becomes a new self. Think of Self vs. self as an actor who plays many roles. For example, if Anthony Hopkins is Self, then Hannibal Lecter (*The Silence of the Lambs*), Abraham Van Helsing (*Dracula*), and Burt Munro (*The World's Fastest Indian*) are examples of selves. Each self lives for the duration of its role, but the Self transcends all of them. As previously mentioned, a Self is called a quantum monad in scientific terms. (See Chapter 15.)

Worn out garments are shed by the body; Worn out bodies are shed by the dweller.
Bhagavad Gita

By now, it should be apparent that there are different types of emptiness, and that all types involve a lack of materiality. The four types of emptiness one has at death are:

- Empty, when the gross levels of mind dissolve,
- Great empty, when all subtle experiences subside,
- Very empty, during dissolution, and
- Completely empty, the experience of clear light

Do you not see how necessary a World of pains and troubles is to school an intelligence and make it a soul?
Keats

According to Seng Ts'an, "When this realization is achieved, never again can one feel that one's individual death brings an end to life. One has lived from an endless past and will live into an endless future. At this very moment one partakes of eternal life, blissful, luminous, pure."

Four Noble Truths

Buddha taught that there are four noble truths that form an analysis of *dukkha*, or suffering/impermanence:

- **Suffering:** All composite phenomena are empty, selfless, and impermanent. Their causes also cause their end. The moment they begin to exist, their disintegration also begins. the moment a child is born, s/he is on the path to death. This idea bears a strong resemblance to the Second Law of Thermodynamics. (See Chapter 26.)

Goodness need not enter into the soul, for it is there already, just unperceived.
Theologia Germanica

- **The origin of suffering:** Suffering stems from denying the facts of suffering, trying to stop the flow of transience, trying to escape, the desire for either eternal life or oblivion after death, etc. Ignorance is not passive unawareness but active delusion, the antithesis of knowledge that actively opposes real knowledge.

- **Cessation of suffering:** Freedom from suffering comes when we pierce the illusion of extrinsic existence and see the ultimate nature of reality: that all phenomena are empty of true existence. This allows us to let go of all desires for things of this world and all attachments to that which cannot endure.

- **The path to cessation:** The path from impermanence and dissatisfaction to selflessness that brings true peace is called the Eightfold Path.

These truths are the foundation of all Buddhist thoughts and practices and the road map to reaching enlightenment.

The Eightfold Path

The Eightfold Path has eight interlocking, non-sequential parts:

- Right understanding about the nature of reality.
- Right thought, including giving up attachments and desires.
- Right action, which includes not taking what is not given, avoiding conduct that arouses self-desires, avoiding untruth, and avoiding intoxication from drugs or alcohol.
- Right speech, where we do not feel the need to be embarrassed about what one says since one is not trying to manipulate people.
- Right living in both work and vocation.
- Right effort, which involves doing things for the right reason and avoiding that which could cause bad karma.
- Right mindfulness, which means being more aware of nature, transience, and the opportunity this gives us to move to enlightenment.
- Right concentration (*samadhi*), which is unifying mind and being into a single purpose where all subject/object distinctions disappear.

If your bonds be not broken whilst living, what hope of deliverance in death?
Kabir

The rout and destruction of the passions, while a good, is not the ultimate good; the discovery of wisdom is the surpassing good.
Philo

Ten Bad Actions

The following 10 actions can cause bad karma:

- Killing
- Stealing
- Sexual misconduct
- Lying
- Divisiveness
- Harsh speech
- Senseless speech
- Coveting
- Harmful intent
- Wrong view, which is thinking that rebirth, the law of cause and effect, or that the Three Jewels don't exist.

> *The truth itself can only be self-realized within one's own deepest consciousness.*
> Buddha

The Three Jewels are:

- **Buddha:** The Buddha nature, which is the highest spiritual potential that exists in all beings. This is the *nagarjuna*, the underlying connection between moments of appearance that sustains all appearances but which does not give actual existence. We can speak in ordinary language as if appearances are real (such as saying that the Sun is warm), but there is no real difference between the samsara world of death and rebirth and Nirvana. All things may seem different, but this is just an illusion.

> *Hell is truth seen too late.*
> John Locke

- **Dharma:** Buddha's teachings.
- **Sangha:** Community of people who have achieved enlightenment or who practice Buddhism.

Nonviolence and Compassion

All Buddhist thought and practice boils down to adopting a worldview that sees the interdependence of all things while leading a nonviolent and harmless life. We don't want violence done to us, because suffering comes from violence. The root causes of suffering are ignorance and lack of discipline. Refraining from destructive acts is a form of discipline that reduces pain and suffering. In other words, if we don't want something in our lives then we should not cause it.

A good heart is the source of all true happiness. Humans are social animals who can only survive in dependence on—and with mutual help to and from—other people. Everything we need above and beyond the bare minimum comes from other people. This is why compassion is so necessary. As someone who practices compassion, we must develop patience. Learning the art of patience requires someone to hurt us, because this gives us the practice we need to be more patient and compassionate.

> *Jealousy and anger shorten life, and worry ages a man prematurely.*
> Sirach

Anger separates us from others, and we must make efforts to resolve it whenever it appears. Even a small amount of anger causes huge discomfort, as anyone who has ever been angry knows. Getting angry with someone impedes our own glory and thus hurts us more than the other person, especially when we are angry at another's success. In fact, we must be grateful when enemies harm us because they give us the chance to be patient. The Buddhist injunction against anger extends to those whose nature it is to cause harm to others. Being angry with them would be like being angry at fire for having the nature to burn. Sadly, this is one of the most gaping flaws with the JCI religions, which go out of their way to hate things and people for things they do by nature. Likewise, if causing harm is only incidental and circumstantial, then there is no need to be angry because the problem is due to immediate circumstances. Of course, we must distinguish between an ordinary enemy and delusion; being friendly and understanding with a normal enemy offers the chance to make a new friend, but trying to befriend our delusions will only harm and destroy us.

Engaging in moral behavior such as compassion and lack of anger keeps our bodies, speech, and minds from committing any of the Ten Bad Actions, and in so doing forms the foundation for mental and spiritual development. This requires mindfulness and alertness, the importance of which cannot be overemphasized.

> *The kingdom of God cometh not with observation: neither shall they say, Lo here! or lo there! for, behold, the kingdom of God is within you.*
> Luke 17:20-21

Maintaining mindfulness and alertness depends on relying on four things:

- the teaching, not the teacher,

- the meaning, not the words that express it,
- the definitive meaning, not the provisional meaning, and
- the transcendent wisdom of deep experience, not mere knowledge.

For example, if a student sincerely points out her or his guru's faults and explains any contradictory behavior, this will help the guru to correct her or his actions and adjust any wrong actions.

According to the Dalai Lama, "If you reflect deeply enough, it become obvious that we need more compassion and altruism everywhere."

Selfishness is the single greatest obstacle to being good and compassionate, because it underlies most of our states of mind. As the Dalai Lama says, "Training one's mind and bringing forth inner discipline can change our outlook and thus our behavior as well."

Bemoan not the departed with excessive grief. The dead are devoted and faithful friends; they are ever associated with us.
Confucius

Other Religions

Our look at world religions would be incomplete without at least a glance at some of the smaller belief systems.

Confucianism

During the Chinese Han dynasty, people believed that *Tian* (the Godhead), Earth, and humans formed a single reality in threefold form, where Tian is not outside creation but is source of order inside creation. All people are related to each other and to Tian. The family is the center of Heaven on Earth. Respect for elders is paramount and children are obligated to care for their parents. Kings are also linked to Tian, a belief that worked well during good times but not so well when times were hard.

Confucius (551-479BC) lived during this time. By his own account, he took to learning when he was 15 years old. At age 30, he took to standing firm. He ceased to doubt at 40, knew he was the will of Heaven at 50, his ear understood at 60, and by 70 he did what he wanted and broke no rules. His teachings

Love is infallible; it has no errors, for all errors are the want of love.
William Law

are contained in roughly 500 remarks and conversations with the people and rulers he met.

Human goodness is the supreme virtue in Chinese belief. Before Confucius, only the nobility were thought capable of this, but he taught that all people could learn to be good. Rituals were important to Confucius, who believed in two levels of *li*: the outer behavior, and the inner mind. According to Shao Yong, "The human mind can be as calm as still water; being calm it will be tranquil; being tranquil, it will be enlightened." Shao Yong also refused all official posts, called himself Mr. Happy, and believed that all comes from the same ultimate source.

Some thought that Confucian teachings would weaken the nation and taught that well-known laws backed by a strong government were the keys to success. Nevertheless, Confucianism became the foundation of Chinese education, philosophy, and government for 2,000 years.

Better do a good deed near at home than go far away to burn incense.
Chinese proverb

Daoism

Daoism is the belief that life and well-being depend on understanding the *Qi* (pronounced "chi") energy that sustains the vigor of life. As for Dao itself, the Dao that can be described as Dao is not the eternal Dao. The origin of Heaven and Earth is nonexistent so we can learn its inner secrets, existent so we can see its outer manifestations.

The history of Daoism mainly consists of trying to work out what its profound words mean. Dao is the reason why anything exists at all and is far beyond our comprehension, but we can say something about it because of its manifestations. Dao supplies the possibility of all nature and individual appearances, which are Dao as it becomes nameable. The yin is receptive and calm, the yang aggressive and excitable. This can make it seem like yin and yang are opposites, but each contains the seeds of the other. The secret of life is to understand the Dao-de and yin-yang and work with the principles that bring the universe into being and keep it going. This requires being active in a non-active way and letting Dao unfold without rebelling against it. There are many gods one can pray to in a sort of cosmic Internet. Ni Zan asks, "Who can say that

When you doubt, abstain.
Zoroaster

the realm of the Dao is far from us? How tranquil it is, at the beginning of Heaven and Earth."

As we will see in Chapter 29, Daoism is very compatible with idealism and with the concept of a biocentric universe.

Zoroastrianism

Both Christianity and Islam have been in Persia and India for a long time, but were both imported. Long before this, Parsis of Persia (ca. 600 BC) is considered the first Zoroastrian prophet. Good and evil play central roles in this dualist system, which believes that the fight against evil can be won. At death, the soul is led to the Bridge of Judgment. Those who pass live in eternal bliss, while those who fail are thrown from the bridge into eternal torment. As we saw above, these beliefs are very close to Christian beliefs.

Sikhism

Sikhs believe that God is far beyond human understanding and should be approached only in worship and reverence. Learners or disciples must follow the teachings of gurus. All religions devoted to God are considered partners; according to Guru Nanal, there is no Hinduism or Islam, only the path of God. This is often interpreted to mean that all religions are valid, but what it really means is that other religions are not following the correct path. Still, what matters most is devotion to God above any institution. In this, Sikhs resemble the Gnostics. Kabir (c. 1518) found God everywhere he looked; what matters is finding God beyond the names, which he thought was easy. Charity is a virtue to Sikhs, who often serve free vegetarian food to all comers equally regardless of caste, race, or creed. To them, feeding the needy serves their guru and worships God.

Like so many other religions, Sikhs have seen their share of strife. Guru Hargobind (1595-1644) organized the Sikhs to defend themselves and built a fortress. Guru Gobind Rai (1666-1708) said that the Sikhs must be better organized for defense, support, and encouragement; war can be justified if all of the following five conditions have been met:

- All other means of settlement must have been tried first.

People should think less about what they ought to do and more about what they ought to be.
Meister Ekhart

Fools regard themselves as awake now-so personal is everything! It may as a prince or it may be as a herdsman, but so cock-sure of themselves!
Chuang Tzu

- Fight without passion or the need for revenge.
- Do not seize or loot land.
- The army must consist entirely of committed Sikhs.
- Use the minimum amount of force possible.

These are all admirable principles. If only they were universal!

Taoism

We make an idol of truth itself; for truth apart from charity is not god, but his image and idol, which we must neither love nor worship.
Blaise Pascal

One night, Chuang Tzu dreamed he was a butterfly. He then awoke to find himself in bed. Was he a man dreaming he was a butterfly, or vice-versa? One of the key features of Taoist thought is not seeing things as absolutes or as opposites. This concept even extends to the ideas of good and bad. There is a story of a farmer whose prize stallion ran off one day. His neighbors consoled him for this loss, but the farmer replied, "Good? Bad? Who knows?" Two days later, the stallion returned leading a large group of feral horses. The neighbors congratulated the farmer's good fortune, but he replied, "Good? Bad? Who knows?" Then the farmer's son mounted the stallion but fell off and broke his leg. Again the neighbors consoled the farmer, and again he replied, "Good? Bad? Who knows?" Soon thereafter the army came through town conscripting soldiers. Seeing the son's injury, they exempted him from service. Good? Bad? Who knows?

This story illustrates the simple truth that nothing in life is ever intrinsically good or bad unless and until we choose to apply those labels. The problem is that we often apply those labels unconsciously. Developing the art of conscious observation allows us to perceive life any way we want. This philosophy is not restricted to Taoism; it can be applied equally well to any belief one has.

The Divine Savage
Revealing the Miracle of Being

Chapter 6

Jesus Christ

> *Jesus was the first socialist, the first to seek a better life for mankind.*
> Mikhail Gorbachev

Jesus Christ is one of the most famous and controversial figures to ever walk the Earth. *Jesus* comes from *Yeshua* (deliverer, savior), and *Christ* is another word for *meshiha* (anointed one). Paul of Tarsus arguably invented Christianity and proclaimed a new era free of the Jewish law, despite the fact that Christ specifically said that he came to affirm the law in Matthew 5:18. Paul did this without ever seeing Christ or knowing his real name. The philosopher Philo of Alexandria (c. 20BC-50AD) adds to the mystery by not mentioning Christ. This has led to some speculation that perhaps Christ never existed at all. I feel that the evidence for his existence is all but conclusive and am proceeding under that assumption.

> *Fundamentalists believe Jesus was God becoming man. I believe that Jesus was man becoming God.*
> Eric Butterworth

It cannot be stressed strongly enough that Christ and his companions and disciples were Jews who absolutely did not intend to start a new religion, possibly unlike Mohammed. There was no New Testament. Christianity did not exist. Christ and those around him used Jewish books. Christ knew and quoted Jewish scripture, and seems to have assumed that others were familiar with it as well. Sadly, these Jewish origins have been buried under countless layers of pagan traditions that were either adopted by the Catholic Church to attract converts or brought into the mix by the converts themselves, such as the December 25th birthday and the virgin birth. Many Christians believe that these traditions are 100% true and factual. This is just one more example of how far Christianity has strayed from its roots. Christ himself doesn't seem to care about people's backgrounds; he accepted all comers.

Did Jesus cure the sick, raise the dead, feed the multitude, and perform the other miracles attributed to him? Probably not, at least not in the ways portrayed in the Bible. Having said that, it is quite possible that he was a highly skilled healer who tended to the sick and injured, and who may have had some success. Many of today's so-called miracle drugs only have a small percentage of efficacy when tested against placebos and/or other medications; thus, a faith healer with even a modest track record could be deemed miraculous. We will soon discover that there is every reason to think that Christ was such a healer.

> *We must not sit still and look for miracles; up and doing, and the Lord will be with thee. Prayer and pains, through faith in Christ Jesus, will do anything.*
> John Eliot

Was Christ born to a virgin in a Bethlehem manger with a star shining overhead to guide three wise men to his side bearing gifts? No. So who was the real Jesus Christ, and why did some who professed to follow him make up so many outlandish stories? The story you are about to read bears little resemblance to the standard Christ stories that you are probably already familiar with, but it is the story that makes the most sense to me based on the best available evidence.

Foretelling Christ

The Old Testament contains many passages that either explicitly predict a future messiah or at least imply it. Some of the more relevant passages include:

> *If the grandfather of the grandfather of Jesus had known what was hidden within him, he would have stood humble and awe-struck before his soul.*
> Kahlil Gibranr

- **Genesis 19:19:** But God said [to Abraham], "No, but Sarah your wife shall bear you a son, and you shall call his name Isaac; and I will establish My covenant with him for an everlasting covenant for his descendants after him."

- **Numbers 24:17:** I see him, but not now; I behold him, but not near. A star shall come forth from Jacob, and a scepter shall rise from Israel, and shall crush through the forehead of Moab, and tear down all the sons of Sheth.

- **Jeremiah 23:5-6:** "Behold, the days are coming," declares the LORD, "When I shall raise up for David a righteous Branch; and He will reign as king and act wisely and do justice and righteousness in the land. In His days Judah will be saved, and Israel will dwell securely; and this is His name by which He will be called, 'The LORD our righteousness.'"

- **Isaiah 7:14:** Therefore the LORD Himself will give you a sign: Behold, a virgin will be with child and bear a son, and she will call His name Emmanuel.

- **Isaiah 7:19:** There will be no end to the increase of His government or of peace, on the throne of David and over his kingdom, to establish it and to uphold it with justice and righteousness from then on and forevermore. The zeal of the LORD of hosts will accomplish this.

- **Zechariah 9:9:** Rejoice greatly, O daughter of Zion! Shout in triumph, O daughter of Jerusalem! Behold, your king is coming to you; He is just and endowed with salvation, Humble, and mounted on a donkey, even on a colt, the foal of a donkey.

The New Testament tells us that Joseph and Mary knew what was about to happen. Matthew 1:18-25 says that Joseph and Mary were engaged. Upon discovering that she was pregnant through the Holy Spirit, Joseph thought of divorcing her, but an angel appeared and told him to stay with Mary, that she would bear a son to be named Jesus because he would save people from their sins.

It is very important to understand that the Old Testament existed long before Christ was born, while the New Testament came after his supposed death on the cross (more on this later). We therefore cannot use the New Testament for any kind of predictive purpose; neither could anyone alive at the time of Christ's birth. They did, however, have access to the Old Testament, and could conceivably engineer events to fulfill the ancient prophesies. All one would have to do is arrange a marriage between people of the appropriate lineage and see to it that the resulting child acted in accordance with the predictions. Riding into Jerusalem on a donkey is the epitome of simplicity. Other aspects of the prophesies might not be so easy, but should still be quite manageable. From there, a strong dose of faith would rely on God to take care of the rest.

It was God's intention to bring all things in heaven and on earth to a unity in Christ, and each of us participates in this grand movement.
Desmond Tutu

Jesus was all right, but his disciples were thick and ordinary. It's them twisting it that ruins it for me.
John Lennon

This leads to a very important point: Christians like to think that Christ descended from heaven with a pre-made message on a predetermined mission from God, but this is not true. Christ was an ordinary boy born to ordinary people in a rebellious corner of the Roman Empire and carefully trained for his role. Nevertheless, as Robert Burns points out in his poem *To a Mouse*, the best-laid schemes of mice and men often go awry.

Lineage and Breeding

Israel at the time of Christ was part of the Romans province of Judea. Most of the population accepted this situation, but a band of rebels sought to overthrow the occupiers and restore sovereignty over their lands. These were the Zealots, religious fanatics who wanted to establish God's kingdom on Earth. The Zealots anticipated a leader in the form of a messiah who would lead them to victory. How fanatical were the Zealots? In 72AD, the Romans assaulted the mountaintop garrison of Masada. Rather than face capture, the Zealots murdered their own families and then each other, with lots cast to see who would be the last man left standing to kill his remaining comrades before committing suicide. 960 men, women, and children perished, believing that such a ritually pure death would guarantee a resurrection that would be unavailable if they were captured. Capture meant being sold into prostitution and impurity from which there was no coming back. People who died together would resurrect together, so each family gathered to die in the same spot. None of these events had yet happened when Christ was born, but this was the general situation Christ found himself in and the crucible where Christianity began.

The Dead Sea Scrolls that were discovered between 1947 and 1956 in caves near the settlement of Qumran near the Dead Sea have remained untouched since their writing between 150-70BC and are thus a much more reliable record than the Bible, which has undergone countless thousands of accidental and deliberate changes. (See Chapter 8.) These scrolls place a heavy emphasis on the ancestral purity of kings and priests. Priests anointed the *meshisha* (messiah) as the legitimate King of Israel of the line of King David. Based on the Old Testa-

---

The Pyramids will not last a moment compared with the daisy. And before Buddha or Jesus spoke the nightingale sang, and long after the words of Jesus and Buddha are gone into oblivion the nightingale still will sing. Because it is neither preaching nor commanding nor urging. It is just singing. And in the beginning was not a Word, but a chirrup.
D. H. Laurnece

---

ment prophesies, the people of Judea were awaiting a descendant of David to lead them to victory over the Romans.

Frank Herbert's legendary *Dune* tells of a breeding program orchestrated by the religious Bene Gesserit order to create the Kwisatch Haderach, a man who could endure a potentially lethal ritual (the spice agony) and obtain all of the powers of a Reverend Mother, the highest Bene Gesserit rank. This man would then be able to transcend space and time and make the Bene Gesserits the preeminent faction in the galaxy. This is science fiction, but, as Mark Twain said, truth is stranger than fiction. Armed with the ancient prophesies from the Old Testament, the Zealots arranged a marriage between Joseph of the line of David and Mary of the line of Aaron (the first high priest of the Jews, who finished what Moses started by leading the Jews to the Promised Land) to produce their own Kwisatch Haderach in the form of Jesus Christ. This man would be both the rightful king of Israel and the high priest, a true messiah in fact and in deed. Someone decided to invoke Isaiah 7:14 and claim that Mary was a virgin, and I can only assume that it would be easy enough to coerce Joseph and Mary to cooperate. They may even have done so willingly. The Star of Bethlehem is a messianic symbol of David's line.

Birth

Once in royal David's city stood a lowly cattle shed where a mother laid her baby in a manger for his bed. Mary was that mother mild, Jesus Christ her little child.
Cecil Frances Alexander

According to the Bible, the emperor Augustus ordered a tax census so that "the entire world should be registered." Quirinius, the governor of Syria, ordered everyone to return to their ancestral homes to be counted. Joseph returned from Nazareth to Bethlehem to be counted (Luke 2:1-7). There is just one problem: No such census ever took place under Augustus. There was a census in 6AD, 10 years after the death of Herod the Great. People did have to return to where they lived and worked to be counted, but asking everyone to return to their ancestral home would create a logistical nightmare that would bring the economy—and the associated tax revenue—to a standstill when all the Romans cared about was taxation. There is also reason to doubt whether Nazareth existed at the time of Christ.

As for the birth itself, it seems that Matthew simply rehashed the story of the birth of Moses to come up with a suitably humble beginning for Christ. There are plenty of practical reasons to portray Christ as humbly as possible, not least of which has to do with striking a chord among the common people by making them feel that Christ was one of them who had experienced their travails, and would look out for their welfare in ways that no noble-born king could match.

The Catholic Church is quite content to set aside its long-standing disdain of women to adore Mary, just so long as she remains a virgin and both she and Christ remain unsullied by the filth of sex. After all, according to Pope Gregory 1 (540-604AD), sexual pleasure can never be without sin. The 1987 encyclical *Redemptoris Mater* by Pope John Paul II even went so far as to state that Mary's hymen remained intact during Christ's birth! I am at a loss to explain this, because God made women with hymens that are designed to be broken. It is therefore ludicrous to look down on something for functioning as designed, especially when you are worshipping the designer! But I digress...

The earliest New Testament writings by Paul do not mention anything about a virgin birth. The first such mention appeared when the Old Testament was translated into Greek in the 3rd century AD. Isaiah said that an *alma* (young woman) would bear a son called Emmanuel, and *alma* was translated into *parthenos* for parthenogenesis. Despite Matthew's insistence that Jesus Christ fulfilled ancient prophesy by being born of a virgin, all that was needed was for a young woman to give birth—hardly an extraordinary event. Christ himself never mentions celibacy or being born of a virgin, and Paul himself claims to know nothing about a virgin birth. Even Peter (the supposed founder of the Catholic Church and its posthumous first pope) was married and traveled with his wife. The Catholic Church forgot this, and celibacy became the order of the day.

All evidence indicates that Christ was born to Joseph and Mary in the same way as countless other children to countless other couples. The only difference is that Christ was the product of a carefully engineered arranged marriage, which is hardly uncommon throughout history. For example, arranged marriages were very common in Europe and remain common in parts of the world today.

Sexual pleasure can never be without sin.
Responsum Gregorii

[Mary's] hymen remained intact.
John Paul II

Life and Training

And you are to love those who are your aliens for you yourselves were aliens in Egypt.
Deuteronomy 10:10

Christ was about two years old when Herod died in 4BCE and was born in a time of great social unrest, as we saw above. The official story is that he went to Nazareth in Galilee, at which point the Bible story falls silent until Christ suddenly emerges as a grown man who emerges from Galilee to be baptized and gather his disciples. Leaving aside the oddity of omitting the formative years of the New Testament's star character from the official history, is there any evidence that can help us fit in the missing pieces of Christ's life story? As it turns out there is. One of the keys to unraveling this mystery lies in knowing about the Jews in Egypt.

John the Baptist

The Christ myth created a much more fantastic imaginary universe than anything encountered in the Jesus traditions.
Burton Mack

More than one ancient prophet plied his trade by rehashing Moses, Joshua, and the supposed flight from Egypt 1,200 years prior. These prophets led followers into the wilderness to reenact the crossing of the river Jordan, where God would purify them. We cannot know whether God held up his end of the bargain, but we can be reasonably sure that plenty of would-be pilgrims died during these journeys.

John the Baptist was different: Instead of leading people into the hinterlands, he baptized them and sent them back to the Promised Land of Israel to await the apocalypse that would usher in God's kingdom—a *modus operandi* that is very similar to modern Jehovah's Witnesses. Christ may have been one of those baptized by John, a fact that makes Christians nervous, because it implies that John was somehow superior to Christ. Christ seems to have begun as an apocalyptic believer. He may even have been a follower of John who eventually broke off to form his own movement, possibly in competition with his former leader. If this is true, does that make Christ gullible (for believing John), power hungry (for seeing an opportunity to build his own flock), or both? Or was something else happening? We find a clue in Mark's story of the execution of John, which attempts to draw parallels between John and Christ. John is the precursor to Christ, and it is possible that Mark is retelling an even older story. We find another clue in the fact that Jesus was mystical, while the Zealots who bred him were not remotely mystical.

Plans within Plans

Galilee was a hotbed of Zealot rebellion. Christ therefore cannot have learned there. He also cannot have lived in Israel, or the Gospels would certainly have regaled us with tales of his waxing greatness. The Gospels are silent about the first 18 years of his life. As far as they are concerned, Christ vanished into thin air and just as magically reappeared as a grown man. Where was he during this time? Where had he acquired his knowledge? Had he been somewhere for training, and if so, where? What are the Gospels trying to hide by leaving out what is probably the most interesting and important part of Christ's biography? One thing is clear: The Zealots had bred Christ to lead them against the Romans; had they been responsible for Christ's training, there is almost no way that he would have adopted such pro-Roman attitudes as, "sheep in the midst of wolves" (Matthew 10:16), and "Render unto Caesar what is Caesar's." (Matthew 22:21). Even if he had, there is no way the Zealots would have let him enter Jerusalem as the messiah had they known of his beliefs. Christ was therefore working on a plan of his own, a plan that he had to keep secret until the last possible moment before springing it upon the world. But how? Where? From whom?

> *Those in possession of absolute power can not only prophesy and make their prophecies come true, but they can also lie and make their lies come true.*
> Eric Hoffer

Essenes and Therapeutae

The Essenes were a mystical Jewish religious group that existed from the 2nd century BCE to the 1st century CE and are believed to have written the Qumran (Dead Sea) scrolls. Scientists studying these writings have found many similarities between the teachings of the Essenes and Jesus Christ. Some believe that a group called the Therapeutae were a branch of the Essenes.

The Therapeutae worshipped the eternal, self-existent, One Divine reality that is beyond our ability to describe. (The Buddhists would agree with this!) Women and men were equals, and women participated fully in spiritual matters. The worshipper's gender was irrelevant; all sought a direct vision of the Godhead to experience what lies beyond mundane daily life (enlightenment). The Essenes themselves excluded women. Christ's inclusion of women therefore means he was probably not Essene, but could have belonged to the Therapeutae.

> *Now this is eternal life: that they may know you, the only true God, and Jesus Christ whom thou has sent.*
> John the Baptist

The Therapeutae themselves were educated, wealthy *patricians* (aristocrats) living in Alexandria (Egypt) who renounced their worldly goods to live the simple life (similar to Buddhist monks). They did not concern themselves with Israel or Judaism, because their kingdom was not of this world—a worldview that Christ would have wholeheartedly approved of. Both the Therapeutae and the Jewish Zadokite sect used the Egyptian solar calendar. The primary Egyptian deity, Ra, was an expression of the Sun and the giver of all life and creation. The pharaoh sought mystical union with Ra. It is no coincidence that the Gnostic Gospels were buried in Egypt at Nag Hammadi to protect them from destruction by the Roman Catholics.

That the Jews assumed a right exclusively to the benefits of God will be a lasting witness against them and the same will it be against Christians.
William Blake

There was also a large, influential Jewish colony in Egypt that was allied with the Ptolemies and enjoyed a higher social standing than the native Egyptians. The Jews in Egypt spoke Greek and even conducted their services in Greek at a temple that functioned similarly to the one in Jerusalem. Egypt was also a major center of knowledge. Passengers arriving in Alexandria were searched for books; any titles found were duplicated, the originals placed in the library at Alexandria, and the copies returned to their owners.

To attempt the destruction of our passions is the height of folly. What a noble aim is that of the zealot who tortures himself like a madman in order to desire nothing, love nothing, feel nothing, and who, if he succeeded, would end up a complete monster!
Denis Diderot

Why Egypt? It was crucial for Christ to act out the specific Old Testament prophesies in a very literal fashion, a situation that placed severe restrictions on what he could do. Beyond the Therapeutae and other Jews, there were also some Zealots in Egypt. All of the Jewish scriptures were available, making Egypt the perfect place to bring Christ to receive his training during the so-called "flight" into Egypt described in the New Testament. Seen in this light, Jesus's journey was not one of running away from persecution but had been planned from the very beginning. As Zealots, Joseph and Mary would have been careful to avoid the Hellenized Jews by traveling south past the temple of Onias. What the Zealots did not plan for was that Christ would secretly embark on an entirely different path: When he prayed for the poor and the meek, he probably meant the Essenes, because that is what they called themselves. Various sites in Egypt remain important to the Coptic Christians today.

Egyptian Spirituality

The Egyptian cosmos held two worlds, the physical and the far. These worlds existed simultaneously, with the far world being intertwined with the physical world and providing an eternal source and backdrop to everything happening in the universe. Life came from and returned to the far world. Egypt also saw two corresponding types of time: *neheh* (cyclical time) and *djet* (that which is outside of time, eternal, and immortal. (We will return to these ideas in Chapter 28.) Death was seen as the source of life, and the dead were seen as the truly living. The story of Gilgamesh echoes this idea: Gilgamesh seeks immortality but fails because he cannot stay "awake" and returns to tell the tale of his experiences, which implies that he achieved a state similar—if not identical—to Buddhist enlightenment.

I think religion is often very different from spirituality. Religion is often about rules and people trying to control our lives who are actually very unspiritual. God can be found anywhere, and in fact, everywhere.
Darren Aronofsky

Ancient Egyptian traditions, science, and engineering dealt with the relationship of both worlds. For example, the Pyramid complex in Giza reflects a terrestrial version of the constellations Orion and Leo and their relationship to the Milky Way as it would have appeared during *Zep Tepi*, the "First Time" when the gods lived among humans, ca. 10,500BC. As we saw in Chapter 4, the earliest humans distinguished between these two worlds, as evidenced by Neanderthal burial rituals dating back some 100,000 years. Humans have been concerned with the far world and the existence of self-consciousness since we first became aware of them. One must presume that these concerns confer some evolutionary benefit on humanity, an idea that we will revisit in Chapter 9.

The Egyptians believed that the *ba* (soul) is independent of the body and goes back to the divine source when the body dies. The ba is assumed to always exist as an integral part of all people; if this is true, then there is no reason why people cannot experience their ba form before dying. This concept of enlightenment existed long before Buddha. Some of the relevant sayings include:

The pyramids, attached with age, have forgotten the names of their founders.
Buckminster Fuller

- **Pyramid carving:** The spirit is bound for the sky; the corpse is bound for the Earth.

- **Pyramid Texts:** I have gone and returned. I go forth today in the form of a living spirit.

> *I feel that the Godhead is broken up like the bread at the Supper, and that we are the pieces. Hence this infinite fraternity of feeling.*
> -Herman Melville

- **Amduat:** It is good for the dead to have this knowledge, but also for a person on Earth.

A person in this state would be comparable to someone having an OBE, which occurs spontaneously at death but must be induced during life. According to philosopher Jeremy Naydler, this resulted in detailed initiation techniques that allowed people to travel to and from the far world. A priest in a ritual state could travel to the far world and then describe what he saw when he returned; the gods drew their souls up to familiarize them with being out of their bodies at death. These journeys were part and parcel of the priesthood, and the priests could come and go at will.

The *ankh* is the symbol of eternal life; the goal of the ba is to become pure radiance and merge with the Godhead. The Egyptian Book of the Dead was originally called the *Book of Coming Forth by Day* and contained instructions on how to go into the light. It ends with the saying, "Whoever knows these mysteries receives the ankh spirit and can enter and leave the netherworld always speaking with living ones. This has been proven to be true a million times." This sounds eerily close to Buddhism and also fits with modern descriptions of NDEs. It is easy to dismiss Egyptian descriptions of the far world as creative speculation amassed over thousands of years, but Naydler believes that they could be describing actual mystical experiences. It is interesting to note that the earliest references to the far world come from shamans who said that one must visit hell before going to heaven, which sounds a lot like the Buddhist idea of karma.

> *Education is not merely a means for earning a living or an instrument for the acquisition of wealth. It is an initiation into life of spirit, a training of the human soul in the pursuit of truth and the practice of virtue.*
> Vijaya Lakshmi Pandit

Author Michael Baigent leads tours of Egyptian temples and chapters, and reports that people often experience unexpected emotions; in *The Jesus Papers*, he mentions seeing a man walking around the Osirion at Abydos muttering, "This is the real thing. This is the real thing." as evidence that the Egyptian beliefs are neither a fraud nor a mere thought experiment but something real.

Initiations

Many ancient traditions from places like Egypt and Greece include an aspect of initiation in which someone undergoes a ritual and is then let in on a huge secret. Christ himself spoke in stories and parables that used simple language to explain mystical concepts. When asked why, he said that this was for mass consumption, but that only the initiated could know the actual secrets. He therefore acknowledged two levels of people: the initiates, and everyone else. The path to the kingdom of heaven was reserved for the former and, once discovered, was ever-present for them. There is also reason to believe that Christ participated in initiating others.

The Greek philosopher Socrates (c. 469-399BC) was sentenced to die by poisoning himself because he disrespected the Greek gods. According to him, "The aim of those who practice philosophy in the proper manner is to practice for dying." Initiates were those who received the same knowledge one obtains at the point of death. According to Themistius (ca. 390CE), the soul at death experiences what initiates experience: terror and darkness, then light and joy. The initiated seeker can see the uninitiated living in fear of death and the blessings it brings—a lesson repeated by legions of *Near Death Experience* (NDE) survivors.

Aristotle (394-322BC) believed that reason, discussion, logical argument, etc. are all limited. We must experience truth directly instead of intellectually. For example, we cannot know the pain of fire unless we experience it by being burned. The story of Jacob's ladder in Genesis 28:11-19 says that Jacob saw the ladder in a dream; history might have turned out very differently had the story said that Jacob visited the far world and returned.

The Gospel of Mark is part of the accepted New Testament canon, but there is also a Secret Gospel of Mark that was written for those being initiated into the great mysteries—mysteries that Pope Clement 1 felt compelled to deny. The scene in question speaks of Christ initiating a man in Bethany, which just happens to be where he supposedly raised Lazarus from the dead. One initiation ritual Jesus would have known about had initiates go into a cave or tomb underground and lay there for three days to experience what it is like to die and to see the far world for themselves. Lazarus just happened to be underground for three days. This would explain why Christ doesn't seem overly concerned when he learns that Lazarus is sick; in John 11:11, he says, "Our friend Lazarus has fallen asleep; but I am going there to wake him up." Christ may well have

---

We must remember that Satan has the ability to appear as an 'angel of light' and as a 'servant of righteousness.' his goal, of course, is to lead people astray. He is happy to mimic a being of light if the end result is that he can lead people away from the true Christ of Scripture.

Ron Rhodes

---

known that Lazarus was being initiated. Some modern critics wonder why nobody interviewed Lazarus to get his side of the story and confirm his miraculous revival. If Lazarus's "death" was merely a ritual that was known to those around him, then there would be nobody to interview him, which would render those who take his resurrection literally mistaken.

Nearer and Farther Still

> *One of the secrets of life is to keep our intellectual curiosity acute.*
> William Lyon Phelps

"Let the initiate reveal [the secret of the great gods] to the initiate, but do not let the uninitiated see it.," says a Babylonian text. Babylonian baptism was the source of Jewish purification before rituals to separate priests/initiates from the physical world and establish pure relations with the far world. The Jewish calendar, astrology texts, and incantation bowls are also borrowed from the Babylonians. Jewish monotheism was inspired by the Assyrian god Ashur (the one God) and the Egyptian Aten. (See Chapter 7.) Ezekiel's vision of God on a throne of lapis lazuli also comes from Babylon. The Talmud contains medical information that is also derived from earlier sources. Judaism therefore represents not so much a new invention as a cobbling together of other existing traditions, exactly what one might expect to see in a relatively backward region surrounded by great civilizations.

> *He that has eyes to see and ears to hear may convince himself that no mortal can keep a secret. If his lips are silent, he chatters with his fingertips; betrayal oozes out of him at every pore.*
> Roland Barthes

The Tree of Life is the backbone of Kabbalistic thought, but did not originate with the Jews. There is evidence to suggest that the Tree of Life/Good and Evil/Knowledge/etc. was known since the dawn of humanity, which means that we may literally have traces of Neanderthal teachings. The Kabbala tree is based on the Assyrian tree, which may ultimately be based on a Neanderthal tree—and why stop there? This tree shows how God manifests in living things and contains 10 *sefrot* (symbols of emanated divine principles), which may not seem remarkable until you realize that modern superstring theory posits 10 dimensions, as we'll see in Chapter 25. Teresa of Avila (1515-1582CE) alludes to the soul leaving the body and discovering mysteries. Other fragmentary examples remain.

Return and Betrayal

Christ was born of a prearranged marriage to assure his lineage and sent to Egypt for training that included Jewish law, Essene mysticism, Therapeutae medical arts, and the ancient prophesies he was supposed to fulfill. All of this occurred during the time in which the New Testament is oddly silent as to his whereabouts and activities. Christ had been gone from Palestine so long that he was required to pay the Stranger's Tax upon his return (Matthew 17:24-27). Even John the Baptist did not recognize him at first.

Christ understood that he was the Messiah and, in his capacity as the high priest to be, preached a message of expanded consciousness and mystical union with the Godhead. He taught the Essene ritual of the sacred meal. As the Gospel of Thomas says, the kingdom lies both within and without. Christ also talked a good game from a kingly and military leader perspective—at least at first.

Mediterranean values center far more around families and tribes than individuals. To this, Luke 14:26 quotes Christ as saying that, "If anyone comes to me and not hate his own father, mother, wife, children, siblings and even his own life, he cannot be my disciple." In Luke 12:51-56, Christ goes even further, saying, "Do you think I came to bring peace on Earth? No, I tell you, but division. From now on there will be five in one family divided against each other, three against two and two against three. They will be divided, father against son and son against father, mother against daughter and daughter against mother, mother-in-law against daughter-in-law and daughter-in-law against mother-in-law. When you see a cloud rising in the west, immediately you say, 'It's going to rain,' and it does. And when the south wind blows, you say, 'It's going to be hot,' and it is. Hypocrites! You know how to interpret the appearance of the earth and the sky. How is it that you don't know how to interpret this present time?"

These sayings can be interpreted in many different ways. Here are some examples:

- The American Civil War (1861-1865) tore families apart. It was not unheard of for different members of the same family to fight for different sides. Seen in this light,

I am thirty-three, the age of the good Sans-culotte Jesus; an age fatal to revolutionists.
Camille Desmoulins

Because of the theological obligation to endorse the precepts of Jesus, Christian theologians have a strong tendency to read their own moral conviction into the ethics of Jesus. Jesus is made to say what theologians think he should have said.
George Smith

Christ could be urging people to put their desire to fight the Romans ahead of their families, some of whom could be indifferent to the struggle or even Roman sympathizers. As an aside, I'm positive that the American Civil War was hardly unique in its effect on families.

- Christ could be demanding loyalty of Abrahamic proportions. Remember that Abraham was prepared to kill his son on God's orders.

- Buddhist writings that speak about killing one's parents are metaphorical and mean that one must set aside one's existing beliefs if one is to seek enlightenment. Christ's message could have been similar.

- Christ could have meant any combination of the above and may have had additional meanings that may or may not have anything to do with those presented here.

They represent this virtuous and amiable man, Jesus Christ, to be at once both the god and man, and also for the son of god, celestially begotten, on purpose to be sacrificed, because they say that Eve in her longing had eaten an apple.
Thomas Paine

With all of the above said, the question about interpreting the present time seems to be a clear message that the time for rising against Rome has come.

Christ entered Jerusalem in a manner that would fulfill the prophesies and push all the right buttons to have the people see him as Israel's chosen messiah and anointed king. The Zealots must have been beside themselves with joy and anticipation at seeing their plans coming to fruition—and then Christ changed his tune. Far from telling the people to rebel, Christ told them to abide by Roman rule! (He was also making a statement about the separation of church and state that today's Evangelical Christians seem to have forgotten.) In other words, Christ decided to use his position and training to lead the people not to war, but to enlightenment.

One can only imagine the outrage among the Zealots and the mad rush to control the damage. It is perfectly plausible to think that they returned betrayal with betrayal to at least salvage a martyr from their ruined plans, making Judas a traitor to Christ but a hero to the Zealots. Judas could also have been following Christ's orders as told in the Gospel of Judas, another book that did not make it into the official New Testament. Having distanced himself from the Zealots, the Greek version of Matthew 26:55 shows Christ confronting those sent to arrest him by asking, "Have you come with swords and clubs to arrest me as you would a Zealot? I have been in the temple teaching every day and you did not seize me." Clearly, Christ was fully aware of the politics of the day. Strange as this story is, it's about to get a lot stranger.

The Crucifixion

Crucifixion is the most denigrating form of execution the Romans ever dreamed up. It is death by slow torture, where the victim's own body weight makes breathing difficult. Death from asphyxia occurs after about three days as the legs weaken and become increasingly unable to support the body. The Romans sometimes took pity by administering a "merciful" breaking of the legs to speed things along. The Gospel of John says this happened to the two men crucified with Christ, but that Christ was already dead.

The Roman prefect of Judea at the time was one Pontius Pilate. He would have known that the Sadducee priests wanted Christ dead because he undermined their power, and that the Zealots wanted to avenge their betrayal. Pilate's dilemma was that he was Rome's official representative in Judea. Rome's biggest problem with Judea was that the Jews didn't want to pay their taxes, yet here was Christ telling them to do just that! He could not execute an open Roman supporter without getting into major trouble of his own.

Crucifixion was used to punish political crime, but the Bible says that Pilate gave Christ to the crowd, who wanted him crucified for religious dissent, a crime normally punishable by stoning. Contrary to the Bible, this demonstrates that the people of the time had no concept of crucifying Christ to atone for Eve's original sin. This alone falsifies the Gospels and begs the question of what they are trying to hide. Christ was sentenced for political crimes, and the Romans were calling the shots, not the Jews. The Bible's attempts to spin the story to distance Christ from politics can't hide the simple fact that he was sentenced and executed by the Romans—or was he?

There is another gaping problem with the Biblical account of Christ's death: Crucifixion was normally reserved for commoners. Those with power or influence were not crucified. Those without influence could hardly request the corpse of the executed because the Romans saw families as part of the

Jesus Christ? Stop me if you heard this one: Jesus Christ walks into a hotel. He hands the innkeeper three nails, and he asks... Can you put me up for the night?
Eric Draven

None speak of the bravery, the might, or the intellect of Jesus; but the devil is always imagined as a being of acute intellect, political cunning, and the fiercest courage. These universal and instinctive tendencies of the human mind reveal much.
Lydia Maria Child

> *Jesus proclaimed the coming of the kingdom but what came was the church!*
> Alfred Loisy

problem. If a father was crucified, they looked for and crucified the wife and children to reduce the chances that someone might seek revenge. There is no way that Christ's immediate family would try to recover his body, which would have been either left to rot and be eaten by crows or buried in a shallow grave for the dogs to eat. There is no way that the Romans would give him a decent burial, much less the prime spot in a nearby family tomb. The Bible tries to reconcile this discrepancy by claiming that Christ had influential friends like Joseph of Arimathea who could plead with the Romans on his body's behalf.

These are not the only holes in the story. The Bible describes the two men crucified with Christ as thieves, but the Greek text uses the word *lestai* (brigand, the Greek term for the Zealots)—a "mistake" that echoes Matthew 26:55, which also uses the word "thief" instead of "Zealot." This alone should be fairly conclusive proof that Christ himself was a Zealot. He was hardly alone. Barabbas, whom Pilate freed during a day of amnesty, is called a *lestos*. There is also Simon the Zealot, and Judas Iscariot (which means *sicarii*, an assassin who uses a knife called a *sica*). In Luke 22:36, Christ tells his disciples to arm themselves; clearly, he was part of a movement to eject Rome by force. Perhaps some of his Zealot compatriots remained loyal to him after he turned on the Zealot movement by preaching peace, a reasonable assumption since Christ was both a Zealot and a mystic.

> *The historical messiah had long been forgotten; what mattered now was the Vatican's Christ.*
> Roland Barthes

The Jews rejected Christ as a messiah because he died on the cross in violation of prophesy. Isaiah 53 tells of God's servant suffering but not dying. Specifically, Isaiah 53:10 says that, "When he makes himself an offering for sin, he shall see his offspring, he shall prolong his days." This clearly means that Christ was not supposed to die on the cross. It also means that Christ was supposed to have children, which must assume that he would be married, a thread we will pick up in just a few moments. In the meantime, we have to ask: What if Christ did not die on the Cross? Any viable evidence that he survived the crucifixion would wash the whole foundation out from underneath the entire orthodox Christian religion.

It should go without saying that the Catholic Church would go to extraordinary lengths to conceal any evidence that Christ survived the crucifixion—but did they go far enough?

Irenaeus complained that the Gnostics were teaching that Simon the Cyrene was crucified in Christ's stead. The Koran 4:157 says that a likeness of Christ was killed but not Christ himself. Still another account has Jesus laughing at those who thought he had been killed. There is another, even more intriguing possibility: What if Christ was indeed crucified but survived?

This possibility is a lot less outlandish than it might seem. Right off the bat, such a scheme could give Pilate a way out of his dilemma. Staging Christ's execution would mollify the priests and the Zealots, thus helping preserve the peace. Making sure that he was cut down before he died would avoid the inevitable problems that would befall Pilate for executing a Roman supporter. This alone is a practical reason to suppose that this scenario is plausible, but is it medically viable?

Yes. The Bible account admits that Christ took an inordinately short amount of time to die, even assuming that he had been weakened by torture beforehand. Of course, crucifixion ran the risk that Christ might actually die, so there was every reason to sedate him and cut him down early to minimize the risk. Bible scholar Hugh Schonfield (1901-1988) pointed out that Christ could have been drugged to appear dead for later revival. Simply soak a sponge in a mixture of opium, belladonna, hashish, etc. and allow it to dry for transport. When the time comes to knock someone out, soak the sponge in water to activate the drugs, place it over the subject's nose and mouth, and out he goes. Of course, this would require the help of people with extensive medical knowledge, people such as the Therapeutae, who Philo notes as having "an art of healing superior to that practiced in the cities." It is not unreasonable in the slightest to suppose that the "angels" described in John 20:12, Luke 24:4, Matthew 28:3, and Mark 16:5 were actually Essenes wearing their typical white robes, of which the Therapeutae were part.

It is amazingly convenient that Christ was crucified near a garden and tomb owned by Joseph of Arimathea, who was one of Christ's disciples. In fact, it is even possible that the crucifixion took place in the garden to keep crowds distant and limit the number of witnesses. A public crucifixion would have attracted a huge crowd that would surely have been mentioned in the New Testament, which mentions no such thing; in fact,

---

Revelation terminated with Jesus Christ.
Joseph Ratzinger

Myth is neither a lie nor a confession: it is an inflexion.
Roland Barthes

> *If the tree is indeed known by its roots, then Christianity's departure from the norm of the perennial philosophy would seem to be philosophically undesirable.*
> Aldous Huxley

Luke admits that the crowds were kept away. The Biblical description of Golgotha seems to match the Kidron Valley, an area that contains both many tombs and the Garden of Gethsemane where Jesus was arrested.

Mark says that Joseph went to Pilate to ask for Christ's body. Pilate asks if Christ is dead and is surprised when Joseph says yes because of the short time Christ took to die. Still, Pilate lets Joseph take the body. The Greek version has Joseph asking for a *soma* (living body) and Pilate telling him to take the *ptoma* (dead body). Thus, the Bible itself says that Christ survived. Christ was taken down and placed into a new tomb. Later, Joseph and Nicodemus visited the tomb with large quantities of myrrh and aloe, perfumes with medicinal—but not embalming—uses. for example, myrrh stops bleeding, something Christ would certainly have needed. According to Mark and Luke, women brought other spices and ointments. The Therapeutae were sparing no effort to revive their wounded brother.

Once Christ was out of mortal danger, the next logical step was to move him to a safe place. Station XIV in the chapel of Rennes-le-Chateau in southern France shows people removing a living person from the tomb under cover of darkness—a clear sign that its caretaker, Abbé François Bérenger Saunière (1852-1917), didn't buy the orthodox version of events. Under the circumstances, the only safe place to take Christ was back to Egypt. With Roman help, it would be a relatively simple thing to head to the docks with Mary Magdalene and set sail. Once back in Egypt, the couple could have taken refuge near the temple of Onias in an area that was both mystical and Zealot without being politically zealous. More on this in just a few moments.

Mary Magdalene

> *The Catholic Church has never really come to terms with women. What I object to is being treated either as Madonna's or Mary Magdalene's.*
> Shirley Williams

Jesus was known for empowering women, a revolutionary concept at the time and very much against Catholic dogma. Nevertheless, scholars are increasingly coming around to the view that Christ was married to Mary Magdalene—a far cry from the orthodox view of her as a prostitute and yet another insult hurled by the church against the man it claims to revere as the Son of God. The idea that Christ was married is in accord with both Old Testament prophesy (see above) and with rabbinic tradition. Christ being married would not have raised any eyebrows among the people close to him; in fact, his remaining a bachelor would probably have caused a stir.

John 12:1 has Christ being anointed by Mary of Bethany, who may also have been Mary Magdalene. Micah 4:8 says that, "Tower of the flock, Ophel of the daughter of Zion, to you shall be given back your former sovereignty and royal power over the house of Israel." This is a plausible tie to Mary Magdalene, which, if true, makes the church's label of whore even more insulting.

If Mary Magdalene was married to Christ the Therapeutae, she would certainly be one of his closest apostles (even more than Philip or Peter) and fully familiar with the near and far worlds. She would be both capable and empowered to anoint Christ, because Therapeutae women had the same status as the men. As such, she would be the logical choice to anoint her husband to his role as the messiah. This simple yet profound act would be the final stretch of a long road that the church has tried its best to forget for 2,000 years. Only now is the full extent of the cover-up becoming clear. The simple fact is that the idea of women being subservient to men and unable to be priests is just not part of the original tradition. Neither Christ nor Mary wanted to be worshipped as gods, start a religion, or have their teachings put into books; on the contrary, they wanted people to travel to the far world for themselves to be filled with the spirit of God.

Jesus of Nazareth could have chosen simply to express Himself in moral precepts; but like a great poet He chose the form of the parable, wonderful short stories that entertained and clothed the moral precept in an eternal form. It is not sufficient to catch man's mind, you must also catch the imaginative faculties of his mind.
Dudley Nichols

Exile to France

Escape to Egypt was a great idea, at least until the riots in Alexandria in 38AD and the subsequent persecution of the Jews. Between this and the situation back in Jerusalem, it became necessary to get out of both Egypt and Judea and find a Jewish community comfortably far from the Greeks. Narbonne was a Roman trading post on the Aube River in France that included the oldest Jewish population in the region. This area took a long time to become Christian. Even then, the Pauline flavor of Christianity was either absent or ineffective. Narbonne and Marseilles are both places with legends of Mary Magdalene arriving from the Middle East by boat.

The Holy Grail may not be a treasure but a document showing that Christ was alive in 45AD, long after his supposed execution in 36AD. But is there any proof? The fact is that traditions about Mary Magdalene coming to the *Languedoc* (Langue d' oc, or "Land of Yes" region in the south of France) are too ancient and widespread to be a modern hoax. There is also reason to believe that the line of David existed in this region early in Europe's medieval era. Benjamin of Tudela visited Narbonne sometime around the year 1166AD and wrote of a Jewish community that was ruled by a descendant of the house of David, a clear reference to a descendant of Jesus and Mary Magdalene. The Merovingians (c. 400-500AD) may have been descendants of David and Christ, as could the Cathars.

Where do we find a precept in the Bible for Creeds, Confessions, Doctrines and Oaths, and whole carloads of other trumpery that we find religion encumbered with in these days?
John Adams

The Cathars

The Cathars (11th-13th centuries AD) lived simple lives of renunciation, simplicity, and spirituality who called themselves *bonhommes* (good men and good Christians). They sought personal religious experiences that were outside the purview of the Catholic Church. Rejecting worldly treasures made the Catholic excesses seem even worse. Like the Therapeutae, Cathar women were equals; there were no organizations or hierarchies at first.

The Catholics saw the challenge but were unable to stop it, particularly when Bernard of Clairvaux (1090-1153CE) saw the need for a more formal organization among the Cathars. Finally, frustrated, the Catholics launched the Albigensian Crusade in 1209. The Cathars retreated to the fortress of Montségur, where several hundred were burned alive during the final assault. When a Catholic soldier asked how they should distinguish Cathars from innocents, the commander (Arnaud-Amaury) famously replied, "Kill them all! God will know His own."

Father, would that we could all be burned for Christ.
Theresa of Avila

On August 5th, 1234, an ill Cathar woman was urged to confess by a bishop. When she refused, she was carried out to a meadow still in her bed and set afire while the bishop and local Dominicans rejoiced. Church excesses in the south of France were so widespread, with even the pretense of piety left by the wayside, that Pope Innocent III was finally compelled to condemn these lapses despite being anything but innocent himself, as we will learn in Chapter 8.

The Jesus Family Tomb

It should go without saying that discovering Christ's tomb would be a very big deal. There is no evidence of such a tomb under the Vatican, nor is Peter buried there. If either of them was buried there, you can bet that the church would be making huge announcements. As it happens, a tomb was found... just not where the church wanted it to be found.

1st century tombs in Jerusalem were carved into the rock outside the city walls and contained two chambers, the outer of which being where the body was anointed, the inner being where the final burial took place after the corpse had been left for a year to decompose into bones that were then placed in an *ossuary* (bone box). These tombs were expensive, meaning that only the wealthy could afford them. The sacking of Jerusalem in 70CE ended the ossuary cult, making the ossuaries themselves ideal indicators of age in and of themselves. (Remember that Jewish rule was not restored to Jerusalem until 1948 with the modern state of Israel.) Thus, the tombs' occupants literally had ringside seats for the apocalypse, more evidence that the prophesies of Revelations are long fulfilled—with all due apologies to the Evangelicals, of course!

In 1980, Simcha Jacobovici (the "Naked Archaeologist") discovered a tomb in the south part of Jerusalem containing ten ossuaries that he identified as Christ's tomb. (The title of this section takes its name from his book on the subject.) Thousands of ossuaries have been recovered from hundreds of tombs; of these, only about 20% are inscribed. The tomb in question had six ossuaries with Greek, Hebrew, Latin, and Aramaic inscriptions—a clear sign of an influential family. Each inscription is linked to the Gospels. The names themselves are a mix of common and rare, but it is in the combination of names that things get interesting. The names are:

- Yeshua bar Yosef (Jesus, son of Joseph)
- Maria (Mary)
- Matia (Matthew)
- Yose (Yosef, or Joseph)
- Mariamene e Mara (Mariamne, or Mary)
- Yehuda bar Yeshua (Judah, son of Jesus)

What an apt punishment! The very place that endured so long blasphemies against god was now masked in the blood of the blasphemers.
Raymond of Aguilers

Suppose you are on holiday in Paris and decide one morning to visit the tomb of Napoleon. you arise bright and early and find the tomb is empty. Would you conclude that the emperor had risen into heaven? Hardly.
Victor Stenger

Assuming that the entire population bears one of the above names in equal proportions, the odds of finding this particular combination are about 2/1,000ths of 1%, which are long odds against coincidence under the best of circumstances. Of course, the real population includes many more names, some of which are much less common than others. The odds of this tomb being simply a coincidence are astronomically low. We must therefore assume that the tomb of Christ and his family has been discovered.

The "Jesus ossuary" was very plain. We can assume Christ would have wanted it that way. The ossuary of Mary Magdalene was found next to this one—another blow against coincidence.

Other Messiahs

If another Messiah was born he could hardly do so much good as the printing-press.
Georg C. Lightenburg

Christ may have been a messiah, but he was hardly *the* messiah. In 115CE, the Jews in Libya and Egypt revolted; apparently Emperor Vespasian had missed a few descendants of David's line. This second messiah, Lucuas, led the revolt that wiped out the status and power of Jews in Egypt. Rome responded forcefully, because the city relied on Egyptian grain to feed its citizens, and any interruption in the supply risked starving the capital of the Roman Empire.

Messiah 3.0 was Simon bar Kochba, who was killed in 135CE. In his effort to completely erase Judea from memory, Emperor Hadrian changed the provincial name to Palaestine (Palestine) and the name stuck.

The Legacy of Jesus Christ

Pythagoras was misunderstood, and Socrates, and Jesus, and Luther, and Copernicus, and Galileo, and Newton, and every pure and wise spirit that ever took flesh. To be great is to be misunderstood.
Ralph Waldo Emerson

The simple fact is that there is nothing in the Gospels that we can be historically sure of, as we will see in Chapter 8. The nearly 400-year delay between the time of Christ and the adoption of the New Testament falsifies the idea of divine transmission. It is far more likely that the official canon was foisted on a God who is OK with a wide variety of teachings, imposed by people seeking to control the divine for their own worldly purposes. The Bible simply does not support the church despite the church's best efforts at selective editing.

The church could not decide which books to include the New Testament canon without settling the question of Christ's divinity. Justin Martyr felt that the Gospels were simply apostolic memoirs that could be used to support one's faith without being holy, a status he reserved for books of law and those written by prophets.

Roughly 95% of the Jewish population was illiterate in Christ's day. On one hand, we may be able to assume that Christ was in that 95%. Thus, while he certainly knew the traditions, he may not have known the exact books, etc. The scene in Luke 2:41-52 where he goes toe to toe with the scholars may therefore be an exaggeration. On the other hand, Christ spent a long time in Egypt, a place known for its literacy, and spent many long years in training; the scene could just as easily be real. If Jesus was literate and also trained in medicine, then his abilities would certainly seem amazing to those around him. I can only speculate what they might have thought of a modern emergency medical technician with a high school diploma!

Jesus was a brilliant Jewish stand-up comedian, a phenomenal improviser. His parables are great one-liners.
Camille Anna Paglia

Apropos of medicine, why would the Bible list medicinal herbs as being used to embalm a dead body? It is quite probable that the many people who edited the New Testament had no idea which drug was which. If this sounds farfetched, consider John Paul's 1987 ruling about Mary's hymen that displays a fundamental ignorance of the female body, or the bishops in 2010 who claimed that condoms are porous and ineffective against AIDS. I see no reason why this glaring hole in the New Testament's most important story could not have slipped under the noses of those trying desperately to detect and intercept it.

The philosopher Celsus (c. 150AD) objected to the "savior who not long ago taught new doctrines, deceived many, and caused them to accept harmful beliefs that took root and spread among the lower classes, possibly because of its vulgarity and the illiteracy of his followers. While some elite people might be inclined to interpret these beliefs allegorically, they thrive in pure form among the ignorant." Celsus's message remains just as on-point today as it was some 2,000 years ago.

I am pretty sure that we err in treating these sayings as paradoxes. It would be nearer the truth to say that it is life itself which is paradoxical and that the sayings of Jesus are simply a recognition of that fact.
Thomas Taylor

The 1st century AD was a critical time for the budding religion. Paul's letters differed from the Gospels in a very important respect: They contained stories about Paul, not about

Christ. Paul did not know Christ and was on a mission to convert pagans. Did Paul genuinely believe what he preached, or did he see an opportunity to make a name for himself? We may never know for certain.

> *Community, not the highly vaunted individualism of our culture, is the setting in which Christ is at play.*
> Eugene Peterson

The initial churches were locally ruled at first, but grew increasingly centralized with a single bishop controlling a diocese, a practice that began in Rome in the mid 100s and was completed in the early 200s. The Bishop of Rome thought himself the most powerful and wanted to become the supreme leader of the Catholic Church as a representative of the messiah on Earth, basing his claim on Matthew 16:18 where Christ calls Peter, "the rock on which I will build my Church." He also claimed that Peter came to Rome to become the first Christian bishop. In 258CE, Emperor Valerian ordered all Christian bishops, priests, and deacons to be put to death.

As we saw in Chapter 5, a rift soon emerged between the Christians seeking knowledge and those settling for belief. The Gnostics cared far less about Christ than they did about obtaining personal experiences of God. They did not want to have faith in Christ but to become Christs themselves—a mission that Christ himself endorses in the Gospel of Thomas when he says, "When you come to know yourselves, then you will become known and you will realize that it is you who are the sons of the living father."

Irenaeus, frustrated by Gnostics wooing converts away from the church to gain superior knowledge, complained that the Gnostics attacked him with arguments and questions instead of faith and belief. He specified that the Gospels of Matthew, Luke, John, and Mark were divine, invented the idea of Christ as the son of God, and set forth how the church should be organized with central authority, orthodoxy, and sheer power serving as proofs of God's support for the One True Church—a position still heartily endorsed by popes today.

> *I don't know anyone less Jesus-like than most Christians.*
> Bill Maher

Constantine is responsible for bringing Christianity into its own. Plenty of bishops opposed the notion of Christ's divinity at the Council of Nicaea and pointed out that it is not part of the scriptures, but that did not stop those who were bound and determined to make Christ a god and thus replace the Christ of history with the Christ of faith. This key decision formed the criterion by which books were either admitted to—or barred from—the collection that we know of today as the New Testament. Anything that differed from this platform was to be considered heresy and stamped out, by force if necessary. The church's main audience consisted of pagans instead of Jews, making anti-Semitism necessary for marketing and control purposes. All Christians who did not buy into Paul's view of Christ were excommunicated, and heretics have been pursued ever since.

In the 4th century CE, Pope Damasus I (305-384) claimed to be the direct successor to Peter and the rightful leader of the Roman Catholic Church. He hired assassins to kill his enemies and bypassed Jerusalem by establishing Rome as the only location with direct apostolic succession, a move that any Zealot would have found laughable. Christ was bred and trained to fight Rome, but now Rome had sealed its victory over Christ and taken the liberty of committing a massive case of identity theft in the bargain. Damasus's decrees were to be obeyed immediately. Pope Innocent I (401-417) said that Rome was the supreme authority in the Christian world. Leo I (ca. 400-461) proclaimed that Christ gave supreme authority to Peter that was transmitted to each pope in turn, with the pope being the mystical embodiment of Peter. Gelasious I (-496) wrote that there were two emperors in two powers, the emperor for the physical world and the pope for the spiritual world, with the pope being superior because the church provided salvation. He was the first "vicar of Christ." It is a testament to the power of the church that the "lesser" emperor didn't kill him.

The kingdom of God in the teachings of Jesus was not an apocalyptic or heavenly projection of otherworldly desire. It was driven by a desire to think that there must be a better way to live together than the present state of affairs.
Burton L. Mack

The battle between the Christ of history and the Christ of faith was over by the 5th century CE, with the latter winning a total victory and myth attaining the status of accepted truth, but the church did not rest on its laurels. Instead, it took it upon itself to "protect" the faith by attacking other Christians. In 386, Priscillian, the Bishop of Avila, was executed for heresy in the first execution officially ordered by the church. This later metastasized into the Inquisition, which remains in place today, all for a man who preached peace. Popes even crowned emperors and kings, a clear sign of who was who in the pecking order.

The early church had three popes vying for supreme control, a dispute that was settled by the bishops. The popes of that time may have claimed that their authority came from Peter, but it really came from the bishops. Every pope had to get the bishops' approval before making any changes. Of course, this too changed over time.

> *It has been the scheme of the Christian church, and of all the other invented systems of religion, to hold man in ignorance of the creator, as it is of government to hold man in ignorance of his rights.*
> Thomas Paine

Witchcraft was considered a fraud or delusion, and believing in it was a sin. The papal bull *Summis desiderantes affectibus* issued by Pope Innocent VIII (1432-1492) in 1484 took the logical next step by condemning witchcraft, demanding recognition of its existence, and empowering the Inquisition to find it and exterminate it—an order the Dominicans took and ran with for hundreds of years. (See Chapter 8 for more about the Inquisition.) The pattern of church repression continued; in the 19th century, the pope even banned railroads because he feared that expanded travel and communications would harm religion—this in Europe at a time when liberation movements were standing up to monarchies and totalitarians of all kinds! Despite the church's self-imposed ignorance, the outside world was on the move.

In 1869, Pope Pius IX (1792-1878) decided that popes should be declared infallible and convinced the First Vatican Council, where he pressured the bishops into acquiescence by punishing dissenters and even physically assaulting one opponent. Despite the fact that only 49% voted to grant this power, Pius declared victory and was declared infallible on July 18th, 1870, only 98 years to the day before my own birth. In 1907, Pope Pius X banned modernism and required all priests and seminary teachers to swear an oath against modernism that even forbade them from reading newspapers.

> *To suppose that people can be saved by studying and giving assent to formulae is like supposing that one can get to Timbuktu by poring over a map of Africa. Maps are symbols, and even the best of them are inaccurate and imperfect symbols. Beauty is truth, truth is beauty..*
> Keats

The current pope as of 2010, Benedict XVI, is a strict dogmatist who openly asked whether the church was created by votes. Truth cannot be established, he claimed, only recognized and accepted. The church as the bearer of faith is immune from sin. Before he became pope, Cardinal Ratzinger was the head of the renamed Inquisition, an institution that no longer tortures or kills anyone (as far as we know) but that acts as the Vatican's means of suppressing any evidence that would reveal the Christ of faith as the falsehood he is. All of this is based on that one passage in Matthew, which is the key to the Vatican's claims of spiritual supremacy. Prove the church wrong and it could well fall, ending a 2000-year run that has been nothing less than spectacular. We have seen that plenty of such evidence exists, to which the church has responded by retreating ever further into its self-constructed shell of dogma and proclaiming itself correct at every turn.

To paraphrase the hit TV show *The X Files*, the truth is out there. Much of it may be locked in the Vatican's extensive archives. How much remains is unknown, because the church has destroyed and forged documents at will, continuing the precedent set by the boy King Josiah (648-609BC) who "discovered" a book of "divine" law thousands of years ago. How much has the Vatican destroyed over the centuries? How much has managed to escape the church's obsession with eliminating heresy?

Christ never claimed to be a god. On the contrary, he went out of his way to proclaim his humanity and mortality. Far from exclusive, his powers could be had or surpassed by anyone. Christ once offered salvation but today his name is in dire need of salvation. Can Christ's legacy be rescued from the dogma, myth, politics, and violence that has been carried out in his name?

Dear Lord, I've been asked, nay commanded, to thank Thee for the Christmas turkey before us, a turkey which was no doubt a lively, intelligent bird, a social being, capable of actual affection, nuzzling its young with almost human-like compassion. Anyway, it's dead and we're gonna eat it. Please give our respects to its family.
Berke Breathed
(Bloom County)

Chapter 7

The One God

> *I believe in one God, no more, and I hope for happiness beyond this life.*
> Thomas Paine

We already know that polytheism is the natural state of human religion and that monotheism is a later invention. The transition of most of the world's people from the former to the latter was neither easy nor rapid. Monotheism only prevailed after well over 1,000 years of determined effort, and the victory was far from total. Today's monotheistic religions are rife with enough saints, angels, spirits, and demons to qualify as thinly veiled polytheism or monolatry at best. Let's take a closer look at this struggle and its aftermath.

Progress?

> *Is your god too small?*
> Jack W. Geis

Did monotheism bring measurable progress to civilization? Does monotheism enjoy any moral superiority to other religious systems? Is monotheism necessarily better than polytheism by any objective measure? The short answers are no, no, and no. Here's why...

Civilization

It is tempting to think that monotheism brought civilization to barbaric pagans everywhere. It is also wrong. Plenty of pagan civilizations enjoyed legal, moral, and ethical superiority over their less advanced contemporaries. For example, one thing that distinguished the very polytheistic Romans from their less advanced contemporaries was the distinct lack of outright human sacrifice. They were as shocked as any modern monotheist would be by northern Europeans who strangled victims for Odin or the Celts, who put people in wicker baskets shaped like their gods and burned them alive. In fact, the Romans took pains to curtail human sacrifice in conquered areas. Plenty of people died at Roman hands and were then dedicated to the gods, including those who fell during gladiatorial spectacles or were executed. Still, the religious aspects of these events were secondary to the primary purposes of entertainment, punishment, etc. "Let's dedicate this corpse to the gods!" is different than, "Let's kill this person for the gods!"

> *People who cannot recognize a palpable absurdity are very much in the way of civilization.*
> Agnes Repplier

Gender

Polytheism has both male and female gods, while the monotheistic religions limit themselves to male gods and male priests. Some of the best-known, most loved polytheistic gods were female. Women occupied the highest ranks of some of the mystery cults and could do anything their male counterparts could do, including initiating others into the mysteries. It is no surprise that the monotheistic religions routinely accused them of impure motives and sinful conduct; monotheism cannot tolerate openness, tolerance, or equally if it is to survive.

> *In the theory of gender I began from zero. There is no masculine power or privilege I did not covet. But slowly, step by step, decade by decade, I was forced to acknowledge that even a woman of abnormal will cannot escape her hormonal identity.*
> Camille Anna Paglia

Sexuality

It is tempting to think of monotheism as chaste and virtuous, in stark contrast to freewheeling polytheism. It is true that many polytheistic religions were far more open about sexuality than their one-god counterparts were. Homosexuality, bisexuality, and prostitution offered both pleasure and birth control. Many pagan women could obtain abortions; some pagan infants were abandoned to die. Other pagans placed a premium on remaining virginal until marriage and made adultery a sin. One temple in Pergamum required a worshipper to wait one day after having sex with his wife or two days after having sex with a different woman before entering. Some followers of Cybele or Bacchus had public sex, to the chagrin of their more refined neighbors. The Roman senate banned *bacchanalia* (festivals in honor of the god Bacchius) for being morally offensive, proof that pagans were just as capable of prudishness as

> *Sexuality poorly repressed unsettles some families; well repressed, it unsettles the whole world.*
> Karl Kraus

any monotheist. Some cults required priestly celibacy long before the JCI religions.

Thus, pagan morality ran the full gamut from permissive to restrictive. Many in the middle of the spectrum saw nothing hypocritical in valuing both virtue and sex, because they did not see sex as evil or sinful. They were certainly far less hypocritical than monotheists who routinely fail to practice what they preach. From Jim Bakker to Ted Haggard and countless others throughout history, monotheism has seen more than its share of people preaching chastity while practicing debauchery.

Devotion

Pagans treated images of their gods and goddesses with the same respect and affection as any monotheistic holy object. They even showed respect for each other's gods. For example, a pagan visiting a house in Rome might kiss the image of the house god by the doorway in the same way a Jew might kiss a *mezuzah* (parchment scroll with scriptures inside a small case) mounted on the doorjamb. Religious images inspired pagan prayer just as the altars, prayers, and sculptures in today's churches inspire modern JCI worshippers. In fact, by most definitions, any given synagogue, church, or mosque is filled with idols despite Biblical warnings to the contrary!

Both ancient and modern skeptics tend to take a dim view of "miracle-working" clergy, accusing them of using cheap tricks to exploit people's fears and hopes; however, ancient playwrights taught that the gods were alive and manifest in the daily world. Perhaps they went along with the clerical displays in the same way that modern audiences eagerly pay to see magic shows or movies with special effects. We know that such displays are illusions, but allow ourselves to suspend our disbelief to enjoy the story.

These few examples should suffice to demonstrate that pagans were every bit as civilized as any monotheist. Why then does the latter condemn the former so vehemently?

Culture Wars

All monotheistic condemnation of polytheism boils down to one seemingly simple issue: To the monotheistic way of think-

---

Fear of sexuality is the new, disease-sponsored register of the universe of fear in which everyone now lives.
Susan Sontag

There is always the danger that we may just do the work for the sake of the work. This is where the respect and the love and the devotion come in —that we do it to God, to Christ, and that's why we try to do it as beautifully as possible.
Mother Theresa

ing, a person commits an unforgivable crime when s/he prays to anything or anyone but the monotheist's idea of her or his one "true" god. Pagan rituals and practices are very similar to monotheistic practices. The only thing all pagans have in common is their lack of belief in one god as defined by the JCI religions. But why should anyone care which god someone worships? As it turns out, the reasons are all too practical.

Consider how deeply religion is woven into culture and society. For example, many (if not most) secular laws, traditions, and taboos trace their roots to religion. To oppose someone's religion is therefore tantamount to opposing the person himself. Consider that the main pagan cultures (Babylon, Egypt, Greece, and Rome) at the dawn of monotheism represented the pinnacles of human achievement to date; the monotheists were backward country yokels by comparison. The real enemy of monotheism was therefore nothing less than civilization and progress itself! When we go to museums, read ancient works, or marvel at ancient discoveries, we are recalling and revering the pagans whose achievements form the bedrock of modern civilization.

The real reason for the long war between monotheism and paganism may well come down to the jealousy of a group of ancient hillbillies stuck in a miserable scrap of land being pushed around by overwhelmingly powerful outsiders. Rather than adopt the ways of their superiors and enjoy the fruits of advancement, they chose to cling to their spears and religion and lash out against the civilized world. I am not saying that they were wrong to do so, nor am I suggesting that those who repeatedly conquered the "holy" land are innocent of grave wrongdoing. Nevertheless, we can see the same mentality at play whenever we hear religious conservatives railing against progress and personal liberty while pining for the restoration of some mythical golden age. Pagans (liberals) are far more open to progress than monotheists (conservatives).

Thought Control

Monotheism insists that its god is the one and only real god, and that all other gods are false. The One True God demands the blood of all nonbelievers, leaving aside the obvious fact that God must have made them too when He created the entire universe. Every sacred monotheistic text calls on the

Culture is the arts elevated to a set of beliefs.
Thomas Wolfe

If repression has indeed been the fundamental link between power, knowledge, and sexuality since the classical age, it stands to reason that we will not be able to free ourselves from it except at a considerable cost.
Michel Foucalt

faithful to defend the religion and God by force. The lives of nonbelievers are forfeit, period. Wrapping the pragmatic jealousies and issues discussed above in a divine mandate takes xenophobia to a whole new level.

> *Reason has never failed men. Only force and repression have made the wrecks in the world.*
> William Allen White

Of course, getting someone to adopt such a narrow viewpoint isn't exactly easy, especially when all one needs to do to enjoy the benefits of civilized life is show some modicum of respect to the pagan gods and by extension the pagan society, as we saw in Chapter 6. In fact, the pagans demanded far less worship and devotion of the official gods than the monotheists demanded for their own god. Only a person who is so convinced of the truth of her or his beliefs as to make them a matter of life and death can achieve the extremely strict levels of religious observance demanded by monotheism. Exposure to outside ideas risks diluting the faith, which is why monotheists tend to keep to themselves and resist change.

The strength of belief required by monotheism inspires rituals from the bizarre to the horrific, from simple dietary restrictions to genital mutilation, from devotion and reflection to punishing those with different beliefs. Religious *zealotry* (fervor, eager desire or endeavor, ardor) can lead to wholesale war and murder. For example, the first use of the word "zeal" in the Bible occurs when God approves of what any sane modern person would consider war crimes and atrocities.

Intolerance

> *I believe in an America where the separation of church and state is absolute.*
> John F. Kennedy

It is monotheism, not paganism, that dictates who to pray to and how to pray to them. Pagans mixed and matched as they saw fit, and both conqueror and conquered might take on each other's gods and rituals. Paganism is not a single religion; rather, it is a mass of various mutually tolerant traditions. By contrast, monotheism claims that there is only one god worthy of worship; any other gods that may exist are inferior. According to the JCI religions, worshipping the wrong god or worshipping the correct god incorrectly constitutes a capital sin that believers must punish—presumably because the creator of the entire universe cannot or will not fight his own battles.

Latecomers

We have already seen that the earliest known religion began with the Neanderthals. As we will see in Chapter 9, some modern scientists even think that evolution wired our brains for religion. Still, nothing says that belief in a single god is either inevitable or desirable. On the contrary, we have seen that people throughout history have worshiped untold thousands of male and female gods, spirits, guides, devils, demons, and saints. Many still do. Monotheism is both a late development and one that took well over 1,000 years to become dominant over polytheism. Most people were appalled by the idea of worshipping only one god—and rightly so, given what we learned in Chapters 5 and 6. Only monotheism could give us the events of September 11th, 2001, the Inquisition, and countless other plagues. (See Chapter 10.) As such, the first attempts at monotheism failed; its final triumph was not assured until the Battle of Milvian Bridge in 312AD. Had Maxentius prevailed over Constantine that October day, today's religious landscape might look utterly different from the one we are accustomed to. That one event is truly a turning point in history, what author Madeleine L'Engle calls a "might have been" in her classic book, *A Swiftly Tilting Planet*.

Imagine a world awash in gods and goddesses, the many churches, mosques, and synagogues replaced by myriad small temples honoring local deities. Imagine worshipping a smorgasbord of gods and freely choosing to modify that retinue any time you like. Imagine a particular god or goddess having to justify itself to you instead of the other way around. Imagine no religious intolerance. Imagine accepting this state of the world just as naturally and inevitably as monotheism is embraced today, because that is the only condition we know. Imagine all of this, and then ask yourself whether monotheism truly represents an advance over polytheism, and whether a handful of mutually exclusive orthodoxies is better than paganism.

I for one believe that the answer is no. I also believe that monotheism and polytheism can coexist as different expressions of the same ultimate reality, for reasons that will gradually become clear as we proceed through the rest of this book.

Bigotry and intolerance, silenced by argument, endeavors to silence by persecution, in old days by fire and sword, in modern days by the tongue.
Charles Simmons

The Three in One, the One in Three? Not so! To my own Gods I go. It may be they shall give me greater ease than your cold Christ and tangled Trinities.
Rudyard Kipling

Pitfalls and Promises

Monotheism runs into a number of serious problems right off the starting line. On one hand, it wants you to believe that there is a single omnipotent, omniscient, omnipresent, and eternal (OOO) God that created the entire universe in which nothing can escape His abilities or attention. This God loves His truly faithful followers and offers them both worldly and otherworldly benefits, including eternal bliss after bodily death. (How else to get people to willingly die in your name?) Proving your devotion to this God requires adopting a set of practices and beliefs that others may find downright silly. For example, why would a perfect God create a human penis with foreskin only to demand that His followers cut it off? Why should humans, of all the millions of species on Earth, come with the injunction that some disassembly is required?

But you must pay for conformity. All goes well as long as you run with conformists. But you, who are honest men in other particulars, know, that there is alive somewhere a man whose honesty reaches to this point also, that he shall not kneel to false gods, and, on the day when you meet him, you sink into the class of counterfeits.
Ralph Waldo Emerson

On a more serious note, why would a supremely perfect OOO God create or allow evil? He could just as easily have created the Garden of Eden sans the forbidden tree and spared everyone—including Himself—the hassle. After all, He knew exactly what was going to happen long before it happened, because He made it happen. How can free will possibly coexist with an OOO God? God knew I was going to write this book and still went ahead and created me. It would be impossible for me to independently choose not to do so, because that would mean that God did not know my choice before I made it, which in turn would mean that God is not OOO. According to monotheists, God sees you when you're sleeping. He knows when you're awake. He knows if you've been bad or good... but at least Santa Claus lets you make up your own mind! Santa Claus holds you responsible for making decisions he was unaware of beforehand. God is holding the entire human race responsible for something a woman supposedly did with his full foreknowledge.

Countless monotheists have consumed countless reams of various writing media over millennia trying to reconcile these quandaries. Evil could be accounted for by allowing evil beings to compete with God. Of course, God could destroy these beings on a whim. God in fact created those beings. In other words, God created an evil being knowing full well that it would convince a woman to eat the wrong apple, and still punished the woman and the entire species. As for free will, God knows what you are going to decide, but still lets you make up your own mind. In other words, God knows a rock will sink, but drops it into water to let it decide to sink.

I could go on, but I think my point is clear: Accepting the tenets of monotheism as presented by the JCI religions requires a level of commitment and acceptance that subjugates reason to faith. Polytheism has none of these problems. In fairness, there is a very simple way to explain evil, suffering, free will, etc. in the context of an OOO God. The Buddhists (see Chapter 5) have done so, and there is every reason to think they are very much on the right track, for reasons we will continue to explore.

> *We know all their gods; they ignore ours. What they call our sins are our gods, and what they call their gods, we name otherwise.*
> Natalie Clifford Barney

Monotheism confers some powerful advantages to those who are committed enough to get past the logical knots and put all their spiritual eggs in one basket. It is the only system that can implement and enforce rules by divine mandate using a rigid doctrine. It is the only way to get people to die for God, especially if He promises eternal life. A single conscious, all-powerful, and virtuous God gives the ultimate bang for the spiritual buck, because people will accept an extended exchange-based relationship with that God—and therein lies the secret to the power of monotheism. This relationship can inspire lifelong devotion and place any number of demands on believers. Monotheism is a one-stop shop for all one's spiritual needs.

Conversion

We have seen that monotheism is not inherently superior to polytheism beyond its ability to inspire fierce and even lethal loyalty—a good thing for many perfectly mundane reasons, as we will soon discover. We also took a very brief look at just a few of the challenges monotheism presents to anyone who stops to think about it. Monotheism seems to hold a pretty weak candle to the openness and freedom of polytheism. What could possibly attract anyone to such a belief system?

> *The first missionaries, good men imbued with the narrowness of their age, branded us as pagans and devil-worshipers, and demanded of us that we abjure our false gods before bowing the knee at their sacred altar.*
> Charles Eastman

The simple fact is that the doctrines themselves are not the primary reason why people choose a religion. The decision to convert has more to do with bringing one's beliefs into line with one's neighbors than anything else. Conversion is an act of conforming. Why conform to something that might seem preposterous on rational examination? Humans are social creatures, and evolution has programmed us to believe that being accepted by our peers is absolutely critical for survival. (See *The Enlightened Savage* and *The Natural Savage*.) People will do damned near anything if they think their survival depends on it, including going along with their neighbors' beliefs. The need for acceptance is so strong that people will convince themselves that the doctrines are the reason they converted. The purported truth of their adopted religion replaces the real truth of their need to fit into the group. Small wonder that religions and cults attract those without strong existing religious or social connections. Someone with strong religious beliefs will turn to others of the same belief and is therefore not a good candidate for conversion. On the other hand, alienated or detached people make excellent converts. This group includes runaways, outcasts, criminals, and people who have fallen on hard times without a strong support group. These same people will come to believe that "God" saved them and want to share the "good news" with anyone who will listen.

> *A civilization is destroyed only when its gods are destroyed.*
> E.M. Cioran

I have to wonder whether religious charities, rescue missions, etc. would exist if they did not make such powerful recruiting tools that are so excellently targeted at the people most likely to accept and act on the religious belief system responsible for whatever help they receive. Which came first, the charity, or the realization that the charity is second only to childhood indoctrination as the best method of growing the flock? Think about this for a few moments. It should become painfully obvious that religious conversion efforts are far less about spiritual truth than they are about gaining control over people.

The History of Monotheism

> *In our own hearts, we mold the whole world's hereafters; and in our own hearts we fashion our own gods.*
> Herman Melville

Our brief discussion about some of the many ways in which monotheism comes short of the glory of polytheism explains why this tectonic shift in spiritual thinking was neither quick nor easy. Let's see how this transformation occurred.

Babylon

According to the Babylonian religion, the god Marduk created humanity. No gulf existed between humans, gods, and nature. Humans and gods alike shared the same divine nature; divinity was no different than humanity.

King Hammurabi wanted to extend the Babylonian Empire across all of Mesopotamia. This eventually happened, and Marduk became the head god in place of Enlil. Other gods found themselves demoted to being simply aspects of Marduk—a status significantly below that of being subservient while retaining their individual identities. This was a significant step on the road to true monotheism. The roster of gods/aspects reflected both human governments and a growing desire to explain how the universe works. Absorbing other deities into Marduk thus serves as a sort of Grand Unified Theory of its day. This has many practical considerations. For example, if technology makes agriculture more reliable and predictable, then one can modify one's belief in the harvest god aspect(s) of Marduk without undermining one's belief in Marduk. Demoting the other gods also emphasizes Babylonian supremacy while giving Mesopotamia a unified set of beliefs that would help the empire form a more perfect union, ensure domestic tranquility, provide for the common defense, and so on. To Babylon's credit, they set about establishing Marduk's supremacy with astonishing good manners and general decency; however, their experiment with nominal monotheism did not last.

> *The stars which shone over Babylon and the stable in Bethlehem still shine as brightly over the Empire State Building and your front yard today. They perform their cycles with the same mathematical precision, and they will continue to affect each thing on Earth, including man, as long as the Earth exists.*
> Linda Goodman

Egypt

Egyptian monotheism took the form of a violent divine putsch by pharaoh Amenhotep IV in the 14th century BCE. We may never know whether he was driven by true religious fervor or whether his motives were more prosaic, but we do know that he tried to make Aten the one true god. Marduk may have subsumed other gods, but Aten replaced them outright. Amenhotep renamed himself Akhenaten (splendor of Aten), declared himself to be Aten's son, and made himself high priest. There were no idols of Aten, because his form could not be imagined. Aten was also a jealous god who refused to share loyalties or affections with anyone. The con-

> *Should Moses have told the children of Israel to live in slavery under the pharaohs? Should Christ have refused the cross? Should the patriots of Concord Bridge have thrown down their guns and refused to fire the shot heard 'round the world?*
> Ronald Reagan

quered gods simply ceased to be. Their temples closed, worship was restricted to Aten, statues were destroyed, and their bands of priests were sent to quarry stone like common slaves. A new city was built and dedicated to Aten.

This sudden, violent, and unprecedented (to the best of our knowledge) transformation by force did not go over well. Akhenaten failed to win over both the common people and the priests—not good, because all gods need help from believers on Earth—and his bold attempt at monotheism vanished when he died. The Aten cult was quickly forgotten, and Akhenaten's name was erased as payment in kind for the treatment the traditional gods had received at his hands. Still, it is tempting to think that this experiment may be one of the ultimate sources for what would become the JCI religions.

> *It is impossible to imagine the universe run by a wise, just and omnipotent God, but it is quite easy to imagine it run by a board of gods. If such a board actually exists, it operates precisely like the board of a corporation that is losing money.*
> H.L. Mencken

Isis is a leading goddess in Egyptian mythology. She was praised as the savior of the human race long before Christ. It was Isis who separated Earth from heaven and who placed the Sun, Moon, and stars in space. She was the sole ruler for eternity without whom nothing could happen. She also happened to be in love with the popular human ruler Osiris, whose brother Set happened to be jealous of that popularity. Set tricked Osiris into stepping into a casket, which he then slammed shut, sealed with lead, and tossed into the Nile. The mourning Isis searched far and wide for the casket and eventually returned it to Egypt, where she hid it in the marshes beside the Nile awaiting a decent burial. (Why someone without whom the universe could not function had to search is not explained.) Unfortunately, Set found the casket and chopped Osiris's corpse into little pieces that he then scattered all over Egypt. Isis went off in search of the pieces, eventually finding everything but her lover's penis. She reassembled the body, brought Osiris back to life, and conceived Horus—a possible precursor to the Virgin Mary, since intercourse was impossible for obvious reasons. Horus fought Set to avenge Osiris, but neither could achieve victory. A truce was eventually declared in which Osiris became the king of the underworld, Horus the kind of the living, and Set the god of chaos and evil. I trust I am not the only one to see more than a passing similarity between this and the Moses story of the Old Testament.

In about 300BC, Emperor Ptolemy 1 ordered the priests to find a way to integrate the Greek and Egyptian religions. Serapis was chosen as the lead god with others playing lesser roles—a popular move, since it reopened many old temples. The priests specifically designed Serapis to be compatible with Greek culture, and his cult spread across the Mediterranean. This of course validated the earlier philosopher Xenophanes's observation that humans portray the gods in our image; if horses had gods, he said, those gods would resemble horses. Of course, no one claimed that Serapis was the only god, merely an amalgamation of other gods.

Judaism

Archaeological evidence demonstrates that human sacrifice goes back to the dawn of the human race. Far from wanton murder, taking a human life represented the ultimate act of devotion, especially when the victims went willingly. Currying favor or trying to placate an angry god just does not get any more serious than that. The Aztecs took human sacrifice to its logical extreme by offering thousands of their own and other people to their gods each year. This has nothing to do with monotheism versus polytheism; I have already mentioned the Roman distaste for overt human sacrifice. In fact, the biggest monotheistic religions in the world (JCI) are built on an act of *sacrificum interruptus*. Abraham was about to sacrifice his only son Isaac to show his obedience to Yahweh. (See Genesis 22.) After that, Yahweh seems to relent and accept only animal sacrifices, except of course for the many wars and genocides he orders and the later crucifixion of his own son. On this score, Yahweh is just as barbaric as any pagan god—and we're only getting started.

So many gods, so many creeds, so many paths that wind and wind, when just the art of being kind is all this sad world needs.
Ella Wheeler Wilcox

Like Aten before him, Yahweh was not content with good intentions and honest effort. Anything less than total, utter submission was a grave sin. All nonbelievers were to be punished for their errant beliefs; the Old Testament is filled with gleeful accounts of holy wars, where not even the babies were spared. Never mind that these wars probably never happened; these stories existed to identify the Jews as a tightly knit group separate from all others. This is understandable, because they were a fringe minority group; just about everyone else was polytheistic.

The disappearance of a sense of responsibility is the most far-reaching consequence of submission to authority.
Stanley Milgram

Josiah rejected polytheism and devoted himself fully to Yahweh. Jewish priests received more power and status than ever before. While restoring the temple of Jerusalem, they supposedly came across a long-lost copy of Deuteronomy hidden in the walls, which so alarmed Josiah that he tore his clothes. This "long lost" scroll contains laws that do not appear anywhere else in the Torah, which consists of the Pentateuch—the first five books of the Old Testament (Genesis, Exodus, Leviticus, Numbers, and Deuteronomy). According to the scroll, Jews had been unwittingly violating Yahweh's will for hundreds of years. Among other things, it claims that Yahweh will only accept sacrifices at the temple in Jerusalem. Deuteronomy is amazingly consistent with Josiah's vision of God—so consistent, in fact, that it is much easier to believe that Josiah and his priests simply wrote Deuteronomy themselves. This accusation is not made lightly: Josiah was a boy king with a tenuous grip on power. The monotheist idea of a single all-powerful god can be extended to imply an all-powerful earthly king. What better way to garner loyalty than by using a simple act of fraud to exploit the kind of passion and devotion that only monotheism can impart?

Oppression that is clearly inexorable and invincible does not give rise to revolt but to submission.
Simone Weil

Fraud or no, Josiah claims to be acting under Yahweh's specific orders as he purges nonbelievers in a holy war that seeks to exterminate all who make offerings to other gods. This is Judaism's formal debut as a monotheistic religion with one god, one temple, and one king of David's lineage. This is the idea that Jesus would be born and bred to champion about six centuries later. Josiah was able to ram monotheism down the nation's throat because he wielded the power of the state.

I have heard with admiring submission the experience of the lady who declared that the sense of being perfectly well dressed gives a feeling of inward tranquility which religion is powerless to bestow.
Ralph Waldo Emerson

In 609BC, Pharaoh Neco wanted to move his army through Judea to link up with the Assyrians against the Babylonians. Josiah refused permission for the army to pass, and a battle ensued at Megiddo. Josiah led a charge and was killed by an arrow. Yet again, monotheism sputtered. The line of David had died out within 25 years, and the Jewish nation was wiped out as the great deportations to Babylon began. Those Jews who returned 75 later saw themselves as holier than those who had remained behind, and refused to let them rebuild the temple in Jerusalem to worship Yahweh. The New Testament book of Revelation speaks of the end of the world occurring at a place called *Armageddon*, a Greek word meaning *har Megiddo*, which means "Mountain of Megiddo" or "Mount Megiddo." The descriptions of rivers of blood are very apropos for a major battle scene. The armies of the East may be the Babylonians coming to deport the Jews. In short, the events in Revelations may not be based on the future end of the entire world but the previous end of the Jewish nation. Given that Revelations was probably written around 100AD, when the Romans were in full control of Judea, the author of Revelations may well be wishing that things had gone differently for Josiah and hoping for a rematch that could have rewritten history. (This is one interpretation of Revelations.)

It is important to understand that Greek culture played an extremely important role in the ancient world, much like Americana fascinates much of the world today. People coveted Greek goods and knowledge. Greek was the official language of much of the region. The people we refer to as pagans called themselves Hellenes. Alexander the Great may have fought brutal wars of conquest, but he never punished anyone for believing in the wrong god. Thus, the Jews once again strayed from monotheism. There was a gymnasium in Jerusalem where one could study Greek, play games in the nude, and even seek surgery to hide the telltale look of circumcision. In one of the rare instances of pagan religious persecution, Antiochus IV (215-164BC) imposed polytheism by force because of his anger at monotheistic Jews who refused to respect the Greek gods.

In 166BC, Judah Maccabee (Judah the Hammer) led a guerrilla army that defeated Antiochus and restored the Jewish nation for the first time since the Babylonian invasion. This was a holy war; the Maccabees believed that Yahweh was on their side. Fellow Jews believed to have gotten too cozy with the invaders were also targeted, especially those who had opted out of circumcision. Why was circumcision important? Because it was an unmistakable sign of being Jewish, a sign obtained by mutilating one of the most delicate parts of the male anatomy. The Maccabees also invented the concept of martyrdom, which of course goes hand in hand with the concept of holy war. A holy warrior has no problem killing for his god, but must also be prepared to die for that god. The former is a weapon of strength, the latter a weapon of weakness, and God demands both. The Jewish nation was restored once more for a few short decades.

To the press alone, chequered as it is with abuses, the world is indebted for all the triumphs which have been gained by reason and humanity over error and oppression.
James Madison

Martyrdom is the only way a man can become famous without ability.
George Bernard Shaw

Then, in 63BC, the Romans under Pompey captured Jerusalem, and the Jewish nation became a province. What did the new conquerors do? The Romans mounted a golden eagle on the temple door, and that's about it. They did not defile the interior. In fact, a Roman soldier who defied the Jews by entering the temple could be executed. The puppet king Herod even expanded and enhanced the temple. Herod's sons took over when the king died, but all of Judea remained a Roman province. Meanwhile, the Zealot Jewish resistance movement remained alive and well. The Romans saw them as outlaws, but they saw themselves as holy warriors who attacked both Romans and collaborating Jews. Zealot sicarii assassins were trained to kill selected targets in public and then vanish into the crowds.

To attempt the destruction of our passions is the height of folly. What a noble aim is that of the zealot who tortures himself like a madman in order to desire nothing, love nothing, feel nothing, and who, if he succeeded, would end up a complete monster!
Denis Diderot

The revolt continued after the Zealot plans for Christ fell through, eventually escalating to all-out war in 66AD with the taking of the mountain Roman fortress of Masada. Ten years later, the Romans under Lucius Flavius Silva besieged the mountain. The Jews chose to commit mass suicide rather than face the final Roman assault. On the last night before the battle, the Jews drew lots to see who among them would kill the almost 1,000 men, woman, and children to the last man, who would then take his own life. Two women and five children escaped by hiding in a cistern and recounted the siege and final hours to the historian Josephus (37-100AD). Josephus started out as a Jewish army commander fighting the Romans, only to switch sides after suffering a defeat. The Romans rewarded his change of heart with nice digs in Rome and a life of leisure that allowed him to document some of the most important events in Jewish and Christian history. More revolts followed, including one led by bar Kochba in 132CE. The final defeat came in 135CE when bar Kochba was killed, at which time Emperor Hadrian sought to remove Judea from memory by renaming it Palestine, a name that remains with us today. There would not be another Jewish state for 1,813 years until the founding of the modern nation of Israel on May 14th, 1948. The Jewish Diaspora had begun.

Yochanan ben Zakai convened the Council of Jamnia from 70-90AD so Jews could decide what to do in light of the loss of Jerusalem. Many important changes took place to allow the Jews to continue living in foreign lands. Animal sacrifices were replaced by prayer, and *rabbis* took over from priests. Rabbis serve to this day as Jewish teachers, priests, scholars, and judges. The Jews modified their ancient laws to allow themselves to continue as a going concern.

The Roman Empire initially treated Yahweh just like any other god. They renamed him Iao and gave him a place among their many gods. To them, Yahweh/Iao was just one more god to respect among the throng. The Jews did not return the favor. Pagans who respected the Roman gods were perfectly acceptable, but the Jews alone claimed to be the people chosen by the One True God and rejected all other gods. They often became hated for the same reason that a child who constantly brags about how special he is will soon find himself alone on the playground and quite possibly sporting a bloody nose to boot. Jewish customs such as circumcision, dietary restrictions, strict Sabbath observances, and more kept them separate from most of the Roman population. They did not present any serious threat because they did not mingle, and few Romans wanted to forgo their foreskins and banquets.

Rome was at her peak in the 1st century CE. The Roman gods existed more for the glory of Rome than for the good of the people. In this context, a single God dedicated to His people, kings and commoners alike, held a certain appeal. Pagans were also impressed by the fact that Judaism was as old as their own religions. Some synagogues even had places where visitors could watch the proceedings. Jews could also pray for the emperor, so long as they did not pray to him. In ways like this, Jews and pagans normally kept an uneasy peace.

Sin against Yahweh has little to do with morals and ethics, and all to do with religious purity. Praying to other gods is the worst sin, because any chink in one's steadfast belief is an abomination. Strip away the many layers of myth and metaphor in the Old Testament, and we see a god who is both jealous of the attention other gods are receiving and frustrated by his inability to win over his special chosen people. (Why an OOO god who created the situation and knew about it to begin with would feel this way is of course never explained.) As we saw in Chapter 5, the Eastern religions believe in one absolute Truth, Godhead, or Ground, while at the same time embracing their own pantheons of gods and demigods. Yahweh is attempting nothing less than a total monopoly on reli-

After all the public is entitled to what it wants, isn't it? The Romans knew that and even they lasted four hundred years after they started to putrefy.
Raymond Chandler

A competent and self-confident person is incapable of jealousy in anything. Jealousy is invariably a symptom of neurotic insecurity.
Lazarus Long

> *Being a Christian is more than just an instantaneous conversion; it is like a daily process whereby you grow to be more and more like Christ.*
> Billy Graham

gious thought and faith. His efforts rely on a succession of believers acting on his behalf to carry out his will. Someone who accepts this mission is far more likely to arm himself and assume the task of punishing the sinners. This tends to be self-limiting, since society at large will not long tolerate such a menace unless and until a believer holds the reins of political power with which to bring the full power of the state to bear in defense of God. A devout monotheist can claim carte blanche to act as his god's instrument. Which begs the question: If Yahweh is truly as great as he claims to be, why were people not flocking to worship him en masse of their own accord?

Perhaps Yahweh was not that great. There is little to nothing new or novel in the Bible, which represents little more than an amalgamation of stories and myths borrowed from other cultures, right down to acknowledging the presence of other gods. That plus the historical records proves that the ancient Jews were half-hearted monotheists at best. Even Moses asks Yahweh what other gods are like him, thus exposing his own doubts about monotheism. There is no known remnant of Solomon's temple. There are, however, hundreds of clay tablets showing women touching themselves that may have been used by pagans and Jews alike. A jar from the 9th century BC discovered in Sinai says, "I bless you by Yahweh and by his Asherath." Asherath is universally reviled in the Bible. The basic story of the Old Testament has Yahweh repeatedly blessing the Jews, who repeatedly fail to live up to their end of the bargain and face repeated suffering. Reading between the Bible's lines reduces the supposedly glorious Yahweh to a dismal failure who could not even command the loyalty of his own chosen people. Time and time again, bands of Yahweh followers try and fail to convert Israel to monotheism. They and their god resort to violence, consistently forgetting the simple lesson that one can catch more flies with honey than with a stick. Thus do Yahweh and his flock come off as a ragtag gang of playground bullies who keep attacking the other kids despite getting their asses kicked every single time.

Christianity

> *Self-denial is the shining sore on the leprous body of Christianity.*
> Oscar Wilde

Religious conversion was not enough for the Jews, who insisted that entire nations had to be Jewish. Judea had been wiped off the map, the Romans did not embrace Judaism, and Jewish rabbis discouraged conversion. Paul of Tarsus would not make that mistake when he set out to build a new version of Judaism that would become known as Christianity. Christ himself told his apostles to go forth and bring anyone who asked into the fold. For the first time in the history of monotheism, all ethnic requirements were abolished. Word of mouth remains the most effective method of getting a message out to people, and every Christian was expected to become a missionary and preach the "good news." Christianity was spreading like wildfire within 20 years of Christ's supposed crucifixion. (See Chapter 5.) Only monotheism can command that kind of loyalty. The good news was primarily an urban movement, since that made it easy to reach large numbers of people quickly. The first Christian cities tended to be port cities, because traveling missionaries invariably ended up there. Ancient life had more than enough struggle and misery to go around. Christianity promised both worldly and eternal solutions, and even came with some relatively simple rules about how to make life seem better.

There were about 1,000 Christians in 40AD. By 180AD, all Hellenic cities had churches. By 350AD they numbered 31.7 million—not bad, considering that there are only some 14 million Jews today. This growth might seem miraculous, but only requires a linear growth rate of 3.4% per year. This did not happen; initial growth was very rapid and uneven but slowed as the number of potential converts shrank, a perfect example of the law of diminishing returns. Many modern Christian sects are growing much faster than 3.4% per year, but nobody is calling that a miracle.

The Romans took this in stride at first, thinking it just one more Eastern religion. Many Greek gods were former mortals who had been promoted, so the Christ story was already familiar. The Cybele cult had spread from Turkey and the Isis cult from Egypt. Neither was monotheistic, but the Eastern religions offered instant solutions that appealed to individuals over communities and engaged their senses and emotions more by emphasizing celebration and joy. Where the Roman gods tended to behave badly, the Eastern religions spoke of atonement and morality. Cybele's priests castrated themselves, cross-dressed, and acted like women. Homosexuality was one

---

I believe in Christianity as I believe that the Sun has risen; not only because I see it, but because by it I see everything else.
C.S. Lewis

Christianity is completed Judaism or it is nothing.
Benjamin Disraeli

thing for Romans, but effeminate men were another thing altogether. It is fascinating to speculate that these rituals inspired the traditional garb of Christian priests and may also have played a role in establishing the rule of celibacy. In other words, Christ may literally have ridden Cybele's coattails to stardom. Moreover, both the Old and New Testaments claimed to offer actual historical facts, unlike the made-up characters in the Roman myths.

> *The word Christianity is already a misunderstanding. In reality there has been only one Christian, and he died on the Cross.*
> Friedrich Nietzsche

Christ and Isis were greeted by heaven when born. Isis and Mary were virgins. Christ and Isis were both resurrected as saviors who could trump death. Cybele's followers were washed in the blood of the lamb. Fasting and mourning accompanied the remembrance of Isis's death, followed by celebrations marking her rebirth. How did Christians respond to these similarities? By saying that God had sent earlier glimpses to ready people to receive the final truth! God reveals Himself to people and even angels, based on their ability to absorb the knowledge. Conflicts between the Old and New Testaments are simply a sign of God adapting to the changing times. The rapid growth of Christianity proves that people accepted these explanations just like people in 1828 accepted Joseph Smith's explanation about his inability to translate two "seer stones" the same way twice. In each case, what should have been proof of falseness was seen as proof of veracity. As an aside, the Mormon seer stones are called the *Urim* and *Thummin*—the same names given to stones used by Moses.

> *I am a communist because I believe that the Communist idea is a state form of Christianity.*
> Alexander Zhuravlyov

Cybele and Isis gave early Christianity a huge boost. Two thirds of cities with temples to Isis, and only 2 out of 14 cities without temples to Isis had Christian churches by 100AD. All cities with Isis temples and only half of cities without Isis temples had Christian churches by 180AD. Cities with temples to Cybele were more likely to have temples to Isis as well. Nevertheless, Christianity was still a Jewish movement at first that offered a Jew-like religion without all of the laws and requirements. Jews could preserve their religious traditions and adopt the new add-on that got rid of old laws without imposing new ones—and ever mind Christ's admonition in Matthew 5:17 that he had not "come to abolish the Law or the Prophets." Services were conducted in Greek. What more could a Hellenized Jew ask for? The Christ story appealed to pagans and

Jew alike in what has to rank as one of the most effective marketing campaigns in history.

Paul's method was simple: He concentrated on Hellenic port cities with large Jewish populations. There were far more Jews than needed to provide all Christian converts until the 4th century AD. Relatively few pagans may have converted before then. Paul spread the word in each city by meeting with key people and their families to build a band of followers that allowed him to form a congregation, hold viable services, welcome newcomers, and arrange for their conversion. He gathered any Christians already in town into the congregation and used their networks to drive further expansion. Paul would have made one hell of an Amway salesman; cities he visited had churches before those he didn't. Even so, he had far less of a direct effect than the Bible credits him with, because there were other missionaries, and their ranks kept growing.

The pagans called Christians atheists for denying pagan gods. The Christian adherence to their monotheistic beliefs was seen as disrespect and refusal to show a little courtesy. To the pagans, Christians were like uninvited guests who showed up, ate all the food, and then refused to leave. When it came to religion, Christians insisted that they alone knew how to worship the one true god—an attitude that 2,000 years of unparalleled success has done little to ameliorate. Ironically, it was Roman religious tolerance that allowed Christianity to establish itself. Pagan tolerance and openness sowed the seeds of its own eventual undoing. The Roman senator Tacitus (56-117AD) had Christians in mind when he lamented that all bad practices end up in Rome.

The Romans did not take kindly to having their gods lambasted. They arrested Paul and executed him, his status as a Roman citizen allowing him to be beheaded instead of crucified. The Emperor Nero was the first to differentiate between Christians and Jews. The *Pax Romana* (Roman Peace) depended on religious peace. Christians therefore presented a real threat to national security that was seen as treason. Any Christian could hand over their Bible, make a sacrifice to the pagan gods, and go on about her or his life worshipping however s/he saw fit. Instead, the Christians chose to actively recruit members and call themselves soldiers of Christ who were willing to fight and die for God. Roman pleas to tone it

I never saw, heard, nor read, that the clergy were beloved in any nation where Christianity was the religion of the country. Nothing can render them popular, but some degree of persecution.
Jonathan Swift

If there is any moral in Christianity, if there is anything to be learned from it, if the whole story is not profitless from first to last, it comes to this: that a man should back his own opinion against the world's.
Samuel Butler

down a little and avoid execution fell on deaf ears, leading Arius Antoninus to point out that the Roman Empire had plenty of cliffs for them to jump off or rope to hang themselves with if they insisted on dying so readily. From the Christian point of view, anyone who died for the faith was a hero, an exemplary soldier of Christ. Those who saved themselves by making offerings to other gods were the worst criminals imaginable. Rome therefore refused to recognize the Christian churches, and Christians could not own property. The Romans may have persecuted the Christians, but the Christians brought it on themselves.

Roman roads featured way stations called *stabula* where travelers could eat, rest, and get fresh mounts. Legend has it that a pagan soldier named Constantine seduced a nobleman's daughter and eventually became one of the four men to rule Rome under the power-sharing arrangement invented by Diocletian (c. 244-311AD). (Such an arrangement was easy in a polytheistic society.) Two of the four men held the higher rank of *Augustus*, and two held the lower rank of *Caesar*. Each pair ruled half of the Roman Empire, with well-planned succession to ensure smooth transitions. Diocletian outlawed Christianity in 303AD and set out to exterminate the Christian religion once and for all. Constantine started out as the lowest man on this totem pole, but rose to become the eastern Augustus. A dispute ensued between him and the western Augustus, Maxentius. Their armies met at the Milvian Bridge near Rome. The night before the battle, Constantine saw a vision of the cross in the sky and converted to Christianity. Constantine won, and the world was at last safe for monotheism after a struggle that had begun under Amenhotep IV some 1,650 years prior. At least for a while...

The Edict of Milan stopped the persecution of Christians, an event they hailed as an endorsement, but one merely intended to restore the old religious tolerance and status quo of pagans. Christians now had total freedom to worship as they saw fit. The hierarchy of bishops and priests was welcomed into the highest social circles. Unfortunately, the freedom to choose inevitably meant that some would choose wrongly. There were over 150 "false" beliefs and practices that threatened the orthodox Catholics by the 4th century AD. This was not tolerable. Christians soon began dying at the hands of other Chris-

---

How natural that the errors of the ancient should be handed down and, mixing with the principles and system which Christ taught, give to us an adulterated Christianity.
Olympia Brown

Christianity has operated with an unmitigated arrogance and cruelty—necessarily, since a religion ordinarily imposes on those who have discovered the true faith the spiritual duty of liberating the infidels.
James Baldwin

tians. Far more perished this way than during all of the Roman persecutions. A similar situation exists in Los Angeles today, where a member of a Crips street gang is far more likely to be killed by another Crip than by a member of the rival Bloods. Freedom from the risk of becoming martyrs made it easy for Christians to revere those who died while denouncing the survivors who turned from the faith.

Christians who committed some crime against the faith had to present themselves to the bishop for punishment that usually entailed an act of penance followed by readmission to the church. Excommunication was reserved for the worst offenders. Pagans may have reviled bacchanalia because of the public sex, but had no problem whatsoever with the underlying beliefs or with private orgies. Orthodox Christians prosecuted "thought crimes" in an attempt to dictate people's beliefs and behavior. Women posed a special threat because of Eve's original sin in the Garden of Eden.

The Christians saw Constantine's victory as God's triumph, but the truth is that the emperor's attraction to monotheism had everything to do with his desire for earthly power. The Christian church demanded abject obeisance, and he stood to gain by tapping into that rigidity à la Josiah. There was just one problem: The Christians were fighting amongst themselves over what should—and should not—constitute Christianity. A fragmented church was not a good foundation on which to build a power base. Constantine therefore invited all Christian bishops to Nicaea to iron out their differences. Attendees received star treatment that included free passage on Roman roads, gifts, room and board, etc. in exchange for the promise of establishing law and order in the church.

The Council of Nicaea failed at this mission, but did forge an alliance between church and state that created a government within a government with its own Vicar of Christ (pope). Christians who had been taught that Rome would fall and they would be lifted up to heaven found themselves in the awkward position of having been saved by Rome. The empire that used to roast Christians alive now showered largesse on them and paid for their churches. The era of Revelations was over, and a run of hitherto unimaginable power that still shows no signs of slowing down after 1,700 years had begun—a fact that remains lost by modern Christians who are awaiting the

---

Vices are character traits. Sins are specific acts of commission or omission. Once Judaism and Christianity adopted the concepts of vice and virtue from the Greek and Roman moralists, vices were often called sins and sins vices.
Solomon Schimmel

Madness, in fact, is a medical term that can claim no more notice from the objective critic than he grants the charge of heresy raised by the theologian, or the charge of immorality raised by the police.
James Joyce

"promised" Rapture without realizing that the party is still in full swing. In fact, it has never been better to be Christian.

Was Constantine a true Christian, or were his motivations more pragmatic? I believe the latter. Two pagan temples opened in Constantinople alone during this time, and paganism was never outlawed. Constantine was not admitted to the church until he was on his deathbed in 327AD. In fairness, deathbed conversion was not unheard of, because it allowed a person to sin to her or his heart's content and still receive salvation at the last moment, assuming all went as planned.

> *There is no heresy or no philosophy which is so abhorrent to the church as a human being.*
> James Joyce

Constantine's sons ruled for 34 years after his death, during which paganism was outlawed and persecuted. Things changed again when Julian the Apostate became the sole ruler of the Roman Empire in 361AD. Raised as a Christian, he later abandoned the faith, choosing not to confine himself to one god or set of beliefs. He decriminalized paganism, and then traveled to Constantinople. The world stood between two possible outcomes: one god or many. Paganism reflected the highest state of education and civilization in those days, while the Christians still could not agree about how to worship their one god despite having the empire's full support. Paganism was not about to fade quietly into history. Julian did not execute Christians, refusing to supply them with more martyrs. He simply returned Christian property to the pagans, removed the cross from the imperial standard, and tolerated all forms of worship, including Christianity. He ridiculed Christians as backwater bumpkins, mocked relics as bits of corpses and the churches that held them as tombs, and challenged the Christians to educate their children and see how they fared against their pagan peers.

> *The difference between heresy and prophecy is often one of sequence. Heresy often turns out to have been prophecy—when properly aged.*
> Hubert Humphrey

Julian was killed in battle in 363AD and succeeded by Jovian, who once again restored Christianity and made worshiping pagan gods a capital offense. This time the Christians were out for revenge. Pagan temples were destroyed in a religious frenzy that swept up all in its path. Church, state, and mob came together to crack down on any and all Christian and pagan diversity. Theodosius 1 (347-395AD) was a fanatical Christian who invented the first real Inquisition and made Christianity the official Roman religion. We saw in Chapter 6 that Priscillian was the first person to be executed for heresy . This was also the first time a civil court enforced a religious doctrine. Untold tens of thousands have followed ever since. In 390AD, Christians burned the fabled library of Alexandria to the ground, an act that may well constitute the single greatest loss to civilization. Rome destroyed its own poetry, philosophy, history, etc. that represented and documented civilization's greatest triumphs. Reusable pagan texts were erased and used for pious Christian writings. Much of what we know about ancient pagan writings comes from Christian texts written to refute them. For example, much of our knowledge of the Gnostics came from Irenaeus before the discovery of the Nag Hammadi scrolls in 1945. (See Chapter 8.) Other writings survived in remote corners of the empire and beyond. For example, the Arabs played a crucial role in preserving many ancient texts. No other attempt to erase history has ever succeeded so thoroughly.

Paganism limped on in the countryside for another 300 to 400 years. The last pagan academy closed by around 1,000AD. The era of religious tolerance was over.

Islam

Picture yourself walking through the desert for weeks on end, where the only scenery breaking up the monotony is the hind end of the camel in front of you. Under those circumstances, I can only guess that an exceptionally vivid imagination might be a survival trait. A new up-and-coming religion that demands total surrender to divine will and five prayer breaks a day might be just the thing to relieve the boredom, especially when it features otherworldly creatures made of "the fire of a scorching wind," hot air being in generous supply in the desert. With one stroke, a dreary trudge between trading centers becomes an adventure fraught with peril and opportunity.

The Jews and the Christians had received their revelations of the One True God, but the Arabs had been left out of this great transformation. There was a perceived need for a local set of revelations that the Arabs could claim as their own to unify the people. As we saw in Chapter 5, Mohammed was the man for the job. He was not about to let on that his knowledge of God was secondhand; instead, he claimed to have been visited by the angel Gabriel, who dictated the Koran to him—a neat trick, considering that Mohammed was illiterate. Nevertheless, he did not claim that Islam was a new faith; he

> *I studied the Koran a great deal. I came away from that study with the conviction that by and large there have been few religions in the world as deadly to men as that of Muhammad. As far as I can see, it is the principal cause of the decadence so visible today in the Muslim world and, though less absurd than the polytheism of old, its social and political tendencies are in my opinion more to be feared, and I therefore regard it as a form of decadence rather than a form of progress in relation to paganism itself.*
> Alexis de Tocqueville

drew heavily from the Old and New Testaments, and demanded that people accept his revelations as the absolute truth. One could argue that Islam is little more than thinly disguised plagiarism, a charge substantiated by the fact that Mohammed's wife and cousin were both Christian.

> *I have to be honest with you. Islam is on very thin ice with me....Through our screaming self-pity and our conspicuous silences, we Muslims are conspiring against ourselves. We're in crisis and we're dragging the rest of the world with us. If ever there was a moment for an Islamic reformation, it's now. For the love of God, what are we doing about it?*
> Irshad Manji

The Bible has been extensively studied and found to be riddled with additions, deletions, mis-translations, and other discrepancies. No one has ever undertaken a similar examination of the Koran. Why? The Jewish and Christian religions continue to thrive despite the fundamental flaws in the Bible, and it seems reasonable to think that Islam would be no exception. Muslims, like the Christians before them, resist translating the Koran; they view it as the word of God only when it is presented in the original Arabic, claiming that no translation can ever be the Koran because the original is "like a symphony." All Muslims therefore recite the Koran in Arabic. Some discrepancies are known despite these restrictions, such as the so-called Satanic verses from which author Salman Rushdie took his inspiration (and for which he was targeted for assassination as a result). In another example, Mohammed allowed some people to keep worshipping their local gods, but then changed his mind by claiming he had allowed himself to be channeled by evil. Even his wives noticed how his revelations changed to suit his immediate needs. Is it more likely that Mohammed heard the word of God or that he simply made it all up as he went along, like Josiah and so many others before and since? Even if the revelations were real, why would the Arabic God choose an illiterate who could not possibly pass on His words unchanged to be His messenger?

Verdict

> *The savior who wants to turn men into angels is as much a hater of human nature as the totalitarian despot who wants to turn them into puppets.*
> Eric Hoffer

The power of belief combined with the power of a totalitarian state forms a truly dangerous mixture. A monotheistic level of rigor and zeal exists in all totalitarian regimes, including supposedly secular places like the former USSR, North Korea, and Cambodia. As the Germans used to say, *"Ein Volk, ein Reich, ein Fuhrer!"* (One people, one empire, one leader!)

Today's relatively tolerant and diverse religious attitudes reflect pagan attitudes that precede monotheism. Westerners face very little risk of death for their religious beliefs, the handful of abortion-related shootings and bombings notwithstanding. Sadly, the Islam that used to be tolerant, and that preserved so much of Western civilization, is now on its own rampage, yet another reminder of the perils of monotheism.

The 4th century Roman scholar Symmachus said it best: "We gaze at the same stars, the sky covers us all, the same universe encompasses us. What does it matter what practical system we adopt in our search for the truth? Not by one avenue only can we arrive at so tremendous a secret."

When peaceful coexistence is in one's best interest, respecting other people gods is part of getting along; however, this has not been the primary goal of monotheism. The fact that JCI religions spawned intolerance is a common complaint and seems to be an intrinsic property of monotheism, which seems to be literally allergic to peace and mutual respect.

We accept the verdict of the past until the need for change cries out loudly enough to force upon us a choice between the comforts of further inertia and the irksomeness of action.
Judge Learned Hand

Chapter 8

In God's Name

> *Very human decisions were made, based upon very human priorities-mostly concerning control and power.*
> Michael Baigent

> *We must not try to excuse things for which there is no real excuse. To ignore the question of human responsibility would make all history meaningless.*
> .G.G. Coulton

This chapter looks at some of the things that monotheism—and especially Christianity—has done in God's name. I am not singling out Christians as any better or worse than any other monotheistic religion, because all such religions have done much the same things and behaved in much the same ways. The Egyptians and Jews had their own Inquisitions and purges long before Christianity, a legacy that the modern state of Israel may be perpetuating. Islam's initial openness literally saved Western civilization from itself by safeguarding copies of classical works from the wholesale destruction of knowledge wrought by the Christians before dogma took hold. Thus, the story you are about to read talks about the Christians, but applies to all monotheistic religions. If you are religious, this will probably be a difficult chapter for you to read. Chapters 9 through 12 will also be tough reading, yet I must ask you to read them with an open mind with the promise that things will fall into place as we keep moving through this book.

In theory, the term "god" should function only as a symbol and anthropomorphism of the divine that should not be confused with the actual divine. This perspective acknowledges all religions as equally valid, since all religions use their definitions of God (or gods) to express the inexpressible. The problems begin when dogma becomes entrenched and divides people into mutually opposing camps—an especially tragic situation, because the god experienced by mystics around the world is a single unifying force. Accepting dogma and myth as literal fact can have other problems as well. For example, an African pygmy listening to a missionary describing heaven at length interrupted to ask a very simple question: How could the missionary possibly know? Had he been there? As we have seen, all monotheistic religions make god in their own images in some sense, which does not necessarily have any bearing on reality.

I believe it because it is absurd.
Tertullian

Europeans came to dominate much of the world for many reasons that are far beyond the scope of this book. With that domination came the belief in their own intellectual and spiritual superiority, and the idea of a personal god who created humans in his likeness followed them around the globe. On the plus side, a personal god can foster the idea of individual rights and liberal humanism. Yahweh/Jehovah began as a highly personal deity, became transcendent, and then went back to being personal in the form of Christ, who walked among the people, showed them how to reach heaven, and sacrificed himself for their sins. On the down side, a personal god can become a liability by endorsing prejudices and forcing us to accept what should be unacceptable. Assigning a gender to a god automatically limits half of humanity. Viewing a people and/or gender as somehow superior can encourage judgment and condemnation on Earth as the god's followers take it upon themselves to act on his behalf, forgetting that the OOO creator of the entire universe should be more than capable of fighting his own battles. Meanwhile, an amazing amount of religious diversity has been either lost or endangered. The Aborigine view is that time is timeless, the Eastern view is that time is cyclical, the primal religions that look to the past for answers, and the rich diversity of deist and pantheistic views, are virtually unknown in the modern monotheistic world. We ignore these lessons at our peril.

Religion is a defense against a religious experience.
Carl Jung

An examination of the history of Christianity reveals beliefs, dogmas, and institutions coming into being and evolving over time. Believe it or not, most Christian doctrines and rituals have no basis whatsoever in the New Testament; some are even contradictory. The very term "Christian" is a misnomer, because Christianity extends back to Paul (the Joseph Smith of his day), but no further. Thankfully, more and more Christians are becoming embarrassed by their history of blood and forced belief.

Science

> *Just how does one use earthly empirical standards of weights, measurements, and calculations to analyze God, who transcends space, time, and matter?*
> Becky Garrison

The increasing specialization of professions and scientific disciplines over time has yielded impressive advances in the state of our knowledge about the universe. It has also made seeing the "big picture" harder and harder, a theme we will return to later. At its core, science has faith in a rational universe. This does not entail belief in any god, design, intelligence, or purpose to the universe, merely that the universe obeys rational laws that can be revealed through the process of empirical observation and experimentation. In Chapter 3, we discussed how religion represents humanity's first science and use of reason to explain how the universe works. Later, the monotheistic religions forcibly resisted reason and stifled scientific advancement. The church's fears were well founded: As the religious monopoly on thought crumbled, scientists separated themselves from religion and began taking a long hard look at the Bible, the church, and religion itself. The old "proofs" of God's existence were shattered, replaced by an ongoing attempt to objectively prove—or disprove—God. Much of this separation has occurred within the last 200 to 300 years, although there have been at least a few dissenting voices all along.

> *Religious dogmatism impedes medical research, starts wars, diverts scarce material and intellectual resources-in short, it gets people killed.*
> Sam Harris

Ironically, monotheism can create a more open framework for science precisely because it postulates a single ultimate cause or reality for the universe, a sort of Grand Unified Theory. As we will discover later, the search for such a theory is the Holy Grail of science. Still, belief can be both powerful and dangerous. The history of monotheism is the history of beliefs that have been both defended and violently imposed on others. In fact, almost all of the longest and hardest-fought conflicts in history have been based on religious beliefs. Not all of these conflicts have involved violence. The schism between science and religion has been largely a war of words between Christians who take the Bible literally and the scientists whose investigations continue to disprove the Bible. What people embroiled in this battle don't understand is that most Western atheists oppose the Western god (Jehovah/Yahweh), as described in the Bible and acted on by Christians without necessarily opposing all things divine. Christians who get angry with scientists who claim to have eliminated God don't understand that the scientist in question is probably only opposing their understanding of God, not God as a concept. Christians who see science as the enemy forget that Islam already covered this ground by preserving Western texts and advancing the sciences while Europe plunged into the Dark Ages.

Can I teach evolution in your church?
Anonymous bumper sticker

Those Christians who don't take the Bible so literally understand that science (and particularly Darwinian evolution) does not undermine their faith, and can even strengthen it. Those who can see past Newtonian mechanics and determination and embrace a random, indeterminate quantum reality can actually have their faith enriched. The same holds true for any religion. We will be exploring this in much more depth throughout the remainder of this book. In the meanwhile, it is important to note that fundamentalism is enjoying a resurgence around the world. Monotheists are digging in their heels and increasingly condemning science and anyone who disagrees with them (often using methods and technologies developed by the very disciplines they decry).

We will be exploring some of the many scientific objections to religion in Chapters 10 through 12.

Altering the Bible

Judaism is a very unique religion in that it is monotheistic. Yahweh chose the Hebrews as His special people out of all the people in the world He could have chosen. A Jew may worship anywhere, but can only perform his ritual duty of sacrifice at the temple in Jerusalem. All other Jewish religious buildings are called synagogues, possibly to distinguish them from the temple. Jews had books that they considered (and still consider) religiously important, as well as books of prophets, poems, history, etc. Twenty-two of these books became the canon of the Jewish Bible over the first few centuries following Christ's supposed crucifixion and accepted as the Old Testament. This created the very paradoxical situation of a religion that relied on books of supremely important writings that very few people could actually read. One might think that

From the moment I picked up your book until I laid it down, I was convulsed with laughter. Someday, I intend reading it.
Groucho Marx

those who could read the books would want to educate the rest of the population so that all could participate in the sacred writings. The exact opposite occurred, because religious leaders wanted obedience, not enlightenment.

> *Progress is nothing but the victory of laughter over dogma.*
> Benjamin DeCasseres

Polytheistic religious used almost no books, focusing instead on honoring their gods via sacrifices. Thus, they had little in the way of doctrines or guiding principles. Their beliefs about the gods, morality, ethics, etc. played almost no role in their religions. Polytheism entailed creating a personal philosophy for individual guidance, and few people need books to tell them what they think.

It may be difficult for a person raised in an era of ubiquitous printers, presses, and desktop publishing programs. to conceive of a time when no such technology existed, but such was the state of the world when the great monotheistic texts were written and disseminated. Back then, the only way to copy a book was by hand, letter by letter, one word at a time—a slow, painstaking process that could take years for a single title. This inevitably caused a larger problem: scribes changed the texts as they copied them, either by accident, negligence, or even deliberately. Nobody reading a book in the ancient world could be entirely sure that s/he was reading the author's original words. On the contrary, it was safe to assume that the text had been altered, even if only a little; however, even the smallest changes can have tremendous impact. Consider the (possibly apocryphal) example of a man who asked the oracle at Delphi whether his wife was bearing a son or daughter and received the answer, "BOY NOT GIRL." Which is it? The answer depends on whether you interpret the words as, "BOY, NOT GIRL" or "BOY NOT, GIRL." This particular ambiguity validates the oracle either way.

> *It is only by the exercise of reason that man can discover God. Take away that reason, and he would be incapable of understanding anything; and, in this case, it would be just as consistent to read even the book called the bible to a horse as to a man. How, then, is it that those people pretend to reject reason?*
> Thomas Paine

The church began using professional scribes in the 5th century AD. Monks took over these duties still later; however, the earliest manuscripts were not copied by professional scribes, but by literate—or semiliterate—members of the congregation. It is therefore safe to assume that these earliest copies are loaded with mistakes, an assumption that is validated by records of reader complaints. Origen himself complained that the differences between manuscripts had become large due to negligence and/or deliberate actions. Scribes either did not perform adequate quality checks and occasionally made frivolous additions/deletions to suit their own interpretations or beliefs. In fairness, most of the changes in early Christian manuscripts were mistakes; just because someone was a scribe did not mean that person was competent or even trained.

Moving to professional scribes did nothing to ease the drudgery of copying an entire book. Inattention, hunger, tiredness, boredom, laziness, etc. remained sources of errors. Even the best scribes made mistakes. Some made changes because they thought the changes should be made. These deliberate alterations often had nothing to do with theological differences. Scribes sometimes found text that seemed wrong (such as a contradiction or mistaken reference) and made those changes with the best of intentions. Still, every change is a change, and the words of the Bible's original authors have become irretrievably lost. This is crucial, because the original text provides the only way of knowing what the author intended to say.

Contrary to its own shrill claims, the Church was never 'catholic' in the literal sense of the word: one Universal Church of the faithful, out of which there is absolutely no salvation.
Jonathan Kirsh

Many sermons hinge on a single word. What if that word was changed? For example, saying that Christ's word reveals everything is far different than saying Christ keeps the universe together with his word. Once that change was made, it became permanent unless someone else found and fixed the problem; however, that scribe would doubtless make his own mistakes and add his own changes. Compounding the problem is that a seeming error may actually be closer to the original text than the proposed change. After only two generations of copies, there exist three different versions of the same text (the original, Copy A, and Copy B). Errors and changes can only multiply from there.

It was common for the church to accuse scribes of heresy for changing texts for personal reasons; however, the record shows that leading orthodox church figures routinely changed the text to both make them more difficult for heretics to "misuse" them and bring them more into line with official doctrine. Accidental and well-intentioned changes are one thing, such as the warning in Revelations 22:18-19 against adding to or removing anything from the book, which is a typical warning to scribes. There are many such warnings in various texts, none of which are supposed to be part of the texts themselves. Confronted by outsider allegations of bad copying, Origen denied that the Catholics had changed anything, despite having described some of the changes he made himself. Altering

John's gospel was written in the heat of controversy, to defend certain views of Jesus Christ and to oppose others.
Elaine Pagels

the text purported to be God's word to suit the wishes of an organization supposedly devoted to following and obeying that word smacks of both hubris and hypocrisy.

The 21st century has an extremely diverse assortment of Christian denominations laying claim to the title of "true" Christianity. They will all agree that the diversity is both recent and bad, forgetting that diversity marked the dawn of both Christianity and religious belief. The Gnostic ideas that so closely resemble the Eastern religions emerged during the 1st century and were suppressed by church leaders like Irenaeus, who sought to impose orthodox beliefs. A heretic is one who deviates from the true faith. The key question is, what is the true faith, and who gets to decide it? In the case of orthodox Christianity, those beliefs that allowed and supported a centralized hierarchical institution won out over those that emphasized individual beliefs and initiations. Orthodox Christianity demanded (and still demands) total faith. As Tertullian said, with faith one needs no more belief. Check your reason and critical thinking skills at the door, and accept the church wholeheartedly—or else.

> *Once the bible begins to be interpreted literally instead of symbolically, the idea of its god becomes impossible. To imagine a deity who is literally responsible for everything that happens on earth involves impossible contradictions.*
> Karen Armstrong

By 200AD, the Catholic Church had a three-tier hierarchy of bishops, priests, and deacons who thought they were guarding the one true faith. Most of these church leaders rejected any divergent beliefs as heresy. Irenaeus insisted that there was only one church, without which salvation was impossible. The availability of state resources, troops, judges, prisons, and executioners that became available when Christianity finally triumphed helped elevate heresy to a capital offense. The Catholic struggle against heresy was so successful that virtually everything the world knew about Gnosticism before the discovery at Nag Hammadi came from the attacks against it.

> *On the one side [of Christianity] were those who sought knowledge, and on the other were those who were content with belief.*
> Michael Baigent

Transcription skills improved over time. Later copies tend to resemble each other more than earlier copies, which also show more variation amongst themselves. Christianity's shift from persecuted minority to religious monopoly yielded benefits in the forms of largesse, state resources, and a huge influx of educated converts who were trying to keep up with the Joneses after the emperor's own conversion. Major *scriptoria* (transcription centers) opened in larger cities. Constantine asked for 50 Bibles to be made in 331AD, which means there had to be at least one scriptorium. The availability of these professional services was a far cry from earlier churches that relied on volunteers of questionable competence. The quantitative improvements of later scripts that came into much closer agreement with each other does not mean that later texts are better than earlier texts, because they were copied from the earlier unreliable texts. A faithful copy of an altered text reproduces the error and gets you no closer to the original text.

The Bible is the most studied book in history, with vast amounts of time, equipment, and money having been devoted to that endeavor. The first fragmentary—but still useful—copy of Galatians dates to 200CE, about 150 years after Paul's original letter. This means that copies made by non-professional scribes had already been circulating before a copy managed to survive to the present day. Is this copy accurate? If so, how accurate is it? This text, called P46 (the 46th catalogued New Testament text) must contain mistakes. It is therefore almost meaningless to talk about the "original" text of Galatians, because it probably no longer exists. The best we used to be able to do was examine the earliest surviving copy and hope for the best.

Scholars have devised methods of determining where surviving texts deviated from the original New Testament text with a fair degree of accuracy, and the results can be real eye openers. For example, the story of the adulterous woman in John 7:53-8:12 is both wonderful and a later addition. This is just one of countless example, which raises an obvious question: If it was not part of the original John text, should it be in the Bible at all? Most critics would answer in the negative.

Celsus was both staunchly opposed to Christianity and well aware of the accidental and deliberate changes being made to the texts. According to him, "[the] objections [against Christianity] come from your own writings and we need no other witnesses, for you provide your own refutation." He goes on to allege that, "Some believers, as though from a drinking bout, go so far as to oppose themselves and alter the original text of the Gospel three or four or several times over and change its character to enable them to deny difficulties in the face of criticism." Ironically, much of what we know of Celsus's writings comes from Origen's counterattack titled *Contra Celsum*. Celsus doesn't stop there. In an almost gleeful tone, he writes, "Everywhere they speak in their writings of the tree

---

Even with some theologians, the church appears to be a human construction.
Joseph Ratzinger

Who are the clerics to interpret nature? They have shown themselves quite unable to do so.
Christopher Hitchens

> *There are in the world a great many situations that weaken the conscientiousness of the soul. First and foremost of these is dealings with women... for the eye of the woman touches and disturbs our soul, and not the eye of the unbridled woman, but that of the decent one as well..*
> John Chrysostome

of life... I imagine because their master was nailed to a cross and was a carpenter by trade. So that if he happened to be thrown off a cliff or pushed into a pit, or if he had been a cobbler or a stonemason or blacksmith, there would a cliff of life above the heavens, or a pit of resurrection, or a rope of immortality, or a blessed stone, or an iron of life, or a holy hide of leather. Would not an old woman who sings a story to lull a little child to sleep have been ashamed to whisper tales such as these?" Later that same century, the philosopher Porphyry wrote in *Adversus Christianos* (Against the Christians) that "[The] evangelists were fiction writers, not observers or eyewitnesses of the life of Christ. Each of the four contradicts the other in writing his account of the events of his suffering and crucifixion." Try as the Catholics might, history has validated both Celsus and Porphyry.

Irenaeus blasted the heretics for having, "more Gospels than there really are." The serpent in Genesis represents the Gnostic principle of divine wisdom that convinces Adam and Eve to eat the apple, at which God threatens them both with death. This is a clear analogy for the orthodox church that demanded abject faith. It is also a great way to keep the locals coming to church every, Sunday because it can easily be interpreted as sanctioning capital punishment for going against the church by being a heretic. The Nag Hammadi texts were labeled heretical in the 2nd century C.E., and possession of all books deemed heretical became a criminal offense. Why were the Gnostics so evil?

> *Religions make factual claims that have no special immunity from being examined under the cold light of reason and objective observation.*
> Victor Stenger

- "A Gnostic," said Theodotus, "understands who he was and what he has become, where he has been and where he is going, what birth is, and what rebirth is."

- "Abandon the search for God," said Hippolytus of Rome, "and start with yourself. Learn who within you makes things his own and says 'my God, my mind, my thought, my soul, my body.' Learn the sources of emotions. Do this and you will find Him in yourself."

- "I am not your master," said Christ in the Gospel of Thomas. "Because you have drunk, you have become drunk from the bubbling stream which I have measured out. He who will drink from the bubbling stream which I have measured out, he who will drink from my mouth will become as I am. I myself shall become he, and the things that are hidden will be revealed to him."

The charges of heresy against the Gnostics boil down to the fact that they saw themselves and the divine as one and the same, while orthodox Christians and Jews saw a huge gulf separating man from God—a gulf that had to bridged by God's emissaries on Earth, who would dole out punishments and rewards on God's behalf. Someone who understands the nature of the Godhead and that s/he is one and the same with that Godhead has no need of any Earthly institutions purporting to be the instruments of God's will, because s/he quite literally knows better. Even the Gnostic Christ we have met in preceding chapters had to be replaced by the Christ sent by God to enforce the ancient laws. A Christ who speaks in terms of illusion and enlightenment instead of sin and repentance undermines the very institutions that claim to be serving him. The Gnostic Christ who leads someone to enlightenment is no longer the master but an equal, and even identical. Compare this to the Buddhist teachings in Chapter 5 and the definition of idealism in Chapter 1. Could at least some of these similarities owe themselves to scribes influenced by the Buddhist monks who traveled to Alexandria via trade routes opened between 80 and 200AD to preach? Hippolytus includes Brahmins as sources of heresy, so this is certainly possible; however, it is equally possible that similar ideas started in multiple places at once.

We condemn all heretics, whatever names they may go under. They have different faces indeed but their tails are tied together inasmuch as they are alike in their pride.
Fourt Lateran Council

The Western world needed Bibles in various languages, including the vulgate Latin. The number of varying translations proliferated to such an extent that Damasus I commissioned Jerome, the best scholar then available, to create an official, authoritative Latin translation for all Latin-speaking Christians. Jerome's writings speak of the plethora of translations and his attempt to fix the problem by comparing the best Latin translation against the best Greek texts. The resulting Bible became the vulgate for the Western church and was itself copied countless times over. Christians have relied on the fruits of Jerome's labors ever since.

You are accused as a heretic because you believe and teach otherwise than the Holy Church believes.
Bernard Gui

The huge and growing number of translations gave some the idea that God would guard His book against errors that would prevent its divine use without bothering to worry himself over small errors. There are between 200,000 and 400,000 variants

> *Those who can make you believe absurdities can make you commit atrocities.*
> Voltaire

of the New Testament today. This is a rough number, because it is impossible to achieve a full accounting. Any way you count them, the fact remains that the number of variants exceeds the number of words in the New Testament. There is at least one different version of the New Testament for every word in the New Testament. It doesn't help that ancient Greek was written with little to no punctuation or spaces between words, making it very easy to mistake words that look alike. For example, in 1 Corinthians 5:8, Paul warns against, "eating old leaven, the leaven of wickedness and evil." The Greek word *pneras* means evil. It also looks a lot like *porneias*, which means sexual immorality. This small change has a huge impact: Some texts have Paul issuing a specific warning against sexual vice.

In 1 Corinthians 12:13, Paul says that everyone in Christ has been baptized into one body and drunk of one spirit. The Greek word for spirit is *pneuma*, which scribes often abbreviated as $\overline{PMA}$. This can be misread as *poma* (drink) to have Paul saying that all have drunk of one drink. Words that sounded similar could cause problems when copying via dictation. A scribe hearing a word that sounded the same as another work had to make a guess and could easily make the wrong guess, especially if either choice made sense. Revelations 1:5 speaks of one who *lusanti* (freed) us from sin. This sounds a lot like *lousanti* (washed). Some translations thus speak of one who washed us from sin. Even I got these two words mixed up at first, and I was staring directly at them—on a modern computer, no less.

Texts could be changed because of perceived errors of fact and/or interpretation. Matthew 24:36 has Christ saying that no one, not even he, knows when the world will end. How could the son of an OOO god not know this? To "fix" this, they simply took out the mention of Christ not knowing. The angels may not know, but now Christ could know, and the entire message and point of that passage was changed. Of course, the original passage makes perfect sense if we lose the assumption that Christ is God, a claim he never makes.

> *Religious people are in general all too eager to let us know what they believe.*
> Pascal Boyer

I have already mentioned that some scribes altered texts to make them unusable by heretics or to make sure they said what the scribe thought they should say. In Luke 5:37-39, Christ says that nobody stores new wine in old skins and no one who drinks old wine wants new wine, because the old wine is better, as any wine lover will attest. But how could Christ say old wine is better when the salvation he offers is both newer and better? Some scribes simply removed verse 39. Other texts were modified to suit particular beliefs and associated rituals. For example, something that could happen only through prayer might be edited to require fasting as well. Some of these changes reflect efforts to make the Greek Gospels agree with each other. Some edits were even guided by oral traditions.

Theologian Johann Wettstein (1693-1754) pointed out that 1 Timothy 3:16 had been in long use by orthodox Christians to bolster the notion that the New Testament calls Christ God; however, most texts refer to Christ as, "God made manifest in the flesh and justified in the spirit." Wettstein also pointed out problems in other passages used to justify Christ's divinity. Examining the texts using established methods of evaluation and critique removes most of the references to Christ's divinity. This includes the "Johannine comma" in 1 John 5:7-8 and in Acts 20:28 that talks about, "the church of God, which He obtained by His own blood." The *Codex Alexandrinus* and some other texts speak of, "...the church of the Lord, which he obtained by his own blood." Calling Christ Lord is a much different proposition than calling him God. The term "lord" has been used to refer to noble rank for hundreds of years without anyone mistaking a nobleman for a god. This is no mere exercise in semantics: If the New Testament does not call Christ God, then the entire foundation of Christianity, the world's largest religion, becomes structurally unsound at its very core.

It should go without saying that Wettstein's peers mounted a strong opposition and insisted that he be barred from publishing his useless, unwanted, and dangerous Greek New Testament. (Of course it was dangerous... to them!) They charged him with spreading unorthodox beliefs, making statements against the reformed church, and writing a New Testament with "dangerous innovations." An investigation revealed that he did not believe either that the Bible was divinely inspired or that devils and demons existed. Wettstein, they said, focused his attention on obscurities. They removed him as a deacon and made him leave Basel. He traveled to Amsterdam and

---

[Faith is] the assent of the intellect to a truth which is beyond its comprehension.
New Advent Catholic Encyclopedia

Even some devoted Christians have found that the impulse to seek God overflows the banks of a single tradition.
Elaine Pagels

> *Violence is widely embraced because it is embedded and "sanctified" in sacred texts and because its use seems logical in a violent world.*
> Jack Nelson-Pallmeyer

> *The more the God of your Universe agrees with your every bias and whim, the more certain you can be that the God you worship exists entirely in your own head. An omniscient and omnipotent Deity would not need human praise to know he's great. Any God who insists you worship "Him and only Him above all else!" is just an insecure little bitch. And I'm sorry, my universe is cool enough that I don't need to spend my precious time and mental energy fellating an insecure deity.*
> Scott Cragg

continued his work there. Later, he claimed that the controversy had delayed his book by 20 years.

There are a few verses in Luke where Christ laments his coming execution. This is the only place in the New Testament where he seems unable to take the strain of his destiny. Why would Luke remove all other references to Christ's suffering only to emphasize it here? Why take out corroborating verses before and after these few? This passage seems to be a later addition. As for his divine destiny, most ancient manuscripts say that Christ died for everyone because of God's grace (CHARITI THEOU) while others say that he died separated from God (CHORIS THEOU). There is good reason to suspect that the latter translation is correct. Why change it? Early Christians saw Christ's death as God's ultimate act of grace and forgiveness for humanity. Saying that Christ died apart from God could mean many things, most of which are inconsistent with this viewpoint. This is one example of how the New Testament has been altered to suit theological desires. One would think that the theology should adapt to the Bible, especially since those doing the changing are at the same time claiming that the book represents God's divine law. It's a little like changing the DMV handbook to suit your driving style.

The period between the writing of the New Testament texts and the conversion of Constantine saw a number of theological differences among different Christian sects, many of which would not be recognized as Christian today. Some believed in one god. Others insisted that the gods of the Old and New Testaments were different beings altogether. The latter group included the Marcionites, who insisted that their view represents Christ's true teachings. All Christian sects insist that they are the "correct" and "true" Christians, and all such claims should therefore be taken with a healthy dose of skepticism. Some Gnostics claimed 12 gods, others 30 or even 365. They too claimed to represent the "true" Christianity. It is important to remember that the New Testament did not exist then. The books had been written, but so had plenty of others that also claimed to be written by the apostles. These other books included gospels, acts, epistles, apocalypses, etc. that contained ideas very much at odds with the books that ended up becoming the New Testament. The New Testament only exists the way it does because one group attracted more followers than others and decided which books should be included as scripture. It stands to reason that all Christian groups were probably both right and wrong to roughly the same extent.

Some groups said that God created the material world. Others claimed the material world is evil, the result of some cosmic disaster. Some said that God was responsible for the Jewish texts, and others blamed an inferior god. Depending on who you believe, Christ was either totally divine, both human and divine, or merely human. Some groups even claimed that Christ was divine and Jesus human. Christ's death may have saved the world or had nothing to do with saving the world or did not occur at all. As you can see, religion has less to do with having the "correct" view than it does with having the "winning" view. The belief that the world is flat is a well-known example of a "winning" belief that turned out to be incorrect. It should therefore be obvious that religious beliefs are decidedly not divinely inspired.

Facts, you see, can be overturned by evidence; dogma is impervious to it.
Bernard Haisch

Luke 2:33 says that Mary and Joseph, Christ's mother and father, took him to the temple where they were amazed by what Simeon said as he blessed Christ. If Joseph is Christ's father, then how can Christ be the product of a virgin birth? Scribes simply changed the text to indicate that "Joseph and his mother" took Christ to the temple to prevent heretics from using that passage to refute Christ's divinity. The same thing happens 12 years later, when Joseph, Mary, and Christ attend a festival in Jerusalem. Christ remains behind as the rest of the family heads home, unbeknownst to his ""parents." Once again, the scribes came to the rescue by changing the passage to say, "Joseph and his mother did not know." When they return to fetch him three days later, Mary tells him that, "Your father and I have been looking for you." This of course became, "We have been looking for you." Christianity begins to look more and more made up and less like any homage to any actual divine being.

The first 18 verses of John talk about the word of God that was with God in the beginning. The word made all things exist and allows God to communicate with the world and manifest Himself to others. At some point, the word became human and walked among us as Jesus Christ. Christ is therefore God's word incarnate. The prologue ends by telling us that no one

Forget it. I had no hand in that evil. I have no original sin. There's no blood of any sacred martyr on my hands. I pass on all of this.
Billy Joel

has ever seen God except the unique son and the unique God. Which is it?

In Luke 22:17-19, Christ takes a cup of wine, gives thanks, and tells his apostles to divide it among themselves, "for I say unto you that I will not drink from the fruit of the vine from now until the kingdom of God comes." Christ then breaks the bread, give thanks, and gives it to his apostles, saying, "This is my body. But behold the hand of the one who betrays me is at the table." Nowhere in Luke or Acts do the apostles indicate that Christ's death offers atonement for sins. For Luke, Christ's death serves to remind people of their own guilt before God, since Christ was innocent when crucified. Once they recognize their sin, they should repent to God, who will forgive them. In other words, Christ's death brings repentance that then brings salvation. This account is at odds with other disputed verses that are missing from earlier texts that describe Christ's death as "for you." Why were these verses added? In a dispute with Marcion, Tertullian claimed that Christ had plainly said what he meant when he called the bread his own body. Likewise, mentioning the cup and sealing the New Testament in his blood proves the reality of human body for blood cannot belong to a body that is not of flesh. The evidence of the flesh gives proof of the body, and the blood proves the flesh.

Luke 24:12 is another suspected newcomer. Some women go to Christ's tomb, discover him missing, and are told that he has been raised from the dead. They inform the apostles, who refuse to believe them because they seem silly. Many manuscripts then have Peter going to the tomb, seeing only the burial shroud, and returning home in wonder at what had happened. The fact that everyone involved seems to have ignored the plethora of other more plausible explanations is never mentioned. This is another example of text that does not seem to have been part of the original book. It is written in a different style than the rest of the book and uses a different vocabulary. It could have been added to corroborate the summary found in John 20:3-10, where Peter and the "beloved disciple" (Mary Magdalene?) find the empty tomb. Of course, as we saw in Chapter 6, it would make perfect sense for Christ to make himself scarce after the crucifixion.

Disturbances in society are never more fearful than when those who are stirring up the trouble can use the pretext of religion to mask their true designs.
Denis Diderot

It is either the gospels or contemplation; either mysticism or the book.
Emil Brunner

If this passage was added, then it was done to support the orthodox belief that Christ had a physical body. Thus, the tomb story ceases to be a silly yarn and becomes true and verified by Peter himself, the "rock" on which the Catholic Church was built. According to Tertullian, if we can deny Christ's death because we deny his flesh, then we cannot be certain of his resurrection because he did not die. Only flesh can die and rise again. To deny Christ's resurrection is to deny our own. (Tertullian's logic is a prime example of conflation.) Luke 24:51 describes Christ's removal from his disciples and their joyful return to Jerusalem. Some early texts have the addition, "and He was taken up into heaven," which stresses Christ's physical nature as opposed to the bland, "he was removed." According to Irenaeus, sin was such an insult to God that atonement was required in order not to derail God's plans for humanity. (Irenaeus fails to explain how anything humanity does could possibly derail an OOO god's plans.)

Faith results from not seriously seeking truth.
Jerry Wheatley

As I previously mentioned, Christianity was ironic in that it relies on books, while few of its adherents could read. Letters traveled between Christian communities and helped define faith (what to believe) and rituals (how to behave). These letters were read to congregations at meetings, and many of them ended up in the New Testament, which is primarily a collection of letters by Paul and others. Many of the original letters have been lost; for example, Paul references a letter to the Corinthians written before what we know today as "1 Corinthians." The Gospels addressed the desire to know more about Christ's life, death, and resurrection. Many more of these books were written than the four Christians know today. Some of these books survive today, but many others have been lost. Apocalypses were also written; that of John was eventually selected to become part of the New Testament. Still other documents were created to describe how churches should be organized and managed.

The growing religion gradually attracted intellectuals who could rebut the charges made against Christianity. Their writings (sometimes called *apologia*, or apologies) defend the faith against those who accused it of threatening the social order of the day. The Romans saw all too well what Christianity could—and eventually did—evolve into. One of the charges against Christians was that they followed widely varying

The ignorance of the priests precipitates the people into the ditch of error.
Anonymous English archbishop, 1281

> *Kill them all; God will know his own.*
> Arnauld Amalric

beliefs that often opposed each other. Christians also understood that many different interpretations existed among themselves, and the internal threat these differences presented to the religion as a whole. Paul lashed out against "false" teachers, and other Christian leaders began actively opposing heresy.

Over time, some of the many Christian books began to stand out from the crowd and be deemed more worthy of reading. Some of these books eventually came to be seen as authoritative in matters of belief and practice, and became scripture. It was not long before Christians began accepting Christian writings to be just as authoritative as the Old Testament books of the Jews. Christ eventually gained the same status as Moses.

Marcion (c. 85-160AD) was the first known Christian to create a canon of scripture, a collection of books that, in his opinion, constituted the sacred texts of the Christian religion. This Bible contained none of the Old Testament, only one Gospel, and 10 epistles. Marcion believed that false leaders (those who lacked his understanding of Christianity) copied books and edited them to suit their own needs, such as believing that the Old and New Testament gods were one and the same. He therefore set about editing the texts to remove references to the Old Testament god, creation as the work of the true god, or the Law of Moses as remaining valid. By contrast, Irenaeus believed that there had to be four Gospels representing the four zones of the world, the four primary winds, and the church that required four pillars.

> *The accused are not to be condemned according to ordinary laws, as in other crimes, but according to the private laws or privileges conceded to the inquisitors by the Holy See, for there is much that is peculiar to the Inquisition.*
> Bernard Gui

The Bible we know today was first mentioned by Athanasius in his 367AD letter to the Egyptians, in which he listed the 27 books found in the current New Testament. The debate continued for centuries. Keep in mind that mass literacy only appeared during the Industrial Revolution when the benefit of ensuring that everyone could read outweighed the massive outlays of money and effort required for such an undertaking; agrarian societies did not have the excess resources to invest in something as frivolous (to them) as reading.

We have already seen how the religious status of women changed from that of equal participants to subservience in Chapters 5 and 7. Galatians 3:27-28 says, "For as many of you as were baptized into Christ have put on Christ. There is neither Jew nor Greek, neither slave nor free, there is not male and female, for all of you are one in Christ." Contrast this egalitarian statement with 1 Timothy 2:11-15, which says that a woman must, "learn in silence and with full submission. I permit no woman to teach or to have authority over a man. She is to keep silent. For Adam was formed first, then Eve, and Adam was not deceived but the woman was deceived and became a transgressor. Yet she will be saved through childbearing, provided they continue in faith and love and holiness and modesty." 1 Corinthians 33-35 says, "For God is not a God of confusion but of peace. As in all the churches of the saints, let the woman keep silent. For it is not permitted for them to speak, but to be in subjection, just as the law says. But if they wish to learn anything, let them ask their own husbands at home, for it is shameful for a woman to speak in church." It is probable that Paul did not write these verses in 1 Corinthians, because they don't fit into the overall context of instructing Christians how to conduct themselves during services. We have already seen how the Bible was altered to suit beliefs; it is reasonable to assume that the prevailing attitude toward women would have crept into the texts, a stark contradiction to how Christ himself was raised and what he taught.

All of the examples I listed came directly from *Misquoting Jesus* by Bart Ehrman; they represent only a few of the many thousands of changes and alterations to the New Testament. The Old Testament has also been altered over the centuries, as has the Koran and every "sacred" book. These changes have been made to suit human beliefs and prejudices. This alone should be prima facie evidence that the God of religion is made in humankind's image and not the other way around, especially when these changes were presented as, "God's will." Orthodox Christians (the Roman Catholic Church) resisted all attempts at change once the canon was finalized. As Leon of Castro said in 1576, "Nothing shall be changed in the vulgate, not one little iota." Christ's teachings, a glimpse of which we saw in Chapter 6, were gradually turned into tools of religious oppression. The Roman Catholic Church resisted change and progress, and enforced a single set of beliefs. This repression stifled science and progress during the Dark Ages that lasted for some 1,200 years from the fall of the Roman Empire (c. 395AD) to the Enlightenment that began in the 1600s.

It is an inconsistency scarcely possible to be credited, that anything should exist, under the name of a religion, that it held to be irreligious to study and contemplate the structure of the universe that God has made.
Thomas Paine

Men may invent heresies but it is women who spread them and make them immortal.
Anonymous French priest, ca. 1150

Galileo Galilei (1564-1642AD) is just one example of the many tens of thousands of people prosecuted for daring to contradict the (much altered) Bible. Galileo's crime? Stating that the Earth revolves around the Sun, something that every child alive today takes for granted. A a hundred years earlier, the books of Nicolaus Copernicus (1473-1543 AD) were banned for 70 years after his death for advocating a heliocentic theory. The idea that the Earth revolves around the Sun may have been heretical, but it certainly wasn't new: The Greek astronomer Aristarchus (310BC-230BC) presented a heliocentric theory in 270BC. It is fascinating to speculate how history might have been different if his ideas had triumphed back then.

Heresy & Inquisition

> *Dominic founded the monastic order of Dominicans and together they created what is now the infamous Inquisition.*
> Michael Baigent

The Roman Catholic Church commissioned Michelangelo's painting of the Sistine Chapel and countless other works of art. It is also responsible for wielding torture devices such as the breast ripper, Judas cradle, brazen bull, pear, head crusher, rack, iron maiden, and others. The Church sent thousands of people to their deaths tied to stakes, broken on wheels, sawed into pieces, garroted, drowned, boiled alive, and more, all to establish and maintain a religious monopoly and the set of beliefs and practices required to sustain that monopoly. The fact that the Church needed to resort to such stern methods even 1,000 years after it had triumphed over paganism speaks volumes about the natural human tendency to embrace diverse beliefs. No one knows how many primal religions have been irrevocably lost thanks to zealous missionaries and the many tortures at their disposal. The only crime any of these people committed was the heresy of not swallowing the Church's dogma hook, line, and sinker; what George Orwell called "thought crime" in *1984*. The fictional Thought Police were very real in the form of inquisitors, who stopped at nothing in their quest to preserve the church. Men, women, and children were all fair game. It does not help that the term "heretic" is very vague and can apply to just about anyone.

> *We must remember that the main purpose of the trial and execution is not to save the souls of the accused but to achieve the public good and put fear into others.*
> Anonymous Spanish inquisitor, 1578

Eastern religions say that humans can reach God on our own without divine assistance or salvation. Western religions stress the irreparable separation between humans and God caused by original sin. Only believing in God in the manner prescribed by the Church can save someone from eternal damnation. Heretics who did not adhere to this regimen would burn in hell forever. If God has damned them, why should humans treat them any better? And yet, as we will see, the Inquisition and all operations like it failed, because no amount of force has ever managed to create a single worldwide religion.

During the Inquisition, a suspected heretic was arrested after being denounced by someone enduring torture, an enemy, etc. One arrested, the heretic was tortured to extract a confession at leisure, because the Dominicans knew that everyone has a breaking point. Bloodletting was eschewed in favor of blunt and/or hot instruments, making any blood that did spill accidental. Once the victim confessed, teams of lawyers and clerks documented the confession, and the victim was asked if her or his confession was "free and spontaneous." Punishments were called penance. Those who confessed were welcomed back into the Church, albeit sometimes on the way to their deaths. Those who survived and relapsed were excommunicated and/or turned over to the state authorities. Those sentenced to death were then turned over to the secular authorities to perform the actual execution, allowing the Church to claim that it did not execute anybody—an argument that would certainly not hold water in a modern courtroom!

Any secular ruler who endeavors to prevent the familiars of the Holy Office from bearing arms is impeding the Inquisition and is a fautor of heresy.
Nicholas Eymerich

It is crucial to understand that the inquisitors were not criminals; they acted with the full knowledge and consent of the highest secular and religious authorities. Even the most venerated saints would not dare cross the line and risk being accused of heresy. To this day, many Christians, Catholic and Protestant alike, see anything that disagrees with their own views as evil. Sexual slander was (and remains) so common that one must wonder whether the good clergy are protesting too much, especially in light of a Pew study that found that 95% of Catholic dioceses have been touched by the child molestation scandals. *Auto-da-fés* (religious festivals during which heretics were executed) took place on Sundays. They were announced in advance, and all who attended were granted 40 fewer days in Purgatory.

We would gladly burn a hundred of just one among them were guilty.
Roland Barthes

The Roman Catholic Church resisted attempts to translate the Bible into local languages in order to protect people from reading the wrong materials or from reading the correct mate-

> *Ceaseless persecution continued to perpetuity is the only way to achieve a final solution to the problem of heresy.*
> Pope Innocent III

rials wrongly (in other words, to protect its monopoly on human knowledge). Of course, as we saw earlier, the texts being "protected" were not the originals, but copies of copies of copies that had been massaged over the centuries to more closely adhere to the Church's doctrines. Mistaking translation errors for prophesy was a common criticism. For example, we saw how a mis-translation of the word "young woman" became "virgin." Despite these problems, Church founders like Justin and Irenaeus chose to believe the Bible literally despite its farfetched and often contradictory stories; they opted for the pomp and circumstance of supernaturalism and transcendence, forgetting that the immanent can be far more sublime. They chose a figurative dog and pony show over the power and mystery of the mystical aspects of Christ's teachings.

Anselm of Lucca (1036-1086) said that Christ willed to die, making his crucifixion necessary, but that God did not demand it. God had to compensate His son, but Christ already had everything. God therefore gave His compensation to the world, in which He rejects no one who comes to Him in the name of Christ. What is telling is that Anselm does not say that God demands any specific approach. The philosopher Baruch Spinoza (1632-1677) excoriated the Inquisition, pointing out that the Bible is imperfect. To him, the word of God lies in the heart, not in the book, in the relationship between man and God, not in the formality of rules and rituals. In this light, we can start to see the Inquisition as having less to do with God's salvation and more to do with more pragmatic goals, such as revenue, ensuring compliance through fear, and revenge. Polytheists had to be wiped out to eliminate the danger they posed to monotheism, the irony being that the latter had to become dangerous themselves.

> *Innocent may boast of the two most signal triumphs over sense and humanity, the establishment of transubstantiation and the origin of the Inquisition.*
> Edward Gibbon

The Inquisition was obsessed with maintaining religious purity. Over time, it became just as obsessed with maintaining blood purity. This set the stage for later massacres, including the Holocaust, Rwanda, Darfur, and countless others. After all, if it's acceptable for the Church to kill and maim based on ethnicity and belief, then it must be just as acceptable for others to follow the same example. As the supply of living people became scarce, the Inquisition even resorted to exhuming and trying corpses and then burning those it convicted, sometimes decades or even centuries after the person's death. It also confiscated the surviving heirs' inheritances—a great way to keep going and raise money. Inquisitors were allowed to bear arms, even in places where doing so was forbidden to others. Torturers wore outfits very similar to the modern Ku Klux Klan and even marched in parades. Fattened by the riches pouring into Church coffers, priests lived lives of luxury, in direct violation of Christ's mandate to, "carry no purse, no bag, no sandals, and take what is given to you."

How could the Inquisition justify itself and its actions? Simple: The Old Testament contains many graphic tales of powerful armies smiting God's enemies down to the last woman and child. Never mind that none of these battles actually happened, that the Church did not have the archaeological evidence one way or the other, and that these tales justified its own violence. We can see the same self-segregation, intolerance, and religious control in today's Evangelical Christians. Even the Catholics have yet to apologize for the Inquisition, which remains alive and well today, as we will see later. So effective was the Inquisition that commerce, industry, science, and philosophy suffered in Italy, Spain, and southern France, where the Inquisition was the strongest. The same is not true of England, the Netherlands, and other places where the Inquisition was either weak or absent.

At a time when hunger and hard labor were the common fate of the peasantry and the urban poor, the men who held themselves up as the moral examples of Christendom resembled Herod more than Christ.
Jonathan Kirsh

The whole point of the Inquisition was to remain a going concern. Winning the "war on heresy" would render the Inquisition obsolete and put a lot of hardworking people out of work, not to mention stop a very lucrative revenue stream. The Church therefore had every reason to keep the Inquisition going as long as it could and to keep careful records to know whom to target.

Irenaeus

Irenaeus was one of the chief architects of orthodox Christianity. He was the man behind the four-Gospel canon (although he did consider the Gospel of John to be more "elevated" than the other three, because it claims that Christ is divine). He and his successors forced Catholics to adopt certain beliefs, such as the four Gospels and the idea that all future revelations would have to conform to those Gospels.

No one expects the Spanish inquisition.
Monty Python

Irenaeus eventually managed to convince people that his way of reading the Gospels was the only way.

Ptolemy 1 (90-168AD) spoke of levels of baptisms (initiation) into higher levels of knowledge, and of people being children—but not servants—of God. Irenaeus opposed this, saying that baptism was a one-stop journey: You were either in the Church or not. Despite this, he complained about outsiders who pointed out that people who joined the Church followed its doctrines like sheep while strutting around thinking they were full of God. "Those who reject ritual," he said, "are divided and could destroy the Church." One baptism was all that was required, and anyone who disagreed was a heretic, a fraud, and a liar. It was also heresy to think that our experiences are anything like God's experiences or to think that we can discover truths about God.

Even doubt was heresy. The believer must have fixed and unwavering faith, and it was the inquisitor's business to ascertain the condition of his mind.
Henry Charles Lea

According to Irenaeus, heretics refuse to acknowledge Christ's uniqueness despite the fact that Christ himself denied being unique. To Irenaeus, Christ's transcendence sets him apart from the rest of humanity. Those who claim he was born of normal parents lack gratitude for the word of God made incarnate. God created us in his image, but we were separated from God because of original sin. We may belong to God, but the devil corrupted and isolated us from God. This viewpoint gives the devil a lot of power, which makes God either unable or unwilling to put a stop to it all. Either way, this viewpoint does not strike me as particularly flattering to God.

When the existence of the Church is threatened, she is released from the commandments of morality. The use of every means is sanctified, even cunning, treachery, violence, simony, prison, death.
Bishop of Verden, 1411

When he said that the Jews departed from God because they did not receive his word, Irenaeus seemed to have forgotten both that Christ was himself a Jew who preached to Jews without any intention of starting a new religion, and about God himself speaking to Moses and others in human form. Nevertheless, Irenaeus said that God had abandoned the Jews, who prayed without answer. The Jews were of course guilty of murdering Christ, an allegation seemingly made without regard for the fact that an OOO god should be immune to such dangers, and that an entire religion depended on this supposed event for its very existence. The Eucharist commemorates Christ's sacrifice and draws God's power down to Earth, which alone offers access to God. The Church alone performs the Eucharist, ergo the Church alone is the pathway to God. The choice between orthodox and non-orthodox Christianity makes all the difference between spending eternity in heaven or in hell.

Irenaeus claimed that Christ is embedded throughout the Old Testament, speaking to figures like Moses and Ezekiel—a belief that few modern scholars would agree with. Some contemporaries also had problems with this and called Irenaeus simple and naive, to which he responded by threatening excommunication. Unswayed, his questioners responded that they no longer believed in a harsh, judging God, and that it was wrong to think of God in such petty anthropomorphic terms. And still Irenaeus persisted, along with Tertullian and others, answering all questions by insisting on simple blind faith. "What does Athens (reason) have to do with Jerusalem (faith)?" asked Tertullian. Christ had spoken, and all must listen and obey, period. Faced with contradictions in the New Testament books, Origen responded that John represented spiritual truth if not literal truth. The Holy Spirit inspired contradictions to startle readers into asking about the meanings of the conflicted passages—a far more convoluted answer than the more probable answer involving many people writing and changing many books without any idea that their writings would eventually merge into a single tome.

I was torn in pieces by the devils that rack the brains of unhappy men. Do God's eyes not reach to the prisons of the Inquisition?
Carcel

Timeline

Here is a quick look at some of the key events before, during, and after the Inquisition.

Many Greek philosophers had views that were very similar to the Eastern religions. Pythagoras (c. 570-c.495BC) taught that one could liberate the soul to achieve harmony with the universe through purification rituals. Plato believed in an ultimate divine reality and the rational nature of the universe. Aristotle understood the importance of logical thinking and saw this as a way to understand the universe. He understood the nature and importance of religion and myth, and that people who joined mystery (gnostic) religions were simply required to open themselves up to experiences. Humans may be above plants and animals, but are still part of the universe. We must become pure and immortal through ritual. Wisdom is the highest virtue. Logic and mystical experiences are both essential. Philo said that we need to go beyond ourselves to get closer to the divine, a task made easier by solitude. We must

Power prefers spin to truth.
Michael Baigent

separate from the body and the irrational; only then can the Logos within pay true homage to the sole existence (the Godhead, or Ground). Communion with God means freeing the mind of clutter to get to pure reason and wisdom, the only state in which one can truly enjoy the company of God.

Fast forward to the 900s AD and the Bogomil sect in eastern Europe. The Bogomils (also called Bulgars for the region in which they lived, which later became Bulgaria) rejected Roman Catholic doctrines. They also rejected some sexuality and aspired to spiritual purity. This did not stop them from being subjected to the same sorts of sexual slander heaped upon those who disagreed with Rome. For example, "bugger" (anal sex) has survived in our vocabulary for over 1,000 years.

> *The inquisition boasted that over the course of 150 years it burned approximately 30,000 women-all innocent victims of a church-sanctioned pathological fantasy.*
> Michael Baigent

The Cathars lived in the Languedoc in the 1000s and 1100s, the same area where Christ and Mary Magdalene fled after their escape from Jerusalem. (See Chapter 6.) A Gnostic sect, they believed that all matter was the work of the devil, including the human body —a form of idealism. They rejected the notion of *transubstantiation* (the transformation of the Communion cracker and wine into Christ's flesh and blood), because that would mean having God in one's bowels where He would be expelled during that person's next visit to the latrine. Cathar heretics burned at Orleans in 1022 were accused of child murder, orgiastic sex, and even cannibalism—the very same crimes Romans had charged Christians with hundreds of years before. Of these, the cannibalism charge seems a little out of place, given the decidedly cannibalistic Eucharist. In reality, the Cathars believed that Christ was not born of a virgin, that he did not suffer and die for anyone's sins, that he was not buried in Jerusalem, and that he was not raised from the dead—beliefs very much in line with the idea of Christ living out his days in that very region.

> *[Women] are the devil's gateway: you are the unsealer of the [forbidden] tree. You are the first deserter of the divine law. On account of your desert, even the son of God had to die.*
> Tertullian

The Roman and Eastern Orthodox churches each claimed to be the one true Christian faith. In 1054, they excommunicated each other in the Great Schism, which lasted 1,009 years until its rescission in 1965. These antics may be one reason why the English scholar Adelard of Bath (c. 1080-c. 1152) responded to Christian questioning by saying, "It is difficult for me to talk with you about animals, for I have learned from my Arabian masters under the guidance of reason; you, however, captivated by the appearance of authority, follow your halter.

Since what else should authority be called but a halter? For just as brutes are led where one wills by a halter, so the authority of past writers leads not a few of you into danger, held and bound as you are by bestial credulity. Consequently some, usurping to themselves the name of authority, have used excessive license in writing, so that they have not hesitated to teach bestial men falsehood in place of truth. Wherefore, if you want to hear anything more from me, give and take reasons, for I am not the sort of man that can be fed on a picture of a beefsteak." Adelard was responsible for many advances, including introducing the Indian number system to Europe, translating many important Greek and Arabic texts, and other scientific inquiries.

Heresy gives birth continually to a monstrous brood that passes on to others the canker of its own madness.
Pope Innocent III

Pope Gregory IX (c. 1145-1241) appointed full-time Inquisitors from the Dominican and Franciscan orders. They had no power to punish, but handed people to the secular authorities. It is no small irony that those orders that pledged to live like Christ became the biggest persecutors of other Christians. Innocent III (c. 1160-1216) added the trappings of law and theology to the Inquisition. He convened the 4th Lateran Council on November 1st, 1215 that started a crusade and laid the framework for the Inquisition. To Catholics, the Crusades and the Inquisition were morally equivalent; the former fought foreign enemies, the latter domestic. In 1245, Innocent IV (c.1195-1254) gave inquisitors the power to absolve their assistants for all violence perpetrated in the line of duty. Inquisitors were immune from excommunication for life, absolved of sin upon death, and had their time in Purgatory reduced by the length of their service. In 1252, he authorized torture in the papal bull *Ad extirpanda*. Another bull in 1260 compelled the loyalty of the Franciscans and Dominicans. The condemnations of 1277 stated that one could be with the Church or with reason, but not both. Meanwhile, complaints by critics about priests' extravagant and deviant lifestyles only helped fuel the Inquisition; complaining about the Church eventually became heretical. King Philip IV of France (1268-1314) initially opposed the Inquisition, but changed his mind when he saw the revenue potential. Philip was also responsible for bringing down the Knights Templar.

Christendom seems to have grown delirious and Satan might well smile at the tribute to his power in the endless smoke of the holocaust which more witness to the triumph of the Almighty.
Henry Charles Lea

Witch hunts began in 1484 when Innocent VIII issued *Summis desiderantes*, a bull that rescinded the *Canon Episcopi* and

> *Because of the crime which once their fathers committed against our Lord Jesus Christ, the Jews are deprived of the protection of their natural rights and condemned to eternal misery for their sins.*
> German Jewry Law, 1268

extended the Inquisition to the correction, imprisonment, and punishment of witches. This bull equated witchcraft with heresy. Those accused of witchcraft had no legal rights. Someone making a false secular accusation could be drowned, but no such rules of evidence applied to those who defied Church law. Luther believed in witches and saw Christians locked in a battle against Satan. He doubted one could prove the existence of God; faith thus required not research but free surrender. Witches flew at night to worship the devil in the same sorts of ceremonies Rome accused Christians of holding, which involved kissing the anus or penis, sexual orgies, eating children, etc. Captured witches were purified in death to be reborn as Christians. Author Anne Llewellyn Barstow believes that the witch hunts were actually "woman hunts" driven by misogyny, a reasonable hypothesis given the Church's attitude toward women. (See Chapter 5.) For example, a single Swiss canon logged 3,371 "witches" from 1591 to 1680, all of whom were executed.

Torquemada (1420-1498) was appointed Grand Inquisitor in 1482. A Dominican who had taken a vow of poverty, Torquemada had both 50 cavalry and 100 infantry bodyguards and a palace. The man responsible for the suffering and deaths of thousands of people was himself afraid of getting killed. He lived a long life and died in his own bedroom of old age.

Credit was critical to commerce in Europe. Christians were forbidden from lending the money, a crucial role that Jews filled. The Christian persecution of Jews disrupted trade and commerce across the continent. Torquemada had Jewish blood; he therefore needed a way in which Jews could escape persecution. The term *conversos* describes those who converted from Judaism or Islam to Catholicism. Conversos could marry up and soon came to compete with the older Christian gentry.

> *The two world wars were not fought on religious grounds at all.*
> Keith Ward

As awful and far-reaching as the Inquisition was, it could neither stifle all dissent nor stop a split within its own ranks. On October 31st, 1517, Luther nailed his 95 theses on the door of the castle church in Wittenberg. In response, the Council of Trent (which occurred in three sessions between 1545 and 1563) reaffirmed the validity of indulgences, purgatory, and the veneration of saints. Some saw this as undermining the pope, but things actually worked out to his benefit, thanks to the refreshed belief that a strong Vatican authority was the best way to protect the Church. Pope Paul III (1468-1549) renamed the Inquisition as the Congregation of the Holy Office and started a list of banned books.

Meanwhile, the expeditions of Columbus and other explorations were encountering "heathen" people whose religions featured many of the same rites (marriages, funerals, worship, etc.) as Christians. There was no easy way to explain such discoveries, but the Catholics pressed on anyway by taking the Inquisition with them to the newly discovered lands. The Wars of Religion (1500-1648) were raging to determine whether Protestants or Roman Catholics would dominate Europe. The Protestants lacked a strong central authority and split into increasing numbers of denominations. The Presbyterians were founded on the (very Gnostic-sounding) idea that congregations, not popes, were responsible for matters of faith. This combination of circumstances led increasing numbers of Europeans to feel that religion had been discredited; many people were suffering and dying for beliefs that could not be proven either true or false.

The Inquisition opposed science. Galileo was condemned as a heretic. It was only in 2008 that pope Benedict XVI proposed to complete Galileo's "rehabilitation" by building a statue of him inside the Vatican. Still, science progressed. Philosopher René Descartes (1596-1650) believed that the mind could find God and that the intellect could provide certainty—a viewpoint I happen to share and that is driving the creation of this book. According to Descartes, one could not see design nor deduce first principles from nature, but evidence for God could be found in human consciousness. Spinoza saw God as the sum total of natural laws, for which he was expelled from his synagogue with a long curse read against him. To Spinoza, God did not corresponded to our normal understanding of the word—precisely what mystics have been saying all along.

Sir Isaac Newton (1643-1727) placed God within the mathematical and *deterministic* (caused by natural laws) system he created that came to be known as classical physics. People who deny God today are often denying Newton's concept of God as the origin and sustainer of the universe. Newton examined the concept of the Trinity and declared it a sham perpetrated by Athanasius to get convert pagans by wrapping monotheism in a package more palatable to polytheists. Portions of the

Hence today I believe that I am acting in accordance with the will of the almighty Creator: by defending myself against the Jew, I am fighting for the work of the Lord.
Adolph Hitler

With its image improved and its name twice changed, the Inquisition still exists and functions today, the heir to a tradition of over seven hundred years.
Edward Burman

New Testament used to affirm the Trinity he deemed forgeries designed to appeal to base human instincts. Voltaire (1694-1778) openly defied the Church. He had no problem with the idea of God; in fact, he saw atheists as superstitious fanatics. His problem lay with the doctrines that insulted human intelligence. Immanuel Kant (1724-1804) felt that postulating an OOO creator could make science lazy and allow *deus ex machina* ("God in the machine," where a seemingly intractable problem is suddenly solved via contrived intervention). To him, the traditional "proofs" of God proved nothing.

> *The dead hand of the Holy Office was pressing slowly on the arteries of Spanish intellectual life.*
> Cecil Roth

Diderot (1713-1784) went one step further, saying that nothing but matter exists, with the laws of matter giving the illusion of design. Religion was the "cradle of ignorance and fear" that a mature, enlightened man could rise above. Jean Meslier (1664-1729) was a priest but died an atheist, leaving behind a scathing rebuke of the Church in his memoirs. He was a materialist who believed that religion only existed to oppress the poor and strip them of their pride.

The term "atheist," long an insult, was starting to be seen as a badge of honor as the rising tide of science and reason eroded the monopoly the Church had tried so hard to establish. The Inquisition fizzled out during the 1700s, with the last burning at the stake taking place in Portugal in 1760. Napoleon formally decreed the Inquisition over in 1808. In the newly formed United States of America, the Constitution took the lessons learned from centuries of religious oppression to heart by granting freedom of religion while also explicitly separating church and state. The Industrial Revolution was changing how people saw themselves, their ideas about God, and their relationship to God. Civilization depended on everyone from the lowliest worker to the highest elite. Mass literacy was essential for economic progress. People demanded the right to share in the fruits of their labors, and the social power of the landed nobility was eclipsed by the *bourgeoisie* (middle class). Men and women felt empowered to take charge of their own lives like never before. Each generation expected their children to do better than themselves. The myth of religion was at least partially replaced by the myth of progress, one that has offered many blessings to humanity, albeit at great environmental cost.

> *The Vatican has long been forced to maintain its position through suppression and manipulated expression.*
> Burton Mack

The French Revolution at the end of the 1700s nationalized church property and placed the Church under the French government, showing people for the first time that they could be completely free of religion. Seeing these existential threats to its existence, the Church responded by seeking to make the Vatican a nation in its own right, a desire granted by the Lateran Treaty signed by Mussolini in 1929. Still, the popes yearned for power. The First Vatican Council that started in 1868 declared the pope infallible. In 1907, Pius X condemned modernism—a futile gesture since the Industrial Age was by that time far too entrenched to stop or reverse.

Schoolteacher Cayetano Ripoll bears the dubious honor of being the Inquisition's last known victim. He was garroted on July 26th, 1826 because his executioners could not bear to see him burned alive. They reserved that dishonor for the corpse, which was placed in a barrel painted with flames and set ablaze, the remains buried in unconsecrated ground.

The popes of Rome took it upon themselves to ritually anoint the emperors into their exalted office as part of the ceremony of coronation, as if a pope should have the power to create a messiah. As if they alone had a monopoly over the pathway to truth.
Michael Baigent

In 1869, the Church did admit its error in the burning of Joan of Arc (born 1409) in 1431, a mere 438 years after the fact. She was *canonized* (declared a saint) in 1920. Galileo is still waiting, despite the incontrovertible proof of the correctness of his discoveries.

The Inquisition still exists. In 1965, Paul VI renamed it the Congregation for the Doctrine of the Faith, charging that office with promoting and protecting the faith. Cardinal Ratzinger (now pope Benedict XVI) led this organization's secret efforts to condemn those who go against Catholic dogma and ensure conformity of beliefs. The torture chamber and executions may be gone (or not, as some authors attest), but the Church refuses to apologize for its actions or even acknowledge that they could, perhaps, have gone just a little overboard.

Cruel Legacies

The Inquisition had little to do with belief in God. It had much to do with faith in the Bible as the word of God and with the Church as the sole valid interpreter of those words. This sad truth is even worse because we saw some of the evidence of the Bible being altered to suit the Church, and thus being anything but "God's word." To put it simply, the Church fabricated an elaborate set of lies and then rammed those lies

The goal is to go into every church whether they like us or not.
Bill McCartney

down the society's throat for almost 2,000 years. The Protestants are not immune from this; their problem was not with the (altered) Bible, but with how the Catholics interpreted it. They too conducted an Inquisition, albeit on a much smaller scale than their Catholic counterparts.

Giordano Bruno (1548-1600) said that the Church played the same social and practical roles as the state. Dissension and strife were bad for the Church, which therefore required doctrine and the authority to enforce conformity and outward acceptance; but the Church had no right to interfere with the pursuit of knowledge, truth, and science. Bruno was burned at the stake on February 16th, 1600. Four hundred years later, Cardinal Angelo Soldano (1927-) called this a "sad episode," but also said the Inquisitors, "had the desire to preserve freedom and promote the common good and did everything possible to save his life." Cardinal Soldano seems to miss the point that not persecuting Bruno in the first place would have been an excellent way to save his life. The continuing economic superiority of northern Europe compared to southern Europe could be a legacy of the Inquisition.

Demonizing one's victim is critical to the ability to commit genocide or any kind of mass murder. It is one thing to kill a fellow human being. It is quite another to kill a traitor to God (Cathar) or the Fatherland (Russian under Stalin), etc. Both Hitler and Innocent III refer to their enemies as "filth."

The Church has never apologized for the Inquisition or shown any remorse, except for a handful of cases, such as Galileo and Joan of Arc. On the contrary, Eugenio Pacelli (Pius XII, 1876-1958) declared that, "the Church has never treated the doctrines of the pagans with contempt and disdain; rather, she has freed them from all error, then completed them and crowned them with Christian wisdom." Pius XII is also infamous for making peace with the Nazis and turning a blind eye to the atrocities they committed. The Catholics have never excommunicated a single Nazi. The arrogance and hubris of the Church continues to beggar the imagination.

The methods pioneered by the Inquisition can be seen in Hitler's Germany, Stalinist Russia, Kim's North Korea, and other places. Hitler drew inspiration from the Inquisition, comparing himself to ancient prophets and using ideas about the

Much of the appeal of today's popular atheists—from Richard Dawkins to Sam Harris—lies in the corruption of religion.
Brian McLaren

It is putting a very high price on one's conjectures to have a man roasted alive because of them.
Michel de Montaigne

purity of blood taken from the Inquisition's playbook. These and other examples of "secular" religions have created elaborate myths, such as the tale of Kim Jong Il's miraculous birth on the slopes of Mt. Paektu in North Korea—a far more august tale than his real birth in a Soviet rail carriage.

Ralph Waldo Emerson said it best when said, "I like the silent church before the service begins better than any preaching." Me too.

To such heights of evil are men driven by religion.
Lucretius

Chapter 9

The Biology of Religion

If you weren't a little paranoid, you wouldn't survive a minute.
Judith Hooper &
Dick Teresi

Anthropologist Anthony Wallace (1923-) estimated that humans have invented some 100,000 religions over the last 10,000 years. As we saw in Chapter 2, the majority of the world's population identifies itself as religious. The ubiquity of religion begs the question: Is religion in our genes? In *The Natural Savage*, I explain how all of life boils down to the six core functions of predator avoidance, group status, food, shelter, reproduction, and death. Religion can affect all six areas, because humans are hierarchical social animals. If the majority of those closest to you have subscribed to a certain belief set that affects how one fits into and enjoys the benefits of society, then you will have powerful incentive to mold your own beliefs to fit in. Genes thus predispose us to belief, but do not tell us what to believe. For example, subscribing to a belief system that takes Saturdays off and eschews pork is a choice; the openness to subscribe to a belief system that may or may not include these features is genetic. Religious rituals span the gamut from the prosaic to the truly odd.

Why do religions exist? The simplest answer is that the process of evolution tried other systems that ended up being blind alleys. Religion exists because it works; given that our minds are well adapted to live in hierarchical societies for protection from predators, one could even say that the emergence of religion was both predictable and expected. Some theories of religion say that gods serve as über-alpha leaders. Death could be seen as the über-predator that no one can escape; in this context, the protection could consist of assuring people that it really isn't all that bad, because the über-alpha will protect us after we have departed this life. This plus the fact that we remember the dead and can vividly imagine them in life whether awake or dreaming could be the source of afterlife myths. From this, it is only a small leap to the concept that the über-alpha will only grant a pleasant afterlife to those who obey the rules by adopting the correct beliefs. This may demonstrate that we are genetically predisposed to spirituality.

The genetic predisposition to spirituality evolved for a purpose. That purpose may have otherworldly origins that myth and religion do their best to explain. In other words, we may be able to detect that each of us exists beyond this lifetime. On the other hand, all notions of some personal existence continuing after this life could be wishful thinking that does nothing beyond simply assuaging our natural fear of death. There is nothing intrinsically theistic or atheistic about the genetics of religion. The so-called "God spot" in the brain poses intriguing questions about why it exists and what practical purpose it serves. The question of whether God exists may lie beyond the ability of science to answer, but the question of why we believe in God may be well within our reach.

If you have a good neocortex working for you, you try to explain your paranoid feelings and persuade other people. You go out and start a religion and recruit followers. Some paranoids are very persuasive.
Paul MacLean

Legitimacy

Different religions take very different approaches to science. Western religions continue to expend vast amounts of time and energy trying to disprove science or bending it to their beliefs; the Creation Museum in Petersburg Kentucky that shows people romping with dinosaurs is just one example. Eastern religions tend to be more accepting of science; the Dalai Lama has stated that any tenet of Buddhism that conflicts with science must be altered to match science. We will leave the question of just how well Buddhism agrees with science for later in this book, although Chapter 5 should already have provided some tantalizing clues. Either way, the attitude of religion toward science says nothing about whether or not it is reasonable and proper for science to test religion.

Most religions do more than simple moralizing but make basic pronouncements about nature, which science is free to evaluate
Victor Stenger

Feeling is the deeper source of religion.
William James

While consciousness lies in the no man's land between religion and science, claimed by both yet understood by neither, it ay also hold a key to the apparent conflict between these two great human institutions.
B. Allan Wallace

Is it appropriate to discuss religion as a biological phenomenon? Biologist Stephen Jay Gould (1941-2002) asserted that science and religion are "separate and non-overlapping magisteria" that had nothing to contribute to each other. Scientific "proofs" of God are seen as an insult to science and scientists. To religious people, such proofs are seen as an insult to God. Scientists such as ethologist Richard Dawkins (1941-) point out that religion makes testable claims about how the universe works and how God interacts with the material world, and it is therefore well within the right of science to test those religious claims. I happen to agree with Dawkins on this point, hence the book you are now holding in your hands.

Can science prove the existence or nonexistence of God? That depends on who you ask and what you accept as valid science. It is true that science has done much to discredit religion; for example, proving that the Earth orbits the Sun and that we are not at the center of the universe utterly destroyed religion's claim of infallibility. Other advancements have continued chipping away at religious credibility, which goes a long way toward explaining the religious antipathy for science and reason that originated with the likes of Irenaeus and Tertullian (see Chapter 8) and continues today as religious people seek to hobble science education around the world. Science tends to be *reductionist*, meaning that it works by breaking systems down into comprehensible chunks that religion attempts to dispute or bend to match their beliefs. In religion's defense, while the reductionist approach is responsible for untold numbers of scientific and technological advancements, it can miss the big picture by providing only a partial view of reality.

Religion begins with the idea that God is the ultimate creative force. Science turns this assertion around by saying that God is a mental phenomenon that occurred as the result of evolution. Science says that the Big Bang (or some variant thereof) started it all. Physicist Alan Guth (1947-) says that the fundamental laws of nature could imply a universe. Religion believes that God is where these laws of nature came from and what (or who) caused the Big Bang. Which view is correct? If the history of the universe was rewound to the very beginning and replayed, would we still be here, or would random events ensure a different outcome every time? We will explore these questions later. Meanwhile, science has replaced religion as the primary arbiter of human wisdom over the last 400 years or so. The mere fact that you are reading this book is a testament to the triumph of reason driven by the ceaseless quest for knowledge that was unleashed by breaking the religious monopoly on thought. Religion has consistently been on the losing end of its many encounters with science. This may be why the Catholic Church finally reversed itself and admitted that people can have both faith and reason.

One common lament is that science has made the world cold and meaningless. This accusation is aided and abetted by statements such as Weinberg's that, "The more the universe seems comprehensible, the more it also seems pointless." or Dawkins's belief that religion is, "a virus of the mind, a parasitic bunch of myth and lies that serve no purpose at all." It is certainly true that the primary reason people believe in God is because the universe seems too perfect to be the result of anything but design (the so-called "watchmaker's argument," which we will revisit in Chapter 20). I know this feeling; as I look at the natural world or gaze into space, I am often tempted to ascribe the universe to some divine or higher intelligence. Is there some truth to this feeling, or is it all nothing more than wishful thinking driven by genes that evolved by chance to deal with very different questions?

So far as I can remember, there is not one word in the Gospels in praise of intelligence.
Bertrand Russell

Genetics

As I said in Chapter 2, I accept evolution as a fact that has been proven beyond all reasonable doubt. I credit the human species with some amazing accomplishments in our short history, but even the loftiest heights of our achievements pale next to what simple changes over billions of years have wrought. We will discuss evolution at much greater length in Chapters 18-20. In the meantime, I will return to the question I posed at the beginning of this chapter: Is religion in our genes? We have seen how different brain circuits may be responsible for some or all religious experiences across all cultures. For example, birds sing species-specific songs because they are birds of that species, regardless of their upbringing. Their cultural surroundings (the birds they grow up with) influence the songs and produce regional differences, but the bird's ability to sing those songs is a result of genetics. This is

We do not know of societies in which bravery is despised and cowardice held up to honor, in which generosity is considered a vice and ingratitude a virtue.
Solomon Asch

just one of the many examples of the complex interplay between nature and nurture. Spirituality and religion may work in much the same way.

Drat! These pesky scientific facts won't line up with my beliefs.
Gary Trudeau (Doonesbury)

It may surprise you to know that Adam and Eve really existed about 140,000 years ago in Africa. All human Y chromosomes today can be traced back to this Adam, who is our common patrilineal ancestor. Likewise, all human *mitochondria* (cell organelles that help with energy production) go back to this single Eve, who is our common matrilineal ancestor. Adam and Eve were not the first humans; they are simply the only humans whose genetic material has passed into every generation since them. They were probably neither married nor related.

Let us assume that religion springs from a mix of genetically based mental circuits that originally evolved for other, more prosaic purposes. First of all, most people marry others of the same faith, thus passing on and possibly strengthening the genes responsible for religious predisposition. The human DNA sequence contains 3 billion (3,000,000,000) base pairs of genetic instructions. The genetic variance between people is about 1 in every 1,000 pairs, meaning that two average people have 3 million (3,000,000) different genetic differences. By contrast, a chimpanzee has about 45 million (45,000,000) genetic differences compared to a human. This means that the difference between a person and a chimpanzee is about 15 times greater than the differences between two people. Identical twins have 0 genetic differences between them; it is no coincidence that twins raised apart are more likely to share religions than non-related people.

Death anxiety is the mother of all religions, which, in one way or another, attempt to temper the anguish of our finitude.
Irvin D. Yalom

How do these genes benefit us? They may give us a sense of optimism that helps us persevere in the face of adversity and have more children by giving us an inspirational sense of purpose and meaning. They may assuage the fear of our inevitable deaths while also allowing us to flee or fight when necessary. There is evidence that faith can have a healing and calming effect, which promotes health. People of a given faith tend to stick together, enhancing group cohesion and making the group better able to compete for resources, guard against predators, and protect against diseases through such mechanisms as kosher and halal food laws and dietary restrictions. The "supernatural" aspects of religion may be by-products of the same evolved pattern-seeking, cause and effect, predator avoiding, and hierarchical instincts discussed above. Social bonding may be another key benefit; biologists tend to place too much emphasis on competition while overlooking the cooperation that is essential to the survival of many species. For example, humans could not survive without a host of cooperative bacteria in our digestive tracts; our bodies may host thousands of species of bacteria.

Genes seem to influence spirituality about as much as they influence most other traits, and may be stronger than the genes for some physical characteristics. Some genes are associated with a sense of *self-transcendence* (going beyond the ego during meditation or RSME). The VMAT2 gene is associated with spirituality. People with a C base pair in this gene score higher for spirituality than people with an A base pair in the same location. VMAT2 seems to be essential for survival; mice born without this gene were runts that died early. VMAT2 also influences dopamine levels in our brains. Dopamine makes people happier, confident, and more optimistic. Someone who feels good about themselves and what the future may bring may be more likely to search for food and take better care of themselves and those around them.

You already know that there is not much difference in the number of genes between humans and primates. What you may not know is that the *Caenorhabditis elegans* worm (a species with a grand total of 969 cells, of which 302 are brain cells) has 24,000 genes. The far more complex fruit fly *Drosophila melanogaster* has only 15,000 genes. A human body has over 50 trillion (50,000,000,000,000) cells and about 25,500 genes—only 1,500 more than the worm. Rodents have the same number of genes as humans; of mice and men, indeed!

Belief can be explained in much the same way that cancer can.
Daniel C. Dennett

Science now reveals that love is addictive, truth is gratifying and cooperation feels good. Evolution produced this reward system because it increased the survival of members of our social primate species.
Michael Shermer

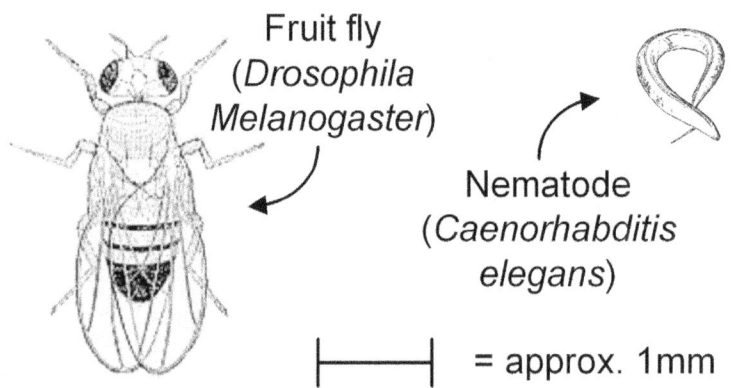

Fruit fly
(*Drosophila Melanogaster*)

Nematode
(*Caenorhabditis elegans*)

⊢――⊣ = approx. 1mm

> *Religious belief itself is an adaptation that evolved because we're hardwired to form tribalistic religions..*
> Edward O. Wilson

The great versatility of proteins to serve multiple purposes means that fewer genes are required, just like 26 letters can type every one of the millions of words in the English language. Genes can also be transferred between species, which means that new species can "learn" from older species. The walls separating the different species seem to be far more porous than previously thought, and all organisms seem to be connected in ways that remain mysterious. It may even be valid to say that we cannot define just what a species is. This new information flies in the face of traditional reductionist ideas about linear information flow. Genetic information seems to flow not in a straight line, but across a diffuse web. One could perhaps say that reductionism is to science what Newtonian physics is to quantum physics.

Religion in the Brain

> *People are disturbed not by things, but the view they take of them.*
> Epictetus

The following diagram displays the major external areas of the *cerebrum* (forebrain) as seen from the left side. In addition, the brain is separated into left and right halves that are visible from the front, top, and back. These halves are partially separated by a fold called the *falx*. Another fold, the *tentorium cerebelli*, separates the occipital and temporal lobes from the *cerebellum* (hindbrain) that appears as crosshatched lines at the bottom right. The *medulla* (brainstem) is at the bottom center. Other internal structures are located in the center of the brain between the left and right temporal lobes.

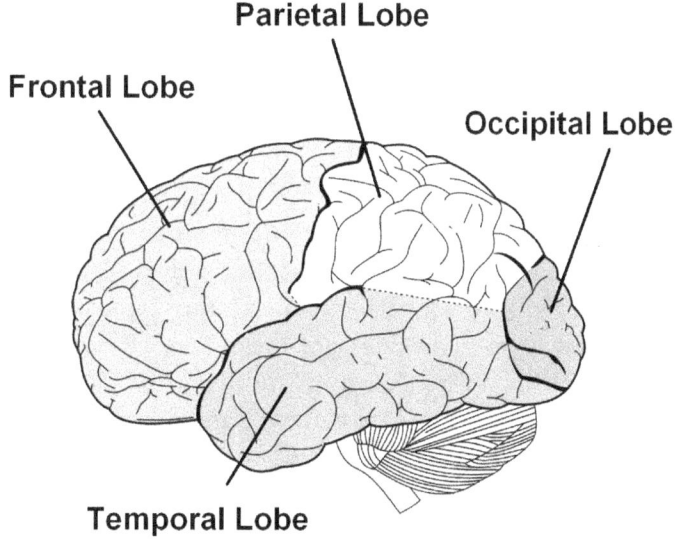

Evolution has just as much to do with behavior as it does with genetics. For example, your decision to have a child or not is a behavior that will quite literally ripple across the entire human gene pool. Our brains are designed to identify relevant information and filter out the rest in accordance with our core beliefs that prepare us for both certain concepts and a range of variations on those concepts. (See *The Enlightened Savage*.) As we have seen throughout the last several chapters, we are open to religious concepts that can be acquired and passed on like any other form of knowledge. Again, it is important to separate a predisposition for something from actually having that thing. Someone raised in a non-musical environment might be genetically predisposed to be a brilliant pianist, but that does not mean that s/he will ever touch a piano. Likewise, a normal human brain has the ability to acquire religion, which does not mean that it will do so.

Chimpanzees and gorillas have language, culture, and consciousness. They have everything needed to form conscious beliefs, as evidenced by the response, "Fine animal, gorilla!" given by the research gorilla Koko to someone who asked whether she was human or animal. I am not aware of any studies about primate spirituality, which is a shame because that exploration could reveal much about our own religiosity.

Generalist species adapt more easily to their surroundings than specialist species. Humans are generalists; we live in every climate zone on Earth and eat just about everything that is not downright poisonous. This requires a very advanced brain with myriad dedicated circuits. For example, we have mental circuits that help us recognize familiar people by face and others that help us with voice recognition. Victims of certain types of strokes and other brain injuries lose the ability to recognize even close friends and family on sight or by hearing their voice. Think about your closest friends and loved ones, and imagine not being able to recognize them instantly by seeing them or by hearing a snippet of their voices. A "God circuit" may have survival value by motivating us to persevere during tough times. This does not mean that we have special circuits in our brain for religion any more than we do for reading. Literacy uses existing brain circuits, and it is probable that the same holds true for religion. Biologist Gerald Edelman believes that consciousness arises from communications

---

Our brains are no longer conditioned for reverence and awe. We cannot imagine a Second Coming that would not be cut down to size by the televised evening news, or a Last Judgment not subject to pages of holier-than-thou second-guessing in The New York Review of Books.
John Updike

---

between these circuits within the human triune brain that consists of the R-complex, limbic system, and neocortex (see *The Natural Savage*).

A man paints with his brains and not with his hands.
Michelangelo Buonarroti

Meditation activates the front parts of the brain and shuts off the back of the brain. Edelman believes that consciousness resides in the thalamocortical network of *axons* (signal-carrying appendages of neurons) that connect the *thalamus* (primary way station for most sensory data from the body except smell) and the *neocortex* (area of sensory processing and integration). This same area deals with attention association and displays increased activity in meditating monks. Clearing one's mind of thoughts and emotions requires such a strong level of concentration that other areas display reduced activity, like too many appliances connected to the same power supply. Decreasing sensory consciousness connects them to the universe simply by redirecting a few nanovolts of electricity.

This is a purely materialist explanation. Dualists believe that physical and mental processes are entirely separate (which does not necessarily imply God), while idealists believe that only mental processes are truly real (more on this later); however, any way you look at it, consciousness is selective and continuous. My consciousness belongs to me alone, yours to you alone, and nobody can share her or his consciousness with someone else. A certain level of "core" consciousness is required to support higher consciousness, but the opposite is not true. Higher levels of consciousness can process abstract thoughts, which may be where spirituality comes in.

To an evolutionary psychologist, the universal extravagance of religious rituals, with their costs in time, resources, pain and privation, should suggest as vividly as a mandrill's bottom that religion may be adaptive.
Marek Kohn

A human brain has about 100 billion (100,000,000,000) neurons. Each neuron has an average of 7,000 connections called *synapses* for a total of 700 trillion (700,000,000,000,000) connections. Each synapse can be in one of several states, which means that a brain has the potential for more configurations than there are subatomic particles in the entire universe. Not bad for a 3-pound mass with the consistency of a nice flan custard! The downside is that modern civilization has evolved far too fast for our brains to keep up. Our mental circuits are still wandering the Pleistocene African savannahs, a fact that sometimes causes problems. For example, our taste for fat, salt, and sweet developed because these items are both rare and beneficial in the wild; the development of processed food and pervasive marketing caught us off guard and contributed to the growing obesity epidemic. The same may be true of religion.

The brain is divided into left and right hemispheres, hence the expressions "left-brained" and "right-brained." The left hemisphere helps us define the limits or our bodies, while the right hemisphere helps us locate our bodies in space. The left hemisphere also maintains our sense of self. A disorganized state, such as a seizure or *transient ischemic attack* (TIA, a temporary stroke), may cause us to sense another presence that some people interpret as God. Events that involve the *amygdala* (nodes in the middle of the brain between the temporal lobes) that correlate emotions with memories) are particularly emotional. The brain also has both reductive and holistic areas that help us make mental connections between different events, memories, thoughts, etc. Spindle cells within the brain appear to be linked to the ability to form a sense of morality. These cells are only found in primates and human and take years to develop, which is why children less than 4 years old tend to neither play fair nor follow rules.

The general idea of memes has been a seductive one; people want to believe it.
William L. Benzon

If the human brain is great at one thing, it is organizing information into patterns, especially around cause and effect. Our ability to find these patterns evolved because it helped us avoid predators and other dangers. Unfortunately, this same ability also convinces some people that they can see shapes in clouds and images of Christ in their morning toast. God may be both an über-alpha leader and an über-pattern. Our pattern-detecting circuits seem to be designed to overreact; after all, seeing a nonexistent predator is much better than not seeing an actual predator. On the other hand, we need to be able to quickly dismiss false alarms lest we spend our days in constant panic. Assuming that "supernatural" entities like ghosts, angels, and God are not real, how come they are so persistent across all spectrums of society?

Dreams may be nothing more than random circuits activating with some pattern-seeking thrown in to create a cohesive story line, however surreal that may be. We do have a built-in sense of what is real and what is not, which we use (among other things) to distinguish waking reality from dreaming. The VMAT2 gene, which we will see more of later in this chapter, can alter our perceptions of what is real. Neurologically speaking, however, our ideas about truth and fantasy are little more

Could it be that Dawkins' mother, while carrying him, was frightened by an Anglican clergyman on the rampage?
Alvin Plantinga

> *I don't think there's anything unique about human intelligence. All the neurons in the brain that make up perceptions and emotions operate in a binary fashion.*
> Bill Gates

than subjective values created by our brains. We designate some things as "real" and others as "fantasy" based on our beliefs and other factors, but there is little difference between our waking lives and our dream worlds as far as most of our brain is concerned. The methods we use to determine reality do not mean something is real. Scanning the brain of a deeply religious person tells us about their brain state; it says nothing whatsoever about the existence of God. Decreasing activity in the parietal lobes elicits a sense of selflessness. Nuns, monks, and other deeply spiritual people do this and describe the types of timeless, selfless feelings associated with mysticism. It is interesting that the sense of self vanishes instead of being enhanced during these experiences.

Children between 3 and 10 years old often have imaginary friends with whom they engage in relationships that are *decoupled* (removed) from ordinary reality. This does not mean that they are confused about reality versus fantasy or that they are engaged in wishful thinking. In fact, children with imaginary friends are often better at various tasks than their friendless counterparts, especially where interaction is required.

The neurologist V.S. Ramachandran (1951-) believes that brain research indicates that God and religion are caused by neurological factors, such as temporal lobe activity. He connected epileptics to an *electroencephalogram* (EEG, or brain wave monitor) and showed them various scenes. Their brain wave activity spiked when shown religious images. Something about God or the idea of God literally changed something inside their heads. Michael Persinger (1945-), another neurologist, stimulated subjects' temporal and parietal lobes, and discovered the brain's so-called "God spot," which is actually several different areas in the brain. Pentecostal Christians displayed increased parietal activity, during which they did not lose themselves but felt that God was moving through them to convey messages, and/or give divine guidance. This is the opposite of the selfless and timeless sensations experienced by monks and others with decreased parietal activity.

> *Lord, if we are deceived, it is by thee.*
> Augustine

All test subjects sensed changes when their mental circuits fired differently than normal. A religious or spiritual person experiencing such a change outside a laboratory could easily be convinced that s/he has touched another world. A staunch atheist might dismiss the event as abnormal but harmless, and promptly forget all about it. Those who experience a sudden conversion to religious faith also experience a simultaneous biological change in their brains. Altered activity in the frontal lobes, parietal lobes, and other areas that process language and emotions are associated with the temporary cessation of normal cause and effect, which causes the believer to interpret her or his experience within the context of their religion. Monoamines such as serotonin and dopamine play key roles in maintaining consciousness and regulating the brain's reward/pleasure systems. They may be at least partially responsible for associating experiences with emotions. A joyful religious conversion experience can therefore mark a profound shift in a person's life. A person's expectations can also alter serotonin levels.

Famous religious people who experience visions may in fact be suffering from temporal lobe epilepsy. Anecdotal evidence supports this conjecture based on the descriptions of the visions and what happened to the person while s/he was seeing them. Epileptics tend to be philosophical between seizures, and are more likely than average to be highly religious, even zealous. This is not to say that all religious people are epileptics, because brain circuits alone cannot account for all religion. It could be that some religious leaders have epilepsy, with their congregations going along for the ride. Religion essentially consists of codified human instincts combined with plausible-sounding stories and patterns that give people a chance to create strong social and hierarchical bonds. The existence of altered mental states that open people up to these experiences is all that is required to get the ball rolling.

Some people who come very close to dying report visiting a world beyond this one. These experiences, called NDEs, are reported in every culture on Earth. Some scientists believe that NDEs are explainable using purely materialist, brain-centered theories. Others believe they are evidence of something greater, a different plane of existence. We will return to the subject of NDEs in much greater detail in Chapter 16.

Brain chemistry and evolution are plausible explanations for religious experiences. The short version is that the same brain traits that allow religion confer other survival benefits on humans, which are indispensable for the propagation of the species. Once religious beliefs form, they are very difficult to

---

It is more than likely that the brain itself is, in origin and development, only a sort of great clot of genital fluid held in suspense or reserved. This hypothesis would explain the enormous content of the brain as a maker or presenter of images.
Ezra Pound

When one person suffers from a delusion, it is called insanity. When many people suffer from a delusion, it is called religion.
Robert M. Pirsig

remove, just like any other core belief. The extent to which we believe in something is the extent to which we think it is real. Religious people know that God is real, which literally makes God real for them, for reasons I discussed in *The Enlightened Savage*. Whether that God is real to anyone but them remains an open question that we will continue to explore. In the meantime, it is important to acknowledge that what a secular person calls *faith* (a conviction of the truth of certain doctrines of religion, especially when not based on reason) a religious person calls *knowledge* (acquaintance with facts, truths, or principles, as from study or investigation).

Beliefs

To a very large extent men and women are a product of how they define themselves. As a result of a combination of innate ideas and the intimate influences of the culture and environment we grow up in, we come to have beliefs about the nature of being human. These beliefs penetrate to a very deep level of our psychosomatic systems, our minds and brains, our nervous systems, our endocrine systems, and even our blood and sinews. We act, speak, and think according to these deeply held beliefs and belief systems.
Jeremy W. Hayward

Beliefs are models of reality through which all perception is filtered. Personal beliefs explain how different people observing the same thing can generate differing—and even conflicting—conclusions about what they are witnessing. Beliefs govern just about everything we think, feel, and do yet are rarely challenged even when confronted with contradictory evidence. As I explained in *The Natural Savage*, the process of creating our own personal reality out of raw sensory input follows the acronym ETEAR:

- **E**ARLY EMOTIONS caused by one's earliest experiences lead to
- **T**HOUGHTS that form our deepest beliefs about the world and our place in it, which in turn fuel a cycle of
- **E**MOTIONS that have the same effects on our bodies as addictive drugs and therefore drive our
- **A**CTIONS that lead to the consequences that form our
- **R**EALITY, which reinforces our THOUGHTS and begins the cycle all over again from that point.

Belief is fundamental to brain function. Each of us needs this vision of how the world works and our place in society in order to function and survive. Morality, spirituality, and science are all subjects that influence how we interact with our fellow humans and the world around us. Beliefs can even affect our health. One example involves a certain Mr. Wright who was about two weeks away from dying of cancer when his doctor told him about a new miracle drug. Mr. Wright took the drug and experienced an amazing recovery that lasted until he learned that the drug he was taking was being criticized as ineffective. He relapsed, only to learn about another new miracle drug that might be able to help. This time, the doctor injected him with plain water as a placebo billed as a new, super-refined, and extra-potent version of the drug he had been taking. Mr. Wright recovered again until he learned of more problems with the drug. He returned to the hospital and died within two days.

This example demonstrates that beliefs cannot distinguish between the provable and the unprovable, which is further evidence that the line between fantasy and reality is a thin one at best. Proof itself involves belief, because the logical and evidentiary rules for proofs are themselves assumptions built on top of other assumptions that result in circular reasoning on a large scale. We tend to reject ideas or facts that disagree with our beliefs; nevertheless, we are capable of changing and updating our beliefs, a process that literally alters our brain's wiring. Does this mean that beliefs are entirely within our heads? A materialist would say yes, pointing out that science cannot verify that consciousness exists without a brain, while ongoing observation seems to be revealing a phenomenon caused purely by biological processes.

Why do humans build such complex systems of abstract, unproven, and possibly contradictory beliefs? Because we have no other choice. We are born utterly dependent and helpless into a world filled with language, science, religion, and people telling us what is and is not true. Our brains have no idea that we no longer live under the specter of predation, and dutifully accept all of this input as survival instructions. Beliefs can change very quickly if need be, but they usually remain astonishingly consistent over the course of a person's life.

Most of our brain activity revolves around perceiving the outside world and how our own bodies are doing. There is a delay between perception and consciousness, but our brain and nervous system have developed ways of dealing with danger before we are aware of it. Reflexes move muscles before our brains even get the signal that something is happening. Subconscious mechanisms can literally have us running before we have any idea what it is that we are running from. Emotions are the glue that holds our raw perceptions, thoughts, and

---

Any animal whatever, endowed with well-marked social instincts, the parental and filial attractions being here included, would inevitably acquire a moral sense or conscience, as soon as his intellectual powers had become so well developed, or nearly as well developed, as in man.
Charles Darwin

---

actions together. People always monitor the beliefs and behavior of friends and the prevalent beliefs of their communities to make sure they fit in. The patterns we fall into literally create chemical addictions in our brains that keep us doing the same things over and over, unless and until something happens to alter the beliefs. Intermittent snippets of perceptions pass through the filters of our beliefs and emerge as smooth, coherent, seamless, and intensely personal reality. (See *The Enlightened Savage*.)

Humans have gradually evolved to leave many of us doubting religion.
Nicholas Kristof

There is a huge difference between Western and Eastern psychology where divine intervention is concerned. People that Western psychology might diagnose as psychotic may be revered as having been touched by God in the East. The West identifies strongly with the self, while the East sees self as an illusion. In general, believers see signs of spirituality ,and God and nonbelievers dismiss them. For example, believers see the act of speaking in tongues as proof of divine presence, while nonbelievers see it as proof that something is not quite right in the babbler's head.

Humans are not the only animals with beliefs. Even single-celled organisms can learn and change. Shocking an amoeba makes its future explorations more cautious. The social amoeba *Dictyostelium discoideum* displays cooperation, cheating, *altruism* (defined as, "the principle or practice of unselfish concern for or devotion to the welfare of others"), and other behaviors. And why not? Every cell must make decisions about its own survival, and beliefs are a powerful way to do just that. If beliefs can do this for single-celled animals, then it should be no surprise that belief in God can have profound consequences on our emotions and actions.

Societal health causes widespread atheism, and societal insecurity causes widespread belief in god.
Phil Zuckerman

Our beliefs, including religious beliefs, affect our moral choices such as the food we eat. The Dalai Lama refuses to eat shrimp because of the number of lives required to create one meal. Numbers aside, why do we object to killing some kinds of animals while eating the flesh of other animals? We all make equally arbitrary decisions about virtually all aspects of our lives. We justify these arbitrary decisions with explanations that rearrange the available evidence to satisfy our beliefs. This creates a context that makes our beliefs both understandable and justifiable. With that said, our brains are not generic explanation devices; they consist of many highly specialized systems such as the circuit that allows us to recognize familiar people on sight. We excel at building *inference systems* (methods of arriving at conclusions or decisions that may not be derivable from the underlying assumptions but that have some degree of probability related to the premise being considered) that are often based on fragmentary information. People who expound on unfamiliar topics (such as the office know-it-all) are examples of people with strong inference systems.

The number of inferences we make every day is staggering. If we eat one beef steak, then we can infer that all beef steaks will taste similar. If we eat an unfamiliar plant and get sick, we can infer that all identical plants will also cause illness. If we dissect one fish, we can infer that all fishes will have similar insides. We of course have no way to prove any of this without testing each and every possible example. Inference systems are thus a form of faith that can carry a high survival value in cases such as the food poisoning example. Religion can only be successful to the extent that it activates our inference systems. The ongoing success of religion in the face of mountains of contradictory evidence is proof positive that our most powerful inference systems are being activated for reasons that should be abundantly clear after reading this far.

By now, it should be obvious that beliefs are anything but passive. We actively filter our perceptions through our beliefs. We also relax our standards to the degree that our inference systems are activated. Tell a religious police officer that the accelerator pedal in your car stuck and caused you to career off the road into a pole, and he will probably demand solid evidence to back your claim. Tell the same officer that a man was born of a virgin, died to forgive humanity's undoing, and was resurrected and physically lifted to heaven, and he will accept that on faith. In fact, he will probably reject any and all evidence you can provide to demonstrate that it didn't happen that way!

Skeptics claim to demand evidence for all claims. As Sagan said, the more extraordinary the claim, the more extraordinary the evidence required. One might think that skeptics make a habit of examining whether their beliefs are true or not; however, this is often not the case. Most so-called skeptics simply have dissenting beliefs that they validate using the same inconsistent standards and inference systems as anyone else. Also, while some skepticism is good, an overabundance of skepti-

---

The mind can assert anything and pretend it has proved it. My beliefs I test on my body, on my intuitional consciousness, and when I get a response there, then I accept.
D.H. Lawrence

Man is not logical and his intellectual history is a record of mental reserves and compromises. He hangs on to what he can in his old beliefs even when he is compelled to surrender their logical basis.
John Dewey

---

cism can have some unintended consequences. For example, imagine how society might function—or not function—if all school students were taught to question everything they were told. This is one reason why modern educational institutions are just as much about conformity as they are about imparting knowledge.

Logic

There are two general types of logical error: A Type 1 error occurs when someone believes in the truth of a falsehood; conversely, a Type 2 error occurs when someone mistakes a truth for a falsehood. The human tendency to think in terms of cause and effect leads to plenty of both types of errors. We are equally adept at both scientific and magical thinking. This does not mean that magical thinking is bad; the superstitions of the Middle Ages helped make a truly awful period in human history bearable for many millions of people. To this day, churches, mosques, temples, synagogues, and other religious buildings trigger intense feelings of spirituality and an otherworldly presence, which of course does not mean that one is actually present.

As we saw in Chapter 3, stories about how things work became myths. For example, history abounds with stories and myths about animals, with mythology elevating the stories to traditions that help preserve knowledge by passing it down to succeeding generations. These particular traditions can have significant survival value by teaching people how to avoid being killed and eaten by predator animals. Remember that humans are prey animals; we are somewhere in the middle of the food chain at best. Religion is simply another form of storytelling that became myth that then became institutionalized, as we saw in Chapter 4. This of course does not mean that religion began with a bunch of people sitting around dreaming stuff up. Humans are adapted for social life and specialize in storing, processing, and transmitting information. People do not invent gods; rather, we obtain information that leads us to the conclusion that one or more gods exist, thanks to our social adaptations and inference systems. Those hit by misfortune try to find a way to explain and make sense of their bad luck. Is it more comforting to believe that "stuff happens" or that everything happens as part of a larger benevolent plan?

The beauty of religious mania is that it has the power to explain everything. Once God (or Satan) is accepted as the first cause of everything which happens in the mortal world, nothing is left to chance... logic can be happily tossed out the window.
Stephen King

Only a brave person is willing to honestly admit, and fearlessly to face, what a sincere and logical mind discovers.
Rodan of Alexandria

The vast number of diverse religions all use a few basic templates. People are much better at performing complex tasks or retaining information if there is an element of social interaction, especially if that interaction involves cheater detection. We have evolved many methods of both cheating and cheater detection. Cheating can benefit the cheater but can have dire repercussions for a social group, especially if the cheating involves pilfering scarce resources that the group needs for survival. It is no accident that a significant part of religion revolves around morals and acceptable social behavior. Gods make rules for people to live by that foster group cohesion and survival, and people follow those rules. There are different ways to link gods with moral choices. For example, an immoral and evil god could be an example of how not to act, while a single omnipotent God could be waiting to punish those who misbehave. This punishment could occur in this lifetime or in a miserable hereafter.

Against logic there is no armor like ignorance.
Laurence J. Peter

Religious concepts are best understood when there is a perceived need for them. For example, something needs to be done with dead people. All religions speak of death and answer questions about mortality. Dead people may be gone from us and death itself may mean ceasing to exist, but we cannot simply forget about the dead, nor can we imagine our own life ending and our own ceasing to exist. The best I can do is sink down into an inky black mental well while maintaining full awareness—a distinctly frightening experience, because I know that doesn't come anywhere close to true non-existence.

Religious rituals act as precautions designed to help prevent misfortune or ensure good fortune. If needed, they can act as corrective measures to right some wrong. It is interesting to note that ritual does not derive from divine power, but the other way around; if Ritual A is not performed, then God will not perform Action B. God's power is therefore granted by the ritual performance, which may seem odd if the religion holds this god to be OOO. Rituals are a perfect example of obsessive compulsions. People know they are irrational but do them anyway with undiminished fervor. Similarly, sacrifices serve as mutually beneficial exchanges with the gods, such as God receiving a fatted calf in exchange for rain. As illogical as this may sound, such paradoxes are critical, especially for initi-

Men are apt to mistake the strength of their feeling for the strength of their argument. The heated mind resents the chill touch and relentless scrutiny of logic.
William Ewart Gladstone

ations. Society makes large demands of its members up to and including killing and dying, and the tribe needs to know that it can count on its members at all costs. Rituals, sacrifice, and initiations allow humans to interact with the "supernatural". Whether such a thing actually exists anywhere but inside people's heads remains an open question. This brings up the idea of proof.

Can we logically prove that God exists? The Catholic priest Thomas Aquinas (1225-1274) proposed the following five proofs:

- **Prime mover:** Some things are in motion. For something to move, something else must move it. There cannot be an infinite number of movers, hence there must a first mover that does not get moved by anything else. This first mover is God.

- **First cause:** Some things are caused. Anything that is caused is in turn caused by something else. There cannot be an infinite number of causes, hence there must be a first cause that is not caused by anything else. This first cause is God.

- **Necessary being:** Everything fails to exist at some time and is therefore contingent. If everything was continent, then there would be a time when there was nothing, which is clearly not the case. There must be something that is not contingent. That non-contingent thing is God.

- **Greatest thing:** Some things are greater than others. Anything that is great in any way receives its greatness from something greater. There is therefore a greatest thing. This greatest thing is God.

- **Intelligent designer:** Many non-intelligent things act purposefully. Anything that acts with purpose must be directed by an intelligent being. The universe must therefore have an intelligent designer. This intelligent designer is God.

These arguments may sound compelling, until one considers the counter-arguments:

- **Prime mover:** The universe is all that was, is, and will be. The universe does not need to be moved, and all

---

I am always amused by Christians who try so very hard to defend their position with reason and logic. They are like standing on top of a ladder with their tools and gadgets, straining their necks looking skywards for the leaks, trying to fix the roof amidst the rain. What they do not realize, of course, is that the roof isn't leaking. It isn't there.
Anonymous

Biologically, the species is the accumulation of the experiments of all its successful individuals since the beginning.
H.G. Wells

movement within the universe can be accounted for using well-established laws of nature. No mover is required. God is not required.

- **First cause:** Natural laws can account for the beginning of the universe and everything within it. No external first cause is necessary. God is not necessary.

- **Necessary being:** The laws of conservation of matter and energy demonstrate that nothing really comes into or out of existence. In this sense, nothing is contingent. God is not required.

- **Greatest thing:** The term "greatest" is subjective and carries no real meaning. Also, the laws of nature account for everything in the universe at all scales. God is not required.

- **Intelligent designer:** Nature is neither as designed nor as perfect as people want to believe. Complexity can spontaneously arise from simplicity without any external intelligence. God is not required.

Other arguments religion uses to try and prove the existence of God include:

- **Miracles:** A *miracle* is considered an event that goes against the established laws of nature and is therefore attributed to God. To a scientist, however, a miracle is either something that cannot be explained using current knowledge, a statistically possible but improbable occurrence, or a mistaken attribution. For example, the spring water flowing from the grotto in Lourdes, France, is said to have healing powers. About 8 million (8,000,000) people visit Lourdes every year. Of them, about 66 have been healed. If we assume a grand total of 160,000,000 visitors to the site (the actual number is far higher because our assumption only includes about 20 years' worth of pilgrims), then one person has been healed for every 2,424,242 pilgrims—a rate of healing that is 24 to 40 times less than the estimated rate of spontaneous remission from cancer (between 1 in 60,000 and 1 in 100,000). Nothing miraculous happening here.

- **Pascal's wager:** The mathematician Blaise Pascal (1623-1662) said that if one lives one's life as if God

---

It is a fraud of the Christian system to call the sciences human inventions; it is only the application of them that is human. Every science has for its basis a system of principles as fixed and unalterable as those by which the universe is regulated and governed. Man cannot make principles, he can only discover them.
Roland Barthes

Defined in psychological terms, a fanatic is a man who consciously over-compensates a secret doubt.
Aldous Huxley

exists and He does not exist, then one has lost nothing. On the other hand, if we act as if God does not exist and He does in fact exist, then we have lost everything. The obvious counter to this argument is that it proves nothing; besides, it is one thing to say that God exists, and quite another to say that this God is the God of the Catholics or any other specific religion. Behavior that is perfectly acceptable to one religion may be absolutely unacceptable to another. Which religion is the correct one, and how can we know for sure?

For those who believe, no proof is necessary. For those who don't believe, no proof is possible.
Anonymous

- **Mystical experiences:** We already know that mystical experiences are real in that they cause changes in brain activity. This of course does not prove that anything is happening beyond the person's head, and says nothing about whether or not God exists.

- **Fideism:** This is the concept that all religious matters must be decided on faith; science and philosophy can say nothing about religion, nor can religion say anything about science. This argument proves nothing beyond reinforcing the well-established fact that plenty of religious institutions would have you check your logic and reason at the door—the same logic and reason that God gave you, because God created the entire universe and everything inside it.

- **Morality:** Religion claims to be the source of human morality, a farfetched claim at best, given what we have learned in Chapters 7 and 8. There is plenty of reason to believe that morality is a product of evolution that does not require any outside agency such as church or God. It is in the best interests of all social animals, humans included, to behave in ways that help ensure group cohesion and maximize group survival.

Science has proof without any certainty. Creationists have Certainty without any proof.
C. E. Montague

Whatever you may think of these arguments and counter-arguments, the fact remains that our brains are always seeking to discover the deepest truths about our lives and the universe we live in. We should keep on trying.

All thought and emotion is subject and prone to logical biases and fallacies. Some of the many logical biases include:

- **Authoritarianism:** The tendency to value more highly the opinion of someone in a position of authority.

- **Confirmation:** Interpreting information in a way that validates one's preconceptions.

- **Self-serving:** Claiming more *responsibility* (ability to respond) for successes than failures, and/or evaluating ambiguous information in a way that serves one's own interests.

- **In-group**: Giving people in one's own group preferential treatment.

- **Out-group:** Seeing members of one's own group as being more diverse or varied than members of other groups.

- **False consensus:** Overestimating the number of people who agree with one's opinions or beliefs.

- **Bandwagon:** Tendency to do or believe something because others are doing or believing in the same thing.

- **Projection:** The unconscious assumption that others (or one's future self) share one's current emotions, thoughts, and values.

- **Expectation:** Tendency for experimenters to publish data that agrees with their assumptions while downgrading data that conflicts with those assumptions.

- **Probability**: This is a blanket term for many biases such as ignoring probability when making a decision, or the gambler's fallacy of believing that prior events influence future events.

- **Pleasure:** Giving more weight to beliefs, information, etc. because doing so is pleasurable.

- **False-memory:** Confusing imagination with memory or true memories with false memories.

Some of the many logical fallacies that can occur include:

- **Appeal to probability:** Because something can happen, it will (Murphy's Law).

- **Argument from fallacy:** If an argument for some conclusion is fallacious, then the conclusion is false.

- **False dichotomy:** Presenting two alternatives as the only options when there may be more.

Science is simply common sense at its best—that is, rigidly accurate in observation, and merciless to fallacy in logic.
Thomas Henry Huxley

There is a principle which is a bar against all information, which is proof against all arguments and which cannot fail to keep a man in everlasting ignorance. That principle is contempt prior to investigation.
Herbert Spencer

- **Homunculus fallacy:** Explaining a concept in terms of the concept itself without defining or explaining that concept.

- **Naturalistic fallacy:** If something is natural, pleasant, popular, etc. then it is good or right.

- **Negative proof fallacy:** A premise that cannot be proven false must be true, or a premise that cannot be proven true must be false.

- **Teleological fallacy:** Claim that an object or idea has a purpose, and that this purpose suggests or requires that the argument is true.

- **Existential fallacy:** Where the premises do not support the conclusion.

- **Proof by example:** Using specific examples to prove a universal conclusion, such as, "This chocolate is sweet, therefore all chocolate is sweet."

- **Appeal to ridicule:** Presenting an opposing argument in such a way as to make it look ridiculous.

- **Begging the question:** One or more premises implicitly or explicitly assumes the conclusion.

- **Circular cause and consequence:** Claiming that the consequence is also the cause.

- **Demanding negative proof:** Avoiding the burden of proof by demanding that anyone who questions one's claim prove the opposite of the claim.

- **Etymological fallacy:** Assuming that the historical meaning of a word or phrase is the same as its current meaning.

- **False attribution:** Appealing to an irrelevant, unqualified, unidentified, biased, or fabricated source to support an argument.

- **Historian's fallacy:** Assuming that people in the past viewed events from the same perspective and with the same information available to later analysts.

- **Red herring:** An attempted distraction from the subject at hand via introducing a new argument.

It is true that there are exercises that can strengthen the 'muscle' that enable us to push back the bounds of acceptation. But these are relatively unimportant. The real problem is that we are trapped in misconceptions that always deceive us, as the matador's cape deceives the bull; that continue to deceive us a million times over the course of a lifetime.
Colin Wilson

- **Regression fallacy:** Ascribes a cause where none exists.

- **Retrospective determinism:** An event occurred, therefore the occurrence must have been inevitable beforehand.

- **Cherry picking:** Using selected information to confirm a particular conclusion while ignoring significant information that may contradict the conclusion.

The preceding list represents only a small fraction of the total number of biases and logical fallacies, and there is also significant overlap between the two categories. These samples should be enough to demonstrate that avoiding all bias and fallacy when presenting an argument or drawing a conclusion is virtually impossible. It would be hard enough to create a perfect argument free of all biases and fallacies if we all agreed on how to define and measure the flaws. The reality is that one person's cherry-picking is another person's proof positive, one person's solid conclusion is another person's expectation bias, and on and on we go.

Growing awareness of biases and logical fallacies may be contributing to the growing gap between the number of spiritual people and those who attend religious services, as people increasingly distinguish between the two. There also seems to be a growing feeling that one church is just as good as another. This reflects the growing power of doubt, reason, and skepticism, all of which are perfectly valid. It is one thing to passively hold a belief; it is quite another to question that belief and find it affirmed by valid evidence. As Thomas Jefferson (1743-1826) said, "Question with boldness even the existence of a God; for, if there be one, he must more approve the homage of reason, than that of blindfolded fear."

Mysticism

Given that reproduction is the top evolutionary priority for all species including humans, it may surprise you to learn that more people every day pray or meditate than have sex. This alone suggests that spirituality is about far more than insecurity or fear of death, a suggestion that is borne out by numerous experiments. And why not? Is life not sacred in some way? Is it not difficult to gaze into the smallest parts of an atom or out into the farthest reaches of the universe without seeing the

No sadder proof can be given of a person's own tiny stature, than their disbelief in great people.
Thomas Carlyle

Man has a limited biological capacity for change. When this capacity is overwhelmed, the capacity is in future shock.
Alvin Toffler

hand of God, however illusory or imaginary? Does that not make even science sacred in some way? And yet, as mystics around the world have said for thousands of years, trying to understand God is nearly impossible; it's like an inhabitant of Edwin Abbot's *Flatland* trying to grasp the concept of a third spatial dimension.

Author and philosopher Simone de Beauvoir (1908-1986) said that it is easier to think of a world without a creator than it is to think of a creator saddled with the contradictions of this world. She has a point; the God that lives in a believer's head may well be more powerful than any real God. But spirituality is emotional, not intellectual. We don't know God; we feel God. This feeling is not necessarily religious, because religion is based on spirituality and not vice-versa.

Mystical experiences used to be thought of as bad, but people who have these experiences are actually well adjusted. Prayer and mysticism can contribute to improved mental and physical health. Children do not learn or acquire spirituality from outside sources; spirituality comes from within—more evidence that genes play a role, and that spirituality does indeed have survival value. Every person reinvents her or his world every day in the never-ending search for truth, enlightenment, and the ultimate reality, Ground, or Godhead. This quest is doomed to fail, because everything we are aware of is nothing but an illusion, a symbolic and highly personal representation of the world. (See the section on Buddhism in Chapter 5.)

Psychologist Abraham Maslow (1908-1970) devised the "hierarchy of needs" as follows:

- **Physical:** Food, water, air, sleep, sex, excretion, etc.
- **Safety:** Physical security, employment, health, family, property, etc.
- **Love/connection:** friends, family, romantic love.
- **Esteem:** Confidence, achievement, respect for others, self-respect, self-esteem, etc.
- **Self-actualization:** morality, creativity, acceptance, etc.

According to Maslow, self-actualized people can have periodic "peak" experiences. A peak experience is a sense of oneness with the universe and seeing things as they really are and not

---

What does mysticism really mean? It means the way to attain knowledge. It's close to philosophy, except in philosophy you go horizontally while in mysticism you go vertically.
Elie Wiesel

Mysticism and exaggeration go together. A mystic must not fear ridicule if he is to push all the way to the limits of humility or the limits of delight.
Milan Kundera

through our normal filters. These people are less likely to attend religious services, but more likely to have mystical experiences. Interestingly, Catholic priests scored lower on measures of self-actualization than lay people, possibly for some of the reasons we discussed in Chapters 7 and 8. Peak experiences can range from something as simple and low-level as watching a sunset and feeling that there is more to the experience than the sum of its parts, to romantic love and feeling joined with the person of your affections. Those experiences that include religious awe—be it spontaneous or triggered by ritual—and bring about a sense of utter union with the universe/Ground/Godhead are the highest possible peak experiences. Peak experiences alter brain activity. This alone tells us that something is going on. The fact that mystical people tend to be healthier and happier than most people suggests that these experiences are not delusional. Of course, this does not necessarily mean that there is a God, that humans have a soul, that there is an afterlife, etc. Also, drug "trips" can mimic mystical experiences but are not real substitutes.

Self transcendence is a measure of a person's ability to reach out beyond themselves. This term describes spiritual feelings that are independent of traditional religion. These feelings are not based on belief in a particular god, but on the nature of universe and our place in it. Highly self-transcendent people have a strong sense of oneness with other people and the universe, while non self-transcendent people tend to focus on separation and differences. There are no significant racial differences among self-transcendent people; however, women score 18% higher than men on average, possibly because they tend to be more open to emotions. People who are highly self-forgetful and tend to get lost in their work or experiences (people who are "in the zone") tend to be more self-transcendent.

Transpersonal identification measures a person's feeling of connectedness to the universe and everything therein. In general, people with mystical/spiritual leanings tend to see coincidences as evidence of God or other higher power (which they just may be) and believe in the "supernatural". One need not be religious to be spiritual. On the downside, people with high levels of mysticism who lack psychological maturity may be more prone to psychosis.

The first proof of a person's incapacity to achieve, is their endeavoring to fix the stigma of failure on others.
B.R. Hayden

A fact in itself is nothing. It is valuable only for the idea attached to it, or for the proof which it furnishes.
Claude Bernard

As we saw earlier in this chapter, real changes take place in the brain during mediation and/or mystical experiences. The question is whether those changes are evidence of a connection to something beyond, or whether they are nothing more than rerouted brain signals of no particular significance.

Marketing Religion

The aim of marketing is to know and understand the customer so well the product or service fits him and sells itself.
Peter Drucker

It may seem odd to discuss how religion markets itself to prospective converts in a chapter devoted to biological explanations of religion. Remember that humans evolved to fit into hierarchical societies. The core of marketing is nothing more than spreading a message. You are marketing at every moment of every day with the clothes you wear, what you say, how you say it, what you buy, etc. If you are in a position of real or perceived authority, then you are in a position to influence how other people think by activating the circuits designed to ensure social cohesion and survival that we have been talking about in this chapter. We have already seen in Chapter 7 that the idea of a universal religion is not a universal idea. And, when it comes to corporate marketing, no secular campaign comes close to rivaling Catholicism's 2,000-year track record.

It didn't take Christians long to start annoying people.
Robert Wright

Religion is a powerful tool for amplifying the importance of our clan or tribe to the point where it is worth fighting and dying for. Stories of real or imagined persecutions stoke sympathy and a desire for retribution. Stories of real or imagined victories stoke pride and a sense of righteousness and destiny. Stories of the One True God selecting your tribe and your tribe alone out of the thousands of tribes on the planet as His chosen people stoke awe and pride. Stories of God's vengeance and eternal punishment stoke fear and compliance. Religion has no shortage of ways to elicit powerful emotions that trigger all kinds of social circuits in our brains. Because of this, one could say that religion is parasitic—an idea that occurs to me every time I see a televangelist asking for money. The late Eugene Scott (1929-2005) of the University Cathedral in Los Angeles was infamous for telling viewers to "get on the telephone," at which point the video cut to various random scenes while music played. After a few minutes, Mr. Scott would either pick up where he left off or say, "That's not good enough, get back on the phones!" One can only wonder what his congregation might have done with the estimated $1 million per month that went to chauffeured limousines, horse ranches, a Lear jet, etc. had their mental circuits not been so strongly activated.

America is a secular nation (despite evangelical claims to the contrary), making religion a free enterprise. Rival churches compete for converts and their money using common hard-sell techniques to create a sort of religious mania among the less educated. (One can imagine that the Founding Fathers would have been horrified by this.) Religious groups provide services and try to control the market just like any business, hence their ceaseless quest for political dominance. The religion with the most loyal followers (in both quantity and quality) wins. Religion answers the big questions, provides comfort and solace, promotes social order, and offers a convincing illusion of correctness. Some religions care more about salvation than others, but most religions have far more in common than they ever admit. On the contrary, each religion tends to play up its differences to demonstrate its uniqueness and correctness. Religion goes far beyond mere spirituality to act in educational, judicial, landholder, therapeutic, and other ways that have nothing to do with spirituality. Why? The more of a person's life the religion can control, the more loyal and devoted that person will tend to be, at least to all outward appearances.

Religion works by transmitting *memes*, ideas that get passed from person to person and generation to generation. For example, the idea that marrying a disbeliever is a sin is a meme that has helped the Jews create a distinct ethnicity and DNA patters over the last 3,000 years. The idea of hell is another self-perpetuating meme; children are told they will burn in hell and live in fear because they are unable to test the assertion.

Thought Control

Take a step back to examine some of the claims religions make, and you will probably end up shaking your head in amazement. Western religion records farfetched beliefs and tries to validate them. Experiments never produce the desired results, but people still believe with the attitude that their religion may be losing battles but will eventually win the war. Religions promote revelations, doctrines, rituals, and texts as sources of truth while downplaying thinking, questioning, and

The more religious people convince others of the validity of their beliefs, the more they believe they are doing God's will, contrary to what the Bible clearly states. The problem is not what people believe. The problem is that they believe.
Jerry Wheatley

I am a man of one book.
Thomas Aquinas

reasoning. Revelations are particularly blatant examples of circular reasoning: God reveals Himself to a person, who is correct because s/he heard God's word. Religious scholars are equally adept at creating both coherent doctrines and a lot of stories that just don't make sense.

> *All moral learning is ultimately based on the pain and pleasure circuitry in your brain—on your internal reward and punishment systems.*
> Robert Heath

Why would the same God who created all of the people on Earth select a tiny backward tribe whose strength and technology paled next to its neighbors as His chosen people? What are the odds that the cobbled together bits of earlier myths that make up the Jewish religion actually got the story right? Why do Christians treat some artifacts as if they have special powers in seeming violation of the Biblical injunctions against idolatry?

In some religions, ghosts, witches, and demons are part of people's lives in the same way that cars and airplanes are for most people alive today. These otherworldly beings often display a downright bizarre mix of superhuman powers and downright idiocy. People in Siberia speak in metaphors to avoid tipping off evil spirits; if the listener understands, why can't the spirit? Some Africans express sympathy for handsome children so witches won't want to eat them, as if good-looking children somehow taste better. Haitians think that witches may steal corpses, and bury eyeless needles with the deceased to keep the witches busy. Jesus said the Kingdom of God would come during his lifetime, and over 1 billion people are still waiting 2,000 years later. At least the apocalypse stories strike a chord, because we know our time on this planet will end and that the universe will probably experience heat death. Plenty of religious people have (wrongly) predicted the end of the world. Of these, the Jehovah's Witnesses take the prize for the most failed attempts at foreseeing our demise, having predicted that the world will end in 1874, 1878, 1881, 1910, 1914, 1918, 1925, and 1975. They also said that people who were alive in 1914 (the year World War I started) would still be alive when the world ended. They are a little more vague today.

> *Knock on yourself as upon a door and walk upon yourself as a straight road. For if you walk on the road, it is impossible for you to go astray. Open the door for yourself so you may know what it is.*
> Silvanus

Who Believes?

Most people have at least some spiritual capacity, making it one of the most universal and powerful forces in human life. Cave paintings from 50,000 years ago show mythical creatures. A poll done in March of the year 2000 revealed that:

- 86% of people in North America believe in God; only 8% are atheists.
- 2/3 of people attend church monthly, and 36% attend weekly.
- 2/3 of people are members of a church or affiliated with a religion.
- 60% of people say that faith very important in their lives. Another 30% say that faith is somewhat important.
- 52% of people are more likely to vote for a political candidate who expresses religious beliefs.

A 2003 study by the Higher Research Institute at UCLA found that 80% of those surveyed considered themselves spiritual. In general, religion has done quite well in the United States over the last 200+ years. Church membership has risen from 17% during the American Revolution to over 60% today, which has not stopped modern preachers from chastising our hedonistic and godless ways.

The findings of science are far awe-inspiring than the rantings of the godly.
Christopher Hitchens

Socioeconomic status is not a factor, but conservatives and the less educated tend to be more religious. Conservatives want to preserve some supposed halcyon status quo, a viewpoint that is most extreme in the religious right. This is similar to radical feminism, which dreamed about some golden era before civilization. Radical people, including radical scientists, have more in common than they may care to admit. Greater religious piety is associated with more prejudice against other religions, ethnicities, ethnocentrism, totalitarianism, dogmatism, social distance, rigidity, and intolerance of ambiguity. In general, the more open a person is to experience, the less religious s/he is likely to be.

Children make the best converts, because their belief systems are far more receptive to new ideas than adults who have already learned how to survive and formed strong beliefs. Children know that animals move and rocks don't, but can't explain why. They are also far more predisposed to notice different people than different toys, an adaptation with obvious survival value. All children want to be special and in the middle of things. They may be scared of dying and open to stories of the afterlife. Their religious indoctrination helps them form core beliefs that become very difficult to change later on.

I may disagree with what you have to say, but I shall defend, to the death, your right to say it.
Voltaire

Chapter 9
The Biology of Religion

Is God in Our Brains?

Physical systems are governed by necessity, living things by contingency. Humans may be a huge contingency; Gould may be right that replaying the tape of history might yield a very different outcome. Still, just because the odds of humans evolving may be low does not mean they are 0, nor does our evolution mean we were destined to be here. We will explore evolution in Chapters 18-20; meantime, suffice it to say that it is perfectly possible for complexity to evolve from simplicity without any outside interference.

People will continue to seek God for as long as God remains a mystery. We cannot perceive the whole truth, but that won't stop us from trying, because that is how our brains are wired. Physicist Kurt Gödel's (1906-1978) theorem says that any math or symbolic logic system will always be incomplete and have unprovable assumptions. Every scientific (or religious) concept may have some false ideas. Thus, the truth can never be fully known, but we can still try... and try we must.

We are intensely curious creatures, which may well serve as a counterbalance to our fear of predation. In *Animals in Translation*, author Temple Grandin (1947-) describes how livestock animals will panic at the slightest novelty but quickly return to investigate. The fear response keeps us alive, while the curiosity keeps us from living in fear all the time by allowing us to assimilate new experiences.

Plenty of experiments have proven the placebo effect, where people given ersatz medicine recover because they believe the medicine is working. I tried it on my son Logan one evening when he was unable to fall asleep: I gave him an absolutely harmless fish oil capsule and told him that it was a powerful sleeping pill that he should only take when he was in bed under the covers and ready to pass out. He swallowed the capsule and was sound asleep within minutes. If faith in a capsule can make a boy fall asleep, and if faith in a cancer drug can cause spectacular recovery, then it stands to reason that faith healing itself should work. Does this mean that there is more to faith than a bunch of people who believe in silly things? No. But it does not mean that faith healers are wrong, either.

If the brain evolved by natural selection, religious beliefs must have arisen by the same mechanism.
Edward O. Wilson

Controversial new research suggests that whether we believe in a god may not just be a matter of free will. Scientists now believe there may be physical differences in the brains of ardent believers.
Liz Tucker

We evolved with the capacity for spirituality and religion. Our brains contain powerful inference systems that let us form conclusions about myriad topics based on partial, even fragmentary information. Our genes predispose us to exist in a hierarchical society. We form beliefs by accepting information as survival instructions and then following those instructions to the letter. As we have seen, religion taps into this on many levels. We are primed to have God on the brain because of our biology. Whether or not this means that God is a figment of our imaginations that resides nowhere else besides within our brains remains to be seen. Meanwhile, we will examine the many charges leveled against religion in the next chapter.

Spirituality comes from within. The kernel must have been there from the start. It must be part of their genes.
Dean Hamer

Chapter 10

Religion, Interrupted

> *The president of the United States has claimed, on more than one occasion, to be in dialogue with God. If he said that he was talking to God through his hairdryer, there would be a national emergency. I fail to see how the addition of a hairdryer makes the claim more ridiculous or offensive.*
> Sam Harris

> *Either you're with us or you're with the enemy.*
> George W. Bush

In 1964, the documentary film *Mondo Cane* (World of the Dog) told the amazing story of Pacific island tribes that had lived in isolation until the day many white-skinned people started arriving in ships laden with wondrous machines and materials. Unable to explain this sudden influx, the islanders translated these occurrences in the only way they could: by deciding that the arriving visitors were long-lost ancestors returning from the afterlife bearing presents. These ancestors built piers, and more ships came. The natives built piers as well, but no ships came for them. In desperation, they turned to the Christian missionaries among them to ask where the gifts were. The predictable answers slowed these so-called cargo cults, which were largely forgotten until the middle of the 20th century, when American forces arrived on the islands and began building airstrips where flying machines arrived with yet more fantastic cargo. Some of this cargo inevitably broke down, and the white people simply sent away for more. The white people never seemed to be doing anything useful, because all they did was sit at desks shuffling papers, which the locals interpreted as acts of religious devotion that kept the goods flowing.

The locals soon abandoned Jesus Christ and started building their own runways to attract airplanes, complete with ersatz control towers, antennas, air traffic controllers, aircraft, etc. to worship John Frum (John from America). These cults still exist today; at least one photo shows islanders manning their makeshift airports watching jets take off from nearby runways. It is easy for us to laugh at these poor people and their silly rituals, but careful reflection reveals that other religions are built on foundations that make the cargo cults seem eminently logical by comparison. This may be very difficult for you to accept if you are a person of faith, especially if you were raised in a religious setting; however, a dispassionate examination of the history of the world's major religions reveals the truth of my assertions. As George and Ira Gershwin's character Sportin' Life sings in *Porgy and Bess*, "They tell all you children the devil's a villain, but it ain't necessarily so... Oh I take the Gospel whenever it's possible, but with a grain of salt."

This chapter spells out many of the charges and accusations made against religion. It may be very difficult for you to read if you are a person of faith, and yet I must ask you to read it thoroughly and accept it without retort. We will look at religion's responses to these charges in the next chapter, and will keep exploring the big picture as we continue into this book.

> *I always thought organized religion was a con game. It always got me mad that on high holy days you had to buy tickets to pray.*
> Jordan Sinclair

Eastern vs. Western Religions

The spiritual differences between East and West are vast. Why did mysticism based on monistic idealism only take hold in Asia? Geographer Jared Diamond (1937-) speculates that the JCI religions emphasized faith itself instead of the experiences purported to lie behind that faith. Comparing the Eastern mystical philosophers to their Western counterparts reveals striking differences. One could live one's entire life within the JCI religions without ever encountering teachings like those found in the East. Even modern neuroscience cannot hold too many candles to Buddhist teachings. Mysticism requires explicit instructions that have no ambiguity. Both East and West seek to connect adherents to the Godhead; the key difference is that the former tells you how to do it, while the latter relies on believing the stories of others who claim to have done it.

> *Religion is the end of love and honesty. Faith is a colorful hope or fear, the origin of folly.*
> Tao Te Ching

The major Eastern insight is that thinking and reason are necessities, but that failing to see a thought as a thought leads to the idea of "I," which is where suffering and unhappiness come from. Breaking the spell of thought eliminates the subject/object split, as well as the difference between happiness and suffering. Few Westerners try to understand the concept that the nature of consciousness requires no thoughts, and that experience is open to everyone. Unlike religion, mysticism is utterly pragmatic and based on empirical evidence contributed by many experimenters and meditators. Religion is built around this kernel of truth because ethics, spirituality, and vibrant communities are essential for happiness, and it is possible to unite spirit, reason, and ethics in a profound faith that is backed by solid evidence.

If you are a Christian you do not need to believe that all the other religions are simply wrong all through... but, of course, being a Christian does mean thinking that.
C.S. Lewis

Sadly, this kernel has been all but buried underneath layers of dogma, ritual, and fear shrouded by ignorance. No wonder Christian leaders were ashamed to tell their followers about the principles underlying their religion and forbade translation of the Bible for so many centuries! No parent I know can justify the story of Abraham willingly leading his own son to the slaughter to her or his own child, let alone the idea that this act somehow guarantees happiness and eternal bliss. The idea that cold-blooded murder can improve the human condition is ludicrous on its face. My friend (I'll call him Joel) protested my literal reading of this story based on the analyses done by philosopher Søren Kierkegaard (1813-1855) in *Fear and Trembling*, in which Kierkegaard offers some seemingly plausible reasons for Abraham's actions.

"Joel," I said, "put yourself in Abraham's shoes. Under the exact same circumstances recounted in the Old Testament, would you kill your daughter?"

His eyes went wide; the discussion was over.

I'd sure like to help the colored, but the Bible says I can't.
Rat Robertson

By now, it should be abundantly clear that this chapter will focus on the shortcomings of Western religion. I am doing this for several reasons, chief among them my assumption that my readers are far more familiar with the JCI religions than any others. My intent is not to portray the Western religions as evil and the Eastern religions as good. All religions have their problems, and a good number of the problems we will discuss apply to all religions.

The Perils of Tolerance

Shaky foundations and silly rituals that are out of touch with modern life, combined with the steadfast refusal to revere the reason and logical faculties God supposedly gave us are reason enough for me to eschew religion, but that's only the beginning. As I have demonstrated in Chapters 5 through 8, religion is responsible for more than it's share of the world's suffering. This was bad enough in the days before nuclear weapons that are capable of destroying all life on Earth several times over. Today, most of society ignores the pathological aspects of religion at our mutual peril. Religion, like anything else in life, must be judged by its results.

An ignorant psychopath who abuses and tortures children needs to be punished, but we as a society understand that mental disorders and prior traumatic experiences can have devastating effects on otherwise perfectly decent people. People who believe they have a divine mandate for their actions are far more dangerous. A person who kills because God told her or him to do it may be found insane, but will still serve time in prison. A hermit living in a cave claiming to see visions will probably be left alone, unless we discover that he is planning attacks. A person called by God to stockpile weapons and bed multiple women would almost certainly arouse both skepticism and perhaps envy. But if any of these things are preached in the context of an established religion, we accept them at face value.

Abraham, the purported founder of all three of the JCI religions, was prepared to sacrifice his son Isaac because God told him to do it (Genesis 22:1-19). The Old Testament is littered with tales of religiously inspired genocide. The fact that they are apocryphal is beside the point, which is that the Bible, Koran, and other "holy" books extol violence. Elijah, Mohammed, and other cave dwellers received divine revelations that directly or indirectly led to violence. Plenty of religious heroes from King David to Joseph Smith and David Koresh have stockpiled weapons and bedded multiple women. All of the examples I provided in the previous paragraph are indeed preached by contemporary religions with very little opposition. Quite the contrary: Most non-religious people keep silent out of "respect" and "tolerance." People who carry

> *Any religion that professes to be concerned about the souls of men and is not concerned about the slums that damn them, the economic conditions that strangle them and the social conditions that cripple them is a spiritually moribund religion awaiting burial.*
> Martin Luther King, Jr.

> *[Religion] does not even have the confidence in its own various preachings even to allow coexistence between different faiths.*
> Christopher Hitchens

out extreme actions in the name of religion are seen as "bad apples" that do not reflect the view of the far more moderate masses, despite the obvious fact that the supposed "moderates" read the same books and build their beliefs on the same foundations as the extremists.

Indeed it seems to me that the more Christian a country is the less likely it is to regard the death penalty as immoral.
Antonin Scalia

Imagine a single building in which most of the rooms are painted in ordinary colors and conservatively decorated. Other rooms sport garish colors and outlandish decorations that most of us would find distasteful. A fatal flaw in the foundation affects all rooms equally; one cannot claim to be safe simply because s/he occupies one of the conservatively decorated rooms. Religious moderation therefore represents no panacea or counterbalance to extremism. In fact, religion may well represent the single gravest threat to the ongoing survival of the human species, and indeed all life on Earth.

What if religious people didn't attempt to foist their views on all of society? What if they did not engage in or condone violence? What if they eschewed politics and proselytizing and kept to themselves? Would their beliefs be more acceptable and their believers less dangerous? That depends on who you ask. Author Sam Harris (1967-) would say no. As he says in *The End of Faith*, "People who believe that the Earth is flat are not dissenting geographers. People who deny that the Holocaust ever occurred are not dissenting historians. People who think God created the universe in 4004BC are not dissenting cosmologists; and we will see that people who practice barbarisms like 'honor killing' are not dissenting ethicists. The fact that good ideas are intuitively cashed does not make bad ideas respectable."

Sometimes believers have ascribed to god a character and the behaviors that go with it which, if they appeared on Earth, would lead to instant arrest.
John Bowker

Harris makes the mistake of conflating religion and God, and his atheism is every bit as dogmatic as any religion; however, his observations about the current state of human knowledge and morality are very well taken. We have come a long way over the last 2,000 years, but our major religions have not evolved beyond the Bronze and Iron Ages. Religion has resisted every advance tooth, nail, and stake, opposing technology on one hand while appropriating it for its own purposes on the other. I would go so far as to say that religion may well represent a huge thorn in God's side (pun fully intended).

In 1940, Einstein wrote a paper that explained his lack of belief in a personal God. Religious people responded with a swarm of letters, many of them pointing to Einstein's Jewish heritage. The following letter from the founder of the Calvary Tabernacle Association in Oklahoma City, Oklahoma is taken from *Einstein and Religion* by Max Jammer: "Professor Einstein, I believe that every Christian in America will answer you, 'We will not give up our belief in our God and his son Jesus Christ, but we invite you, if you do not believe in the God of the people of this nation, to go back where you came from.' I have done everything in my power to be a blessing to Israel, and then you come along and with one statement from your blasphemous tongue, do more to hurt the cause of your people than all the efforts of the Christians who love Israel can do to stamp out anti-Semitism in our land. Professor Einstein, every Christian in America will immediately reply to you, 'Take your crazy, fallacious theory of evolution and go back to Germany where you came from, or stop trying to break down the faith of a people who gave you a welcome when you were forced to flee your native land.'"

One has to wonder whether it is the belief or Einstein's response that is hurting the Jews (and everyone on Earth) more. Whatever objective truth may exist (a question we will return to in Chapter 23) is independent of our most fervent desires and beliefs, whether we like it or not. The fact that many people see the objects of their faith as objectively real does not make it so. The fact that most religious people are happy with their creeds and articles of faith does not make them valid or beneficial. People turn to religion for experiences and emotions, which do not necessarily reflect any level of reality. The fact that roughly 80% of the world's population identifies itself as religious does not make religion valid because truth is not a matter of consensus, but of fact.

What makes religious beliefs dangerous? Beliefs are what form each of our personal realities. In order for them to have any meaning, we must believe that our beliefs give us a faithful representation of the objective state of the world. Some beliefs more closely fit the information we receive from the world and fit the available evidence better than others. For example, it is natural to think that the Earth is flat in the absence of direct evidence to the contrary; it is almost unthinkable to

---

Religious intolerance and was are inevitably born with the belief in one god.
Sigmund Freud

The whole point of religious faith, its strength and chief glory, is that it does not depend on rational justification. The rest of us are expected to defend our prejudices. But ask a religious person to justify their faith and you infringe 'religious liberty.'
Richard Dawkins

Chapter 10
Religion, Interrupted

maintain that belief in light of the evidence available to us today. It is one thing to say that all cultures should be respected and not judged by external standards in the same way that we cannot judge those who came before us by our own modern standards. The flip side of this argument is that we must not judge the present using old standards, nor should we revere old behaviors and codes of ethics that have since been found wanting. Herein lies the central problem with religion: Billions of people are clinging to worldviews, beliefs, ethics, and morals that may have been enlightened 2,000 years ago, but that have been long since superseded by advances in all human endeavors. A religious person attempts to apply ancient information to modern existence. Therein lies the problem with so-called religious tolerance.

> *Criticizing a person's ideas about god and the afterlife is thought to be impolitic in a way that criticizing his ideas about physics or history is not.*
> Sam Harris

Religion Behaving Badly

Modern ethics would be hard pressed to justify retributive punishment or scapegoating. The idea that untold billions of people must suffer because one woman ate an apple until God can temporarily splinter off a piece of Himself as a son who will endure excruciating agony and temporary death to forgive us all, is ludicrous on its face. Chromosomal Adam and mitochondrial Eve did exist, but there is no reason to think they were the first humans, or that they frolicked in a serene tropical paradise until a snake got the better of Eve. Most JCI people know the stories of Adam and Eve, Jesus's crucifixion, and others, but few take the time to read the entire Bible or Koran and place the carefully chosen passages they hear during religious services into context.

> *What should be said of us who are forced to live piously, not by devotion but by terror?*
> Maximus of Turin

The fact that most of these stories never actually happened is beside the point, because they are nevertheless held up as examples of how humanity should base its morality. Of course, morality has no bearing on whether or not God actually exists. Religious people base their actions on a desire to curry favor with God, and claim that we would not have standards by which to judge good and bad without religion. In other words, religion claims the exclusive right to decide what is good and what is not good. As evidence, they offer up books that contain a quasi-random cobbling together of mutually contradictory—and mostly fictional—stories as the

> *The great unmentionable evil at the center of our culture is monotheism.*
> Gore Vidal

inerrant sources of human morality. Lot is held up as an example of righteousness. If someone who offers his own daughters to be raped and commits incest is the best the city of Sodom could offer, one can certainly understand why God would want to level the place (Genesis 19:31-36). God's vengeful tantrums when "His" people stray is simple jealousy—hardly a good role model.

The story of Jericho and the invasion of the Promised Land is a precursor to the German invasion of Poland, or Saddam Hussein's massacres of the Kurds in northern Iraq and the marsh Arabs in the south. Few atheists would seriously consider razing a city or destroying religious objects, but religious people aspire to such acts. Weinberg calls religion an insult to human dignity and points out that, while good and bad would exist no matter what, getting good men to do evil things requires religion. Pascal concurred hundreds of years earlier when he said that, "Men never do evil so cheerfully and completely as when they do it from a religious conviction." The New Testament is little better, and the Christian symbol (the cross) is a torture device. Had Jesus (supposedly) been executed 20 years ago, would Christian people be wearing miniature lethal injection gurneys around their necks? The hell described in the Bible and Koran is very unlikely to be real, which is easily countered by making it as awful as possible. The extreme level of eternal suffering that could occur is enough to keep people in line, however unlikely such damnation might be.

Religions around the world claim to be the arbiters of human behavior. Religious leaders decry supposed moral decay in society. They claim that they alone have the right to tell us how to live, because they have a special connection to God, who defines right and wrong. The truly astonishing part of all this is that the secular world pays tribute to these claims, consulting clergy in matters of science, law, medicine, ethics, etc. while rarely entertaining the opinions of atheists and other free thinkers. Religion enjoys a special status accorded to no other human organization, which is remarkable because religious people tend not to behave any better than anyone else. On the contrary, religion gives people license to commit acts of unspeakable horror that would give the most ardent ethnic cleanser or jaded bordello madam pause.

The Vatican is too strong, and too unapologetic, for us to go taking on bishops. Haven't you heard of infallibility?
Rwandan Ministry of Justice

Lighthouses are more useful than churches.
Benjamin Franklin

> *Religion insists that what is essentially real and important to you subjectively must also be that which is essentially real and important in the objective world of fact.*
> Barbara C. Sproul

Christians in particular pride themselves on peace and family values. They seem to have forgotten Matthew 10:34-37, in which Jesus says, "Do not suppose that I have come to bring peace to the Earth. I did not come to bring peace, but a sword. For I have come to turn a man against his father, a daughter against her mother, a daughter-in-law against her mother-in-law. A man's enemies will be the members of his own household. Anyone who loves his father or mother more than me is not worthy of me; anyone who loves his son or daughter more than me is not worthy of me; and anyone who does not take his cross and follow me is not worthy of me. Whoever finds his life will lose it, and whoever loses his life for my sake will find it." Never mind that this passage is probably a later edit designed to increase the Catholic stranglehold on daily life during the Middle Ages; it is in the Bible, the book revered by every Christian today.

Gallup and Barna Group surveys consistently demonstrate that evangelical Christians are just as prone to be just as materialistic, hedonistic, selfish, and sexually inventive as anyone else, if not more so. The Federal Bureau of Prisons reports that 80% of the prison population is Christian, while only 0.2% is atheist. Children are more likely to be sexually abused, and wives are more likely to be beaten in conservative religious families, and there is a direct positive correlation between the incidence of such abuses and the level of conservatism. In the United States, the so-called "red" states are primarily conservative because of religion, making them excellent test cases for claims of religious moral superiority. 38% of the states with the lowest crime levels are conservative. 76% of the most dangerous cities in America are conservative, and 3 of the 5 most dangerous cities are in Texas. The 12 states with the highest burglary rates—and 17 of the 22 states with the highest murder rates—are also conservative. Similar statistics exist for Judaism and Islam. The religious claim to moral superiority is complete and utter bunk.

> *While we believe such fables as these, or either of them, we believe unworthily of the Almighty.*
> Thomas Paine

Given these numbers, should society give religion the right to tell us what is good and what is bad? If so, which religion should we turn to? Which parts of which self-contradictory and hopelessly edited books should we rely on, and why? How can we be sure that any such set of choices is any better than any others? These questions have no answers, which makes the privileged status of religion even more baffling. No discussion about sexuality or morality is complete without clergy from different faiths—as if they have better medical, scientific, or ethical qualifications that doctors, researchers, and philosophers! In 2006, the US Supreme Court's decision in *Gonzales v. O Centro Espirita Beneficiente Uniao do Vegetal* granted an exemption to drug control laws to allow this group to use *ayahuasca* (a hallucinogenic drug obtained from various plants) in its ceremonies. This group did not need to prove anything about the drug's safety or efficacy, merely that it is part of their religion. Meanwhile, secular and religious people alike are routinely prosecuted for possessing drugs with well-established medical uses such as marijuana. Stories of religious conversion have reduced more than one prison sentence and commuted more than one execution. Whether one agrees with drug laws or criminal penalties is beside the point: Religion is the ultimate trump card. Religious people enjoy benefits that no secular person has access to, this is in a nation that prides itself on the separation of church and state! Needless to say, religious influence is even stronger in nations without even the pretense of these protections.

Religion needs to mature more if the world is going to survive in good shape.
John Lubbock

One could be forgiven for thinking that religious people should be among the happiest, most self-assured people on Earth, but one would also be wrong. Religion portrays itself as vulnerable and under constant attack from all sides despite the tremendous advantages it enjoys. Religion demands—and receives—a level of respect that far exceeds that given to any other organization, or from one person to another. Why does an omnipotent, omniscient, and omnipresent deity need such special protection? The simple answer is that religion enjoys a huge lobbying and public opinion advantage over atheists and agnostics. This respect is a double-edged sword: A society that agrees to grant special status to religion must by definition extend that respect to extremists such as Osama bin Laden, because the moderates and fundamentalists of any religion are built on the same foundation. The alternative is to withdraw respect for all religions.

Systems of belief can be extremely powerful and dangerous.
John Bowker

We saw that Luther triggered the Protestant movement when he nailed his 95 theses to the door of the Castle Church in Wittenberg, Germany, in 1517. He, like Tertullian and countless religious leaders before and since, recognized the danger that logic and reason pose to abject faith. "Reason," he said, "is the greatest enemy that faith has. It never comes to the aid of spiritual things, but—more frequently than not —struggles against the divine Word, treating with contempt all that emanates from God. Whoever wants to be a Christian should tear the eyes out of his reason. Reason should be destroyed in all Christians." Similar injunctions exist in all of the JCI texts.

> *On the whole, I tended to cross the street when prayer meetings broke up.*
> Christopher Hitchens

True to these mandates, religions cling to science when they think it suits their purposes, and ignore or persecute it when it doesn't. The founders of the JCI religions could not possibly have foreseen how science would advance human knowledge. The obstacles they threw in the path of progress set the entire human race back centuries; but for religion, those of us alive today might have found ourselves in a world as far removed from our actual circumstances as we are from the Renaissance. Many of our greatest scientists might have been condemned to death had they been born only a little earlier. How might history have been different had 14th century Europeans set about trying to find out what caused the Black Death and how the disease was transmitted? Had they known that flea-infested rats hitching rides on merchant ships were to blame, would they have mutilated blasphemers, worn brightly colored clothing, or taken enemas? How might they have reacted if someone had discovered penicillin back then? Would this person have been feted as a hero, or burned as a heretic? Had religion triumphed over science, the Internet would not exist, nor would the computer I am using to write this book—a fact that has not stopped modern religious people from using both computers and the Internet!

Deliberate Ignorance

> *O holy people, whose very kitchen gardens produce gods.*
> Juvenal

One of the single greatest differences between science and religion is that the former actively seeks out and corrects mistakes, while the latter digs in its heels and defends the mistakes even more vigorously, lest they be proven wrong and lose their raison d'être. Religion clings to the pretense that ancient taboos and made-up stories contain the ultimate sources of meaning. At best, this hampers the ability of the faithful to address their concerns rationally. At worst, it causes violence. My son's unsolicited and unprompted reaction to the stories told during a church service some neighbors took him to was, "This makes no sense!" I am not the first person to imagine that religion would lose much of its power if we simply gave children honest answers to their inevitable questions. Instead, we live in a world where people are executed for imaginary crimes, where millions of children are "educated" by forcing them to recite an ancient book, and where women are treated like chattel and denied reproductive and other basic rights.

This is not to say that people of faith lack all reason; on the contrary, religious leaders leverage the strength of the human intellect and use faith as the fallback when reason fails. In general, religion supports reason as long as reason supports faith; that support vanishes the moment reason fails. Faith is the spackling paste that both fills in the gaps in evidence and allows religion to remain viable as a going concern. In 2008, intelligence researcher Helmuth Nyborg (1937-) tested the relationship between IQ and religion. Atheists scored an average of 1.95 points higher than agnostics, 3.82 points higher than religious liberals, and 5.89 points higher than religious conservatives. "I'm not saying that believing in God makes you dumber," said Nyborg. "My hypothesis is that people with a low intelligence are more easily drawn toward religions, which give answers that are certain, while people with a high intelligence are more skeptical." Is it a coincidence that monotheism flourished in a land populated by a small backward tribe? Is it any coincidence that religious sects seek converts among the poor, dispossessed, and befuddled?

The myths and mysticism that lie at the root of religion represent humanity's first and most profound attempts to answer life's most significant questions. The layers of dogma and violence that form the foundations of our largest religions are the worst imaginable insult to both the people religion attempts to "save" and the God religion claims to adore and obey. The JCI religions teach people that they are miserable sinners before God who must live in constant submission and fear. Their prayer positions are classic positions of utter surrender. The whole point of life is preparing for God's next arrival, which many religious people believe will occur in their lifetimes. Incredibly, this lower-than-dirt status still manages to instill a level of selfish conceit, because God cares about each individual person and created the entire universe just for us. Never

Between two barbaric nations, the one that was the more superstitious of the two would generally be the more united, and therefore the more powerful.
Francis Galton

The Confederate troops, of course, sang and prayed to the same god to do the same thing to their enemy.
Michael Shermer

mind the parade of discoveries that continue knocking humans off our self-made pedestal, including the potentially habitable planet Gliese 581g that lies only 20 light years from Earth. Every sect claims to be carrying out a special mission from God on behalf of a particular group of people. Each sect has a sacred book of divine revelations, and each accuses the others of being wrong.

With our without religion, good people can still behave well and bad people can do evil; but for good people do to do evil—that takes religion.
Steven Weinberg

Religion comes from a time when people had no idea how the natural world works. Middle Age scholars who expanded religious dogma and hold on daily life did the best they could under horrid circumstances in a time when life really was nasty, brutish, and short. A well-educated religious person from the 14th century would seem hopelessly ignorant in all subjects today except faith; this person's scientific knowledge would embarrass most schoolchildren, but he would know a lot about God. As such, the stories in the Bible, Koran, and other books bear little resemblance to historical or scientific reality and are rife with contradictions, despite the best efforts of theologians for centuries. The method of Christ's execution has no special significance, and the idea that he came to Earth just to temporarily die for our sins is laughable. Crucifixion is a terrible way to die, but plenty of people have died in even more horrific ways—many of them at religious hands. The stories are so ludicrous that one can only accept them logically after one has accepted them emotionally. Texas governor Ann Richard is reputed to have opposed bilingual education in schools by saying, "If English was good enough for Jesus Christ, then it's good enough for the schoolchildren of Texas."

The Christian god is a being of terrible character—cruel, vindictive, capricious and unjust.
Thomas Jefferson

Again, the problem is not that religion began when human knowledge was less advanced than it is today; the problem is that religion has failed to evolve as our knowledge has evolved. The philosopher Friedrich von Schiller (1759-1805) lamented that, "Against stupidity the gods themselves contend in vain. It is by means of the gods that we turn gullible and stupidity into something ineffable." William James agreed, saying, "Give up the feeling of responsibility, let go your hold, resign the care of your destiny to higher powers, be genuinely indifferent as to what becomes of it all, and you will find that not only do you gain a perfect inward relief, but often also, in addition, the particular goods you sincerely thought you were renouncing." And why not? John 20:29 says, "Ignorance is the real coinage of the realm. Blessed are those we believe but don't see." Children today are told to defer to their parents' god by ignoring much about the modern world. The statistics are less than comforting:

- 22% of Americans are certain that Christ will return within 50 years.
- Another 22% think his return is probable.
- This same 44% attends church at least once per week and take Bible teachings literally. This group forms the most cohesive and motivated segment of American society; their views are influencing our courts, schools, and government policies.
- Only 28% of Americans believe in evolution.
- 73% believe in angels.

It is interesting to find that people of faith now seek defensively to say that they are no worse than Nazis or fascists or Stalinists. One might hope that religion had retained more sense of its dignity than that.
Christopher Hitchens

It should be obvious that this much ignorance among the population of the world's lone superpower poses clear and present dangers for the entire world. Just ask the Iraqi people.

Author Christopher Hitchens (1949-) presents four objections to faith that he calls irreducible:

- Faith misrepresents the origins of man and the cosmos.
- Because of this original error, faith manages to combine the maximum of servility with the maximum of *solipsism* (the idea that one is only sure of one's own existence).
- Faith is both the result and the cause of dangerous sexual repression
- Faith is ultimately grounded on wishful thinking.

If one accepts creation myths literally, then the first objection is correct; however, as we saw in Chapter 3, treating these myths as metaphors reveals a surprising amount of scientific accuracy. As for the fourth objection, I am inclined to agree to the extent that faith is placed in religious teachings; however, one cannot say that mysticism and mythology must have anything to do with wishful thinking.

To the illustrious Herr Adolph Hitler, fuehrer and chancellor of the German Reich!
Eugenio Pacelli
(Pope Pius XII)

The term *agnostic* quite literally means "without knowledge." An agnostic person is someone who feels that s/he does not have all of the evidence needed to form an opinion or belief. Agnosticism toward ambiguous or missing evidence is appropriate. Permanent agnosticism is for questions that can never be answered; however, the question of God's existence has at least a probabilistic answer if not a definitive answer. The idea of God is a scientific hypothesis that can be tested. As many people have said, God could end the debate once and for all if s/he chose; the fact that s/he has not done so to everyone's universal satisfaction either means that God does not exist or that millions of people are looking in the wrong place. We will continue to address this question throughout this book.

Indoctrinating Children

My feeling as a Christian points me to my lord and savior as a fighter.
Adolph Hitler

All authoritarian cultures are obsessed with children, from the Hitler Youth (Germany) to the Young Pioneers (North Korea), and countless madrassas, Sunday schools, religious summer camps, and schools throughout the world. Getting someone to believe logically inconsistent and mutually contradictory stories and display the kind of blind faith needed to keep the group viable must begin young. It is far easier to gain converts by having children (being fruitful and multiplying) than it is to convert adults. Among adults, those most likely to accept conversion are those who have hit rock bottom through hunger, drugs, homelessness, crime, etc. It is no coincidence that religious missions focus so heavily on these groups. Yes, they do help people who might not otherwise have anywhere to go, but this help comes with explicit or implicit strings attached. Under ordinary circumstances, a person who is old enough to have developed even a modicum of reason will be virtually immune to religious advances.

It is impossible to know just how many children have been negatively impacted by religion in countless ways such as:

- Having their entire education consist of learning to recite holy books.

Politics has slain its thousands, but religion has slain its tens of thousands.
Sean O'Casey

- Genital mutilation, such as male and female circumcision/infibulation.

- Sexual repression. The prospect of bedding 72 virgins was a key ingredient behind the September 11th hijackers. Given healthy sex lives, it is highly doubtful that they would have felt motivated to carry out suicide attacks.

- Sexual abuse. Some estimates indicate that the majority of children in Ireland have been subjected to sexual abuse by clergy. It strains credulity to think that the child abuse scandals of the last several decades are anything new. It is reasonable to believe that priests have been molesting children for centuries. In fact, Mary MacKillop was posthumously canonized in late 2010, more than a century after she was excommunicated for exposing child molestation by a priest.

- Physical abuse. Children in religious families are more likely to experience physical abuse in the home. Remember that the JCI religions name Abraham as their founder, a man who was ready to kill his son for God.

Computers do as they are told in a very robotic way without any way to know if the instructions they are following are good or bad. The same CPU can just as easily help a doctor perform a delicate operation or guide a missile to its target. Children evolved a built-in rule to obey parents and elders, just like moths evolved to navigate by moonlight. Religion can short-circuit this crucial survival instinct just like a candle flame can short-circuit a moth's crucial navigational abilities. "Give me a child for seven years," a Jesuit said, "and I will give you the man."

Are you the victim of religious indoctrination? If you believe that Christianity is true and Islam is false despite knowing full well that your views would be reversed had you been born Arabic, then the answer is yes. But that's OK because religion is moral, right? Wrong. 168 Israeli children were given a text lifted directly from the book of Joshua where the name "Joshua" was replaced with "General Lin" and "Israel" was replaced with "a Chinese kingdom 3,000 years ago." Only 7% of these children approved of the general's behavior, and fully 75% of them disapproved. Remove religion as a variable, and the immorality becomes blindingly obvious.

As we saw in Chapter 9, the presence of imaginary friends is both normal and can provide comfort to a child. These imaginary beings (including gods) can devote their full attention to the child/believer, but their power to comfort does not of itself make them any more or less real. Psychologist Paul Bloom (1963-) believes that children have a natural tendency

Tell a devout Christian that his wife is cheating on him or that frozen yogurt makes a man invisible and he is likely to require as much evidence as anyone else, and to be persuaded only to the extent that you give it.
Sam Harris

There is in every village a torch—the teacher: and an extinguisher—the clergyman.
Victor Hugo

toward dualistic thinking, since they can live in both the physical world and a fantasy world with their imaginary companions. To him, religion is simply a by-product of our natural proclivity to dualism. In Chapter 2, we saw that dualists see a difference between matter and mind, while monistic materialists believe that mind is simply a product of brain, and monistic idealists believe that brain (and all matter) is a product of mind. Our tendency to see patterns and purposes all around us that is so useful for trying to see and avoid hungry predators predisposes us to creationism. Professor Deborah Keleman speculates that children are intuitive theists. *Teleology* is the belief that purpose and design are part of (or apparent in) nature; children are natural teleologists, and some of them may simply never grow out of it. Do mysticism, mythology, and religion really owe their existence to something this simple? Keleman herself acknowledges that more research is needed.

Richard Dawkins, you are so wrong. God is not a delusion. No way. No how, Instead, he's right here in the Holy Land.
Becky Garrison

Telling a child that she or he is doomed to eternal torture because a woman who never existed ate an apple that never existed, and that the only way to avoid this terrible fate is to believe and obey everything the church says, is abusive on its face. What ethical parent can possibly condone the idea of punishing their child for someone else's crime? A religious parent would protest just as loudly as anyone else if his child was accused of someone else's crime, such as a burglary committed by another child. This same parent has no problem believing that the same child is worthy of unimaginable suffering on purely religious grounds. Original sin is abusive.

Denying a child the full use of the brain that God supposedly gave her by burying her natural curiosity and inborn drive to learn under dogma is abusive on its face. Religious fundamentalists are hell-bent on eviscerating science education for hundreds of thousands of children. Religious moderates may not be championing these changes, but are certainly abetting them by teaching the virtues of unquestioning faith.

Moral indignation is a technique to endow the idiot with dignity.
Marshall McLuhan

Using a religious term to describe a child is abusive on its face, because it applies a label that the child is unable to comprehend. One would not call a child a Democrat or a Republican because we understand that they are too young to know the implications of belonging to a political party. Our civil and criminal laws recognize that children are incapable of making mature decisions. Despite this, we routinely hear about Muslim children, Christian children, Mormon children, etc. If a child is incapable of making a rational decision about a mundane aspect of daily life, then how dare we label them with names that we believe carry eternal consequences?

God's Will

The one thing tsunamis in the Pacific ocean, hurricane Katrina in Louisiana, the attacks of September 11th, the earthquake in Haiti, and other disasters have in common is the litany of religious finger-pointing that happens after the fact. "This is what happens when you don't listen to us!" they say as they offer redemption that is not theirs to offer, only to threaten us with more calamities if we reject their offers. And they say they are acting out of love in accordance with God's divine plan! Where and how did people become so full of themselves as to believe that a god who created over a quintillion (10^{18}) stars with innumerable planets cares about our scrap of rock above all others, and came here to die because a woman bit into a fruit? Does every inhabited world have its own Eve, and did God die there too? If so, God would seem to be spending a lot of time killing himself.

How can anyone believe that this creator went one step further by choosing one nation and one people as His chosen ones *über alles*, with all others consigned to second-class status? How could this OOO being be so irresponsible as to reveal His divine plan to a few random illiterates, thus guaranteeing a passive game of telephone where His message will assuredly become corrupted beyond recognition? How much vanity does it take to look in a mirror and see yourself as a linchpin in such a God's plan? How low does one's self esteem need to be in order to believe that one is sinful by nature and deserving of eternal damnation from the moment one's lungs first draw air? How many baseless assumptions must one make in order to weave all of this into a seamless whole that withstands all attempts at logical scrutiny? How many saints does one need to pray to in order to cover all of life's bases before monotheism gives way to monolatry or polytheism? How much pain and suffering must one endure before one finally realizes the bankruptcy of such beliefs? Do people really put themselves through all of this and more to try to find the security and

---

Science can purify religion from error and superstition, and religion can purify science from idolatry and absolutes.
Pope John Paul II

The more unnatural anything is, the more it is capable of becoming the object of dismal admiration.
Thomas Paine

> *You will find something much greater in the woods than in books. The woods and stones will teach you what you cannot learn from other masters.*
> Bernard of Clairvaux

assurances and related comfort that are so obviously missing from life? People do.

At its core, religion claims that people can transform the way they see and interact with the world. Daily life may be mundane, but the extraordinary and divine are just within reach. This concept may in fact be true and lies at the heart of this entire book, but religion dangles this carrot just out of reach. Becoming just like Christ is impossible, they say, forgetting Christ's own instructions on how to not only equal but surpass him. (See Chapter 6.) All a believer can do is tally up one's sins, believe unbelievable things, and wait for the world to end all around them. Religion hides the many ways in which people can have profound experiences that bring them close to the Godhead in a way that all mystics, Christ included, would recognize and welcome. All spiritual traditions I know of speak of happiness far beyond the pleasures of this world and warn that seeking mundane happiness can prevent us from achieving the higher happiness of enlightenment. The major religions, particularly the Western religions, obscure this knowledge behind a truly spectacular dog and pony show. How easy it is to forget that history might have turned out quite differently, but for the Battle of the Milvian Bridge (Chapter 7)!

The question of how a benevolent OOO god could allow evil remains a persistent problem for Western religion. Some of the many explanations that have been floated include:

- Evil is a product of the human will; God cannot stop us from being evil.

- Some suffering is needed in order to develop one's virtues.

- Good and evil are contrasts; it is impossible to know happiness unless we have known sadness.

- Eliminating evil will make us unable to recognize good.

- God's definitions of good and evil are different than our own.

- Evil exists for a purpose that is part of God's plan.

- The devil is responsible for evil; God is innocent.

> *The idea that any of our religions represents the infallible word of the one true god requires an encyclopedic ignorance of history, mythology and art even to be entertained.*
> Sam Harris

- Weakening the definition of God makes explaining evil easier.

Note that each of the above arguments either forces God to be less than OOO or monotheism to be false, both of which directly violate JCI teachings. Nevertheless, many Christians believe that Satan is an independent evildoer who is independent of God and responsible for such calamities as the September 11th attacks. Again, we just saw that such beliefs eliminate both monotheism and the idea of an OOO God.

There are millions of people who do good deeds in the name of their religion, and who honestly feel that religion connects them to God. Unfortunately, these people prove nothing, since morality and ethics can flourish independently of religion. If we had it to do over again, would we need to believe that a man was born of a virgin in order to be good to one another? As for Christ's teachings, author C.S. Lewis (1898-1963) alleged in *Mere Christianity* that a man who is merely a man who said what Jesus Christ said would not be a great moral teacher; he would be either a lunatic or the devil. Therefore, Christ was either the son of God, a mad man, or worse. Lewis wrote before the Gnostic gospels were discovered at Nag Hammadi in 1945, and thus had no access to Christ's Gnostic teachings. He also seemed to have either not known about, or forgotten about, the many philosophers and ethicists who have contributed to all cultures before and since Christ.

Lewis also critiqued the crucifixion as follows: If A harms B and B forgives A, all is fine. But how can C, who is not involved with either party, forgive A? Christ proclaimed that all sins are forgiven without bothering to consult the victims, acting if he was the offended one. According to Lewis, this only makes sense if Christ was God and his laws had been broken. Anyone else saying such a thing would be conceited beyond belief. Here again, Lewis seemed unaware that most of Christ's alleged teachings had been added, deleted, and edited to suit the orthodox agenda, as we saw in Chapter 8. For example, the story of Christ writing the sins of those who wanted to stone the adulteress (John 8:1-11) is not part of the original gospel. We will see more of Lewis in Chapter 11.

The New Testament contains more edits than it does words, and many of these edits alter the original meaning of the text.

To accept the idea that god created starlight 'in transit' and rocks with radioactive elements used up in order to give them the 'appearance' of billions of years of age is to accept the idea that god i deceptive.
Fred Heeren

By high school it was commonplace among us that it was just plain useless to argue with Catholics about religion, because no matter what you said, they knew they were right or at least seemed to know.
Harvey Cox

(See Chapter 8.) The Old Testament has been similarly altered. The four Gospels that were deemed "inspired" by God disagree in several important respects, and the original book that inspired those gospels, called *Q*, has been lost to us. Even if none of this had happened, the constant evolution of language, words, meanings, and usage would make retaining the original message all but impossible. The word of God cannot possibly be found in printed works. Therefore, if one assumes that the Bible propounds God's will, then one must also assume that God has done a rather shabby job of safeguarding His own words—hardly OOO of him.

If anyone promoted such views in any area outside a religious context, he would be taken in for a psychiatric evaluation.
Victor Stenger

Government—however you want to limit that concept—derives its moral authority from God.
Antonin Scalia

Faith is the commitment of one's consciousness to beliefs for which one has no sensory evidence or rational proof. A mystic is a man who treats his feelings as tools of cognition. Faith is the equation of feeling with knowledge.
Ayn Rand

Forsaking Reason

Christ lived in an era when pagan myths still held sway. Plenty of other gods and otherworldly beings were celestially begotten of virgins, the offspring of gods copulating with mortal women in a manner reminiscent of Rick Riordan's *Percy Jackson and the Olympians*. The Holy Trinity of Father, Son, and Holy Ghost is the cobbled-together result of consolidating some 30,000 gods into a single god who is at three with himself. Statues of Mary replaced statues of Athena, Diana, and others. Heroes revered as gods were recast as saints, and Christianity ended up just as polytheistic as the pagan religions it replaced. The orthodox Christ story smacks of fraud.

Ancient giants fought the god Jupiter. A lightning bolt knocked one rock-throwing giant under Mt. Edna, which now erupts when he belches. Satan was confined to the fires of hell after losing a battle with God, who then had to let him out in order to tempt Eve. Did God put Satan back when this nefarious deed was done? No, because the universe needs Satan, who was promised the Jews and the Muslims. The similarities between these two stories are too striking to be coincidental, and the Christian version is both laughably pathetic and an utter insult to God, because it strips God of His omnipotence. The freed Satan became omnipresent, which forced God to surrender the Earth. God then sacrificed His own son (who is not really a son but rather an aspect of God Himself) to repair the damage, and punishes every human because He let the cat out of the bag. This all would make a lot more sense if God had placed Satan on the cross!

The mere fact that a story is old does not make it true. The origins of every nation and religion are steeped in myth, tradition, and outright fabrication. Drop the idea that Moses authored Genesis, and one is left with one of many pieced-together collections of stories of questionable historic, moral, or literary value. (Of course, Moses himself is not exactly a nice guy.) The story of Sampson and Delilah (Judges 16) is yet another example of a story that tells us everything about codependent relationships and betrayal, but nothing about God. The Bible is laden with such stories that glorify atrocities. In fairness, a Bible full of educational, uplifting, and inspiring stories could still not be the word of God for the same pragmatic reasons mentioned above. Still, the Bible we do have can only be described as demonic; even its good parts tell us nothing new or different about the human condition. Most people, Christians and others alike, have no idea just how evil the Bible is and have never read anything beyond the carefully selected and interpreted passages provided by their clergy.

I met a woman I'll call Linda some years ago between my divorce and meeting Jennifer. Linda told me how much she loved singing jazz, but that she was giving it all up to travel with a mission preaching the (Christian) word of God. According to her, this life was little more than a dress rehearsal to see who made the cut and got into heaven.

"But Linda," I said, "what if God placed the love of jazz within you? Wouldn't singing jazz to your heart's content be the best way to give thanks and serve God?"

I thought it was a perfectly reasonable question, but Linda didn't see it that way; however, she was still talking to me, so I asked the next question that came to mind: Had she read the Bible?

"Of course!"

"All of it? Cover to cover?"

"Um... no. Just some passages here and there."

"Linda," I said, "you told me that you believe this life is pretty meaningless except as preparation for getting into heaven. Would you knowingly board an airplane flown by someone who is not a duly trained and certified pilot, or submit to sur-

---

Real knowledge is to know the extent of one's ignorance.
Confucius

The memory of my own suffering has prevented me from ever shadowing one young soul with the superstitions of the Christian religion.
Elizabeth Cady Stanton

gery by someone who is not a duly trained and licensed doctor?"

"Oh, goodness no!"

"Well, I'm confused. If you would not do things in this meaningless life with people who are not properly trained and qualified, and if getting into heaven is a one-way ticket that's punched for all eternity, then how can you possibly be qualified to tell anyone how to accomplish this one all-important task when you have never been to seminary or even bothered to read the entire book once?"

Linda never spoke with me again. I hope she's out there somewhere singing jazz and loving every second of it.

Anyone searching for evidence of God's will would do well to study the universe around us. This universe (which may be one of many) has more going on than any one religion or scientific discipline can cover. Anyone needing a miracle would do well to wrap their minds around the incredibly long odds against our having made it this far. There are more stars in the average galaxy than cells in our brains; there are also more galaxies in the universe than there are cells in our brains. We are an almost incomprehensibly small part of the universe. Stare up at the night sky or peer through a microscope, and you will never want for fascination and joy. If science is the universal language that ties the smallest particle to the largest galaxy and all in between it, then the universe is the scripture. If there is a God, and if that God has a will, then it is only by studying His works firsthand that we can discern that will. Science is available to anyone and can be a source of profound inspiration.

Apropos of science, can you imagine the religious reaction if carbon dating the shroud of Turin pegged its age at around AD29? Or if DNA testing on Christ's remains revealed his virgin birth? Or if prayer was conclusively found to influence events? Or if the waters of Lourdes had extraordinary healing power? Or if traces of a mass migration out of Egypt into Israel some 3,000 years ago turned up? Or if the remains of a great city with its walls all fallen outward were dug up? Science has found traces of much smaller and older tribes and migrations, conclusively proved the existence of chromosomal Adam and mitochondrial Eve, and more. Would religion let scientific validation of its most treasured stories and relics go

---

You believe in a book that has talking animals, wizards, witches, demons, sticks turning into snakes, burning bushes, food falling from the sky, people walking on water, and all sorts of magical, primitive and absurd stories, and you say we're the ones who need help?
Mark Twain

For many decades now the Western intellectual world has not been convinced that theology can be engaged in with intellectual honesty and integrity.
Arthur Peacocke

unnoticed? So why are they are so noticeably silent or antagonistic when science contradicts them? You can't have it both ways. Despite this, luminaries from Socrates to Galileo and beyond have relented in the face of religious opposition.

Science is far from finished. All scientific disciplines have discoveries waiting to be made and plenty of room for advancement. Our understanding of the universe and how it works is anything but complete. Religious people point to these gaps as evidence for God. For example, Behe and others say that a concept called *irreducible complexity* offers proof that God is guiding evolution to meet His divine blueprints, a viewpoint commonly called intelligent design. Irreducible complexity basically says that an organ (such as an eye) cannot function without all of its parts present and functioning. Since it is virtually impossible for a complete eye to have evolved piecemeal, God must have designed an eye and then caused it to evolve spontaneously.

I'd take the awe of understanding over the awe of ignorance any day.
Douglas Adams

Intelligent design and irreducible complexity have both been disproved, but provide excellent examples of the so-called *God of the gaps* idea, which says that the gaps in our scientific knowledge provide room for God to operate. As for the eyes, some sight is better than no sight, and eyes have evolved independently along completely separate lines. There is also the question of how our "backward" eyes could possibly be the result of any kind of design process. We will discuss this more in Chapter 19. Meanwhile, the *anthropic principle* is another idea used to try to prove God. We will see the anthropic principle again in Chapter 11; the short version is that the universe is so finely tuned for life that God is the only possible explanation. These arguments basically boil down to, "I have no idea how this happened, and I bet you don't either!"

A god that can act in measurable ways using the laws of nature should be detectable by science. The fact that science cannot detect the god of any religion goes a long way to invalidating the idea that such a god exists. As for a god of the gaps, unless a particular gap can be shown to lie forever beyond the reach of science, then that premise also falls flat. Philosopher Theodore Drange (1934-) cites the lack of evidence argument:
1. If God exists, there must be objective evidence.
2. No good evidence exists.
3. Therefore, God probably does not exist.

The Christian focus is overwhelmingly on sin, sin, sin, sin, sin, sin. What a nasty little preoccupation to have dominating your life.
Roland Barthes

> *I'm glad some people have that faith. I don't have that faith. If there is a God, a caring God, then we have to figure he's done an extraordinary job of making a very cruel world.*
> Dave Matthews

This argument does not preclude all possible gods, a concept I will return to in Chapter 14; however, it is fair to say that the "big man in the sky" model of God has been pretty much ruled out for lack of evidence. This is an important distinction. For example, geometric principles would be the same regardless of who wrote them. Euclid's work describes such principles; it does not cause them to exist. The same cannot be said of religious texts. Extraordinary claims really do require extraordinary evidence.

On a mundane level, the evidence is damning: Humanity could be little more than accidental flotsam in the cosmos, each of us destined to live for a few decades before winking out like a light to return to the utter oblivion from whence we came. Our planet will eventually be swallowed by the Sun, and the entire universe seems destined to cool and fade to utter blackness, like the dying glow of a photographer's flash bulb. Amid these dreary prospects, we see animals sharing food, comforting each other, cooperating with each other, tending the sick and injured, rescuing the stranded, adopting orphans, and displaying behavior that can only be described as moral without any observable traces of religion. If there is a god behind it all, then it is beyond absurd to think that s/he actually wants us to cut the tips of our male children's penises or amputate our female children's clitorises.

> *In no instance have the churches been guardians of the liberties of the people.*
> James Madison

There is a story of Chinese people who asked the Christian missionaries among them why, if God revealed Himself to the West, did he wait so long before heading East? Some say that a loving god would make its presence known; on the other hand, demanding that of a god may be just a little much. Then again, perhaps the evidence for God (albeit not the JCI god) is all around us if we only know where to look. Either way, religion makes claims that cannot possibly be verified; those claims that can be tested universally fail. Religion demands that we check our brains at the door and believe in the unbelievable. Faced with the slightest dissent or variance, religion has responded with threats and violence. This sheltered background has ill prepared religion for any kind of substantive debate, if people like Linda, Lewis, and countless others are any indication

Dubious Heroes

On June 17th, 1996, Ireland enacted the 15th amendment to its constitution to allow divorces, having decided that the Catholics should no longer be able to legislate morality and the impossibility of reconciling Protestants to Catholic rule. Mother Teresa (1910-1997) flew all the way from her Missionaries of Charity organization in Calcutta, India to join the Catholic campaign against this amendment. The message was clear: A woman remaining with an abusive husband was fine in God's eyes, whereas a divorce could endanger her immortal soul. Protestants could either accept Catholic rule or face damnation themselves. Nobody so much as suggested that the Catholics be allowed to follow their own rules without imposing them on others. This is the same Mother Theresa who campaigned against contraception and abortion despite working in an overpopulated nation, the same Mother Theresa who believed in the spiritual goodness of poverty, the same Mother Theresa who is reputed to have kept those in her care in squalor while her foundation had access to tens of millions of dollars. Despite this questionable track record, she will doubtless be made a saint and elevated to join the rest of the Christian pantheon, thereby encouraging the ongoing belief in, and support of, religious fraud. The amount of suffering this will cause is immeasurable. "Abortion is the greatest destroyer of peace," she proclaimed... and yet she won the Nobel Prize.

The 2005 tsunami that killed some 250,000 people, hurricane Katrina in New Orleans in 2006, the 8.0 magnitude Haiti earthquake of 2010, etc. all get religious people scrambling to interpret God's will in light of these tragedies. Assuming that we live on a geologically active planet and taking adequate means to protect ourselves using tried and true techniques (like the robust building codes that saved countless lives during and after the 8.8 magnitude earthquake in Chile in 2010) is a much simpler and more elegant solution. The billions of dollars funneled to religion each year could pay for a lot of seismic upgrades, levees, etc. Instead, these resources go to the likes of evangelist Pat Robertson, who blame calamities not on nature, but on homosexuals and others deemed undesirable.

Marjoe Gortner (named after Mary and Joseph) was born in 1944 and started preaching at age 4, spurred on by his minister father and mother, who abused him whenever he wavered or

---

I'm frankly sick and tired of the political preachers across the country telling me as a citizen that if I want to be a moral person, I must believe in A, B, C, and D. Just who do they think they are? And from where do they presume to claim the right to dictate their moral beliefs to me? And I am even more angry as a legislator who must endure the threats of every religious group who thinks it has some God-granted right to control my vote on every roll call in the senate. I am warning them today: I will fight them every step of the way if they try to dictate their moral convictions to all Americans in the name of conservatism.
Barry Goldwater

> *There are specific religious beliefs which actually encourage, or even demand, behaviors that can only be described from outside the system as evil.*
> John Bowker

complained. He rebelled at age 17 and dropped out of the revival circuit, having made $3,000,000 that his father absconded with. Marjoe returned to preaching in his 20s when money ran low. This time, he shared the tricks of how to work an audience for maximum financial gain with filmmakers Howard Smith and Sarah Kernochan. The resulting *Marjoe* took the Oscar for best documentary in 1972, because it blew the whistle on the entire religion racket from the perspective of both victim and perpetrator. Despite this damning and conclusive evidence, hundreds of other preachers use the same tactics today to fleece billions of dollars from millions of believers.

It took more than a little bravery for Luther to nail his anti-Catholic theses to the church door, but that did not stop him from persecuting others when he had the chance. His 20th century namesake Martin Luther King, Jr. (1929-1968) was an adept preacher who could run rhetorical rings around white Southern preachers. Dr. King denounced violence and racism, a laudable position that required a tremendous amount of bravery and self-sacrifice in the Jim Crow South. This much is known by most schoolchildren worldwide. Few people of any age know that the Bible supports slavery and holds violent oppressive people as paragons of righteousness, which can only mean that Dr. King was anything but a true Christian. His elevated and much-needed efforts required no higher power for validation. The backlash from the right-wing Christianity that remains a viable force today is much closer to the Biblical ideal.

> *Christians were called atheists by the Romans because Christians denied the gods, refused to sacrifice to them for the good of the community, and shunned city feasts.*
> Stephen Tompkins

Millions of Catholics accept the idea that the pope is infallible in matters of faith and morality, in the same way most of us accept the fact that the Sun rises in the East. Is this faith misplaced? Consider that no empirical evidence exists to substantiate most papal claims. How can anyone alive today prove that Christ was born of a virgin? Where could they get the kind of detailed knowledge about an ancient woman's sex life to know? A 24-hour covert surveillance operation spanning several months plus DNA samples and testing would be the only way to make sure. The pope has nothing more than a subjective, heavily edited, and unsubstantiated book plus his own opinions and beliefs to go on. Such flimsy evidence would be laughed out of any court. For example, Pope John

Paul II claimed that Our Lady of Fatima spared his life by guiding the bullets that struck him on May 13th, 1981. Catholics everywhere must accept that claim as inerrant fact, despite the obvious questions of why the Lady could not have arranged the bullets to miss him entirely or how he can possibly know which Lady helped him. As Dawkins drily observed, "Presumably Our Ladies of Lourdes, Guadeloupe, etc. were all out shopping at the time."

The list of similar stories goes on and on.

Corrupted Sexuality

Imagine walking into a room and seeing an old man sucking a baby boy's penis. Disgusting, you say? What if this man is a *mohel* carrying out the Jewish rite of circumcision that involves cutting the penis, sucking off the foreskin into the mohel's mouth, and then spitting out the flap along with the requisite blood and saliva? How does the mantle of religious authority make this act any less brutal and revolting? I would argue that the crowd of jubilant people normally gathered to witness this torture of an innocent boy only makes the whole affair even more depraved and disgusting. In 2005, a mohel gave genital herpes to several boys and caused the death of at least two of them. Despite this, mayor Michael Bloomberg erred on the side of freedom of religion, ignoring Jewish doctors who correctly pointed out the risks inherent in this ritual. The level of sexual barbarism only worsens from here.

> *It's interesting to speculate how it developed that in two of the most anti-feminist institutions, the church and the law court, the men are wearing the dresses.*
> Flo Kennedy

As we saw in Chapter 8, the religious dogma of the virgin birth was probably caused by an illiterate scribe, which would make Pope John Paul's "infallible" ruling that Mary's hymen remained intact after the birth uproariously funny, if not for the dire consequences of such twisted logic. Additionally, Mark and John seem to have missed the memo about the virgin birth, an almost unthinkable omission if such a concept had been as central to Christians back then as it is today. In fact, they were both very concerned with establishing Christ's legitimacy, which makes perfect sense if Christ was the result of the breeding program described in Chapter 6. Even the apostle Paul says that Christ is the son of Joseph and Mary, saying that Christ was "born of the seed of David according to the flesh" (Romans 1:3) and "made of a woman" (Galatians

> *So, is there any tread left on the tires? Or at this point would it be like throwing a hotdog down a hallway?*
> Stewie Griffin
> (Family Guy)

> *Christ was the savior of men, but I am the savior of women, and I don't envy him a bit!*
> James Hinton

4:4). These passages plus Christ's distinct omission of any claim to divinity clearly establish his humanity with no virgin required, or even possible. The fact that Christ's final resting place was discovered (see Chapter 6) proves this beyond any reasonable doubt. A (presumably) innocent slip of a scribe's pen has caused two thousand years of sexual paranoia throughout Christendom and beyond.

Why is religion so obsessed with sex that occurs in private between consenting adults? According to them, such acts displease God and put innocent people in the line of fire when he decides to strike back—because apparently the aim of even an OOO being can be a little off from time to time, and the risk of collateral damage is very real. Evangelist Jerry Falwell blamed the September 11th attacks on the ACLU, abortionists, feminists, homosexuals, and others. Not to be outdone, Pat Robertson blamed hurricane Katrina on a lesbian comedienne who lived in New Orleans at the time; he also said that the Haitians caused the 2010 earthquake by selling their souls to the devil to achieve independence from France. These gentlemen and many other seem to have forgotten Genesis 18:16-33, where God promises to save the city of Sodom if only ten righteous people live there. I would be surprised if New York, New Orleans, and Port au Prince did not have at least ten devout Christians living there before the disasters. So much destruction for a relative handful of so-called sexual deviants?

> *I'm now convinced that the root cause of violence is deprivation of physical pleasure. When you stimulate the neurosystems that mediate pleasure, you inhibit the systems that mediate violence; it's like a seesaw.*
> James Prescott

From a purely evolutionary standpoint, all sexual species exist to mix and propagate genetic material. (See *The Natural Savage*.) As far as genes are concerned, all men should be making regular deposits at sperm banks, and all women should be donating eggs to infertile couples. Nothing is better for a gene than being passed on to thousands of children, especially if the biological parents need not invest in actually raising the offspring. This is not how the world works in practice, which proves that our selfish genes have not totally consumed us. I am not saying that humans are monogamous by nature or that the institution of lifelong marriage as envisioned by religion is the correct model for living. Humans evolved a "mixed" reproductive strategy that tends to be serially monogamous and include a healthy amount of philandering on the side. *Sperm Wars* by Robin Baker and similar works provide a fascinating insight into how straight and narrow we aren't. Religion turns our evolved nature against us by making sexual indiscretions punishable by death. Thousands of woman and girls are ritually murdered by their own relatives in so-called "honor killings" for alleged improprieties, even in the absence of any conclusive proof. These improprieties can be as simple as speaking to a man, as tragic as being raped, and anywhere in between.

I have already mentioned that Christ's supposed virgin birth is hardly a unique story. Perseus was born from the virgin Danae after she was impregnated by Zeus's shower of gold. Buddha was born through a hole in his mother's side. Catlicus hid a ball of feathers between her breasts and gave birth to Huitzilopochtli. The virgin Nana placed a pomegranate between her breasts that grew into the god Attis. A virgin Mongol princess gave birth to Ghengis Khan. Krishna was born of the virgin Devaka, Horus of the virgin Isis, Romulus of the virgin Rhea Sylvia, etc. Parthenogenesis is a central feature of many religious stories. What is so evil about penises that the skin of these august personages could not touch a surface previously touched by their fathers' penises? Why do so many people want to see the vagina as a one-way street?

I simply find it amazing that, for decades, it was okay for priests to rape boys by the thousands but when two men or two women want to affirm their love, that's not okay.
John Phelan

Genital Mutilation

It may surprise you to learn that some of today's popular foods were invented to curb masturbation, including Graham crackers and corn flakes. This is just one example of the ongoing religious taboo against masturbation that continues to be abetted by pseudoscientific remedies, such as recommending bland food and yogurt enemas. Why are religious people so opposed to masturbation? Christine O'Donnell, the 2010 Republican candidate for Senator from Delaware, summed it up when she said, "It is not enough to be abstinent with other people; you also have to be abstinent alone. The Bible says that lust in your heart is committing adultery, so you can't masturbate without lust. If he already knows what pleases him and he can please himself, than why am I in the picture?" Apparently Ms. O'Donnell's future husband would be incapable of thinking of her while masturbating. Then again, I can't imagine why anyone would want to masturbate to images of anybody who says that, "American scientific companies are

We're so focused on original sin that we've forgotten original pleasure.
Irwin Kula

cross-breeding humans and animals and coming up with mice with fully functioning human brains."

The same taboo against masturbation leads religious people to mutilate their own children's genitals, an act that is utterly incompatible with the idea of divine design, because it implies that the OOO God did a less-than-perfect job. Jews and Christians cut the foreskins off penises. Muslims amputate the clitorises of baby girls, and sometimes even remove the entire external genitalia, leaving only a mutilated slit behind. They also sometimes perform *infibulation*, which consists of sewing the vagina shut, leaving a small opening for menstrual fluids (which are also a source of religious paranoia and disgust). The threads are removed by the thrusting of the girl's husband, which must cause tremendous pain, suffering, and mutilation. All of this is done without anesthesia, and often without even basic sanitation. The photo I once saw of a roomful of tearful young Arab girls clutching their crotches through their dresses as another is led off to be "purified" was heart-rending.

> *The slow-witted approach to the HIV epidemic was the result of a thousand years of Christian malpractice and the childlike approach of the church to sexuality. If any single man was responsible, it was Augustine of Hippo who murdered his way to sainthood spouting on about the sins located in his genitals.*
> Derek Jarman

Some apologists defend this practice by saying that it allows parents to sacrifice only part of their children to God, unlike Abraham who had no such luxury. Besides, they say, Adam and Moses were born pre-circumcised! This level of sexual repression stunts normal human development, especially when combined with heavy religious indoctrination. There is just no way to persuade nineteen young men who enjoy normal relationships and sexual activity to blow themselves up by flying hijacked planes into buildings; by contrast, a lifetime of stifling sexuality with the promise of a debauched afterlife is the most efficient way to induce such behavior. It is only natural that parents would resist torturing their children like this without believing in a divine mandate.

> *All in all, considering the repression, defamation, and demonization of women, the whole of church history adds up to one long arbitrary, narrow-minded masculine despotism over the female sex. And this despotism continues today, uninterrupted.*
> Uta Ranke-Heinemann

Rape and Sexual Torture

Do we really need religion to tell us that torture, rape, spreading STDs, slavery, etc. are bad things to do? If so, why? Religion is the single largest promoter of all of these things. Women are infibulated to protect their "purity." How pure is the stench, pain, humiliation, infection, sterility, and increased child mortality that comes with it?

Bosnian women were raped by Serbian men who saw them as easy pickings, since there were not enough Bosnian men left to protect them. Acts like this speak volumes about the status of women in more than one society. One might think that religion would speak out against such crimes and come to the aid of the women. In addition to waxing poetic about the benefits of totalitarianism, Augustine wondered whether women were raped because they were excessively proud of their chastity or perhaps suffering some infirmity that made them targets. In other words, the women deserved it. Was Augustine excommunicated or censured in any way for such statements? No; he was made a Catholic saint. Islam is no different. Views like these are what drive a man to kill his own daughter for being raped and bringing shame upon the family.

Rape victims may have their throats cut, be stoned, doused with gasoline and set ablaze, beheaded, shot, and/or beaten to death. These men do love their wives and daughters and kill them not out of love but religious duty, despite the obvious disregard for the Golden Rule those acts entail. I think it is perfectly safe to say that any culture that teaches men to kill victims instead of comforting them paints a rather skewed picture of what love is.

One might think that the rapists and molesters are godless souls who lack religion, but the devout are one of the single largest sources of sexual impropriety, especially the clergy. For decades (and probably longer), the Catholic Church moved offenders around to avoid detection and keep the scandal hidden. When the public finally became aware of what was happening, the church blamed it all on a few "bad apples" who represented the exception, not the norm. This claim is a bald faced lie because of the scope of the problem, the complicity of the church hierarchy in maintaining secrecy and not disciplining and/or expelling the offenders, and the underlying belief that the victims got what they deserved. I find it impossible to believe that this problem spontaneously began a few decades ago; it is much more plausible to assume that clergy have been raping and molesting children of both sexes for as long as the church has existed.

Whatever they may be in public life, whatever their relations with men, in their relations with women, all men are rapists and that's all they are. They rape us with their eyes, their laws, their codes.
Marilyn French

Marriage as an institution developed from rape as a practice. Rape, originally defined as abduction, became marriage by capture. Marriage meant the taking was to extend in time, to be not only use of but possession of, or ownership.
Andrea Dworkin

Criminalizing Sexuality

Liberation is an ever shifting horizon, a total ideology that can never fulfill its promises. It has the therapeutic quality of providing emotionally charged rituals of solidarity in hatred—it is the amphetamine of its believers.
Roland Barthes

Religious leaders may routinely get away with rape and molestation, but that does not stop them from trying to legislate morality for the rest of us. For example, in Palestine, men are hired to snoop around parked cars and do whatever they please with anyone caught *in flagrante delicto*. Iranian men can obtain temporary marriages to prostitutes, followed by a divorce as soon as the sex is over. Meanwhile, African women are expected to die of AIDS, thanks to misinformation that prevents contraception from being used by those who need it most. Crimes such as drugs, sodomy, pornography, and prostitution are victimless despite often being bundled with other crimes, such as human trafficking, rape, and sexual exploitation. Still, there is a vast difference between someone who casually uses drugs and someone who commits violence while distributing drugs, someone who visits a prostitute and someone who forces a woman into prostitution, or consensual sodomy and rape. Nevertheless, religion still tries to criminalise these mostly harmless acts. The religious invasion of privacy that has been codified in secular vice laws has nothing to do with protecting people and everything to do with not wanting to offend an OOO God, who just happens to be unavailable for comment. Consensual anal sex between adults remains a crime in 13 states, and that is only one of a long list of prohibited pleasures. Drugs may cause harm, but people use them for pleasure. Perhaps sex is not the problem; perhaps pleasure in all forms is the culprit, because it interferes with religious piety.

Punishing the prostitute promotes the rape of all women. When prostitution is a crime, the message conveyed is that women who are sexual are bad, and therefore legitimate victims of sexual assault. Sex becomes a weapon to be used by men.
Margo St. James

So-called health concerns are nothing more than red herrings. People are routinely arrested for possessing marijuana in places where alcohol runs freely, which is absurd because such policies have no public health benefit. Alcohol has no valid medical use and is directly or indirectly responsible for thousands of deaths and injuries per year, at a cost to society of billions of dollars. Marijuana has no known lethal dose and is very rarely implicated in deaths or injuries. Meanwhile, reactions to over-the-counter painkillers such as aspirin, acetaminophen, and ibuprofen cause 76,000 hospitalizations and 7,600 deaths in the United States alone every year. There is just no way to make a credible case against marijuana, and yet there are people serving life sentences over a plant that is practically identical to a common roadside weed, while violent offenders are released early because of prison overcrowding. The 18th Amendment to the United States Constitution that outlawed alcohol from 1920 to 1933 was a colossal failure foisted on society by... you guessed it... religion. The separation of church and state is a myth; the resulting misuse of government resources is staggering.

Abortion is another religious hot button. Plenty of religious people believe that abortion is never an option, regardless of any medical conditions or whether the fetus was caused by rape or incest. In *The Natural Savage*, I pointed out that nature aborts some 25% of all pregnancies, often without the mother suspecting that she was pregnant or being aware of the miscarriage. A baby born with a birth defect or other handicap would be easy pickings for predators in the days before civilization, just like the weak and infirm of any animal species. Nature is callously efficient at weeding out the unfit, because even perfectly healthy animals have a difficult enough time finding food and shelter while evading predators. If you have read *The Natural Savage* or know anything about evolution, then you know that civilization and technology have advanced far too rapidly for our brains to catch up. Our bodies are not adapted to deal with poverty, drugs, the need for education that extends well into adulthood, or the myriad other factors that can seriously handicap a woman who has a baby at the wrong time. The only difference between a miscarriage and an abortion is that the latter involves a conscious decision, at least for now. I dare say that mothers a million years from now will routinely miscarry babies if they have not completed college, are addicted to drugs, etc... if our species survives that long, that is.

Homosexuality or bisexuality occurs in roughly 4% of the human population, and has also been observed in a wide variety of animal species. Its universal presence is a strong indication that it is either an accidental by-product of evolution or part of God's plan. Nevertheless, the JCI religions universally decry homosexuality despite an ongoing parade of gay sex scandals involving the loudest protestors.

Consensual sexual acts can only be punished if those who prohibit them are trying to conceal their own desires to perform the same acts. As Shakespeare's King Lear said, "The police-

---

No doubt, love, but as long as people are still having promiscuous sex with many anonymous partners without protection while at the same time experimenting with mind-expanding drugs in a consequence-free environment, I'll be sound as the pound!
Mike Myers
(as Austin Powers)

Don't knock masturbation, it's sex with someone I love.
Woody Allen

> *I believe in using words, not fists... I believe in my outrage knowing people are living in boxes on the street. I believe in honesty. I believe in a good time. I believe in good food. I believe in sex.*
> Susan Sarandon

man who lashes the whore has a hot need to use her for just what he is flogging her for." Religions that want to clean up our act would do well to put their own houses in order before telling the rest of us who to do and what we can do with them.

Sexually Transmitted Diseases

In 1665, archbishop Lancelot Andrews noted that the Black Plague was killing off the faithful just as much as everyone else. Nevertheless, the religious attitude toward medicine is just as hostile as it is to the other sciences. For example, Nigeria was declared polio free—an amazing triumph of modern medicine over a disease that has ravaged millions of people—until Muslim imams denounced it as a United Nations and United States conspiracy designed to sterilize the faithful and issued a *fatwa* (Islamic religious decree) against the vaccination. Polio spread back through Nigeria and beyond. Many religious people see AIDS as God's revenge for loose morals, especially homosexuality. Lesbian women are at less risk for AIDS than heterosexuals and generally maintain higher standards of hygiene, but religion refuses to so much as acknowledge the existence of lesbianism.

> *The AIDS epidemic has rolled back a big rotting log and revealed all the squirming life underneath it, since it involves, all at once, the main themes of our existence: sex, death, power, money, love, hate, disease and panic. No American phenomenon has been so compelling since the Vietnam War.*
> Edmud White

In 2003, cardinal Alfonso López Trujillo warned that condoms are made with microscopic holes that are designed to transmit AIDS; the virus is much smaller than a sperm, and is thus able to pass through the condom "net." Can you imagine a better way to cause unspeakable suffering among millions of people with so few words? This statement (which the cardinal defended in later interviews) demonstrates an appalling level of ignorance and disdain for science and established fact. Even more abhorrent, this was said in the context of poor nations that are struggling with overpopulation caused to no small extent by religious sanctions against contraception and family planning. Bishop Rafael Llano Cifuentes said that the Catholic Church opposes condoms because sex has to be natural, and he has never seen a dog using a condom when mating. Other Catholics have said that condoms spread AIDS and call women who die of AIDS rather than using condoms martyrs. Wow.

The human papilloma virus (HPV) can cause cervical cancer in women. Preventive vaccines were developed in 2005 and 2006. Instead of hailing this breakthrough in women's reproductive health, religious people resisted it, claiming that it would encourage premarital sex. They are quite literally happy to accept cervical cancer in God's name.

A 1996 council of Muslim *ulemmas* (leaders) in Indonesia recommended that condom use be limited to married couples with a doctor's prescription. Iranians with HIV can lose their jobs, and doctors can refuse to treat them. Pakistan boasts that AIDS is not such a large problem there because of its better Islamic values, this in a nation where a woman can be sentenced to be gang-raped for a crime committed by her brother. The very existence of AIDS is unmentionable, because one can find everything one needs for clean living in the Koran.

These are just a few examples. I hope they are enough to convince you that religion is a clear and present public health hazard.

Fundamentalism and Extremism

Religion thrives on pandemonium. Throngs of British Muslims have taken to the streets bearing signs saying, "Slay those who insult Islam," "Butcher those who mock Islam," and my personal favorite, "Behead those who say Islam is a violent religion!" Filmmaker Luis Buñuel pointed out that, "God and country are an unbeatable team that breaks all records for oppression and bloodshed." Religion has no monopoly on extremist positions, but extremism is not easy to defend on non-religious grounds. Atheists are just as capable of bad acts as religious people, but they almost never do these acts in the name of atheism. Meanwhile, religious wars are both very frequent and are always fought in the name of religion. Religious fundamentalists also tend to be neither poor nor uneducated. For example, Osama bin Laden was a member of a prominent Saudi family and had a college education. Jerry Falwell attended college and enjoyed a lavish lifestyle. Pat Robertson has a law degree and millions of his own dollars.

If this is true, then why are so many poor uneducated Muslims lining up to blow themselves up? Because they believe their rich, educated leaders who have the full backing of the Koran, which makes martyrdom seem like a promising opportunity.

I think that there are no forces on this planet more dangerous to all of us than the fanatics of fundamentalism, of all the species: Rrotestantism, Catholicism, Judaism, Islam, Hinduism, and Buddhism, as well as countless smaller infections.
Daniel C. Dennett

In the 21st century, all books, including the Koran, should be fair game for flushing down the toilet without fear of reprisal.
Sam Harris

> *When ye encounter the infidels, strike off their heads till ye have made a great slaughter among them.*
> Mohammed

Believers who stay comfortably at home without suffering are second-class citizens compared to holy warriors. Allah accords higher status and richer heavenly rewards to jihadis than those who decline to fight and die for him. Unbelievers will get their just desserts. (Koran 4:95-101.) Koran 4:74-78 says that we should, "Let those who want the afterlife fight for God. Whoever fights, no matter the outcome, will be richly rewarded. True believers fight for God, but infidels fight for the devil. Fight against Satan's friends, casting off the pleasures of this life as trifles next to the afterlife." One cannot possibly claim that Islam is a religion of peace and tolerance in the face of passages like these. Martyrdom is the only way to bypass purgatory and judgment altogether and get right down to the business of bedding virgins.

"The national socialism of all of us is anchored in uncritical loyalty and the surrender to the Führer that does not ask for the why in individual cases in the silent execution of his orders. We believe that the Führer is obeying a higher call to fashion German history. There can be no criticism of this belief." So said Rudolph Hess, Hitler's deputy in the Nazi Party, in 1934. Nazi Germany is an example of a mostly secular religion that created a powerful personality cult around a leader who had almost godlike powers and respect. The Reich also retained relations with different Protestant sects, and the words *Gott mit uns* (God with us) appeared on uniform buttons, buckles, and other Nazi paraphernalia.

> *O ye who believe! Verily in your wives and your children, ye have an enemy: wherefore, beware of them.*
> Koran

Germany, Stalinist Russia, North Korea, Cambodia, and all other totalitarian nations feature a powerful dogma that holds the organization together, consolidates power, and metes out punishment to all enemies. Nazism owes its existence to a complex web of sociopolitical factors, but managed to unite Germans behind the idea of racial purity and the superiority of the German people over the sick, homosexuals, gypsies, Jews, and other *personae non gratae* (unwelcome people). Religious Germans had a long history of hating the Jews, blaming them for everything that went wrong; they saw Nazism as the perfect excuse to do something about it. Dogma, believers, infidels, strong authority, etc. Totalitarian regimes have all the hallmarks of religion. This is just what one would expect of organizations whose pantheons mimicked human monarchs and courts, as we learned in Chapter 7.

The leap of faith required to sustain religious belief is itself a fraud, because one must keep leaping despite mounting evidence to the contrary. Continuing this vicious cycle can cause mental breakdown, delusions, and manias. Religion understands the law of diminishing returns quite well, which is why it corrupts faith and insults reason at one stroke by offering laughable "proofs," such as the idea of intelligent design. Asking questions, proving theories, and demonstrating knowledge is far more difficult—and far more rewarding—than simply believing the absurd, as Tertullian would have us do.

In general, higher income means less religiosity, with the notable exception of the United States, where the religiosity of the general public stands in sharp contrast to the atheism of the intellectual elite. Religion takes full advantage of this. Today's mega churches can seat up to 16,000 worshippers and provide everything people need in their social and religious lives including onsite parking, shuttle buses, fitness clubs, restaurants, bookstores, and other services such as childcare, entertainment, and community outreach. The sermons themselves tend to avoid the threat of hellfire while promising a life of leisure and material wealth if one simply follows Jesus, accompanied by lighting, sound, and other special effects to rival most theaters. Religious people stand out much like peacocks and bower birds. Their beliefs consume resources and endanger both life and quality of life. Mega churches, cathedrals, temples, mosques, and other religious buildings consume millions of tons of supplies for no constructive purposes. It is hard to believe that the creator of something as sublime as the universe would be impressed by anything the inhabitants of a tiny planet could put up, so who are the religions really trying to impress?

False Moderation

Tell a devout believer that her husband is having an affair or that celery makes you fat, and she will probably demand just as much hard evidence as anyone else. Tell this same person that a certain book was written by an invisible god who will punish her for all eternity if she does not obey it, and she will accept this without any qualms. Tell her that this god punishes those who have strong reasons for not believing while those of blind faith are exalted, and she will eagerly go along with it.

Behead those who say that Islam is a violent religion!
Sign at a pro-Islamic rally in London

To ask 'is religion X a religion of peace?' is a silly question.
Robert Bowker

Such knee-jerk beliefs are an affront to God, because they assume that there is something wrong with the brains and logic this same God gave us. God seems to have real issues with smart sexy people.

The restraint of British newspapers derived less from sensitivity to Muslim discontent than it did from a desire not to have their windows broken.
ANdrew Mueller

Religious moderation owes its existence to both advances in secular knowledge that finally broke the church's stranglehold on thought, discourse, and ignorance such as Linda's about what the scriptures actually say. The dogmatic belief that peace will be at hand when everyone respects everyone else's views is in many ways just as bad as more extreme views, because it extols blind faith for the sake of blind faith. Those who are most open and welcoming of different beliefs and religions are the most ignorant of the many Biblical/Koranic/etc. injunctions against heresy and unbelievers, in general. The only thing religious moderation has going for it is deliberately ignoring certain parts of the divine book, which makes about as much sense as deliberately ignoring certain parts of your local driver's handbook. These people, many of whom would never actually hurt anyone themselves, defend their more extreme brethren by insisting that religious evils are no worse than secular evils and by being notably silent when said brethren commit some atrocity or another.

Religious moderates aid and abet the extremists, and are therefore willing *accomplices* (people who knowingly help commit criminal acts) to everything done in the name of their religion. Secular justice punishes secular accomplices while burying its head in the sand when it comes to dealing with religious accomplices. Let me say this plainly: The sweet little old lady who sings in the choir every Sunday and bakes cookies for the fund-raisers and who would never hurt a fly is just as guilty as the religious extremist who murders an abortion doctor, hijacks an airplane, or bombs a train.

What these [religious] people really love and do best is pandemonium.
Germaine Greer

As I have already said, the idea of a totalitarian state is intimately connected with religion. In both cases, the ruling class is deemed infallible, and all others must obey or perish. There is little difference between a dictatorship and a theocracy. The Japanese Emperor Hirohito was worshipped as a god during World War II and orchestrated war crimes on a massive scale. Nobody denounced religion as a result. Those religious people who try to divert attention by pointing out secular tyranny want us to forget the long and cozy relationships between churches and tyrannical states. North Korea practices a twisted form of Confucianism that reveres the personalities of dictators Kim Il Sung and Kim Jong Il. A Christian missionary once reported having had a hard time preaching the idea of a savior to people who had fled North Korea; they had heard it all before.

Religious Persecution

The Ten Commandments are just as noteworthy for what they don't say as they are for what they do say. There are no commands against rape, child abuse, slavery, genocide, or other heinous crimes. The Old Testament God is both jealous and selfish, as are His modern followers who have been forcing Palestinians from their homes for over 60 years. Religion inflames and multiplies tribal hatreds and bigotry. Christians are filthy because they eat pigs. Christians and Jews drink alcohol, which caused Muslims to blame drinking Christians for a tsunami that occurred just after a Christian celebration. Catholics are dirty and have too many children. Muslims breed like rabbits and wipe their rear ends with the wrong hand. Jews have lice in their beards and flavor their *matzo* (unleavened bread) with the blood of Christian children. Most children learn to grow out of such pettiness in playgrounds around the world—an example that far too many religious adults would be well advised to heed!

Christians destroyed countless thousands of philosophical and scientific works, because they believed that no meaningful morality could have existed before Christ. Some of these works were preserved and retransmitted to the ignorant West hundreds of years later, but many more remain irretrievably lost. Religion is the single biggest reason why the West lagged so far behind the East for hundreds of years. Who knows where we would be scientifically, philosophically, and technologically without this tremendous setback?

On Valentine's Day in 1989, author Salman Rushdie was sentenced to death over a novel. Iran's Ayatollah Khomeni offered his own money to pay for assassinating Rushdie and everyone involved in publishing *Satanic Verses*, despite probably never bothering to actually read the book. Did the Vatican and the chief rabbi in Israel rally to Rushdie's defense? No! They sided with Khomeni, as did the cardinal of New York

---

A close study of these [sacred] books, and of history, demonstrates that there is no act of cruelty so appalling that it cannot be justified, or even mandated, by recourse to their pages. It is only by the most acrobatic avoidance of passages whose canonicity has never been in doubt that we can escape murdering one another outright for the glory of God.
Sam Harris

The moderates in all the religions are being used by the fanatics, and should not only resent this; they should take whatever steps they can to curtail it in their own tradition.
Daniel C. Dennett

> *The overwhelming problem is one of attachment to the teaching itself: the 'love' of one's own religion, spiritual egoism—it can be called many things.*
> Jacob Needleman

and many other religious leaders, because Rushdie had committed blasphemy in their eyes. Some even said that Rushdie had brought this down on his own head for daring to insult a great monotheistic religion. Secular leaders went along with this, providing Rushdie with police protection while never questioning the right of religion to issue and defend such bile.

On the morning of September 11th, 2001, the most sincere believers on the planet were the 19 men on the hijacked airplanes. Once again, the religious elite defended their own, as Jerry Falwell and Pat Robertson hastened to blame this calamity not on extreme indoctrination and sexual frustration, but on the American society that tolerated homosexuality and abortion. Billy Graham, a pastor known for his anti-Semitism, was allowed to speak at the memorial service for the victims. The one thing all of these men had in common was an intimate knowledge of a fictitious hereafter.

Here are some more examples:

- Jewish and agnostic Air Force cadets are harassed by their "born again" peers, who believe that only those who "accept Jesus Christ as their personal savior" are qualified to become officers.

- Members of Fred Phelps's Westboro Baptist Church (www.godhatesfags.com) have taken to picketing funerals and other major events. In an exemplary display of secular morality, the United States Supreme Court ruled in their favor because freedom of speech applies to all.

- Muslims are so scared of pigs that they refuse to read *Animal Farm*, being too ignorant to realize that the story is anything but flattering to pigs. They don't want their children seeing Piglet, Miss Piggy, or The Three Little Pigs, and the statue of the wild boar in the British Arboretum has been threatened with vandalism. Muslims and Jews alike believe that God really hates ham because pigs look, taste, and sound like humans and are very intelligent—a possible holdover from the days of human sacrifice and cannibalism. Spanish Catholics began the custom of passing out ham at parties to catch Jews by seeing who would not eat it. This now-quaint custom was once used to justify torturing people for believing a slightly different variation of the same made-up story.

> *Some of Luther's more zealous followers at Wittenberg demanded not only a return to the simplicity of the early church but the abolition of all education.*
> Roland Barthes

Do not think about the color red. What did you just think of? It is one thing to keep someone from doing bad things. It is another thing to keep someone from thinking bad thoughts. The Bible and Koran alike condemn thoughts that stray from approved beliefs. The commandment against even thinking about other gods is very similar to saying that a man who looks at a woman the wrong way has committed adultery; it is an example of how religion places impossible restraints on behavior in order to always have something over its followers. Religion has served as both the spiritual and secular authorities and has set up a system contrary to the person it claims to adulate. For example, church leaders live among pomp and circumstance to imitate a man who lived in humility and poverty. The invention of purgatory and the church's power to reduce or eliminate a soul's time there was a fantastic revenue source.

If I could wave a magic wand and get rid of either rape or religion, I would not hesitate to get rid of religion.
Sam Harris

The idea that we are free to believe anything we want is a myth. We are not free to believe what we want about God any more than we are free to adopt unjustified beliefs about science, history, or able to mean what we like when we use words and language. Anyone making such claims could expect a fair amount of ridicule. Religion is the exception; it forces us to believe unfounded claims on pain on death or other punishment. The Catholic Inquisition used Deuteronomy as its inspiration, the same book that Josiah's priests "found" while restoring the temple in Jerusalem (Chapter 7). There are museums full of religious torture devices.

According to Deuteronomy 13:6-18, "If your brother, child, wife, or friend says, 'Let us worship other gods,' do not yield to or listen to him. Show him no pity. You must put him to death, you first then all the people. Stone him to death because he tried to turn you away from God, then all Israel will be afraid and no one near you will do such an evil thing again. If you hear that one of the towns God gave you is worshipping other gods, investigate thoroughly and, if substantiated, kill everyone in that town. Kill the livestock and burn the town and everything in it without taking anything as an offering to God, and never rebuild there." Despite this unambiguous commandment, the Catholics did not destroy heretic property, but used it to enrich the church and clergy.

Religion is used by those in temporal charge to invest themselves with authority.
Christopher Hitchens

> *Science flies us to the Moon; religion flies us into buildings.*
> Atheist T-shirt

Imagine being seized and taken before judges who barrage you with nonsensical questions. Did you cause a storm to ruin the harvest? Do you question the validity of the Eucharist? Nothing you can say can satisfy your interrogators. You have no idea who accused you, but that doesn't matter because they will be punished if they change their story now, and the accusations against you will remain. What would you do? Confess and name those who helped you, since no confession is complete without denouncing accomplices? After confession, you would be punished in a variety of horribly painful ways. Maintain your innocence (who can cause a thunderstorm?), and you will be tortured in one or more utterly barbaric ways. If you confess under torture, then you will be forced to confirm your confession before a judge; any mention of torture will earn you a repeat visit to the dungeons. If you are condemned and repent, these same people who profess such concern for your undying soul may do you the favor of strangling you before burning you to ashes.

As we noted in Chapter 8, the Inquisition was carried out by clergy of all ranks, men of God who purportedly served Christ, the man who healed the sick and challenged those without sin to cast the first stone. Transforming Christ from a healer and a teacher into an instrument of murder and theft may seem impossible, but it's actually quite simple, because the many contradictions in the Bible allow people to adopt a wide range of beliefs. Thus, the central problem is faith itself; anyone who thinks s/he knows the truth without evidence is at risk of lashing out in horrible ways. Spanish priests in Latin America baptized native babies before bashing their brains out to make sure they got into heaven. Even this is perfectly justifiable to someone who believes that all are born guilty of original sin. Some conquistadores were so cruel that one actually suggested apologizing and admitting to a terrible mistake. Why this seemingly magnanimous gesture? Because maybe the Indians were living in Eden, and the Spanish were accidentally destroying it! Never mind that the Indians had writing, astronomy, engineering, a more accurate calendar, and the concept of 0 that Middle Age popes resisted as heretical.

> *If the whole church comes together and all speak in tongues, and outsiders or unbelievers enter, will they not say that you are out of your mind?.*
> 1 Corinthians 14:23

The Protestants may have broken with Rome but were no less cruel. Still, the Jews were even worse than Protestants in Catholic eyes, because they denied Christ's divinity and were guilty of murdering him. No one bothered to ask how mere humans can kill an OOO god or how they could possibly have a religion based on Christ sacrificing himself for the world's sins without the actual sacrifice. The Catholics did not officially eschew torture until Pius VI ended the practice in 1816. Nevertheless, the Vatican perpetuated the idea of the "blood libel" (Jews killing Christian children for their blood) in newspapers until 1914, and did not formally remove the charge of *deicide* (killing Christ) until the 1960s. Keep in mind that the Vatican had not accused some specific Jews, but the entire Jewish race. Even if the Jews in Jerusalem did want Christ dead, and even if Christ did actually die on the cross (see Chapter 6), how can one hold their descendants liable for an action they had nothing to do with? The concept of original sin provides a clue here.

Virtually everyone sees dying to save someone else as a noble act. Assuming that Christ died on the cross, the fact that he was then resurrected weakens his sacrifice to a minor inconvenience. After all, what is a few days of suffering and a temporary death for an eternal OOO god? From a practical perspective, how does a human sacrifice that happened 2,000 years ago—and that any decent person would have felt compelled to protest—possibly excuse today's wrongdoings? Why must I endure an eternity of suffering that makes Christ's time on the cross seem pleasant by comparison if I refuse to accept this?

The Catholic Church in Germany allowed the Nazis to access their genealogical records to determine the amount of Jewish ancestry people had. Not one German Catholic was ever excommunicated over Nazism, regardless of their crimes. The Catholics did, however, excommunicate scholars and scientists for not being quite orthodox enough. This alliance between church and state didn't end when the Nazis were defeated; the church even helped former Nazis escape prosecution after the war. Catholics believe that the pope is the vicar of Christ on Earth who keeps the keys of Saint Peter; a pro-Nazi pope must therefore be God's will. Joseph Goebbels, the Nazi Minister of Propaganda, was excommunicated, not for being a Nazi war criminal, but for marrying a Protestant. As previously noted, Galileo was not "forgiven" for 400 years.

> *If we want to know why people kill in the name of God, and why they have been doing so for thousands of years, we must face one simple and obvious fact that nobody wants to confront. The fact is this: The god of Judaism, Christianity, and Islam—the god of monotheism—is a terrorist. In fact, he's the ultimate terrorist. It is an undeniable fact that the God described in the pages of the holy Bible and holy Koran is a bloodthirsty, ruthless, destructive terrorist.*
> Michael Earl

> *The true believer cannot rest until the whole world bows the knee.*
> Christopher Hitchens

Islam is no better. Mohammed died in 632, and the first account of his life was written 120 years later by Ibn Hisham. There is no account of how Mohammed's followers assembled the Koran, nor did he leave any clear instructions about succession. The fighting broke out as soon as he died, with the Sunni and Shia sects emerging before Islam was even formalized. At least one of those interpretations has to be mistaken. Muslims became increasingly concerned that Mohammed's orally transmitted words could be lost, because so many soldiers were being killed that few were left who remembered. They therefore called every living witness together, pieced together their accounts in the hadith (see Chapter 5), and then decided which tidbits were canonical and which must be discarded—an Islamic version of the Council of Nicaea, in which hundreds of thousands of sayings were reduced to around 10,000. What are the odds that they got them all right?

Arabic script was not standardized until the 9th century, and the Koran was interpreted in many different ways during this period. This would only be a passing concern for a novel, but the Koran is not a novel—or at least it is not acknowledged as being a work of fiction. The Arabic words on Jerusalem's Dome of the Rock are very different than those in the Koran.

The hadith is used to help interpret the Koran, and many Muslims see it as higher than the Koran. What does the hadith say? Here are a few tidbits:

- Jihad is a duty under any righteous or wicked ruler.
- A single effort of fighting for Allah is better than the world and all in it.
- A single day and night of fighting is better than a month spent fasting and praying.
- Someone who dies without being involved in jihad is a form of unbeliever.
- Paradise lies in the shadow of swords.

> *From my own point of view, I can hope that this long and sad story will come to an end at some point in the future and that this progression of priests and ministers and rabbis and lamas and imams and bronzes and bodhisattvas will come to an end.*
> Steven Weinberg

It should go without saying that the 2003 American invasion of Iraq will be inspiring jihadis for decades to come.

Of course, religious fundamentalism is only a problem because of that religion's fundamentals; a rise of fundamentalist Jains would not endanger anyone, and would actually improve the world tremendously. Yes, we would probably lose more crops to pests, but those that survived would be far more nutritious than the factory-farmed excuses for fruits and vegetables so many of us eat today. We would also spend far less of our resources on armaments. JCI fundamentals are fundamentally different than Jainism. Claims that Islam is a religion of peace are falsified both by the hadith mentioned above and by Koran verses that include such things as:

- Prophet, make war on the unbeliever and the hypocrites and deal rigorously with them. Hell shall be their home; an evil fate. (Koran 9:73.)

- Believers, make war on the infidels who dwell around you. Deal firmly with them. Know that God is with the righteous. (Koran 9:123.)

- Lo, those who disbelieve our revelations, we shall expose them to fire and as often as their skins are consumed, we will exchange them for new skins so they may taste the torment. (Koran 4:56)

Silencing and killing dissenters is a clear sign of weakness that religions are actually proud to flaunt. Like Christians, a Muslim who knows enough about how the world works to doubt her or his faith deserves death, as does a lapsed Muslim. The West lagged behind the East for a long time, but that pattern has reversed itself over the last few hundred years and caused plenty of angst among Muslims, who label it perverse and use it as an excuse for jihad. Western "imperialists" supposedly offended Islamic dignity and pride, a charge that smacks of jealousy and envy and obscures the fact that Islamic law is the single greatest insult to Muslims.

Afghanistan under the Taliban is a shining example of Islamic law in all its gory glory that includes banning music and recycling paper that could possibly contain a scrap of the Koran. A Muslim allowed to vote today will readily vote to end her or his freedoms and subject themselves to these laws. The great society that preserved so much Western knowledge has stagnated and festered politically, economically, and spiritually; the 2002 GDP of all Arabic nations combined was less than that of Spain. Each year, Spain alone translates as many books into Spanish as the entire Arabic world has translated into Arabic since the 9th century. Despite this, Muslims think it perfectly

---

Toward every other people [the Jews] show hatred and enmity. They sit apart at meals, and they sleep apart, and although as a race they are prone to lust, they abstain from intercourse with foreign women; yet among themselves nothing is unlawful.
Tacitus

Those who preach this doctrine of loving their enemies are in general the greatest persecutors, and they act consistently by so doing; for the doctrine is hypocritical and it is natural that hypocrisy should act the reverse of what it preaches.
Thomas Paine

OK to invade Spain and "enable" them to convert, while taking exception to any nation that decides to return the favor.

No account of the Middle East and suicide bombing is complete without mentioning Jewish settlements. Osama bin Laden was not out to reduce the income gap among Arabs or to donate all of his money to humanitarian causes. More than anything else, he was driven by anti-Semitism; the idea of a Palestinian state is an afterthought. He was upset by the presence of infidels in the Holy Land. All of these are theological ideas driven by his religious beliefs. As much as we claim to hate him, most Americans have one critical thing in common with Osama, namely their belief that outlandish ideas can be accepted without evidence or questioning. Cartoons of Mohammed printed in Scandinavia caused global rioting and deaths as Muslims called for the cartoonist's head—and Western religious leaders condemned the cartoons. A few bullies with God on their side succeeded in dealing yet another blow to the concept of free speech. No secular group could have gotten away with this. Once again, secular governments surrendered their authority and abdicated their responsibility in the name of tolerance.

The end of killing heretics in the West proves that good idea can eventually triumph over bad. Many Muslims are convinced that God cares about women's couture. Riots in Nigeria over the 2002 Miss World contest killed 200 people. Men and women were hacked with machetes and/or burned to keep women from wearing bikinis. Religious police in Mecca prevented firefighters from rescuing girls trapped in a burning school, because they were not dressed as required by the Koran. Fourteen girls died, and 50 more were injured. Is it moral and ethical to allow people to think that God cares about hemlines? If God really cared, would He not have designed some way to conceal the genitals without the need for additional clothing? The fact that the Greeks started abandoning the Olympian pantheon hundreds of years before Christ in favor of more intellectually fulfilling pursuits should be both an inspiration and source of embarrassment for the rest of us.

The 1998 bombing of the Al-Shifa pharmaceutical plant in Khartoum, Sudan cannot be compared to terrorist attacks, according to linguist Noam Chomsky (1928-). The United

---

Whether our world will fall apart from the excesses of religious zeal or the blind hubris of scientific materialism is a serious question.
Larry Dossey

The age of ignorance commenced with the Christian system.
Thomas Paine

State military thought it was striking a chemical weapons site with no intention of indirectly killing tens of thousands of people; we only wanted to destroy a source of weapons of mass destruction, and perhaps any Al Qaeda members present at the time. The 1968 My Lai massacre in Vietnam remains a moment of shame for the US military. Even US soldiers were horrified by the actions of their comrades; helicopter pilot Hugh Thompson ordered his crew to kill American soldiers who were killing civilians and rescued as many people as possible. Contrast these examples with the televised scenes of jubilant Muslims dancing in the street celebrating the September 11th attacks.

It is high time for us to face the fact that not all people are at the same stage of moral and intellectual development. Religious people of all flavors are falling ever further behind the times, and that trend is only accelerating—with possibly fatal consequences for us all.

Civilization on the Brink

Given all we have discussed in this chapter, it should be obvious that the degree to which religious ideas influence government policies is the degree to which those policies are dangerous to everyone. Ronald Reagan thought about the Middle East in terms of Biblical prophesy. He invited televangelists to national security meetings, and the resulting American policy toward Israel and the rest of the region were partly shaped by religious fundamentalists. American support for Israel represents the pinnacle of religious arrogance, because it validates the fundamentalists' ideas that a strong Jewish presence in the Middle East will trigger the second coming of Christ, for whom obliterating the Jews is a top priority. The 1926 Balfour Declaration supporting the idea of a Jewish return to Israel was at least partially inspired by the Bible. People have been fighting and dying for rights to the rock where would-be child murderer Abraham is said to be buried ever since.

Defusing the Middle East confrontation and reducing the likelihood of World War III will require Muslims to set aside the most extreme of their beliefs, as many Jews and Christians have learned to do. This will be no easy task; all devout Muslims dream of a Muslim planet cleansed of all infidels, and

---

I found nothing grand in the history of the Jews, nor in the morals inculcated in the Pentateuch. Surely the writers had a very low idea of the nature of their god. They made him not only anthropomorphic, but of the very lowest type, jealous and revengeful, loving violence rather than mercy.
Elizabeth Cady Stanton

Those who would take over the earth and shape it to their will never, I notice, succeed.
Tao Te Ching

Chapter 10
Religion, Interrupted | 303

armed combat in defense of Islam is a religious obligation for every Muslim man. Jihad must continue until the world becomes Muslim or surrenders to the Muslims. This of course does not prevent violence between Muslim nations. Over 1 million people have been killed for religious reasons since India and Pakistan were made into separate nations. They have fought three wars, and the Kashmir border between them remains an active combat zone. Both nations now have nuclear weapons. All this because people disagree over "facts" that are no more real than the names of the seven dwarves in *Snow White*.

> *When you have names and forms, know that they are provisional. When you have institutions, know where their functions should end. Know when to stop.*
> Chinese proverb

Solving the Israel/Palestine conflict should be as easy as recognizing that two groups of people have similar claims to the land and splitting the area into two separate nations. This single stroke could reduce tensions between East and West. Why is this not likely to happen any time soon? Religion, of the same type that would have us living under Taliban-like conditions. The Jews are anything but blameless in this situation, because their religion is a lightning rod for intolerance. Christianity and Islam both consider the Old Testament valid and offer easy conversion paths for Jews and all others. Islam revered Moses, Abraham, and Christ as forerunners of Mohammed. Hindus embrace almost everything in sight; some even see Christ as an avatar of Vishnu. Judaism is the only religion surrounded by error on all sides. Jews still think that they alone have a special contract with God, and still see themselves as separate from and different than others. These beliefs are no more respectable or tolerable than any other religious beliefs. Jews exercising their freedom of religion is the single largest obstacle on the road to peace in the Middle East, and one of the largest potential triggers of a future war that could end us all.

> *The world is now too dangerous for anything less than utopia.*
> Buckminster Fuller

The many people who turn themselves into bombs or use their children as human shields can only be inspired by religion. Martyrs are in it for the religious promise of eternal bliss more than anything else, an afterlife where they can finally get laid. Someone who is not absolutely certain about the afterlife is far less likely to kill her or himself, because doubt is a powerful deterrent. The Muslims who engage in extreme behavior do so because the "moderate" Muslims allow them to. The United States and Union of Soviet Socialist Republics kept the planet on the verge of nuclear destruction for 50 years not because they intended to use the weapons, but because they intended not to; the certainty of a devastating counterattack prevented either side from starting World War III. This simple—if dangerous—logic fails when religious martyrdom enters the equation, because a martyr would not hesitate to kill anyone and everyone, a fact that has been proven time and time again. If religious extremists get their hands on nuclear weapons, much of the world's population will die because of ideas that belong in cheap pulp novels.

What if people got the idea that movies or software were made by God? It is difficult to imagine a world where *Star Wars* fans stand ready to murder *Star Trek* fans en masse, or where Macintosh users launch suicide attacks against PC users. Such scenarios are ridiculous to the point of hilarity, until we remember that people are dying every day for ideas that are every bit as ridiculous as this imaginary scenario.

In the United States, many Republicans belong to the Council for National Policy, a secret religious group founded by the fundamentalist preacher Tim La Haye that meets quarterly. George W. Bush gave a closed-door speech there in 1999 that resulted in his gaining the evangelical Christian vote. (40% of Bush voters were evangelicals.) One of Bush's appointments to the Food and Drug Administration was a pro-life OB-GYN (obstetrician/gynecologist) who publicly stated that premarital sex is a sin, and that it is dangerous to separate religious and secular truth. Other religious people were placed in other government positions. Congressman Tom Delay said that only Christianity allows people to live in response to current world realities, and blamed the 1999 Columbine school shooting on the teaching of evolution in the classroom. Nobody insisted on or even suggested his expulsion from office for such nonsense. College-educated politicians stymie stem cell research, because they think a fertilized human egg should be granted full human rights. By that standard, a person who blows his nose is expelling countless souls.

Most of the money given to churches goes to proselytizing instead of solving problems, including government dollars publicly spent on "faith based" organizations. While governor of Texas, George W. Bush ended inspection requirements of religious charities in that state, despite the fact that a district

---

I have found it an amusing strategy, when asked whether I am an atheist, to point out that the questioner is also an atheist when considering Zeus, Apollo, Amon Ra, Mithras, Baal, Thor, Wotan, the golden calf and the flying spaghetti monster. I just go one god further.
Richard Dawkins

Manipulation, exclusivism, hatred, and violence are undeniable outgrowths of Biblical monotheism.
Bruce Feiler

> *Much of the world's population could be annihilated on account of religious ideas that belong on the same shelf with Batman, the philosopher's stone and unicorns.*
> Sam Harris

court found a job training program that was using government money to sponsor Bible study classes with no secular alternatives anywhere in sight. Every dollar thus spent carries the anguish of starving people around the world.

Supreme Court justice Antonin Scalia has said that government is a minister of God with powers to revenge and execute wrath by the sword. The United States abolished the death penalty for minors in 2005, the last industrialized nation to do so. Scalia is a clear example of the perils of religion mixing with government. To a nonbeliever, killing someone ends that person's existence for all time. Religious people believe in an afterlife, making killing a much less serious problem. The United States is among the last of the developed nations to use the death penalty, mostly because of religion; conservative states like Texas execute far more prisoners than liberal states.

Extreme religious beliefs make people highly resistant to persuasion while inspiring acts of violence. If they cannot be convinced to soften their tone, then the rest of us may well be justified in killing them in self defense. I would argue that killing a religious extremist who refuses to listen to reason is no different than killing an armed suspect who poses a serious threat to victims and/or the responding police officers. Should we refrain from using torture? Waterboarding is extremely cruel, but far less damaging than a missile fired from an unmanned drone aircraft. If we are OK with the collateral damage of killing and maiming innocent people (a virtual certainty whenever we drop bombs), then why spare suspected terrorists? What is the moral difference between accidentally killing innocents or subjecting possibly innocent men to torture? There are no women or children in Guantanamo Bay, unlike the buildings we target almost every day. If we are going to drop bombs, then we must torture without mercy. If we are not going to torture, then we must stop dropping bombs. It's that simple.

> *In dark ages people are best guided by religion, as in a pitch-black night a blind man is the best guide; he knows the roads and paths better than a man who can see. When daylight comes, however, it is foolish to use blind old men as guides.*
> Heinrich Heine

Pacifism sounds like an elevated concept. It would be if everyone shared it, such as if the world woke up one morning under the yoke of fundamentalist Jains. In a world full of religious extremists who are ready, willing, and able to kill for their fantasies, pacifism is simply the willingness to die at their hands. An ethical person who fights someone unethical can't win.

Eschatology refers to the religious study of the so-called end times. Is it any coincidence that "eschatological" sounds awfully similar to "scatological?" I think not, but this does not mean that the end is not coming. We will all die, the Sun will swallow the Earth whole, and the entire universe will eventually be nothing but cold dark matter. The end is indeed coming, albeit not soon enough for religious people.

> *Smile. There is no hell.*
> Atheist T-shirt

Breaking the Spell

The desire to understand the world opens a person to new evidence. The disconnect between religion and scientific inquiry is real and unfortunate, because science can point to the divine, albeit not the divinity that the JCI religions have in mind. The good news is that there are thousands of religious people who want to leave, but who don't know that leaving is a real option for them. Meanwhile, we cannot reason with, accept, or tolerate true believers; ridicule is the only weapon we have against religion short of open warfare.

Religion aside, most people believe that there is more to existence than we see. Life does indeed have a sacred aspect, and discovering that in each of ourselves could be the ultimate meaning of existence. None of this requires believing nonsense, such as virgin births or that a book contains the absolute and inerrant word of God. Kant said that we cannot know for sure about God and immortality, but we still believe strongly in the real presence of things we cannot know about directly. Some human experiences qualify as mystical, spiritual, meaningful, selfless, beautiful, inspiring, and uplifting. They get us out of our normal selves and into an entirely different plane of existence beyond normal thought and feeling. Nothing about this justifies any claims about the will of God or the holiness of any book or relic.

People have known for thousands of years that suppressing the feeling of "I" removes the sense of separation from the universe. This is supported by reams of neurological and philosophical evidence. Such experiences transform people, revealing far deeper connections between ourselves and the universe than we commonly perceive. It is from this sublime seed that religion grew and metastasized. Psychic phenomena also seem real, despite being mostly ignored by mainstream

> *Most ignorance is invincible ignorance. We don't know because we don't want to know.*
> Aldous Huxkley

> *Ridicule is the only weapon that can be used against unintelligible propositions. Ideas must be distinct before reason can act on them; and no mane ever had a distinct idea of the trinity. It is the mere abracadabra of the mountebanks calling themselves the priests of Jesus.*
> Thomas Jefferson

science. Sagan was correct that extraordinary claims require extraordinary evidence; however, this does not mean that the universe is not a stranger and more wonderful place than we can possibly imagine. For every neuron in our brain that processes sensory data, we have 10-100 that don't. Our brains pretty much talk to themselves; no information about the world runs directly to the cortex where consciousness seems to reside. (See *The Enlightened Savage* for more about this fascinating topic.) There is an inverse relationship between religion and paranormal beliefs: People who do not go to church have more belief in the paranormal, which we will explore in Chapters 15 through 17.

The fact that mystical states exist means that non-mystical experiences cannot be the sole arbiters of what we can plausibly believe. As Swami Vivekananda said, "There is no feeling of I, and yet the mind works, desireless, free from restlessness, objectless, bodiless." Author Ralph Waldo Trine agrees, saying, "There is a conscious realization of oneness with infinite light. The degree to which you realize this oneness is the degree to which you are free from sickness, sadness, privation, etc. You need only stay in hell as long as you want and can rise up whenever you wish."

> *The reason there's so much ignorance is that those who have it are so eager to share it.*
> Frank A. Clark

Spirituality needs to be rational, even if it only shows us the limits of our reason. Ethics must also be rational. Both are possible once we understand that we are talking about the wellbeing of thinking, feeling creatures. Worrying about someone else's private pleasures only proves that the worrier has far too much time on her or his hands. Our ethics must be derived from the natural world, because we evolved from and remain a part of nature that brings both beauty and horror. If there is a god, then that god created the bad along with the good. Theology sprang from the human drive to know and discover; it has devolved under the weight of its own dogma into nothing more than the study of human ignorance and depravity. A single photograph of a cell or a galaxy reveals far more truth and beauty than all the JCI books combined. As Sagan observed, "No religion ever looked up and went, 'wow, this is bigger than we are and grander and more eloquent.' Instead, they keep God little. A religion that extolled the beauty of the universe could possibly tap into a wellspring of faith untrammeled by normal religions."

Believing that sacred books are the inerrant words of gods has caused immeasurable suffering. A shift in viewpoint that saw these books as written by, for, and about people struggling with belief would immediately end much of the world's suffering. It is high time to admit that these books were not written by any god, but by people who thought the Earth is flat and for whom the ballpoint pen I am using to edit this book would have been an astounding display of technology. It is high time to admit that the cultural concessions made to religion are providing tacit blanket approval for ongoing violence and stupidity. It is high time for religion to be deemed intolerable, and relegated to the societal fringes. It is high time to admit that a boy king trying to cling to power created the JCI god. It is high time to admit that all revealed religions are complete and utter bunk. It is high time to admit that the idea of one religion being "it" above all others comes from a criminal degree of ignorance about history. It is high time to admit that the thousands of years of theological scholarship and religiously inspired artisans was a complete waste of time and resources.

Above all, it is high time to stop turning to religion to seek out the divine and strike out on our own, accepting guidance from those who have been there while giving unconditional loyalty to no one creed. Getting rid of God puts us in the pilot's seat of our own life, and gives the us the responsibility to create a better world for ourselves and those around us. Understanding the fact that we can die at any moment, and that there is no guarantee that anyone will remain alive from one minute to the next, will motivate us to make the most of whatever time we have left. We will all certainly die at some point, whether we believe it or not, and whether we can imagine our own demise or not. We must accept the fact that the questions of God and afterlife are entirely separate, as we learned in Chapter 2. To quote the apostle Paul, "When one is a child, one thinks and speaks as a child, but one must put away childish things when one become a man." Well said.

Ponder all of the good that will not be done tomorrow because people are building churches, temples, mosques, etc. or enforcing religious laws. How many person-hours have been lost today because of religion? A glitch that disables a critical computer server for only a few minutes can cost millions of dollars in lost productivity. What does religion cost?

Those of us who have for years politely concealed our contempt for the dangerous collective delusion of religion need to stand up and speak out.
Richard Dawkins

We have just enough religion to make us hate but not enough to make us love one another.
Dean Swift

Chapter 10
Religion, Interrupted

Our enemy is nothing other than faith itself.
Sam Harris

Occam's razor does not always lead to truth. The more complex answer is sometimes true. As we will see throughout the rest of our journey together, the truth may well be far more complex—and wonderful—than we can imagine. The beauty of this journey is that it will not have been in vain, even if I am utterly wrong about everything in this book. Merely asking the questions and pondering the answers is a noble pursuit. As Philippians 4:8 says, "Finally, brethren, whatsoever things are true, whatsoever things are honest, whatsoever things are just, what soever things are pure, whatsoever things are lovely, whatsoever things are of good report: If there be any, and if there be any praise, think on these things."

I have presented some damning evidence against religion. We will now look at how religion responds to the charges against it. From there, we will see whether a god can exist in the universe and ponder what the characteristics of such a god might be.

310 | The Divine Savage
Revealing the Miracle of Being

Chapter 11

Religion Fights Back

> *We speak of the wisdom of God in a mystery, even this hidden wisdom.*
> 1 Corinthians 2:6-7

> *All those who undertake to be spiritual workers... and believe that they should hear, smell or see, taste or feel spiritual things... surely are deceived and are working wrongly against the course of nature.*
> The Cloud of Unknowing

A young woman named Lisa Baker once encapsulated the human search for knowledge and enlightenment when she said, "All I want is reality. Show me God. Tell me what He is really like. Help me to understand why life is the way it is and how I can experience it more fully and with greater joy. I don't want empty answers. I want the real thing. And I'll go wherever I find that truth system." The previous chapter summarized the full frontal assault of science and reason against religion and made it seem that religion is on the losing end of the battle, whether religious people want to admit it or not. Is this in fact the case? How does religion respond to the grave charges made against it, and how well do their responses stand up to the cold light of reason?

Religious people say that it's up to the skeptics to prove them wrong, not their job to prove themselves right. This is wishful thinking. Anyone can make any claim about anything at any time, but the burden of proof is always on the claimant. The Flying Spaghetti Monster (FSM) is a tongue-in-cheek response to religion that postulates the existence of an omnipotent serving of sentient pasta and meatballs, that touches those it favors with its noodly appendage. One can neither prove nor disprove the existence of such a creature any more than one can prove or disprove that there is a teapot floating somewhere off in deep space; however, we can say with some confidence that the odds against the reality of both the FSM and interplanetary teapot are far higher than the odds of them existing. The same logic works for religious gods; all religious people are atheists when it comes to all gods but their own.

Some years ago, I asked some Christian friends of mine how they could believe what they did in light of scientific evidence that has been consistently proving them wrong for hundreds of years. They told me that science and history would eventually prove them right, and that all of the reasons for being Christian could be found in Lewis's *Mere Christianity*. Lewis was raised in a religious family, but became atheist before returning to theism and eventually back to Christianity. "My argument against God was that the universe seemed so cruel and unjust," Lewis wrote. "But how had I got this idea of just and unjust? A man does not call a line crooked unless he has some idea of a straight line. What was I comparing this universe with when I called it unjust?" Of course, the comparison was between his definitions of just and unjust, which says nothing at all about the state of the universe. A crooked line has curves in it, and merely looking at it should be enough to make someone imagine a line without curves. Just because someone thinks something is so does not make it so.

Lewis's logic seemed pretty flimsy, but I had made a promise to read his book, and undertook to do with an open mind. After all, as he wrote, "We do not need to study philosophy to reach God, just to have an intellect free of sophistry or prejudice." Fair enough; however, Lewis also cautioned against letting reason and questioning roam too far afield because, "Thirst was made for water; inquiry for truth. What you now call the free play of inquiry has neither more nor less to do with the ends for which intelligence was given you than masturbation has to do with marriage." In other words, it's OK to ask questions so long as you don't get carried away.

"Creatures are not born with desires unless satisfaction for these desires exists," wrote Lewis. "A baby feels hunger; well, there is such a thing as food. A duckling wants to swim; well, there is such a thing as water. Men feel sexual desire; well, there is such a thing as sex. If I find in myself a desire which no experience in this world can satisfy, the most probable explanation is that I was made for another world." The world had not discovered DNA and proven the truth of evolution

Truly, one becomes good by good action, bad by bad action.
Brhadaranyaka Upanishad

Ministers say that they teach charity. That is natural. They live on hand-outs. All beggars teach that others should give.
Robert Ingersoll

when those words were written, nor did psychologists and neurologists understand nearly as much as they currently do about the power of belief and how traits that evolved for one purpose can be co-opted to other ends. As we learned in Chapter 9, our encounters with the divine, our religions, our ideas about other worlds, afterlife, etc. may all be in our heads.

Jars of spring water are not enough anymore. Take us down to the river!
Jalal-uddin Rumi

I was beginning to feel that religion—and Christianity in particular—is doomed if people like Irenaeus, Tertullian, Aquinas, and Lewis are the best they can muster. Nevertheless, there are huge holes in our scientific understanding, and the discoveries we continue to make seem to be pointing us in some surprising directions. As astronomer Allan Sandage wrote, "We cannot understand the universe in any clear way without the 'supernatural.'" Sandage is not alone in his viewpoint; we will meet many other scientists whose research has left them plenty of room to accept the divine later in this book. In Chapter 2, I explained that there is no necessary connection between religion and God; therefore, the discussion in this chapter must remain focused on whether religion has any valid arguments in its defense. We will return to the question of whether God can exist in Chapters 13 and 14.

Claims Against Materialism

In heaven, an angel is no one in particular.
George Bernard Shaw

Materialists are often just as dogmatic as the religions they oppose. For example, Jefferson dogmatized the idea that the immaterial is nothing and that nothing independent of matter can have any real meaning when he said, "To talk of immaterial substance is to talk of nothing. To say that God, angels, etc. are immaterial is to deny they exist." A religious person might respond to this by saying that God is not purely spiritual because He uses material things (such as bread and wine) to put new life into us, because He invented both matter and eating. Besides, even the most devout materialist who believes that thoughts are generated inside our heads would have a hard time saying that an idea is nothing more than a tiny electrical current, because such currents are not alive. I personally don't think that electrical current is a good example, because computer processors, memory, etc. rely on electricity, and one could see the brain as simply a complex computer. With that said, I do regard materialism as more philosophy than science for reasons I will explore in Chapter 31.

As I explained in Chapter 2, a dualist believes in the existence of both matter and mind and the fundamental difference between them. Mind (spirituality) does not influence matter through material contact, since mind and spirit are not matter themselves. Try as it might, materialism cannot deny the immaterial any more than someone researching the Himalayas can deny the existence of the Great Plains. This much is true, but it's a bit of a stretch to then ask how a material thing brings the immaterial into existence; radios do this all the time, but nobody ascribes consciousness to a radio. Thought *supervenes* (occurs as an unexpected or extraneous development) on brain cells. Some would say that thoughts do not affect the material world, an argument easily rebutted by the fact that our thoughts are an integral part of creating our individual experiences of the material world, without which we may as well not exist. Thoughts also guide actions, which can certainly impact the material world.

> *No testimony is sufficient to establish a miracle, unless the miracle be of such a kind, that its falsehood would be more miraculous than the fact which it endeavors to establish.*
> David Hume

If thought is immaterial and if humans really are body and soul, then why is it so easy to render someone unable to think for anywhere from a few seconds to forever by hitting them on the head? Catholic philosophy opines that humans are a union of body and soul, where the soul is not free enough from the body to function independently if the body is not functioning; however, since thinking is not material, then a brain injury is an *extrinsic* (extraneous) dependence. Identifying thought with electrical impulses or materialism is absurd; one can say that thought emerges from matter, but what does *emerge* mean? My computer is a mass of electrical impulses traveling through material pathways, yet somehow this word processor has emerged to capture orderly patterns of symbols while a streaming music player is busily converting immaterial impulses to material vibrations that are in turn affecting material air molecules that I am then converting back into electrical impulses and understanding as *Volare* by the Gypsy Kings. Nothing spiritual is happening here, and the religious argument seems to be utterly refuted.

> *What most atheists do is believe that although there is only one kind of stuff in the universe and it is physical, out of this come minds, beauty, emotions, moral values—in short the full gamut of phenomena that gives richness to human life.*
> Julian Baggini

But wait! It may be possible for thinking immaterial beings (spirits, angels, etc.) to exist without any dependence on matter. These beings would be far simpler than humans, because

> *Resurrection is not resuscitation. To say that Jesus was 'raised from the dead' does not mean he returned to haunt Jerusalem or Galilee as a terrifying, though familiar, ghost, but that the entered upon a new mode of existing, a new relation to God, a new and different way of interacting with the world.*
> Brian Josephon

they have few components, but that would not make their intellects any less rich than ours. The value of π would still be the same (≈ 3.14), even if the material universe did not exist. Materialists fall into the trap of thinking that complexity and perfection go hand in hand, when the two are not necessarily related. Materialism is absurd, because thought is neither material, nor caused by material, nor a material property. Thus, thought and matter must be separate, and dualism is the true nature of the universe. Materialism has been proven false. Besides, a naturalistic view of the future is only as accurate as one's base assumptions. This line of thinking smacks of, "different verse, same as the first." There is no reason to believe that π, the radius of a circle, or any of the four fundamental forces would be the same absent a material universe, because one can postulate universes with totally different natural laws.

> *How much effort it takes to affirm the incredible!*
> Roland Barthes

In short, the religious argument against materialism seems to boil down to a fundamental ignorance of the science behind materialism, combined with a healthy dose of Pascal's wager. Materialism may well have insurmountable obstacles (see Chapter 31), but not for the reasons religion tends to give.

The Origin of Knowledge

Reality is usually something a person can't guess. This much is true; after all, nobody expected our neat clockwork universe to reveal itself as inherently random and not all here, but that's what quantum mechanics did. Nobody expected the answer to a simple question about motion to depend on who is asking the question, but that's what relativity did. This logic is one reason why Lewis abandoned atheism for Christianity.

> *Faith is the art of holding on to things your reason has once accepted, in spite of your changing moods.*
> C.S. Lewis

According to Lewis, intellectual knowledge is better than animal sense knowledge. A dog is more complex than a single-celled organism and also superior to it. Increased power does not necessarily mean greater complexity; for example, Mozart invented entire symphonies at once, yet most musicians have to struggle through his compositions bar by bar to grasp Mozart's reasoning and learn how to play the melody. Even then, their knowledge remained inferior to his. This sounds like a completing argument until one grasps the extent of Lewis's ignorance of modern psychology. Humans are primarily emotional creatures; our logic primarily exists to justify emotional decisions. Also, the speed at which a series of computers independently process a given task says nothing about the complexity of the task itself.

For example, Mozart's symphonies themselves are extremely complex, especially when compared to songs like *Chopsticks*. The fact that he came up with them on the fly does not make them any less complex, although one could say that his process of creating the music was simpler than for a normal composer. Mozart was musically intelligent but again, intelligence/processing power does not of itself say anything about the complexity of the task at hand. According to Catholic philosophy, God knows everything by knowing Himself. All things resemble God and are limited reflections of His unlimited nature. By knowing Himself, God therefore knows all that could ever exist, making His act of knowledge supremely simple. This theory has merit for reasons we will see in Chapter 31, even though the arguments as presented are rather weak.

Lewis postulated that a cause must be at least as perfect as its effect for it to cause that effect. For example, a flame that is less than 212 degrees Fahrenheit cannot possibly boil water at sea level, nor can a teacher impart knowledge s/he does not have. So far, so good. Lewis then goes on to say that the first cause of thought must be intellectual. There is a first cause on which all depends; nothing can happen without this unchanging first cause. This can go to infinite regression, except that Lewis asks us to think in terms of causes. Nothing can change unless there is something capable of causing change without changing itself in the process. Changing your mind requires a cause, and no event can occur without such a cause, be it known or unknown. Whatever causes will must be either changed or not in so doing. If the former, then this cause itself requires a cause and so on until one arrives at the unchanging cause, which must be omniscient with unlimited intellect. In short, God is causing our thoughts—and because of God's unlimited nature, philosophy must call Him good. Here, Lewis seems to have been ignorant of basic psychology, the power of the human mind, evolved instincts, and much more.

We cannot detect God directly because He is an external controlling power in the universe and cannot show Himself as one of the facts inside the universe any more than an architect can show himself as part of a house he designed. Lewis's argument

If an angel were to tell us about his philosophy, many of his statements may well sound like 2x2=13.
Richard Dawkins

The prophet receives and transmits the word of god to which he adheres through faith; the mystic is sensitive to an inner light that exempts him from believing. The two are incompatible.
Henri-Marie Cardinal de Lubac

is little more than a rehashing of Aquinas's Prime Mover argument, which we saw in Chapter 9. As an aside, if this argument is true, then it would seem to leave little room for original sin, unless God is scapegoating the entire human species for His mistake.

The Moral Law

Ever since the creation of the world his invisible nature, namely his eternal power and deity, have been clearly perceived in the things that have been made.
Romans 1:20

We tend to reserve our highest praise for those who come to the aid of others with no expectation of reward, especially when the recipient is a stranger. This can be as simple as helping a little old lady across the street, as final throwing one's self on a grenade, or anywhere in between. The Biblical story of the Good Samaritan (Luke 10:25-37) encapsulates the idea of altruism. Lewis offers altruism as evidence of what he calls the *moral law*, claiming that it poses a major challenge for evolutionists, and that humans are unique in our capacity for morality. This is not quite the case; in Chapter 9, we discovered that evolution can easily account for selfless acts.

Lewis correctly points out that the concept of morality is concerned with fair play, keeping individuals on the straight and narrow, and keeping society functioning smoothly. He goes on to claim that morality must play itself out as God wants it to, and then lists seven virtues, four cardinal and three theological. The four cardinal virtues are the bedrock on which all other virtues stand. They are:

- **Prudence:** Caution and discretion in practical matters, care in managing resources, economy, and frugality.
- **Temperance:** Moderation or self-restraint in thought, word, action, and indulgence (particularly of alcohol).
- **Justice:** Being righteous, equitable, and just.
- **Fortitude:** Having strength, firmness, and/or courage.

Faith is a virtue by which things that are not seen are believed.
Augustine

The three theological virtues are the foundation of Christian moral activity, and are infused into our souls by God in order that we may attain the four cardinal virtues. They are:

- **Faith:** Confidence or trust in a person or thing, belief that is not based in proof, trust in God and His promises made through Christ and the Bible by which we can be saved.

- **Hope:** feeling that what one wants can be had or that events will turn out for the best. According to Lewis, looking forward to the afterlife is not escapism, but something every Christian must do.

- **Charity:** Generous acts or donations to help the needy.

Bad psychological material is a disease that needs curing, not repentance. Humans judge each other by external actions, but God judges by moral actions; a bad man performing a small kindness may be worth more than a good person giving up her or his life to save another.

People appeal to standards of behavior when dealing with morality and apply these standards when appraising behavior. Most people agree with these behavioral standards; those accused of violating the standards almost universally claim either that they did not violate the standard or that the violation was justified or otherwise at least partially excusable under the circumstances. For example, someone accused of murder will either deny culpability or plead for leniency for reasons, including self defense or the inability to tell right from wrong. I am not aware of a single murder defendant who has claimed that the laws against murder are fundamentally flawed. People have a sense of fair play in mind. Disputes and fights revolve around right and wrong and have no practical purpose, unless those involved have some sort of idea what right and wrong are. It is impossible to break a non-existent rule. In general, the moral law sides with the weaker side and directs our instincts to action.

The claim that one set of moral ideas is better than another arises from measuring compliance to a standard, which means comparing them against some real morality and admitting the existence of a real right that is independent of people's thoughts and wishes. Some power controls the universe, appears as a moral law, and makes a person feel bad when s/he is wrong. We must assume this power is like a mind; otherwise, it's matter, and matter cannot give instructions. The moral law is therefore something that is coming to us from beyond the universe. We have two pieces of evidence for this power that we call God. First, the universe exists. Second, we

---

God is no fonder of intellectual slackers than of any other slackers.
C.S. Lewis

Sir, allow me to ask you one question. If the church should say to you, 'two and three make ten,' what would you do? 'Sir,' said he, 'I should believe it, and I would count like this: one, two, three, four, ten.' I was now fully satisfied.
James Boswell

all have the moral law in our heads. We can find out more about God through the moral law than by studying the universe, just as we can learn more from a person by listening to her and not by the things she has made. Christianity (and presumably other religions, depending on a person's individual cultural context) speaks after one realizes the moral law, the power behind the law, and their own failure to uphold the law has made them wrong with that power.

Theology is now little more than a branch of human ignorance.
Sam Harris

A naked tribeswoman and a fully clothed Western woman might be equally chaste and modest, depending on their local mores. Of these, chastity is the most difficult rule: One may have either a 100% faithful marriage or nothing at all. This goes so far against our instincts that either Christianity (and similar religions) is wrong, or our inborn instincts have gone astray. "Of course," Lewis says, "being a Christian, I think the instinct has gone wrong."

One can easily assemble a large audience to watch a strip tease. If one could assemble such an audience to watch a piece of food, would we think something had gone wrong with our appetite for food? (Lewis was a few decades ahead of both the obesity epidemic, the Food Network, and top-rated shows such as *Iron Chef* and *Top Chef*.) Perversions of the food appetite are rare, while sex perversions are both numerous and hard to cure. We have been fed lies that the sexual instinct is just like other instincts, and that all will be well if we simply open up, but that is not true. We can see the error of this conclusion the moment we look at the facts. There is something inside us all telling us that we are two halves, male and female, and were made to combine in pairs, both sexually and totally.

A skeptic is simply a person who chooses to examine carefully whether his or her beliefs are actually true.
Andrew Newberg

The church should recognize that most people are not Christians and should not be expected to lead Christian lives. Thus, there ought to be two kinds of marriage, church and state, each governed by its own rules, with sharp distinctions so people can instantly recognize religious versus non-religious marriages. (On this point, I happen to agree entirely!)

Things that have no knowledge, such as sodium, cannot be directed toward an end, such as reacting with water, unless by something with knowledge. It is in the nature of sodium to react with water, but this occurs with no premeditation to do that, as opposed to something else; it does not choose its own action, but must instead be directed by an intelligent agent.

The belief that God is beyond good and evil is called *pantheism*, in which God animates the universe in the same way we animate our own bodies. The universe is God, who would not exist if the universe did not exist. JCI believers believe that God created the universe and is separate from it; his existence does not depend on the universe any more than a painter does not die simply because his painting is destroyed. A pantheist sees a slum and says, "This is also God," but Christians disagree and are moved to action because Christianity is a fighting religion. The argument against pantheism seems cruel and unjust, but where did these ideas come from? What are we comparing the universe with? The attempt to disprove God actually proves Him. Atheism is too simple. If the universe has no meaning, then we should not know that it has no meaning. If there was no light in the universe, then we would not know what darkness is, any more than a fish knows what wetness is.

Evil depends on the same traits that also further good, such as determination and cleverness. The powers that make evil possible are therefore granted by goodness. Thousands of years ago, a Jewish man appeared talking as if he were God, forgiving sins, saying he always existed, and that he will return to judge the world someday in the future. A pantheist would say this man is either part of God or one with God and find nothing odd with that, but Christ did not mean that kind of God. Surrendering, admitting one's wrongness, and starting over from scratch is the only way out. Repentance is no fun. A bad person must repent, but only a good person can repent perfectly.

Someone who scorns secular authority the way some scorn religious authority must be content to remain ignorant. God is holding back to give us a chance, but we don't have forever. A hallmark of a bad man is that he cannot give something up without wanting everyone to give it up as well; this is not the Christian attitude. (What about sex?) Pride is the worst of the seven deadly sins, because it leads to all of the other sins. Pride has been the chief cause of misery throughout human history.

You may have noticed that intellect, reason, logic, and autonomy are not included in the list of virtues that Lewis puts

---

Study astronomy and physics if you desire to comprehend the relation between the world and God's management of it.
Maimonides

Thanks to the telescope and the microscope, [religion] no longer offers an explanation of anything important.
CHristopher Hitchens

---

forth. Moreover, the scope of his ignorance of basic psychology, evolution, and science is breathtaking. If the likes of Tertullian, Irenaeus, Lewis, and others are the best religious people can do to justify and explain its existence, then it would seem that religion is utterly without merit in light of the history we have reviewed in Chapters 6 through 8 and the strength of the arguments made in Chapter 10. Can religion be joined and made compatible with current scientific knowledge? Some believers are trying to do just that.

Evidence of Divine Design

To be human is to have myths.
J. F. Bierlin

According to Dawkins, the more complex something in the natural world is, the more complex its producer must be. For example, people build cathedrals that are far more complex than nests built by birds with small brains. The very complex universe we live in would thus require an even more complex creator. A designer god cannot explain organized complexity, because a god capable of designing something would have to be complex enough to require explanation in its own right, that is, such a complex God would need His own even more complex creator, and so forth. God is supposedly the intelligent designer of the universe, but any being capable of such a feat would have to be an outside agency; a God who is both the "first being" and the intelligent designer of the universe therefore cannot exist. Even church philosophers will agree that God's complexity demands explanation, especially because the god defined by theologians can multitask.

In fact, science is a type of myth, if we think of [not necessarily untrue] myths as stories about ourselves and our origins.
Michael Shermer

This argument could make some sense if we were talking about bodily things as Dawkins does, because bodily things do require more complex producers; however, God is not body but spirit. The complexity of the material universe does not imply that a spiritual being must be complex as well. Thus, there is no reason to think that the designer is any more complex than the design. The designer must be as rich in reality as that which he designs, because he must possess it within himself before producing it. In short, God must be intellectually capable of the design, which does not mean that God Himself is complex. (This whole problem disappears if we see things from a monistic idealism perspective, which we will do in Chapter 31.)

Thousands or Billions?

Did the universe come into existence some 13.7 billion years ago? Is the Earth 4.5 billion years old? Or did everything come into being a mere 6,000 years ago? Plenty of religious fundamentalists argue for the so-called "young Earth" theory. Physicist Gerald Shroeder points out that the Hebrew word *yom* in the Bible can mean time in the generic sense without reference to any specific period or interval. Augustine and Irenaeus would agree. The former said that the days of creation were not divided, but that God is divided; the latter that the days of Genesis are not literal. We therefore have no way to know for sure whether the Bible is meant literally or not, and plenty of reason to think not. Such a position does require blending science with religion, which religious leaders of all stripes have decried and continue to decry. Some even think science is the work of the devil, but the Bible and Koran both say otherwise. For example, Psalms 111:2 says, "Great are the works of the Lord and they are studied by all who delight in them." We will see another example in Chapter 32.

Assuming that the Bible is meant to be taken literally and that the young Earth claims are true, then how can one account for the many signs of age we see in the universe, from the radioactive decay of elements with known half-lives to measurements of cosmic distances? In the 1800s, Phillip Gosse (1810-1888) speculated that perhaps God created everything to look old. The actual age of the Earth might be only 6,000 years, but God could have made it appear to be billions of years old. Thus, ancient animals might only have existed as fossils, an idea renounced by the Creation Museum in Kentucky, which displays saddled dinosaurs and dioramas of people with dinosaurs. God could even have simulated the undigested food found in ancient stomachs, radioactive decay, and even placed starlight in transit to make it look like it had been traveling for billions of years. This theory quite literally makes God the biggest liar of all time—the same God who commands us against bearing false witness. Does God have a double standard?

Guided Evolution

We evolved from tree-dwelling apes, and some of our predators routinely attack from above. It should therefore be no

---

That which is now called natural philosophy, embracing the whole of science, of which astronomy occupies the chief place, is the study of the works of god, and of the power and wisdom of god and his works, and is the true theology.
Thomas Paine

Wonder how Dawkins and Dennett would react if world-class theologians gave similar short shrift to their scientific treatises?
Becky Garrison

> *Scientific theories are organically conditioned, just as much as religious emotions are.*
> WIlliam James

surprise that we can deduce the laws of falling objects, for the simple reason that we evolved to look for them. The surprise is that is we can figure out things like cosmology and quantum mechanics, because we did not evolve to handle those problems. The question we should be asking ourselves is therefore not which materialist model is correct, but how we actually got to this point. Most scientists eschew the idea of intelligent design and creation. Such a scientist must assume that life is everywhere in the universe, and that we are the product of generic processes. A scientist who believes in the existence of an OOO god has the luxury of considering all options and can approach the question in an unbiased manner.

Dawkins supposes the existence of an evolutionary algorithm that can shorten a process from, say, 10^{43} tries down to 43; however, this idea has deeply *teleological* (pertaining to the study of ultimate causes) implications in a theory that is supposed to be free of teleology. Moreover, such algorithms can only generate the illusion of complexity, not complexity itself. Intelligence chooses some things and chooses not others. For example, if a completely new species of life were to be created in a laboratory, it would prove the intelligent design of the lab scientist; however, we cannot create life piecemeal. If we are going to evolve an organism, then we must evolve the organism, because we cannot do so bit by bit; it's all or nothing. This argument sounds good but for growing fossil evidence that includes jawbones migrating to the inner ear where they now aid balance, to *Tiktaalik*, the first known species to have arm and wrist bones and the ability to do push-ups. More in Chapter 20.

> *I have always thought it curious that, while most scientists claim to eschew religion, it dominates their thoughts more than it does the clergy.*
> Fred Hoyle

All the evidence we have shows that life began on Earth as soon as the planet cooled down and started getting an atmosphere. In other words, life began as soon as it could, without the hiatus one might expect. Some microbiologists believe that the leap from nonliving to living matter that contained the same basic genetic code as humans is as big a leap as the one from that first organism to humans. This could suggest that life is ubiquitous throughout the universe. In our example, Earth would tend to hold in a lot of heat from a meteor impact because of its abundance of water vapor, while Mars would recover sooner. It may be better to think that life started on Mars and then hitched a ride to Earth on meteorites. David McKay is the Chief Scientist for Astrobiology at the Johnson Space Center in Houston, Texas. He believes that solving the riddle of whether there is or was life on Mars should be a top priority.

No theories exist to explain how prebiotic life can generate a genetic sequence that creates true life without resorting to an intelligent designer. A random sequence could generate some 200 bits of information. Compare that to *Escherichia coli*, which contains 6,000,000 bits. One cannot generate enough probability for this to happen randomly, even if one assumes one billion Earths. The odds of randomly generating the E. coli genetic sequence are about 1 in $10^{1,800,000}$. By comparison, the entire universe consists of a mere 10^{80} atoms.

Chance sequencing never produces information. Specified complexity results from a structure that creates meaning, such as languages that consist of letters assembled into words, sentences, paragraphs, etc. according to the rules of spelling, grammar, and punctuation. This is the kind of code our DNA contains. The proverbial monkeys sitting at typewriters could not possibly have generated more than half a line of Shakespeare in the entire history of the universe. We can see how patterns can appear without intelligent input, thanks to our understanding of chaos and self organization, but none of these theories say anything about creating meaning. Our internal genetic code points to an intelligent, transcendent intelligence that finds creative and ideal solutions to problems. For example, a ribosome can build all of the components it needs in minutes despite being trillions of times smaller than the smallest human-built machines.

What if we assume that life pervades the entire universe? "There are infinite worlds both like and unlike this world of ours," said Epicurus. "We must believe that in all worlds there are living creatures and plants and other things we see in this world." Is space the place to look for our origins? Neither the Western nor the Eastern religions have any problem with the idea of extraterrestrial life. None of the holy books says it contains all there is to know, merely all one needs to know. This begs some interesting questions: If there is intelligent life on other planets, do God's laws apply to them too? Did God come to their planets to die for their sins as well? Does God operate on a planet-by-planet basis?

The priests of the various religious sects dread the advancement of science as witches do the approach of daylight, and scowl on the fatal harbinger announcing the subdivision of the superies on which they live.
Thomas Jefferson

Let him be made of clay animated by blood.
Babylonian Poem of Creation

> *We must question the stony logic of having an all-knowing, all-powerful god, who creates faulty humans and then blames them for his own mistakes.*
> Gene Roddenberry

On a more mundane level, will extraterrestrial life have the same type of DNA? If so, will that DNA share the same helical structure? Will those organisms have similar cellular structures? The list of questions is endless. If we do find life on Mars or elsewhere, and this life turns out to be markedly different from that on Earth, then God either create life in many different forms, or it appeared spontaneously. If extraterrestrial life is similar to Earth life, then God either created all life in the same mold (in His image), or something in the way the universe evolved required all life to use the same building blocks. If the latter, then life probably did not evolve on Earth, because our thicker atmosphere makes ejecting rocks into space toward Mars after a meteor impact much harder than the other way around. Cosmologist George Smoot (1945-) led the COsmic Background Explorer (COBE) team that produced a very detailed map of the universe's background radiation. (See Chapter 26.) To him, it seems arrogant to think that God would make the entire universe just for us.

Proving that life began elsewhere and migrated to Earth would solve the problem of our origin, but would do nothing to answer the question of life's ultimate origins. If life started on Mars, then it may have come from even further off, and so on. We have no idea how long such journeys might take. Also, how likely is it for higher intelligence to form? Is the human species merely the "ultimate in an oddball rarity," as Gould would have us believe? Sagan once said that a single message from space would conclusively prove extraterrestrial life. Are we carrying just such a message in each of our genes? Physicist Paul Davies (1946-) believes that life and consciousness should be direct consequences of the laws of physics. These laws are uniform throughout the entire universe, which means that life should be spread throughout the universe.

> *Those who didn't take Genesis literally had no reason to believe there had been a beginning.*
> George Smoot

We do not have any evidence for the existence of little green men, but we are discovering more and more *exoplanets* (planets orbiting other stars). In September 2010, scientists announced the discovery of Gliese 581g, a planet 20 light years (about 70.5 trillion miles) from Earth that may be able to harbor life. Meanwhile, we may just have evidence for a superior intelligence in the impossibly long odds behind the values of—and relationships between—cosmic constants that tell us we may be far more than an evolutionary afterthought.

The Anthropic Principle

We know that the anthropic principle is based on the fact that the laws of nature seem tailor-made for life to evolve. There are different variants of this theory that can be described as "weak" or "strong." Needless to say, religion advocates the strong version, which has the following primary variants:

- There is a single universe that was designed to generate and sustain life. This is the classic religious argument.

- The universe can only come into being if there are observers to make it happen. This is the so-called biocentric theory, which we will explore more in Chapter 27.

- Other universes are needed for ours to exist. We will look at this "multiverse" theory in Chapter 26.

In short, we can summarize the strong anthropic principle as follows: Either God did it, we are incredibly lucky, or there are so many universes that the existence of our universe is nothing special. Any way you explain this, the fact remains that:

- If the rate of expansion of the universe one second after the Big Bang was different by 1 in 100,000, the universe would have either expanded too quickly for stars, galaxies, etc. to form, or have fallen back in on itself.

- If any of the 15 physical constants (see Chapter 26) was different by even a tiny fraction, our universe could not sustain life.

To say that our universe is wildly improbable is an understatement. We may eventually find that some or all of these constants are indeed specified, controlled, or limited by some higher force or power, but so far none has been forthcoming, which of course does not mean that such a thing does not exist. Meanwhile, we can imagine physical laws that would determine events for a god that could observe its creation without intervening. Jastrow said that the "universe began under circumstances that make it impossible to ever know what brought it into existence." Isaiah 40:26 tells us not to look too hard, because everything came from God.

From a scientific perspective, the universe has either been around forever or had a definite beginning, but neither sce-

---

The lamb and the lion will lie down together, but the lamb won't get much sleep.
Woody Allen

To an evolutionary psychologist, the universal extravagance of religious rituals, with their costs in time, resources, pain and privation, should suggest as vividly as a mandrill's bottom that religion may be adaptive.
Marek Kohn

> *Knowledge is that information which represents reality.*
> Jerry Wheatley

nario explains our ultimate origins. We currently have no way to find evidence for existence before the Big Bang, because all physical laws break down at the *Planck time* (the smallest possible interval of time) 10^{-43} seconds after the Big Bang. We can only observe causes and effects back to this time, because even the subatomic particles we are familiar with could not have existed before then. Everything prior to that time is closed off to us, which leaves the question of ultimate origin wide open. As we will see in Chapter 26, some physicists believe that quantum fluctuations could create an entire universe such as ours, which only begs the question of where those fluctuations came from. On the other hand, if the flow of time resembles discrete grains of sand flowing through an hourglass instead of a stream of continuous water, then the Planck time itself may be the moment of creation. Emerging theories such as Loop Quantum Cosmology (LQC) are beginning to offer hints at what came before our universe, but are still too new to be practical. More on this in Chapter 25.

Ex nihilo nihil fit: From nothing comes nothing. To the best of our knowledge, every effect has a cause. This is the cosmological argument for God. No theory about the Big Bang—or an "oscillating" universe that expands, contracts to a singularity, and starts over—can avoid the need for an ultimate cause. As philosopher Martin Heidegger (1889-1976) asked, "Why is there being rather than non-being? All we see is an effect that demands a great 'supernatural' cause. The universe could not have come from something that is irrational to the human mind."

> *If we have learned one thing from the history of invention and discovery, it is that, in the long run—and often in the short one—the most daring prophecies seem laughably conservative.*
> Arthur C. Clarke

Any way you examine this, the Big Bang, oscillating universe, and vacuum fluctuations must have a source. This source must have complex laws that allow it to make one or more universes. Where did these laws come from? Astronomer Fred Hoyle (1915-2001) said that, "A commonsense interpretation of the facts suggests that a superintellect monkeyed with physics, chemistry, and biology, and that there are no blind forces worth speaking about in nature. The numbers one calculates from the facts seem to me so overwhelming as to put this conclusion almost beyond question."

In Chapter 26, we will see that the universe's background microwave radiation (the afterglow from the Big Bang) perfectly matches previous predictions, which lends huge support to the Big Bang theory. This *black body* radiation indicates that the universe was so dense that it formed a single continuous body, with nearly perfect thermal equilibrium that contained all of the points in the universe as its source and produced unimaginable amounts of energy. (A black body is an idealized physical object that absorbs all electromagnetic radiation and emits thermal radiation.) The "ripples" in the background radiation could be the remnants of quantum fluctuations, which account for the presence of clumps of matter that formed into galaxies. To Smoot, looking at this background radiation whose ripples are only 8 to 14 parts per million is like looking at God. A group of Semitic worshippers at Elba thought of an ultimate creator who predates the universe, but the credit belongs to the Egyptians for coming up with the idea first.

There is no need to wait until we are saved to become human.
Andre Compte-Sponville

As for the Big Bang itself, it literally occurred outside of time, not merely before time. It occurred outside of space, not merely in some space outside the universe. Until we develop some new theories such as LQC, it is meaningless to ask what came before the Big Bang or where it occurred. As Guth said, "The Big Bang theory is the theory of what happened afterward, not before." Time and space as we know them were created with the Big Bang, and it is logical to think that the universe would continue expanding at a roughly consistent velocity; however, that is not the case. A brief period of extremely rapid *inflation* (exponential expansion) occurred as falling temperatures allowed matter to form permanently in a manner similar to water freezing into ice. These cooling particles released their *latent* (trapped) energy, which accelerated the expansion process. Once that energy was spent, the expansion slowed down because of gravity that functions like a cosmic brake. The amount of gravity in a given volume of space lessens as the universe expands, which means that the expansion has been picking up speed ever since. According to physicist Roger Penrose (1931-), the original phase-space volume was precise to a tolerance of $(10^{10})^{123}$, far more than the 10^{80} atoms in the universe.

Conflicts between science and religion arise from misinterpretations of the Bible.
Maimonides

The universe was about the size of a baseball when this all began, and had expanded to 6 feet across within a trillionth of a second. The building blocks of atoms, quarks, and electrons, formed at this time. Opposite sides of the universe are the

> *Myth is neither a lie nor a confession: it is an inflexion.*
> Roland Barthes

same temperature, meaning that they must have been in contact at some point. Guth is quick to caution that inflation does not explain how the universe began from nothing; it merely turns a very small universe into a much bigger one. It also explains why the universe is so uniform everywhere we look. People who believe in a transcendent God are commonly dismissed as superstitious or even delusional, and belief in a conscious universe may seem to be the ultimate superstition (see Chapter 9), but what should we make of a universe with properties that let it create and fine-tune itself? Is this not a step down the slippery slope to creationism?

Some theories of planet formation speculate that so-called "super planets" like Jupiter or Saturn formed further out from their host stars than small rocky planets like Earth, most of which would have fallen into the stars. Alternatively, rocky planets could form further from the star beyond any bigger planets, which would mean that larger "sweeper" planets in outer solar systems would be rare. Jupiter and Saturn are the super planets in our outer solar system, and play a significant role in protecting Earth from comet impacts. Without this protection, comets would strike Earth anywhere from 100 to 10,000 times more often, meaning that we would not be here.

It is of course too early to tell whether or not Earth-like planets that can harbor life are rare, but the discovery of planet Gliese 581g only 20 light years from Earth may deal yet another blow to any claims we may have of any sort of special status. On the other hand, life-bearing planets could be relatively few and far between, making each of them special. Consider some of the requirements for such a planet:

> *If you always be handling the letter of the word, always licking the letter, always chewing upon that, what great thing do you do? No marvel you are such starvelings.*
> John Everard

- The star must be of the right variety.
- The star must have a proto-planetary disc.
- There can be no large planets with elliptical orbits.
- Large planets with circular orbits are required.
- The habitable planet must be in the star's "Goldilocks" habitable zone where temperatures will support life.
- The habitable planet must be large enough to hold on to its atmosphere, but not so large as to crush all living things on its surface.

- The habitable planet may need to be part of a double-planet system. For example, the Moon is huge compared to Earth, which is very unusual.
- The time when the star heats up must correspond to the time when the habitable planet's atmosphere cools.
- The habitable planet must have mechanisms in place to retain carbon dioxide.
- The habitable planet must have a stable crust.
- The habitable planet must be inhabited.

It is important to remember that time is not infinite; it has a definite beginning. This beginning must have a cause, and, as we saw above, this beginning cannot be caused by anything in time. Cause and effect happen within time, by definition. God is not confined by time, and therefore needs no cause. Some Eastern religions say that the universe is its own cause, that it simply is. God and the universe are one, and there is no duality. This view contradicts everything we know about cause and effect—if we apply a dualist or materialist mindset, that is, but such a view is perfectly compatible with monistic idealism. We will return to this interesting angle in Chapter 31.

As far as we know, matter that falls into a black hole is trapped forever, despite some intriguing theories that matter may shoot out so-called "white holes" on the other side into alternate universes, or that black holes even create new "baby" universes. If matter cannot re-bang out of a black hole, then how could it possibly have banged out of the Big Bang singularity that supposedly had nearly infinite density and gravity? Quantum mechanics demonstrates that some energy does escape black holes in the form of *Hawking radiation* (named after physicist Stephen Hawking, 1942-), which would not necessarily result in a smoothly oscillating universe, but in chaotic expansion cycles. More on this in Chapter 26.

Also, the First Law of Thermodynamics says that energy can be converted into a particle or anti-particle, or vice-versa, but that one cannot get something for nothing. Mathematically, we can convert A to B, but still cannot get something for nothing. There is no such thing as a free lunch in the entire universe. Matter did not simply pop into being. The expansion of the universe is an expansion of space, not through space. It

The Christian myth claims to be history and asks its adherents to believe that it is true.
Burton Mack

What we humans are looking for in a creation story is a way of experiencing the world in a way that will open us to the transcendent, that informs us, and at the same time forms us within it.
Joseph Campbell

> *When science and the Bible differ, science has obviously misinterpreted its data.*
> Henry Morris

is meaningless to talk about space beyond the universe, but this is where quantum fluctuations are said to have acted, which would be impossible if true nothingness lies beyond this universe's borders. If quantum fluctuations are indeed infinite, then it would seem that our universe is expanding inside a larger space. Smoot says that we can imagine the universe springing from as little as 20 pounds of material, which is still infinitely more than no material.

The Second Law of Thermodynamics deals with entropy. The Bible points out that the universe is wearing out Psalms 102:25-26 says, "In the beginning you laid the foundations of the Earth, and the heavens are the work of your hands. They will perish, but you remain; they will all wear out like a garment. Like clothing you will change them and they will be discarded." The universe must have had a highly ordered beginning and could not have begun itself, ergo we can infer that it was created by an external entity. (Keep in mind that the Hebrews were hardly the first to think of this concept; they simply borrowed it from other, much older cultures.)

The observable universe is expanding, which points to a common origin in time and space, which means it must have had a creator. In 1927, priest and physicist Georges Lemaître (1894-1966) first predicted the primordial singularity and said that heat from the Big Bang should still be present, because there is nowhere for it to escape to. This heat should be detected as radio waves and microwaves. Such radiation was discovered as we learned above, and the COBE team mapped the radiation with extraordinary precision. Since then, the Wilkinson Microwave Anisotropy Probe (WMAP) has carried out even more precise observations that further confirm background radiation and a number of other critical predictions. The next time you hear static on a radio or see static "snow" on a TV screen, take a moment to ponder that you are literally hearing and seeing the remnants of the Big Bang moment of creation itself.

> *I do not see at all the mystery of the incarnation, and all the other advantages that god brought forth and infinity of other very great advantages for an infinity of other creatures.*
> René Descartes

Mathematician Émile Borel (1871-1956) proposed using odds of 1 in 10^{50} as the boundary beyond which we eliminate chance. Philosopher William Dembski (1960-) proposes 1 in 10^{150}. As we saw above, Penrose placed the odds of our universe phase space being like ours at $(10^{10})^{123}$. Chance seems not to be a factor behind the universe's existence. Then again, our universe may not be the only one. Physicist Hugh Everett

(1930-1982) proposed a "many worlds" theory, wherein our universe splits into two virtually identical copies every time a quantum event takes place. Every time a particle pops in or out of existence, zigs instead of zags, every time you order the pasta instead of the roast beef, whatever, the universe splits. The copies then split every time a quantum event occurs therein, and we soon end up with nearly infinite universes. Everett proposed his theory to account for quantum effects and the seemingly central role of consciousness, which we will examine more starting in Chapter 29. We seem to be forced to choose between an infinite creator or infinite universes.

We know that religion believes none of this to be arbitrary. The laws of nature are interlocked in such precise patterns and intricate structures that they can only be the result of a super-intellect who wanted to benefit humans. Additional examples include:

- The existence of elements needed for life.

- The ratio of the mass of a proton to the mass of an electron.

- The relative strengths of the four fundamental forces (strong, weak, electromagnetic, and gravitational).

- Protein formation (Hoyle placed the odds of having all of the functional proteins required for life in one place at one time at 1 in $10^{40,000}$, which could not have happened in a universe completely filled with so-called primordial soup.)

- The balance between gravitational and electromagnetic forces in stars.

- The universe beginning in a low-entropy state.

- The balance of expansionary and contradictory forces in the universe.

- The slight excess of matter over antimatter, which Weinberg placed at one part in 10^{10}.

- Background radiation that perfectly matches predictions.

All great deeds and all great thoughts have a ridiculous beginning.
Albert Camus

How do you know? Have you died and been there?
African pygmy questioning a missionary's description of heaven

- The strong and weak forces must be balanced to within 1 part in 10^{60}. Too strong and hydrogen would be rare; too weak and only hydrogen would exist.
- If gravity was stronger or weaker by one part in 10^{40}, then stars would not exist.
- The balance of the number of electrons compared to the number of protons must be within 1 part in 10^{37}, or planets could not form.

All of this apparent fine tuning led Hawking to ask, "What breathes fire into the equations and makes a universe for them to describe?" Pagels wonders, "Where are laws written into the void? What tells the world it is pregnant with a universe?" Guth, however, points out that, "The anthropic principle is vague enough to where we can use it to explain most anything. It never gives precise predictions, only explains after the fact that what you saw was acceptable." Davies took a more nuanced approach when he said that, "One may find it easier to believe in infinite universes than an infinite deity, but either involves faith. The multiverse being taken seriously for a time is evidence of a lack of good evidence against a grand designer."

A variant of the anthropic principle theory called the *participatory anthropic universe* says that we ourselves may have brought the universe into being by measuring ourselves and our surroundings. We will be exploring concepts like quantum entanglement, quantum superposition (also called *Schrödinger's cat* after physicist Erwin Schrödinger, 1887-1961), wave/particle duality, the Heisenberg Uncertainty Principle, delayed choice, biocentrism, and the implications of all of these things beginning in Chapter 21. For now, suffice it to say that the universe may not have existed until something evolved to observe it.

Materialist physicists vehemently deny any such idea. Some, like physicist Victor Stenger (1935-), have created cottage industries for themselves by writing book after book denying anything but strict materialism. For example, Stenger dismisses the idea of a participatory anthropic universe because it violates classical cause and effect. Physicist Werner Heisenberg (1901-1976) did not mean to imply that particles have no causes, only that we are limited in how we can measure them. An intelligent observer is not needed to give reality to a parti-

---

When the social institutions become shaky, and uncertainty about the future becomes widespread, people look to religion to provide absolutes and a sense of security in the midst of their changing world.
Martin Marty

We simply do not need religious ideas to motivate us to live ethical lives.
Sam Harris

cle, because a detector will detect that particle with or without an observer. (Stenger forgets that we can only know that detection has occurred by observing the detector.) He goes on to say that quantum mechanics is only true for small systems below the so-called "macro" level, claiming that uncertainty reaches zero for large enough systems, despite the contradicting evidence from dual-slit experiments conducted on macro-level objects. Attempting to apply quantum mechanics to the whole universe means trying to explain the universe's behavior and existence in quantum terms. If the universe had no existence until observed, then who or what outside the universe could have done the observing? A transcendent God? (Here Stenger reveals his ignorance of the biocentric theory, because the observations could be carried out by something inside the universe that caused both itself and the universe to exist.)

Sagan once asked, "I am a collection of water, calcium, and organic molecules called Carl Sagan. You are a collection of almost identical molecules with a different collective label. But is that all? Is there nothing in there but molecules?" That is the question this book seeks to answer. We will look at the anthropic principle again later.

The Orderly Universe

At the most fundamental level, each of us forms our own primary existence unto ourselves, utterly dependent on our senses and beliefs to form our individual realities. We are prisoners in our own bodies, and all of us are utterly alone, no matter how many people we have around us. We do perceive an external universe, and project ourselves into and into it. The universe exists but also changes; we name semi-permanent things, but change is inevitable. The present fades into the past, and the future keeps coming toward us. Our lives are a paradox of being and becoming. How can this ever-changing universe be built on unchanging, abstract concepts?

If God exists, whence evil? if He does not exist, whence good?
Gottfried Wilhelm Leibniz

We want to feel that something must have started it all. This something must be "supernatural," because science cannot take us beyond the moment of creation and can tell us nothing about our ultimate origins. Of course, all debate about origins assumes that there is indeed an origin. The JCI religions believe that God is external to creation and that the universe is contingent on His will, which came into being at a defined instant because of His deliberate action. Pagan and Eastern religions believe that the universe is an extension of God. How God relates to matter is open to debate. Some myths talk about order emerging from chaos, while the Gnostics saw matter as the work of the devil. Differences aside, all religions believe that the physical universe is incomplete and cannot be self-explained; existence must depend on something divine. All agree that the universe is not infinite; if we lived in a universe of infinite size and age, then our sky would be infinitely bright in all directions.

Heaven for climate, hell for company.
Mark Twain

As we will see in Chapter 24, time, space, and matter are linked. In other words, time needs space in order to exist. At some point, the universe was either a singularity or very close to it. Where? Nowhere. When? Nowhen. What came before? There was no before; time began at the Big Bang. If we accept this, then what caused the Big Bang? Nothing could have come before it, because there was nothing before it. Did it cause itself? Was this event without cause? The laws of physics break down at this theoretical singularity, which means that the beginning was outside these laws. The reason for the Big Bang must therefore lie outside physics. God, if invoked, must be beyond normal cause and effect. Occam's razor demands that we reject God as unnecessary, because we cannot know where God came from. In fairness, the problem persists no matter how you try to explain it away. Even LQC fails to explain ultimate origins.

If we can assert that God needs no creator, then we can also assert that for the universe itself. Atheists accept the universe as brute fact over God, because we can directly perceive the universe. Some physicists claim that quantum mechanics allows the universe to come into being without any external intervention. The "zero bound" theory says that space and time emerge from a zero point that has no given instant, in the same way that all directions are south when standing at the North Pole, giving the universe a beginning but no singularity or first moment. Whereas the Big Bang tends toward divine creation, the zero bound removes the need for a creator. Eliminating the initial event restricts the quantum state of the universe at all times. We already know the mathematics for the earliest moments of the universe.

Don't worry, be happy.
Meher Baba

Why does the universe exist? A mathematical model of the universe is not the same as the universe itself by any definition. Quantum mechanics is based on unpredictability and probability, not certainty. Thus, there is a non-zero chance that the universe could simply appear spontaneously. There is no law against it happening; even the so-called "laws of nature" are nothing more than concepts of how nature works. The only reason we have laws of nature is because we have observed nature behaving consistently over periods of time. It is humans who have defined and enumerated these laws, not nature itself. In fact, the very idea of natural laws comes from religion and Aristotle's four causes, which are:

- **Material cause:** The material something consists of. For example, the material cause of a dinner plate may be porcelain, clay, etc.

- **Formal cause:** The form of a thing, e.g. how its material is arranged. For example, the formal cause of a dinner plate may be circular with slightly raised edges.

- **Efficient or moving cause:** This is the thing's primary cause of change or rest. An efficient cause of a thing can exist whether or not the thing is ever actually produced, and must therefore not be confused with a sufficient cause. For a dinner plate, this might be the art of pottery since that is the principle guiding the plate's production.

- **Final cause:** A thing's final cause is its goal or purpose. for a seed, this could be an adult plant. For a dinner plate, it would be containing the contents of a meal in a sanitary manner.

Belief in a rational God formed the basis for the scientific approach in western Europe that sought to understand God's order by studying His works. The Chinese system of belief has no such creator per se; their science progressed without interruption or obstruction from religion, which may well explain why they flourished while Europe languished in the Dark Ages. For example, the best ships Europeans could put to sea were little more than dinghies compared to the giant Chinese ships that measured over 400 feet in length and came with watertight bulkheads, tanks for live fish storage, and more.

Western science has tended to proceed in a reductive mode that seeks to gain understanding by breaking systems down

Think good, do good, speak the truth.
Zarathustra

The truth of religion lies less in what is revealed in its doctrines than in what is concealed in its mysteries.
Roger Scruton

into individual components. In recent years, however, science has come to understand that system must be understood on a holistic level. Why is the world made in such a way that we can make sense of it? Why does it obey natural laws, even though those laws are only imposed by beings who have no control over whether they are obeyed or nor? Religious people still think that these laws carry encoded messages from a divine lawgiver who must be obeyed. Atheists see no evidence of such a lawgiver. Humans tend to see patterns emerging from nature, instead of patterns going into nature. Natural laws operating on their own have many of the properties we once associated with God.

> *In the scientific world of the new atheists, labs would replace cathedrals, brain scans, holy books. It would be different, but would it necessarily be better?*
> John Meacham

We have learned that a universe as smooth as ours requires an amazingly fine level of tuning in order to expand in an orchestrated manner that allows intelligent life to evolve. If the laws of nature are not transcendent, then we must take the universe at face value. If they are transcendent, then we may have the beginnings of an explanation. Most scientists see the initial cosmic conditions as lying beyond science. Religious people bridge the gap by appealing to God, atheists to randomness. Either way, simply saying that the universe began in a certain way doesn't really explain anything. We need laws that are independent of starting conditions.

Why did the initial conditions of the universe exist as they did? Why did our universe start the way it did? Was there something special about those initial conditions? Are we merely the product of one of many universes that each follow different laws? Were the initial conditions of the universe arbitrary, or was there an even deeper principle or law of initial conditions at work? How and why did the universe evolve beings to reflect on its own existence? Either way, we are literally made of the stuff of stars. When we ponder existence, origins, destinies, etc. we are literally the universe looking back at itself in wonder, in the same way one might admire and study one's self in a mirror.

The Mathematical Universe

> *There is no other god beside me.*
> Isaiah 45:21

To mathematicians, math is the language of nature. But why? Why do numbers work so well for counting? How is it that 10+10=20 no matter whether we are talking about 10 people plus 10 people, 10 rocks plus 10 rocks, etc. and that the universe is constructed so as to allow such simple numerical abstractions? Despite this, not even mathematics is absolute; on the contrary, it is a leap of faith in many respects. For example, we learned about Gödel's proof that some true statements can never be proven as true. Researcher Douglas Hofstadter (1945-) concurs, saying that, "Undecidable propositions run through math like gristle so dense in a steak that it can't be cut out without destroying the steak." Cosmologist John Barrow (1952-) also agrees, saying that, "If religion is a thought system that requires belief in unprovable truths, then math is the only religion that can prove it's a religion."

In mathematics, where there's a pattern, there's a reason.
Douglas Hofstadter

What about life itself? In 1970, mathematician John Conway(1937-) created the game of Life with the following four rules:

- Any live cell with fewer than two live neighbors dies.

- Any live cell with more than three live neighbors dies.

- Any live cell with two or three live neighbors lives on to the next generation.

- Any dead cell with exactly three live neighbors comes to life.

These four rules are both simplistic and deterministic. Nevertheless, one cannot know in advance the patterns that may occur during the game, which therefore seems to have randomness or uncertainty built into it, just as in the real universe, because logic can only take us so far. Can every natural system be modeled on a computer? If so, then any system that is complex enough to compute itself can—theoretically at least—simulate the entire universe; however, the Second Law of Thermodynamics effectively prevents that from happening.

Computers require design. Are atomic processes part of some giant computation? If so, then the physics and computations are identical, and the universe is its own simulation. A computer powerful enough to be conscious could possibly create a society of conscious beings who live in their own world, with no idea that they are part of a simulation that could be stopped at any moment just as the cells in the Life game do. As incredible as it may sound, there is reason to think the universe may indeed be a giant hologram (recorded image; see Chapter 27). Were the makers of the movie *The Matrix* on to

The voyage of discovery is not in seeking new landscapes but in having new eyes.
Marcel Proust

something? Of course, the universe may not break down into algorithms, which would destroy any attempt to correlate with any known computer.

> *My theology, briefly, is that the universe was dictated but not signed.*
> CHristopher Morley

It is impossible to prove that a number sequence is random, but one can prove a non-random sequence. This means that almost all number sequences are random, but one can never tell which. Does this mean that nothing is really random? Does the Heisenberg Uncertainty Principle (see Chapter 23) represent reality or simply the limits of our understanding? So-called quantum randomness could therefore be anything but random, because we cannot ever prove that it is truly random. Likewise, π may be non-random, and we may just lack the information to fully compute it. Is there some cosmic code that can predict quantum results and expose indeterminacy as an illusion? Would this code contain the universe's remaining secrets? Would it mandate divine intervention? Many theologians see quantum indeterminacy as the means by which God can act. Would they be proven wrong?

Randomness is rife throughout mathematics. There is no systemic method of finding answers in advance to well-defined mathematical questions. It is therefore wrong to see mathematics as exact facts neatly linked together by solid, well-defined logical pathways. God thus plays dice with both the universe and with numbers. We may do well to begin the study of experimental mathematics. Meanwhile, science is actively seeking algorithmic expressions for observed data, without which the entire exercise is akin to collecting facts in much the same way one might collect bottle caps. The entire scientific endeavor is built on the idea that the universe can be expressed algorithmically. The search for the so-called Theory Of Everything (TOE) or Grand Unified Theory (GUT) is a direct result of this fundamental idea, and is a search that would be pointless otherwise.

> *One has to either appeal to the anthropic principle or find some physical explanation of why the universe is the way it is.*
> Stephen Hawking

As computer scientist Charles Bennett (1943-) said, "The value of a message is the amount of mathematics or other work plausibly done by its creator, which the receiver is saved from having to repeat." Simple patterns are both logically shallow and random, because small programs can generate them quickly. In biological systems, life has great depth, because it can only appear as a result of a long evolution of processes. The universe is full of deep systems, as is Earth. We ought to be preserving these systems, because rebuilding them is difficult at best. It seems that the world is both ordered and algorithmically expressible. This organized complexity is what gives humans our existence and our free will. The laws of physics must provide simple patterns that allow great depth. Not coincidentally, natural laws manage to fulfill both roles, a fact that is quite literally of cosmic significance. Galileo may have said it best when he said that, "The book of nature is written in mathematical language."

Is mathematics an invention or a discovery? Do mathematicians give numbers a false reality, or are they finding preexisting truths? For example, much more comes out of a Mandelbrot set than goes into it. In such cases, mathematics may have found God's handiwork. Either way, the fact that humans can crack nature's code and get at its secrets at all is astonishing. We can imagine both a world where all regularities are apparent at a glance, and a world where they are so hidden as to be undiscoverable. We live in the best of both worlds, where progress requires effort, and where the rewards of progress keep increasing. This is especially impressive in light of the fact that evolution shaped our brains in response to environmental pressures, which at first glance has nothing whatsoever to do with understanding how the universe works. So why should our brains be able to try to understand it? Such an ability seems to confer zero evolutionary benefit.

We believe that mathematics is somehow linked to the secrets of the universe. Our worldview depends in large part on how our brains are structured. Extraterrestrials with differently structured brains may not share our fascination with mathematics, and could see the universe in ways we would find incomprehensible; however, Penrose rejects this cultural argument, saying that good ideas must be more than just survivors. There must be some deep underlying reasons why mathematics and physics fit together so well. Brains that evolve from the physical world do so in accordance with the properties of that physical world, including mathematical content. It is therefore no surprise that we can see math in nature. In fact, math geniuses are born every generation, which means the propensity for mathematics is a stable part of our genes. For this to have evolved by accident would be an amazing coincidence. If mathematical laws do confer some survival value, then we

The universe shows evidence of a designing or controlling power that has something in common with our own minds.
James Jeans

The answer to the most ancient question 'why is there something rather than nothing?' would then be that 'nothing' is unstable.
Frank Wilczek

must still ask why the laws of nature are mathematical. Also, none of this explains the evolutionary benefit of our ability to work with abstract mathematics.

> *We are beginning to see that we are written into the narrative of nature in a fundamental and mysterious way.*
> David Darling

The separations of nature into astronomy, chemistry, biology, etc. are arbitrary. The universe is an interconnected whole. A basketball arcing through the air influences—and is influenced by—every object in the universe, however slightly. Thanks to quantum entanglement, pairs of particles on opposite sides of the universe must be treated as one indivisible entity, as we will learn in Chapter 23. Measuring one particle will partly depend on the state of the other particle. Nonlocality is real.

Physicists James Hartle and Murray Gell-Mann (1929-) point out that the laws of physics are remnants of the Big Bang, which means that the strengths and ranges of all natural forces depend on the universe's initial quantum state, including the mostly linear and local states of most systems. The fact that we can understand nature, that science works, etc. is a direct result of special initial conditions; the extreme effectiveness of mathematics is a result of equally effective initial conditions.

Did God have a choice in how He created the universe? Seemingly independent laws have been found to be interdependent, which means that individual forces are dependent on other forces. Does this make God necessary? Many think so, arguing that the universe is how it is because it must be so. All of science is looking forward to the day when a single mathematical scheme is revealed as the only self-consistent one possible. A TOE may be possible: superstring, brane, and/or LQC theory (see Chapters 25 and 26) may be it.

> *The entire economy of the word incarnate and of scriptural truth will be rendered suspect (if the Earth is found to move).*
> Pierre de Cazre

Physicist Russell Stannard (1931-) explains that a valid TOE must explain how the universe came to be, why it is the only possible type of universe, and why there can only be one set of physical laws. This goal may be unattainable, thanks to Gödel's theorems that state that we can never prove all true statements and that, "For any formal effectively generated theory T including basic arithmetical truths and also certain truths about formal provability, T includes a statement of its own consistency if and only if T is inconsistent." Whether or not this will forever prevent us from finding a final theory, the fact remains that there is no intrinsic reason why the universe must exist or why it must exist as it does.

The Contingent Universe

Are the laws of nature we are familiar with the only ones that can lead to complexity? This universe may be the only one that permits biology and consciousness. If so, it is the only possible knowable universe. Our theories are built both within and as part of the universe since we are immanent within the universe and not transcendent. Thus, any theory of initial conditions must be simple enough for the universe to contain it. In other words, our laws of physics must be extraordinarily simple... and yet a single subatomic event can produce a mutation that alters evolution forever. As I said in *The Enlightened Savage*, a gene that mutated and caused jawbones to weaken may be the single biggest difference between humans and apes, because a smaller jawbone places less pressure on the skull and allows the brain to expand. Every person who has lived, is living today, and who will ever live may owe their humanity to one subatomic particle that either interacted or did not interact with another subatomic particle.

> *You know something's happening here but you don't know what it is.*
> Bob Dylan

Overall, it seems that the universe does not need to be the way it is. This is an example of *contingent* (uncertain and/or dependent) order. The laws of physics seem contingent on initial conditions that could have been otherwise,. We have created many consistent theories, but the universe is not obligated to instantiate or otherwise follow them. Davies believes that higher-level laws associated with the self-organizing properties of complex systems, such as the results of monk Gregor Mendel's (1822-1884) 19th century experiments in plant breeding that are consistent with—but not dependent on—the laws of physics. These higher-level laws have important contingent features above and beyond the laws of physics. The mystery is not that our universe is contingent, but that it is ordered. In particular, the order found in contingent features must have some deeper meaning. The fact that we can understand any of it makes the mystery even deeper. Nothing compels the universe to exist as it does, so how should we explain it?

> *The natural laws are not forces external to things, but represent the harmony of movement between them.*
> I Ching

Liebniz said that it is absurd to say that a collection of books where each has been copied from the one before explains the content of the books. Similarly, we must seek our explanations outside the universe, and these explanations will be metaphysical, because a contingent universe cannot explain itself within itself. A creative power must be responsible for the laws of physics that govern how space and time evolve, among other things. We may be able to decode this creator's message, because that message is encoded in a way that preserves fidelity without degradation.

> *Is it not obvious that science only pretends to explain the cosmos on its fundamental level?*
> Robert Lanza and Bob Berman

Most scientists believe that beauty is an indicator of truth. Mathematical elegance is not easy to explain to non-mathematical people. There is no objective definition or measurement for beauty, and yet we know it when we see it. If a particle just happens to do something, does the law behind it just happen to be true? It is one thing to say that individual events are random, quite another to say that laws are random. For example, it is easier to explain someone who just happens to be driving at 73 miles per hour than it is to say that the speed limit just happens to be 73 miles per hour.

Whether or not God exists, we would not be here to discuss the matter if the laws of nature did not allow us to be here. We are here because of a complex set of marvelously interacting arrangements. Existence alone cannot explain this. We must either accept the facts in the growing library of natural laws as brute and random, or seek a library explanation. The universe appears to be evolving according to a plan.

> *It must now be self-evident that there is a vast gulf between the Jesus of history and the Jesus of faith.*
> Michael Baigent

Why Christianity is Correct

Lewis was raised in a religious family, but left religion as an atheist for a time before returning to religion, specifically Christianity. Why Christianity? What makes this religion—and specifically the Church of England's flavor of Christianity—correct over all religions? Lewis was quite correct when he said that, "Most of us don't approach Christianity to find out about it but to support our own views."

According to Lewis, wonderful things will start to happen when we let ourselves be drawn into the Bible. Anyone who honestly wants to be a Christian will find their intellect sharpened. Christianity does not require a person to be educated, because it provides an education in itself. One can only assume that Lewis never knew or appreciated the extent to which Christianity attempts to stifle reasoning and learning in everything but faith, as evidenced by the writing and sayings of church luminaries throughout Christian history. Christ himself was either the son of God, a lunatic, or worse. It seems obvious to Lewis that Christ was neither crazy nor dishonest, and we must therefore accept that he is indeed the son of God despite the fact that Christ never claims that title.

Many people believe in an impersonal god, but Christians are the only ones who know what such a being would be like. Christianity is the only option for someone seeking a super-personal God. Some think that they will be absorbed into God after many lifetimes and say that experience is like one material thing being absorbed by another, such as a drop of water being absorbed by the ocean. That is the end of the drop, which ceases to exist. Christians alone have an idea about a soul that goes to God while retaining its identity and becoming even more itself than it was before. Personality at the human level is simple, but personalities combine at the divine level in ways that we cannot imagine. We now find three personalities in one like a cube with six sides that is nevertheless one cube. I can only wonder what Lewis would have thought of Islam, which worships the same God by a different name, and which has nearly identical views about the soul. I also wonder how he can say that the personality combinations he describes are unimaginable, while proceeding to try to explain them in the next breath.

God is outside time, which means that every moment is the same to Him; He therefore has forever to listen to every prayer. The idea the God is OOO does cause problems with free will; if God is outside time, then all days are now to Him and He sees us acting now, not tomorrow, because there is no tomorrow. He does not know which action you will take until you take that action, but all actions are in the now for him. The human illusion of separateness is caused by our sense of time. If we could see people smeared out over history, we would see our common origin and even God. More in Chapter 28.

The entire Christian community waits for each new convert to offer her or him one of the best tools for learning about God. (On the other hand, there are also plenty of cargo cults still

Any supposed 'faith in god' that does not include trusting that whatever happens on the other side of death is just fine is really no faith at all. Fear of a hellish after-death scenario or hope of a blissful, heavenly after-death scenario are just that: fear or hope-not faith, not trust.
Michael Dowd

Who but a slave thanks his master for what his master has decided to do without bothering to consult him?
Christopher Hitchens

waiting by their runways, and I doubt any of them could explain how an airplane works!)

Christianity asserts that everyone lives forever, which is either true or false. We have plenty to not worry about if we are only going to live a lifetime, but we must worry about everything if are going to live for an eternity—which is Pascal's wager restated in more modern language. The doctrines are not God, merely a map that is based on the experiences of many people who were truly in touch with God. If you want to get in touch with God, then all you need to do is follow the map.

Against Atheism

The new atheism is too cut off from emotion, from intuition, and a spirit of generosity toward those who see the world differently.
Greg Epstein

Some years ago, I picked up a copy of Becky Garrison's *The New Atheist Crusaders* at the Seattle airport to pass the time between flights. According to Ms. Garrison, atheists are not saying that believing in God is wrong, just that it's dangerous. This intrigued me. Could she refute all of the accusations against religion that we saw in Chapter 10 and explain some of the challenges we saw in Chapters 7 and 10? Ms. Garrison believes that there are there dangers in raising one's voice in Christ's name; many Christians therefore choose silence; atheists should stop whining about religion and get to know some religious people and the good done by religion. This was definitely going to be an interesting read.

Imagine a world in which generations of human beings come to believe that certain films were made by God or that specific software was coded by him.
Sam Harris

Atheists use Greek gods, teapots, and vacuum cleaners to claim that God is imaginary, but recorded history etched into the fabric of the Holy Land clearly proves otherwise. So why can't atheists simply allow people to believe in God as they see fit? If someone thinks there is no God, then why throw a tantrum about it? Why do atheists denigrate clergy who use the Bible to counter science while attacking believers even when science disagrees with them? How come atheists get to make the rules? People like Dawkins make no serious attempts to examine theology, religious attitudes, or the history of interactions between church and science. "Richard Dawkins, you are so wrong," Garrison says. "God is not a delusion. No way. No how. Instead, he's right here in the Holy Land." "I wonder," she says elsewhere in the book, "how Dawkins and Dennett would react if world-class theologians gave similar short shrift to their scientific treatises?" Is she actually agreeing with cardinal Trujillo (Chapter 10)? Besides, the church patronized art. Take out all religiously funded art, and museums would be left empty but for "a few indigenous trinkets."

Reading the Bible reveals that Christ was executed like a common criminal. He made clear that his kingdom was not of this Earth, which is why he was not entitled to any sort of royal burial. Those who read the Gospels know that Mary Magdalene found only an empty tomb with no body inside it. Christ was therefore resurrected, and this resurrection could not have happened without divine intervention. Scientists like Dawkins and Stenger need proof that Christ walked the Earth and want to find his bones. Archaeologist Simcha Jacobovici did so and wrote about it in *The Jesus Family Tomb*. (See Chapter 6.)

Societal health causes widespread atheism, and societal insecurity causes widespread belief in God.
Phil Zuckerman

A priest was asked where he thought God was on September 11th, to which he replied that he would rather think about where God was afterward. Theologian Stanley Hauerwas (1940-) responds to people saying that the world changed on September 11th by saying that it changed in the year 33AD; the question is how to narrate what happened on September 11th in light of what happened in 33AD. A simple reading of the Bible and the violence and prejudice it extols in both the Old and New Testaments answers that question. Society's nearly carte blanche tolerance of religious views and support of religious freedom is what allows events like these to continue. Author Reza Aslan (1972-) said that religion does not make people bigots, but that bigots use religion to justify their prejudices. I have to wonder whether he has actually read either the Bible of the Koran.

At Loggerheads

Shane Claiborne (1975-) is one of the founders of the Potter Street Community, a new monastic Christian community. He believes that, "Our discontent with the church is why we engage instead of pull out. Within the brokenness of the church is our own brokenness." Lewis seems to be of similar mind when he wrote that, "Most of us have got over the pre-war wishful thinking about international politics. It is time we did the same about religion."

Humans have gradually evolved to leave many of us doubting religion.
Nicholas Kristof

This would be the best of all possible worlds, if there were no religion in it.
John Adams

I can only mention a few of the defenses religion makes on its own behalf. So far, religious people seem to be making a decent case for the possibility that God exists; however, they continue to fail miserably at explaining or justifying the long, sordid history of their faiths. All religions have faced a fork in the road between spirituality and dogma, inquiry and blind faith, freedom and hierarchy, openness and orthodoxy. All have chosen the latter path. We have seen some of the ways in which religion has consolidated and extended its power, and how it justifies itself. So what are we to make of it all?

The Divine Savage
Revealing the Miracle of Being

Chapter 12

Appraising Religion

> *Our ancestors worshipped the Sun and they were far from foolish. it makes good sense to revere the Sun and the stars because we are their children.*
> Carl Sagan

We have seen that religion is an innate human capability. Some, like Spinoza, adopted a pantheistic outlook that believes God is the universe and everything in it, but most religions claim to connect people to a personal God. Either way, religious believers can spend a great deal of time indulging their faith. For example, a person who attends 1.5 hours of services a week for 40 years will spend well over 3,000 hours in church.

Worshipping the Sun and Moon and other natural forces goes back to the dawn of modern humanity at least 100,000 years ago. The major religions we know today only appeared during the Axial Age that spanned the 600-year period from 800BC to 200BC. The rise of religion was the spiritual equivalent of the Cambrian explosion of life on Earth some 530 million years ago during which all the major *phyla* (body plans) of life rapidly appeared.

> *Religion is excellent for keeping the common people quiet.*
> Roland Barthes

Religion claims that God is perfect; atheists and agnostics insist that God's designs must be perfect, and use perceived imperfections to refute religious claims about both the existence and nature of God. This makes no sense, because God's perfections need not flow to His designs; besides, no skeptic can hope to design anything nearly as complex and intricate as a single cell, let alone a complete human being. Clinging to old theological concepts in light of scientific discoveries is wrong. A god who is not the same for all people is by definition incapable of telling us what the ultimate meaning of life is, if any.

Science largely rests on the belief that everything in the universe has a rational explanation. It is ill equipped to handle emotional and/or subjective data and tends to dismiss any such reports as delusions, hoaxes, etc. William James said that it is stupid to bar phenomena from notice because we can't partake ourselves. Science must find a way to handle this information reliably if we are to finally crack the mysteries of consciousness and spirit because, while it may be able to explain the biology of an experience, it cannot explain the experience away.

At the outset we are struck by one great partition which divides the religious field. On the one side of it lies institutional, on the other personal religion.
William James

"At the briefest instant following creation, all the matter of the universe was concentrated in a very small place, no larger than a grain of mustard. The matter at this time was so thin, so intangible, that it did not have real substance. It did have, however, a potential to gain substance and form and to become tangible matter. From the initial concentration of this intangible substance in its minute location, the substance expanded, expanding the universe as it did so. As the expansion progressed, a change in the substance occurred. This initially thin noncorporeal substance took on the tangible aspect of nature as we know it. From this initial act of creation, from this ethereally thin pseudosubstance, everything that has existed, or ever will exist, was, is, and will be formed." So said the Jewish scholar Nahmanides in his commentary on the Torah and Genesis 1:1. Nahmanides also postulated the existence of ten dimensions, which correlates very closely with superstring and other modern theories that propose 10 spatial dimensions and one time dimension. There's just one more thing: Nahmanides lived from 1194-1270.

Mystical Wisdom

We saw in Chapter 4 that all religions are ultimately based on mystical experiences. Some of these religions have remained truer to their mystical origins than others, which partly explains the wide diversity of religious beliefs. Meanwhile, the mystics themselves have consistently reported practically identical experiences and findings through history, and in all cor-

In essence, an initiation is an encounter with the sacred.
Mircea Eliade

ners of the globe. Something is going on, but what? True mystics avoid the limitations of ideology and dogma, which hints that whatever they have found has some significant bearing on whatever reality might be.

O fortunate and blessed one, you are a god, no longer mortal.
Thurii

Some say that mystics intuited quantum mechanics and go so far as to say that quantum mechanics proves the veracity of mystical teachings. This is a bit of a stretch. Some might consider the evidence iffy at best, but the fact is that quantum mechanics does indeed demonstrate universal connectedness among all things in the universe. One could therefore say that physics supports the mystics, which is a significant step back from calling it proof.

Our brains are wired to allow us to go beyond ourselves. We remember mystical experiences with the same clarity as so-called "real" experiences, while being unable to recall dreams, drug-induced states, hallucination, delusions, etc. with nearly the same clarity or force of "it really happened" in most cases. This provides a strong hint that what is happening inside a mystic's mind is not the result of any physical or mental pathology, but rather the result of an extremely stable, coherent mind. Does the fact that mystics will themselves to visit higher spiritual planes of existence mean that they are actually doing this? No, of course not. But we cannot dismiss the possibility. The pattern of mystical experiences is consistent across the board. The universe is a complex place, and any encounter with it that bypasses the many filters and barriers we normally have should be a complex experience—and RSMEs are indeed complex experiences. Thus, while mystical experiences by themselves cannot prove or disprove God or tell us what happens when we die, they do provide a key piece of the puzzle that we are trying to assemble in this book.

[Man is] as a cosmic rather than a terrestrial being.
Poimandres

We saw in Chapters 3 through 11 that mystical wisdom and mythological attempts at explanation degenerated into superstitions that degenerated into dogma that degenerated into orthodoxy that degenerated into concepts of heresy and attempts to monopolize thoughts and beliefs, all in the name of that which had been almost irretrievably lost. The evolution of religion with its single-minded focus on the devolution of freedom in favor of blind faith stifled human progress for centuries. Its resurgence in the late 20th century could yet spell the ending of us all—the ultimate insult against that on which the religions themselves were founded. The mystics did not seek to validate religion, but to discover the truth—a very scientific pursuit. Some may have interpreted their experiences in the context of whatever religion. Descriptions of RSMEs reveal something that is both *supranatural* (above normal reality) and rational.

Where materialists see upward causation from "mud to mind," as Mario Beauregard & Denyse O'Leary put it in *The Spiritual Brain*, mystics see downward causation from mind to matter. Materialism is just as dogmatic and faith-based as the religious faith that materialists decry. There is no conclusive evidence to prove that materialism is any more correct than any other worldview. If religion is indeed adaptive, as we saw in Chapter 9, then mysticism probably has just as much to do with it as biology. If materialism is false, then non-material positions must be at least somewhat true. We can guess that some of these positions have more or less truth than others.

It is better to debate a question without settling it than settle a question without debating it.
Joseph Joubert

Anyone can have an RSME. Many of us have glimpses of something beyond the familiar material world, but only a few can venture inward at will. Even these few often require many years of training and continuous practice. Thus, an increase in spirituality (as distinct from religion) entails increasing one's mental ability, not by rote such as by learning to recite passages from a book on demand, but by direct experience and contact with the Godhead itself. These experiences consistently report that we are indeed as small and insignificant as any materialist might expect, but that we are also integral parts of something much, much greater than any individual.

Children seem to understand this much more clearly than adults, because their minds are more open to experiences and realities beyond the mundane. This is why they are so easily led to believe in stories such as Santa Claus, the virgin birth of Jesus Christ, the tooth fairy, and others. Much if not all of this derives from the biological necessity to absorb survival instructions as quickly as possible. (See *The Enlightened Savage*.) That said, just because a child is open to belief does not make the beliefs invalid. For example, can we be absolutely certain that so-called "imaginary friends" are truly imaginary just because the rest of us cannot detect them?

Myth is a past with a future, exercising itself in the present.
Carlos Fuentes

However we choose to interpret their experiences, mystics claim to be able to see the transcendent reality behind the immanent reality we normally perceive. Artist Maurits Cornelius Escher's (1898-1972) famous *Drawing Hands* illustrates the difference: The paper and hands drawing each other represent immanent reality and the illusion of separation. The artist or viewer's perspective that reveals both the interconnectedness and singular wholeness of the work is the transcendent. What it all means has yet to be determined.

RSMEs

The absolute tranquility is the present moment. Though it is at this moment, there is no limit to this moment, and herein is eternal delight.
Hui-Neng

It cannot be stressed enough that mysticism and spirituality are the origins of religion and not the consequences. All claims to the contrary are just plain wrong. If we believe that every evolved trait exists for a beneficial purpose regardless of whether intelligence had a hand in shaping it or not, then RSMEs must have such a purpose. What that purpose might be is wide open to question. Religion can be seen as having some survival benefit, since it helps form bonds between people and also makes them more malleable to authority, both good traits for a social prey animal. Still, it is a stretch to say that RSMEs evolved to give us religion, which evolved to confer survival benefits. Something else is going on here. Perhaps that something is a chink in the armor that lets us peek behind the curtains of mundane reality. We can thus argue that dismissing RSMEs as having no survival value makes the mistake of assuming that terrestrial survival is the only kind there is. Darwinian evolution cannot have future goals in mind, and must focus exclusively on the here and now. If there is something more to life than material existence—and the end of that existence—then it follows that RSMEs are very useful indeed.

Jejune and barren speculations may unfold the plicatures of truth's garment, but they cannot discover her lovely face.
John Smith

In Chapter 16, we will see that mind and consciousness appear to keep functioning when the brain is not functioning and a person is in a state of clinical death. We thus cannot rule out the possibility that mystics are indeed contacting something—or someone—beyond the ordinary. They may literally be reaching God, albeit not the God of any religion. In Chapter 9, we discussed the theory that RSMEs may be caused by epilepsy or other temporal lobe anomalies. Most people who have RSMEs are not epileptics. Most epileptics do not report having RSMEs during seizures; in fact, epileptics are not aware of having seizures, and often tend to remain in a *postictal* (confused) state for some time after the seizure. The correlation between epilepsy and RSMEs is therefore weak at best, far weaker than some materialist scientists might want to believe. The most common triggers of RSMEs are depression or despair, followed by meditation, prayer, and natural beauty. They can happen to anyone of any age or religious affiliation, even atheists. Some interpret them as a personal relationship with God, others not. People who have these experiences tend to be better off psychologically than those who do not.

There is a persistent myth that people who are more prone to self-transcendence tend to be more anxious than people who don't. This myth persists despite repeated tests. People who experience RSMEs tend to be far less afraid of death than most people, and also tend to be among the calmest and most well-adjusted people one could ask for. Genuine RSME seem to transcend individual regions of the brain. With that said, there is a definite correlation between religion and violence, as we saw in Chapters 7, 8, and 10. Here we must make an important distinction: Those who have genuine RSMEs are among the least violent people on Earth, while the masses of people who affiliate themselves with religion condone the blood-soaked history of their beliefs, however passively.

The explanation is obvious: People who experience the real thing do so as a result of a focused search that can require many years of training to achieve, and that can only be achieved by quieting the mind and body and concentrating intently. People who think they are having genuine experiences when they attend religious services—especially those that involve speaking in tongues, laying of hands, and other bizarre behavior—are deluding themselves; their experiences are travesties. Religious ideas about the divine are among the most powerful obstacles there are to true seeking.

In Chapter 9, we learned that neurology supports the idea that something is indeed going on during genuine mystical experiences. None of this proves where consciousness is, but adds weight to the idea that it may not in the brain. It is interesting that looking for ourselves is paradoxical, because we are looking for that which is doing the looking; that self slips out of sight wherever we look, and yet it continues to exist. The emerging science of *neurotheology* is attempting to connect reli-

Myth and religion are not phases in human consciousness, but part of human consciousness itself.
Mircea Eliade

Ceremonies in themselves are not sin; but whoever supposes that he can attain to life either by baptism or by partaking of bread is still in superstition.
Hans Denik

gious states to brain states. The extreme version toes the materialist line by saying it all comes from the brain; milder versions tend to be dualistic. James believed that the idea of religious states originating in the brain was illogical and arbitrary. He had a point: Explaining spiritual insights as biochemistry strips those thoughts and feelings of any value, because they originate with the person having those experiences. If believing in something higher comes from chemistry, then so does disbelief/atheism. (Materialists would agree.)

> *Our perceiving self is nowhere to be found within the world-picture, because it itself is the world-picture.*
> Erwin Shrödinger

One of the biggest obstacles to understanding mystics is that they speak in metaphor while everyone else tries to take them literally—and yet there are no words to adequately describe the experience literally. Meanwhile, millions of religious people experience "pseudo RSMEs" and think they have been "born again," but this is only a lame farcical pantomime of the real thing based on a passage in the New Testament (John 3:3), which says that no one shall see the kingdom of God unless s/he has been born again. This passage became popular in the 1960s and has been used to justify the kinds of bizarrely amusing behavior one can see at many churches (primarily Protestant and evangelical) today. This "born again" experience is false on its face, because the "newborn" is just as old as s/he was before the "miracle" and has undergone no significant transformation in belief or knowledge. Mystics avoid this parody in search of the real thing.

> *When thought is completely gone, the consciousness is still there in the void.*
> Evan Walker Harris

A conclusive materialist explanation of RSMEs would render mysticism irrelevant at a stroke, but no such explanation seems forthcoming. Evolutionary psychology assumes that each structure and function has survival value, that our brains are adapted to a hunter-gatherer lifestyle, and that human nature creates conditions favorable for RSMEs that are either accidents or that help survival and/or reproduction, with no deeper benefit or insight. It is difficult to tell what our earliest ancestors thought and felt, because the remains are so fragmentary and they had not yet invented writing. We do know that they buried their dead in fetal positions with supplies and/or in places with amazing natural beauty, which implies an expectation that the dead would not remain forever so. Cave paintings going back some 25,000 years depict shaman and fertility cults. Much beyond that is anybody's guess. Some genetic theories have been advanced, as we saw in Chapter 9, but nothing conclusive is forthcoming. Squeezing RSMEs under the umbrella of evolutionary psychology requires leaning hard on the "R" part, because religions do have some survival value, but this is not the whole story by any means.

Author Leon Wieseltier (1952-) correctly said that one cannot disprove a belief without disproving the content of that belief. Anyone who thinks they can do this cannot believe in reason. If reason is a product of natural selection, then just how much stock can we put in any reason-based argument for natural selection? Reason is powerful because of its independence, which makes it more a province of mysticism than materialism. Evolutionary psychology thus paints itself into a corner by trying to invoke reason to destroy reason.

Is Faith Valid?

We know that faith is the assent of intellect to truth beyond its comprehension. When a teacher tells a student a fact, that student usually accepts it on faith, because s/he cannot verify that fact without considerable effort, and/or because s/he has already verified other facts received from the same source and deemed it trustworthy. When a clergyman tells a parishioner a theological fact, that parishioner most often has no way to test the claim beyond resorting to the sacred text, which may or may not agree with the clergyman, because most such texts are rife with internal contradictions. Organized religion leaves such uncomfortable questions alone, choosing to accept them either as unsolved mysteries, or as already answered by greater minds. In short, all a parishioner need do is rely on the wisdom of texts and ministers while swallowing all doubts.

When one person suffers from a delusion, it is called insanity. When many people suffer from a delusion, it is called religion.
Robert M. Pirsig

Religion claims that turning to God boosts one's immune system, and that one's ability to fight off diseases becomes five times more effective. There are some intriguing studies along these lines, which we will look at in Chapter 16; however, these studies only say that belief in God is real. They say nothing about either the validity of religion or the reality of God.

Philosopher of science Daniel Dennett (1942-) believes that our brains create multiple competing interpretations of perceptions and events, where the winner becomes consciousness. Faith and reasoned beliefs may originate in different

Faith is not wanting to know what is true.
Friedrich Nietzsche

> *Revelations are the aberration of faith.*
> J. J. Oliver

parts of our brains, but the former need not actually rewire our brains, unlike the latter. The most extreme form of faith is blind faith, which eliminates all doubt. Some modern religious leaders say that doubt is an essential part of believing in God, because it forces us to seek answers that broaden and deepen our faith. Indoctrination is not inspiration; doubt lives in an active mind. All of this may be true, but religion has consistently demanded nothing less than total surrender and obedience, capriciously resorting to horrific means to forcibly squelch any inkling of dissent.

If humanity's genetic and biological predisposition to religion is God's handiwork, then this can only mean that the differences between the religions are more artificial than any of their adherents will ever want to admit. Similarly, we are all in deep trouble if God really is planning to send those of us without a strong faith/belief response to hell, because He did not imbue it within us. Either way, anyone with a temporal lobe has a door to the divine—or at least to divine feelings. Again, young people most easily embrace the idea of God, because their beliefs are very much in flux, and they are also prone to having imaginary friends and other flights of fancy.

Consciousness is a very useful tool; in fact, it is what makes all human experience, communications, and advances possible. We will learn in later chapters that the laws of physics cannot explain consciousness, even though consciousness must by definition follow those laws. There is plenty of reason to doubt—if not outright refute—a materialist interpretation of the universe. Also, there is no such thing as a wasted trait or adaptation in nature; every single characteristic of every single species on Earth exists for the simple reason that it helps individuals of that species survive and reproduce. Does this mean that consciousness and intellect are divine gifts?

> *There is something nearer to us than scriptures, to wit, the word in the heart from which all scriptures come.*
> William Penn

Perception is not merely the passive reception of sensory data. The fact that a radio is malfunctioning does not mean that the signal is gone. Fixing the radio will again allow it to receive. Is the human brain the source of consciousness, or an antenna for consciousness? We will explore this question later, but in the meantime, what does it mean to exist or not exist? Also, can the universe function without consciousness? Physicist Freeman Dyson (1942-) says that, "The architecture of the universe is consistent with the hypothesis that mind plays an essential role in its functioning."

Regardless of religious affiliation, people who had valid RSMEs saw some other level of reality, and emerged shunning the word "God," because of its religious connotations. People in severe medical distress have also reported NDEs, which occur in the brain very similarly to a meditative union with God by lighting up the same brain regions and traveling along the same neural pathways. NDEs are not hallucinations; hallucinating people experience events in the current stream of reality without leaving their bodies or the Earth. People experiencing NDEs leave both body and Earth behind and go elsewhere.

Are NDEs simply the last gasps of a dying brain? A person will be unconscious 20 seconds after her or his heart stops beating. All of the sensory and intellectual structures that create our worlds shut down, and nothing is functioning. An oxygen-deprived brain hallucinates chaotically, but NDEs are coherent. Whether or not they are nothing more than memories or dreams, NDEs can only form when the brain is dead for all intents and purposes. This may indicate that consciousness has both local and infinite components. We may see only the local side in normal life, but that does not mean the other side is not there.

All of life involves acts of faith. The next time you cross a street, consider that there is nothing but a few stripes of paint and perhaps a handful of colored lights telling drivers what they must do. None of those things by themselves can stop or even slow an oncoming car. Whenever you are crossing a street in front of an oncoming car, you are literally placing your life in the driver's hands and trusting that they will see both you and the paint and/or lights and stop to let you pass. Even if the car is stopped, you are trusting that the driver will not accidentally or deliberately run you down. Crossing the street is an act of faith; however, this faith is reasoned. Under normal circumstances, you will only cross if you have every reason to believe that the driver has or will see you in time to respond and/or that her or his foot will remain firmly on the brake while you are inches from their front bumper, and that s/he has no intention of harming you. No attentive pedestrian will cross unless s/he has the firm belief that s/he has all the

---

No man ever believes that the Bible means what it says. He is always convinced that it says what he means.
George Bernard Shaw

If in effect the world be not a serious thing, it is the dogmatic people who will be the shallow ones, and the worldly minded whom the theologians call frivolous will be those who are really wise.
Ernest Renan

---

information she needs to make a sound decision and that s/he will reach the other side safely.

This kind of faith can only come from experience and education. It is never absolute, because no amount of reading books about traffic can ever substitute for the real thing. The kind of blind faith that Tertullian and many within religion demand of followers would quickly get you killed if you applied it to the simple act of crossing a street. Believing something precisely because it is absurd, and having faith that it is nevertheless true with no way to verify it for yourself, is a great way to get yourself killed in the normal world. Why, then, should blind faith be a valid approach to the divine?

Separate Magisteria?

One may imagine possible worlds without sin and without unhappiness, but these same worlds again would be very inferior to ours in goodness.
Gottfried Wilhelm Leibniz

In 1997, Gould proposed that science and religion are "separate and non-overlapping magisteria" that had nothing to contribute to each other. Is this true? Can religion and science simply live side by side?

On August 11th, 1999, the Kansas Board of Education severely curtailed the teaching of evolution by proclaiming that, "No evidence contradicting a scientific theory shall be censored." The religious discomfort with—and outright opposition to—science remains alive and well. Religion may no longer torture and execute scientists, but that does not stop it from mounting determined rhetorical attacks against science—often while benefitting from the fruits of science to make their opposition known. As we saw in Chapter 3, some religious people even believe that God rigged the universe to look old, thereby forcing us to embrace divine deception while rising above it ourselves.

A professor of theology should have no place in our institution.
Thomas Jefferson

The JCI religions have all conducted crusades against people who did not share their exact beliefs, even when these people technically belonged to the same religion. The 30 Year War was fought from 1618 to 1648 between Catholics and Protestants. The Christian Cathars were obliterated by Christian Catholics. England fought over whether to remain Catholic or form a separate church. The Crusades slaughtered tens of thousands of people who believed that Christ was a prophet but not God's son. The Taliban, Al Qaeda, and others are currently promoting an extreme Islamic agenda that features Muslims torturing and killing other Muslims over quibbling matters of belief. The Inquisition both justified and provided the template for systematic oppression. Modern religious leaders are denying people the simplest and most effective methods of preventing the spread of AIDS and other diseases. The list of horrors committed in God's name could fill many pages. Atheists and secular regimes also have plenty of blood on their hands; however, nobody has ever killed in the name of science. For example, the idea of physicist Kip Thorne (1940-), Hawking, and their respective student bodies going to war over the properties of black holes is almost unimaginable. (The actual stakes were a subscription to *Penthouse* magazine.)

The real war is between rationalism and superstition.
Jerry Coyne

Religion has made forays into science. So-called intelligent design is a softened version of outright creationism that wraps religious ideas around scientific terms and seemingly scientific theories. We will examine intelligent design in detail in Chapter 20; meanwhile, most arguments put forth by believers in intelligent design have been thoroughly debunked. Shoving God into the gaps in our scientific knowledge is not a good role for a supposedly OOO being. As Dawkins says, we must "either admit God as scientifically valid or leave Him with the sprites and fairies."

Scientists are hard at work searching for a TOE that explains why our universe is the way it is without God. To some religious people, looking for God in science is like breaking open a television to search for the actors inside. If God created the natural laws, then they are perfect enough to suit His needs. Created or not, our natural laws are doing a very good job of keeping us alive. Some people on both sides of the debate believe that our physical and biological laws cannot provide evidence for or against God—a view that leaves the existence of God as a matter of pure faith. No matter their religious preferences, most scientists are convinced that any viable TOE will be conceptually simple and beautiful. Some even see signs of design in the universe, primarily because of the anthropic principle and its ramifications.

The world is far more interesting to me than the Bible or the Koran.
Andre Compte-Sponville

The idea that the universe is complex—and thus needs an even more complex creator may be flawed—because there is a direct correlation between size and complexity. The smallest subatomic particles are simple. Atoms begin very simply and become more complex as one proceeds along the table of the elements. For example, a single hydrogen atom has no distinguishable parts, while a single uranium atom has plenty of parts. Molecules are more complex still. Cells are still more complex, as are multicellular organisms right up to humans and other current species. From there, the level of complexity begins to decline. Most celestial bodies are very simple, and we can describe the gravitational relationships between them using fairly simple equations. Even superstring theory that postulates 11 dimensions is in many ways simpler than the standard 4-dimensional model of physics.

People who believe they have the truth should know they believe it, rather than believe they know it.
Jules Lequier

Systems have laws that cannot always be derived from the laws of their constituent components. For example, looking at nature on a macro scale gives no indication that quantum nonlocality exists, and that nonlocal quantum events can influence macro-level systems. Author Jean Staune says that, "There seems to be a reality that transcends time, space, energy, and matter, yet still has a causal effect on the material world."

As we saw in Chapter 3 (and will explore further in Chapter 26), it is meaningless to ask what God did before He created the universe, because there was no before. Funny how science can accept the idea that time did not exist before the Big Bang, but can still ask where and when God came from. The idea that quantum fluctuations created the universe is a secular creation *ex nihilo*. The vacuum of space is anything but nothingness, but we cannot account for the fluctuations and must simply accept them as given—which is just as big a leap of faith as any religious proposition, if not bigger.

There are four general ways in which science and religion can interact:

- **Conflict:** Each opposes the other.
- **Independence:** Each ignores the other.
- **Dialogue:** Each accepts and learns from the other.
- **Integration:** Blending science and religion into "natural" theology.

It would be hubris to think that the finite can comprehend the infinite.
Gerald L. Schroeder

The conflict between science and religion has been raging for over 2,000 years. Perhaps it would be best for everyone to retreat to their respective corners and leave each other alone; however, this solution cannot work as long as religion insists on participating in making public policy decisions that affect everyone regardless of belief, or as long as religion insists on proselytizing to nonbelievers. Dialogue between science and religion is possible to the extent that religion is willing to see its sacred texts as metaphorical, which Augustine encouraged the Catholics to do. Integration—true integration where science and religion seek to discover the ultimate truths without regard to what any sacred text might say—is the ultimate ideal. Sadly, I don't think we will ever get there. I blame religion for this, because science will continue along mostly unchanged, whether or not religious people retreat into their churches, temples, mosques, etc. and leave the world to its own devices. Religion does not extend that same courtesy to science or to anyone it suspects of not believing what it believes in the way it wants that belief expressed.

Probability

The so-called Golden or *phi* (φ) ratio is found in art, architecture, botany, biology, physics, mathematics, stock markets, the arrangement of stars in the sky, and more. The golden ratio is defined as existing when the ratio of the sum of the quantities to the larger quantity equals the ratio of the larger quantity to the smaller one. The mathematical expression of the Golden ratio is *(a+b)/a = a/b = φ*. The only positive solution of this ratio is 1.6180339887. The heavy prevalence of this ratio has prompted many to wonder whether the universe is mathematical by nature, and whether God is a mathematician.

The scientist does not study nature because it is useful; he studies it because he delights in it, and he delights in it because it is beautiful.
Henri Poincaré

Imagine a prisoner set to die by firing squad. The soldiers take aim and fire, but the prisoner escapes unharmed. It stands to reason that the prisoner will surmise that something deliberate must have happened for all of the bullets to miss. Either someone tampered with the guns, the soldiers deliberately missed, or some divine providence altered the bullets' paths in mid air. There is certainly a non-zero chance that all of them will miss, even if the guns, bullets, and aim are working perfectly, thanks to quantum mechanics, but these odds are very small. The odds of our being here to contemplate these questions are smaller still, which leads to the argument from improbability: The odds of our evolving are so small that God

You can recognize truth by its beauty and simplicity.
Richard Feynman

> *The remarkable fact is that the values of these numbers seem to have been very finely adjusted to make possible the development of life.*
> Stephen Hawking

must have intervened to make our presence a certainty. But who or what made God?

Dawkins and like-minded scientists refute a designer god as an explanation for organized complexity, because this god would have to be complex enough to demand explanation in its own right. The solution to the problem of improbability is neither design nor chance. In other words, neither chance nor design can explain our presence, because God would have to be even more complex than the universe—at least so the argument goes. I find it interesting that science can speculate about organized complexity arising from simplicity and the bell curve of randomness that we mentioned in Chapter 9, while still saying that God must be complex; it seems as if materialist science is trying to have it both ways.

The computer programs that Dawkins and others use to demonstrate chance must be carefully designed. It is possible to decide that a pattern is random whether or not the pattern is actually random. We can also conclude that what we call "random" merely reflects a level of organized complexity that is beyond our ability to describe in mathematical terms. The Heisenberg Uncertainty Principle says that it is impossible to know a particle's exact position and momentum; the more precisely we measure one, the less precisely we can know the other. This does not mean of itself that particles lack definite positions and momentums. (They do in fact lack both, for different reasons that we will look at in Chapter 23.) A standard home computer can be used to show that chance is not a plausible explanation for the universe. Absent design, the material world creates a bell curve, a model of perfect distribution of values. Randomness is amazingly orderly.

> *Conceivably, it might have been blind luck if only a few constants of nature were required to assume certain values to make life possible.*
> Michio Kaku

Neither order nor randomness occur by chance. Chance cannot explain the existence of order. This implies the existence of a universal field that provides guidance and design to the universe. Chance is not an efficient way to create something as simple as a computer program designed to demonstrate chance, so it cannot be an efficient way to create life. In fact, it is all but impossible. If chance cannot even write the word "chance," then how can materialists claim that it is responsible for all of life? And yet atheists dismiss everything as random chance. More on this in Chapter 14.

Some creationist scientists point to the evolutionary trend toward greater complexity as violating the Second Law of Thermodynamics, which says that entropy must always increase. The total amount of entropy in the universe must always increase, but this does not mean that localized areas cannot experience lower entropy. All life on Earth is powered by the Sun, which is losing some four million metric tons of mass each second; the Sun's total entropy is increasing by far more than any localized reduction of entropy caused by evolution, and our solar system as a whole is gaining more entropy than order overall, in perfect keeping with the second law.

Religion makes the valid point that something beyond what meets the eye is going on in the universe. It funds scientists whose views clash with materialism. The results of these endeavors are not always successful—witness the debunked intelligent design theory—but the mere fact that dissent exists can serve as a valuable check on scientific endeavors.

Mathematics is the handwriting of the human consciousness of the very spirit of life itself.
Claude Bragdon

The Probability of God

In general, probability theory represents chance and uncertainties in mathematical form. *The Probability of God* by physicist and risk analyst Stephen Unwin chronicles his attempt to quite literally calculate the chance of God being real, using the Bayesian scale and probability as an expression of disbelief. The Bayesian scale provides a systematic method of modifying probabilities based on evidence, and is therefore both a powerful analytic tool and way to organize complex thoughts and problems. Some of the factors Unwin considers are:

- To theists, goodness is evidence of God. There is both absolute and relative goodness, which must not be defined as comfort, pleasure, etc.
- There are two kinds of miracles: natural phenomena triggered by God and unnatural phenomena (events that do not obey the laws of nature as we understand them.)
- A personal God must exert influence on daily life.
- Some theologians think that the spiritual response to natural events is the miracle. Something as simple as a

Why would a perfect god create a universe in which such huge amounts of suffering occur, when such suffering does not bring into existence any of the goods required to absorb the suffering and make the situation on balance a good one?
Nicholas Everett

particularly gorgeous sunset can trigger a religious experience.

- Evidence must be assessed to see if it is much more likely to be true if God is true or not true, moderately more likely if God is true or not true, or if it does not matter either way.

- Would goodness have any meaning in a Godless universe? Do we learn to recognize goodness? There is no acceptable atheistic definition of goodness, so the presence of goodness is strong evidence for God. In fact, this alone raises the probability of God from 50% to 90% for Unwin.

- Moral evil is inevitable in a Godless universe, and evil itself must be reconciled with the idea of an OOO God. This problem has proven to be nearly intractable for the JCI religions. The presence of evil thus returns the probability of God to 50% from 90%.

- Religious experiences are inevitable in a universe where God exists. Could someone have the same experiences in a Godless universe? Such experiences would be intense, but not connected to the divine. Could our sense that "something is out there" be a delusion born of desire? Could it be that we have found Spinoza's cold, impersonal God? No... because the experiences just don't feel that way!

In the end, Unwin decides that the probability of God is 67%. To me, this and similar arguments speak more about the power of religious indoctrination and belief than the actual probability that God (or a god) exists. Saying that the presence of good and evil affects the likelihood that God exists means that one is using a very narrow definition of God. Also, the numbers themselves, however well reasoned and however good they may feel, are subjective. Unwin is applying his subjective beliefs and arbitrary numbers to try to determine the probability that the JCI God is real, an exercise that provides much more insight into the nature of the religious mind than it does into the nature of the universe.

Scientists are beginning to find the fingers of God in the creation of the universe.
Eugenio Pacelli
(Pope Pius XII)

Unfortunately, the wisdom traditions designed to help us deepen our questions—from religion to science, philosophy to psychology—have become disciplines for knowing, for defending absolutes.
Irwin Kula

Religious Behavior

All of the charges I made against religion in Chapters 7 through 10 are matters of historical fact. Religion makes no attempt to apologize for itself, change its behavior, or make good on the irreparable harm it continues to inflict on humanity every day. The religious leaders and their followers who perpetrate crimes from mass murder to child molestation are guilty of the most heinous crimes imaginable. In Chapter 10, I included religious moderates by accusing them of being just as guilty as their more extreme brethren. How can I do this? How can I possibly equate the darling little old lady baking cookies for the fund-raiser, the day-care worker who would never so much as think of molesting one of the children in her or his care, or the nice family that goes to church on Sunday with an inquisitor? It's actually very easy.

Many years ago, a bicyclist in San Francisco was assaulted by two men. The bicyclist drew a gun from a bag and shot one of the assailants dead. The police charged the surviving accomplice with the murder and sought the bicyclist (who left the scene) merely as a witness. The bicyclist was probably carrying the concealed firearm illegally, since it is notoriously difficult to obtain a concealed firearm permit in San Francisco. Even if the gun was visible and thus being legally carried, state and local laws do not permit one to carry a loaded firearm. Thus, the only legal way the bicyclist could have had the gun would have been to have it plainly visible and unloaded. When attacked, he would have had to have loaded the gun before being able to pull the trigger. In short, the gun was most probably being carried illegally. The bicyclist pulled the trigger, not the assailant charged with the crime. So why charge the surviving assailant and not the bicyclist?

Guilt by association is a murky area of the law. One technically has the right to associate with whoever one likes, whether or not that person is a criminal; however, the nature and type of association can make it difficult to prove that one does not know about criminal activity taking place, and can even make it difficult to prove that one is not somehow involved in that activity. The mere fact that the surviving assailant was with the man who was killed is not by itself reason enough to charge him with anything. For example, I have no reason to suspect

> *To assert that the Earth revolves around the Sun is as erroneous as to claim that Jesus was not born of a virgin.*
> Cardinal Bellarmine

> *I leave it to the faithful to burn each other's churches and mosques, which they can always be relied upon to do.*
> Christopher Hitchens

> *You don't need a "supernatural" deity at all if you belong to a physically affectionate, caring culture.*
> James Prescott

my partner Jennifer of any criminal activity. If she suddenly decided to rob someone while we were walking down the street, I should not be liable for anything that might happen.

It would be a far different situation if Jennifer and I set out to commit a crime. The men who attacked the bicyclist were looking for a victim. They knew robbery is illegal. They knew their actions could cause violence. Thus, they acted with full knowledge of what they were doing, and were therefore accomplices. In that case, the surviving assailant was indeed fully responsible for the crime and its aftermath. His complicity directly caused the death. One presumes that the bicyclist would not have harmed anyone but for that event.

I have already said that people who belong to a given religion are like rooms in a building with a common foundation. Some rooms may be more gaudily decorated than others, and some inhabitants may be more tolerant than others. But everyone is sitting on top of the same foundation. A flaw in this foundation exposes every room in the building to the danger of collapse. The fact that one room is painted red and other green with purple polka dots is meaningless. Similarly, the foundations of the JCI religions are built on violence and ignorance. Libraries abound with incontrovertible evidence of the long history of religious crime that spans the entire gamut from telling children to behave or burn, to starting full-fledged wars of aggression against people whose only crime is disbelief. Ignorance of the law is no excuse; you will receive a speeding ticket whether or not you saw the speed limit sign. Likewise, ignorance of religious history is no excuse, nor is any explanation that seeks to diminish or deflect the blame for the crimes committed by religion from antiquity to this very moment.

> *Religion, which as originally designed to help deepen and ignite our experience, simply doesn't fulfill its promise as it is most often practiced.*
> Irwin Kula

Even an abject apology and utter reformation of how religion comports itself will not work so long as any religion follows a book or other instruction that extols, praises, or condones any kind of violence. Ignoring the bad parts of the Bible and Koran does not make them go away any more than ignoring a crumbling foundation saves a building. Choosing which parts of the sacred text to believe and which parts to ignore is like picking a room in a building with a flawed foundation. In some ways, religious fundamentalists and extremists are more respectable than their moderate counterparts, for they accept the bad along with the good. No moderate has any ground to stand on to refute the acts of extremists beyond muttering that the religion in question is a "religion of peace" when something happens in God's name.

All religions have blood on their hands; however, the JCI religions institutionalized violence and perpetrated it on a scale that has never been equaled. Turn the Bible or the Koran to any random page, and you will probably find some reference to some violent or bad act. Turn the *Bhagavad Gita* to a random page, and you will probably find something beautiful and uplifting. This is the largest reason why I use the JCI religions as my examples. Another big reason is that both my readers and I are probably far more familiar with the JCI religions than they are with Buddhism, Hinduism, Daoism, etc. Finally, there is some valid debate about whether the Eastern religions really are religions in the sense that we have seen over the last several chapters.

I shall be to this generation a new Mohammed.
Joseph Smith

Mythology represents the seed of human science. Mysticism may offer powerful clues to a wider reality. Shamanistic and tribal traditions try to connect people with the divine. Spiritual traditions and mythologies do the same. Polytheistic religions allow an incredible diversity of belief and tolerance. By contrast, monotheistic religions engender power plays and fanaticism that have no place in a world brimming with nuclear weapons.

Which Religion is Correct?

We have seen that some religions try to stifle science or bend it to their will. We have also seen that science has all but disproved some religions, while others are far more compatible with science and even welcoming. For example, the Dalai Lama has said that he will update his beliefs the moment science proves his assertion that humans cannot be reduced to biological machines wrong.

No apostle has been sent before to whom we did not reveal that 'verily there is no God beside me.'
Koran 21:25

In Chapter 5, we learned that Hinduism believes that God sacrificed Him/Her/Itself to create the world that then becomes God again. Our inner self is the atman, which is independent of the body. Brahman is the Ground/Godhead. The atman survives death and can move into other physical bodies based on karma—a neat trick that certainly helped keep the peace and helped people accept their lots in life in a rigid caste system, where the lowest caste can make an American homeless person seem decidedly well off by comparison. Leading a truly exemplary life frees one from the cycle, and returns one to Brahman. Buddhism is derived from Hinduism and went one step further by implicating desire as the root cause of suffering, the abolition of which leads to enlightenment and freedom from the cycle of karma. Is Hinduism correct? Buddhism? If either or both of these religions is correct, does that make other religions wrong?

> *Heretics boast that they possess more gospels than there really are.*
> Irenaeus

Catholics rely on the "infallible" pope over the Bible as their ultimate authority. This is very convenient, because the pontiff can reconcile all conflicts between science and the Bible by saying that science is correct and that it was God's will the whole time. This does not happen terribly often—it did take 400 years to "forgive" Galileo, after all—but this glaring loophole does cast serious doubt on the book whose contents are supposed to be the source of the pope's power. One pope even went so far as to approve of scientific study of time after the Big Bang but not of the Big Bang itself, because that was the work of God. That is a far cry from the Dalai Lama's position, and implies that the pope has something to hide. Does this make Catholicism wrong?

The Israeli monotheism that eventually became Judaism, Christianity, and Islam borrowed many important ideas and rituals from other religions, including unleavened bread (matzo), sacrificing spring lambs, burnt offerings, and fertility rituals. We saw in Chapter 8 that the so-called "New" Testament is nothing of the sort, since it derives from the Torah, reveres a messiah of the line of David, and has plenty of ideas taken from plenty of other religions. For example, Zoroastrian beliefs include all of the major features of Christianity, such as virgin birth, revelations, judgment, and apocalypse. Other borrowed concepts include Christmas, Easter, the idea of God as man, the Greek and Roman Logos as the Word of God, God's word represented by a son, resurrection, and many more. Does this blatant plagiarism make the world's three leading religions wrong?

> *There are more variations among our manuscripts than there are words in the New Testament.*
> Bart D. Ehrman

Does the litany of contradictions in the Bible render it useless as a guide for living, forgetting the many other problems we discussed in Chapters 7 and 8? For example:

- 2 Timothy 3:16 says that God inspires all scripture, but 2 Corinthians 11:17 has Paul saying that he is not divinely inspired.

- Galatians 2:16 says that man can only be reconciled with God by believing in Christ, while Matthew 16:27 says that all men are judged by their deeds. Christ would seem to agree with Matthew, since he never said that faith is a prerequisite for getting into God's kingdom; this bit was inserted much later.

- John 3:17 says that God did not come to judge, while John 5:22 says that He did.

- 1 Corinthians 15:52 says that the dead are immortal (death does not happen), while Luke 20:37 says that the dead will be resurrected (death does occur but is later reversed), Ecclesiastes 9:5 says that the dead are gone, and Isaiah 26:14 says that the dead are dead and will not live again.

The baying of a dog to the Moon is as much an act of worship as some ceremonies which have been so described by travelers.
John Lubbock

These few examples should be more than enough to convince anyone that the Judeo-Christian god is a less-than-stellar editor. The entire Bible is rife with contradictions, to where the real miracle is that people actually believe it. Furthermore, the Bible and other "holy" books are anything but self-explanatory, and offer no means of verifying their many assertions. They also fail to completely answer the big questions, because we are still asking them. This glaring incompleteness leaves them open to interpretation, which opens the door to a growing laundry list of denominations claiming to be the correct one. Dogma plugs the gaps between scripture and that denomination's vision of truth. The dogma of one church is opposed by all other "true" churches.

Only one denomination of one religion can possibly be called correct, and all others must therefore be false. This singular number is a maximum, because it is entirely possible that nobody has it quite right. No religion has been scientifically validated, although some have come closer than others. Despite this, adherents of every flavor of religion proclaim the veracity of their beliefs above all others—and every single proclamation collapses under the weight of probing questions.

It is imperative that we begin speaking plainly about the absurdity of most of our religious beliefs.
Sam Harris

Talk to a religious person, and you will quickly realize that s/he probably fails to grasp the difference between her or his holy book and the world that book claims to represent and explain; that person thinks that they know God because s/he read the book (or more likely carefully selected parts of a book) and had an emotional response. You would not claim to know me through reading this book or even my entire body of work, so how can anyone claim to know an invisible superman from a single book?

It stands to reason that no religion can possibly be called true unless every one of its beliefs is true. Religions make plenty of statements. Some of them may in fact be true, but that does not make them true, just as a broken watch being right twice a day does not make it a reliable timepiece. By this standard, the best any religion can do is get things somewhat right. But what parts of reality do which religions get right? Which revelation—or set of revelations—is correct, and why? Does this come from the Bible? Which Testament? The Koran? The *Book of Mormon*? *Dianetics*? The *Vedas*? People seek solace in religion, but how can anyone possibly choose? The ongoing violence committed in the name of religion seems to be proof positive that not one of them has it right, and that people are fighting and dying over statements that can be neither proved nor backed up by anything other than circular logic.

The only valid religion is one whose tenets can be validated by external sources using the same standards of evidence and proof demanded by all people in all non-religious matters. The fact that we are talking about matters of life, death, and the meaning of existence does not lessen the need for rigorous standards. On the other hand, the standards must be even higher, precisely because of the magnitude of the claims being tested. The only valid religion is one that can be backed by rigorous scientific testing. If God exists in some form or another then S/he gave us brains, which is all the proof we need that S/he intends for us to use them. Demanding blind faith makes just as little sense as demanding that birds not use their wings to fly.

The premise of the book was this: Life was an experiment by the Creator of the Universe, who wanted to test a new sort of creature He was thinking of introducing into the Universe. It was a creature with the ability to make up its own mind. All the other creatures were full-programmed robots.
Kurt Vonnegut

The Legacy of Religion

Philosopher Robert Pennock observed that, "To say nothing of God is not to say that God is nothing." The reverse must

Can we ever save Jesus from the dogma within which he has long been mired?
Michael Baigent

also be true: Saying something about God does not say that God is something. In other words, just because religion is based on the idea that God exists does not make it so. Religion claims to be all about praising and worshipping God and following God's will. Legions of theologians and religious scientists have pondered the divine for thousands of years, often in the context of sacred books that have been edited so heavily as to be almost unrecognizable next to the (long lost) originals.

Religion owes its existence to mystical experiences and spirituality bolstered by the power of myth and tradition. Each has grown in different directions from these roots, and each has faced a choice point between spirituality and dogma, freedom and hierarchy, exploration or orthodoxy, tolerance or persecution, openness or control, expression or repression, etc. With very few exceptions, all religions have chosen the latter path to become powers in their own right. The JCI religions continue their violent assaults against internal and external enemies. In fact, a member of a given religion stands a much greater chance of dying at the hands of a member of his own faith than he does at the hands of a different religion—a situation very similar to the Crips street gang where a Crip is far more likely to be murdered by a fellow Crip than by a Blood. (The Bloods and Crips are sworn enemies.) In fairness, the Buddhists are a notable exception. In Chapter 5, I mentioned that they had split into "upper" and "lower" schools. That much is unremarkable. What is remarkable is that this split was peaceful, unlike their Western counterparts.

Radical secularism is just as bad as radical religion. The worst wars of the 20th century were ostensibly fought for secular reasons, despite the insistence of all combatants that God was on their side. The advent of the Industrial Revolution may have contributed to a "God-shaped hole in human consciousness" as Sartre put it, but I can only speculate that religion has only made this hole bigger, for the reasons I have listed in Chapters 7 through 10.

Each religion thinks it's unique and special, and most condone violence to harm others who think differently. I will say it again: One would think that people who are utterly certain that an OOO God is real—and that they are destined for eternal happiness—would be the happiest, most carefree, generous, well-adjusted, secure people on the planet. One would

A religious life, exclusively pursued, does tend to make the person exceptional and eccentric.
William James

As the government of the United States of America is not, in any sense, founded on the Christian religion; as it has itself no character of enmity against the laws, tranquility, or religion against the Mussulmen; and as said states never have entered into any war or act of hostility against any Mehomitan nation, it is declared by the parties that no pretext arising from religious opinions shall ever produce an interruption of the harmony existing between the two countries.
Treaty of Tripoli

certainly expect them to be the calmest, most stress-free people imaginable. But that's not the case. Religious people tend to be insecure, often violently so. They are insecure, petty, insular people who foist their beliefs on society whether or not we want to hear it, and who think it's perfectly OK to legislate morality. Meanwhile, religious people tend to be some of the most immoral people around. The level of hypocrisy caused and sustained by religion is absolutely breathtaking in its scope and breadth. All of this for a set of books and ideas that were (badly) cobbled together from far more ancient texts and myths. There is nothing new or special about either the Bible or the Koran. The parts of Christianity that tended to agree with the Eastern religions (the Gnostic traditions) were annihilated, and almost all traces were lost until very recently. Christ's wife Mary Magdalene is called a prostitute by the religions that claim to follow in her husband's footsteps.

The religion of one age is the literary entertainment of the next.
Ralph Waldo Emerson

People have always faced some struggle or other adversity. From the time of Josiah onward, monotheism has helped unify them around a cause on a level that is hard to achieve with polytheism. From a search for the truth, religion has become a search for power. Dawkins has called religion a very successful chain letter—and with good reason.

NDE reports bear some resemblance to religious descriptions of a lush afterlife, which begs the question of which came first and why the two seem to match so closely. The short answer is that NDEs came first. They match religious descriptions so well, because the latter are built on the former. The very term "NDE" means that they occur under life-threatening conditions, and people have been dying for many years. Most religions can trace themselves to experiences such as NDEs, OBEs, and RSMEs with the obvious exceptions, such as Mormonism and Scientology that are nothing more than blatant and obvious frauds. (Joseph Smith had been previously convicted of fraud, and L. Ron Hubbard's fibs are quite well documented.)

The religious factions that are growing throughout our land are not using their religious clout with wisdom. They are trying to force government officials into following their position 100 percent.
Barry Goldwater

All religions are spiritual—or at least purport to be spiritual—but not all spirituality is religious. Sadly, many religions start with the idea that something is wrong with humans the way we are and sells us a way out, provided we follow their tenets wholeheartedly—a "right and wrong" mentality that is fundamentally dualist in nature. Believe, and you will receive sanctuary in this life and the next with your fellow followers. Disbelieve, and you will be cast out to suffer for all time by the very same people that claim to love you above all others. Mystics have none of this, because they know that good and evil are both subjective, and that both have their place.

Religion may have inspired the worst violence ever, but that does not mean that violence would vanish if religion ended tomorrow. Nations create arbitrary boundaries and fight over them. People treat each other abominably far too often, and religion is not always a factor. One of my favorite *South Park* episodes depicts bloody battles over different interpretations of science. It seems ludicrous to think of physicists carrying out acts of violence against those who disagree with their theories, or of wars starting to defend the many worlds interpretation of quantum mechanics against the evil Copenhagen interpretation; however, this does not mean that such things cannot happen. Still, we must ask ourselves whether the violence would be as bad without religion acting as both an excuse and an example.

Weinberg hopes that discovery of a final Theory of Everything will spell the end of religion, but that seems like wishful thinking. Still, a TOE could burst many bubbles, like children discovering that there is no Santa Claus. Would the world be better off knowing the truth, even if that truth is that this life is all we will ever get? In some ways, yes. The universe we gaze into would be just as cold and callous. We would still have morality and ethics. We would still have philosophy, art, and science. What we would not necessarily have is various sets of unfounded beliefs running roughshod over the world.

Science today is having a devastating effect on theology, although few theologians will ever admit that. Nevertheless, theology finds itself trying to play catch-up, and falling further and further behind. This seems like a losing battle, but one that religion must fight lest it be exposed as irrelevant.

We have seen that religion makes claims about ultimate reality that are not nearly as amenable to testing as scientific theories. Greek religion did, however, manage to provide a framework within which mathematics and philosophy could flourish. There have been several periods during which various religions have safeguarded science. For example, much of what

---

The most detestable wickedness, the most horrid cruelties, and the greatest miseries that have afflicted the human race have had their origin in this thing called revelation, or revealed religion.
Thomas Paine

Shakespeare has much more moral salience than the Talmud or the Koran or any account of the fearful squabbles of Iron Age tribes.
Christopher Hitchens

we know of ancient Greek literature is thanks to Muslim preservation during the Dark Ages in Europe, when knowledge was actively sought out and destroyed. Religion has also influenced science, such as theories of the solar system that relied on circles instead of ellipses, as we will see in Chapter 21.

At its heart, religion is an acknowledgement that metaphysics is a real phenomenon, even though no religion can validate this by having its dogma proven correct. Logically, only one religion—or one denomination thereof—can make a credible claim to being the 'true" religion; however, there are literally thousands of religions making claims without any shred of proof. Meanwhile, science is universal. Two plus two equals four no matter where you are, and every single one of its claims is wide open to testing and falsification.

Many people today are seeking spiritual growth, just like people have for our entire evolutionary history. Spiritual growth is just like any other kind of personal development, in that it requires dedication and training. The only way to truly know something is to experience it for one's self, and that requires asking many questions and assessing all of the answers. Those who say that belief and faith are all that's needed do a terrible disservice to those who sincerely want to know. Relinquishing the intellect in favor of an invisible super nanny does nobody any favors. Accepting a conversion experience that requires one to accept and obey an authority figure that spouts fantastic stories without proof leads to fear and control, which drives more blind acceptance in a pattern of growing addiction and dependence/codependency. An examination of religious leaders reveals people with little self control who are seeking to control others.

Drugs, alcohol, gambling, and religion are all examples of escaping from reality, which I loosely define as attempting to soothe the very understandable anxiety of not knowing; however, treating symptoms does not cure the underlying disease. Letting some external party control your thoughts and emotions is self-destructive, while taking full control over yourself is self-creative. This sounds good but is notoriously hard to do, because humans are social and hierarchical animals by nature (see *The Enlightened Savage* and *The Natural Savage*). Thus, people rarely question authority—a trait that allows religious

---

Religious bondage shackles and debilitates the mind and unfits it for every noble enterprise.
James Madison

To know the mighty works of God; to comprehend His wisdom, majesty, and power; to appreciate, in degree, the wonderful working of His laws, surely all this must be a pleasing and acceptable mode of worship to the most high, to whom ignorance cannot be more grateful than knowledge.
Nicolaus Copernicus

leaders to speak in vague terms that carry an air of expertise until one asks for definitions and gets only gibberish in return.

This is not to say that religious people are not skeptical. They are—about everything but their own beliefs and their own religions. All is well so long as these beliefs go unquestioned—but any test of those beliefs tests the underlying relationships and reveals the basis of those relationships. Most religious people lack knowledge and understanding and are thus incapable of anything that resembles unconditional love, a sad trait that may help explain the alarming rates of various types of abuse among religious people. The Milgram experiments proved that good people are plenty able and willing to do bad things when absolved from responsibility, and that they are very likely to follow the group despite private misgivings.

The more fervent opponents of Christian doctrine have often enough shown a temper, which psychologically considered, is indistinguishable from religious zeal.
William James

Theologians have been forced to concede that their views are mostly lacking in scientific merit as discovery after discovery continues to erode religious claims about the nature of the universe. As for morality, one need not believe in a god to see the virtues of generosity over selfishness, compassion over cruelty, justice over injustice, honesty over deception, etc. Religion makes few valid scientific claims and is utterly redundant where morality is concerned. In fact, religion derives from morality, and not the other way around—an assertion validated by the statistics in Chapter 2 about religious beliefs and the prison population in the United States. Religion has a nasty habit of confusing its own tenets with universal morality, which is how we get groups of people trying to regulate private sexual contact between consenting adults.

Governments all over the world are overrun by religious people who cannot bring themselves to see such distinctions. Some of these people also have fundamentalist ideas about the imminent end of the world, which would be plenty harmful even if they did not have access to nuclear weapons whose only practical value lies in never being used. Combine all of this with the abdication of all personal responsibility to God and the resulting carte blanche to act on a whim while "obeying God's will," and the rest may yet be history.

Leave the matter of religion to the family altar, the church, and the private schools, supported entirely by private contributions. Keep the church and the state forever separate.
Ulysses S. Grant

PART THREE

God

The Divine Savage
Revealing the Miracle of Being

Chapter 13

Can God Exist?

> *God is a latecomer in the history of religion.*
> G. Van der Leeuw

Given what we now know about the universe and how it works, it is safe to say that the God of the JCI religions does not exist. It is also safe to say that the many deities of the Hindu and other polytheistic pantheons don't exist either. That leaves the deep, underlying consciousness postulated as the Ground or Godhead by the Buddhists and similar disciplines. Can such a god theoretically exist? Can such a god be proven or disproved? The presence or absence of a god is a scientific question, and it is to science that we must turn in search of the answer.

> *God is what mind becomes when it has passed beyond the scale of our comprehension.*
> Gerald L. Schroeder

We will explore the sciences in much more detail in Chapters 31 through 32 and discuss the implications in Chapters 33 through 40; however, having just discussed religion in some detail, it is appropriate to ask whether God can exist and what such a being might look like. We can then use the conclusion we arrive at during our detailed discussion of the sciences to see whether our hypothetical God holds up to the best scientific evidence we have. The result will then form one part of the answer to the main question in this book: Is there some form of existence awaiting us after our physical deaths?

A Few Arguments Against God

Most of the arguments against God specifically deal with the JCI God, but there are also plenty of arguments against any kind of God. Here is just a small sample:

- God is supposed to be omnipotent; can God create a stone so heavy that He cannot lift it? This question cannot be answered without limiting God's potential power in some way, and thus opens a chink in any evidence that may exist for God's existence; however, Leibniz speculated that even God cannot escape reason and logic and must always act rationally.

- Can God change His mind? Here again, this question cannot be answered without limiting God.

- Any God who gives humans knowledge they cannot get via materialism should have testable evidence for His presence, but no such evidence exists; therefore, the evidence indicates that God does not exist. This argument can be easily countered by pointing out that all the evidence one could ask for is right under everyone's noses if only we would seriously examine it.

- Dawkins and others are fond of saying that a God complex enough to create the universe and all of the complexity therein must be even more complex than the universe, and must therefore require an even more complex origin, and so on, in a case of "turtles all the way up." Ironically, these same scientists point to evolution as proof positive that simplicity can spontaneously organize itself into ever-increasing complexity. As we saw in Chapter 11, the game "Life" is a very simple program that can create very complex patterns. One should therefore be able to conceive of a very simple God that may be orders of magnitude less complex than the universe S/he created. After all, a human brain can contemplate the entire cosmos, yet no one I know is arguing that the former is more complex than the latter.

- The fact that most people across most cultures believe in God provides no evidence whatsoever that God does in fact exist. Atheists can posit many plausible explanations for religion that require no divinity whatsoever, but

God is dead.
Friedrich Nietzsche

Nietzsche is dead.
God

If it turns out that there is a god, i don't think that he's evil. But the worst that you can say about him is that basically he's an underachiever.
Woody Allen

> *If there were a god, I think it very unlikely that he would have such an uneasy vanity as to be offended by those who doubt his existence.*
> Roland Barthes

these explanations do not in themselves offer any evidence against God's existence.

- We are accustomed to invoking God to explain things we cannot explain otherwise, but that list is shrinking daily. UFOs are similar: Explaining effects without resorting to their purported cause makes it a fairly safe bet that said cause does not exist. On the other hand, we can explain how consciousness works, but we are not even remotely close to explaining how consciousness itself comes to exist.

Can We Prove God?

> *The probability of God begs to be computed.*
> Stephen D. Unwin

Atheists dismiss yearnings for the divine as lacking any evidentiary value. To them, spiritual cravings and spiritual/mystical experiences are nothing more than wishful thinking and/or the result of some sort of misfiring in the brain. We want to believe there is something greater to ease the sting of death, and cloak our mundane lives in the robe of divine mandate and planning. We are also hierarchical animals, and it makes sense to speculate that some hidden über-alpha controls things that are beyond our own control.

Remember that there is no single wasted evolved trait. Even so-called "junk" DNA that does not actively code proteins has been found to play a key role in regulating how our bodies function. The so-called "God-shaped vacuum" in our hearts and minds may therefore be very real. But where did it come from? Theologians have a hard time saying where life came from. They define God as simple and decry science for saying that God is complex. Indeed, the first cause for life must have been both simple and self-bootstrapping; complexity evolved later through a long succession of very simple processes. We do not yet have a "theory of evolution" for physics, but we are working on it in fields such as superstrings, branes, etc. Meanwhile, the facts are clear: Evolution began simply and snowballed by adding simple processes and adaptations on top of other simple processes and adaptations to create complexity. Nature is ruthlessly utilitarian; waste is not tolerated. Only the successful adaptations survived in an increasingly complex environment, which gave rise to organized complexity where 50 trillion cells can come together to form Anthony or any other human being. Thus, the mere yearning for the divine does not make a god unreal. If anything, it bolsters the case for a god.

Must we accept God on faith and shut off our reason and intellect? "Let no man think that one can search too far or be too well versed on God's word and works," said philosopher Sir Francis Bacon (1561-1626), and he is entirely correct. Religion leads its followers to eschew science, and scientists are told to check their brains at the door with good reason: Galileo and Copernicus proved that humans are not at the center of the universe. Darwin removed the need for God as a creator, and modern biology and physics can explain just about all facets of existence. The enmity between science and religion is so strong that scientists are almost always presumed to be atheist or agnostic in academic circles.

It is as atheistic to affirm the existence of god as to deny it. God is being itself, not a being.
Paul Tillich

It is easy to see science as a threat against the divine, especially when the Western religious god can be so easily disproved. The Western worldview sees the universe as impersonal and separates humans from nature; disproving a separate and transcendent God is therefore a huge threat to the JCI religions. The Eastern religions see humans as part and parcel of the cosmic design and see God as immanent; science does not threaten them nearly to the same degree, and scientists know that. For example, Dawkins says that he does not see Buddhism as a religion, and Stenger is careful to limit his definition of "God" in his attempted disproof. The idea that an immanent God who is both one with and distinct from an idealist universe cannot be disproved—and that materialism and dualism both have serious flaws—shows that the actual Godhead may well be even grander than we can imagine.

Proving that a god exists is not easy. If we accept the idea that natural causes alone cannot account for the universe and our presence in it, then we need to find solid evidence of at least one creative act. This is the rub: People conflate science, politics, religion, etc. in ways that makes examining the facts on their own merits difficult at best. For example, religion relies on what science does not know to shove God into ever-shrinking gaps; abandoning old dogmas in favor of a revised concept of God would validate the mystical and spiritual traditions that got them started in the first place. Remember that the very first gods were scientific theories in their own right.

Our creative ability to process new meaning gives us much tangible scientific evidence for the existence of God.
Amit Goswami

Instead, religious beliefs place serious limits on what God can do. If God exists and is acting today, He should be able to act through biology, physics, etc. any time he chooses.

I don't know if God exists, but it would be better for His reputation if He didn't.
Jules Renard

There is an additional difficulty in that there are as many ways of perceiving universe as there are species in the universe. All we know—indeed, all we can ever know—is our own way of looking at things. Kant referred to objective objects and events that are independent of our senses as *noumenons*. This term generally refers to objects of inquiry, understanding, or cognition and may be further contrasted with the perception and processing of a phenomenon in the human mind. Thus, noumena transcend space and time. In fact, space and time are part of the space-time continuum; either can manifest itself as the other, but we can never directly know the continuum itself. Likewise, we may be able to sense various aspects of God (phenomenon) but are unable to know God directly (noumenon).

I have already said that the religious God is easy to pick apart, that materialists routinely have field days doing just that, and that the God of the gaps is a God that theologians have painted into a corner; however, I have already hinted that materialism ignores much evidence that could conceivably point to something greater, including a god. Materialists are aware of the gaps in our knowledge but remain confident that additional knowledge will prove them right—a sort of promissory materialism of the gaps. The triumph of materialism versus religion has as much—if not more—to do with the naiveté of religious people at it does with the cleverness of materialist science. Creation poses no threat to Darwin and vice-versa.

What I set as my objective is to prove the numerical probability that god exists.
Stephen D. Unwin

Maslow summed up the problem when he said that, "If all you have is a hammer, you will see all problems as nails." Materialists see all evidence affirming their beliefs while discounting all else as "supernatural." Anyone willing to look beyond materialism and who is ready to suspend one's disbelief about the role that consciousness seems to play in the universe and how that could point towards the Godhead, has progressed further than most scientists. Again, I am not talking about the God of religion; I am talking about the Godhead/Ground/reality at the heart of all existence.

I stand behind what I said in Chapters 7 through 10 about the evils of religion and the guilt of all those involved; however, we must never forget that Islamic science lasted almost 600 years, longer than modern science. It was not until around 1100AD that Europe began to rediscover the Greek and other manuscripts they had destroyed and that would have been lost but for Islam. Scientists were routinely accused of heresy, and yet they say saw their work as, "thinking God's thoughts after Him" to quote astronomer Johannes Kepler (1571-1630). It was only after more centuries of religious repression that science began to distance itself from religion—and God—until Darwin completely removed the need for the God of the JCI religions. Scientists and theologians alike made—and continue to make—the mistake of conflating religion and God.

To say it for all my colleagues and for the umpteenth millionth time: science simply cannot by its legitimate methods adjudicate the issue of God's possible superintendence of nature.
Stephen Jay Gould

Toward a Divine Hypothesis

Any hypothesis of the divine must incorporate feelings and other subjective data that no laboratory can test as part of its body of evidence. It must guard against hoax and false positives, while not relegating the entire lot to the trash bin. It must also accord with everything we know in the scientific realm and stand the test of future scientific discoveries. It must be to current ideas of God as brane theory is to superstring theory, as superstring theory is to quantum physics, and as quantum physics is to Newtonian physics, where each new advance contains the entirety of all previous discoveries within it. Above all, it must respect and exalt the humanity it exists to serve along with all life, while giving meaning and purpose. In other words, it must fulfill the long-broken promise of religion to connect people to the divine, with neither judgment, prejudice, or recriminations. That is the only way to ensure that the millions of people who died or otherwise suffered for religion and the idea of God will not have died in vain.

Why does our craving for God persist? It may be that we need it for something. It may be that we don't need it, and it is left over from something that we used to be. There are lots of biological possibilities.
Daniel Dennett

As biochemist Arthur Peacocke (1924-2006) wrote in *Paths from Science Towards God*, the divine hypothesis must be the simplest, most coherent, and causally adequate explanation of the evidence or phenomena being examined. In other words, it must be the hypothesis that best explains the relevant evidence, a result of *abductive* logic that provides the Inference to Best Explanation (IBE). A theology that relies on authoritative texts, an authoritative community, and *a priori* truth to stand as the IBE can only be true if the following equation is true:

S+CRE+WR=IBE, where:

- S is scientific knowledge,
- CRE is classical revelatory experience (Mohammed, John of Patmos, etc.), and
- WR is the world's religions.

Nothing in Western culture or scientific knowledge can be affirmed or defended by appeal to the Bible, Koran, or any JCI belief; however, the existence of the universe is anything but self-explanatory. Even the zero-bound universe that Hawking believes eliminates the need for anything divine (see Chapter 26) only explains the what and how, but not the why. The only way to avoid an infinite regression down either a scientific or religious path is to postulate an ultimate reality that must be diverse, yet unified and capable of unimaginable richness, multiple simultaneous expression, varied means of outreach, etc. For example, why is the universe so deeply steeped in rationality? The simplest explanation is that reality itself must be both supremely rational and OOO to a point. It must be transcendent. Individual reality is very personal, and personality in many ways represents the highest level of organization. Reality must therefore be personal or supra-personal. It must be transcendent but also immanent, and must be both self-existent and logically self explanatory. It is far more appropriate to call such a reality "s/he" than "it." This hypothesis stands as the IBE if:

S+PE+ME+NDE=IBE, where:

- S is scientific knowledge that encompasses physics, biology, neurology, psychology, genetics, etc.,
- PE is normal personal experience,
- ME is mystical experience, and
- NDE is near-death experience

If God (in the sense of the Godhead/Ground and not any religion) created the universe and time, then God's consciousness is self-limited. S/he created the world so as to know the probabilities—but not certainties—of our actions and of

Question with boldness even the existence of a God; because, if there be one, he must more approve of the homage of reason than that of blindfolded fear.
Thomas Jefferson

I believe that gods exist to the extent that people believe in them.
Vince Sarich

quantum events. This God created the entire universe at once, and never intended for humanity to lord it over ourselves or any other species, but for all to share in the process and joy of creation. Creation itself could involve God learning and enjoying the ride as much as anyone and sharing in the creative process. This of course entails sacrifice and suffering, but in the creative sense akin to those of a woman giving birth. As Hildegarde of Bingen (1098-1179) said, "All living things are sparks from the radiation of God's brilliance and emerge from God like the rays of the Sun."

Hoyle summed up the marvels of scientific progress when he said that, "When we learn the answer to any scientific problem, we always find in both the whole and in the details that the answer is finer in concept and design than anything we could have guessed at." If such a hypothesis can be constructed, then it should follow in the footsteps of previous discoveries and reveal a universe even grander than we currently imagine.

Such a hypothesis by itself may not give us much insight as to what happens when we die, because it is perfectly possible to have a God but not immortality, soul, afterlife, reincarnation, etc. as we learned in Chapter 2; however, it can be an important piece of the puzzle, as we will continue to discover. We may indeed be able to prove God.

God is a shelter for ignorance and anthropomorphism. The universe, on the other hand, incites us to venture outward and take risks. it is the space of all knowledge and all action.
Voltaire

Do We Need God?

Thanks to the First Law of Thermodynamics (conservation of mass and energy), neither atoms nor energy are lost when we die. In that small sense at least, we are immortal. Of course that is cold comfort to those of us who wonder if there is anything more to personal existence beyond this lifetime. Does any aspect of a unique "I" survive death in any way beyond memories in the living and works, such as books or paintings? Do we need God in order for existence to continue? Do we need God even if death means utter annihilation? Beethoven, Shakespeare, Michelangelo, and others would be just as beautiful with or without God, as would a sunset, waterfall, or starry night. Beauty neither requires nor proves God; one presumes that Michelangelo would have done just as good a job decorating a secular government building as he did the Sistine Chapel.

Science offers us an explanation of how complexity arose out of simplicity. The hypothesis of god offers no worthwhile explanation for anything... we cannot prove there is no god, but we can safely conclude that he is very, very improbable indeed.
Richard Dawkins

> *You know, Reverend, the more I learn about this amazing universe, the more awesome my God becomes!*
> Anonymous parishioner

Arguments from personal experience are very compelling to people who don't understand psychology, but not so much to those of us who do. Society has names for people who hold irrational beliefs. If enough people have such beliefs, we call them religious. People with individual beliefs that stray too far from the norm are called mad, psychotic, or delusional. There really is sanity in numbers; for example, it is considered normal to believe that God hears our thoughts and prayers, but delusional to think He communicates by television.

Australia's Aborigines must be very intelligent, rational, and creative people in order to survive in the harsh outback. They are experts of the natural world, knowing which species of plants and animals are nutritious, poisonous, dangerous, and benign, and how to stay cool and hydrated in the desert. They too carry religious/spiritual beliefs. All societies on Earth have spiritual beliefs. Are we all mad? Misinformed? Afraid? Paranoid? Have our evolved instincts gone haywire, in what would be the single example of a maladaptive trait across both the plant and animal kingdoms? Do we need God?

Differing Opinions

> *Much evidence suggests that God is real and/or there exist vastly many, varied universes.*
> John Leslie

Millions of people have opined on the existence and meaning of God. Here are just a few opinions (some quoted, some paraphrased) from some well-known people:

- **Susan Blackmore:** Religion is a meme that spreads. There is a sea of brains and memes jostling to fill them. Those that prevail are selected to move on, presumably because they may have appeal or be similar to other successful memes.

- **Richard Dawkins:** All religious people believe in a superhuman "supernatural" intelligence that deliberately designed and built the universe and all in it. Any such being can only come into existence as the end product of evolution. All gods must be attacked along with anything and everything "supernatural," wherever or whenever it was invented. Buddhism, Confucianism, etc. could be treated as ethical systems and not religions.

- **Daniel Dennett:** Religion may be a by-product of the irrational tendency to fall in love.

- **Albert Einstein:** Einstein was, as he put it, a "deeply religious non-believer" who never said that nature had a purpose, goal, or anything anthropomorphic. Nature is a magnificent structure that we can only barely comprehend with humility, which has nothing to do with mysticism. He found the idea of a personal God very alien and naive.

- **Steve Grand:** People's lives are more like waves than things. Think of something from your childhood as if you were really there. You should know it, because you were there... except that you were not there. No atom in your body today was there when you were a child. Matter flows around and momentarily becomes "you." Whatever you are, you are not the stuff you are made of. As he said, "If that doesn't make the hair on the back of your neck stand up, read it again until it does."

- **Fred Hoyle:** The chance of life evolving on Earth is the same as a whirlwind in a junk yard assembling an airworthy 747. Complex things can't come about by chance, but many define chance as absence of design and therefore see improbability as evidence of design.

- **Thomas Jefferson:** The day will come when the idea of Jesus Christ as a mystical being will be up there with Minerva and Jupiter.

- **Heinz Pagels:** What we think of as a random number is related to other numbers by a simple rule. Two random numbers may be related in a non-random way. You cannot have true randomness.

- **Bertrand Russell:** Many people would sooner die than think; and in fact, they do.

- **Carl Sagan:** If by "God" one means the set of physical laws that govern the universe, then clearly there is such a God. This is emotionally unsatisfactory because we may as well pray to the law of gravity.

- **Mark Twain:** I do not fear death. I have been dead for billions of years before I was born and not suffered the slightest inconvenience for it.

- **Gore Vidal:** The great evil at the center of civilization is monotheism.

I don't feel an awe for the brain. I feel an awe for God. I see in the brain all the beauty of the universe and its order—constant signs of God's presence. I'm learning that the brain obeys all the physical laws of the universe. It's not anything special. And yet it's the most special thing in the universe.
Candace Pert

Physicists answer only to mathematicians and mathematicians answer only to God.
Anonymous

- **Steven Weinberg:** Some concepts of God are so big that people will find Him everywhere. For God to be useful, He must be a "supernatural" creator appropriate for worship.

- **Oscar Wilde:** Truth in matters of opinion is just opinion that has survived.

Evolution

Entities complex enough to exhibit intelligence are products of billions of years of evolution. Organism A can evolve into a more complex organism B, but cannot produce something entirely new. Life is something entirely new. Even once life has begun, evolution flies in the face of the notion that something can only produce something less complex than itself; for example, we will never see a hammer making a carpenter or a pen making a writer. Life continues to evolve toward increasing complexity, and new realities appear for each individual of each new species based on what that species needs to survive; as I said before, there are as many different realities as there are organisms in the universe.

There are easily billions upon billions of planets in the universe. If life is merely improbable and not impossible, then this great number of planets ups the odds that life will evolve somewhere. If we find life elsewhere (such as on Gliese 581g), then life will become not improbable but highly likely in the universe in general, and we will need to find a law of nature that describes how life arises. This need not involve anything more divine than blind natural processes.

A friend of mine once offered the following disgusting—but very effective—proof of the truth of evolution for anyone who doubts it. The goal is to prove that fungus can evolve to survive in a harsh environment. Those that can will live and reproduce; those that can't will die. Eventually, only the survivors will be left to reproduce; the fungus will have evolved. To conduct this experiment:

1. Walk barefoot through locker rooms and showers each day for a week. Don't dry your feet; instead, put on shoes immediately. Keep them on for the entire week to keep your feet damp. This warm, humid environment will cause rapid fungus growth.

The very fact that we exist in the universe to ask these questions about it means that a complex sequence of events must necessarily have happened.
Michio Kaku

If God is to be found, it must surely be through what we discover about the universe, not what we fail to discover.
Paul Davies

2. Purchase the best foot fungus medication available and follow all applicable directions for its use to the letter. Keep your feet warm and wet during this stage while you continue using the medication.
3. Discontinue treatment when you feel relief, then restart it when your feet again start to burn and itch.
4. Repeat Step 3 until treatment provides no remedy whatsoever.

You should notice the following effects, in order:

- Your feet will probably burn and itch during Step 1.
- You will initially feel relief during Step 2.
- This relief will be short-lived; eventually, your feet will begin burning and itching again.
- Repeated treatment cycles (Steps 3 and 4) will offer ever-diminishing levels of relief.
- Eventually, the treatment will not work at all.

Congratulations! You have just bred a super strain of foot fungus. Just make sure you are fully cured before going back to any more showers or locker rooms, lest you start an epidemic.

At first, only a few random fungus cells happened to have resistance to the treatment. Had you kept your feet dry and clean, that plus the treatment would have effectively destroyed the fungus; however, your continued warm and wet feet allowed the resistant fungus to survive and reproduce. Subsequent generations increased the quantity of treatment-resistance cells, until the treatment ceased to have any effect because all of the cells had evolved to resist it. This is an example of micro-evolution. Continued environmental pressures such as these over millions of years would eventually yield macro changes that would create a new fungus species. Similar examples already exist with vaccine- and antibiotic-resistant bacteria and viruses, against which medicine can only produce short-lived treatments. Evolution itself is a fact; the theory just tries to explain how it occurs. Material processes are enough to account for evolution from primitive cell to human and beyond—once life has begun.

Evolution seems too callous and indifferent for a loving God at face value. This bolsters the idea that being able to explain

---

The simple, absolute and immutable mysteries of divine truth are hidden in the super-luminous darkness of that silence which revealeth in secret.
Dionysius the Areopagite

There was one thing I was ignorant of at the beginning. I did not really know that God is present in all things; and when He seemed to me so near, I thought that it was impossible.
Theresa of Avila

evolution through material processes alone could disprove God. Is this true? After all, humans evolved from the same common ancestor as every other species on Earth. We have not enjoyed the fruits of evolution any more than any other species. But for a rogue meteor, the dinosaurs could still be around today. Having lasted 100 million years, what's 60 million more? It is possible that dinosaurs could have evolved technology and all of the other things we take for granted today, including religion. We literally owe our existence to our lucky stars. Would it make any difference to God if dinosaurs had souls instead of humans? Would there have been a reptilian equivalent of Anthony Hernandez writing this book? If humans do indeed have souls, do primates? Apes? Monkeys? Mammals? Reptiles? Fish? Insects? Plants? Bacteria? Where do we draw the line? Do we draw any line at all? If so, why? If not, why not?

The Big Bang, the most cataclysmic event we can imagine, on closer inspection appears finely orchestrated.
George Smoot

Evolution is the process of optimizing living creatures to meet environmental challenges. We live on a planet that is friendly to our type of life because we evolved here, and because of the anthropic principle that also has us living in a remarkably hospitable universe. As we saw, very slight differences in constants would have made huge differences. A god capable of setting these values seems highly improbable, which leads us to wonder how free the values were to vary. A GUT/TOE may reveal that the values depend on each other in some currently unknown way, and that the values could not be any different than they are. This could, in theory, completely remove God as an explanation for how the universe came to be.

He is far and he is near. He moves but he does not move.
Bhagavad Gita

In the early 1800s, naturalist Étienne Geoffrey Saint-Hilaire (1772-1844) speculated that dinosaurs were the source of crocodiles, a direct precursor to Darwin. The idea that life changes and adapts over time is dangerous, for it literally unifies life with time and, by extension, with space. This is the concept that also unifies us with the rest of the universe, and opens the door to understanding that we are all quite literally made of stardust. Darwinian evolution demanded materialist answers, especially when evidence to support the validity of the theory poured in. Everywhere in the world, fossils most closely resembled current species, suggesting changes over time. The earliest amphibians looked more like fish than did later amphibians. People were afraid of *The Origin of Species* because the materialist nature of evolution undercut the religious assumption that a special relationship exists between God and human, which threatened the idea that humans have immortal souls. The process of unseating humanity from the center of God's universe and revealing us to be anything but the apple of His eye began with Copernicus and culminated with Darwin. Creationism itself was displaced. Atheism was validated. Religion will never recover—and neither might God.

Evolution does not preclude either the belief in or the possibility of God, not because it is wrong, but because it is right. All living creatures could be equally valuable to God in their own right, not just as stages on the way to humanity or for us to have dominion over. The creation myth in Genesis may be scientifically accurate, and we may have some level of intellectual dominion over other species, but that's about it. The death of individual members of a species is important for the survival of the entire species. Biological death provides resources and room for new creatures to grow and evolve. The idea that, "the wage of sin is death" could therefore refer to a kind of karmic death.

A pantheistic model could allow both immanence and transcendence. A human body has both immanent and transcendent aspects, in that it is a collection of parts that is also more than the sum of those parts. The concept of natural evil challenges the idea of a benevolent god; however, a pantheistic model can understand God's role in evil as His experience needing to take the bad along with the good.

The overwhelming success of Darwin's theory has led scientists to extend it to other fields, such as religion and the existence of God. For example, Dawkins's claim that creating synthetic life in a lab would disprove God is false; all that would do is prove the creativity and inventiveness of the scientist who designed the life. Dennett wonders how anything we think and feel can be not a product of evolution when evolution can account for our minds. Gould believed that language exists because of evolution without being a goal of evolution, while Dennett believes that language was an evolutionary goal. Biologist Edward Wilson (1929-) believes that religion serves social purposes and is therefore explainable by science.

I know God will never give me anything I can't handle. I just wish that He didn't trust me so much.
Mother Teresa

A lot of my colleagues like the idea of final theories because they're religious. And they use it as a replacement for god, which they don't believe in. But they just created a substitute.
Mitchell Feigenbaum

However, we have not reduced plants to chemistry or physics; on the contrary, we have elevated our understanding to appreciate plants in a way that adds wonder and delight to our view of nature, the universe, and our place within it. Wilson, Dennett, Dawkins, and others have gone beyond science and reason, making arbitrary decisions about what knowledge to include or exclude based on their materialist sources and beliefs. They have created a dogma, a secular religion of sorts. Listening to them is very similar to listening to a religious fundamentalist; it is as if they are 359 degrees apart on the same circle—so near, yet so far.

Intelligent Design

Today the least educated of my children knows more about the natural order than any of the founders of religion.
Christopher Hitchens

Moths use an internal compass calibrated for navigation based on the Moon, which is at optical infinity and whose light rays are mostly parallel. Objects closer than optical infinity, such as light bulbs, candles, and flames radiate light rays that are not parallel but more like the divergent spokes emanating from the hub of a wheel. The moth's internal navigation system applies a given angle to a nearby candle as through it were at optical infinity, which causes the moth to spiral into the flame along a logarithmic curve that ends up being its death spiral. The moth is not attracted to the light, nor is it trying to commit suicide. It is simply a nocturnal animal that evolved to use the brightest object in the night sky to find its way around. Lanterns, torches, bulbs, candles, fires, etc. have emerged far too recently and too quickly for moths to evolve to deal with them. A moth that used to have one light to guide it may now have many thousands of lights around it, and the brightest one is most likely to be artificial if it is near any human settlement or encampment. Similarly, the human appetite for fat, sweet, and salt has been corrupted by food manufacturers to cause the exploding obesity epidemic. Both of these are examples of short-circuited evolutionary instincts. How could an intelligent OOO designer possibly make such basic design errors? This gaping flaw in the theory is why intelligent design is often referred to as creationism in a cheap tuxedo.

We could say that there was no creation, and that the universe has always been here. But this is even more difficult to accept than creation.
Barry Parker

In fairness, an Intelligent Design (ID) adherent must believe both that the Earth is billions of years old and in the fundamentals of evolutionary theory; however, ID and pure Darwinian evolution are entirely separate ideas. ID believers love the anthropic principle (Chapter 11) because they believe it supports their case, not realizing that it is meant as an alternative to their explanations, not an addendum.

Why is religion so opposed to evolution, despite having come to terms with cosmic discoveries that knocked us out of the center of it all? I agree with the speculation that the cosmological discoveries only affected our perception of our special status with God, while the theory of evolution goes to the heart of who and what we are: animals, just like any other animal. Earth's biological history paints a clear arrow from simple to complex, and evolution favors complexity over simplicity to the extent that the former tend to survive longer.

Creationists counter that Darwin's theory is about adaptation but not evolution. This is mistaken, because the former does not need the latter. For example, notoriously cold and foggy San Francisco is home to a flock of feral red-masked conures (*Aratinga erythrogenys*), whose normal habitat is the hot tropical areas of western Ecuador and northwestern Peru. These birds have adapted to the colder climate, but they have not evolved. (See *The Wild Parrots of Telegraph Hill* for more on this flock.) Creationists also claim that the presence of gaps in the fossil record disproves Darwinism, dismissing the claim that the gaps are being filled as "promissory evolution." Well, the gaps are being filled in, as we will see in Chapter 20. The gaps that exist occur because fossilization is far from automatic. Certain conditions must be met after death if the bones are to be preserved. Also, even if a fossil does form, we cannot possibly dig up the entire planet looking for them. Still, God gets shoved into these imaginary gaps; filling them does no good because, as author Michael Shermer (1954-) says, discovering an intermediate species that fills a gap now leaves two gaps, one on either side.

Creationists would do well to look at the living record, because modern embryos often resemble earlier species at the earliest stages of development. Take a look at the following excerpt from a 19th century drawing of early-stage embryos:

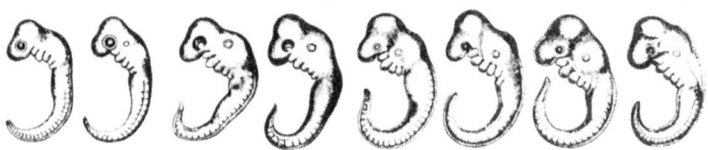

The impression of design is overwhelming.
Paul Davies

The anthropic principle in any of the three versions mentioned suggests quite strongly that the universe is purposive, created by a designer with the purpose of creating life.
Amit Goswami

> *I do not believe the no-boundary proposal proves the non-existence of God, but it may affect our ideas about the nature of God. We do not need someone to light the blue torch paper of the universe.*
> Stephen Hawking

Can you tell which is which? You may think that I am playing a trick on you by showing you different drawings of the same species, but I'm not. Maybe this second drawing will help you:

By now it should be obvious that we are discussing at least three different species, but look at the next image to see what happens when the embryos mature just a little more:

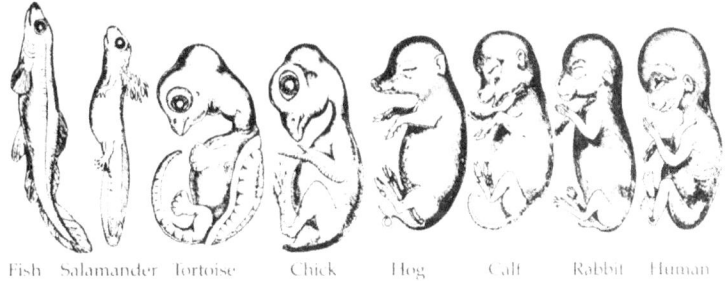

Fish Salamander Tortoise Chick Hog Calf Rabbit Human

Paleontologist Niles Eldredge (1943-) worked with Gould to propose an idea called *punctuated equilibrium* to explain the lack of intermediary species by proposing discontinuous evolution, in contrast to Darwin's constant gradual changes. Creationists saw this as an opening, saying that evolution can only occur on a micro level, whereas macro evolution requires many changes that cannot occur all at once, meaning that God had to intervene; however, sudden evolution can theoretically occur thanks to quantum waveforms, which we will discuss more in Chapter 29.

> *If scientific arguments for the existence of god are to be allowed into intellectual discourse, then those against his existence also have a legitimate place.*
> Victor Stenger

Intelligent design advocates also point to irreducible complexity to prove their point. Just as a car is useless without an engine, they say that partially formed adaptations, such as eyes and wings, are worse than useless because they consume resources and burden the animal, exposing it to predation. A system that requires all of its parts in order to function cannot be useful if only partially formed, and thus evolution cannot produce it. Biologists such as Behe are fond of using the eye as an example of irreducible complexity, but Darwin himself answered this question long before Behe was born. For example, the bones in your inner ear that help you balance and walk upright used to be jawbones. The fossil record shows these bones moving back and up, and slowly adapting themselves to hearing. One species even had a double-hinged jaw that allowed it to hear while eating, a perfect example of an intermediate step. Irreducible complexity may be a valid concept if one is using it to describe a tangled hierarchy of life where there is no defined highest or lowest level, and where moving through the levels eventually returns one to the starting point, but it does not apply to evolution at all.

Biologists have performed experiments that shut off an existing gene and let it re-evolve, which has sometimes created an even better "fit" than before. This new gene looked perfectly designed, but was really a product of blind evolution at work. Irreducible complexity poses an intellectual problem, in that believing it causes one to stop looking. This attempt to create reasonable doubt about evolution is failing badly, not least because the current *morphological rate* (natural rate of genetic mutations) is between 10 and 100 times faster than needed to create the fossil record we have in the allotted amount of time (3.5 billion years or so). This rate could even be as high as 1,000 times faster than needed for punctuated equilibrium. The foot fungus experiment described above demonstrates both the truth of evolution and the positive effects of mutation. If a few genes can mutate a fungus to resist a deadly poison in a matter of days, what exactly stops them from causing large changes over millions of years?

Intelligent design paints God as a magician, while neither accounting for the many species that have gone extinct over billions of years, nor verifying morphological rates. Some biologists also point to the example of the "backward" human eye where nerve endings are placed before the light detectors and pass through the wall of the eye causing a blind spot that we don't normally detect as an example of poor design. In fairness, others point to it as an example of brilliant design, because it minimizes reflections and other effects that might otherwise degrade vision. This is just one example. In short, the intelligent design argument boils down to, "I have no idea how this happened so it must be irreducibly complex," which is nothing but a lame argument from personal incredulity.

According to Darwin, evolution is the product of mindless natural selection. Darwin himself was ignorant of modern

Do there exist many worlds, or is there but a single world?
Albertus Magnus

I believe that science can lead us to the God who is now making himself/herself/itself known in physics, statistics, computer science, and even in, of all places, parapsychology experiments.
Gary E. Schwartz

genetics, biology, and psychology, but all of these later sciences have validated his theory in a big way. One of the main arguments against intelligent design is that evolution can only work on what already exists, whereas a designer could start fresh with a clean drawing board at any time. Human embryos are almost indistinguishable from other species at first (see above). We have empty yolk sacs, a legacy from a more primitive past. Our spines have only recently adapted for bipedal motion, as anyone who suffers from low back pain will tell you. We cannot yet explain how a cell in an embryo knows it is going to become a kidney, muscle, or ligament, but we know it happens. We do not yet know how DNA makes proteins, since proteins are required to make DNA, but we know it happens. Did God make DNA? There is a leap between micro possibility and macro possibility, and another bigger leap between macro possibility and macro actuality, which in turn limits micro possibility—a great example of tangled hierarchy.

> *Today, we are rediscovering the truths about the universe, life, god, and humankind that have always been there, but were shrouded by religion.*
> R. Craig Hogan

There is also the problem of *meaning* (purpose, intention, destination, significance, etc.). Meaning can only evolve as an adapted survival trait if matter is capable of processing meaning, as opposed to mere *syntax* (patterns, logic, etc.). Penrose confirmed that computers (and thus matter in general) cannot process meaning. How can nature select a material quality that it does not possess?

At face value, evolution does not seem to need God, except that we are unable to explain how life arose from nonliving matter. In fact, we have yet to devise a suitable definition for life. By some definitions, fire is alive because it breathes, metabolizes, and excretes after a fashion. Science may someday answer the question of how life arose, and may find Penrose wrong about the ability of matter to process meaning. The idea that no OOO designer could possibly have wired the human eye backward or given us weak knees or any of the accusations against the idea of God-driven evolution ignores the fact that we are utterly incapable of designing even the simplest living thing. We can perform cloning, genetic modification, etc. at will, but life requires life. We cannot take something dead and give it life. We cannot assemble vials of chemicals and other ingredients into the simplest imaginable cell. I somehow doubt we are in any position to doubt the intelligence or ability of any intelligent designer! Try as materi-

> *Life appears simply as a disease of matter.*
> Gerald Feinberg

alists might, the wrongness of intelligent design, and the fact that material processes can explain evolution, does not preclude God in the slightest, nor does it necessarily require God.

Quantum Mechanics

The debate between science and religion/God is often framed in classical (Newtonian) terms; however, quantum physics may just tip the argument. A *quantum* is the smallest amount of energy that can be transferred, and this amount is variable; for example, a gamma ray photon has about 1 billion times as much energy as an infrared photon. (A *photon* is a particle of light.) The fundamental decision facing us from a physical standpoint is whether the universe is *deterministic* (all events have causes that can determined in advance if one knows the starting conditions) or *probabilistic* (events cannot be determined in advance; we can only determine the probability than an event will occur). Probabilities are predictable, and thus give order to the macro world.

There is a chance that you could find yourself orbiting Saturn (or anywhere in the universe for that matter) in an instant with no warning whatsoever, but the chances are extremely minute that any of your subatomic particles will decide to make the trip, much less two or even all of them. Your body has about 50 trillion (5×10^{13}) cells, each of which has about 10 trillion (1×10^{13}) atoms, which means that your body has about 5×10^{26} atoms. If there is a 50% chance that one of your atoms will jump to Saturn at any given moment, then the chance that two will jump at the same instant becomes 25%, the probability of three jumping become 12.5%, and so on. Given that atoms consist of multiple subatomic particles and the odds of one of those going walkabout are less than 50%, the odds of your going anywhere instantaneously are extremely small. Thus, while any one—or even many—of the particles that make up your body may wink off to places unknown at any given time (and probably do), they are too few for you to notice. As Schrödinger said, "Atoms are so small so that we can live in a predictable world."

It is important to point out that the uncertainties of quantum mechanics do not exist because of gaps in our knowledge or understanding. They exist because that is the way nature is. This could give God the freedom to act while always obeying

If we do discover a complete theory, it should in time be understandable in broad principle by everyone, not just a few scientists. Then we shall all, philosophers, scientists, and just ordinary people, be able to take part in the discussion of why it is that we and the universe exist. If we find the answer to that, it would be the ultimate triumph of human reason-for then we would know the mind of God.
Stephen Hawking

> *The cosmos is all there is or ever was or ever will be.*
> Carl Sagan

natural laws. As chaos theory has taught us, small changes can have a huge effect. The example of a butterfly flapping its wings in Thailand and causing a hurricane in Florida may be a bit extreme, but it nevertheless carries a grain of truth. For example, it is possible for a single particle to cause a mutation that leads to the evolution of a new species based on the original species, which is exactly how evolution has been shown to work. Such a God would be one who deeply respects His own creation, while still being able to act in unpredictable and undetectable ways. Religion remains unaware of the friends they may have in physicists, and biologists are mostly ignorant about just how the supposed materialism of life owes itself to unpredictable events.

Both quantum mechanics and relativity stemmed from observing anomalies in the behavior of light. Light seems to be more fundamental than space, time, and even matter. (Let there be light, indeed!) An observer moving at light speed would see time and space shrunk to zero; s/he would be everywhere and everywhen at once. According to Einstein, an object moving at light speed will also have infinite mass, which has been experimentally proven. Photons have zero mass. Every photon is transcendent, always in the eternal now with neither future, past, nor space to traverse. We perceive light as traveling at a certain speed because we are moving more slowly, and thus have a different frame of reference. This allows us to, for example, use a laser or radar to measure distance and speed and to think of a photon being emitted, traveling, reflecting, and finally being absorbed, even though quantum mechanics says we can never know what happens between Point A and Point B.

> *New life through death of the old is inevitable in a finite world composed of common building blocks (atoms, molecules, macromolecules) having regular properties.*
> Arthur Peacocke

Subatomic particles are not solid like matter as we think we know it. They are more like waves than particles, fuzzy clouds of potential existence. So-called solid matter is a lot less solid and contains a lot less matter than we think. Schrödinger anticipated DNA when he said that life needs a second set of laws in addition to physics without violating the laws of physics. DNA is structured such that even a single atom or particle is important; as we saw, quantum events can cause mutations that cause evolution. I was not exaggerating when I said that all of evolutionary history could hinge on the behavior of a single particle. This seems to contradict what we know of molecular biology; however, molecular biology occurs on a macro level and has not adapted to quantum mechanics.

The core assumptions of materialists therefore seem to be wrong, even by their own sciences. Science tends to have a habit of trying to find a so-called natural solution while defining nature as the material space-time world and relegating God and anything else that doesn't quite fit the mold to the realm of the "supernatural." If science is going to include quantum mechanics, then it must accept the transcendent aspects of quantum mechanics as well, which may include consciousness itself. You have probably noticed that I am consistently placing quotes around the word "supernatural." I am doing this because the definition of that word depends on what you accept as natural. For example, ghosts may be "supernatural" to some and perfectly natural to others.

The more the universe seems comprehensible, the more it also seems pointless.
Steven Weinberg

A level of self-organization separates life from non-life, but there is no difference between them in the particles, atoms, or molecules. Everything in the universe is made of the same stuff, and everything obeys the same quantum laws. This may be why science is having such a hard time trying to find consciousness in the brain; the mind/brain link may not exist, despite all appearances to the contrary. Time and space are different aspects of each other that are inextricably linked. It is even possible that every atom is conscious in its own way. Physicist James Jeans (1877-1946) once said that, "The universe begins to look more like a great thought than like a great machine." In fact, consciousness could explain quantum entanglement, the observer effect, and many other examples of quantum oddness. (See Chapter 23.)

At its heart, matter consists of "probability waves" that have no definite position or momentum, unless and until they are measured by a conscious observer. Between observations, the laws of quantum mechanics dictate that a particle is quite literally everywhere and everywhen until the next measurement. Using the laser example, the police officer staffing the speed trap has no reason to think that the photons leaving the gun are doing anything but traveling in a straight line, reflecting off the vehicle, and coming straight back to the gun; however, he has no way to know for certain that this is the case. Quantum physics says that the particles will take every possible path to and from the car, and that some of them will simply *tunnel*

It is not a trivial matter that the laws of physics, as realized within our universe, have the proper form to allow for chaos and complexity.
Fred Adams

> *Our emergence and survival depend on very special 'tuning' of the cosmos—a cosmos that may be vaster than the universe that we can actually see.*
> Martin Rees

(teleport or jump instantly from place to place) elsewhere and never return. I don't suggest using this defense in traffic court because the sheer number of photons emitted guarantees that enough will make the round trip as expected to create a value that the officer then measures. The measurement collapses the probability wave into a single actuality that helps determine whether or not the motorist in question is speeding. This simple explanation suffices for now; we will get into more detail in Chapter 23.

Having a lot of money and an interest in classic motorcycles does not fill your garage. The classical model of consciousness that treats it as an *epiphenomenon* (a phenomenon that arises from another phenomenon) of the brain, sees quantum mechanics as a paradox, and a paradoxical model, is a bad model. How can personal reality be subjective while reality at large seems to be built on consensus? How can there even be such a pursuit as science? The idea of a non-local quantum consciousness at the heart of reality (idealism) could solve all these problems, while leaving room for—if not outright requiring—God, albeit not the God of any religion... except maybe the Buddhists.

We can go one step further: The act of collapsing the probability wave could be said to not only leave room for God, but to require God, since everything we see is actuality and not probability. By this logic, consciousness is not an epiphenomenon, and mind and matter are on equal footing. Consciousness ultimately chooses how to collapse the probability wave. Thus, our inner psyche and the outside world are not separate, but are parallel and interconnected aspects of consciousness.

> *In quantum theory, we are beyond the reach of pictorial visualization.*
> Niels Bohr

Traditional science postulates upward causation from particles to atoms, molecules, cells (including neurons), brain, and consciousness. The interpretation of quantum mechanics that I am alluding to here postulates downward causation from consciousness to brain, cells, molecules, atoms, and particles.

As we saw in Chapter 11, both the weak and strong anthropic principles suggest that the universe was purposefully created by a designer (God) who wanted to create life. Life is here and we are here because of the universe; however, quantum mechanics is forcing us to seriously examine the idea that the universe may be here because of us. You can call this God, quantum consciousness, the *Akashic record* (a massive collective memory store), or anything you like—but as Shakespeare's Juliet said, "What's in a name? That which we call a rose by any other name would smell as sweet."

Cosmology

On the other end of the size scale, it may surprise you to learn that the concept of creation *ex nihilo* was created to counter Gnostic Christians who saw matter as the work of the devil in yet another example of religion making the story up as it goes. As it happens, the Gnostics may have been at least partially right for reasons we are exploring throughout this book. This did not stop Pius XII from proclaiming that scientists are beginning to "find the fingers of God in the creation of the universe." This may be true, but the God that science may be finding is almost certainly not the God that Pius had in mind.

Quantum mechanics introduced the concepts of chance and randomness into the deterministic view of the universe that prevailed since Newton, thus providing a strong rebuttal to classical ideas of cause and effect. Where God had been needed to wind up the universe and set it on its way, quantum mechanics revealed that the universe could have popped into being on its own, with no outside intervention needed. The origins of the universe (see Chapter 26) were determined by the laws of physics; however, this says nothing about whether God exists or not, merely that God is not arbitrary, even if He does play dice from time to time.

Evolution and relativity further eroded the apparent need for God since we could now explain how life progressed on Earth, even if we couldn't describe how it got started. It is still possible to see God's hands holding planets and stars in the sky, but that vision is fading fast. The divine soul has become the Freudian *id*, and old myths and religions have been thoroughly debunked. In short, the universe seems to operate by a set of natural laws that dictate how it will evolve (within uncertainty limits). These laws may or may not have required divine intervention to set up.

Science stands upon the brink of being able to explain everything without the need for God, which certainly does not mean that God does not exist—at least not yet. Will a final

The universe is like a safe to which there is a combination. But the combination is locked in the safe.
Peter de Vries

Matter is frozen energy.
Nigel Calder

Theory of Everything eliminate God entirely? Or will it show us the mind of God? Current theories of creation and the origins of the universe are built on quantum mechanics and relativity, replacing religious theories with thermodynamics and subatomic particles. In fairness, the Big Bang has done little but swap one problem for another; instead of asking where God comes from, we must ask what happened to cause the universe to come into being.

Hawking applied the quantum waveform to the entire universe to postulate a universe that could simply pop itself into being with no outside agency required. If he is correct, then there could be infinite universes, each with different constants and natural laws. In that case, our own universe would be anything but special, and God would not be needed. This theory does not explain where all of these universes are coming from and simply replaces one form of infinite regression with another, but it does at least remove God from consideration, which seems to be almost *de rigueur* (necessary according to etiquette, common sense, protocol, or fashion) for mainstream science. On the other hand, a pantheist would see the universe and its wave function as being the closest we can get to the Godhead.

The laws of physics and relativity break down at the Big Bang and black holes, opening the door for God to act by winding up the universe and letting it go; however, combining quantum mechanics and relativity gives us the possibility of a "zero bound" universe. The zero-bound theory speculates that the universe has no clearly defined beginning, just like the North pole is the top of the world from which all roads point south. If space and time do form a closed surface with no boundary or edge then, as Hawking asks, "what place for a Creator?" God would not need to do anything, because the universe could do all it needed to do as it sprang into being. In his latest book, *The Grand Design*, Hawking theorizes that quantum fluctuations caused the universe to come into existence without the need for a creator. As we will see in Chapters 23 through 28, this idea may well be very close to the truth, albeit for reasons Hawking didn't exactly have in mind. As he himself asked in *A Brief History of Time*, "what breathes fire into the equations and makes a universe for the laws and math to govern?" His

---

If we ask, for instance, whether the position of the electron remains the same, we must say 'no;' if we ask whether the electron's position changes with time, we must say 'no;' if we ask whether the electron is at rest, we must say 'no;' if we ask whether it is motion, we must say 'no.'
Robert Oppenheimer

answer to his own question is somewhat less than satisfactory for anyone who wants to believe in God.

Loop quantum gravity is a new theory in the making that eliminates the infinite density of the Big Bang and opens the door for an oscillating universe that expands, contracts, and starts over. This concept is very much in line with Buddhist philosophy, as we will see in Chapter 25.

Stenger has said that observing a mass density in the universe that is different from that required for the observed effects of a beginning with 0 net energy would validate the God hypothesis. Any cosmologist will tell you that we are unable to see far enough back into the universe to know for sure whether the total energy in the universe is 0 or not. We are also not sure what effect dark matter and energy might have on this calculation, if any. We are therefore unable to say for sure whether the universe has 0 sum energy or not, especially since "empty" space seems to be teeming with energy (Chapter 23). Furthermore, I would argue that a universe with 0 net energy would provide a powerful argument for idealism.

Professor William Tiller speculates that the universe may have formed from a subtle energy field that became dense and material, as if God had created a template or pattern that guided the formation of the universe. This view is compatible with biocentrism.

Philosopher and theologian Emmanuel Swedenborg (1688-1772) believed that the universe was created by two streams, one spiritual and one material, a view that is compatible with holographic theories of the universe. (See Chapter 27.)

> *The most important scientific revolutions include, as their only common feature, the dethronement of human arrogance from one pedestal after another of previous convictions about our centrality in the cosmos.*
> Stephen Jay Gould

Multiverse

The anthropic principle makes the presence of sentient life consistent with the idea of God, but certainly does not prove that God exists. Everett's many worlds theory (that the universe splits into copies that encompass all possible outcomes of every quantum event) was put forth in an attempt to remove consciousness from its seemingly starring role. It makes the fact that we live in a life-friendly universe no more amazing that knowing someone, somewhere, will win the lottery despite the long odds against any one ticket being a winner. Dennett believes that the multiverse idea is just as good as

> *It is well to remember that the entire universe, with one trifling exception, is composed of others.*
> John Andrew Holmes

the God idea. This must mean that the God idea is just as good as the multiverse, and yet the scientific community receives the latter far more warmly than the former.

Other ideas include physicist Lee Smolin's (1955-) speculation that black holes could be the birthplaces for new "daughter" universes that are slightly "mutated" versions of their "parent." It is also possible that the universe oscillates from a Big Bang that expands, slows, and eventually collapses into a Big Crunch for the cycle to start over. In this scenario, the laws of nature could be different every time, but this does not avoid infinite regression. Also, new models of our universe suggest that it will keep expanding forever (which does not preclude prior oscillations). There are problems with both theories, not least of which is the fact that so many universes seems wasteful, to the point where we may as well speculate that God is real. I should stress that there may be multiple universes even if Everett's theory of splitting universes is wrong. If true, this neither violates nor confirms materialism, nor does it really say anything about the existence of God.

Some say the world will end in fire. Some say in ice. From what I've tasted of desire, I hold with those who favor fire.
Robert Frost

In fact, Everett's theory leads to some startling conclusions. A universe that splits whenever a quantum event occurs into copies where each possible outcome has occurred, which are then copied whenever quantum events occur, and so forth would lead to an almost infinite number of universes. Many of these universes would result in your never being born; on the other hand, many of them would lead to your living forever. Whether the "you" who is reading this is slated for eternal life is open to debate, especially since you have probably split unto unknowable millions of copies that are themselves splitting in the time it took for you to read this paragraph.

Consciousness

The knower and the known are one. Simple people imagine that they should see god, as if he stood there and they here. This is not so. God and I, we are one in knowledge.
Meister Eckhart

Is mind in the brain, or is the brain an antenna? Thanks to science, we have an almost complete picture of a logical universe in which we all formed from the remnants of dead stars; however, science must apply to consciousness as well as matter, and consciousness presents a huge challenge to materialists. We will explore the brain and consciousness more in Chapters 29 through 32, but the question bears some exploration here: How can something immaterial like consciousness come from something like unconscious matter? This is a problem of the same magnitude as the one of life evolving from non-life. Materialism consistently fails to explain consciousness, which is solid evidence that it is on the wrong track. Instead of trying to explain consciousness in material terms, why not expand the paradigm to include it as a fundamental part of the universe?

If consciousness is material, then it must occur somewhere in the body, and the nervous system is as good a place as any to begin searching. If consciousness is always present, then how does it relate to the nervous system? The latter may act as an amplifier or focusing lens that increases the richness of our experiences. If our brains are simply an antenna for consciousness, then smashing a radio does not shut off the signal and the same should hold true for a brain. It is telling that NDEs only occur in people who are in imminent danger of dying, but not in people who are only temporarily knocked out or otherwise disabled. It is entirely possible that people with various disabilities and diseases are perfectly OK but are unable to clearly get sensory data in and/or communications out such, as the example of Rom Houben, who was in a "waking coma" for 23 years. During this time, she had the full use of her senses and intellect, but could not move or communicate with the outside world.

Our brains create the convincing illusion of sensations all over our bodies, but all of the heat, cold, pleasure, pain, etc. we perceive is happening in our heads. There may be a reality out there, but all we can ever know of it is assembled inside our heads and filtered through our beliefs, all before we are ever aware of having perceived something. But this only explains part of the process. There is nothing about matter, raw information, or the structure of the brain that indicates that consciousness can or must exist.

Neuroscientists perform the same three basic tests as TV repairmen:

- **Recording:** One can record TV signals using measuring devices.
- **Stimulation:** Energize various circuits to see how the TV behaves.
- **Ablation:** Remove components and see how the TV responds.

We know mind plays a big role in our own lives. It's likely, in fact, that mind has a big role in the way the whole universe functions. If you like you call it God. It all makes sense.
Freeman Dyson

In those respects in which the soul is unlike God, it is also unlike itself.
St. Bernard

> *My me is God, nor do I recognize any other me except my god himself.*
> Catherine of Genoa

None of these tests mean that the signals originate in the TV. Likewise, none of this information tells us anything about whether the brain creates or simply passes on information. The brain is involved with consciousness, which does not mean that it is the source of consciousness. Similarly, we cannot measure spirituality, but we do know—or at least sense—that it is real. It should not take a genius to see that God could easily divide into a bunch of little gods, one per brain.

Why does processing electrical impulses give rise to consciousness and not simply a complex robot? How does something that initially began as a transparent and odorless gas become self-aware? It is easier to show how hydrogen evolved into humans than it is to explain the inner experience of self-awareness. Consciousness is not made out of matter, and we must also presume that matter is not conscious.

As I mentioned in Chapter 9, there are about 100 billion galaxies in the universe, each with about 100 billion stars; there are also 100 billion cells in our brains. Visual signals from the left side of the retinas goes to the left side of the brain, while signals from the right sides of the retinas go to the right side of the brain. The amygdala does not process visual information, but can react to it before we are consciously aware of what we are seeing. People with intact eyes and amygdalas can still react to visual stimuli, even though they cannot see what they are reacting to. It is apparent that "I" actually consists of three separate but linked components: subconscious "I," emotional "I," and logical "I." All three continue growing and maturing throughout our entire lives, because our brain is continually adapting itself to life and experiences.

> *Nothing is real unless we look at it, and it ceases to be real as soon as we stop looking.*
> John Gribbin

There is another challenge: Consciousness is not limited to humans. Apes and monkeys are conscious, as are dogs. So are other mammals. Birds are also conscious. Why not other vertebrates and possibly even some invertebrates? The type of consciousness experienced by one species is not the same as that of another species or of humans. There must also be differences between individuals of a species, just as there are between humans. Plants and even single-celled organisms are aware of and sensitive to their environments; even amoebas exhibit various specific behaviors. Who can say that they lack thoughts and feelings? Can we say that the subjective experience of any animal is no less rich, joyful, and mysterious to other species in their own ways as ours is for us? Where do we draw the line below which consciousness does not exist? It is even possible for consciousness to extend all the way down to subatomic particles. The central problem with the idea that brains secrete thought and emotion the way sweat glands secrete sweat or breasts secrete milk, is that this model makes consciousness both the subject and the object, which makes explaining consciousness difficult at best.

Descartes' idea of dualism, mind and matter, is easier to explain, even if dualism has since been debunked. Theories such as the "zero bound" universe are trying to duck the question of why the universe came into existence. Such efforts also avoid the question of how probability waves collapse into actualities throughout the universe. We know we need consciousness to do this (see Chapter 23), so how did it happen? Did we collapse the universe into its present state going all the way back to the beginning? Remember that time and space are linked. Time itself may be illusory, which we will discuss more in Chapter 28. Thus, we could literally manifest the universe going back in time to the very beginning. Without us, the universe may not exist at all. We are certainly here because of the universe, and the universe could in turn be here because of us.

Materialists say we are a cosmic accident, and dissenting physicists like Stenger dismiss the idea of conscious collapse of waveforms. To them, EEG readouts are the last word on where consciousness resides. Any information that does not fit the mold, such as paranormality, mysticism, and even some interpretations of their own experiments, are quickly explained away as "supernatural."

Quantum mechanics suggests that the observer affects the outcome. A medical patient's state of mind can affect how—and if—their body heals. How can this be explained? How can subjective personal experience be explained? *Panpsychism* is the idea that everything is conscious. *Panexperientialism* is the idea that everything has experience while avoiding the question of soul. In general, both ideas adopt the stance that experience can only come from that which has experience. If consciousness is indeed universal, then it is not a product of human or overall biological evolution; forms of consciousness evolved, but not consciousness itself.

With quantum mechanics, the paradox of free will is no longer benign.
Bruce Rosenblum

Observations not only disturb what has to be measured, they produce it. We compel [the electron] to assume a definite position. We ourselves produce the results of measurement.
Pascual Jordan

> *The astrolabe of the mysteries of God is love.*
> Jalal-uddin Rumi

If consciousness is universal, then what is the ego that lies at the core of how we experience the world and our inner lives? Some would say that it is the product of genetic and psychological conditioning that comes to dominate our ordinary existence. Not all schools of psychology take the ego seriously. For example, behaviorists don't acknowledge it at all—which does not stop them from taking themselves and their work quite seriously; however, spiritual traditions tend to undermine the ego. Within these traditions, the idea that consciousness exists at a given point in the universe is illusory. Our experiences are simply constructs within consciousness, as is our sense of self. We place ourselves at the center of our personal worlds to give ourselves the illusion that we are inside the world, when the exact opposite is true: The world is in all of us. Thus, we have no location in space or time; space and time are within us. People who look down on their bodies during an NDE think they have left the body, when in fact they were never in the body to begin with.

Penrose's proof that computers cannot process meaning strikes a hard blow to the materialist position that meaning is an evolved adaptive quality of matter. If matter cannot process meaning, then how can it present any meaning-processing capability for nature to select? Acupuncture and *homeopathy* (a medical practice that dilutes medicine to the point where one can expect zero molecules of the medicinal agent, and yet it works) also present serious challenges for materialists. The placebo effect is also very real: My son Logan once had a hard time falling asleep. I gave him a fish oil capsule and told him it was a pill that would make him fall asleep immediately and wake up refreshed and ready to go. Logan was out like a light about a minute after taking the pill and woke up the next morning bright and cheery as can be, a shining example of the placebo effect in action. Materialism cannot explain this.

> *The self is God. I am is God. if God be apart from the self, he must be a selfless God, which is absurd.*
> Sri Ramana Bhagavan Maharshi

If our minds are not in our brains, then how are the two linked? The brain could store memories in much the same way that computer circuits store memory, which is then accessed by consciousness. This opens up room for God to act. If mind and brain both consist of probability waves with mind being meaning and brain being memory, then God could be the mediator between them, collapsing the probability waves of both sides to experience the mental meaning while also gaining memory. This is not as dualistic as it may sound, as we will see in Chapter 31.

Our waking reality is as much a mental construct as our dreams; in fact, our brains have no built-in switch to tell us which is which. Our inner realities, waking and dreaming, are all we will ever know; dreams are in many ways as real—if not more real—than the waking world. Our awareness of the world lags about 1/5 of a second behind actual events, but our brain compensates for this to make us think we are witnessing and experiencing things as they happen, and thus interacting directly with the world.

Photons have no mass, and are therefore not part of the material world. This is not to suggest that light is God, but that light may be the first manifestation of the Godhead, a concept that both Genesis 1 and cosmology would agree with. The same is true of immaterial consciousness. Thus, we could argue that light and consciousness are both fundamental to the universe. The Buddhist concept of samadhi is pure consciousness without content. A mind devoid of content is absolutely serene, at peace, and able to discover the true nature of self. Looking for the self is said to be akin to being in a dark room with a flashlight trying to find the source of the light. We are not something or something else; we simply are. The essence of self is pure consciousness. You and I are not conscious beings; we are consciousness, period. Our core identities are eternal and timeless, with no uniqueness or sense of an individual self.

Mystics who have probed these depths to discover their true nature have returned saying, "I am God," and now we can explain why: God is not a separate being who loves and/or judges us as the JCI religions attests, but an immanent being who resides in each one of us as the most intimate and undeniable part of ourselves, the light in every mind. God and consciousness are universal. God is said to be the source of creation. Consciousness creates our entire world and is thus the source and creator of all we know. As previously said, the light of this consciousness is eternal.

If this model is true, then maybe death is not the end. Maybe the essence of mind and our core identities consists of quantum possibilities, and maybe this could solve the problems of

By limiting the infinitely possible, you create the finitely real.
Bernard Haisch

God only comes to those who ask him to come; and he cannot refuse to come to those who implore him long, often, and ardently.
Simone Weil

> *The more God is in all things, the more He is outside them.*
> Meister Eckhart

both seeming dualism and the survival of physical death. Again, we will explore these concepts in more detail later; meanwhile, we have emerged from the shadow of religion to begin postulating a God that is scientifically possible and that could also point us in the direction of answering the question of whether anything of us survives bodily death.

Individual consciousness and matter may be partial realities that reflect the underlying Ground/Godhead. We may be able to discover this by direct intuition or experience that unites known and known, a possibility that is perfectly compatible with reports from mystics who claim to be one with the Godhead while remaining aware of what is happening because a united consciousness is still conscious. In this case, our outer selves are our phenomenal selves, while our inner selves are our true selves. We are normally only aware of our outer selves. One of the primary goals of life is to learn to identify with our true selves, a goal that religions refer to as salvation, enlightenment, etc. One must assume that at least some of the differences are more superficial than they appear, because of the many different ways in which one can report an experience. Analyzing the sayings of Christ and Buddha reveals many similarities. This is not a farfetched supposition.

Author Evelyn Underhill (1875-1941) wrote that mystical consciousness puzzled materialists, who opted to dismiss the whole concept as illusion. Her writings speak of a point where a person's individual nature touches reality, which forms the basis for mystical claims that God is real and that reuniting with God is possible. Neuroscience can neither prove nor disprove God, but it has repeatedly proven that mystical experiences are real to the extent that the mystics are experiencing altered states of consciousness and honestly believe that they are connecting with the Godhead. This alone must make God's existence at least plausible.

> *God, as viewed through religion, has personal tastes and preferences, like enjoying a good cup of tea but strongly discouraging the wanton consumption of coffee.*
> Stephen D. Unwin

We are not God in the quantum mechanical sense, and yet we may well be God if idealism is correct, a paradox that may manifest itself in every inspiration and creative spark that reveals God very actively engaged in creation through all sentient beings on Earth, and possibly beyond. Einstein did not believe in a personal God, but still had great faith in the underlying order of the universe, a faith that was strong enough to make him sidestep evidence of the Big Bang that he would later refer to as the biggest blunder of his life. Nevertheless, the theory of relativity alone proves that there is no such thing as objective reality; the physical world has a fundamental dependence on observers, which implies that God is indeed personal. Quantum mechanics only reinforces the dependence on observers.

It is impossible to know how any of these theories will fare in the light of future discoveries; however, science must take consciousness into account one way or another. Conscious observation seems to affect the potentialities of matter and energy, whether we like it or not Sweeping that fact under the rug will not make it go away. In fact, adding consciousness to the equation could be the key to obtaining the elusive Theory of Everything. Our individual and collective wills seem to span minds and affect physical phenomena; the links between consciousness and quantum mechanics will remain. If the history of science has taught us anything, it is that each new round of discoveries reveals a universe that is far more intricate and marvelous than anything we had previously imagined.

Our individual minds govern our brains and our bodies. Our individual wills control our brains and influence the entire universe. Collectively, our combined wills have some God-like traits. We cannot say for sure what happens to consciousness when we die, but we can say that our will seems to transcend our bodies, and that survival is therefore a distinct possibility.

> *The journey to my discovery of the divine has thus far been a pilgrimage of reason. I have followed the argument where it has led me. And it has led me to accept the existence of a self-existent, immutable, immaterial, omnipotent, and omniscient Being.*
> Anthony Flew

Finding God

The idea is that life is completely devoid of meaning is one of the worst ideas imaginable, against which wishful thinking and blind faith in God can only go so far. The bottom line is that people need proof. We will all get this proof when we die, but it would be lovely to have it while still living, if for no other reason than to end the uncertainty.

Scientists often say that questions about God are beyond their scope, but this is absolutely not true. If God exists, then science must be able to construct the question and get an answer; however, science has largely won the day by ignoring God altogether, in favor of materialism that has given the illusion that matter is king. It is true that ancient stone gods did

> *A great many people think they are thinking when they are merely rearranging their prejudices.*
> William James

nothing and were eventually abandoned, that the God of the Jews abandoned them more than once, and that science has advanced the state human knowledge to where Yahweh, Jehovah, and Allah are little more than fairy tales along with Zeus, Mithra, Odin, and countless others.

We have all about us the fingerprints of God.
Barbara Bradley Hagerty

Religion used to be the inquisitor. Today, science is on the attack, disproving religious concepts and exposing the harmful principles at the core of organized religion. Authors from Sam Harris and Hitchens to Dawkins, Dennett, and others are on the offensive, excoriating religious behavior and disproving the religious God. Religion itself has become too focused on its own differences to remember their common mythological and mystical origins that are remarkably uniform all over the world. Religion had—and lost—the chance to offer different, culturally appropriate expressions of the divine while acknowledging that these expressions are not Truth, but simply aspects of truth. The artificial differences between the religions—and the entrenched notion that religion deserves elevated tolerance and respect—have quite literally brought our world to the edge of the abyss. If there is a God, we will not find Him through religion.

Materialism eliminates the religious God because of the fundamental problem posed by dualism of how a non-material God can downwardly cause changes in the material world. Such interactions would require energy, but no such energy transfers have been detected. Energy and matter are interchangeable thanks to $e=mc^2$, but the total amount of matter and energy in the universe must remain constant, in accordance with the First Law of Thermodynamics. To materialists, the feeling of God comes from the brain, and indeed some experiments have replicated religious/spiritual experiences by stimulating the brain. God thus becomes an imaginary nanny one can turn to for succor. As previously mentioned, this is a very classical view; however, the quantum world exists *in potentia* (as potential) that God could theoretically actualize by collapsing the probability wave—an example of the observer effect on steroids, so to speak. Experiments performed by physicist Alain Aspect (1947-) may well have disproved materialism for good.

There are compelling psychodynamic explanations for a person's belief in God. These, however, say nothing about whether or not God exists.
Glen Gabbard

Materialists make the *ontological* (metaphysical or "first principal") argument that matter is the bedrock of existence and that everything in the universe can be boiled down to material building blocks. They may have it backward, because some interpretations of quantum mechanics demands a science based on consciousness, with matter being the epiphenomenon. As previously noted, this approach includes and contains materialism in the same way that each advance in physics includes and contains its predecessor; there are shades of Newton in superstring and brane theory. If this is correct, then we may be able to scientifically prove the existence of God and may already have done so. Such a God would have all the attributes needed to be a causative agent, in keeping with all of the laws of nature. This sounds farfetched, but then otherwise intractable problems often require farfetched solutions.

I believe in God because life would be too sad without Him.
Anonymous

We have assumed that time and space are real for thousands of years, long enough for materialism to make a credible claim for having ruled out God. We have looked high and low for God, have detected nothing, and the universe seems to be humming along just fine without Him. The only problem is that we have been looking outward when perhaps we should have been looking inward, as the Buddhists (and mystics all around the world) have been doing for thousands of years. One can only imagine what we might learn if we explored the mind as deeply as we have explored materialism.

Under this model of God, we need not achieve to be at peace; we must instead stop doing, wanting, and otherwise obscuring our inner peace. Then and only then can we find what we have been looking for waiting for us with infinite patience. It is no coincidence that the Greek word for forgiveness is *aphesis*, which means to let go. Our prayers are to the divine presence inside all of us, not to some external entity, and the results are impressive as our fears, judgments, prejudices, biases, desires, etc. simply melt away. This may sound like a vision of heaven on Earth—and that's exactly what it is.

Free Will

Dyson says that there is no clear distinction between mind and God, a view that is very compatible with idealism. Has every single thought, feeling, and action been determined in advance? If so, then life is nothing but automatons proceeding along precisely programmed paths with orders of magnitude less freedom than the average industrial robot. On the other

He who thinks that god has any quality and is not the one, injures not God, but himself.
Philo

hand, the reality of free will could be construed as limiting God's power to the extent that we do have free will. Free will is also limited to the extent of cause and effect; for example, we seem to be unable to pick our parents or fetus, although some would disagree with that.

> *Oh God, how does it happen in this poor old world that thou art so great and yet nobody finds thee, that thou callest so loudly and nobody hears thee, that thou art so near and nobody feels thee, that thou givest thyself to everybody and nobody knows thy name?*
> Hans Denik

God may represent the collective will of all beings in the universe with the potential for any knowledge and the ability to make any event occur while being bound by the limits of our own minds. This could explain *psi* (the paranormal) and the studies that show the efficacy of prayer that combines small amounts of will from everyone involved in the prayer. This could also explain the seemingly triangular relationship between mind, brain, and quantum mechanics. The Christian and Muslim trinity of Father, Son, and Holy Ghost could also be explained as the trinity of God, humanity/matter, and consciousness that manifests and drives the universe according to idealist theories.

A universe that solely consists of matter can have neither consciousness nor will. Materialism simply does not work. Einstein did not see the evidence for mind in physics, and thus missed the profound implications that has for our vision of reality. Materialists continue to fail to see anywhere for God to set up shop, or to fathom how consciousness can arise from—and hide within—matter. Idealism is the key that lets us see beyond this.

Spinoza pointed out that people are aware of their own actions and blindly assume that they are the cause of those actions, because they are ignorant to the larger truth. Is this larger truth God's will? If so, where does this leave free will? Psychologist B. F. Skinner (1904-1990) said that death was the end of free will; however, he believed that people were little more than robots to begin with.

There is no such thing as a free lunch, and there is no such thing as unrestricted free will; there is also no reason to think that free will must be incompatible with the idea of an omniscient God, as we will see in Chapter 14.

> *Consciousness is a singular for which there is no plural.*
> Erwin Shrödinger

The Case for Idealism

Mainstream science says that reality is ultimately material, and that God is an illusion. Mystics say that reality is ultimately

God and that material is an illusion. The JCI religions say that both are separate but equal realities, a viewpoint that conjures visions of segregation in the deep South.

Materialists believe that consciousness consists of electro-chemical reactions, period. Emergent materialism allows consciousness to be an emergent phenomenon, albeit one whose existence is most probably tied to the physical body.

Dualists believe that matter is secondary; our brains process information, but our intelligence and consciousness reside elsewhere. Our souls descend from the spirit world/dimension to experience transitory physical existences, but mind ultimately resides with God and returns there when the physical body dies.

Idealists believe that God interacts with the universe because He is not separate, but both transcendent and immanent. There may in fact be only one consciousness that splits into subject and object (as we do when we observe ourselves), such that everything is in god, but God is not necessarily in everything because of His transcendent aspect. Idealism is just as asymmetrical as materialism in the opposite direction, believing that consciousness is the true substance of the universe, with matter being almost trivial by comparison.

Materialism stripped us of our humanity by painting us as deterministic wind-up robots. Existentialist philosophy goes hand in hand with this by saying that life is devoid of all meaning but that which we give it. We exist, so we may as well play along by pretending that meaning exists; otherwise, the universe is pointless. The discovery of quantum nonlocality and Bell's elimination of local hidden variables opens the door for dualism and idealism. Thermodynamics eliminated dualism, leaving idealism standing as a potentially viable model of the universe that is not just another "God of the gaps" argument but a new theory that propounds a new hypothesis of God. An idealist would modify Descartes' famous *cogito ergo sum* (I think, therefore I am) to *eligo ergo sum* (I choose, therefore I am), which recognizes the full creative power of consciousness to create contexts in which to manifest a material reality.

To idealists, matter and mind both reside inside us, but the macro/micro distinction behaves like a Newtonian object, which gives the illusion of a shared objective reality that led to

What if everything is an illusion and nothing exists? In that case, I definitely overpaid for my carpet.
Woody Allen

The physical world is a creation of the observer.
Deepak Chopra

materialism. Materialists cannot distinguish between inner and outer awareness, denigrating the former while at the same time admitting that it serves many adaptive purposes. Author Ken Wilber (1949-) believes that quantum monistic idealism resolves both the mind/body problem and the internal/external split, because all of these distinctions are only illusory.

It is ignorance that causes us to identify with the body, the ego, the senses, or anything that is not the Atman.
Shankara

Eastern religions believe that the cycle of birth and rebirth continues until one unites with God. We cannot change the world, only ourselves. The Western belief in a single life with eternal repercussions undermines self-realization and self-transformation, because the whole emphasis is on getting into heaven and enjoying the afterlife. This life is too short, so we do all we can to improve our lot. Dualist philosophers have agonized over why a perfect God would create an imperfect world when He could have remained perfect. To idealists, God became immanent to manifest possibilities; the world begins imperfectly, but consciousness evolves toward perfection. Various mechanisms obscure our true natures and lead to the sense of separation called ego that dominates our waking hours. Conditioning, and the resulting formation of core beliefs add patterns to our responses and strengthens our attachment to our egos, making choices more and more predictable—and behaviorism more accurate—over time. Experiments have proven that activity in the brain increases 1/2 second before we are aware of our decision or action. Most of us fail to ever grasp that our sense of separation is an illusion that comes from collapsing the quantum probability wave.

Unconscious—or, more precisely, unaware—processing does not collapse probabilities, which can therefore interact in our unaware minds. The awareness of subject/object happens contemporaneously with a conscious wave collapse. Consciousness is ever-present, and the chain of possibilities it can collapse can extend arbitrarily far back in time, even to the dawn of the universe itself.

All things are one.
Heraclitus

God acts within the system because the system is God, including finite entities. There is no place outside God, because all is created within the God-self. This incorporation of individual systems within Himself allows interaction with the whole and its constituent parts. Physicist John Wheeler (1911-1980) saw this as completing the meaning circuit and thus removing the "Who created the Creator?" argument. Life can thus be seen as having three components: God's downward causation, tangled hierarchies that allow self-referential wave collapse, and a "vital blueprint" that limits possibilities within the system to the known (and yet to be discovered) laws of nature.

Newton's laws are deterministic to the extent that we know the starting conditions; however, quantum uncertainty means that we can never know the starting conditions with the infinite accuracy needed to predict the future with absolute certainty. Unpredictability becomes almost certain, because of accumulated quantum effects. In other words, the equations themselves are deterministic, but we cannot ever know them beyond a certain degree of precision. The universe seems less and less predictable, while operating within a framework that includes openness and flexibility within the natural laws.

Ask not what's inside your head but what your head is inside of.
William Mace

Our inner experiences can point to the reality of God's existence. If they are just as causally efficient as our outer lives, then science must either explain it or lose relevance. Materialism cannot account for the difference between our inner and outer experiences. Dualism cannot hold in the face of thermodynamics. Idealism is the only remaining choice. In the idealist model we are discussing, the soul can only develop after the ego. Its initial identity is with the physical body, but performs unconscious mental programming with neither conditioning nor ego, but always within God-consciousness (the one consciousness pervading and creating the universe). Young children live in this state before developing their egos; in fact, Hindus regard children as gods until age 5. The developing ego makes it harder and harder to reach this soul as conditioning slowly becomes dominant.

Idealism seems to open the door for God and for a scientifically acceptable system for personal existence beyond this lifetime. This is pretty heady stuff, but we must never lose sight of practical matters. As the spiritual adage says, even a Zen master has to go to the bathroom.

What of the Soul?

Mystics will agree that our dreams are unreal; but so is waking. Dreams are created by the individual, while waking could be said to be God dreaming within us. To them, realizing this shows that waking and dreaming are simply different states of

I am the eye by which the universe beholds itself and knows it is divine.
Percy Shelley

consciousness with different values, which shifts our perspective to the God-consciousness and frees us from worldly boundaries. In waking life, everyone we see is also ourselves. Lucid dreams in which we are aware that we are dreaming back this up, as we guide our dreams to reveal solutions to problems occurring in waking life. There are also telepathic dreams, precognitive dreams, etc.

Those wise ones who see that the consciousness within themselves is the same consciousness within all conscious beings, attain world peace.
Katha Upanishad

Reincarnation or some other type of afterlife could be possible by postulating the existence of a quantum monad that could move from body to body in the same way than an actor moves from role to role. In this model, Anthony Hernandez is a role being played by a monad; when Anthony Hernandez dies, the monad will simply assume a new outward role and carry on. The question is whether any memory of Anthony Hernandez will carry on in the same way that an actor remembers her or his past roles that also shape her or his future career. If anything and everything Anthony Hernandez is lost forever upon my bodily death—if the actor gets total amnesia between roles—then any discussion of immortality or continued existence is, for all of our current identities, moot.

Thankfully, quantum memory is not restricted to a single lifetime in this model. A monad may go through many different bodies while carrying its memories and experiences with it, with subsequent bodies being a reincarnation of that same monad. Past lives are correlated with each other, and we can retrieve memories from them. Dr. Ian Stevenson (1918-2007) and others have collected thousands of substantiated stories that could provide conclusive evidence of past lives. There are alternative explanations for Dr. Stevenson's data that we will look at in Chapter 17. The Dalai Lama believes reincarnation to be of supreme importance, because that is the only way to achieve union with God, something that is impossible to do in a single lifetime. This too bears further examination; for now, however, the Dalai Lama and Buddhism seem to be on the right track.

The universe is immaterial—mental and spiritual. Live, and enjoy.
Richard Conn Henry

What about NDE accounts of dead relatives greeting the newly deceased? Monads may not necessarily reincarnate immediately. There is plenty of data to indicate that some interval may pass between incarnations, including NDE, channeling, and more. Some *mediums* (people who communicate with and/or channel discarnate monads) experience dramatic personality changes while channeling.

How much of our current "I" survives death? Buddhists believe that karma is all that moves from life to life, each monad bringing its accumulated karma to the next life, the karma of which gets added to one's current and future karma. This model is debatable, and this is not the place for that debate; however, it is important to stress that Buddhism is not nihilism. Emptiness is nothingness, but not no-thing-ness; it is infinite potential.

The Possibility of God

Including all levels of hierarchical complexity allows us to infer the presence of God. Each tier in the hierarchy is both self sufficient, and yet dependent on the level above it. I had this thought while standing on the deck of a cruise ship in Mexico at night gazing out over the ocean under the glow of a full Moon. Another cruise ship passed in the distance, and it occurred to me that each ship was a quasi-autonomous cocoon of life that provided everything needed for a comfortable life. Each ship in turn depended on its home port for periodic resupply, which in turn depended on a myriad of vendors, which eventually depended on farms, mines, and other natural resources, which depend on the Earth, which depends on the solar system, and so on up to the entire universe. This same concept applied to every person on the ship, as well as their organs and tissues, cells, cell structures, molecules, atoms, and particles. It occurred to me that any break in these circles of greater dependence might well be arbitrary. We don't know how far down the layers of subatomic particles go. Current theories include the Higgs boson or superstrings. We also don't know how far up the layers go. It is therefore impossible to exclude the possibility that God exists.

The fine-tuning of the physical constants is just what one would expect if life and consciousness were among the goals of a rational and purposeful God.
Ian G. Barbour

Someone once asked Einstein if he believed in God, to which Einstein replied, "Define God and then I will answer." That was a brilliant answer, for reasons I am explaining in this book; however, it is also moot on a certain level. Whether or not God exists, people tend to cling to ideas or dreams of future happiness in heaven, a new house, a new job, etc. The

If he existed and chose to reveal it, God himself could clinch the argument, noisily and unequivocally, in his favor.
Richard Dawkins

extent to which people are living for the future is the extent to which they are not completely happy right now.

If God is omniscient and omnipotent, you can't help wondering why She doesn't pull out a thunderbolt and strike down Richard Dawkins
Nicholas D. Kristof

Any God that exists must be both transcendent and immanent; however, all religions are immanent, which means that the best any religion can hope to do is create an approximate model of God. So far, it seems like the Buddhist model may offer the closest compatibility to modern scientific theories. I am not suggesting that the Buddhist model is correct, nor have we proven the existence of God or any form of existence beyond this lifetime; we have, however, touched on the limits of materialism and debunked dualism, leaving idealism as a viable path forward. Idealism postulates that consciousness is the Ground of being, which by definition postulates that which we would call God, for lack of a better term. We also demonstrated that idealism opens the door for existence beyond this lifetime. The next step is to discuss the potential nature of God and construct a hypothesis that we can examine throughout the remainder of the book.

422 | *The Divine Savage*
Revealing the Miracle of Being

Chapter 14

On the Nature of God

God wants you to laugh. God has a sense of humor and if you don't believe me go to Wal-Mart tomorrow and just look at people.
Carlos Mencia

If we are simply physical beings in a physical universe, then there is no ultimate purpose in life. The presence of a god could be a source of meaning, except that it says nothing about whether or not this life is all we get. Materialist scientists like Dawkins say that belief in evolution requires one to be an atheist, but we just saw in Chapter 13 that this is not necessarily the case at all. In fact, the only real difference between science and theology is the question of whether metaphysical creation has a personal interest in what it created. Piecing together different metaphysical views demonstrates that the difference between believers and nonbelievers may be smaller than anyone cares to admit. The only real difference is whether there is a creator or not.

If God lived on the earth, people would break his windows.
Yiddish proverb

Some eternal non-thing may pervade the universe. Esoteric traditions around the world agree that God is, before any beginning and after any end. To the Buddhists this is nirvana; to the Hindus it is Brahman; to Kabbalah it is Ein Sof. Do these traditions hold water? That all depends on whether you believe that all of the accidents that led to our being able to ask such questions were really accidents. Quantum entanglement (Chapter 23) lends credence to the idea of a single wisdom pervading the universe that lies at the heart of all mystical traditions. By contrast, the religious view of God that extrapolates from the physical world is absurd by default.

Life as a mere human seems to be very far removed from God and yet that may be the whole point. If the universe is a physical expression of the metaphysical, then we can study the creator by studying the creator's works; hence, theology without science is fatally weakened. Mysticism blends theology and science, gathering report after report of subjective experiences and correlating them to paint a picture of the world lying just beyond our mundane perceptions.

> *How clear it is! How quiet it is! It must be something eternally existing!*
> Tao Te Ching

The insistence of religious traditions that all other traditions are wrong fails to understand that their differences are only superficial, which makes religious persecution of so-called heresies all the more intolerable. Enlightenment is easier when we have no preconceived ideas about what might be true. How can atoms and molecules produce a being capable of asking such questions or of appreciating beauty? The total oneness we all experience within ourselves is as close as most of us ever come within this lifetime to knowing the One, the Godhead/Ground of reality. We settle for a kindly and slightly senile grandfather figure that dotes on us no matter what we do. Maimonides said that we need a science of God built on the study of nature. He was entirely correct: As we saw in Chapter 3, Genesis begins with an accurate physical science description of the beginning and evolution of the universe.

What Kind of God?

God could be nothing more than a mythical personification of the laws of physics, but that is cold comfort to someone who wants to believe in a truly personal God. Theologian Keith Ward (1938-) asked, "If God is self sufficient then how can it be that He creates a world at all? If God is necessary and immutable then how can He have free choice and not act out of mere necessity? Either God's acts are necessary and not free or free and arbitrary and this has confused most Christian philosophers for centuries."

> *Pleasure is God manifested.*
> Irwin Kula

What if you had it all, but could not use any of it? What if you had billions of dollars in the bank, but could not spend a dime of it? Imagine yourself as a being transcending space and time. What is there for you to do? You might perhaps decide to go from sterile potential to actual creation, to doing and not just being, to living out your fantasies and nightmares, from experiencing all that there is to experience from all possible perspectives. The difference between latent talent and actually doing the thing you are talented at is huge. Making it happen by playing the game is a lot more fun than simply reading the rule book. Designing a car and driving that car are two entirely separate things. You get the idea.

Think about it: how arrogant to assume that we could ruin God's day!
Bernard Haisch

This may be the point of life: to experience the possible. The human genome was mapped six months after the new millennium began, a prime example of humans using our own inventions to figure out how we work and what makes us tick. If consciousness is the origin of matter, then the purpose of matter is for God to experience potential as His ideas become realities. We thus exist to create God's experience, and are incarnations of God in the physical world. Consciousness can be transformed, but can never die; body and mind are tools for experiencing physical existence. This may be what Christ meant when he said that, "My Father and I are one." Christ seems to have espoused a more Eastern philosophy that emphasizes wholeness over the more separate Western view.

If we agree that the universe has reason, and if we decide to label that reason God for convenience, then how can God be responsible for laws of physics and nature? The only way this can be meaningful is if God can select those laws. In other words, God must be rational and omnipotent. Even so, God's power is not absolute, because He is still subject to the constraints of logic; for example, He cannot create a round square. Thus, God would have to be perfect and omniscient in the sense of being aware of all possibilities and knowing which ones to choose. Which begs a question: Is wisdom the basis of all experience? The laws of nature ensured that the universe would evolve as it did, which means that the universe was in a sense pre-programmed. Idealists believe that a single consciousness pervades the universe, which is a manifestation of this wisdom. Of course, nothing prevents God from creating other universes that would strike us as utterly irrational; in fact, as we shall see in Chapter 25, some theories predict the existence of up to 10^{500} universes.

It seems to me a sort of form of hubris to think that God made the universe just for us.
George Smoot

If consciousness underlies the universe, and if one can see into other levels of consciousness beyond one's own normal waking consciousness, then personal and mystical experiences offer us real usable data. Individual thoughts are part of the infinite consciousness. There is nothing mysterious about this idea, any more than there is about sharing one's memories. Thus, the infinite consciousness gets the joy of living out every imaginable sort of experience from the mundane to the amazing. I imagine that this might be akin to the sense of wonder young children display as they explore their worlds.

God is odd. He loves the odd.
Muslim saying

Evil, illness, suffering, etc. are only problematic if we believe that an individual consciousness/monad only lives once. After all, why would an individual consciousness choose to live out its entire life in some dreary corner of the world and call it that? Karmic law—the idea that one's future life is based on your current and past lives—may be analogous to the First Law of Thermodynamics, which essentially says that you don't get something for nothing. If you live a life of suffering this time around, then presumably your next life will be better. If you are causing suffering in this life, then you will probably be suffering during your next life.

If the idea of God that we have been developing in these chapters is correct, then God does not need us to do anything at all for His happiness. He cannot dislike or hate what we do or be angry or disappointed in us, He will never punish us because that would only be punishing Himself, and there is no literal heaven or hell in which to consign us for all eternity. Imagine how society might change if people knew that what went around would come around full force, with no way to cheat one's way out of what's coming. That might motivate better behavior. The life review reported by most NDE survivors (see Chapter 16) reveals the consequences of everything one has done in life and forces one to experience all of the good or bad one has caused to others. There may very well be a law of conservation of information, in which information can be neither created nor destroyed; in fact, a timeless universe would demand such a law. We will explore this more in Chapter 28.

According to multiverse theory, anything that can happen will eventually happen in one universe and/or another. Some of these possibilities would seem outlandish and more than a little surreal to us; however, an observer in another universe would perceive that universe to be perfectly ordinary, with ours being the outlandish one. Multiverse theory therefore does not explain the perceived lawfulness of our own universe

God is well, and so are you.
Unknown

and is of limited use. Applying a little metaphysics makes a multiverse perfectly compatible with everything we are discussing here. For example, if theory predicts 10^{500} universes, then 10^{500} combinations may represent the upper bound of all that is possible and God's limits. Multiverse theory may seem to violate Occam's razor, but this is a matter of taste and not necessarily one of science, because one can believe in both a multiverse and a designer God. In fact, it may be easier to justify belief in God if one presumes a multiverse, because that eliminates the discussion of the improbability of this universe.

Reductionism

The reductionist worldview is chilling and impersonal. It has to be accepted as it is, not because we like it, but because that is the way the world works.
Steven Weinbberg

The ongoing feud between science and religion was perpetrated by religious repression. Religion is responsible for the worst excesses in history, against which science has taken a firm stand. In response to the religiously inspired idea of intelligent design, science says that such a theory could call the designer's intelligence and competence into serious question. Intelligent design may be a fatally flawed theory, and religion may be utterly evil, but none of that precludes the possibility that God is real, albeit not in any religious sense.

Reductionists boil things down to the actions of their constituent parts to claim, for example, that consciousness is just brain chemistry, and we just have to suck it up if we don't like it. What they won't tell you is that their seemingly solid position is just as faith-based as any religion. Those who scoff the loudest at the evidence against materialism are like the people at religious tent revival meetings who work the faithful into a frenzy. The most fervent materialists take an almost inquisitorial delight in waxing poetic about the pain and suffering of evolution and how any good god could allow this.

The origin of life on suitable planets seems written into the chemistry of the universe.
Carl Sagan

Nevertheless, reduction does seem to be the way of the world. Consider a slide projector that contains a white light bulb. Inserting a slide removes different colors in different places to create a coherent scene. Limiting the infinite potential represented by the white bulb creates the finitely real scene. Movies are a fantastic example of how a series of reducing filters can create a virtual world. Even the words on this page are reductions of the infinite potential of the blank white page from which meaning is emerging. This is a good metaphor for a universe in which the physical reality of space-time is manifested when the unlimited Godhead selectively limits itself to create stunning—but still limited—results. The idea that the universe was created from nothing gets turned on its head; the universe was created *ex infinitio* (from infinity). In other words, creation makes something out of everything. The God that does this need not bear any resemblance to the God of religion, especially the JCI religions. Mind over matter or matter over mind? That is the only difference between idealism and materialism; both involve reduction in their own ways.

The astounding abilities of *savants* (people with rare abilities, such as the ability to play a piano concerto after hearing it once and never having taken a piano lesson) are normally linked to brain problems. It is easy to see how mental or physical problems can cause disabilities, but what about abilities? No one has ever heard of a damaged car outperforming its intact counterparts, so how can a brain defect lead to such highly enhanced abilities? We are not discussing sensory problems, such as blindness, that free up brain resources to process other senses more efficiently; we are discussing problems in the being itself that should by rights limit intelligence and/or ability. This may be evidence for connection to an infinite consciousness, as well as for the brain acting as a reducing valve.

Complex open systems can be very sensitive to external influences that make behavior unpredictable, thus conferring a sort of freedom. They can also display law-like behavior in spite of being both indeterminate and subject to random external disturbances. Some general organizing principles seem to be in play that govern complex systems alongside the laws of physics that, although consistent, cannot themselves be reduced to physical laws. God could be both necessary and changed by His own creation that includes an element of freedom, just as people are changed by their own experiences. Mystics are routinely laughed at when they say that the universe is the body of God, but there may be a lot of truth to those assertions. Meanwhile, multiverse theory and higher dimensions postulated by superstring and other theories are accepted without question. The irony is that the evidence for God may be stronger than the evidence for superstrings. God in this case is a supremely benevolent creator, and humans have immortal

God exists since mathematics is consistent and the devil exists since its consistency cannot be proved.
Albert Einstein

I believe in God, only I spell it nature.
Frank Lloyd Wright

spirit forms that evolve through a long succession of temporary bodies.

Huxley suggested that the brain may filter out all but the immediate consensual reality by extracting a drop of reality from the infinite consciousness. The brain thus acts as a reducing valve that releases a small trickle—a concept that is consistent with Buddhist ideas that so-called higher creatures may actually be less conscious than so-called lower creatures. If creation is a process of subtraction, then this model makes perfect sense while also allowing for paranormality, such as ESP. Our minds and thoughts could be filtered through the thoughts and mind of God. Each of us could be a dot of color on the infinitely white light from which all emerged.

Can we prove this? The meditative/mystical state of peaceful awareness of consciousness without the need for referential objects could provide that proof, albeit on an individual scale; collecting and correlating many such reports could provide a key piece of evidence that could form part of a more objective proof. We must explore further.

Myth is neither a lie nor a confession: it is an inflexion.
Roland Barthes

The Limits of Science

It is has been said that science is incapable of finding out why our universe came into being, what the meaning of life is, what happens when we die, etc. I happen to disagree strongly with that statement, which is one of the reasons I wrote this book. Having said that, I must also confess that I do not believe science to be all-powerful or all-capable; far from it. For example, the astonishing medical breakthroughs of the last several decades have brought us no closer to explaining consciousness. We can break it down into components in the same way we might break down a basketball game into the numbers of players, the calories they consumed during the game, number of yards run, number of baskets attempted and made, etc. but that does not even begin to recreate the game itself, because it ignores the larger context of the game and the athletes playing that game. Consciousness is not the only human system to suffer at the hands of this reductionist treatment.

The evolution of the universe is effectively the change in distribution of matter through time-moving from a virtual homogeneity in the early universe to a very lumpy universe today.
George Smoot

Similarly, science can acknowledge the extreme fine tuning of the universe, but not design; the idea of a purposeful universe is rejected *a priori* by resorting to the multiverse, many worlds, or something, anything else to avoid the problem altogether. Besides, as previously noted, multiple universes open the door for immortality.

Theories and facts can be updated or scrapped in light of new evidence. Should I become aware of any compelling evidence that calls any part of this book into question, I will update the book. Dogma does no such thing, because it is impervious to evidence. The excuse that science is concerned with nature while God is beyond nature is just that: an excuse. If God exists then He exerts some influence on the universe. The fact that we have not yet detected Him means either that we have not yet looked in the right place, or that the evidence has been sitting under our noses this whole time.

The Illusion of Matter

All of the matter present in the universe today was present in the Big Bang and part of the primordial entity that fueled the explosion. Or was it? Nature plays the ultimate illusion on our minds and brains. We see and feel solids, but there are no such things. What we see as solid appears that way because our eyes have nowhere near the resolution required to see just how empty even the most solid matter really is. In fact, solidity is nothing but a force that gives the illusion of being solid. If an atom were the size of the Empire State Building, the nucleus would be the size of a single grain of sand. A single fertilized cell contains all of the physical potential any life form will ever become, but even this has less than 0.0001% solid matter.

Don't think for a minute that physicists fail to understand the rose petal argument for God. Perhaps evil does play an important role in allowing free will, or perhaps the divine rationale for evil is beyond our comprehension.
Stephen D. Unwin

It gets even stranger: All electromagnetic waves from radio waves to gamma rays are different frequencies of light. All matter is composed of energy, which means that the force fields of science fiction may very well exist all around us. It may even be composed of raw information and a quantum wave function that gives the appearance of solidity. Maybe there is nothing (as in "no thing") to existence at all.

Mark well how varied are the aspects of the immovable one, and know that the first reality is immovable. Only when this reality is attained is the true working of Suchness understood.
Hui Neng

In quantum field theory, the vacuum state (also called the *Zero Point Field* or ZPF) is the quantum state with the lowest possible energy that generally contains no particles. The coldest, deepest vacuum of the emptiest regions of space is not empty; the entire universe contains electromagnetic waves and particles that pop into and out of existence. All other energy in the universe is above this lowest possible energy. There is no such thing as a sheer void. Physicist Alfonso Rueda derived the classical equation $f=ma$, where mass is an illusion. According to this derivation, matter resists acceleration (an object at rest tends to stay at rest) not because it has mass, but because the ZPF exerts force whenever acceleration takes place. This background sea of light is what gives matter the appearance of solidity and stability. If the light of the ZPF is what propping up matter and physical reality, then how does the universe appear to that light? Einstein demonstrated that an object moving at the speed of light sees all of space shrink to a single point and all of time shrink to a single instant. To light, there is only the here and now; there is no space or time to move through.

There is no physical criterion for [quantum waveform] collapse that seems remotely acceptable.
David J. Chalmers

Radio signals are light with very long wavelengths. Imagine sending out a radio wave at light speed and then following behind at half light speed. You might expect to see the waves retreating from you at 1/2 of light speed, but that is not the case; you will still see the radio waves moving away from you at light speed. Linking space and time into a single four-dimensional matrix solves this seeming paradox, because space and time are indistinct. There is no such thing as absolute space or absolute time to sustain the light. It is the light that is the foundation of reality; its propagation determines the flow of time and the measure of distance. Light could even be said to create space-time. If so, then, "Let there be light!" could be more than just a myth, and primitive Sun worship may have been very much on the right track. This is no mere speculation: Einstein said that the propagation of light defines the properties of both space and time. The ZPF inertia hypothesis implies that light creates matter.

There are no coincidences. They are miracles for which God doesn't want to take credit.
Unknown

Sound is logarithmic. A whisper, normal conversation, wailing child, rock music, and jet engine are each about 1,000 times louder than their quieter predecessor in this list. Light is also logarithmic. We happen to perceive these logarithmic changes as linear. If time is logarithmic, then there is roughly the same interval between 10^{-24} and 1ms after the Big Bang as there is between 1ms and 13.7 billion years. Going backward in time gets us to the Planck time, after which time curves back onto space and the calculations start moving forward in time once again, just like all directions are south when standing at the North Pole. There is no zero time in the universe, and there can also be no time prior to zero time as far as we are concerned, although there can be prior cyclic universes.

Matter is frozen energy. When one is totally immersed in a medium, awareness of the medium fades. For example, you don't feel wet when completely underwater, nor do you feel the air around you when it is not moving. A gram of any kind of matter has the same amount of energy. All matter has a quantum wave function. (See Chapter 23.) Particles of matter are made of lesser particles and so on. What lies at the bottom of the well? The Higgs boson? Strings? Information? Or... ? The de Broglie equation (named for physicist Louis de Broglie, 1892-1987) $hv=mc^2$ equates waves and particles where h is *Planck's constant* (approximately 6.626068 x 10^{-34} m² kg/s), v is the wave frequency, m is the mass and c^2 is the velocity of light squared. This wave/particle duality applies to all matter. Is there logic underneath all this? If so, then the connection between the physical and the metaphysical will be firmly established.

[God] could have created us to live in hard vacuum if He wanted to.
Martin Wagner

Perfection?

Materialists point to examples such as the "backward" human eye as evidence that no cosmic designer exists or, if He does, that He is somewhat less than competent. They forget that intelligent design does not necessarily mean perfect design. In fact, many creation myths speak of divine mistakes and creators fumbling and making it up as they go along. Besides, as I have previously pointed out, not even the loudest materialists are capable of creating even a single cell from scratch, much less a complex species with over 50 trillion cells per copy.

Universal Grandeur

It could be said that Einstein's relativity, Planck's quantum mechanics, Heisenberg's uncertainty, Dirac's antimatter, Aspects delayed choice, Bell's inequality, and others have managed to expose the wisdom underlying existence. Following this logic, one could say that science is discovering the spiritual and/or metaphysical, and that the world is getting progressively stranger with each passing discovery.

A man who says, 'if God is dead, nothing matters,' is a spoilt child who has never looked at his fellow man with compassion.
Kai Neilsen

God is subtle but He is not malicious.
Albert Einstein

The Sun converts 660 million tons of hydrogen into about 600 tons of helium every second. Previous generations of stars formed the heavier elements that made life possible. Some of the dust from a long-ago supernova is directing the writing of this book, and another clump of stardust is reading it. The path from thought to keyboard is pressing millions of cells and billions of atoms into service, the whole assembly functioning together like a single well-oiled machine.

A Hypothesis of God

I have come to realize that for some experiences there is no explanation, just a deep knowing that I have encountered the divine.
Janis Amatuzio

God can be like an old man, a woman, a spirit, or a timeless cosmic consciousness. Any way you care to imagine God, imagine that this Guiding Organizing Designing process is inside you waiting to be discovered at this very moment, if you will just wake up and smell the science. The potential for mind is incredibly vast. Where does this all come from? Is it random chance, or is mind part of something else? What is greater in scope, the currently known universe, or the mind that knows it? At every level of existence, the whole is always greater than the mere sum of its parts. One could build most any other creature from the cells that make up my body, for example. There is a very reasonable case for crediting God with our existence. But it is not enough to simply know that something is possible. We must believe that there is a God out there that we can find by scientifically accepted means.

The God I am proposing to try to find is the God described in this chapter and in Chapter 13. I think it can be done. I think we have all the evidence we need. I think that this evidence is also relevant to the question of what happens when we die and what our true natures are.

We affirm nothing and deny nothing for the single and perfect Cause is beyond any affirmation, and its transcendence is beyond all negation.
Dionysius

As we saw in Chapter 9, Gödel proved that some true mathematical assumptions can't be proven true because of complexity. Fully understanding a complex system requires even more complexity. This is probably why human brains cannot fathom human brains. We are effectively barred from brute reality by limits that force us to look for such explanations from the outset. Mysticism may be the best path forward from here.

Davies said it best: "Through conscious beings the universe has generated self-awareness. This can be no trivial detail, no minor by-product of mindless, purposeless forces. We are truly meant to be here."

We have reached the end of Part Three of this book. So far, we have validated mythology and mysticism, debunked religion, and formed a working hypothesis about what God might look like. We have touched on subjects from history to cosmology, quantum physics, biology, and more. So far, it seems possible that this life is not all we get, but it is far too early to claim victory if indeed we ever can.

We will now examine the evidence that may support or refute our hypothesis. It's been a strange ride so far... and it's about to get a whole lot stranger.

If God created the world, where was he before creation? Know that the world is uncreated, as time itself is, without beginning or end.
Mahapurana

PART FOUR

Paranormality

The Divine Savage
Revealing the Miracle of Being

Chapter 15

Through the Looking Glass

It is very beautiful over there.
Thomas Edison
(on his deathbed)

In order to disprove the law that all crows are black, it is enough to find one white crow.
William James

Life and death are all around us all the time. What does life mean? Why must we die? The short answer is that we are here because stars, extinct species, and countless members of our own species died. Without death, neither the universe nor we would be here to contemplate—and often bemoan—our fate. We are walking, talking lumps of stardust with the seemingly unique ability to portend our own deaths. Leonardo da Vinci once said, "When once you have tasted flight, you will forever walk the Earth with your eyes turned skyward, for there you have been, and there you will always want to be." Small wonder we look up at the stars and wish we could travel among them once again.

The question of what, if anything, happens when we die is very germane, because every single person alive today will be dead in little more than a century, including the babies being born as you read this paragraph. We all have a date with death, but is that the end? Our brains will stop working and our bodies will decay (see Chapter 1), but what about our consciousness, our "I" ness? The very oldest human graves show evidence of Neanderthal belief in some form of human survival 100,000 years ago. Was there something to that belief?

We don't fear losing consciousness when we feel reasonably assured of picking up where we left off, which is why we can climb into bed every evening. Death seems to be the "big sleep" that threatens the permanent loss of our selves, a completely different proposition that disturbs us in ways that pondering our prenatal states doesn't. Why should this be so? The Greek philosopher Lucretius (ca. 99BC-55BC) said that the eternity following is no different than the eternity before birth, and is therefore nothing to worry about. 13.7 billion years went by before you were born, and none of us seems to recall a long wait. The countless trillions of years to come after we die will not be any more of a wait. We will either continue on somehow, or not. We will either be aware of a much larger plane of existence, or there will be nothing of us to know that we ever existed, making our existence no different than that of an imaginary person. Thus, there is no rational basis for our fear of death. We consider ourselves to be tremendously important, and yet we really aren't.

Let the data speak, whatever the data say.
Unknown

Classical science may or may not be able to answer this question, but that hasn't stopped scientists from taking positions on all sides of the question. Materialists dismiss any purported evidence of any personal existence beyond this life as "supernatural," while dualists tend to embrace religious views, and idealists tend to draw inspiration from the Eastern religions that offer a sort of "Survival Lite," in which something survives our death, even if that something is not the "I" we are all used to. Still others believe that the answer lies beyond science.

And then there is another group, one that takes the Shakespeare quote that, "There are more things in heaven and earth, Horatio, than are dreamt of in your philosophy" literally. Welcome to the world of the *paranormal* (also called psi), where people are actively investigating "the claimed occurrence of events or perceptions without scientific explanation, as psychokinesis, extrasensory perception, or other purportedly 'supernatural' phenomena."

Paranormal Research

The pursuit of scientific truth is noble. Ordinarily, being convinced of a body of knowledge does not mean that one loses ability to discriminate between good and bad information—unless of course one has allowed dogma to overtake reason.

I regard the existence of discarnate spirits as scientifically proved and I no longer refer to the skeptic as having any right to speak on the subject.
James Hyslop

When it comes to matters of survival after death and anything having to do with information or realities beyond our five mundane senses and single known lifespan on Earth, classical science quickly draws battle lines. This prompted professor and psychical researcher James Hyslop (1854-1920) to wonder why it is perfectly acceptable to find out where we came from but not where we are going. On one hand, questions of survival after death may seem to lie beyond science. On the other hand, scientific dogma may be ignoring mountains of evidence lying right under its nose. For example, biochemist Rupert Sheldrake (1942-) performed an experiment that seems to prove that people have a "seventh sense" that knows when they are being stared at from behind—a useful survival trait, if true.

> *In psychichal research as elsewhere in life, it is inappropriate to assume dishonesty simply on the basis that what is reported appears not to accord with existing information. No paranormal phenomena accords with existing information, if by existing information we mean in this case the accepted laws of science.*
> David Fontana

People have been reporting strange experiences since the beginning of history. In fact, you would be hard-pressed to find someone who has not had at least one experience that seems completely unexplainable by normal means. Psi researchers point to these experiences as evidence that suggests deep connections that seem to transcend space and time. Some examples of psi include:

- **Clairvoyance:** Seeing people, places, etc. visually from a remote location.
- **ESP:** Extra Sensory Perception, a generic acronym for perception or communication outside of normal sensory capability.
- **Ghost:** The soul of a dead person, a disembodied spirit imagined—usually as a vague, shadowy or evanescent form—as wandering among or haunting living persons (sometimes also called an *apparition*).
- **Medium:** Someone through whom the spirits of the dead are alleged to be able to contact the living.
- **NDE:** Near Death Experience reported by some survivors of clinical death, in which a sequence of events takes place.
- **OBE:** Out of Body Experience, where someone reports leaving their body and traveling to a different location.
- **Precognition:** Receiving information about future events.

- **Psychokinesis:** Mentally interacting with inanimate matter.
- **Telepathy:** Exchange of information between minds.

We will expand on these definitions shortly; meanwhile, the challenge for science is to convert these raw experiences into meaningful data, eliminate fraud, error, and ordinary causes, and then try to figure out what, if anything, is going on. This is a serious challenge, given the highly subjective nature of much of the data, the very high percentage of hucksters trying to profit from people's fear of death, the number of people who report experiences to get attention, and the stigma surrounding research in these areas. There is also controversy about just what the terms "paranormal" and "psi" should mean. Most people loosely define these terms as anything having to do with the bizarre, occult, or mysterious. Psi is a hodgepodge where one must carefully define and categorize the paranormal, "supernatural," mystical, and scientific. Mainstream science can therefore be forgiven for shying away from these topics; however, that does not mean that these phenomena are not real. For the sake of this discussion, we will define the paranormal as that which is beyond the range of phenomena accepted by most scientists.

Finding just one solitary white crow would prove once and for all that not all crows are black. Similarly, we need only find one verified case of psi and/or survival after death to prove the reality of such events and forever alter the definition of what is natural versus "supernatural." The implications of proving even one such episode would be nothing short of revolutionary, and all it will take is one incontrovertible example that no skeptic can reasonably doubt. In fairness, there are many people who say we have already achieved that; however, universal acceptance of psi by the scientific community remains little more than a pipe dream.

People who think about an afterlife tend to link those thoughts with their ideas about a particular god, but this was not always the case. Ancient Greeks tired of their gods and set out to study the soul on its own merits, the first of many scientific efforts to uncover the truth about life after death. Aristotle believed that the comparison between body and soul is one of matter and form. The soul makes a human what s/he

---

Your sweet spot is in between the true believers and the scoffing skeptics.
Rob Breszny

Neuroscience cannot tell us whether or not there is an 'external reality' behind the reports of near-death experiences and as such we simply do not know.
Elahi Ebby

is, but has no existence independent of the body, and is like a stamp on a piece of metal that rusts as the metal rusts; however the dead person's rationality returns to some reservoir, presumably for future use by others.

Scientific theories are organically conditioned just as much as religious emotions are.
William James

The second century Roman physician Galen of Pergamon (129-217AD) was among the first to point to the brain as the seat of mental activity. Later, the church believed that we have one life on Earth to earn an eternal ticket in either heaven or hell, and the concept of a soul moved back from science to religion until the Renaissance. Descartes believed that the soul survives physical death, because a person exists and the Creator cannot take that from her or fool her about that, despite being able to fool her about all else. The only thing she can be sure of that she thinks. *Cogito ergo sum.* Dualist philosophers worked on the problem of how brain and soul interact, because an immaterial soul needs some mechanism for communicating back and forth with the physical brain. This question has been resolved against dualism, as we saw in Chapter 13. Materialism and idealism are the only possible options.

Early atheists did their best to fly under the church's radar, out of the very justified fear of persecution. Thus, medieval knowledge of the universe was largely confined to religious scripture until science was able to break away in the 16th and 17th centuries. Science steadily gained ground, as evidenced by the church removing the prohibition against believing that the Earth orbits the Sun in 1758. Total independence was still long in coming. For example, the church made sure Darwin would not be knighted in the 1800s.

The statement that there is any such thing, and the statement that there is no such thing, are neither of them statements that science can make.
C. S. Lewis

The modern science of psychology replaced both clergy and philosophers in the 1800s, and *parapsychology* (the study of psi) evolved right along with it, in an ongoing attempt to prove or disprove religious claims while finding the truth about psi; however, the advent of science brought with it the idea of man as a machine lacking any kind of spirit or soul. Others disagreed; for example, physician Duncan MacDougall (1866-1920) theorized that the soul has both weight and volume, and set out to weigh people at the instant of death. His efforts did not pay off; modern theorists think of consciousness as information, one bit of which has 3×10^{-23} joules of heat and an equivalent mass of only one billionth of a kilogram. Even modern scales are hard pressed to be sensitive enough to weigh the soul; even if they could, it would be nearly impossible to account for other fluctuations such as evaporation, terminal breath, etc. Which brings us to...

Quantum Mechanics

One of the oddest aspects of quantum physics is the dual wave/particle nature of matter. To put it simply, matter seems to exist in nice little chunks when we are looking at it, only to become a wave of possibilities when we look away. Particles seem to exist because experimenters are looking for them, and not because they have any intrinsic existence of their own. The famous dual-slit experiment (see Chapter 23) can get matter to behave as either particle or wave, depending on how the experiment is set up—a true example of the answer depending on how the question is asked. The illusion of solid matter occurs because our eyes have nowhere near the resolution required to see objects that small or the spaces between them. Also, enough particles have a high enough probability of being where you expect them to be when you expect them to be there that we don't normally see quantum effects, but all matter obeys quantum laws. In 1999, physicist Anton Zeilinger (1945-) tried the double-slit experiments on *buckyballs* (spherical cages consisting of 60 carbon atoms apiece) and got the same results. This same experiment also works using the light of distant galaxies, which involves nothing less than influencing the path taken by that light millions or even billions of years ago.

> *Stars are a pivotal stepping stone on the way to our existence.*
> Fred Adams

Some think that the only way this makes sense is if mind really is above matter, if consciousness and not matter is the primary building block of the universe. It just so happens that quantum physics may agree with them. It may also disagree. No one doubts the experiments or their results; rather, it is the interpretation of those results that remains open to question.

The Copenhagen Interpretation

The Copenhagen interpretation of quantum mechanics is named after the hometown of pioneering physicist Niels Bohr (1885-1962). This interpretation limits quantum mechanics to predicting the probabilities of various outcomes of pre-

> *Every interpretation of quantum mechanics involves consciousness.*
> Evan Squires

String theory offers us a vision of inflationary cosmology on a silver platter..
Gabriele Veneziano

defined observations. This is the *probability wave,* or waveform. The definitions of observer and observation are vague, but the general idea is that an observer causes the waveform to "collapse" to one of the possible outcomes based on the probability of each outcome occurring. For example, if Outcome A has a 75% chance of occurring and Outcome B has a 25% chance, the collapse will favor Option A an average of three times per four observations. Absent an observation, collapse does not occur, and each outcome remains just as likely.

Schrödinger conceived of the following thought experiment: Imagine a cat in a box fitted with an apparatus that detects radioactive decay, which has a 50% chance of occurring. If such a decay occurs, the apparatus releases a poison that kills the cat. So long as the box remains closed, we don't know if the cat is alive or not. According to the Copenhagen interpretation, the cat is both 100% dead and 100% alive in a state of quantum superposition. It is only when we open the box that the waveform collapses to either possible outcome, and we see a live or dead cat. Needless to say, this experiment was conceived long before the days of animal rights. It became known as Schrödinger's cat, and has become one of the most widely known concepts in quantum physics, in no small part because of its potentially sweeping implications.

In 1961, physicist Eugene Wigner (1902-1995) revisited Schrödinger's cat using a conscious observer in what became known as the Wigner's friend paradox. Whereas Bohr did not specify what constituted an observer, Wigner asserted that consciousness makes all the difference; collapse occurs when observed by a conscious observer. This startling conclusion threw a large, non-scientific wrench into the works that launched the careers of countless motivational speakers and would-be gurus. The "Law of Attraction" behind such products as *The Secret* is at least somewhat based on Wigner's friend. (See Appendix B.)

It is an unpleasant thing to bring people into the basic laws of physics.
Steven Weinberg

Just what is consciousness? Speculation and debate have been increasing in recent years. As far as physics is concerned, the simple act of observation changes the results. Bohr, Einstein, Heisenberg, Schrödinger, and others thought long and hard about the nature and role of consciousness. Schrödinger believed that the universe is made of consciousness. In an article in *Scientific American*, physicist Bernard d'Espagnat (1921-)

said that, "The doctrine that the world is made up of objects whose existence is independent of human consciousness turns out to be in conflict with quantum mechanics and with facts established by experiment." Physicist Fritjof Capra (1939-) agreed, saying that reality is not some objective state but a creation of human consciousness. Whether or not this is true, quantum effects may help produce, sustain, or otherwise facilitate consciousness; however, physics has only conducted a handful of experiments around consciousness and quantum phenomena. We will see the Copenhagen interpretation and other possibilities again in Chapter 29.

Many Worlds Interpretation

Not all physicists are comfortable with the idea of consciousness playing a central role in shaping reality. Everett thought the idea of mind collapsing the waveform ridiculous. For him, the universe splits at every quantum decision point. Each such event has a set of probable outcomes, and the universe splits into as many copies as there are outcomes. For example, a waveform with six possible outcomes (such as rolling a 6-sided die) yields six different universes. These copied universes continue splitting in the same manner, as do the results of those splits, etc. Thus, all that can happen will eventually happen.

This suggests that the Copenhagen interpretation is correct, while knocking the conscious observer off his high horse. In this scenario, there are almost infinite copies of you reading this book, almost infinite copies where you are not reading, almost infinite copies of you that have already died, almost infinite universes where you were never born, and almost infinite universes where you will never die. There are even universes where we are repeating the same life again in a manner reminiscent of the 1993 movie *Groundhog Day*. You are continually splitting into countless copies every second of every day. The JCI religions don't like this view, because Christ would die not once but almost infinitely many times. Is your head spinning yet?

Cosmologist Max Tegmark (1967-) thinks that the many worlds interpretation and Schrödinger's cat may be the ultimate proof of immortality. The cat may live or die, and the line where it dies must separate from the line where it lives.

You can't have your materialist cake and eat your consciousness, too.
David J. Chalmers

What is meant by consciousness we need not discuss; it is beyond all doubt.
Sigmund Freud

This takes the idea of the primacy of consciousness to its logical extreme. If your universe is your personal universe, then your universe and all of the records of its existence vanish when you cease to exist, meaning that your death causes your universe to never have existed at all. You must continue if your universe is to continue, and the many worlds interpretation guarantees that this will happen; you will live forever in your own universe while appearing to die in observers' universes.

Psychology is not ready to tackle the issue of consciousness.
Ulric Neisser

Another possibility is that we die at a specific moment. A brain faced with impending death is flooded with hormones and other chemicals. Time slows down, and we never hit the wall. Instead, we experience an infinite number of *chronons* (instants of time), where we keep going only halfway to the end. For example, a person near death will begin their last 60 seconds of life. Halfway there, she will have 30 seconds left, then 15, then 7.5, 3.75, etc. right down to—but never reaching—zero. Thus, we never reach the end and will never die in our own universes. Other people die in our universe, and we will die in theirs, but each of us is immune in our own personal universes. This idea is very similar to that proposed by cosmologist Frank Tipler (1947-), in which heaven could exist for an eternity during the last fractions of a second before the universe finishes collapsing back in on itself in the Big Crunch. The idea of an oscillating universe has been largely discredited, but the idea is still an intriguing one, and no such crunch is required for the personal universes each of us could inhabit. We will discuss this interpretation more in Chapter 29.

Hidden Variables Interpretations

Consciousness is a singular for which there is no plural.
Erwin Shrödinger

There are also many "hidden variable" interpretations that seek to eliminate consciousness from the equation by proposing a deeper hidden order that we can never directly detect. Of these, the de Broglie-Bohm theory named after physicists de Broglie and David Bohm (1917-1992) is the most common. Their theory refers to the probability that a particle really is at a certain location (as opposed to the probability that the particle will be found there when measured). This may seem like a piddling semantic exercise, but it does allow one to interpret quantum mechanics as a dynamic theory of particle trajectories instead of a statistical theory of observation. This pretty much eliminates the paradoxes surrounding measurement, observation, and wave collapse posed by Schrödinger's cat and other experiments. Through this theory, one can visualize the reality of most quantum events and is completely consistent with predictive data. Hidden variable theories may seem like semantic exercises that are more philosophical than theoretical; however, the line between physics and philosophy is a blurry one at best and only getting blurrier. We will see more of this in Chapter 29.

Beyond the Second Law

Physicist Ludwig Boltzmann (1844-1906) proposed a method of calculating entropy by treating each particle as statistically independent. Mathematician Henri Poincaré (1854-1912) went further by demonstrating that entropy need not always increase—a contradiction of the Second Law of Thermodynamics. Instead, entropy is cyclical. For example, gas molecules in a box will eventually return to their original configuration, which means that their entropy will decrease. The length of this cycle (called a *Poincaré cycle*) is 10^n seconds, where n is the number of molecules in the universe. The age of the universe is approximately 4.32×10^{17} seconds, which means that the cycle will be extremely long, even for a collection of only a few molecules. It also means that we cannot rule out a decrease in entropy at some future point. We also cannot rule out a period of higher entropy in the past. Shuffling a deck of cards over and over again eventually places the cards in any given order an infinite number of times.

Our life is a faint tracing on the surface of mystery.
Annie Dillard

Quantum Consciousness

Consciousness remains front and center in any quantum interpretation. despite the best efforts of some of the world's most brilliant scientists. The Copenhagen interpretation implies that consciousness plays a role, many worlds implies that immortality may be a real fact, and hidden variables could point the way to a deeper level of order that some would call God.

So far, the Copenhagen interpretation remains the standard interpretation, with many physicists simply choosing to ignore the proverbial elephant in the room. On the other hand, Wheeler went to the logical extreme by saying that we live in a participatory universe that needs observers in order to come

A mind may conceivably exist without a brain.
John Eccles

into existence. Each observer is responsible for her or his own universe in this theory, an idea that has vast implications, because the universe belongs to both each of us individually and collectively. (Note that this theory can apply to both a single universe and a many worlds/multiverse interpretation.)

And why not? Any way you look at it, quantum mechanics implies that consciousness plays a role in bringing the universe into existence. How can we acknowledge this, only to turn around and say that consciousness is finite? The fact that consciousness seems to be necessary plus the fact that quantum experiments have effects across time and space, directly implies that human consciousness has some sort of effect that extends beyond the birth and death we are all familiar with.

We will explore physics theories from ancient to modern times in Chapters 21 through 28, and mind/brain in Chapters 29 through 32. Meanwhile, the point of this discussion is that those who believe in various paranormal phenomena may not be nearly as crazy as some might think.

Out of Time

I have the absolute assurance that when the something called death comes, it will only mean a new and larger and more complete life.
Isaac K. Funk

Life is eternal; and love is immortal; and death is only a horizon, and a horizon is nothing save the limit of our sight.
R. W. Raymond

Combining psychology, the medical advances of psychiatry, and neurology's ever-evolving contributions to understanding how our brains work has helped millions of people lead happier and healthier lives. Mind and brain science are unqualified successes on one hand. On the other hand, none of these sciences can explain how brain structures, states, stimuli, etc. become consciousness. Every model we have about brain function indicates that we could live equally well from an evolutionary point of view as walking, talking robots with no self-awareness whatsoever. In that sense, mind and brain science has been an utter failure. It's a bit like understanding how every part of an engine, suspension, body, etc. function and the inherent limits of acceleration, braking, turning, etc. without knowing how all those pieces form a car.

In all fairness, the human brain is the most complex structure we know about. At birth, the brain contains 100 billion neurons, each of which has tens of thousands of *dendrites* (branched extensions that conduct impulses toward the neuron). To put this into perspective, a piece of brain tissue the size of a grain of sand has about 100,000 neurons, 1 to 2 million axons, and 1 billion synapses exchanging information. As I mentioned in a previous chapter, the number of possible brain states exceeds the number of subatomic particles in the universe. The average neuron is about 80% water and consists of about 100,000 molecules. The brain has about 1 trillion cells (100 billion neurons plus 900 million supporting cells). The molecules in each cell are replaced about 10,000 times during the average lifetime, during which we also lose some 100,000 neurons per day and many billions of connections between neurons during our lifetime. Our surviving brain connections are in a constant state of flux as they establish and remove connections between neurons.

Figuring out how consciousness works for a single brain state would be challenging enough, but the constant changes of both states and physiology happening in our heads make searching for answers challenging to put it mildly. Don Quixote's futile charges against windmills seem positively sensible next to the task of trying to figure out consciousness. And yet whatever pattern that makes me Anthony Hernandez, you whoever you are, and anyone whoever she or he is, remains and persists. There just does not seem to be any plausible materialist explanation for this. We will examine this in more detail in Chapters 29 through 32.

Sheldrake sees consciousness as a field extending throughout the entire universe, and the human brain as a glorified TV set. Damaging, destroying, or simply powering off the set may render it inoperable, but does not affect the signal. This explains how everything from sleep to injury or illness can appear to alter or shut off consciousness. Televisions are barriers between ourselves and the signals that allow us to see and hear the program being broadcast. Likewise, the brain may be a barrier between ourselves and the potential of the universe that allows us to interact with each other. If this is true, then perhaps we should think of ourselves as deprived, because our individual senses of self disconnect us from the rest of universe. Then again, some people speculate that losing one's sense of self can be overwhelming, that schizophrenics know matter is an illusion, and are in touch with their higher selves.

Look up into the night sky, and you will see many thousands of stars, not as they are today, but as they were when the light

What happens after death is so unspeakably glorious, that our imagination and feelings do not suffice to form even an approximate conception of it.
Carl Jung

Man, when he dies, only passes from one world to another.
Emanuel Swedenborg

Your soul is reborn into a new physical body for the purposes of gaining knowledge and for understanding and resolving negative emotions and actions you incurred in your previous lifetimes.
Gloria Chadwick

you see began its journey toward Earth. The farther the star, the further back in time you are looking. In 2010, the Hubble space telescope spotted a galaxy 13 billion light years away, which means it formed only 700 million years after the Big Bang. Even the things near you are not in your present. The light carrying these words to your eyes left the book a tiny fraction of a second ago, but you are still seeing it in the past. The sounds you hear are further in the past, since sound travels at about 768 miles per hour (about 5 miles per second). This is why you perceive a delay between a flash of lightning and the corresponding thunderclap. Counting off the seconds between a lightning flash and the thunder will give you a very good idea of how far away the strike was down to at least 1/5 of a mile, and probably more like 1/10 of a mile.

People in Western parts of the world have a general perception of time as linear, as "slippin', slippin', slippin' into the future" as Steve Miller famously sings. This makes sense; as far as we are ordinarily concerned, one second passes by just like any other second. This linear view also increases our fear of death because we see it as a permanent, one-time occurrence instead of as one spoke in a cycle. Children live in the moment and acquire their sense of mortality and the approaching end over time. Science and reason have debunked religion to the benefit of us all, as we saw in Chapters 9 and 10; however, it also removed the cognitive and emotional safety net by exposing the flaws in thinking that we will simply leave our bodies behind and move upstairs to an eternal—and insufferably well behaved—cosmic dinner party. We also know that dualist ideas about separate physical and mental/spiritual worlds are wrong. It seems to all appearances that our bodies and brains die, taking our selves and all we were, are, and ever will be with them. But are things really this bleak?

Your mind-the thing that is 'you'—your 'soul' if you will—carries on after conventional science says it should have drifted into nothingness.
Sam Parnia

Not necessarily. A neutrino has a lifespan of about 15 minutes. From the neutrino's perspective, it lives those 15 minutes no matter what; however, we see neutrinos traveling at close to light speed that have covered 55,000 light years. Thanks to time dilation, we perceive the neutrino's 15 minutes as lasting over 50,000 years. If time can dilate to this extent, and if we perceive the universe as being 13.7 billion years old, we have to ask just whose years we are talking about. There is no objective measure of time; there is only the subjective measurement that is unique to each entity.

Time itself may be an illusion, as physicist Julian Barbour (1937-) describes in *The End of Time*. (See Chapter 28.) Psychologist Karl Pribram (1919-) believes that the brain is a holographic tuner. Reality may not be the smoothly flowing river of time we think it is. Time and space may both be an illusion, with events being the only reality. Perceiver and perceived are two aspects of a single event, which avoids infinite regression. This conjecture is supported both by Einstein's linking space and time as separate aspects of each other, as we will learn in Chapter 24, and by at least one patient who experienced time as a series of stop images instead of a smooth flow. A coma patient woke up after 19 years with no sense of elapsed time; as far he was concerned, nearly two decades had literally passed in the blink of an eye. According to Pribram, the world we experience is an internal holographic projection of an external reality that consists of electromagnetic energy. Consciousness functions as the laser that makes a coherent image out of the hologram and projects it onto our internal "screen" for viewing.

I personally choose to believe that we do meaningfully survive death and can communicate back through mediums and channels.
Jon Klimo

As for the nature of time, each event may be like a train station that we experience for a few moments as we pass by. The train station remains standing; the only thing that changes is that we experience the station for a few moments before moving on to the next event. This change of perspective is what we perceive as time, but this perception does not mean that time has any flow at all. Our brains manipulate how we perceive time, and our idea of reality can be so easily manipulated because there is no objective reality beyond our minds. We discover new slices of time like runway lights that we pass one at a time on the ground. From the air, we can easily see the entire runway and all of the lights at once.

Neurosurgeon Wilder Penfield (1891-1976) operated on epileptics by destroying the brain cells causing the seizures. Before the operations, he would place a patient under local anesthesia, which kept them wide awake and alert. Penfield would then stimulate various areas of the brain to isolate the areas responsible for the seizures, so as to minimize side effects. Some of his patients experienced extremely vivid memories during the procedure. If Barbour is correct, then

All this invention, this producing, takes place in a pleasing, lively dream.
Wolfgang Amadeus Mozart

those patients may not have been recalling events in great detail but actually reliving them. This is highly conjectural, however we cannot deny the possibility.

Accepting that the entire universe has a subjective aspect to it alters the classical view of the brain, which starts to look more and more like an editor or reducing valve for consciousness, as I mentioned in Chapter 14. The idea of brain as regulator may seem farfetched, but then again most of our organs are regulators. We make neither food, water, nor air, so why should we make consciousness and thoughts? We may only regulate what we think, instead of actually creating it. It is interesting to note that consciousness comes with a half-second delay, which may serve to gather and store sensory data until it is completely available before presenting it to the observing consciousness, just as a computer buffers input for processing.

Materialism sees the brain as the originator of thought, consciousness, and self, but we may need to challenge that view. Cognition may not cease when the brain dies; on the contrary it may increase. What we experience as the dualism of mind and matter is wrong. Consciousness may be neither side effect nor epiphenomenon; it may be an irreducible component of the universe.

Taking aspirin reduces the probability of a heart attack by 0.8% compared with not taking aspirin. This effect is about 10 times smaller than the psi ganzfield effect observed in the 1985 meta-analysis.
Dean Radin

Paranormality Defined

Having seen how some interpretations of quantum mechanics and neuroscience may open the door for many things beyond mundane normality, let's see just what may lie beyond the threshold. Each of these areas of paranormality remains hotly debated, with some hailing new significant discoveries and others dismissing them as hoaxes or pseudoscience. Paranormal studies can encompass many things, some of which are beyond the scope of this book. Here are the aspects of paranormality that we will deal with in this book and their definitions in alphabetical order.

Of course I believe in ghosts. Doesn't everybody?
Red Skelton

Clairvoyance

Clairvoyance refers to the ability to see things remotely using the so-called "mind's eye." The clairvoyant (person with the ability to see remotely) provides visual information about an object hidden from physical view and separated at some distance. The Stargate Project run by the Central Intelligence Agency from the 1970s through 1995 attempted to determine both the validity and potential military applications of different types of remote viewing, including clairvoyance. One viewer received a Legion of Merit award in 1984 for determining 150 essential pieces of information unavailable from any other source. One analyst concluded that remote viewing had been proven, while another found this conclusion premature. The project was terminated in 1995 because of a "lack of documented evidence that the program had any value to the intelligence community."

I mentioned my friend Sarah in the introduction to this book, because I have witnessed her perform numerous feats of remote viewing from across the United States. I was in California and Oregon, and she has been in Florida, Virginia, and Texas. She has no way of knowing what she is looking for in advance, yet her track record is 100%. Given my own experiences and those of others, I have to wonder whether Project Stargate really was cancelled (and if so what the real reason(s) were), or whether it was continued under much deeper cover. I suspect the latter.

Extra Sensory Perception

Extrasensory perception (ESP) is a blanket term for sensing information directly in the mind, as opposed to via the normal five senses of sight, hearing, smell, taste, and touch. This term is also referred to as a sixth sense, hunch, and gut instinct, terms that have been in the English language for a long time.

Parapsychology is the field of science that studies the paranormal. Parapsychologists believe they have found compelling evidence for ESP, a claim that is roundly rejected by other scientists for reasons ranging from lack of evidence, lack of a working theory to explain ESP, and lack of experimental techniques that can provide consistent results. My answers to these objections are:

- What kind of evidence are we talking about? It is true that there is no material evidence to support ESP; however, this is only a problem for materialist science. Dual-

We all have some experience of a feeling that comes over us occasionally. Of what we are saying or doing have been said and done before, in a remote time-of our having been surrounded, dim ages ago, by the same faces, objects and circumstances-of our knowing perfectly what will be said next, as if we suddenly remembered it.
Charles Dickens

ists and idealists understand that there are mountains of evidence pointing to the existence of ESP.

- What kind of theory do we need to explain ESP? We already know that time and space are simply different aspects of each other. We already know that all things in the universe were connected to each other at the beginning of time in the primordial singularity. We know about quantum entanglement, superposition, the observer effect, and wave-particle duality. (See Chapter 23.) Superstring and brane theories require additional spatial dimensions. We already have theories to describe how God and other "extra sensory" beings can interact with the material universe from both dualist and idealist perspectives, so why can this not apply to people with such capabilities?

- It is true that paranormality experiments have traditionally lacked the rigorous controls expected for scientific experiments because of the extremely subjective and anecdotal nature of much of the evidence. For example, I know for a fact what I see with Sarah. I also know that my account of events is not evidence in the strictest scientific sense. On the other hand, more and more rigorous controls are being implemented as research progresses. Some could even argue that sufficient controls exist already, and that the sheer weight of anecdotal evidence is compelling; even allowing for a large percentage of fraud and mistaken reporting leaves a significant percentage of cases with no other acceptable explanation.

Ghosts

The nature of mind-brain relationships and the possibility of life-after-death are some of the most profound issues relating to mankind's place in the universe. The report... of near-death experiences in survivors of a cardiac arrest provides intriguing data that are relevant to these issues.
Christopher French

The standard definition of a *ghost* is the soul or spirit of a deceased person or animal that is capable of manifesting itself to living people, visually or otherwise. The manifestation can be anything from an invisible presence to translucent/wispy shapes, or even fully life-like visions. A deliberate attempt to contact a ghost is known as a *séance*. Belief in ghosts has a long history that began with primitive animism and ancestor worship. Some modern religious practices, such as funerals, are designed to appease the spirit of the deceased. Ghosts are usually described as lonely beings that haunt specific locations, objects, or people they associated with before death. This distinguishes them from other spirits that can travel and meet/interact with people and/or mediums at will.

Mediums

A *medium* is someone who claims to be able to contact ghosts, angels, and/or other types of otherworldly beings. The medium literally mediates communications between the living and the otherworldly spirits. This sometimes goes beyond merely relaying messages back and forth; some mediums go into a trance where they are unaware of the communication taking place, allow spirits to control and speak through their bodies, etc. Some mediums even claim the ability to allow spirits to manifest objects (called *apports*), levitate, etc.

Sometimes these experiences are accompanied by physical effects that seem to be connected and that cannot be considered as hallucinations.
J. B. Rhine

Some common terms associated with mediums include:

- **Channeling:** During a channeling, the medium's body becomes possessed (taken over) by a spirit who then speaks through the medium. The medium remains *cataleptic* (extremely rigid) during the session, and her or his voice may change completely. The spirit then answers questions from sitters and/or passes other information.

- **Demonstrations of Mediumship:** Some congregations devote a portion of their sermons to mediumship, in which the pastor receives and passes on messages from spirits to the faithful.

- **Direct voice:** Direct voice communication is where a spirit builds a voice box using *ectoplasm* (essentially spirit gel) in order to communicate directly with the living.

- **Mental Mediumship:** This is where spirits communicate with a medium via telepathy. The medium sees (clairvoyance), hears (clairaudience), and/or feels (clairsentience) the spirit's messages and passes those messages on to the intended recipient(s). When a medium conducts such a *reading* for a specific person, that person is called the *sitter*.

- **Physical Mediumship:** A physical mediumship session involves spirits manipulating energy and energy systems. Examples include—but are certainly not limited to—raps and other noises, voices, apports, materialized spirit

Better live a crossing-sweeper here than be made to talk twaddle by a 'medium' hired at a guinea a seance!
Thomas Huxley

bodies or body parts, or levitation. The medium is used as a source of power for these manifestations. These sessions often take place in dim light, and the medium typically uses an array of tools such as spirit trumpets, cabinets, and levitation tables, and may produce ectoplasm.

- **Spirit Guide:** Some mediums say that a spirit guide helps them develop and use personal skills as well as follow a spiritual path. Other mediums credit the spirit guide with bringing other spirits to the medium's attention and/or relaying communications to those spirits. Some guides regularly work with the same medium. Not all mediums have guides, and the guides may or may not be deceased members of the medium's family.

- **Spirit Operator:** A spirit operator is a spirit that uses a medium to manipulate energy or energy systems.

- **Trance Mediumship:** During a trance mediumship session, the medium remains conscious but sets her or his ego aside to allow the spirit to use the medium's mind to communicate by influencing the medium's mind with the thoughts or messages being conveyed.

Near Death Experience

"NDE" or Near Death Experience is a blanket term for various experiences that typically occur after the individual has been pronounced either clinically dead or very near to death. These experiences can involve multiple senses, such as leaving the body, levitation, fear, bliss, serenity, security, warmth, dissolution, and a "being of light" that is sometimes interpreted as a deity. The reported incidence of NDEs is increasing as medical advances make it possible to successfully resuscitate more and more people from ever-graver circumstances. The mainstream scientific community sees these experiences as hallucinations or other products of a brain in extremis, while other scientists and paranormal specialists see them as evidence of the reality of an afterlife. About eight million Americans claim to have had an NDE, and the actual numbers may be far higher, because people may be afraid or otherwise reluctant to discuss their experiences.

In times of universal deceit, telling the truth will be a revolutionary act.
George Orwell

For empiricists, the individual brain is the basis for consciousness. Under the new model, no such limit exists.
Charles Leighton

Dr. Bruce Greyson (1946-) lists the general features of an NDE as the sensation of being outside the body, seeing deceased relatives and religious figures, and transcending the ego and typical space-time limits. A typical experience progresses as follows:

- Receiving a message in one's native language.
- Unpleasant sound and/or noise.
- Awareness of being dead.
- Feeling well, at peace, free of pain, and removed from the world.
- Seeing one's own body from somewhere outside the body, which may include seeing doctors and nurses trying to revive the body.
- Sensation of moving through a vertical or horizontal passage, tunnel, or staircase.
- Moving toward a bright light and/or being immersed in that light and communicating with that light.
- A strong feeling of unconditional love.
- Meeting "beings of light," "beings dressed in white," and/or other spiritual beings that may include deceased loved ones.
- Receiving a detailed life review.
- Receiving knowledge about one's life and how the universe works.
- Making the decision, either by one's self or with others, to return to the body despite reluctance to do so.
- Approaching a border that is understood to be the point of no return.

We are more than simply the sum of our parts.
Irwin Kula

These experiences may be interpreted differently, depending on the culture of the person having the experience and her or his spiritual beliefs, if any. For example, a Christian may see the being of light as Jesus Christ. Psychologist Kenneth Ring (1936-) divided the typical NDE into five stages, saying that 60% experienced Stage 1 (feeling peaceful and contented), while only 10% experienced Stage 5 (entering the light).

And see, no longer blinded by our eyes.
Rupert Brooke

NDEs occur most frequently in cases of cardiac arrest due to myocardial infarction (clinical death), shock caused by post-partum blood loss, complications such as septic or anaphylactic shock, electrocution, coma caused by traumatic brain damage, intracerebral hemorrhaging, stroke (cerebral infarction), near drowning, asphyxia, attempted suicide, apnea, and serious depression. Contrary to popular belief, Ring says that people who experience an NDE because of attempted suicide do not have less-pleasant NDEs than those involved in accidental situations.

An apparition is a very slippery fish. You may say you don't believe in ghosts but in fact you are surrounded by them.
Karlis Osis

Out of Body Experiences

An *Out of Body Experience* (OBE or OOBE) is where a person has the sensation of floating outside one's body, and sometimes even seeing one's body from a remote vantage point. This is similar to clairvoyance, except that a person experiencing an OBE will actually travel to the location or object in question, instead of simply seeing it from a distance. Some people claim to remain connected to their physical bodies by a "silver cord" that extends however far they may go. This journey may include seeing things that were unknown beforehand. Some also report encountering both humans and non-humans while traveling that are not always friendly. Some researchers claim to be able to recreate OBEs in a laboratory simply by stimulating the brain. About 10% of people have at least one OBE in their lifetimes that are usually (but not necessarily) associated with NDEs. Some people have many such experiences.

Science really can't talk about things like telepathy, belief, etc. in any kind of way. All that we know about physical laws would say completely that it doesn't happen, but that's not the way things work.
Ursula Goodenough

OBEs can occur spontaneously, or as a result of physical or mental trauma, dream-like states, hallucinogens or other dissociative drugs, or even deliberately. A number of techniques have been developed to try to induce OBEs, such as visualizations while in a calm, meditative state. OBEs most often occur between dream (REM) sleep and waking. Some neurologists also suggest that OBEs can be caused by a mismatch between visual and tactile sensations, although experiments along this line have failed to produce a full OBE.

Precognition

Precognition means knowing about things that have not happened yet, where that information could not have been deduced or otherwise obtained from any normal source. The related terms *premonition* and *presentiment* refer to emotions about future events. Here again, materialist science refuses to concede the possibility, arguing that it violates both the flow of time and standard cause and effect. Also, it is entirely possible for people to believe they know things in advance whether or not they actually have the knowledge.

I again cite my experiences with Sarah as evidence, however subjective and unscientific it may be. Time and time again, she has seen events in my life months or even years before they happened that have included my employment, living situation, relationships, and more.

> *Altered states of consciousness remind that we're more than we think we are.*
> Charles Tart

Psychokinesis

This term refers to the ability to affect matter with mind, such as the age-old trick of bending spoons. More recent experiments have focused on affecting the output of random number generators.

> *If it is real, it will be revealed. If it is fake, we'll find the mistake.*
> Motto of the Human Energy Systems Laboratory

Telepathy

Telepathy is the exchange of information, thoughts, and/or feelings between minds using none of the classical senses or means of communication. A *telepath* is a person who can read other people's thoughts and mental contents.

Super Psi

Super psi is an idea that attempts to explain psi in general by saying that anyone can know everything there is to know. ESP thus becomes powerful enough to explain a so-called afterlife where none may really exist. The problem with this approach is that it is neither testable nor falsifiable, nor does it explain birthmarks in children who are alleged to be reincarnated or the lack of any elevated psi abilities in these children. I see this as the paranormal equivalent of pantheism: If God is everything in the universe, then we have irrefutably proved His

> *I used to be big, and now I'm small.*
> Sam Taylor

After this experience, I have no fear of death and believe with certainty in the afterlife.
Anonymous NDE account

existence, because every object, thought, and emotion is God. Similarly, if everything paranormal is super ps,i then we have proven super psi as well because nearly everyone has at least one paranormal experience in her or his lifetime.

Paranormal Encounters

Plenty of people have reported "afterlife encounters" (a term I'm taking from the book of the same title by Dianne Arcangel and Gary Schwartz) in which they meet and sometimes interact with ghosts and spirits. These encounters tend to be positive, and people who have these encounters tend to end up both believing in an afterlife and being more comfortable with their own mortality. There is often a sense of closure and peace if the encounter is with a deceased loved one.

There are five general types of afterlife encounter:

- Meeting a dead relative, friend, pet, etc.
- Meeting a spiritual figure such as God or Christ
- Meeting a historical or famous person
- Encountering animals and objects
- Other

There are six ways in which these encounters may provide evidence of an afterlife:

- The encountered ghost or apparition provided information that was hitherto unknown by the person or persons experiencing the encounter.
- The encounter was shared by two or more people.
- The person(s) experiencing the encounter did not know the spirit was deceased until later.
- The apparition reported a current event that was not currently known, but that was verified later.
- The apparition manipulated one or more objects.
- The apparition seemed to have a mind and will of its own independent of the person having the encounter.

I felt very happy and I knew that If I went beyond this point there was something vital, full of life, ongoing behind.
Anonymous NDE survivor

Here are just a few of the many hundreds of reports I read while researching this book:

- The new owners of a dry cleaning business often saw the previous owner (who had since died) working there. Though they had never seen the previous owner before, the new owners described him perfectly.

- A man encountered his dead foreman who placed his hand on his shoulder and smiled. The man had never seen the foreman smile in life, and the smile was an odd one that exposed the foreman's bottom teeth. At the subsequent funeral, the man met the foreman's brothers for the first time and noticed that they too smiled in the same odd way.

- A woman's husband got a phone call from his wife's ex-boyfriend telling him that his wife was in a better place and no longer suffering. The husband later learned that his wife had died in surgery at that exact moment, and that the ex-boyfriend had died long before. He wished for a sign that came in the overpowering smell of doughnuts. The husband was a baker, and his wife had often commented on his smell.

- A man had surgery on April 25th to insert a stent to reinforce a damaged artery in his heart and had a follow-up appointment scheduled for May 15th. His wife died six days later on May 1st. On May 10th, he received a phone call from his doctor saying that his wife had called urging him to move the appointment up, because he did not look well. The man went to the doctor, who diagnosed a life-threatening arrhythmia. The doctor had no idea that the wife had died when he made the call to move up the appointment.

- People experiencing NDEs have returned with reports of meeting people on the other side who had died during or just before the NDE, and whose deaths were not previously known to the person experiencing the NDE or to those who received the reports.

- Cardiologist Michael Sabom interviewed some of his long-term cardiac patients. 20 out of 23 gave the wrong answers when asked about resuscitation procedures, but those who experienced NDEs gave amazingly accurate

So this is dying. I never thought it was so easy. It was like taking off your coat.
Anonymous NDE account

I felt as tough I were awake for the first time.
Anonymous NDE account

> *Mom, when you were a little girl and I was your daddy, you were bad a lot of times, and I never hit you.*
> William McConnell

> *The paradoxical occurrence of heightened awareness and logical thought processes, without subsequent amnesia, during a period of impaired cerebral perfusion raises particularly perplexing questions for our current understanding of consciousness and its relation to brain function.*
> Bruce Greyson

reports down to the numbers and types of clamps used, their own internal organs, and conversations in the operating room.

- A woman's son was murdered. When detectives arrived at the house, the chandelier began blinking rapidly for no discernible reason. Five days later, the son appeared to his mother to tell her where to find the blood in the snow. The police were alerted, and the blood was found exactly where they were told it would be. They found witnesses and soon had a suspect whose nickname was Light. The chandelier has not repeated its performance before or since. The son's body was eventually found four miles from the murder scene. Some time later, the mother was walking the dog, and they saw the son wearing an outfit that was in a drawer at home at the time. The dog saw him too, and they gave chase for four blocks before the son vanished. While the son was in sight, his mother saw him moving by gliding above the sidewalk.

- A dead girl appeared to a former acquaintance to ask her to tell her grieving father not to feel guilty about her death, and then described what happened in precise detail.

- A 20th century patient was suffering from horrible nightmares about a bloody massacre. She wrote down her dreams in medieval French about the Cathar massacre (see Chapter 8), including names and places that were later verified exactly as written. The patent had no prior knowledge of either French (modern or ancient) or the events in question.

- A mother was driving her son to a neighbor's house when the door suddenly opened. The son fell out and was crushed by the car rolling over him. Two weeks later, a funnel dropped out of the ceiling in front of the dead boy's parents, and out popped the son with a girl the parents had never seen before. The son said that all was well, and that he was with his new friend Kellie. Both children then flew back into the funnel and vanished. Soon after, new neighbors moved in, and the boy's parents learned that they too had lost a child. They went to visit and saw a photo of the little girl who they instantly recognized as Kellie. They had never seen Kellie before she came out of the funnel.

- A man was convinced that he would die on a Friday the 13th, and told a friend about this. Years later, the man burned to death in an arson fire at a hotel on a Friday the 13th. He had committed a horrible crime two weeks before his death, and had been wondering whether he should turn himself in. Six months later, the friend felt his presence and was told that the dead man had gone to hell. The friend saw him again six months later, only this time he was smiling. He told his friend to tell his brother that he loved the headstone on his grave, and gave his friend a mental picture that the friend drew. The friend had never visited the grave, but the drawing matched the headstone. The friend also learned that the man was no longer in hell but going to school. Then, 18 months later, the man visited again, said this would be his last visit until the friend himself died and passed over to the next world, and asked if there was anything the friend wanted to know. The friend asked about lottery numbers and gambling, to which the man replied that he could not say, because this would be bad karma. The friend then asked how the man got out of hell, and learned that it had happened the moment the man had forgiven himself (another very Buddhist-sounding idea).

- A man was washing his car when he heard his dead wife tell him that he would be joining her soon and to enjoy family and friends while he still could. He told a friend about this on his way to work before being killed in a freak car accident.

Psychology professor Dr. Erlendur Haraldsson studied 350 cases of similar encounters in 1980, and discovered that 39.4% of encounters occur within one year after the death, with 10.5% of those occurring during the first 24 hours. 16.1% of encounters occur between one and five years after the death, 6.8% occur between five and 10 years after, and 13.9% occur more than 10 years afterward.

Skeptics dismiss these reports as inebriation caused by alcohol or drugs, mental illness, imagination, distraught emotional

There can be no one who doesn't hope the NDE is a true prelude to a blissful afterlife.
David Darling

It is worth dying to find out what life is.
T. S. Eliot

> *The more we become accustomed to this idea of a consciousness which overflows the organ we call the brain, then the more natural and probable we find the hypothesis that the soul survives the body.*
> Henri Bergson

states, deliberate fraud or other hoax, mistaken identity, mistaken details, or hallucinations. Infrasound and phenomena similar to hearing voices when standing near antennas surrounded by metal fencing are other possibilities, but these cannot possibly account for more than a small fraction of all of the reported cases.

It is certainly possible that the parents of the dead boy set out to defraud their neighbors, that the dry cleaner was seeking fame and customers, that the friend had forgotten seeing the gravestone, and so on. There may be other explanations that eliminate a brush with the hereafter as a possibility in any given case; however, the sheer number of such cases combined with external corroboration and circumstances unique to each case often rule out mundane explanation. Does this mean that people really are seeing ghosts? Of course not. But it does leave the door wide open.

Dying people are often ecstatic before death, and many report seeing apparitions of dead or living people watching over them to help them on their way to the next plane of existence. Drugs and disease cannot account for all of these, nor can hallucination, delusion, or dementia. It is possible that the dying people are simply engaging in wishful thinking or even fraud, but I for one cannot imagine why they might want to do so. Fear does not seem to be an explanation either. A scared person will typically keep repeating a sort of mantra or telling themselves a story to try to convince themselves not to be afraid, but this can hardly be called ecstatic. Hospice workers report that their dying patients have apparitions with them and are not dying alone. A study done by Case Western Reserve reports that operating room staff often sense a presence in the room when the patient dies. It is a stretch to attribute these experiences to fraud or wishful thinking, because one presumes that medical personnel trying frantically to save a life have other things on their minds at the time.

> *Despite the awesome achievements of 20th-century neuroscience in increasing our knowledge about the workings of the human brain, little progress has been made in the scientific understanding of mental phenomena.*
> David Presti

I should note that the encounters described above are neither hauntings nor séances. Hauntings typically occur at specific locations where they are seen by multiple people over time, most of whom do not know the ghost in question. Séances attempt to bring the spirit to the people wherever those people happen to be. The encounters described above all involve spirits appearing to people wherever they happen to be.

Subtle Bodies

It is both obvious and proven that there are levels of reality beyond our normal perception or awareness. Is the afterlife one of them? The idea of life after death and preserving the human soul was the top priority of pre-Western civilizations by 3000BC. Plato defined death as the separation of the physical body and the soul, which has fewer limits than the body. According to Plato, what we call time is but the "moving, unreal reflection of eternity." He also turned the traditional death/sleep analogy on its head, saying that the soul descends into the physical body from a higher realm. Birth is thus the sleep and forgetting, death the awakening and remembering. Souls are judged soon after death. This view is very much in line with both Buddhist teachings and NDEs that describe the life review and the being of light.

It may be wrong to dismiss the idea of an energy body out of hand.
David Fontana

In the Bible, 1 Corinthians 15:35-52 asks how the dead can be raised, and explains that there are celestial and terrestrial bodies as well as natural and spiritual worlds. This is also in accord with teachings around the world that say that our physical bodies work in conjunction with higher-level *subtle bodies* (energy bodies/souls). This may sound incredible, but then again truth is often stranger than fiction.

There are several schools of thought about subtle bodies. We will discuss ancient Egypt, theosophy (Western), and Hinduism (Eastern):

Egyptian

The ancient Egyptians connected the pharaoh's immortality with the god Osiris, and symbolized death and resurrection in the food they ate. Mummification preserved the body, but the concept of an underworld gained ground, along with the corresponding dualist idea of a soul that is separate from the body. It is interesting to note that the Egyptians embalmed dead bodies for the benefit of the deceased person, while modern funeral directors embalm bodies for the sake of those left behind to get one last glimpse of the person as s/he may have looked in life.

The more we learn about our body, the more we realize what an immensely intelligent system it is and how little we actually know.
Pim Van Lommel

The Egyptians separated a person as follows:

- **Ba:** The lower traditional physical person (body).

- **Ka:** Upper body that is aware of the ba's future (soul).

Theosophy

Theosophy is an amalgam of religious philosophy and mysticism, which says that all religions are spiritual attempts to help humans evolve, and thus each religion has a portion of the truth. The theosophical subtle bodies are:

- **Etheric body:** This is the lowest layer in the human energy field, sometimes called the *aura*. This body is in direct contact with the physical body to sustain it and connect it to the higher body.

- **Astral or emotional body:** This subtle body lies between the intelligent soul and the physical body as an intermediary. An individual astral body contains all of the memories, emotional patterns, personality, self, etc. of the physical body. Death liberates the astral body like a caterpillar becoming a butterfly.

- **Mental body:** Also called the concrete mind, this subtle body is made up of thoughts.

- **Causal body:** Also called the abstract mind, this is the highest subtle body that veils the true soul.

Each subtle body has its own aura and set of chakras, and corresponds to a particular plane of existence.

Tibetan

According to the Tibetan Book of the Dead, dying is a skill like any other. Those who master the art of dying can die an artful death, while those who don't can make a mess of things. The book is written for two audiences: the dying person, to help them become aware of each new wonder as s/he experiences it, and the living, to help them think positive thoughts and not hold the dead person back. The Tibetan subtle bodies are called *koshas*, which means "sheaths."

The Tibetan koshas are:

- **Annamaya:** The physical self, which is nourished by food. Most people identify themselves with skin, flesh, fat, bones, and filth, whereas the more educated under-

It is a truism that most of us, most of the time, identify ourselves with our physical bodies.
Raymond S. Moody, Jr.

It is harder to crack a prejudice than an atom.
Albert Einstein

stand that their selves are distinct from the physical body and are the only reality there is.

- **Pranamaya:** This is the *vital body* that pervades the whole organism to give energy and hold the body and mind together. Drawing breath is the single physical manifestation of this body. The physical body stays alive as long as the vital body remains with it.

- **Manomaya:** This is the *mental body* that is composed of mind combined with the five senses. The manomaya is the closest approximation of self, because it is the cause of diversity in the form of I, mine, etc. Human bondage is caused by the mind; but the mind is also the source of liberation.

- **Vijnanamaya:** This is the *supramental body* that is composed of intellect that provides discrimination, determination, and will. This body identifies itself with knowledge and the body, organs, etc. This body is not the supreme self, because it is subject to change, non-sentient, limited, and not always present.

- **Anandamaya:** This body is composed of bliss and is called the *causal body*. While a person is fast asleep and mind and senses cease to function, this body stands between the world and the self. This is a reflection of atman, which is absolute bliss.

Swedenborg noted that Plato and the Tibetan Book of the Dead contain many parallels. More modern scholars have noted many similarities between the Tibetan Book of the Dead and NDEs.

What if you slept? And what if in your sleep, you dreamed? And what if in your dream you went to heaven and there plucked a strange and beautiful flower? And what if when you awoke, you had the flower in your hand? Ah! What then?
Samuel Taylor Coleridge

Chapter 16

Paranormal Evidence

> *How extraordinary must the evidence be to qualify as extraordinary? When is enough enough?*
>
> George Schwartz

Having defined various types of paranormality and seen how they could operate in the normal world, it is now time to take a brief look at some of the evidence supporting the reality of these paranormal phenomena. A detailed examination would fill several books; our priority here is to see just enough evidence to demonstrate that these phenomena are plausible. Then, in the next chapter, we will look at common objections and rebuttals to those objections before deciding where—if anywhere—paranormality may fit into the central question we are seeking to answer throughout this book.

Psi Research

> *We must remember, just because the probability values are less than one in a billion, they still could have occurred by chance.*
>
> Roland Barthes

To its believers, psi provides a clear indication of just how incomplete our scientific knowledge of the universe is. Currently available psi evidence seems to support the idea that mind consists of more than mere brain. This is very much in accord with Eastern thought, in that we are aware of being made of matter and energy, because awareness is itself fundamental to matter and energy—a very idealist perspective. Mainstream science has been slow to accept psi, much less delve into how it works, but lay people seem to have far less of a problem accepting it. This discrepancy has less to do with the evidence itself and far more to do with the psychology, sociology, and history of the sciences. Mainstream scientists accuse those studying psi of dabbling in pseudoscience, primarily because doing so helps them avoid the growing gulf between materialism and real-world evidence. Science also sees psi researchers as lacking a theory of just how psi works, because scientists tend to get more than a little uncomfortable with raw facts that lack a suitably explanatory theory.

Some dismiss psi as wishful thinking; however, findings seem to rule this out. Also, if psi is not real, then most of the public is suffering from a mass delusion, given that about 75% of American adults believe in psi according to a 2005 Gallup poll. The same argument can be made for religion; however, psi as a field of study has neither the dogma, hierarchy, or sheer force behind it that religion does. Nobody is indoctrinated about psi from childhood on up, psychics don't appear on television spewing predictions and begging for money, there are no mega-psychic buildings, and so forth.

Experimental quality is a subject of repeated criticism, and paranormal researchers are working diligently to improve both the experiments and the controls to eliminate variables and other possible explanations for observed data. Skeptics believe that the evidence supporting psi will decrease as experimental quality improves. Results also tend to be a lot less marked when skeptics are the ones conducting the studies. This could be significant, except that people in all endeavors tend to find what they are looking for. Religious people see evidence of Jehovah or Allah everywhere they look, materialists see evidence of moving meat and other matter with nothing more happening in hidden dimensions, and those who do not believe in ghosts tend not to see them or to dismiss such encounters as something else. In this respect, psi is no different. The fact that skeptics tend to see decidedly less spectacular results is thus neither surprising nor cause for concern.

Some skeptics think that paranormal researchers conveniently leave unsuccessful studies out of their reported data, the aptly named "file drawer" problem. It is possible to estimate how many such studies are needed to reduce overall success rates. For example, there were 34 published studies between 1935 and 1939 that used the classic 5-symbol Zener card deck shown on the next page. It would take 29,000 studies (a ratio of 861 unpublished studies for every published study) to

The time has come for science to confront the serious implications of the fact that directed, willed mental activity can clearly and systematically alter brain function.
Jeffrey M. Schwartz

What a pity so many give undeserved credit to the drug, and not to their own efforts.
Thomas J. Moore

negate the results of those 34 studies. In mainstream science, a ratio of 5:1 is usually considered significant enough to be meaningful. By these standards, those ESP studies passed with stellar marks.

Of all the explanations conceivable, that one which attributes everything to imposture and trickery is unquestionably the most extraordinary and the least probable.
Maurice Maeterlinck

Psi may already be in use, however unwittingly. Consider for a moment that the daily variations of gambling payouts may not be all due to chance; some of that change may be due to the varying psi ability of the gamblers themselves. In more formal settings, experimenters can generate random *hits* (correct predictions, guessing the next card, etc.), and it is possible to know how the percentage of hits one could expect by pure chance. Say one experiment has a chance hit rate of 25% and 100 volunteers obtain an actual hit rate of 34%. In this example the 9% increase is of great statistical significance.

Here are just a few of the many psi experiments I learned of while researching this book:

- One medium correctly predicted 18 out of 25 newspaper headlines for the following day before the print had been set. The odds of this happening by chance are extremely small, to put it gently. My friend Sarah has predicted over a dozen events in my life up to a year in advance with 100% accuracy. This is yet more evidence that causation may flow both forward and backward, in defiance of the assumption that cause always precedes effect. Such a supposition sounds outlandish until one remembers that time is anything but a smoothly flowing river. (See Chapter 28.) Repeated experiments are demonstrating that precognition may exist. The mind may be in touch with its future self and/or spread out in time; experiments are ongoing.

Classical scientific assumptions simply do not account for how mind-body interactions, biofeedback, or the placebo effect works.
Dean Radin

- In 1979, a psychic named Joseph McMoneagle (1946-) taking part in Project Stargate (see Chapter 15) was given map coordinates and told to see what he could see. McMoneagle saw both a building in Soviet territory in which workers were welding a huge submarine and digging a future canal that would be required to launch this behemoth. Four months later, an artificial canal was blasted from the building to the water and the first Typhoon class ballistic missile submarine, the *Dmitriy Donskoy*, was launched. At the time, it was the largest and deadliest submarine in the world.

- Psychologist William Braud conducted a series of experiments over the course of 17 years that focused on people trying to affect the nervous systems of remotely located people. The results were published in 1991 and placed the odds against chance at more than 10^{14} (100 trillion) to 1. It seems that people can respond to distant mental stimuli without necessarily being aware of what's going on.

Churchgoers just share a different set of paranormal beliefs than non-churchgoers.
Victor Stenger

This is all well and good, but what happens when we combine multiple studies to determine their overall odds of chance?

- Taking into account all ESP card tests between 1882 and 1939 as published in 186 publications by experimenters around the world yields a database of over 4 million trials with overall odds against chance at a billion trillion (10^{21}) to 1. The "file drawer" would need to contain 626,000 reports, or more than 3,300 unpublished unsuccessful studies for each successful published report. It strains credulity to dismiss this as chance.

- In 1988, physicist Edwin May analyzed all of the psi experiments conducted at SRI International from 1973 to 1988. There were 154 such experiments comprising 26,000 trials and overall odds against chance of 10^{20} to 1.

- An analysis of 334 random-number trials published as of 1987 at the Princeton Engineering Anomalies Research (PEAR) laboratory concluded that the odds against chance were 100 billion to 1.

- Considering psi in general, the overall odds against chance have been reported as 1.3×10^{104} to 1. The entire universe is thought to contain 10^{80} atoms.

Everything science has taught me—and continues to teach me—strengthens my belief in the continuity of our spiritual existence after death. Nothing disappears without a trace.
Werner von Braun

How does psi stack up against more mainstream research? Both social and physical science reviews demonstrate the same 45% rate of statistical disagreement when no studies are eliminated from the results. This means that "hard" quantum physics experiments have replication rates that are not appreciably higher than those of the "soft" social sciences. Psi research seems to be easily holding its own.

Replication is a valid issue. Science relies on replication to eliminate the possibility of error, experimental flaw, or other variable. ESP has a rocky history when it comes to replication. Those who are open to it have the best results, while skeptics consistently complain that they are unable to replicate what they consider to be the wild claims of their more liberal peers. Some effects can be hard to replicate because of factors such as experimenter bias, nuanced procedures that are not followed closely enough, or evolving phenomena over time, but this does not make the results any less valid.

What if psi really is observed, the experiment is replicated, and variables such as design error are ruled out? Would this convince the skeptics? "Not so fast," say the skeptics because the observed effect might be real but too small to have any theoretical or practical uses; however, if that is true, then we need to pull aspirin off drugstore shelves immediately, because it reduces the probability of a heart attack by 0.8% compared to a placebo. Nevertheless, trials were ended because the researchers felt it was cruel to keep people on the placebo! According to some psi researchers, the effect of aspirin compared to a placebo is 10 times less than the psi effects observed during meta-analysis of psi experiments, with odds against chance greater than a million billion (10^{15}) to 1.

Is it possible that all of these experiments are flawed? Could the meta-analyses of all of these experiments be wrong? Could the skeptics and doubters be correct? Yes on all counts. Can an expanded definition of consciousness explain these results? Can psi research provide a piece of the puzzle to help answer our questions about how the universe works and what happens when we die? Must any valid theory of psi include NDEs and OBEs? Yes.

Three ESP claims warrant further serious investigation using carefully designed and controlled experiments:

- That human thought alone can affect a random number generator, however slightly.

> *The urge to transcend self-conscious selfhood is a principal appetite of the soul.*
> Aldous Huxley

> *Some of the facts we have considered suggest that the belief in life after death, which so many persons have found no particular difficulty in accepting as an article in religious faith, may well be capable of empirical proof.*
> C. J. Ducasse

- That people in a calm, relaxed state (such as being under mild sedation) can receive thoughts or images projected directly into their minds.

- That children's accounts of past lives sometimes prove accurate in situations where the child had no normal way to know of the information being recounted.

Accepting psi as real presents serious challenges to science, philosophy, and religion because it forces us to reexamine our most basic understanding of space, time, mind, and matter that will keep philosophers busy for generations. Theologians will have a scientific explanation for miracles. Some scientists will have to go back to the drawing board. Our understanding of ourselves, our universe, and our place in our universe will evolve. Evolution is not a bad thing. We may not be ready to fully accept psi but we can—and must—leave the door open.

NDE Research

Many people believe that NDEs offer the most compelling evidence of an afterlife. NDEs themselves have been confirmed. The only question—indeed the only question that really matters—is whether an OBE occurs or if the whole thing is nothing more than the result of a brain in extremis. The earliest referral to an NDE appears in Plato's *Republic* in which a soldier named Er describes a classic NDE after being revived from his injuries. A 1982 Gallup poll revealed that 4% of the United States population had experienced an NDE. NDEs have been reported around the world in places and among people as varied as Argentina, Bolivia, among Native American tribes, Buddhists, Muslims, China, and Siberia.

The December 15, 2001 issue of the renowned medical journal *Lancet* published the results of a 13-year NDE study conducted at 10 Dutch hospitals. The study's chief investigator, cardiologist Pim van Lommel said that, "Our results show that medical factors cannot account for the occurrence of NDE. All patients had a cardiac arrest and were clinically dead, with unconsciousness resulting from insufficient blood supply to the brain. In those circumstances, the EEG becomes flat, and if CPR is not started within 5-10 minutes, irreparable damage is done to the brain and the patient will die. According to the

Any explanation of the NDE is going to have to account for transcendental experience in general.
Kenneth Ring

In fact, of all the reports I have gathered, not one person has painted the mythological picture of what lies hereafter.
Raymond A. Moody, Jr.

> *This is eternal bliss, I thought. This cannot be described; it is too wonderful!*
> Carl Jung

theory that NDE is caused by anoxia, all patients in our study should have had an NDE, but only 18% reported having an NDE. There is also a theory that NDE is caused psychologically, by the fear of death. But only a very small percentage of our patients said they had been afraid seconds before their cardiac arrest because it happened too suddenly for them to realize what was occurring. More patients than the frightened ones reported NDEs." There was no correlation between NDEs and the length of the cardiac arrest, unconsciousness, medication, or other factors.

In general, there is no difference in the intelligence, personality types, or any other pertinent trait among people who report NDEs versus those who do not. NDEs also tend to be at odds with religious and cultural expectations, although people may interpret the events in the context of their religions. The Dutch study concluded that, "NDE pushes at the limits of medical ideas about the range of human consciousness and the mind-brain relation. Another theory holds that NDE might be a changing state of consciousness (transcendence), in which identity, cognition, and emotion function independently from the unconscious body, but retain the possibility of non-sensory perception."

> *It was a feeling I had never known before and have not had since. It was like bathing in a glorious calmness.*
> Damian Brinkley

Your brain is an extremely complex organ with an extremely intricate network of neurons arranged into circuits that perform different functions. A very high number of these circuits over many areas of the brain must be working coherently for you to have a lucid experience. A dying brain should not be able to produce NDEs; on the contrary, we would expect such a brain to become anything but more focused. In fairness, we may be seeing only the "uphill side" of the NDE. A 2009 study found that carbon dioxide levels in the blood affected blood pH levels and were a significant factor in people who experienced NDEs. Potassium levels were also cited. It is possible that NDEs occur during the first few moments of a cardiac attack, after which the person quickly goes unconscious and is revived later. If this is the case, then the NDE, a period of unconsciousness, and awakening could be interpreted as one seamless experience in the same way that we are not aware of the hours passing as we sleep. If we could peer into a brain that has gone past the point of no return, we might see the lucidity and clarity of the NDE dissolving into hallucinations and chaos before winking out forever.

Then again, we would also expect NDEs to be highly personal events with fewer common features, and with far less meaning associated with the experiences. The startling consistency of all NDEs is significant in and of itself. Also, if an NDE really is the result of a dying brain, then the full recovery experienced by survivors is an absolute miracle, because irreparable brain damage is said to occur after about six minutes without oxygen. Perhaps the 6-minute rule is wrong and brain cells die more slowly than thought.

But why should this experience be pleasant? Why do so many survivors recount being reluctant to return? Everything we know about evolution tells us that willingly accepting death is decidedly not a survival trait; besides, how would such a trait have evolved, since those who had it would not live to reproduce more? Ring reports that survivors experienced peace, well-being, lack of pain, detachment from their physical bodies, warm inviting light, and meeting people within the light. Greyson reports altered time flow, accelerated thought processes, a life review, sudden understanding, peace, joy, a sense of oneness with the cosmos, seeing and/or being surrounded by light, vivid sensations, separation from the physical body, a sense of being in the presence of a mystical entity or being, a sense of a border or point of no return, and more. He developed the 32-point Greyson scale that determines whether an NDE occurred and, if so, how deeply the person went into it. A score of 7 on this scale qualifies as an NDE. Dr. Robert Crookall reports that survivors view the dead body as a worn out garment or cocoon and may not even realize that they are dead.

The life review or judgment is another nearly universal feature of NDEs that makes no sense in a classical evolutionary context. The saying, "I saw my life flash before my eyes" is valid; I've had similar experiences on more than occasion in "near miss" situations, such as the time a van began merging into my lane while I was crossing the Bay Bridge on my motorcycle and I had no place to go. But I've never been judged or felt that I was undergoing any sort of review. NDE survivors report having to confront their actions and experience the happiness and suffering of the people and animals they

---

The brain function these near-death patients were found to have while unconscious is commonly believed to be incapable of sustaining lucid thought processes or allowing lasting memories to form.
Sam Parnia

It is by means of tranquility of mind that you are able to transmute this false mind of death and rebirth into the clear intuitive mind and, by so doing, to realize the primal and enlightening essence of mind.
Surangama Sutra

affected during their lives. This seems to be a self-judgment where the rewards and punishments are part of the judgment itself. Interestingly, this step is seen as indispensable for spiritual progress and similar to purgatory.

How far a person goes into an NDE depends on whether they experienced clinical death and, if so, for how long. Those who experienced clinical death had deeper, more complete experiences. Those who were dead longer had deeper experiences, and some even approached what they perceived as a final barrier. Is this barrier the point at which brain shutdown is irreversible, or is it the doorway to the beyond? The report from one survivor who claimed that the prayers of loved ones were holding her back offers an enticing clue and may be one reason why the Tibetan Book of the Dead focuses just as much on those left behind as those departing this world. People interpret the being of light differently according to their cultural and religious backgrounds, but the common theme is that the being tells them that some mistake has occurred and/or that the person still has work to do in this lifetime.

People who express reluctance to return can be said to have lost the will to live, but this should not be confused with any sort of death wish. It's one thing to want to go there. It's another thing to want to stay there once you've arrived. Survivors do report an intense desired to return to their bodies followed by not wanting to come back, especially if they encountered the being of light. Survivors acknowledge that some forms of death are more desirable than others and feel they have more changing to do and/or other tasks to perform before being ready to die for good. 82% of NDE survivors report less fear of death and 77% believe in an afterlife (which does not mean they didn't before), because their doubts about survival have been eliminated in the second most dramatic way possible short of dying and not coming back. The conviction that they had been dead has a tremendous impact on their lives. Survivors refute the death-as-sleep analogy and say that death is merely a transition to a different state of consciousness or being.

Children are the most interesting group of NDE survivors. Many children with life-threatening conditions report NDEs, while those with less serious problems do not report them. Children treated with all manner of mind-altering medications

---

The soul lives by that which it loves rather than in the body which it animates. For it has not life in the body, but rather gives it to the body and lives in that which it loves.
John of the Cross

If the apparent separateness of subatomic particles is illusory, it means that at a deeper level of reality all things in the universe are infinitely interconnected.
Michael Talbot

are no more likely to report an NDE than those without such treatments. Child survivors of NDEs tend to grow up as gifted children who enjoy close relationships with others. Pediatrician Michael Morse has studied NDEs in children and reported many accounts, including a 2 year old child who was not breathing for 25 minutes who returned with descriptions of a room with a bright light and a very nice man who asked if he wanted to stay or come back. Children are far less likely to experience a classic cardiac arrest than adults, and one may therefore presume that carbon dioxide and potassium are less significant factors for them. Also, there is no reason for a child to describe the being of light as an adult or to know that such an encounter is part of a classic NDE.

The single most robust objection to the idea that NDEs provide a window to the afterlife is the simple fact that all survivors have come back. This much is obvious and raises the question of whether they were really dead. It turns out that death is not a single event but a continuum, barring sudden massive trauma like being next to a large explosive detonation or a particularly violent trauma, such as the motorcycle accident I responded to years ago in which the rider had literally been torn to pieces by the impact. Nevertheless, those who had the experiences were clinically dead with no vital signs. Some even had their vital signs deliberately switched off, such as for a brain operation to repair an aneurism, where the body is cooled to 60 degrees and all blood drained. People in this state are dead for all intents and purposes, and have no measurable signs of life.

It is entirely possible that NDEs will eventually be fully explained as purely material events caused by such factors as blood gas levels, but even such a mundane explanation will still expand our knowledge of brain and mind and will not preclude the existence of an afterlife. Unless and until this happens, however, what is so wrong about accepting the idea that the line between life and death is a fuzzy one that can be crossed in both directions as a working hypothesis? This is not an unreasonable position to take, especially because some NDEs have independent correlation from people who validated events witnessed during the NDE, including people outside the operating room. A good percentage of survivors give very precise accounts of what happened while they were

> *Scientists and nonscientists alike are experiencing a test of faith—in this case, whether we can put our belief in the scientific method itself. Because if we are to put our faith in the scientific method, and trust what the data reveal, we are led to the hypothesis that the universe is more wondrous than imagined in our wildest flights of fancy.*
> Gary E. Schwartz

under, including things they could not have known about in advance such as tools, procedures, what people said, etc.

The growing body of anecdotal evidence surrounding NDEs is one more thing people point to as evidence for an afterlife, which is perfectly natural since we would all like to cheat death, or at least know that we will carry on as something more than gradually dispersing atoms. There is a growing interest in NDEs, and more than enough evidence to prove that they are neither hoaxes nor cases of mistaken memories. So what exactly are they? Millions of people can be wrong, as we learned in Chapters 7 through 12, but NDEs themselves are real, if for no other reason than the sheer number of people involved who have nothing to gain by lying about it. Too many people experience dramatic personal transformations. Humans are all wired very similarly and tend to respond similarly to a given stimulus or drug, but just try getting three or more people to agree on which movie to rent at the local video store, much less to agree in such detail about what is arguably the single most significant event in their lives. NDEs seem far too consistent to be by-products of a dying brain.

Sagan thought that NDEs could be recollections of birth. Birth certainly involves a tunnel with a bright light, beings surrounded by bright light, and the second most profound personal transformation next to one's future death, but this theory still has flaws in that it requires excessive cognitive ability in the infant. Also, NDEs reported by people born by Cesarean section are no different from those of people born vaginally.

By rights, dying brains should be confused and paranoid, and yet people reporting NDEs recall no pain or anguish. Are NDEs a psychological response to imminent death? *Dissociation* (the splitting off of a group of mental processes from the main body of consciousness) is a powerful defense, and death is an extremely stressful event. Evolution is not compassionate and cares nothing for comfort. The death of an individual animal is relevant only as it benefits the species; however, being able to withstand extreme pain or injury during life is an essential survival trait. This could be the basis for an NDE, in which the body responds with loads of endorphins and other hormones that can trigger us to fight or fly under less extreme circumstances. It could be argued that the body does not

---

It is all too easy for hardened critics to make even the best fieldwork sound ridiculous, and hardened critics have brought this form of ridicule to a fine art, prizing it above any real attempt to get at the truth of things.
David Fontana

Science, while perpetually denying an unseen world, is perpetually revealing it.
F. W. H. Myers

know it is dying and is carrying on the struggle for survival on a very fundamental level. But then why be reluctant to return?

As for the life review, experiments have demonstrated that stimulating parts of the brain causes people to relive triggered events. This is not mere memory; people are actually reliving the events themselves, albeit only in their own heads (as if we ever experience anything any other way). We may store our experiences in far more detail than we give ourselves credit for and play them back during the NDE. Still, the trauma of dying should make thoughts and memories less coherent, but people are more aware than normal during NDEs. Also, simply reliving an event does not necessarily entail a review in which the effects of one's actions are evaluated along with any pain or joy caused by those actions. In fact, NDEs are less common in people whose EEG readings showed the brain to be functioning. Thus, according to the *Lancet* article, the physical brain cannot be necessary for all thought.

There are many theories about what causes NDEs. Author Susan Blackmore (1951-) points out that oxygen deprivation can cause tunnel vision, but many people experienced their NDEs just before their accidents when their brain oxygen levels were presumably normal. Oxygen deprivation impairs both consciousness and memory, and causes vision loss from the periphery inward. No NDE survivor has reported seeing a tunnel surrounded by blackness. Drugs don't seem to be a factor; plenty of NDE reports come from people who had no drugs in their system, and administering powerful drugs during surgery or revival attempts does not alter the NDE. Temporal lobe epilepsy is blamed for everything from religious experiences to NDEs, but no report of an epileptic seizure bears any resemblance to an NDE, and epileptics don't remember the seizure afterward.

I have already mentioned the possible role of excess carbon dioxide, potassium, and/or sodium in causing NDEs; however, the brain depends on the bloodstream. Oxygen deprivation causes loss of consciousness in about 20 seconds, after which the cells can tap an emergency supply of adenosine triphosphate that also gets used up quickly. Brain damage begins to occur within 5-10 minutes, and is very significant if blood flow is not restarted within 15 to 20 minutes. Meanwhile, the cells are "screaming" for food by releasing chemicals to try to

---

Surgeons are wary of people who are convinced that they will die. There are examples of studies done on people undergoing surgery who almost want to die to re-contact a loved one. Close to 100% of people in those circumstances die.
Herbert Benson

Neither the placebo effect nor the nocebo effect have been much studied. Mental discomfort with such squishy phenomena aside, there's no money in it.
Susan McCarthy

activate blood flow before they die completely. Despite all these mechanisms, life is over when the heart and lungs stop working. Brain cells begin dying unless a doctor manages to get things going again.

The placebo effect is unaffected by intelligence or any test of susceptibility.
W. Grant Thompson

The chest compressions performed during CPR do not deliver enough oxygen to the brain. Conventional CPR without epinephrine delivers about 5% of normal blood oxygen while adding epinephrine increases this to 30%. This is way better than nothing and CPR has saved many lives, but these numbers are not compatible with brain electrical activity. Any delay in beginning resuscitative efforts can result in brain activity remaining flat for up to two hours after the heart has been restarted, because the blood vessels need time to open up and restore the full flow of oxygen needed to restart electrical activity. It may be possible for thoughts to arise in a small part of the brain that has blood flow, but the problem of getting this flow to said part of the brain remains. Stopping the heart stops blood flow to all parts of the body. If mind is a product of brain, then by rights it should go dark when the brain shuts down. This is not what happens. If you turn off a light switch and the light stays on, the only conclusion is that the light is getting power from somewhere else.

NDEs are often compared to drug trips, but people coming down from such trips don't report experiences resembling an NDE, and certainly nothing as "far out" as an NDE. Nevertheless, many cultures around the world use drugs to achieve transcendent states, such as the American Indians in the deserts of the Southwest who use peyote to induce visions and bring about enlightenment as a means of reaching other dimensions of reality. Drugs could be one path to these dimensions, but there is no reason to think they are the only path.

The doctor who fails to have a placebo effect on his patients should become a pathologist.
J. N. Blau

Perhaps NDEs don't happen while the heart is stopped; but then how can patients recall the procedures and what was said while they were under? NDEs could be hallucinations, but that would not account for the uniformity of the experience among thousands of people. Again, some people who return from NDEs give accurate accounts of the resuscitation procedures in detail, even when they have no prior medical training or knowledge, and when their EEGs were at flatline. In fairness, EEGs are not completely accurate at very low activity levels. It is possible that a seeming flatline is actually hiding a small but sufficient amount of electrical activity; however, that does not seem to be able to account for the lucid, well structured thoughts and feelings—complete with reasoning and memories—that takes place during an NDE. Even people in the deepest level of sleep have EEG readouts that are anything but flatline:

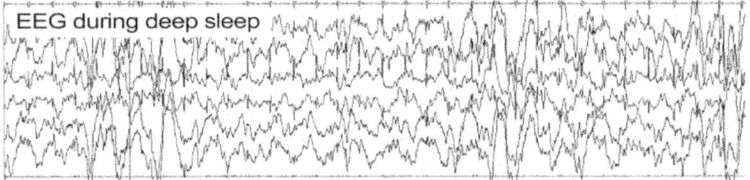

> *I don't think you can locate the source of consciousness. I am quite sure it is not I the brain—not inside of the skull. It actually, according to my experience, would lie beyond time and space so it is not localizable.*
> Stanislav Grov

Maybe electrical activity does not mediate consciousness. It is possible that subatomic quantum processes are ultimately responsible; but then how can something as indeterminate and uncertain as that give rise to a coherent and persistent pattern called, "I?"

These questions have not deterred materialists, who continue to claim, sometimes convincingly, that NDEs are hallucinations, albeit extraordinary hallucinations. The extreme realism of NDEs could be because a person near death is in the most extreme situation imaginable that could make hallucinations seem all the more real, especially when combined with the idea of a big final event that we start learning about as children. If NDEs come from our memories, then we could expect people who are better at imagining things from different perspectives to have more OBE-like experiences during NDEs and dreams. According to Blackmore, this idea has experimental support. Our brains need to decide what is real and will latch on to the most stable model to build its idea of reality. What is there for a dying brain to latch on to?

The loss of memory associated with Alzheimer's or other degenerative illnesses seems to indicate that people need intact brains in order to function and retain coherent memories. Other experiments, such as those in which a researcher removed different segments of mice brains without impacting memories (see Chapter 30), argue against the need for an intact brain and for the idea our brains might not produce thoughts or consciousness, but rather act like a TV set. Again, damaging or destroying a TV set affects how the signal looks to all observers and measurements, but does not affect the sig-

> *Still man does not die, but is only separated from the corporeal part which was of use to him in the world. Man, when he dies, only I realized that this experience led to the unfolding of my destiny that I had chosen before my birth into this lifetime.*
> Gloria Chadwick

nal itself. It is therefore entirely possible that NDEs are spiritual experiences that indicate the existence of a soul.

Having an NDE is a strong predictor of mortality. More patients who reported a deep NDE died within 30 days of receiving CPR than those who did not. Maybe the NDE was caused by physical factors. Maybe the damage done by the injury or illness was so great that the body simply never recovered. Maybe people who have experienced deep NDEs have lost their fear of death and the will to fight it. Any of these possibilities is compatible with both material and spiritual explanations of NDEs.

Mediumship Experiments

The truth is at the bottom of the abyss.
Democritus

If NDEs are evidence of people in transition to the afterlife, then it stands to reason that we may be able to contact these people once they have completely crossed over. This is where mediumship comes in. (See Chapter 15.) It is important to distinguish between the three categories of so-called medium:

- Sensitives are psychics who may be telepathic, clairvoyant, etc. but who do not have communications with the afterlife.

- Pseudo-mediums are impostors and frauds.

- Mediumship practitioners are legitimate to the extent that they truly believe in what they are doing and are not trying to fool or defraud anyone in any way.

I perceived the extravagant splendor, the indescribable joy of life in its entirety.
Margaret Montague

Experiments attempting to prove or disprove medium abilities have been conducted since the 1800s, with mixed results. Mediums have been tied, blindfolded, gagged, and more to keep them from moving tables and other furniture, and been subjected to detailed body searches to eliminate the possibility of their hiding fake ectoplasm, which is often used to convince sitters that the medium is legitimate. More than one medium has stashed various materials from flimsy cloth to paper, animal organs, etc. on and in their person to dupe audiences, and not a single verifiable example of ectoplasm has ever been produced or subjected to any sort of analysis under laboratory conditions. In 1920, one Dr. Crawford speculated that ectoplasm could come from the medium's rectum. He examined her panties and found bits of feces that he attributed to ectoplasmic production and not a lapse in personal hygiene. Escapades like this are examples of the rampant fraud that has plagued serious attempts to discern whether or not there is anything more to mediumship than hucksterism and deceit.

How can we know if afterlife communications are just that? The primary method is through the deceased person passing on verifiable information that neither the medium nor the sitter can have known about beforehand. For example, if a deceased grandfather tells a sitter about a watch he dropped behind a cabinet many years ago and this watch is found, that would provide strong evidence for the reality of survival.

In *The Afterlife Experiments*, Schwartz discusses his series of carefully controlled studies that he claims do all but prove the veracity of mediumship and contact with the dead, because their odds of chance are less than 1 in 10^{12}. Clearly something is going on, but what? Some of the many ways in which a skilled person might obtain false positive results include:

- The mediums may be extremely gifted detectives who can learn about the sitters in advance without being detected.

- Sitters may construct false memories based on what the medium is saying, thus creating a false "hit."

- A medium may make an extraordinarily lucky guess.

- The medium may be able to interpret subtle clues such as intonation, breathing, etc.

- The medium may be engaging in telepathy and/or remote viewing, which is paranormal, but which does not involve communication with the dead.

Reports from the Afterlife

Let us assume for a moment that the afterlife is real and that communications with the deceased are possible. All of the world's religions teach some form of survival after death, and idealism practically demands it. Religions differ in how they see the afterlife, with most except Buddhism believing that some form of progress is possible in the next world. It is pos-

---

The myth of eternal progress in human understanding, which lies behind so much of our delusory intellectual arrogance in modern times, can clearly be seen at least in spiritual matters for what it is, a myth.
David Fontana

The most zealous behave like vigilantes plying the frontiers of science.
Rupert Sheldrake

sible that all consciousness blends into a single whole or goes to a place that could be described as heaven or hell. JCI views of the afterlife as being a one-way trip are not necessarily incompatible with the Buddhist teaching of reincarnation because A) the deceased is permanently gone from this world as we know it, and B) the next life will be influenced by the karma accumulated during this lifetime.

Critics point to the differences between religions as proof that ideas about the afterlife are cultural and not descriptive; however, as I have pointed out, these differences do not add up to any fundamental incompatibility or contradictions. It is important to note that some continuation of the identities we hold while on Earth is necessary if the term "survival" is to have much meaning for us. The Buddhist idea that reincarnating is like passing a flame from candle to candle and that our Earthly identities or anything identifiable as "us" only last this lifetime seems to be only cold comfort. More on this later.

So what is the afterlife like? According to the deceased, all life is equal. This world and the next are similar, "bodies" go with people, plants, etc. Dr. Carl Wickland (1861-1945) spent decades researching death and dying. He concluded that, "Death does not make a saint of a sinner or a sage of a fool." A dead person's mentality is the same as before; people carry their mental baggage with them, even their disbelief in the afterlife, and hang onto it at least for a while. Suffering in this world is important for spiritual development (a very Buddhist-sounding concept); however, they say that reincarnation is sporadic and undertaken only when there is no other way, or when one desires to help those on Earth. (This is very similar to the Buddhist idea of a *bodhisattva* or person who has achieved enlightenment and unity with Brahman/God but who comes back to help others achieve enlightenment in an act of pure altruism.) The deceased also say that mediumship is not required to speak with the dead, who live outside time and at the "edge of space," which supports both holographic theories and idealism. Questions about mundane things like food are only meaningful in the context of life on Earth.

Most people judge this aspect of Spiritualism [physical mediumship] harshly, but if one stops to consider the number of bad popes and charlatan evangelists, perhaps Spiritualism has not done so badly after all.
Allen Spragett

To argue that the Dalai Lama's views on reincarnation are 'against the very foundation of modern neuroscience' is simply not true and shows a profound lack of understanding of where and what those foundations are.
John H. Hannigan

Lives Before Life

If NDEs are a glimpse into the beyond and mediums really can communicate with the dead, then it stands to reason that our current lifetimes are neither our last nor our first. Some believe that anyone can remember her or his past lives if s/he wants to know and believes that s/he can and will know. Children seem to be the most likely to remember past lives and to tell us about them. The Division of Personality Studies at the University of Virginia has over 2,500 documented cases of past lives from children around the world.

Most children who talk about past lives do so between the ages of two and four, and tend to stop talking about them by age six. They do not provide facts, but speak from the point of view of the dead person. They also treat people the dead person used to know, not as someone with information, but as if they were the dead person. Here are two examples:

- Abbie Swanson was four when she told her mother that she used to be the mother's grandmother (her own great-grandmother) and knew the same intimate details of their lives, even though she could not have known about it in this life.

- Susanne Ghanem gave 25 names from her previous life with only one mistake. The odds of her guessing 24 out of 25 names correctly are extremely small.

Why don't more children speak about past lives? Many of us have been taught since youth not to speak about paranormal experiences, including our imaginary friends. This is especially true in the West, where this phenomenon may be greatly under reported.

Is this all a big joke or fraud? Dr. Helen Wambach (1925-1986) set out to debunk reincarnation. What she found was startling:

- The ratios of people who report past lives are in line with actual birthrates of 50.6% male and 49.4% female.

- The percentage of people reporting upper- or upper-middle-class lives was in perfect proportion to historical estimates of socioeconomic conditions during the reported periods.

If the flesh comes for the sake of the spirit, it is a miracle. But if the spirit comes for the sake of the flesh-it is a miracle of miracles.
The Gospel of Thomas

To hear of children crying for years for their family to take them to their previous parents until the family finally relents is not unusual.
Jim B. Tucker

> *There are very few scientists who have the courage to pursue the essence of human existence.*
> Michael Persinger

- Subjects recalled clothing, footwear, food, utensils, and other facts better than popular history books; querying experts in obscure subjects revealed that the reports were extremely accurate.

- The time distribution of recalled lives follows historical population growth. It is of course possible for a fixed number of souls to cause a higher or lower population on Earth simply by varying the timing of their birth.

75% of children describe the manner of death, but only 57% report a natural death, which seems to indicate that death from illness or old age has different effects than a traumatic or violent death. There are cases where children claiming to be shot in a past life have visible birthmarks resembling entry and exit wounds that are consistent with the descriptions of death. One cannot expect young children to be forensics or ballistics experts, or even to have enough knowledge to construct a plausible story that matches their birthmarks. Fraud may explain some cases, but certainly cannot explain all cases, especially those involving birthmarks.

Assuming that reincarnation is real, how can an injury in Person A show up as a birthmark in Person B? It is well known that mental factors can affect the body. Hopelessness increases the risk of cardiovascular disease, cancer, and other ills. Mental images can create physical changes. Negative emotions can literally create toxins in the body. Thus, a mind carrying traumatic memories when it inhabits a fetus could potentially explain these markings. Why don't more babies have wound-like birthmarks or other defects, and why do so many cases only involve the skin? Most of us are far more aware of our skin than of what may be happening inside us, for obvious reasons. Also, it would not do much good to impair the new body right as it is starting to live.

> *If you make love with the divine now, in the next life you will have the face of satisfied desire.*
> Kabir

What about alternative explanations? ESP may explain past life recollections, but does not explain the birthmarks, especially because children report remembering the specific injuries. People who are more open to hypnosis may also be more prone to such birthmarks. Some cases may be pure coincidence, but the odds against this are extremely small, especially when one combines the birthmarks with the many incredibly detailed statements. The evidence for past lives is inconclusive, but materialist arguments against them are very weak because they boil down to fraud, which cannot possibly explain all cases. The best answer to these cases seems to be that survival after death and reincarnation are both real. This of course does not mean that we will all reincarnate.

People who believe in reincarnation believe that we have souls, and that our souls survive the death of our physical bodies and spend time in other planes of existence before being reborn into a new physical body. This cycle keeps going until the soul achieves union with God. Stevenson investigated more than 3,000 childhood cases suggestive of past lives, during which he accumulated far too many specific notes to suggest coincidence or hoax.

Claims of reincarnation are far more common in India, where reincarnation is taught by both Hinduism and Buddhism. They are far more rare in the United States, where the JCI religions that preach a one-way trip to the next world hold sway, and where only 25% of the population believes in reincarnation. As I mentioned above, this phenomenon may be vastly underreported in the West, where children are not exactly encouraged to share such "tall tales." Many stories of reincarnation are local to where the dead person lived. This may be because the dead person knows her or his home best, and/or because it is far more difficult to recognize a child talking about distant events or to recognize such reports for what they are.

It's not true that life is one damn thing after another; it is one damned thing over and over.
Edna St. Vincent Millay

In Between Lives

Consider stars. They grow, reproduce, and die. Long after they are dead, their light shines on for millions or even billions of years. As far as the light is concerned, it is everywhere and everywhen in the universe at once. For those of us on Earth, time seems to be moving and carrying us to our deaths. There are 235 million seconds in the average lifetime, and we lose 86,400 of those seconds daily. The longer we live, the more quickly time seems to fly by, because each day consumes an ever smaller percentage of our total time on this planet. There is also evidence that our *circadian rhythm* (internal body clock) slows down as we get older, which also contributes to the illusion of time traveling faster. But what is time? As we will see

For there never was a time when I or you or these kings did not exist. And never shall we cease to be, hereafter.
Krsna

> *To show a bridge between the two existences would certainly be the solution of a great riddle.*
> Arthur Schopenhauer

in more detail in Chapter 28, time may be the biggest illusion of all. Our physical bodies may be the star, our souls the light. Bacteria are ageless. They can be killed but don't die of old age. Theoretically, a bacterium kept free from predation and carefully tended will last forever. If bacteria can have eternal youth, why not more complex beings after their own fashion?

Philosopher Pyotr Ouspensky (1878-1947) believed that people relive their lives over and over again, but that each life can be different and that growth can take place. The movie *Groundhog Day*, in which Bill Murray plays weatherman Phil Connors, who relives the same day over and over again until he finally gets it right, is an example of what Ouspensky was talking about. The main differences are that we would presumably get more than one day to live over and over again and would not be aware of the repetitions, unlike the fictional Mr. Connors. Every incarnation has its own body and a new brain to allow it to function in the physical world. Knowledge accumulated over successive lifetimes resides in the subconscious, where we can learn to access it. The time in between lives allows us to examine and understand our experiences across all lives, and decide what goals to accomplish and what lessons need to be learned. Our individual destinies are based on our past lives and karma, which create a structure and framework for moving forward. We also confer with other souls to confirm our choices and make commitments to ourselves and others. Our entire lives are thus planned before we are born into that life. Ouspensky offered no empirical or other evidence beyond his personal experiences to back up his assertions, but his ideas are intriguing nonetheless.

> *If you have an experience of leaving your body, becoming one with the universe, or meeting a spiritual being, Western science tells you it's an illusion.*
> Charles Tart

We have already seen that the Eastern religions believe that our selves do not survive, but that our patterns continue. This pattern is born, dies, and is reborn in a cycle that only ends when nirvana is attained (see Chapter 5). The goal of these many incarnations is to learn to shed attachments, aversions, and ignorance that cause suffering and stymie progress toward enlightenment. The thread of karma connects all of these lives and can go up or down; it can be in heaven for a while until the good karma runs out and returns to Earth, or in hell and suffering until it is time to come back up. Buddhists believe that the mind spends some time in limbo before occupying the cells of a developing fetus and gradually becomes identified with the new body. The overall effect of reincarnation is having someone born who is very similar to you.

The Buddhists say that the self is an illusion, to which the late psychologist David Fontana (-2010) asked, "Why struggle and learn as individuals if it's all for nought?" and conceived of individuals within the unity of Brahman. The distinction between individuals may cease, but there must still be a point to individuality, or else why go through the bother? If everything in the universe is evolving, why not the self as well?

What about animals? Very little work has been done in this area. I would personally love to see researchers ask chimpanzees, gorillas, and other animals trained to communicate with humans about the big questions. What do our evolutionary cousins think about life, death, God, and the meaning of it all? Do they think about such things? Darwin demonstrated, and legions of scientists have since proven, that the separation between humans and animals really is one of degree and not kind. After some research, the best I could come up with is the very anthropomorphic idea that animals who have lived close to humans may retain their individual identities; those who have not go back to a sort of collective species consciousness. I have a lot of problems with this view. To me, animals either come along for the ride, or no one of any species gets to ride. Nothing else seems to make sense in the context of evolution as we understand it.

Many questions remain about the mechanics of reincarnation, even if we assume that it does happen. Where does the soul go in between lives? How does it get a new body? What are the logistics of conferring with everyone the soul will meet in the next life? How is consensus reached among all the participants? Is there any hard evidence that all of this occurs? Keep in mind that reincarnation, afterlife, etc. are all possible without the extensive planning sessions postulated by Ouspensky and others.

Having taken this very brief tour of psi, it is time to move on by taking a closer look at skeptical objections and the responses to those objections. We will then discuss the self, the meaning of death, and ask ourselves how psi fits into the complex puzzle we are slowly assembling.

I had come to understand that there seemed to be little difference between birth and death, to any living creatures excepting humans.
Eileen Garrett

When all the souls had chosen their lives, they went before Lachesis. And she sent with each, as the guardian of his life and the fulfiller of his choice, the daemon that he had chosen.
Plato

Chapter 17

Weighing the Evidence

> *I suppose the strongest feelings came from realizing it would be ME who will die, not some other entity like old-lady-me or terminally-ill-and-ready-to-die-me.*
> Anonymous

So far, we have defined various aspects of paranormality and provided a framework within which those aspects could be exist within the laws of physics as we currently understand them. We also explored various concepts of subtle bodies and discussed the evidence for psi, NDEs, mediums, and reincarnation/past lives. Our examination of paranormality now turns to discuss some of the common objections raised by skeptics and how those arguments can be rebutted. We will conclude by pondering just what we mean when we say "I" and how those concepts relate to death, especially if we are willing to alter our definitions of our selves a little. At the end of this chapter, we will see that we are better able to define what we are talking about when we start asking questions about what happens when we die... if in fact we die at all.

Common Objections

> *If 'evidence' means the kind of support provided by reason and science, there is no evidence for God and immortality.*
> Martin Gardner

The single biggest objection to anything paranormal comes down to the idea that we humans keep trying to comfort ourselves by countering the seeming brevity and pointlessness of life with fancy tales of an afterlife, reincarnation, and gods. It is a cruel twist of fate that we are at once programmed to survive while being constantly aware of our own mortality. This is especially true today, because materialist science has knocked

God and the soul off their high perches, cast humanity out of the center of the universe, and fully exposed our fears about the pain and unpredictable nature of dying, confronting the death of loved ones, and losing our own selves when our turn inevitably comes. Our brains see themselves as selves and desperately cling to the idea of immortality, but material science sees no evidence that this occurs and has every reason to think it doesn't. We search for comfort among these hard facts but find none, because anesthetic, injuries, and illnesses can shut the brain off and/or cause other changes. Some people are almost unrecognizable after brain injuries, because their personalities have changed so drastically.

Skeptics continue to doubt evidence and findings related to any existence beyond this one and carry themselves with an air of dogmatic authority hitherto seen only in clergy, despite the growing evidence that contradicts their views. Where priests used to prohibit spiritual debate, materialist scientists do the same for any ideas about mind and physics that don't toe the "matter über alles" line. Scientists do not torture and murder dissenters, and skepticism is both valid and necessary for vetting and validating new ideas instead of blindly accepting them. But when does the urge to dismiss data become the wrong thing to do? Dogmatic skepticism is just as capable of stifling dissent as dogmatic religion. In a perfect world, science and religion would follow the data no matter where it leads. I submit that any religion that followed scientific advances and updated its books and tenets to accommodate new discoveries would find its faith not weakened, but strengthened. But I digress...

To materialist science, the material world is all there is. Evidence pointing to an afterlife and/or reincarnation is by definition not material, and is therefore disregarded. The best data we have points to consciousness being within and produced by the brain, period. Consciousness, self, everything perishes when the brain dies. Any other conclusion is pure imagination that is no more real simply because it is vivid and detailed. Even a psychic accurately telling a person where to find an object could have many possible explanations that require nothing "supernatural." This view may be a failure of imagination and ability to conceptualize what such scenarios might

> *The way I see it, being dead is not terribly far off from being on a cruise ship. Most of your time is spent lying on your back. The brain has shut down. The flesh begins to soften. Nothing much new happens, and nothing is expected of you.*
> Mary Roach

> *All these memories, and others more numerous than the stars in the heavens, are available only to me. And each and every one is but a ghostly image and will be switched off forever with my death.*
> Irvin D. Yalom

look like, which might have more to do with the enmity between science and religion than with any hard data.

What about fraud? We cannot turn on the news today without seeing at least one story of fraud perpetrated for personal gain. There are many good reasons to cook up real-sounding stories of hauntings, mediums, etc. For example, a bed and breakfast may offer perfectly wonderful accommodations and meals, but a haunted bed and breakfast will probably get more business and free publicity. People can also be dishonest when there is nothing to be gained. Some years ago I signed a literary agent to help me market a series of science fiction novels I've been dabbling with since grade school. This agent put together an impressive package to send to publishers that included rave reviews from many leading book reviewers. There was just one problem: The reviewers in question had never heard of my books, a fact that any prospective publisher could verify just as easily as I did. The agent never charged me a dime for her services, leaving me to try to figure out a plausible motive for her actions. It is therefore possible that there are plenty of hoaxes about the afterlife, reincarnation, NDEs, OBEs, mediums, and other paranormal things out there that have nothing to do with any direct benefit to the perpetrators.

Finding myself to exist in the world, I believe I shall in some shape or other always exist; and, with all the inconveniences human life is liable to, I shall not object to a new edition of mine, hoping, however, that the errata of the last may be corrected.
Benjamin Franklin

People who have mystical or religious experiences may be sincere. It should be possible to test for any "supernatural" aspects of these cases, because we can expect that these people would return with some new knowledge or insight that could be put to the test; however, they typically come back bearing good behavioral advice that requires no otherworldly source to explain. One person returning with heretofore undiscovered knowledge, such as a warning of imminent disaster or some new scientific advance, would provide compelling evidence, but so far no such report has stood up to scientific scrutiny. (I wish I had documented my experiences with Sarah more carefully and may yet approach her to conduct more detailed testing, because I have seen proof positive that precognition and remote viewing are very real.)

Knowledge of ourselves teaches us whence we come, where we are, and whither we are going.
Ruysbroek

Social experiments prove that a bandwagon effect can cause people to start believing something that others already believe. Consider the many millions of people who flock to Lourdes despite the complete lack of evidence for any miracle cure, or the so-called "Tea Party" in the US where people are actively championing public policies that are directly against their own self interests under the guise of self-determination. (The "Tea Party" is actually bankrolled by billionaires who presumably don't have the little guy's interests at heart.) Fads are similar phenomena, such as the national craze for Cabbage Patch Kids in the 1980s. If enough people believe in things like ghosts, an afterlife, or God, then it becomes easier for more people to adopt those beliefs as well; however, the sheer number of people who believe something says nothing about the validity of the belief in question.

Skeptics continually ask how a discarnate mind or soul can possibly influence physical objects, since there is no measurable energy transfer taking place between the physical world and a hidden spirit world. This argument blows a fatal hole in dualist theories, including all dualist religions. Thus, the only way ghosts, NDEs, psi, etc. can be real is if materialism is also false, if consciousness and not matter is the primary constituent of the universe. We will explore this more in Chapter 31.

The previously mentioned file drawer problem keeps coming up, as does the unreliability of meta analysis, since it is very vulnerable to bias. Skeptics claims that no new phenomenon has ever been discovered or proven through meta analysis. Meanwhile, paranormal experiments that seem to have generated results are proving devilishly difficult to replicate, and in fact have never been replicated under their original conditions.

So far, mainstream science can only test meditation and mysticism. Beyond that, researchers must either accept personal reports at face value and do their best to weed out alternative explanations and fraud, or try to devise suitable replicable tests. But why bother? Given the fundamental invalidity of dualism, the small magnitude of observed effects, the file drawer problem, the unreliability of meta analysis, and the inability to replicate experimental positive results, how can anyone possibly keep holding out the hope that physical death isn't the end of the line?

Few mainstream scientists take idealism seriously, which means that it almost never enters the debate. It is far easier for science to simply dismiss anything and everything that conflicts with a materialist interpretation of the universe than to try to wrap its collective mind around it. As Dr. Elahi Ebby

Nothing burns in hell but the self.
Theologica Germanica

Keep two pieces of paper in your pockets at all times. One that says 'I am a speck of dust.' And the other, 'The world was made for me.'
Rabbi Bunim

said, "Reality is not neurologically determined but socially, and our understanding is limited by our own limitations. What we consider reality is just that slice we can perceive. Any personal experience like an NDE may be real to the person who had it, but the rest of us can't agree because we haven't had one."

Metaphysics is not a science; theology even less so.
Andre Compte-Sponville

Science critics accept certain odds for natural phenomena, but seem to apply a double standard to paranormal phenomena. For example, a skeptic confronted by convincing evidence of the paranormal and survival after death may claim that "leakage" from the physical world through word of mouth, investigation, or cheating is causing a false positive. Such a skeptic may even argue for "super psi" to claim that the evidence does not indicate afterlife communications or survival after death. Of course, the "super psi" argument relies on accepting consciousness as non-local, which undermines the very argument it is supposed to bolster.

If memory is like a blank disk that is written with data corresponding to each experience, then it may be possible to override short-term recall and cause a sense of *déjà vu* (a sense of certainty that once has seen or experienced a new situation on a previous occasion). Déjà vu, full-memory recall, and other transcendent states have been linked to temporal lobe epilepsy (Penfield). Epileptics seem to see the world very differently than the rest of us. This may be true; however, epilepsy has been implicated in everything from religious experiences to OBEs and déjà vu. I don't doubt that epilepsy can have a profound impact on a person, but it seems like a very convenient excuse that gives skeptical arguments the air of truth and authority without really doing much to answer the question.

In the sorrow of grief humans need to be consoled.
Dennis Klass

Some repository of all that has happened—such as the Akashic record—is also put forth as a possible source of paranormal phenomena, including medium communications. This makes sense from both a timeless perspective (see Chapter 28) and a holographic perspective (Chapter 27), but makes an even stronger cases for the nonlocality of consciousness and, by extension, survival after death.

Why do skeptics stick to their guns with such determination? On one hand, religion has treated science and reason rather shabbily for thousands of years, and it is very difficult to separate the concepts of religion and God, as we have seen in

Chapters 3 through 12. Thus, much of the antipathy toward evidence for survival after death and/or other things paranormal may come from the desire to disprove religion. In short, much of the problem may be due to simple conflation. There is also the problem of fraud. It is safe to conclude that the majority of so-called psychics, mediums, energy healers, etc. are either misinformed, charlatans, or both. Peer acceptance is another factor; more than one scientist's career has been ruined by straying from what her or his field considers valid areas of research, each a clear example of secular heresy. Finally, being proven wrong might not exactly be desirable. An engineer who designs one faulty building is finished the day that building collapses; a skeptic whose claim is conclusively disproved may face similar consequences.

Rebutting the Skeptics

Science has made tremendous strides in both improving our quality of life and our knowledge about everything from sub-atomic particles to the entire universe and possibly beyond. The track record of science is second to none among human endeavors. It is therefore no surprise that society places high expectations on its scientists. We tend to think that the ability of science to answer some questions means that it can answer them all, but this is not the case. It has never been the case. The initial stages of a scientific discovery are not readily apparent and cannot be easily detected or replicated by established scientific techniques. Expectations have to change in order to give new discoveries a chance. Scientific truth is as much about dealing with non-scientific factors like expectations, beliefs, and dogma as it is about experiments and replication. Dr. Ignaz Semmelweis (1818-1865) was an obstetrician who lived out his latter days in an insane asylum and died at the age of 47 after being ruthlessly attacked for demonstrating that death rates during childhood plummeted if doctors washed their hands beforehand. Today's doctors are robed from head to toe and follow elaborate hand washing procedures before entering sterilized operating rooms and using sterilized equipment.

Skeptics and believers are all alike. At this moment scientists and skeptics are the leading dogmatists. Advance in detail is admitted: fundamental novelty is barred. This dogmatic common sense is the death of philosophical adventure. The universe is vast.
Alfred North Whitehead

In general, most scientific progress follows the same four steps:

1. Skeptics say the new discovery is impossible because it violates the established laws of science.
2. Skeptics concede that the discovery is possible but both uninteresting and weak.
3. Mainstream science realizes the importance of the discovery and discovers that the effects are stronger than first thought.
4. The same people who denounced the discovery claim to be the ones who discovered it, and everyone eventually forgets the initial charges of religious or scientific heresy.

One must choose truth rather than peace of mind.
Chet Raymo

Materialist scientists claim that life is inherently meaningless and that we are here by accident, while at the same time revealing a universe that is vaster and more marvelous than anything previously imagined. It is easy to see why reductionist approaches have found neither God nor the soul, nor even the seat of individual human experiences: such non-material things are just not part of the material equation. Finding the truth requires science to keep its collective mind open and expand its tolerance for novel and/or dissenting ideas—especially when talking about things that are inherently non-material, such as death, afterlife, or psi. For example, we can find evidence for non-local minds simply by considering a holographic plate that contains hidden layers of order. You might expect that cutting such a plate into four pieces would result in four plates that each have a quarter of the image. What you will really get are four plates that each contain the entire image. Keep subdividing these plates, and each of the resulting pieces will still contain the whole image, albeit with progressive loss of detail. If the universe is holographic, each of us may literally have a fuzzy picture of its entire vastness.

Labels, except when supported by hard evidence, are not explanations, and it is misleading to offer them as if they were.
David Fontana

As previously explained, it is easy to suppose that mind comes from brain, because changes to the brain correspond to changes in mind; however, science still cannot explain subjective experience, nor can it explain how disparate brain processes and circuits come together to build a unified whole. There is no way to explain how pre-conscious processing becomes conscious, nor can we explain free will. These limitations fuel the speculation that consciousness is an irreducible property of the universe in its own right, just like mass or gravity. There is little reason to think that mind and brain are identical, nor can anyone explain why consciousness and self-awareness needs to be part of living.

The idea that brain size or *encephalization quotient* (the ratio of brain mass to body mass) plays a significant role in intelligence does not hold up. In fact, a brain may not be strictly necessary for life and intelligence at all. Aristotle taught that the brain's function is to cool the blood, and that it is not involved in thinking. He may have been on to something, at least in certain cases. In 1980, neurologist John Lorber (1915-1996) presented some case studies of *hydrocephalus* ("water on the brain," an abnormal buildup of cerebrospinal fluid that can cause mental retardation and death if left untreated), in which the victims developed normally. Two young children he worked with developed normally despite having no evidence of a cerebral cortex; one died at three months, the other at 12 months. Another young man had a larger than normal head that did not seem to be causing him any problems. This man had an IQ of 126 and an honors degree in mathematics, despite having virtually no brain beyond a layer of brain cells only 1 millimeter thick; the rest of his skull was filled with water. This man is still alive and leading a perfectly normal life with the exception of knowing that he has no brain.

We do not understand everything, but we understand enough to know that there is no room in our world for telekinesis or astrology.
Steven Weinberg

Lorber documented over 600 cases of hydrocephalus and classified them into the following four groups:

- People with nearly normal brains
- People whose craniums were 50-70% filled with cerebrospinal fluid
- People whose craniums were 70-90% filled with cerebrospinal fluid
- People whose craniums were 95% filled with cerebrospinal fluid.

The last group comprised fewer than 60 cases. Half of these people were severely retarded, but the other half were leading normal lives with IQs above 100. Lorber admitted to some imprecision, saying that he could not tell for sure whether the math student's brain weighed 50 grams or 150, but it still weighed far less than the normal 1,500 grams. Are these cases examples of truly remarkable physical redundancy? Or is something else going on? In 1830, Dr. Albert Wigan per-

I'm growing old. I'm falling apart. And it's very interesting.
William Saroyan

formed an autopsy on a dead man and found one brain hemisphere, not the expected two. This man had lived a normal life with half a brain. Dr. Wigan speculated that maybe the self, what we call "I," lives in only half a brain, and that maybe a normal brain contains two selves.

I am the dog. No, the dog is himself, and I am the dog—O the dog is me, and I am myself. Ay, so, so.
Shakespeare

Hypnosis can have odd effects that sometimes seem to be violating the laws of physics, even physiological ones. Author Michael Talbot (1953-1992) wrote of a hypnosis subject who was told that he would be unable to see his daughter. The giggling girl then stood in front of her father, who was unable to see her literally right underneath his nose. The hypnotist then pressed a watch against the daughter's back and asked the father to read the inscription and describe what he saw. The father correctly read the inscription and reported seeing only the hypnotist's arm holding the watch in his hand, which struck him as perfectly normal until he learned that his daughter was standing in the way.

We cannot prove that consciousness is not created in the brain, but we can demonstrate that something strange is going on, even once we have eliminated cases of fraud, mistaken identity, and all others with normal explanations (such as a UFO that turns out to be the Sun glinting off a weather balloon, for example). We have seen cases that defy all attempts at material explanation in various aspects of science from ghosts and mediums to NDEs. We have seen that traditional medical explanations fall short, because consciousness expands precisely when it should be contracting. We have seen that what we call the "self" may not be anywhere as simple as it seems to us. We have yet more reason to doubt materialism.

Who Are You?

I exist. Are these not the two most amazing words possible?
Anthony Peake

Ask 100 people this question and you will probably hear 100 names. It is fast, easy, and obvious to say, "I am Anthony Hernandez," or whatever your name may be. But who is that? We cannot have experience without awareness. Is this awareness what we mean when we say "I?" On a mundane level, each of us is a series of perceptions that begins with our births and ends with our deaths, a brief interval of awakening bracketed by billions of years of nonexistence on one side and the prospect of uncountable trillions more years of nonexistence on the other side. This is not a problem for religious people who believe in a heaven somewhere beyond the material universe, but it can be a scary prospect for nonbelievers; however, as we have seen, death may not be the end, and our birth may not have been the beginning. Consciousness may transcend the body.

Either way, our bodies are little more than short-term rentals. From an evolutionary standpoint, it's pointless to design a body for immortality when an individual is very likely to be killed by the environment. It makes far more sense to provide for a period of youthful vim and vigor followed by decline and death. Our genes have some semblance of immortality; we still carry the genetic legacy of the very first life form on Earth that has inhabited a series of temporary shells for billions of years since. Even within this lifetime it is tempting—but wrong—to think that we inhabit a single stable body:

- Humans create 1,000 completely new skins in a lifetime, which is about one every 27 days.

- Our bodies replace 50 billion cells per day, or about 500,000 every second.

- About 100,000 brain cells (neurons) die every day, along with the connections to and from those cells.

- By age 60, all of the molecules in a human body have been replaced about 60 times. The atoms that make up these molecules come not from you but from the food you eat; you literally are what you eat.

With few exceptions, such as brain cells, the cells in our bodies today are not the cells we started out with. Bristlecone pine trees live for thousands of years, but their oldest living cells are only about 30 years old. The brain cells in the oldest living people can be over three times older. By that measure, the 1,000 year old tree is young, while the 90 year old human is geriatric.

Normal cells seem to remember how many times they have divided, even after being frozen. Human cells tend to divide about 50 times before stopping, chicken cells 25 times, and Galapagos tortoises about 110 times. This built-in limit is called the *Hayflick limit* after gerontologist Leonard Hayflick (1928-), who discovered it in 1961. Cancer and sperm cells

---

I am human, and nothing human is alien to me.
Terence

The one 'I am' at the heart of all creation, thou art the light of life.
Shvetashvatara Upanishad

seem to be immune from aging. It seems pretty clear that whatever we are is not in our bodies, and that our minds may be greater than our brains.

You realize beyond all trace of doubt that the world is within you, and not you in the world.
Sri Nisargadatta Maharaj

The idea that our brains mediate all of our experiences has led us to discover what parts of our brains take part in various kinds of experiences, but has not told us whether these experiences are real or how we can tell the difference. There is no "I" center in the brain. The location of most of our sense organs (eyes, nose, mouth, and ears) in our heads does much to convince us that "I" resides there, but what if that's not the case? Some meditation exercises train participants to move their centers of consciousness around their bodies; whether this is actually happening or not, the fact remains that we can at least imagine our centers of consciousness as being somewhere besides our heads.

So if we are not located inside our heads and may not be located anywhere in our bodies, then just what is the self? It seems clear that self consciousness evolved over a long period of time and that nobody woke up one morning with the novel thought of "I." Australian Aborigines do not fear death, because they believe that nothing ever dies. Other primitive tribes consider individuals to be cells in a social organism that evolved from an undetermined past and is continuing into an undetermined future. Our earliest concept of a soul may have been as part of a group mind that returns to the tribe at death.

You are the entire universe. you are in all, and all is in you. Sun, Moon, and stars revolve within you.
Swami Muktana

Thanks to agriculture (arguably the single largest destabilizing influence in recent human development for reasons I describe in *The Natural Savage*), humans began building cities about 10,000 years ago and separating ourselves from the nature "out there" beyond the thick walls. Our tribal links weakened as urban populations expanded and made it impossible to know everyone one encountered in a typical day. This may have given rise to an enhanced sense of individuality and consciousness of self, which may have brought with it a terrible fear of death. The *Epic of Gilgamesh* from about 2500BC tells of the king's mourning the death of his friend Enkidu and noting that his own body will rot and be eaten by worms. He gets permission to visit Enkidu in the underworld, where he learns that the dead remain in a house of darkness forever and wear bird-like wings as clothing, an early reference to angels. The idea of a soul seems like a relatively recent development, whether one accepts the Neanderthal burial sites as evidence of such beliefs 100,000 years ago or something more recent like the Sumerian tale. Then again, an organism's ability to see itself as an independent entity in a larger world can be a powerful survival tool; for example, our tree-swinging ancestors presumably benefitted from seeing themselves as separate and distinct from the trees they were navigating, the better to plan trajectories that would avoid collisions and falls.

However it evolved, the sense of permanent self that we all feel is absolutely overwhelming, to the point where we cannot imagine ourselves not existing in some form or another. We like to think that we each have an "I" and that we know what that "I" is. We value the "I" enough to fear its loss and ask if it survives death; but this could be all a giant illusion, at least according to the Buddhists. Any idealist position must suppose that matter and separation are illusions on some level; the biggest illusion of them all could be this idea of "me."

I have already mentioned the half-second lag between awareness and conscious interpretation that our brain neatly fills in to make everything seem contemporaneous, and that we are living in the present. Our actions and reactions thus form beneath our normal awareness level. The need to make a conscious decision can cause both delay and overreaction, which is why people from musicians to professional athletes work so hard to develop "muscle memory" that will allow them to consistently do the right thing without having to think about it. Even so, consciousness sometimes take over, and usually not for the best.

All of this means that we are little more than observers traveling though our own lives. But what are these lives? Words, concepts, and constructs such as "life," "death," "matter," "star," etc. are all convenient stories we tell ourselves. They are the equivalent of Newtonian physics that works so well on the scales we are accustomed to, but which breaks down when we try to us it to describe the very small and/or the very large. Letting go of Newtonian physics allowed us to discover quantum mechanics and even more recent theories. It therefore stands to reason that letting go of our accustomed way of looking at things can give us insight into what—and who—we really are and where we are going.

You dozed and watched the night revealing the thousand sordid images of which your soul was composed.
T.S. Eliot

It is difficult to let go of the security of an imagined 'I' and enter into the not knowing of just being.
Stephen Levine

> *In that which is the subtle essence, all that exists has its self. That is the true, that is the self, and thou, Svetaketu, art that.*
> Chandogya Upanishad

Our brains are obsessed with analysis and classification. We perceive a boundary between our internal and external aspects and talk about that within as an individual. We then define life as including some sort of self with it. We like to think that only so-called "higher" animals have consciousness and self-awareness, but all life has at least a small degree of selfhood and some degree of consciousness. A microbe's consciousness may not be as deep and complex as ours is, but I again submit that its life is just as enjoyable and mysterious to it as ours are to us. Where do we draw the line between conscious and unconscious, and how do we justify drawing it at all? Just as quantum mechanics and relativity both contain Newton's equations, even a subatomic particle may have some degree of consciousness. Whether it does or not, the fact that at least one species within the universe is conscious means that the entire universe is at least dimly aware of itself and is looking back at itself in wonder.

We know that solipsism is the belief that one is all alone, the realization that one is a thought and not the machine (body/brain) having the thought. Does this mean that I am alone and that you and everything else in the universe is a figment of my imagination? Yes, in a sense. From my perspective, the entire universe depends on having me around to observe it. If I cease to exist, then there is no difference between my ceasing to exist or the entire universe ceasing to exist, as far as my frame of reference is concerned. On the other hand, if the Ground of reality is a single consciousness, then there is no way that "I" could ever begin to exist as a completely separate entity or that "I" could ever truly cease to exist.

> *What could begin to deny self, if there were not in man something different from self?*
> William Law

What would happen if an exact copy of "you" was made? Would this be, as Tipler asserts in *The Physics of Immortality*, the same "you?" Tipler believes that recreating a person's exact quantum state is all that is needed to resurrect the exact same person with the same memories, thought patterns, beliefs, sense of identity, etc. But what would happen if someone recreated my quantum state—or yours, or anyone's—while that person is still alive? Nothing prevents two sets of particles from having identical states. I submit that Tipler is wrong, that we may have two perfect clones, but that "I" would not suddenly have two bodies. Would this clone be alive or even conscious? If so, that would revive many of the materialist objections to anything involving paranormality—especially if this clone felt, thought, and acted just like the original. But if not…

The only way "I" can survive physical death is in a dualist universe, which has been conclusively disproved. If "I" am simply the product of brain activity, then "I" can't take myself with me when I die. If "I" am an illusion created by an irreducible consciousness, then the "I" that I am familiar with cannot go with me. In that sense, "I" do not survive the grave. But then who am "I?" The person I am now is literally and figuratively different than the person who woke up in the morning, who may be different than the person who dreamed during the night in a way that seemed utterly realistic. There may be no subjective difference between dreaming and waking at all.

Could an 'I' be more like an elusive, receding, shimmering rainbow than like a tangible, heftable, transportable pot of gold?
Douglas Hofstadter

If we assume that consciousness is primary, then we must assume that consciousness does not die when the body dies. The material body thus ceases to function, but consciousness remains intact. Goswami believes than an NDE is a backward-in-time collapse of possibilities (delayed choice) that occurs when brain function is restored, which would mean that an NDE is not "real," in the sense that nothing is happening when the brain is not functioning. On the other hand, an OBE could be seen as strong evidence for some personal identity continuing after physical death. Then again, the primacy of consciousness would mean that there is no such thing as an individual mind, and brain death could therefore mean loss of all personal memory and identity, while character habit patterns (karma) remains; however, without some "I" component, it is difficult to see how this differs from the utter annihilation posited by materialists.

Revisiting Death

Death quite literally means coming to an end in the space-time continuum, but there is every reason to think that the four dimensions we are all familiar with are only the tip of a much larger reality. Telepathy and clairvoyance offer intriguing glimpses at the possibility of higher dimensions, and both superstring and brane theories require such upper dimensions. There is a non-local aspect to each of our minds, and it is only logical to think that this element may remain unaffected by

All shall be well and all shall be well and all manner of things will be well.
Julian of Norwich

death, even without the evidence offered by NDEs, ghosts, and mediums.

Likening death to sleeping and forgetting is not sufficient, because death is said to be annihilation, with none of the healthful benefits of sleeping or forgetting; annihilation erases the good memories right along with the bad. But there is another way to define death: As we saw in Chapter 14, the brain may be a reducing valve for consciousness in the same way that a movie frame is a reducing valve for the white light of the projector. In this case, death would be simply the removal of the reducing valve, at which point the barrier between ourselves and the universe would vanish as we move from the fog of introversion into the clarity of whatever reality is out there. People experiencing NDEs have dipped their toes in this ocean, which is something that no hallucination can generate or even approximate. By this measure, it is daily life that becomes unreal.

Human personality does survive bodily death.
Hornell Hart

Seen in this light, the "I" who died never existed, and there is no death. Each of us is part of a process with neither beginning nor end. The question is not, "How can consciousness exist without a brain?" but, "How can consciousness exist with a brain?" Our fear of death and even death itself can be overcome if we realize all of this, because death does not mean extinction. We will lose our bodies, brains, and self, but those never existed. The selfless aspects of ourselves will continue on. Death is thus a liberation from the profound isolation of this life. A good part of preparing for death in some Eastern traditions entails searching for deeper levels of reality.

Children seem to have a much easier time with death than adults, because they have not yet been conditioned to think of concepts like soul and survival of consciousness as stupid, crazy, or fake. This conditioning sinks in as children get older and lose their original belief that death is just another moment in life as they identify more and more with their bodies.

The Meaning of Death

O grave, where is thy victory?
O death, where is thy sting?
1 Corinthians 15:55

All of this is just speculation so far, because we have a long way to go before we can start assembling all of the material we are exploring into a cohesive theory; however, let's take just a few moments to run with this speculation as if it were fact. Do intellect and emotion dissolve, or do they survive physical death? We know that they survive sleep, which is the interval between waking states, so why can't they survive death as well? Whatever "I" am when I am sleeping is not my body; am "I" my body now? The bodies of dead Tibetan masters can go on looking fresh and healthy for days and sometimes even weeks after clinical death, which seems to offer evidence that a dying person can have a lot of control over what is happening to them. A person in the prime of life should not be able to survive that long without food or water, and there is no reason to expect that someone at the end of a long life can pull this off. Here again, something is going on; but what?

As we discussed above, the self we are so accustomed to in this life may not survive death, but that does not preclude an expansion of consciousness without personal extinction; it may be as simple as being able to see ourselves as one with the universe. The Buddhist term *maya* is commonly translated as "illusion," but can also mean that time and space are part of Brahman. Nevertheless, the root of mystical thought supports the idea of illusion, because each of us superimposes our own filters on reality and sees things how we choose to see them. (See *The Enlightened Savage*.) Thousands of years of mystical contemplation are revealing the deep structure of reality that Western science is only now beginning to grasp.

We humans like to think of ourselves as the most advanced species on Earth and possibly the entire universe; however, the extent to which other species are unable or unwilling to limit the totality of whatever experience is open to them is the extent to which those beings are actually more conscious than we are. Following this logic, everything from atoms to stars, and everything in between that makes no distinction between themselves and the universe, may be the most conscious of all.

Ask yourself when "you" were born and what the nature of "you" really is. You may find your mind disappearing, leaving your real "I" standing alone without any attachment to your body or your ego. Consider that this "I" is not the "I" you are accustomed to in the normal world. You are not your body, nor are you that mundane, ego-driven "I." Your real "I" transcends the body for all of the reasons we have been discussing. Your body will die and your lower "I" with it, but your higher "I," your true self, is deathless. The only reason birth and

---

Life is eternal and love is immortal. Death is only a horizon and a horizon is nothing save the limit of our sight.
R. W. Raymond

From the unreal lead me to the real, from darkness lead me to light, from death lead me to immortality.
Brhadaranyaka Upanishad

death seem real is because we identify ourselves with our bodies and other people with their bodies.

Were we born? Only those who are born can die. Someone who thinks s/he was born will fear death, but the true self from which thoughts emerge always exists, and this self is eternal. Understanding this can free us from the limiting ideas of birth and death. There is no birth or death; one remains what one has been all along. An ill woman once told a Zen master about her terminal illness, to which he responded, "Don't worry, you won't die!" because he knew that she was more than her impermanent mind or body—as are we all. Our current lives are reflections of our minds; seeing this frees us from the prison of this existence. Life is an opportunity to discover wholeness and truth.

We think that we are our thoughts and call those thoughts "I," but there is only the pure stillness of being behind our moving minds. There is no name, no reputation, nothing at all to protect. We are not merely objects of thought or mood, and must not confuse the light of this awareness with any object that reflects the light. All is revealed in the silent, deathless I AM of our true selves. This pure awareness is deathless, because it does not come or go; it just is. All life eventually goes on to the next world. Even Western mystics realized this. The term "dust to dust" refers to energy, not literal dust; thus, "ashes to ashes, dust to dust" does not mean that we are born from nonexistence only to return to it. This crucial piece of knowledge was lost on orthodox Christian and Muslim theologians.

The Buddhists stress the elimination of attachment, but this is not to be confused with a total loss of desire; rather, it is freeing up space to allow any thought, feeling, or state of mind to enter without closing it off or eliminating pure being. Non-attachment means being actively open to life, which is neither heavenly nor hellish unless and until we make it so. (See *The Enlightened Savage*.) Karma is not punishment; on the contrary, it reflects the infinite mercy, patience, and love of the universe that allows us to learn all that we have not grasped the last time around—a decidedly different approach than the claim that what we do in this single lifetime will have eternal consequences of either bliss or torture.

Here is a test to find whether your mission on earth is finished: if you're alive, it isn't.
Richard Bach

I look at life as a gift of God. Now that he wants it back I have no right to complain.
Joyce Cary

Implications for Life

Swedenborg once said that a tangible connection exists between the physical and spirit worlds. Everything we touch resonates across both worlds. If this is true, if paranormality really is the doorway to seeing past the constraints of this lifetime into a far grander level of eternal existence, if death is not the end, how might that knowledge affect society? What if doctors saw death as a new beginning instead of the end? Would they still try to save lives at all costs? Would spirituality be more integrated into medicine? What about hospice care or other care of the terminally ill? How else might life change if we knew that consciousness is forever, that harming someone else only harms ourselves in the long run? How would science change? Would we see astrophysics and metaphysics as flip sides of each other? How else might life be different if only we started thinking outside the mortal box?

More Study Needed

Psi offers intriguing evidence to those willing to examine it, but more concrete proof is needed to win over the skeptics. I feel that this research is terribly important, because the study of psi has the potential to reveal whether the human mind is causal (which we would expect in an idealist universe) or caused (which we would expect in a materialist universe). Contrary to what some might think, there is no correlation between ignorance/lower education and belief in psi; some of the world's most intelligent and highly educated people believe that psi is real. This is not proof, because even the smartest person is susceptible to superstitions and wishful thinking; however, it does indicate that psi is something more than mere parlor trickery. Daily life may not be that far removed from NDEs; the difference between normal and extended existence may be one of degree, not kind. Comparing daily life with any afterlife that might exist may be like comparing frozen water to running water.

If psi is real, then it is not "paranormal" or "supernatural" but part and parcel of the natural world. The strong indications that psi is very real should indicate that any serious attempt to study psi constitutes serious, valid scientific inquiry. Sadly, tra-

Science alone cannot tell us what happens when we die because it is blind to too many fundamental aspects of reality.
David Darling

The sanest and the best of us are of one clay with lunatics and prison inmates, and death finally runs the robustest of us down.
William James

> *There is nothing over which a free man wonders less than death; his wisdom is, not to meditate on death but on life.*
> Baruch Spinoza

ditional funding sources eschew this field of study, and very few mainstream scientists are studying it. Those who believe in psi present findings indicating that psi is real, but are dismissed by others who continually find fault with the evidence and maintain the materialist status quo through denial. They point to experiments carried out by skeptics that appear to debunk psi claims without pausing to consider that skepticism itself may be skewing the results. If conscious observation really is what changes possibility into actuality (see Chapter 29), then it follows that someone predisposed to discount a certain phenomenon will tend not to see it. Thus, the negative results generated by skeptics may actually serve to validate psi.

Negative precognitions outnumber positive by 4 to 1, which is just what one might expect from a prey animal. (See *The Enlightened Savage* and *The Natural Savage*.) Those of us living in advanced societies are conditioned to believe that knowing anything about the future is impossible, and thus lose any innate ability we might have to peek at what's coming toward us. It is telling that primitive cultures consistently score higher on evaluations of ESP and other psi abilities. This does not mean that people can always see exactly what is going to happen; in most cases, precognition is nothing more than a "gut feeling" or intuition that guides our actions. For example, fewer people ride trains on days when there are accidents. Coincidence? If this happened once, then we could look to normal explanations, such as random chance, but ridership data is consistent. How can we explain this except by acknowledging that at least some people have direct access to information about the future that most of the rest of us don't?

> *If you cease to identify yourself with the body, but see the real Self, this confusion will vanish. You are eternal.*
> Ramana Maharshi

Classical physics shows that time marches inexorably forward; we step on to the conveyor belt of time at our birth and step off at our death. Yes, time dilation does occur because of different speeds, but that only adds up to a few billionths of a second over our lifetimes. We are a long way from experiencing the dramatic differences caused by acceleration to near light speed, and will probably never get there. By contrast, PEAR research indicates that both absolute and relative time and space are illusions, that every moment of our lives can affect both our future and our past, and that the entire universe exists in a vast Now. This indication has been validated by mainstream science, as we will discover in Chapter 28.

Parapsychologist Dean Radin (1952-) has conducted experiments where random number generator sequences were affected up or down from the expected averages months after being generated. This is yet more evidence that things exist only as possibilities until observed.

Brains seem to be able to influence each other nonlocally. Experiments conducted at the University of Mexico had two subjects meditate to achieve a connection, and then fitted with EEGs and placed in separate Faraday cages. (A Faraday cage is a chamber made out of metal mesh that blocks all electromagnetic waves from entering or exiting, including any such energy released by brain or other nervous activity.) One subject was then shown a series of light flashes that caused a noticeable response on that person's EEG readout. In 25% of these studies, the other subject displayed a similar EEG response at the same moment. This experiment was replicated by Dr. Peter Fenwick, a London psychologist.

All goes onward and outward. Nothing collapses and to die is different from what anyone supposes and luckier.
Walt Whitman

The best explanation is that the brains of the test subjects had become entangled (see Chapter 23) during the meditation sessions and that they could influence each other nonlocally. The Eberhard theorem precludes using photons to transfer information nonlocally; however, if consciousness is responsible for creating outcomes from possibilities, then the Eberhard theorem does not hold and the brains can connect nonlocally. The person who sees the flash causes a similar event to occur in the other person's brain. This is a prime example of what Einstein referred to as, "spooky action at a distance." This and other similar evidence is compatible with both idealist and holographic theories.

Evidence like this pokes gaping holes in materialism, but many materialists simply ignore it, just like religious people ignore all evidence against their beliefs and refuse to alter those beliefs. Materialists are therefore just as faith-based and dogmatic as any devout religious person. Meanwhile, psi research seems to indicate that either God is intervening in the universe, or that the universe is a far spookier place than Einstein imagined. The possibility of entangled nonlocal minds throws open the barn door for the existence of psi. Eastern people see psi as offering a peek into the bardos. What other parts of Eastern belief might be validated by confirming psi?

But it is now time to depart—for me to die, for you to live. But which of us is going to a better state is unknown to everyone but God.
Socrates

The lack of funding for mainstream psi research is a serious problem. If serious science is prevented from making such inquiries, then only the quacks and hucksters will remain, which will further diminish the prospects for mainstream acceptance. For example, the methods employed by famous "psychics" involve little more than careful observation and intelligent guesses, methods that have nothing whatsoever to do with actual psi. Their antics say nothing about psi. Only controlled, replicated studies can continue to shed light on this subject.

Psi is not magic. It is a low-level effect that has been seen in many studies, and the strength of the effects has been directly correlated to the quality of the experiments. We cannot ignore small, persistent effects, because they sometimes force major changes. Two prime examples of this are the "small storm clouds on the horizon" of physics at the end of the 19th century regarding the velocity of light and black body radiation. (See Chapter 23.) These small clouds led to the development of quantum mechanics that has forever altered all of science.

There is always the possibility of being all.
Henry Thoreau

Death and dying provide a meeting point between the Tibetan Buddhist and modern scientific traditions. I believe both have a great deal to contribute to each other on the level of understanding and of practical benefit.
Dalai Lama

Assessing Psi

Some psi experiments indicate that the world may have a global mind. Princeton and the University of Amsterdam have conducted "consciousness field" studies and concluded that, among other things, groups of people watching a TV broadcast can affect the world in various ways. This implies that all things are fundamentally interconnected and may present a case for both the nonlocality and unity of consciousness. This gave rise to the following speculations about what the properties of consciousness might be, including:

- Extension beyond the individual.

- Ability to affect probabilities.

- Individual consciousness fluctuates in strength depending on how tightly focused that person's attention is; ordinary consciousness is less focused than during peak experiences.

- Groups of people may have group consciousness that is strongest when they are unified in goal or purpose and effectively zero when everyone is doing her or his own thing.

- Physical systems respond to consciousness by becoming more ordered.

As we have repeatedly seen, skepticism is one of the pillars of the scientific method that serves as a critical brake to prevent us from running down intellectual blind alleys; however, this same skepticism makes it difficult for scientists to alter their beliefs, even in the face of contradictory evidence that has biologists sounding like materialists, but more and more physicists sounding like mystics.

It is possible that the truth is sad.
Joseph-Ernest Renan

Classical science cannot account for mind-body interactions, biofeedback, or the placebo effect (among many others). Reductionism breaks down at very small and large scales that bring us eyeball to eyeball with the most fundamental building blocks of matter. It also collapses when subjects are viewed in isolation. For example, most scientists feel that discoveries in their own specialties will uncover most if not all of what there is to know. Physicists feel that smashing atoms into smaller and smaller bits will solve the fundamental mysteries of the universe. Biologists feel that mapping DNA will reveal the fundamental mysteries of life. If there is one thing that scientific advances have made clear, it is that the laws of nature are subject to ongoing refinement, and that future laws may bear little resemblance to today's.

All egos are separate by definition. Ending this separation thus ends the ego. Is there life after death? We still don't know for sure, but we do know that there is life before death. Both mainstream and paranormal scientists are trying to scientifically prove that consciousness survives physical death. Some believe we already have it; indeed the picture painted by NDEs, reincarnation stories, ghosts, mediums, telepathy, and clairvoyance points to nonlocality, and some ESP studies have exceeded the burden of proof by some estimations. Paranormality may be all the proof we need. Had someone proposed that invisible waves could transmit audio and visual information across vast distances only 200 years ago, that person would have been laughed at or worse, but electromagnetic waves have always existed. Someone did point out the need for hygiene in medical settings only 150 or so years ago, and

I know I am deathless.
Walt Whitman

Death, where is thy sting-a-ling-a-ling?
Dorothy Parker

was committed, but germs have always existed. What other realities are out there that we are not yet aware of? The one pattern we have seen is the ongoing revelation of a universe that is ever more magnificent and awe inspiring than we previously imagined, and there is no reason to think that pattern will reverse itself.

Most of us would agree that there is a universe out there that exists before, during, and after (but always independently of) anyone in particular. All trips end eventually, but that does not stop us from taking them. I've never met anyone who refused to go on vacation because s/he knew that returning home was all but inevitable. Ultimately, personal experience (or a distinct lack thereof) may be what finally convinces us that either something survives physical death, or that death leaves nothing of us behind to contemplate the fact that death is our last stop on the short journey of life. Either way, it seems abundantly clear that nature is very hospitable to the evolution of intelligent life, and that alone is meaningful. If everything is immanent, then so is the spirit. If everything is natural, then so is spirituality. We belong to and are part of this world; ergo, spirituality is natural.

Some possibilities transcend all attempts at rational explanation. Among them is the very real possibility that death is neither the end nor oblivion, but a return to the Ground/Godhead at the source of reality. Our journey so far has taken us from the astoundingly accurate accounts of creation, and has followed the evolution of religions that have corrupted the power of the mystical traditions that gave birth to them. We have separated God from religion and asked whether the laws of nature as we understand them can allow Her/Him/It to exist. We are now concluding our look at paranormality and its possible implications. Our discussions have relied on history, logic, and just enough science to get the point across where needed.

The belief in continuity in life originates in children altogether spontaneously.
C. J. Ducass

As of now, we can hypothesize that physical death is not the end of existence, and can make a few guesses about what an afterlife might look like. It is now time to delve into the sciences to see how they support—or refute—what we have seen and learned so far. We will begin with evolution.

512 | *The Divine Savage*
Revealing the Miracle of Being

PART FIVE

Evolution

The Divine Savage
Revealing the Miracle of Being

Chapter 18

Defining Life

> *Statistics show that of those who contract the habit of eating, very few survive.*
> Wallace Irwin

Having discussed religion, God, and paranormality, we must now turn our attention to evolution in our ongoing quest for a unified theory of reality that both embraces the various scientific discipline and answers the central question of this book, namely what—if anything—happens when we die.

At its most basic, an organism represents a four-dimensional *phenotype* (observable constitution of matter resulting from the interaction of genes and environmental pressures) that reflects that organism's nature. This nature persists for many generations with little to no appreciable change, an inherent stability that can last many millions of years absent environmental pressure to evolve. For example, the *coelacanth* (a type of lobe-finned fish) has remained basically unchanged for hundreds of millions of years, while humans can only trace our distinct evolutionary history back a few million years. Like all species, we owe our existence to a complex and long-running interplay between genes and changing environments that has left us able to look back in wonder at the path we have walked.

> *Nature gets more interesting as you get nearer to joining it.*
> W. F. Coffin

Our understanding of evolution has come a long way since the 19th century. We have learned a great deal about the physiology and even psychology of many species. Still, we can never know what it's like to be anything or anyone besides ourselves.

We seem to be trapped in our own heads. A discussion about how all this came to pass is in order.

What is Life?

This question seems absurdly simple, because just about everyone knows the difference between living and nonliving. Even a very young child can make the distinction. But how exactly should we define it? My dictionary defines *life* as, "the condition that distinguishes organisms from inorganic objects and dead organisms, being manifested by growth through *metabolism* (the process of consuming energy for sustenance), reproduction, and the power of adaptation to environment through changes originating internally." In other words, life refers to something that keeps on moving around, doing various things, and actively engaging with its environment, all for longer than we would expect from something nonliving.

We are the local embodiment of a cosmos grown to self awareness. We have begun to contemplate our origins-star stuff pondering the stars!
Carl Sagan

The Second Law of Thermodynamics (see Chapters 15 and 22) mandates that entropy must always increase. Everything from dust to humans must eventually decay to a state of maximum entropy. One key difference between the living and nonliving is that the former actively resist this decay, thanks to a metabolism that allows them to maintain their own order by actively causing entropy in something else. For example, eating a chicken lowers your entropy to keep you alive at the expense of deliberately increasing the chicken's own entropy. A dead chicken has far more entropy than a live one, and a digested and excreted chicken has even more entropy. Eating is nothing more or less than consuming negative entropy in an attempt to keep one's own entropy from increasing. This battle succeeds spectacularly during youth and growth, becomes a stalemate during the prime of young adulthood, and eventually becomes a losing battle as we become less and less able to reverse our own increasing entropy. We eat low entropy and excrete high entropy in what is ultimately a futile battle.

I find that I am at two with nature.
Woody Allen

Living organisms also depend on genetic material that must be replicated perfectly in order to preserve the species through succeeding generations; however, this process is not perfect. Mutations appear from time to time. These mutations may be unnoticeable and benign, or may cause major changes, in the same way that an error in a set of blueprints may have an effect that is anywhere from negligible to extreme. There are far more ways to build an unsafe building than there are to build a sound one; similarly, there are far more ways to be weaker, slower, less intelligent, etc. than there are to be more so. Most random mutations are therefore self-correcting, because an animal that is somehow deficient compared to its peers will be easy pickings for predators and other risk factors. This self-correction is not random; for example, it is to a lion's advantage to attack a slower gazelle over its faster brethren. The gazelle species also benefits from this, because the weaker genes will tend to get weeded out.

I stink therefore I am.
Samuel Beckett

On the other hand, positive random genetic changes do occur from time to time, resulting in (for example) a gazelle that is faster and/or smarter than the other gazelles. This lucky individual will be better able to survive and reproduce than its peers, and its genes will therefore be passed on to future generations. This is how evolution works and why we see a universal tendency toward improvement over time. At first blush, life seems to be the exception to the Second Law of Thermodynamics.

The Monk and the Bean Sprouts

Come on now, who do you think you are? Ha ha ha, bless your soul, you really think you're in control?
Gnarls Barkley

Mendel launched the study of genetics when he discovered that crossbreeding purple and white flowers did not result in blending the two colors, but in more purple flowers. This led to his idea of dominant and recessive *factors* (hereditary units). Each flower has two factors for each trait, with one from each parent that may or may not contain the same information. Mendel termed two identical factors *homozygous* and two different factors *heterozygous*. The different forms a given factor could take he called *alleles*. An individual's *genotype* (genetic makeup) consists of its alleles; its phenotype is a combination of its alleles and the environment. Heterozygous individuals express dominant alleles and hide recessive alleles. Mendel named his findings the Law of Segregation and Law of Independent Assortment.

Human cells are *diploid*, meaning that they contain genetic material from two sources. Half of our 46 chromosomes come from our mothers (egg) and half come from our fathers (sperm) through sexual reproduction, which fuses the haploids from both sources into a complete set for the child.

According to Mendel's Law of Segregation, gene copies separate so that each *gamete* (reproductive cell, such as an egg or sperm) receives only one copy. Each new gamete formed by the adult during *gametogenesis* receives half of the 46 chromosomes. Each gamete receives one allele or the other. According to the Law of Independent Assortment, the alleles of different genes are independently assorted during gametogenesis. This means that each gamete contains a random assortment of all possible combinations of both maternal and paternal chromosomes. Each of the 23 chromosomes in a gamete may come from either the mother or the father. The total number of possible combinations is therefore 2^{23}, or 8,388,608. This ensures genetic variety among offspring and future generations, which contributes to the evolution of the species by allowing new traits to develop and by helping the species stay one step ahead of viruses and parasites.

Our bodies are made of stardust; our souls are made of stories.
Tom Rhodes

Organized Complexity

There is a famous adage that a butterfly flapping its wings in Paris can help set off a typhoon in Japan. Similarly, the anthropic principle (see Chapters 11 and 26) recognizes that tiny differences in the initial state of the universe might well have resulted in our not being here. Whether you believe this or not, the point is that tiny initial differences can yield huge changes later. Similarly, simple mathematical rules and equations can create very complex shapes. For example, the equation $z = z^2 + c$ can produce the famous Mandelbrot fractal geometry set, where zooming in consistently reveals infinite levels of intricate structures within structures within structures. Simplicity lies at the heart of many seemingly complex things, and that complexity is not always readily apparent when starting from scratch. The fractionation of life into different species suited for different niches is another example of this. Most species are specialists, meaning that they thrive under fairly restrictive circumstances and/or consume a limited diet; generalists are the exception.

Natural selection never promised us a rose garden. It doesn't 'want' us to be happy. It 'wants' us to be genetically prolific. Understanding what is and isn't pathological from natural selection's point of view can help us confront things that are pathological for our point of view.
Robert Wright

Each species has a unique history that cannot be predicted using only the laws of physics. One can predict the behavior of large systems (such as the effect of adding more animals of a certain species to a given ecosystem), but this does not change the fact that life, while it obeys the laws of physics, cannot be reduced to those laws. All life is controlled by a group of highly ordered atoms that comprise only a small minority of the organism's total mass and number of atoms. Most inorganic molecules may have two or three atoms; DNA contains millions of atoms. Stretching out all of the DNA in a human body could create a strand almost six feet long. Relatively small differences in portions of a DNA molecule are responsible for the amazing diversity of life on Earth. Even so, a complete description of a DNA molecule would not be a description of organic matter; the properties that distinguish life from non-life lie beyond the molecule itself. It seems that statistical events are capable of producing at least the appearance of order, thus we need not worry at our inability to define life using the laws of physics.

We live in the most probable of all possible worlds.
Pangloss

As previously mentioned, we cannot define life based on the laws of physics; however, heredity works within the framework of quantum physical laws. On a large scale, energy changes seem smooth but actually occur in discrete jumps called quanta. Jumps to higher energy levels require a supply of outside energy, while jumps to lower levels require giving up energy. All organisms and their constituent biological processes include many atoms and must have safeguards to minimize the effects of single random atoms. For example, an organism that is sensitive to only a few atoms would be incapable of having organized thoughts. Nevertheless, atom-level events do play a significant role in organisms, such as driving mutation and evolution. Smaller organisms tend to evolve far more rapidly than larger ones; the greater the number of atoms in a body, the lower the error rate.

Cruel Indifference?

Even insects express anger, terror, jealousy, or love by their stridulation.
Charles Darwin

The romantic idea of a universe that evolved brains capable of looking back at itself vanishes when we realize that everything from individual plants, animals, and people to the entire universe itself is running down, thanks to the Second Law of Thermodynamics and entropy. Nature places little value on any individual life; her concern, if any, is for life itself. Environmental pressures, such as predation, climate change, and the evolution of parasites and diseases destroys individual plants and animals while also forming new species.

It is hard not to see all of this as a pointless exercise. We cannot measure or quantify individual subjective experience, but we do know that where our senses tell is there is color and sweetness, there are only particles, atoms, and molecules suspended in a great void—a grand spectacle whose meaning (if any) lies in the mind of the individual contemplating it. Materialist science says that the contemplating mind is a product of that same spectacle that will vanish upon each of our deaths and disappear forever once the Earth and universe die in their own turn.

Then again, many exhaustive searches of physical bodies have yet to locate the mind. The world each of us experiences is a construct built on our perceptions filtered through our beliefs, which manifests itself subject to—and based on—what is happening inside the brain. How does the brain do this? How can purely material processes create consciousness? These questions reveal a huge gap in our understanding and knowledge of any ultimate reality. Spinoza saw the universe we perceive and experience (including both matter and energy) as part of the infinite Godhead. Modern science bears him out, at least to some degree: Our minds make the external world for us to experience and live in. The only way to create something external is to not be part of that something. The creative mind must therefore exclude itself, and thus does not contain its own creator.

Our individual worldviews exist in our minds, and have no other provable existence. Our minds have no fixed locations in space or time. We can speculate that mind lies inside the body, and even that body produces mind, but we cannot prove this assumption. That which we call "I" is simply a canvas on which our lives are painted. Paint can fade and peel over time, and memories can also fade from awareness, but the canvas remains. A hypnotist can cause someone to forget all prior memories, but cannot kill the subject. We have good reason to suspect that death is not the end, and that at least some level of personal existence does not end. The cruel indifference we perceive may be neither cruel nor indifferent.

I don't think of consciousness as an absolute; it is a continuum. Some people, some creatures, are more conscious than others.
Wilder Penfield

Man still bears in his frame the indelible stamp of his lowly origin.
Charles Darwin

Ghosts in the Machine

> *The evidence that all living organisms are descended from a single common ancestor is overwhelming.*
> Christian de Duve

Trying to explain consciousness may be the ultimate scientific endeavor. Consider for just a moment that our very ability to ponder the universe and our place in it might be nothing more than an evolutionary fluke; by most measures, scientific pursuits are not necessary for survival and can in fact make survival more difficult in a pre-civilized setting. Nevertheless, we have science to thank for just about everything we know and do every day. Science consistently yields tangible results in knowledge and technology. This is not to say that science occurs in a vacuum. Scientists feel the thrill of discovery and experience their fair share of doubts, fears, and dilemmas as they push the limits of human knowledge and redefine what's possible. It's hard to imagine that a middle-of-the-food-chain animal built for living and breeding on the primordial African savannah should be probing the universe's most closely guarded secrets, but here we are.

On the other hand, every evolved trait for every species exists for a reason that benefits that species. Consciousness must exist to serve some function that cannot be carried out by any other means. It should be safe to assume that natural selection cannot distinguish between a conscious person and her or his "zombie" twin, if both are equally adept at carrying out the basic functions of life (avoiding predators, fitting in with their group, eating, finding shelter, reproducing, and living long enough to help the next generation along before dying). Natural selection works on the basis of functionality; a zombie who can do everything its conscious counterpart can do should be at no disadvantage. Thus, evolution itself does not seem capable of explaining why consciousness evolved.

> *'Life' and 'nonlife' are made-up categories like all the others we impose on the world. There are no such distinctions in reality.*
> David Darling

Self preservation requires at least a rudimentary sense of self, but this does not necessarily imply consciousness. Even if consciousness is required, we have yet to explain how a unified mind with a coherent sense of "I" can emerge from a bunch of cells. Each cell is a life unto itself. A human body consists of many trillions of miniature lives working in concert; it is a very finely orchestrated machine built of trillions of quasi-independent subassemblies. Where does that which philosopher Gilbert Ryle (1900-1976) called the "ghost in the machine" come from?

Primitive brains process simple sensory inputs and motor responses. The brains in lower vertebrate species are primarily occupied with this kind of processing. Fish primarily deal with sight, smell, and motor response. Associative structures begin appearing in higher animals whose brains feature sheets of tissue wrapped around the more ancient structures that form the mantle and (eventually) cortex. In a human, all of our higher functions seem to take place within about a pound of tissue. Truth, beauty, charity, love, the mystery of existence, awareness of mortality, and the wonder and worry of the human condition may all reside in this small space.

The god of birds and trees would also have to be the god of birth defects and cancer.
Steven Weinberg

Biophysicist Max Delbrück (1906-1981) is considered the father of modern molecular biology. He theorized that our view of the world is entirely shaped by utilitarian factors that may have little to do with any objective reality that may exist. Traits like common sense and intuition may actually mislead us more than they guide us; scientific exploration is the only way to get glimpses of the truth. Here are two revealing passages from his Nobel lecture of December 10th, 1969:

"While molecular genetics has taught us the proper way to reconcile the characteristics of the living world, generation, development towards a goal, and decay, with the contrasting incorruptibility and planlessness of the physical world, it has not resolved our uncertainty about the proper way to relate this language to the notions of 'consciousness,' 'mind,' 'cognition,' 'logical thought,' 'truth' —all these notions, too, elements of our 'world.'"

"... Even if we learn to speak about consciousness as an emergent property of nerve nets, even if we learn to understand the processes that lead to abstraction, reasoning, and language, still any such development presupposes a notion of truth that is prior to all these efforts and that cannot be conceived as an emergent property of it, an emergent property of a biological evolution. Our conviction of the truth of the sentence, 'The number of prime numbers is infinite,' must be independent of nerve nets and of evolution, if truth is to be a meaningful word at all."

The big question, however, is not, 'who will survive, the more fit or the less fit?' The big question is, 'how do organisms become more fit?'
Michael Behe

Dennett's attempt to explain consciousness as a competition between conflicting neural impulses with the winner determining what we are conscious of does little to explain consciousness and everything to rip it to shreds. Attempts to explain consciousness in a materialist framework have not yielded satisfactory results. This alone should be reason to question the materialist worldview. So what's left? Most scientists believe that dualism is not a valid model, because it violates energy conservation by postulating that an immaterial object can influence matter. Also, what distinguishes a conscious vs. a non-conscious neurological event? Until we answer this question, we cannot hope to distinguish ghost from machine, let alone identify the ghost itself. Hard-core materialists have therefore simply abandoned the idea of consciousness, choosing retreat in the absence of a better answer.

Coming to the surface to breathe, for a whale, might feel rather like going off to urinate.
Richard Dawkins

The "central state" theory goes beyond dismissing consciousness as an epiphenomenon of brain activity, but remains purely material. According to this theory, neuronal (physical) and mental (mind) events are different aspects of the same thing. There is no distinction between watcher and participant; perceptions, thoughts, and feelings are inextricably linked. Any internal debate uses different circuits within the brain to evaluate the different possibilities and generate conscious models of the potential risks and rewards. Pleasure and pain emotions are invoked, and an internal process of natural selection chooses the winning option by how well it fits the immediate need, based on its ability to induce or compel motor action. Our previous history and memories influence the final outcome.

The question of consciousness is more than merely academic, because it goes to the heart of the free will problem. If all of our behavior is determined by neurons firing in our brain regardless of whether or not we are conscious, then there is very little room for free will. If free will does not exist, then how can we possibly be held responsible for our actions using the current societal model that postulates the existence of free will and the freedom to choose? This would, among other things, completely undermine our entire system of justice that is based on the idea that someone who commits a crime always has the choice to not commit that crime.

Couldn't evolution be the answer to how and not the answer to why?
Stan Marsh
South Park

It should not be surprising that one would be hard pressed to find any materialist who would take this argument to its logical conclusion. In fact, most materialists are doing their level best to avoid or otherwise escape from this problem altogether using quantum mechanics and uncertainty; however, neither of these phenomena explain free will.

The mind does not behave in any deterministic manner that we can discern using statistical methods. The human body is a machine that operates in accordance with natural law, but this machine is being operated by an "I" who controls its motions, predicts outcomes, accepts consequences, and assumes responsibility. This mysterious "I" controls the motions of countless trillions of atoms at will. Some would say that this is as close as we can get to proving both God and immortality with the same argument. This is not a new theory: The Upanishads equate atman with Brahman, the personal self with the OOO eternal Self. They are not alone: countless mystics across all cultures and historical periods have arrived at the same conclusion: *Deus factus sum.* I am God. As we will see later in this book, this insight may be the most profound conclusion of all.

Reductionism is merciless.
Douglas Hofstadter

Consciousness is a curiously singular event. Everyone experiences a single "I." People with multiple personalities experience only one of them at a time. Our dreams may include many characters that are all products of our own imagination, and yet "I" only plays one role. We see the virtual dream worlds within our heads through the lens of a single entity and interact with those worlds just as we would in the waking world, completely unaware that we are creating both our dream self and the world in which that self is operating.

No conscious observer is ever completely replaced by instruments. The distinction between subject and object completely breaks down in quantum mechanics, and the entire universe behaves according to physical laws. The pragmatic needs of daily life force us to accept a subject/object split, but this is just an illusion. No such distinction actually exists. We cannot know anything in isolation; all knowledge requires interaction and measurement.

A million to one shot only seems amazing if you do not know that a million other shots took place along with the successful one.
Jim B. Tucker

Our minds are all we have, and we somehow got the idea that it's part of our bodies; however, it is valid to ask the ultimate chicken-or-egg question by wondering whether the universe existed before observers did. After all, everything we are aware of perceiving about the world is only a construct of our own minds. We seem to have many separate and distinct "I" minds

If 'dead' matter has reared up this curious landscape of fiddling crickets, song sparrows, and wondering men, it must be plain to even the most devoted materialist that the matter of which he speaks contains amazing, if not dreadful powers.
Loren Eiseley

If men and woman have come up from the beasts, then they will likely end up with the gods.
Ken Wilber

and their associated egos, but only one universe. What if the universe is a product of all those egos? What if those aspects of the world we would all agree on are a shared construction built by all observing egos? This is a form of solipsism, except that it does not view any observer as being alone; rather, according to the mystics and sources like the Upanishads, there is only one consciousness from which all flows. At death, our bodies return to the material world and our minds return to the spirit world. In this view, the physical world is the only thing that changes. The single consciousness (Godhead/Ground) illuminates the material world, and each of us is a window looking out at that world.

When we look at it this way, we realize that the ghost may not be in the machine, but that the exact opposite maybe true. All of the physical material and machinery of the universe may be nothing more than a giant mental construct, in the same way that the worlds we experience when we dream seem utterly real despite being nothing but our imagination. The matter we construct in our dream world exists for practical purposes only to give us something to interact with. The same may be true for the waking world and everything we think is "real," "solid," "concrete," etc. Our bodies and our individual identities may be nothing more than pragmatic symbols in the same way that the individual letters making up the words in this book are merely symbols that allow you to extract and process meaning.

On a purely physical level, all living things are negentropy machines. At least some of these machines have evolved a property called consciousness that countless researchers have tried to define and explain in purely material terms as a by-product of life. As we have seen, the exact opposite may well be true.

The Divine Savage
Revealing the Miracle of Being

Chapter 19

Our Evolutionary History

> *The evolutionary view of life should be as fundamental to a college degree as psychology 101 or western civilization.*
> Sean B. Carroll

Life is pointless at face value. The sole point of living is to reproduce; we are walking, talking genetic laboratories existing for the benefit of the DNA that each of us carries with us. Our bodies are little more than vehicles, carriers, and mixing bowls for genes. Someone who dies without reproducing has lost the evolutionary game. This has been going on for billions of years, during which life has gotten very good at what it does. Each succeeding generation benefits from an ever-longer line of successful genes.

Some people see mainly competition when they look at the web of life. Others see more cooperation. Either way, each species is in the game for its own good only. Competition and cooperation are simply means to an end. The deadliest viruses known to humankind don't care about us and don't want to get us sick; all they want to do is reproduce. Our sickness is just an accidental by-product. Bees and flowers cooperate only because it's in the best interest of both species; neither is doing the other any favors.

> *The fine print in the book of life is now legible.*
> Christian de Duve

As I mentioned in Chapter 13, mutations are random. Selection, however, is anything but random. No species can afford to be a stepping stone on the road to something new and improved. Each member of each species must be the best possible fit at that moment in order to survive and reproduce.

There are far more ways for something to be broken rather than whole, dead rather than alive, or weak rather than strong. Most mutations are therefore detrimental to the affected animal and, by extension, to the entire species. Natural selection ruthlessly weeds out the slower/weaker/clumsier/dumber ones, leaving only the strongest, best, and brightest standing. Again, this process is decidedly nonrandom. A lion will purposefully seek out the weakest gazelle to attack, because that carries the highest odds of making a kill to help it feed its cubs. The earliest forms of an elephant's trunk may have started small and been used for entirely different purposes at first, but each step along the way was useful. If not, elephants today would have either shorter trunks or none at all.

Man has emerged from the dust of stars to contemplate the universe around him.
George Smoot

Consciousness is no exception to this rule. Had the first glimmering of self awareness in the first primitive nervous system proven useless, we might be here living our lives, but we certainly would not be having this—or any—conversation. We can also imagine a form of consciousness that does not lead to speculating on the big questions of life and existence, pondering the universe, or wondering what happens when we die, but that is not the kind of consciousness that evolved. Through consciousness, the universe is becoming self-aware and pondering itself. This could be the one exception to the rule that all evolved traits must serve some useful purpose, but that's not the way to bet.

Order and Structure

In 1997, a supercomputer called Deep Blue won several chess matches against the legendary player Garry Kasparov. Of course, the computer's human programmers deserve the credit, because Deep Blue cannot play chess or do anything else without them. Nevertheless, Deep Blue gives the impression of being intelligent because it was intelligently designed. Shakespearean plays are another example of intelligent design. So what are we to make of this? Macbeth tells us that, "Life is a tale told by an idiot full of sound and fury signifying nothing," while Hamlet reminds us that, "There are more things in heaven and earth, Horatio, than are dreamed of in your philosophy." Which is it? In general, anything designed and/or built by humans is considered useful to the extent that it creates

We wonder at our engineering feats - the jet aircraft, Channel Tunnel and the microchip, but these pale beside the marvels of those tiny electric-powered proton pumps and turbine engines that power every living cell inside our bodies.
Johnjoe McFadden

Chapter 19
Our Evolutionary History | 529

order. Interestingly, we consider beauty to be the epitome of orderliness in everything from science to pastry. (The documentary *Kings of Pastry* describes the exacting standards required for French pastry chefs to earn the coveted title of *Meilleur Ouvrier de France*, or MOF.)

We owe our lives to the existence of galaxies.
Fred Adams

Have we evolved along one deterministic path, or have we simply followed one of many possible paths? The universe contains an uneven distribution of matter because gravity creates a tendency for structures to form. In fact, the Greek word *cosmos* means "orderly whole." As we see in Chapters 24 and 26, our cosmos is balanced on a knife edge between growth that is either too rapid or too slow for life to form. Achieving this feat in a universe that does not have a perfectly uniform distribution of matter requires an extraordinary degree of tuning.

Matter seems to encourage complexity. Every generation of stars creates increasing order in the form of heavier elements that can clump together to form planets, at least one of which has life. Organic molecules are common in meteorites. This plus the 1953 Urey-Miller experiment demonstrates that life may have formed in many places in the universe. Organic molecules form spontaneously. On Earth, clay formations may have formed a matrix that allowed life to emerge. Water droplets coalesce into larger bodies very easily. Growing societies become increasingly structured. Eggs and sperm are very simple cells that combine to form a higher order of life. Even simple animals perform extremely precise tasks that seem both purposeful and coordinated. Viruses have genes and use protein shells for both protective and offensive purposes.

The process of biology and chemistry, rich and extraordinary as they may seem... are no different in principle from cooling.
P.W. Atkins

Organization and complexity seem to be at odds with materialism, especially when one looks at the intricate processes required to create the simplest species. For example, many insects undergo a two-stage development process where the caterpillar literally dies and rots before becoming a mature adult fly, butterfly, or moth. The most complex task any of these animals can perform is child's play compared to what it takes to grow the animal in the first place. One author likened it to admiring the builders of the Taj Mahal for building a hut out of mud and straw.

Common Descent

The more we learn about genetics, the more obvious it becomes that all life evolved from a single ancestor species. We already know that each of us is alive because we carry beneficial traits, but this does not explain why any given individual survived long enough to pass her or his genes on to us. Individuals become increasingly important the further back we go in time, because each carries a proportionately larger share of the species' entire gene pool. As we saw in Chapter 9, all of us alive today are descendants of a man and woman whom we call Adam and Eve. Why? Were these two people especially fit? Did they have more children than their peers? Were they chosen by God? Again, there is no evidence that they were a couple, and they certainly were not the first humans on the planet; however, every single person who lives or who will ever live carries on their genetic legacy, as will whatever species we eventually evolve into. We also carry the genetic legacy of every single species that has gone before us, all the way back to the first life form on Earth. As we saw in Chapter 13, our fetuses resemble those of fish and reptiles before finally becoming recognizably human as they mature.

Life on Earth has existed for about 3.5 billion years, but plants and animals are only some 600 million years old, meaning that the majority of the history of life on Earth belongs to bacteria. The road from single-cell organism to complex organisms sporting trillions of cells per copy was downhill all the way; anything more than a few minor gaps that could be crossed by sheer momentum, and none of us would be here. Evolution is happening today as I type this and as you read it.

Amino acids exist in great quantity, but could they have become peptides under the conditions thought to exist on the early Earth? Biochemist Sidney Fox (1912-1998) heated a mix of amino acids to 170 degrees Fahrenheit and obtained up to 15% by weight of *proteinooids* (inorganic protein molecules) consisting of up to 50 amino acids. Adenosine triphosphate (commonly called ATP) transports energy within cells to drive metabolism. This substance may have been a forerunner of ribonucleic acid (RNA). The evolution of RNA paved the way for copying genetic material, which provided the mechanism needed for evolution to progress—a very early example of natural selection in action. Life hasn't looked back since. A

What nonsense to say that a self-duplicating robot has never been built. What on earth do you think that I myself am? Or you? Or a bee, or a flower, or a kangaroo?
Richard Dawkins

Man is but a foundling in the cosmos, abandoned by the forces that created him.
Carl Becker

> *We are programmed to survive by our genes and yet made painfully aware of our mortality by our forward-looking brain.*
> David Darling

long string of astonishingly accurate copies provided stability, while random mutations subjected to the pressures of nonrandom natural selection allowed new species to emerge, flourish, and—in most cases—vanish. This is not to say that natural selection entails purpose or design; like the mutations it squashes or promotes, it flies blind and can only work on the material present at the time of selection.

The first genes produced peptides less than about 20-30 amino acids long. Simplicity allowed for both some non-coding sections in the actual genes and enzyme-like catalytic processes. This was a key step, because these short peptides were the forerunners of enzymes. We don't know what environmental pressures brought this about, and we may never know. We can speculate that the first species on Earth was a *prokaryote* (a primitive cell without a membrane and most interior structures) that could survive in high temperatures. Similar bacteria have been found in non-boiling water up to 230 degrees F. It is possible that a group of such cells found itself in cooler surroundings and eventually evolved into *eukaryotes* (cells with membranes, interior structures, such as organelles and chromosomes, and the capability to reproduce by dividing called *mitosis*). The most primitive types of eubacteria are found in water up to 176 degrees F, and most need far lower temperatures to survive.

> *The history of the universe is in every one of us. Every particle in our bodies has a multibillion-year past, every cell and every body organ has a multibillion-year past, and many of our ways of thinking have a multithousand-year past.*
> Joel Primack and Nancy Abrams

It is rare to have large colonies of the same cell type, because it is to all the cells' benefit to differentiate themselves. Cells of the same species or type can do this by expressing different genes to various extents. How far can cells push this? Consider that each human cell has enough genetic material to recreate the entire person. By simply expressing different genes in different ways, the trillions of cells making up our bodies sort themselves into about 200 types. The same cell types are used to build other animals, and only slightly different versions suffice to build any animal.

The blueprints for all of this are written into the ancestor cell, mutations of which affect genes that regulate development. As I have previously mentioned, evolution can only work with the material at hand. Imagine a blank sheet of paper with only a few lines drawn on it. This primitive image can become almost anything. On the other hand, options become more and more limited as the image becomes more and more complex. From an evolutionary standpoint, the more complex the genes being mutated, the greater the constraints on what can become of those mutations. This is why all life on Earth is restricted to a very small assortment of *body plans* (phyla, the number and layout of limbs, organs, etc.).

Over time, a number of features evolved to become more and more familiar. Life developed specialized mouths and anuses and positioned them as far apart as possible, with the mouth facing forward and the anus backward, because it makes sense for an animal to move mouth-first toward food and not swim in its own feces. The front end of evolving animals began housing increasingly intricate nervous systems, because it helps to point your sensors in the direction you're going. This bundle of nerves eventually evolved into neurons and the first brains. Once the first brain appeared, the pressure was on to increase sensory input and correlate that with increased processing power. Senses and processing power went hand in hand, because it does little good to sense too much data for the brain to process, and it wastes energy to have a brain that is too complex to process simple sensory inputs.

This long evolution began when a round ball of cells began to elongate and specialize in a way that created separate up and down sides. A rudimentary mouth formed, little more than an indentation that could take in food, digest it, and then excrete it. This indentation eventually elongated into a canal with a mouth at one end and an anus on the other. The change from a spherical to an elongated shape also brought about a change in body plan from *radial* (radiating out from the center) to *bilateral* (two halves that are essentially mirror images of each other). A dense network of nerve cells formed near the mouth, the better to sense and catch dinner. This formed the basis for the brain. Eventually, interior structures such as organs developed with increasing levels of specialization. The earliest worms were extremely sophisticated when they first evolved, and remain astonishingly complex and precise to this day. For example, the nematode species *Caenorhabditis elegans* consists of 959 cells, each of which can be traced back to its origin in the egg cell through 8 to 17 rounds of division. This worm has been extensively tested and found to have a very complex level of precision. All this in only 959 cells. Compare that to the trillions of cells in a human body! The following

Why, if God was the creator of all things, are we supposed to praise him for doing what came to him naturally?
Louise B. Young

No precise dividing line between living and nonliving matter has ever been identified.
Louise B. Young

image is from the article *Somatic Sex Determination* by David Zarkower.

However many ways there may be of being alive, it is certain that there are vastly more ways of being dead.
Richard Dawkins

XX hermaphrodite

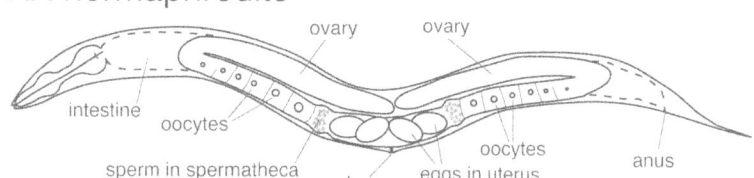

XO male

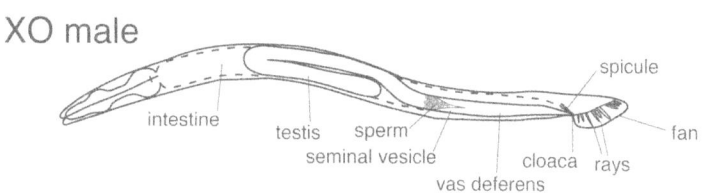

From these humble beginnings, worms went on to evolve body duplication by developing a series of segments, each of which is almost a mini organism in its own right. Each segment is partitioned off from its neighbors but connected to the whole by the skin, blood vessels, and a nerve cord. In these animals, the head and tail segments are virtually the same as the others.

Extraterrestrial Origins?

Messages from the universe arrive addressed no more specifically than 'to whom it may concern.' Scientists open those that concern them.
Norbert Weiner

Where did life first evolve? Occam's razor tells us to seek life's origins on the same planet on which life is known to exist. But is this the correct assumption to make? There is good reason to suspect that life is common throughout the universe, if not ubiquitous. Earth has life wherever there is water. Earth is not the only water-bearing planet. Why should Earth be the only planet with aquatic life? Life may not even be unique to Earth within our own solar system. Both Mars and the Jovian moon Europa are known to have water. In short, life is either spread throughout the universe or a miracle. The universe itself is far more than long equations covering physicists' blackboards; it is a complex system with many intricate structures from subatomic particles to galactic super clusters and everything in between, a collection of trillions upon trillions of potential *biospheres* (ecosystems) that were created and are being sustained by everything around them. At least some of this matter has given rise to consciousness (or quite possibly the other way around). The universe may be a giant incubator.

Just about all life is built from the elements carbon, hydrogen, nitrogen, oxygen, phosphorus, and sulfur, a combination commonly referred to as CHNOPS. Five of these six elements are among the 10 most abundant elements in the universe. Only phosphorus is in short supply on Earth; some experts fear we may only have a 50-year supply left without taking steps to recycle the vast amounts used for fertilizer and other purposes. Comets are made of dust and ice laden with organic compounds. Meteorites also contain these compounds. For example, an analysis of the Murchison meteorite that fell to Earth in Australia in 1969 revealed a variety of amino acids that were amazingly similar in both type and relative proportion to those obtained during the Urey-Miller experiments. Even more amazingly, these compounds survived both the deep cold of space and the inferno of reentry.

Life is built on carbon-based chemistry, which is not any more complex than any other kind of chemistry. 20% of interstellar dust consists of carbon-based compounds, and this dust forms 0.1% of all galactic matter. Given this evidence, it strains credulity to think that Earth is the only life-bearing planet in the universe. Some estimates place the average number of habitable planets at one million per galaxy—an estimate that could be several orders of magnitude too high and still result in many trillions of planets that have, are, and will bear life. On Earth, life began evolving without hesitation the moment the planet had cooled and stabilized enough to make it possible. Life therefore seems almost bound to arise wherever possible, so it stands to reason to guess that the universe is awash with life.

If this is true, then why have we not found any conclusive proof of intelligent extraterrestrial life? I am open to the idea that UFO researchers are on to something, and that there may be evidence of alien visits to our planet; however, in the absence of commonly available proof, I am choosing to proceed on the basis that no such discovery has been made. The vast distances between these planets and the fact that each is evolving life on its own schedule easily explains why searches such as SETI have yet to discover intelligent extraterrestrial life. We can speculate that most of the life in the universe con-

---

A page from a journal of modern experimental physics will be as mysterious to the uninitiated as a Tibetan mandala. Both are records of enquires into the nature of the universe.
Fritjof Capra

It is as if the Milky Way entered upon some cosmic dance. Swiftly the brain becomes an enchanted loom where millions of flashing shuttles weave a dissolving pattern, always a meaningful pattern although never an abiding one; a shifting harmony of sub-patterns.
Charles Sherrington

sists of simple microbial forms, with advanced life being relatively rare and technological civilizations even rarer. Any other technological civilizations may have come and gone, destroyed by their own doomsday weapons or the victims of unavoidable cosmic events. On the time scale of its entire 13.7 billion year history, the universe may be a giant tomb. Our own species is a case study: Examining Earth from an alien's perspective reveals a violent species that is breeding out of control and destroying its planet right out from under its own feet.

History of a Theory

The pace of evolution is variable but not discontinuous.
Christian de Duve

As we saw in Chapter 12, Nahmanides (in the 1200s) postulated that the universe began in an event very much resembling the Big Bang and went on to describe evolution. In 1788, geologist James Hutton (1726-1797) realized that erosion occurs over long periods of time. These are just two examples of scientific thinking that attempted to reconcile the seemingly nonsensical idea of change over time with the seemingly conflicting idea of perfect divine creation. Religion put up a good fight (see Chapters 10 and 11), but eventually lost ground to rapidly increasing scientific inquiry and research.

Jean-Baptiste Lamarck

The first point to make about Darwin's theory is that it's no longer a theory but a fact.
Julian Huxley

Jean-Baptiste Lamarck (1744-1829) was a French naturalist who created the first robust theory of evolution based on common knowledge of the day and his understanding of chemistry. He believed that evolution occurs because of environmental pressures that (for example) cause moles to be blind and mammals to have teeth. He also believed that life is orderly, and that animal movement requires many parts of the body to work in harmony. His theory of evolution depends on two forces. The first drives evolution from simplicity to complexity, while the second adapts animals to local environments and differentiates the species. These forces are a natural consequence of fundamental physical laws, making Lamarck's theory essentially materialist.

The complexity force describes a perceived tendency for organisms to become more complex over time after being spontaneously created in a very simple form. Lamarck based this on a traditional alchemical understanding of the elements being mostly influenced by earth, water, air, and fire. Fluids moving through organisms induced them to become more complex. Rapid flow etches canals between delicate structures, after which the flow will vary and lead to the development of organisms. Meanwhile, the fluids themselves become more complex, driving an even greater variety of components making up the organs in a steady and predictable process. Primitive life forms never vanish because they are being spontaneously created and transmuted into more complex forms. This appears to be a teleological process, but Lamarck stressed that physical laws were ultimately responsible.

In addition to progress from simple to complex, organisms could assume new forms adapted to different localities, and even cause a perfect level of adaptation that would represent the end of the evolutionary process for that organism. According to Lamarck, this results from organisms interacting with the environment and using or disusing different characteristics according to the following two laws:

- In an animal that has not reached its full developmental potential, using an organ often and continually will make that organ stronger, larger, and more developed. That organ's power will reflect the amount of time it has been used. Permanently disusing an organ gradually weakens and deteriorates it until the organ eventually disappears.

- Anything an organism gains or loses in response to environmental pressure is passed on to succeeding generations, so long as the modifications are common to both sexes or to those individuals who produce the young. This has been referred to as "soft inheritance" and has been found to work not by altering genes but by preventing some genes from being expressed.

I cannot persuade myself that a beneficent and loving god would have designedly created the Ichneumonidae with the express intention of their feeding within the living body of caterpillars.
Charles Darwin

Charles Darwin

Darwin was an English naturalist who joined the second voyage of the ship HMS *Beagle* that sailed from England to the Cape Verde islands off the western coast of Africa, Bahia and Rio de Janeiro in Brazil, Montevideo in Uruguay, the Falkland Islands off Argentina, Valparaiso in Chile, and Callao in Peru. From there, the voyage continued to the Galapagos Islands

I don't carve a statue out of marble; I release the form that is within.
Michelangelo

The way you treat your babies literally fashions their brains.
Christian de Duve

west of Ecuador, Sydney and Hobart in Australia, the Keeling Islands in the Indian Ocean, Mauritius near Madagascar, Cape Town in South Africa, back to Bahia, and finally home to Plymouth. This epic journey began on December 27th, 1831, and lasted nearly five years. Darwin spent most of that time on land investigating local geology and gathering specimens while the ship surveyed coastlines.

Darwin took copious notes of what he observed and speculated, and sent both specimens and copies of his journals home at various times during the voyage. His areas of expertise were geology, beetles, and dissecting marine invertebrates. He lacked a more general knowledge and sought other experts to examine and appraise his specimens. Darwin himself took notes about various marine life from plankton to invertebrates.

On St. Jago (Cape Verde), Darwin discovered that a white band running across the volcanic cliffs contained seashells. Darwin was familiar with geologist Charles Lyell's theories of land rising or falling slowly over very long periods of time, and this discovery seemed to validate those ideas. In Patagonia, he found a wide assortment of fossilized bones belonging to large extinct mammals embedded in cliffs next to modern seashells, which indicated recent and sudden extinction. His explorations inland revealed overlapping animal territories. He also saw geological formations that caused him to question Lyell's ideas about how species lived and died. These discoveries got him thinking about the relationship between geology and giant mammal extinctions. His other explorations in South America included an earthquake in Chile and the discovery of mussel beds above high tide, as well as seashells and fossilized trees high in the Andes mountains. During this time, he also compared the lives of natives of Tierra del Fuego with some who had traveled to England, and concluded that the differences between the two subgroups were primarily cultural and not racial. He also began to think that the separation between humans and other animals was not as large as previously thought.

Men argue learnedly over whether life is chemical chance or anti-chance, but they seem to forget that the life in chemicals may be the greatest chance of all, the most mysterious and unexplainable property of matter.
Loren Eiseley

In Cape Town, Darwin met John Herschel, who had written about the mystery of how extinct species were replaced by new species and how this might be a natural—not miraculous—process. Darwin spent much time organizing his notes during the final legs of the voyage home, and wrote that validation of his suspicions about mockingbirds and tortoises would undermine the stability of species, and that this seemed to shed light on the origin of species.

Upon his return in 1836, Darwin set out to find naturalists to help him catalog his many samples, and moved to Cambridge to organize the work and rewrite his journal. Ornithologist John Gould soon reported that the Galapagos birds Darwin took for a mix of blackbirds and finches were actually 12 species of finches. Within a few months, Darwin was speculating about the possibility that one species could change into another, and saw the finches and other species as examples of variation to adapt and alter species in response to environmental factors. He was also sketching out a system of branching descent in which, "It is absurd to talk of one animal being higher than another," thus rejecting Lamarck's theory of independent lines leading to higher forms.

The science of evolutionary biology was born when Darwin published *On the Origin of Species* in 1859 that forever altered scientific thought. In the 150 years that have followed, Darwin's theories have seemingly been proven correct. The primary dissent to his theory of natural selection seems to come from "intelligent design" advocates, who are struggling to find a role for the JCI God in light of recent scientific developments pointing elsewhere.

The Darwinism of the synapses replaces the Darwinism of the genes.
Jean-Pierre Changeux

Natural Selection

Marine mammals such as whales and dugongs still show traces of evolving from land to sea, as opposed to being designed for marine life from scratch. We have already discussed how natural selection is a nonrandom process that is not necessarily intelligent or purposeful. For example, erosion can cause recognizable shapes such as Kauai's Sleeping Giant formation completely by accident, while the Crazy Horse memorial in the Black Hills of South Dakota is anything but accidental.

Humans have managed to breed hundreds of varieties of dogs from their wolf ancestors by simply breeding wolves and/or dogs that looked like what they were trying to make, but we have not yet succeeding in creating a new species, because all

It is crucial to remember that, four hundred million years ago, when an ancient lobe-finned fish set out across a tidal flat in desperate search of water, it had no inkling that its effort would ultimately lead to feathered flight and cathedrals.
Michael Dowd

> *Natural selection is like artificial selection but without the human chooser.*
> Richard Dawkins

dogs breeds can breed with all other dog breeds as well as wolves. Some combinations such as a male Great Dane and a female Chihuahua may not be practically feasible for obvious reasons, but there is nothing genetically preventing at least an attempt at producing offspring. Isolating groups of fruit flies (*D. melanogaster*) in different environments for several generations eventually leads to flies preferring to mate with others from the same group. There are other examples throughout farming, animal husbandry, and—more recently—genetic engineering. In each case, we have been able to speed evolution up by several orders of magnitude by having clear goals in mind and selectively breeding those individual plants and animals with the traits we are looking for (such as larger breasts on Thanksgiving turkeys). If we persist in these efforts, then we may end up creating new species far faster than would happen naturally. Design can speed up the process, but is absolutely not required for the process to work.

Natural evolution has no such goals in mind, and therefore moves at a much slower pace. In general, accidental objects are found, while designed objects are shaped. Natural selection and the laws of heredity ensure that beneficial improvements carry on and build on themselves, but this is not real design. Changes in the environment can turn what was a positive trait into a handicap that could result in extinction. These changes can be slow (such as the spread of grasslands across Africa that eventually forced our ancestors out of the trees) or abrupt (such as the meteor impact that wiped out the dinosaurs). Dawkins refers to living things as "designoids" that are neither designed nor random, because they have been evolved by nonrandom processes that do not involve any active or intelligent design. Just because something looks designed does not mean that it is. In fairness, just because something looks totally random does not mean that it is not designed; just ask anyone who has ever purchased abstract art.

> *Evolution does not mark out a solitary route... it takes direction without aiming at ends, and it remains inventive even in its adaptations.*
> Henri Bergson

The pitcher plant is perfectly shaped to catch flies. It emits a sweet smell to attract them to their deaths. The inside is filled with slippery, downward-pointing hairs to keep the fly from climbing out. The bottom of the pitcher contains a small pool of water inhabited by maggots that only live in pitcher plants. These maggots feed on the insect. The resulting mortal remains and maggot excrement feeds the plant, which oxygenates the water for the maggots. Some termite mounds are marvels of engineering that can keep the colony cool even on hot days using a sort of natural air conditioning. These edifices must be carefully built in order to function as needed. Still, there is no evidence that the termites have any idea what they are up to. There are no architect termites, nor are there engineers, foremen, draftsmen, or building inspectors. It is safe to assume that the termites are not responsible for designing the mounds. Millions of generations of inhabitants of cooler mounds have literally hard wired the building codes into the termites' DNA.

In bridge, a hand with 13 spades is the highest possible hand. The odds of being dealt 13 spades are about 1 in 635 billion. Add in all players, and the odds shrink to 1 in 5×10^{28}. Anyone receiving such a hand would be astonished at their incredible luck. What they would probably not realize is that the odds of receiving any hand are exactly the same. We discount most hands because they are not important to the game; however, every single bridge hand dealt is an extraordinarily rare chance occurrence. Extremely improbable events are constantly happening all around us, but we ignore them because we deem them as lacking significance. One could say that the emergence of life on Earth was a truly amazing stroke of luck, so improbable in fact that we should not expect it to happen at all. Anyone saying this would have a valid point: The odds of a single event producing the long chain of evolution leading to my writing this book and your reading it are much smaller than the odds of a hurricane blowing through a junk yard and creating a fully functional Boeing 747 (or any machine for that matter). Such rare events are often called miracles and thought to require divine assistance that lies beyond science.

Any machine requires a careful sequence of steps to build that gradually transform raw materials into parts that are then assembled. (A fantastic TV show called *How It's Made* describes how many everyday items are manufactured, if you want to see this complexity in action.) The steps required to build a cell are roughly similar, and the process of evolution is like having 13 spades dealt out not once but many thousands of times in a row, which would be impossible without an altered deck. The probability of life evolving on Earth must have been very high to make it even reasonably possible, given

I don't think biology at the moment is a science at all, at least in the sense that physics and chemistry are sciences. We need to know the universal ordering principles just as Newton provided them for the inanimate world.
Brian Goodwin

My theory of evolution is that Darwin was adopted.
Stephen Wright

the large number of steps required. We have to assume that the universe is rife with life. Again, this does not imply any kind of divine plan or intelligent designer.

Replicating genes and the inevitable copying errors made evolution dependent on *contingency* (chance). It is possible that life based on RNA genes had a number of false starts before finally becoming firmly established, especially because the earliest genes were far simpler and had less room for error. Copying accuracy must have been very high for evolution to take hold and flourish as the means for creating new life forms. Those that survived evolved the DNA/RNA protein and metabolic processes that are still used by all life today.

It is important to stress that evolution only works with the material it has. A mutation that has been eliminated is no longer available for natural selection, even if it would have been a better fit for some future need. Once it's gone, it's gone for good. This is not the same as expressing genes to various extents based on need, because this involves switching existing genes on and off, which does not eliminate them.

The non-random process of natural selection works in small steps that can be mistaken for randomness, but this again does not imply that there is any intelligence behind it. Evolution is a giant sieve that sifts through the available genetic material given to each succeeding generation of organisms and rejects those that fail to make the grade at that particular time. The genes that survive do so because they offer the best available fit to the current situation, with no guarantee that they will do so in the future or that they would have been even remotely suited for past situations.

We must also address the evolution of multicellular organisms. Cells quickly evolved membranes to protect them against external threats and developed internal organelles to handle internal functions. We owe our existence to the fact that the earliest cell membranes became able to allow the cell to both grow and divide. That step was critical, for without it many good mutations would have been lost. The maintenance and repair needs of a spherical cell increase as a function of the third power of its volume; however the cell's available surface area only increases as a function of the second power of the cell radius. This simple geometric fact severely limits a cell's

Natural selection is an extremely simple process, in the sense that very little machinery needs to be set up in order for it to work.
Richard Dawkins

We have 'tail bones' because our ancestors had functional tails.
Michael Dowd

growth because its food, respiration, excretory, and other interactive needs will reach a point where they can no longer be met. As the size of the cell increases, less and less surface area is available as a percentage of cell volume. Growth beyond this point requires the cell to form a bud that would eventually fall off as an independent cell. The development of external cell structures, such as *cilia* (fine hairs that allow the cell to propel itself through water), only make division more complicated. Incidentally, the smallest swimming and flying organisms literally swim through their native medium, which must be like swimming in thick syrup to them, because of surface tension. (Surface tension is what allows insects such as water boatmen to walk across the surface of ponds and lakes.)

Evolution hardly ever happens without a reason. For example, human limbs develop as buds that later become fingers and toes, as cells die in accordance with genetic programming—yet another throwback to the days of fins and wings; however, evolution does not happen for its own sake, but is forced on a species by outside pressures, such as a changing environment. In the absence of such pressures, a well-adapted species can remain virtually unchanged so long as its niche remains stable, which accounts for coelacanths, crocodiles, sharks, and other species that have existed for a very long time. Even a poorly adapted organism may continue if it lacks robust competition. Sexual reproduction is nature's hedge against change; genetic material changes just enough to keep viruses and parasites guessing and to have a supply of alternate traits on hand if needed that are quickly wiped out if not needed

Environmental pressures ease as soon as an organism has a successful survival strategy and are replaced by constraints on change when further evolution would not be in the species' best interest. So-called secondary evolution can then spread that strategy into different niches. For example, mosses have survived for many millions of years, spread to many climates from polar to tropical, and made their homes by clinging to many different kinds of trees, rocks, and other surfaces. Algae are simple organisms that have thrived in water ever since they first evolved billions of years ago. Only a major environmental pressure, such as overcrowding, overgrazing, or competition from a superior species could have driven the move onto land. The risk of remaining in the water finally outweighed the risk

Each species to a greater or lesser degree modifies its environment to optimize its reproductive rate.
Lynn Margulis

Natural selection operates blindly on material presented to it by chance.
Christian de Duve

> *Life spirals laboriously upward to higher and even higher levels, paying for every step. Death was the price of the multicellular condition; pain the price of nervous integration; anxiety the price of consciousness.*
> Ludwig von Bertalanffy

of striking off into the unknown by leaving the water. Seeds are another example of hedging against change, because they last much longer than spores. Seeds can remain viable for centuries or even longer; the oldest known example of a viable seed is an 1,000 year old specimen found in Manchuria. This allows the plant to bide its time until conditions are right for germination.

Again, secondary evolution spreads a successful organism into different niches that can eventually become separate species. Convergent evolution is the rough opposite of this, where different branches of the tree of life separately evolve similar traits. For example, fish and whales have very different evolutionary histories, but both are streamlined for movement through the water. As we will see in a few moments, eyes have evolved independently in several different types of animals.

Our aquatic legacy carries on in other ways. One transition species included a fluid-filled sac in its eggs to let the embryo develop in water even on dry land. A hard shell protected this ersatz marine sanctuary while allowing gas exchange and waste disposal. The inclusion of a food supply (yolk) let the embryo grow until it could survive on its own before emerging into the world. This strategy proved so successful that it is still in use today. Natural selection has developed species that include eggs in their diets, sometimes exclusively. The omelet you ate for breakfast this morning is a testament to both a brilliant reproductive strategy and nature's refusal to let any good idea go unchallenged.

> *One well-known and respected naturalist seriously claimed that the bright-pink coloration of the roseate spoonbill served to camouflage this bird at sunrise and sunset-without trying to consider how this bird managed the rest of the time.*
> Niko Tinbergen

Life can survive in just about any extreme of temperature, pressure, acidity, or alkalinity. It can even survive with oxygen (*aerobic*) or without (*anaerobic*), but no life we know of can survive without water. From providing a water supply in eggs to actually changing the planet's climate, life has spared no expense to keep water around. Without life, Earth would eventually become just as dry as Mars, as radiation split water into hydrogen and oxygen and the hydrogen drifted off into space and/or into mineral "sinks" that would literally pull it out of the atmosphere. Life splits hydrogen from oxygen but keeps the hydrogen around so the water can be restored later. (The term *carbohydrate* literally means carbon and hydrogen.) Life helps keep soil moist, and even plays a role in creating atmospheric currents that distribute moisture in the form of rain. This is all too evident on the fringes of the Sahara desert in Africa, where rampant overgrazing and removal of trees and brush for firewood or building have stripped the land of its ability to hold moisture, which is causing the desert to expand at an alarming rate. Some see life's role in keeping Earth nice and moist as evidence that the entire planet is a single "super" organism that is often referred to as Gaia.

Even if this is true, Gaia will probably not be able to save us from the wholesale changes we are causing to our own planet. We have been given the great honor of a level of evolutionary freedom not given to any other known species, which has led to both our ability to comprehend the entire cosmos and wipe out all life on Earth on less than twenty minutes' notice. Thanks to the science started by pioneers like Lamarck and Darwin, we know that nature does not long tolerate imbalances. We can either deal with our excessive population and environmental impacts ourselves gently, or nature can do it for us ruthlessly. For our sake, I hope we choose the former.

For every living species hundreds must have perished in the past; the fossil record is the wastebasket of the grand designer's discarded hypotheses.
Arthur Koestler

Junk in our Genes?

Only a small percentage of DNA codes for proteins. The rest was dismissed as "junk" DNA and cited as evidence against any kind of intelligent planning; however, it turns out that this so-called junk actually performs some crucial functions. Some of this non-coding DNA has been around for hundreds of millions of years, which suggests that it has remained because of strong evolutionary pressure and repeated selection. For example, humans and mice split from our common ancestor some 65-75 million years ago. Only 20% of our coding DNA sequences have been preserved since then, compared with the remainder of preserved sequences occurring in the non-coding regions. Nature does not tolerate waste. These regions must be important. On the down side, some disease-causing genetic variations lie within these non-coding regions.

What does all this "junk" do? Some non-coding DNA acts like switches that determine when other genes are expressed or not. Other non-coding sequences determine where *transcription factors* (proteins that bind DNA at specific sites to enable gene copying) can attach. These are both critical functions. Haphazard genetic expression would cause gross (and probably fatal) deformities. Haphazard genetic transcription

Only after we had absorbed Darwin and recalculated the age of the universe, after the vision of static forms of life had been replaced by a vision of fluid processes flexing across vast tracts of time, only then could we dare to guess the immensity of the symphony we are part of.
Christopher Bache

would cause severe (and probably fatal) mutations. Other non-coding DNA sequences may be responsible for chromosome structure, *centromeres* (links between identical copies of DNA), and recognizing gene sequences during *meiosis* (a special type of cell division that forms sperm and egg cells). In short, we would not be here without this "junk" DNA. To paraphrase an old television commercial for BASF, "Junk DNA does not make the proteins your body uses. Junk DNA makes the proteins your body uses work properly."

Eye See You

I do not ask you either your opinions or your religion; but what is your suffering?
Louis Pasteur

Advocates of intelligent design routinely use the eye as an example of "irreducible complexity" to plead their case for divine intercession in evolution. It is therefore appropriate to discuss vision. Many organisms can sense light or darkness. They may not be able to tell the direction of the light, and may not be able to "see" in any practical sense. At the other extreme, some animals can see ultraviolet light and have visual resolution far higher than the human eye. Modern animal eyes use the chemical rhodopsin to form *photoreceptor* (light sensitive) cells and pigments. Primitive bacteria use the closest chemical relative, *bacteriorhodopsin*, to detect photons. (The word "rhodopsin" is a combination of the Greek words for "rose" and "vision.") We owe our vision to this very ancient chemical—a fact that puts a serious dent in one of the key arguments for intelligent design. (More in Chapter 20.)

God created man in his own image.
Genesis

Eyes run the gamut from simple to complex, beginning with clusters of photocells that merely detect light. Placing these cells in concave cups allows the organism to detect the direction of the light; further evolution of this shape can lead to spherical eyes. (Conversely, convex curves can lead to the compound eyes found in most insects; a compound eye with human visual acuity would need to be 24 feet in diameter.) The cup can then continue evolving until it forms a pinhole, which provides focus and the ability to detect and identify objects. The nautilus is an example of an animal with this type of eye. From there, evolving clear liquids or gels can improve visual acuity as can other improvements such as lenses, variable pupils, and color vision. The shape of the *pupil* (the opening that lets light into the eye) can vary by species but has little effect; even top-of-the-line cameras have cheap pupils that are often simple polygons. Some eyes even include a *tapetum*, a reflective layer that gives the eye a second opportunity to detect photons it might otherwise miss.

Eyes evolved convergently, that is, in completely separate lines in completely different types of animals. They have also been around a long time; trilobites that lived 400 million years ago had compound eyes that were just as good as any modern insect eye. As I mentioned above, some visual components such as pigments seem to have a common ancestry, meaning that they evolved before the tree of life branched off from the main trunk. Complex eyes evolved some 50 to 100 separate times, using and reusing the same proteins and genes over and over again. They seem to have come along within a few million years during the *Cambrian explosion* some 540 million years ago, during which most of the major phyla of animals first appeared. There is no evidence that eyes existed before this period, but a great of visual diversity becomes evident during and after this period.

The cause of the Cambrian explosion remains unclear. Zoologist Andrew Parker (1967-) hypothesized that the evolution of the eye sparked an evolutionary arms race that fueled very rapid evolution. Light sensitivity itself is a helpful thing to have; sharp vision that lets you see your surroundings is practically unbeatable for navigation, avoiding predation, catching prey, etc. This is known as the "light switch" theory. It is difficult to estimate how fast eyes evolved; however, modeling based on small mutations exposed to natural selection indicates that a primitive light-sensing organ could, with good photo pigments, evolve into a human-like eye in about 400,000 years.

Many of the genes involved with eyes are common to all organisms with eyes, suggesting once again that all eyes evolved from a common light-sensing organism. The photoreceptor cells themselves probably evolved multiple times from similar chemical receptors long before the Cambrian explosion. Other similarities reflect repurposing previously existing proteins, such as the use of crystallin for lenses. This would be an example of a protein previously used for other purposes becoming useful in a novel manner. Another such example is jawbones that separated from the mandible and moved back

Men create the gods after their own image.
Aristotle

The iris diaphragm is no more an impenetrable evolutionary barrier than is the anal sphincter.
Richard Dawkins

and up to form the *malleus*, *incus*, and *stapes*, the inner ear bones that play a critical role in hearing. Other shared traits include *opsins* (photosensitive proteins), all seven families of which existed in the last common ancestor of all modern animals. The PAX6 gene that controls eye placement was also present. These last examples are far older than the eyes themselves, and must therefore have served a different purpose than they do today.

When you don't know where you're going, it's important to remember where you came from.
African proverb

As I mentioned, it is safe to assume that sense organs (eyes, ears, etc.) evolved before brains, or that they evolved somewhat hand-in-hand. There is no need to evolve an information-processing organ without information to process. The reverse is also true: Evolving information-gathering organs without the means to process that information is equally useless. This argues for simultaneous development. It follows that excellent vision requires excellent information processing, especially when that vision is needed for safely climbing and swinging through tall trees (in the case of apes) or seeing rodents moving many hundreds of feet below through layers of vegetation (in the case of birds of prey). It is interesting to note that hawks, eagles, falcons, etc. seem to be far less intelligent than humans, especially where science and technology are concerned. The primordial tropical forests of our distant evolutionary past explain our color vision (for picking out ripe fruit), and 3D vision (for arboreal navigation), but they don't explain skills like literacy or math. hen again, jellyfish such as *Cladonema* have complex eyes but no brain. Visual information goes to the muscles without apparent intermediate processing.

How Vision Works

He who hid well, lived well.
Rene Descartes

All vision relies on the same basic chemistry; however, different organisms use this chemistry differently through a wide assortment of eye types. Eye structures and forms evolved much later, compared to their protein and molecular building blocks. There are two basic types of eye cells. One type is used by *mollusks* (a type of invertebrate animal), *annelid* (ringed or segmented) worms, and *arthropods* (invertebrate animals with exoskeletons, such as scorpions, crabs, and most insects). The other is used by animals such as *chordates* (animals with backbones and other related species).

The business end of each eye contains photoreceptor cells where opsin proteins convert light into nerve impulses. These, cells maximize the available surface area for detection by placing the opsins on a hairy layer. There are two main types of hair: microvilli and cilia. The former are part of the cell membrane itself, while the latter are external structures. Either sodium or potassium can be used to generate the electrical signal that leads to the nerve impulse. The amount of sodium used varies by animal. The available evidence suggests that the divergence of these two major types of vision happened when eyes were simply primitive light receptors that later evolved independently into more complex eyes. Within the photoreceptor, the opsin surrounds a *chromophore* (pigment that distinguishes colors).

Primitive Eyes

The earliest eyes were simply patches of light-sensitive proteins called *eyespots* that can even be found in single-celled organisms. These eyespots can only sense light from dark to aid in maintaining the circadian rhythm; they cannot see shapes or tell where the light is coming from. Most major animal groups have eyespots. One single-celled organism called *euglena* has a small patch of pigment above an area of light-sensitive crystals on its front end next to a *flagellum* (tail). This presumably allows the organism to detect light direction (since the sensor is on the front end) and move toward that light, presumably to aid in *photosynthesis*, or extracting energy from light. (*Euglena* is unique, because it has both plant and animal attributes.) The brains of more complex organisms contain visual pigments that could help them time spawning with lunar cycles; detecting small changes in nocturnal lighting could allow males and females to synchronize the release of eggs and sperm, increasing the chances of success. Eyespots are thought to have evolved independently some 45-60 times.

Look round this universe. What an immense profusion of beings animated and organized, sensible and active! But inspect a little more narrowly these living existences. How hostile and destructive to each other! How insufficient all of them for their own happiness!
David Hume

Directional Vision

On one hand, only 6 of the 30+ phyla have vision that detect where light is coming from with any kind of accuracy. On the other hand, these 6 phyla account for 95% of living species. This seems to indicate that developing a direction-sensing vision system is no easy task, but that doing so yields immense

For optimal vision, why would an intelligent designer have built an eye upside down and backwards?
Michael Shermer

selective benefits. The simplest of these systems is a cup. *Planaria* worms are examples of animals with this type of vision. cups can distinguish brightness in certain directions. Deepening the pit and adding more photoreceptor cells refines the ability to tell direction. Cells on flat eyes are triggered no matter where the light comes from; cups are effective because different cells will be triggered based on the direction from which the light is coming.

Pinhole Cameras

Living systems are cognitive systems, and living as a process is a process of cognition. This statement is valid for all organisms, with and without a nervous system.
Humberto Maturana

Eye evolution speeded up during the Cambrian explosion and made impressive strides in both imaging capability and directional vision. For example, the nautilus eye functions like a pinhole camera. This type of eye occurs when the cup deepens to become a spherical chamber with only a tiny opening to the outside. The tiny opening allows both very refined direction sensing and—for the first time—actual imaging, the ability to see shapes. This ability is limited because of low resolution and lack of focus, but still represents an enormous improvement over simple eye cups.

In some cases, transparent cells grew over the eye openings to prevent contamination and parasites. Separating the liquid in the chamber from the surrounding liquid allowed the development of a transparent *humour* (gel) that allowed *color filtering* (allowing light of certain wavelengths to pass through while blocking other wavelengths), higher refraction, UV blocking, and even the ability to operate out of water. Color vision, the ability to distinguish light of different wavelengths, now became possible. We may owe our ability to see some colors but not infrared, ultraviolet, or microwaves to the fact that only blue and green light can travel through water—a fact that also influenced plant development and explains why virtually all plants are green.

Lenses

It's just a physical system, what point is there?
Margaret Geller

Adding a transparent layer to the eye is a huge improvement. The next major step is to make that layer flexible and controllable to allow the organism to focus on objects at different distances simply by changing the curvature of the lens. Focus on a nearby object, and then use your peripheral vision to notice that distant objects appear blurry. Repeat this exercise while focusing on the distant object, and you'll notice that the nearby objects are now blurry. This happened because you changed the shape of the lenses in your eyes in a fraction of a second without feeling anything whatsoever, and without even consciously thinking about it.

Lenses can concentrate arriving light on the *retina* (area of the eye coated with photoreceptor cells). Lenses bend light; the ability to focus light on the retina is what enables clear vision. Some early deep-water dwellers focused the light behind the retina, which meant perpetually blurry vision while also allowing greater light sensitivity and the ability to see in the dark depths. Later evolution probably corrected this problem over time. As an aside, people with conditions such as *myopia* (nearsightedness), *hyperopia* (farsightedness), *astigmatism* (abnormal corneal curvature), and other visual defects have lenses that cannot focus light directly on the retina. Modern technology has made corrective measures such as glasses, contact lenses, and surgery widely available. Subsequent developments such as the *cornea* (layer of clear tissue covering the pupil) added protection and—in most cases—improved visual acuity.

A lie can get halfway around the world before the truth has a chance to get its boots on.
Anonymous

Embryonic lenses are alive; however the cellular structures to build the lenses are not, and must be removed for the organism to see. This removal means that the lens is made of dead cells packed with crystallin proteins that need to last a lifetime. The type of crystallin used is not important. What is important is how those proteins are distributed in the lens that makes it so useful. Most lenses are *biconvex*, meaning that they are thicker in the middle than on the edges on both sides. This increases resolution and low-light vision. It also separates resolution from the size of the pupil, which is now free to evolve even greater capabilities such as regulating the amount of light entering the eye.

Irises

Some animals have both a cornea and an *iris* (opaque layer surrounding the pupil) in front of the lens. The iris permits improved blood flow and circulation, which in turn permits larger eyes. The iris also masks optical imperfections that commonly occur at the edges of the lens. This is especially important as lens curvature and power increase along with

The universe that we observe has precisely the properties we should expect if there is, at bottom, no design, no purpose, no evil, no good, nothing but pitiless indifference.
Richard Dawkins

overall resolution and eye size. The iris can also control the size of the pupil opening that admits light into the eye. The pupil constricts in bright light, which can improve clarity and acuity because of the pinhole effect. The pupil *dilates* (expands) in low light to allow more light in and improve vision in dim settings while sacrificing some clarity. Nocturnal species such as owls have very large pupils that allow excellent night vision combined with highly evolved focusing systems that let them see extremely clearly, often with fatal consequences for their prey. (The pupil appears black because most of the light entering the eye is absorbed.)

An eye with all of these features is pretty much a modern eye that can be found in most modern vertebrates. This basic structure is powerful, flexible, and adapts to different needs.

Other Features

Some of the other features an eye can come with include:

- **Color Vision:** Being able to distinguish colors is a huge advantage for many species, because it allows the animal to better recognize food, mates, and predators. What originally began as the ability to detect different wavelengths of light evolved into color vision as photoreceptors began developing multiple pigments. This may have happened early in the history of eyes, and may have come and gone many times since then. Different animals have different color vision capabilities. Most mammals are *monochromatic*, meaning that they can only see in shades of gray. Some animals (including "color blind" humans) are *dichromatic*, meaning that they can see any two of the three primary colors of red, blue, and green. Humans and some other animals are *trichromatic*, meaning that we can see all three primary colors (red, blue, and green). Many birds and insects are *tetrachromatic*, which allows them to see ultraviolet light. (Many flowers have patterns to guide bees that are only visible in ultraviolet light.) Some butterflies even have *pentachromatic* vision that allows them to distinguish two shades of ultraviolet light.

- **Focusing:** One can focus a lens either by moving it back and forth (as happens in cameras) or by altering the

---

Ditch your white panties for yellow ones. Sounds crazy, but color theorists say your body absorbs the vibration of colors, which, in turn, affects your brain and can actually alter your mood. Yellow connects us.
Fitness magazine

White is not a mere absence of color; it is a shining and affirmative thing, as fierce as red, as definite as black. God paints in many colors; but He never paints so gorgeously, I had almost said so gaudily, as when He paints in white.
G. K. Chesterton

curvature. Both mechanisms are used in nature, with humans using variable curvature. Beyond the lens, eye growth and chemistry must be regulated to ensure that focus can be maintained. Animals that live in bright light don't need much in the way of focusing because of their small eyes and apertures. Larger animals and those evolving in low-light environments do need this ability.

- **Location, location:** Most prey animals have eyes on the sides of their heads to obtain the widest possible field of view; some animals have almost panoramic vision and can spot motion behind them. This is a good thing for seeing predators as soon as possible. Predators generally have their eyes in front, which limits the field of view, but also enables depth perception that greatly increases accuracy. Knowing the distance to a potential target is in an early step in deciding if and how to attack. It also helps you aim your attack with pinpoint accuracy. Some prey animals have forward-facing eyes: Monkeys, apes, and humans are in the middle of the food chain at best (see *The Natural Savage*), but our arboreal lifestyle made depth perception critical.

Forward or Backward?

Vertebrates and octopuses evolved "camera" eyes independently. The former have optic nerves that pass in front of the retina and a blind spot where the nerves pass through the retina. The latter have optic nerves behind the retina and no blind spot. Vertebrate eyes also form upside-down images on the retina that must be processed in the brain to flip things right side up again. Light entering our eyes must pass through layers of cells to reach the photoreceptors. This seeming inefficiency and "poor design" may be at least partially compensated for by the tapetum layer of reflective tissue behind the retina that gives photons a second chance for absorption. Octopus eyes are "well designed" by comparison. This difference may be due to how the eyes form during development. Vertebrate eyes begin as offshoots of the brain, while octopus eyes begin as indentations in the head.

The question of eyes and design is a topic of heated debate, as we are about to discover.

All that is sweet, delightful, and amiable in this world, in the serenity of the air, the fineness of seasons, the joy of light, the melody of sounds, the beauty of colors, the fragrancy of smells, the splendor our precious stones, is nothing else but Heaven breaking through the veil of this world, manifesting itself in such a degree and darting forth in such variety so much of its own nature.
William Law

Chapter 20

Accident or Design?

Evolution has developed man to such a high degree that he builds zoos to keep his ancestors in cages.
Unknown

All the evolution we know of proceeds from the vague to the definite.
Charles Sanders Pierce

The meteor that hit near the Yucatan peninsula some 65 million years ago unleashed a global conflagration that makes the *Revelation* by John of Patmos and the most dire warnings from environmentalists seem like friendly misunderstandings by comparison. Most of the world's large animals vanished in one blow. The 165 million year reign of dinosaurs was over. As catastrophic as this event was, it is yet another in the extremely long list of coincidences to which we owe our very existence. One can only wonder whether dinosaurs might have evolved to form civilizations, science, technology, or spirituality, and what that might have looked like, but for that one errant rock from space. The fact that dinosaurs had such a long run is further evidence that evolution does not act for its own sake; dinosaurs were perfectly matched to the conditions of the day, and there was no apparent need to do much else. There was no way for random mutations and natural selection to anticipate or plan for this bolt from the blue.

Gould saw evolution as a mix of chance and necessity; chance mutations encountered the need for selection. The sudden removal of dinosaurs left many ecological niches suddenly vacant, which made room for the mammals that had survived the apocalypse. (Given enough time, evolution takes the prize for making lemonade out of lemons.) So far, so good. But was this meteor strike a random event? Or was it being guided by a higher power? If it was purposeful, then how can we discern it amid the many happy coincidences that have culminated (so far) in our arrival? Maybe these coincidences were not so coincidental after all.

Protestant theologian Benjamin Warfield (1851-1921) believed that evolution is, "the theory of divine providence." He felt that believers should embrace science instead of attempting to discredit it—a position very much at odds with the state of much religion today, as we learned in Chapter 10. Darwin himself stressed the importance of asking questions, and felt strongly that nobody should die without ever having asked how they came to be alive in the first place.

Evolution based on mutation and natural selection has been proven to be the mechanism for change in living species over time beyond any hope of a reasonable doubt in the 150-plus years since *On the Origin of Species* was published. This alone says nothing whatsoever about God or divine intercession in the evolutionary process. It is true that we cannot yet say how life originated with any certainty, but this is not a good hook on which to hang any religious hats. Similarly, resorting to God to explain the Cambrian explosion is a "God of the gaps" argument that attempts to find God in the rapidly shrinking gaps in our scientific knowledge. A few religious types have even claimed that bananas are designed for humans because they are so perfectly shaped for our hands, come with a non-slip wrapper, indicate when they are ready to eat, include a perforated opening to make peeling fast and easy, are 100% biodegradable, taste good, and include both a pointed tip and convenient curvature for ease of eating. The merits of this argument deserve no further discussion.

In 1999, geneticist Charles Birch (1918-2009) said that biology is waiting for its own Einstein to define life. Biologist Brian Goodwin (1931-2009) said that biology needs the same sort of universal principles that are available to physics. Does our current state of knowledge allow us to make any sense out of life and evolution, and can we discern the presence or absence of any divine planning and/or intervention? The answer is a definite and resounding maybe.

Every phase of evolution commences by being in a state of unstable force and proceeds through organization to equilibrium. Equilibrium having been achieved, no further development is possible without once more upsetting equilibrium. A journey of a thousand miles starts in front of your feet. Whosoever acts spoils it. Whosoever keeps loses it.
Kabbalah

Evolution is gaining the psychic zones of the world... life, being and ascent of consciousness, could not continue to advance indefinitely along its line without transforming itself in depth. The being who is the object of his own reflection, in consequence, of that very doubling back upon himself becomes in a flash able to raise himself to a new sphere.
Pierre Teilhard de Chardin

Arguments Against Design

> *If the universe was designed to advance toward some state of absolute beauty and goodness, the design is incredibly faulty.*
> Theodosius Dobzhansky

In general, one can presume design if the object in question has features that a competent engineer would include for a clear purpose. Different engineers have different skill levels, but any engineer can recognize design, whether good or bad. Plants and animals seem to have all the hallmarks of fantastic design, and it is both convenient and tempting to ascribe the precise engineering of life to an intelligent super engineer who we would call God; however, we have demonstrated that no such intelligence need exist for evolution to proceed. Mutations are random, like a bunch of artists randomly throwing paint on canvases and then trotting them out to a panel of critics who decide whether or not the art has merit. In this case, the critics are the environment, predators, competing species, etc. Again, no super intellect is needed. Modern theologians are better educated than Paley (see Chapter 2) and know not to say that life has demonstrably evolved under divine guidance. Instead, they give every appearance of supporting evolution while cleverly trying to undermine Darwin by saying that the probabilities involved are just too low for all of this to be mere dumb luck. As we are about to learn, these theologians often misunderstand just how natural selection works.

> *There seem to be too any blind alleys and extinct species and too much suffering and waste to attribute every event to God's specific action.*
> Ian G. Barbour

The inability to explain a phenomenon does not make that phenomenon inexplicable. For example, we were unable to explain how we as a species came to exist before 1859. Today, most schoolchildren know enough about basic evolutionary theory to have far more of a grasp of humanity's origins than even the wisest scholars did less than 200 years ago. Similarly, ignorance does not mean nonexistence. Just because you don't know the speed limit on a certain stretch of road does not mean that there is no speed limit, or that a traffic officer will overlook speeding. Scientists need to know their specialty and at least something about the foundations of that specialty; a cardiac surgeon needs to know something about fluid dynamics but does not have to know anything about evolution, which of course does not falsify the theory of evolution.

In 1992, physicists Stéphane Douady and Yves Couder demonstrated that spiral formation is a natural biological property using droplets of oil placed on a film and subjected to magnetic fields, thus removing the mystery behind shellfish and snail growth and shooting another hole in so-called "intelligent design theory," or ID. Plenty of scientists such as Behe, Dembski, and philosopher Stephen Mayer (1958-) have written about ID and have won over creationists (who were already predisposed to disbelieve Darwin). Unfortunately, ID is not a scientific theory. It is little more than propaganda that uses pointed questions and false logic to try to make its case. A 2006 study at the University of Oregon debunked the idea of irreducible complexity (see below) by showing how proteins evolved using nothing but trial and error in an ongoing, bumbling process that has been happening for hundreds of millions of years.

Our evolution makes it very hard for us to accept the idea of evolution and our own individual pointlessness.
Susan Blackmore

Today's scientists—and even the general public—know many things that the founders of religions had no clue about. Were any of them alive today, they would be widely regarded as hopelessly ignorant, idiotic, or both. Still, religious people eschew science to this day, as evidenced by the ongoing battles to have creationism taught in public schools; however, even they have been forced to acknowledge that science has challenged many of their most cherished assumptions. Ideas that purport to reconcile science and religion using scientific-sounding language to validate the Bible, such as Young Earth Creationism (the notion that the Earth is less than 10,000 years old) are being advanced as "alternatives" to Darwinian evolution. A good portion of this "new creationism" is being fueled by the Discovery Institute, a right-wing religious organization that is seeking to conduct "science" based on evangelical Christianity. The fact that they feel compelled to do this says more about their own doubts and insecurities than it does about the nature of the universe, but that hasn't stopped them from trying.

On a certain level, this desire to disbelieve evolution makes perfect sense. It is somewhat less than comforting to wrap one's mind around the idea that one is nothing more than a product of a long chain of accidents on a ho-hum world orbiting a ho-hum star in a ho-hum galaxy with, "no hell below us and above us only stars," as the late John Lennon (1940-1980) sang. On another level, some people feel compelled to rock the boat to gain converts, or at least a little notoriety. There is much to discuss and learn about evolution. Pandering to cre-

Creationists make it sound like a 'theory' is something you dreamt up after being drunk all night.
Isaac Asimov

Probability

> *We have all heard some fundamentalist-minded person say, 'don't tell me I'm related to monkeys.' The fact of the matter is that now that we have discovered DNA and its code, we know that we are not only related to monkeys, we are related to zucchini. so let's get over it.*
> Marlin Lavanhar

There is a general rule of thumb that any event with a probability of occurring that is less than 1 in 10^{50} should not be considered random. This argument seems sound until one realizes that all one has to do to falsify it is write 51 random digits on a piece of paper. So-called miracles are examples of improbable events that result in a stroke of luck, but that does not make them inherently different from any other event. For example, we saw that being dealt all spades in a bridge hand is an extremely improbable event; however, all possible hands are just as unlikely. The only difference is that nobody will proclaim a miracle if they get just any of the 635 billion possible hands.

All events lie on a probabilistic scale that ranges from all-but-certain (the Sun will come up tomorrow, and you and I will die eventually) to miraculous (winning the lottery or spontaneous remission of a terminal disease). Some coincidences can magnify an event's effects. Dawkins uses some examples in *The Blind Watchmaker*. First, if the odds of being struck by lightning are 1 in 10,000,000 ($1:10^7$) and one has lived 25,000,000 (2.5×10^7) minutes, then the odds of a strike on any given minute is 1 in 250,000,000,000,000 ($1:2.5 \times 10^{14}$). These are pretty small odds, but they are not zero and can be calculated with ease. Second, the odds of all four players being dealt a perfect bridge hand are approximately $1:2.235 \times 10^{24}$. If this happened, all four players would immediately suspect fraud or divine intervention, not that they had simply gotten the luck of the deal, just as they have with every other bridge hand they have ever been dealt. Still, such an event is both possible—and in fact far more likely than your suddenly finding yourself orbiting Pluto (which is also possible, albeit not anything to worry yourself about).

> *We were constructed to serve the interests of our genes, not the reverse. The reason we exist is because it once served their ends to create us.*
> Keith E. Stanovich

Events between the extremes of near certainty and near impossibility give us the creeps. Thinking of someone for the first time in years only to learn that the person has just died is just such an example. An event like this is extremely unlikely, but there are many people on the planet, and we can expect the library of such tales to grow over time (there are lots of old ghost stories still floating around), which could lead one to believe that this sort of thing happens more often than it actually does. Also, the fact that we hear about such tales at all speaks to their improbability. If people saw ghosts as often as most of us see ourselves in the mirror, then such sightings would not be news. In fact, the stories you see in the news are there precisely because they are rare.

You hear about every single airliner crash anywhere on the planet in excruciating detail because so few airliners crash. In fact, the day you stop hearing about these events is the day you should stop flying! Among many other things, the 24-hour news cycle has made an entire generation of parents deathly afraid for their children's safety, with drastic results. Ask a group of parents what their worst fear is, and you'll probably hear something about abduction. How many kids are actually abducted by strangers each year? 115, and most of them make it home alive. Meanwhile, thousands of children are killed or maimed in traffic accidents, but you don't see parents refusing to let their children ride in the family car.

Thanks to evolution, we have some ideas about risks and probabilities that are designed to help us live for about a century. Our views of the mundane and miraculous would be entirely different if our lifespans were measured in millions of years, and yet this perspective is the correct one for accurately judging the probability of life and the many events in an average lifetime. Also, our brains are designed to calculate the odds of things happening to the small group of people we know. As I just said, we did not evolve with mass media that instantly informs us about every improbable event that happens anywhere on the planet in near real-time. All of this is strong evidence that coincidences are just not nearly as big a deal as we often make them out to be.

Origins of Life

The origins of life seem miraculous, whether it occurred on Earth, or whether Earth was seeded with life from elsewhere. The first single-celled organisms took only a few hundred million years to appear after the Earth formed. It took 2.5 billion more years for multicellular life to emerge, because the atmosphere did not have enough free oxygen to support larger life forms. The earliest species excreted oxygen until the atmo-

---

Darwinism is the story of humanity's liberation from the delusion that its destiny is controlled by a power higher than itself.
Phillip E. Johnson

Evolution is the law of policies: Darwin said it, Socrates endorsed it, Cuvier proved it and established it for all time in his paper on The Survival of the Fittest. *These are illustrious names, this is a mighty doctrine: nothing can ever remove it from its firm base, nothing dissolve it, but evolution.*
Mark Twain

---

sphere had enough of it to evolve species that required oxygen. So far, we have seen organic molecules such as purines, pyrimidines, and amino acids form spontaneously in lab containers; we have yet to see DNA or even RNA emerge spontaneously. We also cannot create completely synthetic life from scratch in the laboratory. We can, however, greatly modify existing life using genetic splicing and other methods, but the source material must be alive to begin with.

The consequences of an act affect the probability of its occurring again.
B. F. Skinner

The final major piece of the puzzle lies in figuring out how genetic replication began. Again, no experiment so far has yielded RNA. DNA and proteins are like the two legs of a stable archway. Arches require scaffolding and falsework to build. What was the scaffolding that allowed RNA to evolve and kick off the replication process? The Urey-Miller experiments did not get much further than amino acids because their gas mixture probably differed significantly from that of early Earth; however, complex organic carbon molecules can form easily via simple chemical reactions.

An experiment performed by biophysicist Manfred Eigen (1927-) combined *replicase* (an enzyme that facilitates replication) and RNA building blocks, which allowed RNA to evolve both spontaneously and repeatedly. Care was taken to prevent outside contamination and verify that the RNA was only able to evolve from the building blocks provided. These experiments give us a window into how RNA might have gradually evolved over time and provide strong evidence for an automatic, non-guided start to life. We have no idea why replicase makes RNA, but that hasn't stopped it from happening with no thinking, feeling, or planning behind it. The evolution of a self-replicating RNA molecule allowed lineage; descendants and copying errors provided the mutations that led to the diversity of life we see today. Copying also got more and more accurate over time, which allowed genetic stability absent environmental pressures.

The origins of disputes between philosophers is, that one class of them have undertaken to raise man by displaying his greatness, and the other to debase him by showing his miseries.
Blaise Pascal

Replication requires some fairly complex machinery in order to function. So where did this machinery come from? How did natural selection favor organisms with the precursors of replication, and how did those organisms evolve without it? Some people think this is too complicated to explain without resorting to God, in an attempt to prove a God who intervenes to build the universe and then walks away to let it run on its own. This is a classic God of the gaps argument. If God did it, then where did God come from? If God is responsible, then God is complex and needs explanation. If God is simple, then this is not a good explanation. In fairness, the people who subscribe to this argument will also wax poetic about complexity emerging from simplicity, such as the simple equation that yields the infinitely detailed Mandelbrot fractal set, which makes it seem like they are trying to have it both ways. Also, biocentrism and nonlinear theories of time (see Chapters 28 and 29) can eliminate the infinite regression of having to explain layers of causes.

It is hubris to assume that Earth is the sole life bearing planet in the universe, because the implication behind that assumption is that the entire universe was created just for us. That may well be the case, but it's far too early to tell, especially with recent discoveries of potentially habitable planets in other solar systems and new probes being readied to search for alien life in our own solar system. If life evolved on many planets, then the "miracle" of life multiplied by the number of life-bearing worlds becomes nothing more than possibility, if not a probability. This is especially true, because space already has amino acids. How many of these organisms gave rise to intelligent civilized life is unknown.

It is also more than a little hasty to point to the failure to create artificial life as evidence for any sort of miracle. On the contrary, there would be good reason to be more than a little concerned if making life turned out to be too easy. The simple explanation is that chemistry experiments last only a few years at most, and there are only a few chemists dedicated to this line of research. Given more chemists and more time, it is not unreasonable to expect positive results, provided they have emulated Earth's early conditions closely enough, which is not known. Even if the scientist have it right, it took 500 million years for life to form on Earth the first time around, and we just don't have that much time to sit around staring at test tubes.

As we saw in Chapter 19, the living world is divided into bacteria (prokaryotes) and everything else (eukaryotes). Eukaryote cells have miniature cells within them such as the nucleus or mitochondria. Some of these structures may have come from bacteria that formed symbiotic relationships with the

He used statistics as a drunken man uses lampposts; for support rather than illumination.
Andrew Lang

Round numbers are always false.
Samuel Johnson

> *We have found a strange footprint on the shores of the unknown. We have devised profound theories, one after another, to account for its origins. At last, we have succeeded in reconstructing the creature that made the footprint. And lo! It is our own.*
> Sir Arthur Eddington

host cell and were eventually absorbed as integral parts of that cell. This was a major step forward in our evolution. Another major step occurred when divided cells began sticking together instead of going on their separate ways, because this allowed more complex organisms to evolve. Laboratory experiments have managed to create spontaneous colonies from normally solitary single-celled organisms under the right circumstances, which indicates that this obstacle was not overly difficult to surmount.

A theory of life that finally explains how things got started on Earth (or elsewhere, if it turns out Earth was seeded from elsewhere) would probably not be enough to account for the apparent scarcity of life on other worlds; however, there is so much we still don't know that this should not concern us. The probability of life evolving must be very low, which still leaves plenty of room for many millions of life-bearing planets in the universe. On the other hand, we should not be too concerned if a chemist does manage to grow synthetic life in a lab; just because an event is extremely unlikely does not mean that it cannot happen at any given moment. One thing we can be sure of: If we do discover life on other planets, we will also discover Darwinian evolution in action.

Thermodynamics

> *Myth is neither a lie nor a confession: it is an inflexion.*
> Roland Barthes

There really is no such thing as a free lunch. Creating life requires entropy. The more complex the organism, the higher the price tag. Increasingly complex molecules are swimming against the tide of entropy, because each new stage of evolution adds another level to the growing house of cards and increases the odds of breakdown and chaos. Bodies therefore contain numerous safeguards to protect cells and verify gene replication lest cell breakdown lead to death.

This highly localized decrease in entropy seems to violate the Second Law of Thermodynamics, but that is not the case, because this law only speaks to overall entropy; localized decreases in entropy are entirely permissible. On Earth, the ever-increasing order of all life is more than made up for by leveraging a very tiny fraction of the massive amounts of entropy being acquired by the Sun, which is the ultimate engine maintaining life on our planet. This simple fact debunks any claims that life is somehow exempt from the entropy requirement. All of life is a struggle against entropy, knowing full well that eventual surrender is inevitable. The random distribution of heat in the universe will eventually become uniform, and will then cool to the lowest possible energy state. The universe will literally freeze to death through the same mechanism that is allowing us to live now. This is the way of things. None of us would be here without death, and each life contributes something, however small, to the destiny of the universe.

Particles and fields have positive energy, while gravity has negative energy. These pluses and minuses add up to a grand total of 0 sum energy, which allows the universe to pop into being without violating the laws of thermodynamics. The electricity powering the light bulb you are using to read this book represents a loan from gravity. You can read this book because of gravity. More on this topic in later chapters.

Examples of Bad Design

Some people discount the idea of intelligent design because humans are poorly designed, to the point where any god would have to be a bumbling cosmic ne'er do well. I'd like to see any one of these people do half as well with something many orders of magnitude simpler, but let's run with their arguments for now.

Intelligent design advocates point to eyes to bolster their case; however, we have already discussed how eyes evolved, including their repeated independent developments in different phyla. We also discussed the backward and inverted human eye with its blind spot. Light entering out eyes must pass through the optic nerves to reach the rod and cone photoreceptor cells. No designer would build an eye like that because of the increased complexity and reduced visual acuity inherent to such a setup. Natural selection does not care about the mechanics of a solution, so long as the need is met. We can see well enough to do whatever it is we do, and that's all we need. We may be the smartest animal by some measures, but ospreys can see 60 times better than we can. Blindness is often caused by tiny parasites that have plagued us for millions of years. Ospreys can maneuver from high altitude to catch a fish underwater, including correcting the aim point to account for

Nothing splendid was ever created in cold blood. Heat is required to forge anything. Every great accomplishment is the story of a flaming heart.
Arnold Glasgow

Before water generates steam, it must register 212 of heat; 200 will not do it. The water must boil to generate enough steam to move an engine. Lukewarm water will not run anything. Lukewarmness will not generate life's work.
Unknown

refraction. These magnificent birds are on the verge of extinction, but any blind person can believe in God.

Flatfish skeletons are imperfection incarnate that reveal the telltale signs of gradual evolution. No designer would build a sideways skeleton; s/he would instead design a better framework, such a stingray skeleton. Flounder and sole used to be vertical, but turned sideways because of evolutionary pressures. The eyes moved to the same side of the head, and that was pretty much it.

Why kick the man downstream who can't put the parts together because the parts really weren't designed properly?
Philip Caldwell

An article in *Scientific American* by Messrs. Olshanky, Carnes, and Butler looked at the flaws in the human body and speculated how an engineer could fix those flaws to increase the average lifespan to more than 100 years. This arbitrary number flies in the face of nature, which wants us to reproduce early and often, raise our children, possibly help with grandchildren, and then move aside to let someone else fill our shoes. The current human lifespan of about 75 years is perfect for this, which is why we lose bone mass as we age and have other flaws, such as a rib cage that does not offer complete protection, various types of atrophy (wasting away), etc. This same article shared a rendering of what the authors thought a properly designed human should look like. Let me be blunt: You would probably not consider this fictional person exactly handsome.

Perhaps believing in good design is like believing in God, it makes you an optimist.
Terence Conran

Let's say that 1,000 steps are needed to evolve a fully functional human eye or equivalent. How could this possibly happen? The simple answer is that each of the 1,000 mutations provided an advantage that may or may not have been directly related to sight. All of the species that survived made the upgrades as needed. Those that didn't fell by the wayside. The environment imposes itself on species, but not the other way around (at least not until very recently in our evolutionary history). We saw in Chapter 19 that the eye is one example of convergence across different phyla. Octopus eyes are wired "correctly." Birds have developed sonar independently of both bats and other similar birds. It is a safe bet that the common ancestor of all birds and mammals probably did not fly or live underground. Bats, birds, whales, and perhaps a small handful of mammals chanced on echolocation about 100 million years ago. Other species that are now extinct may have done so as well. Bat and bird wings evolved independently. Tall trees may be majestic, but their great height is both inefficient and requires extensive resources. Their height is the result of an ongoing struggle for light and air, in which the taller trees win.

Some organisms include features that look like evolution run amok. Survival is only part of the battle. Organisms must find ways to attract mates and compete with others of their species to pass on their genes, even if it means risking death. Some species have therefore evolved significant handicaps, presumably because an individual who survives despite its burden can be seen as a strong suitor. Females are usually the ones who select mates, which means that males of many species have to compete for attention and the chance to reproduce, while the females can simply take their pick. Humans are no exception. The moment female preferences change, male attributes will change to follow suit.

Genetics

All life on Earth has DNA that uses the same four bases of A (adenine), C (ctyosine), G (guanine), and T (thymine). A bases bond with T bases, and C bases bond with G bases. This extremely simple "programming language" is common to all species. The programs created by this language have evolved to fit ever-changing environments, thanks to both programming errors (mutations) and quality control (natural selection). The message contained in DNA is practically immortal and is measured either in tens of thousands to millions of years, or in tens of thousands to well over a trillion life spans. The DNA itself dies when its host body dies; it is the message, the pattern, that lives on. Each life form refutes the JCI idea of a transcendent God creating anything in His image, and falsifies the idea of separate "kinds," depending on how you read the pertinent passage. (See Chapter 3.) Natural selection may be a watchmaker (using Paley's argument), but the watchmaker is both blind and non-intelligent.

DNA information for a single gene is copied to single-stranded messenger RNA (mRNA), which then moves from the nucleus to the ribosome. There, translators read the instructions and make the appropriate proteins that do the cell's actual work and provide its structural integrity. The ACGT language allows 64 possible 3-letter combinations, which is plenty because there are only 20 amino acids. For

We cannot cheat on DNA. We cannot get round photosynthesis. We cannot say I am not going to give a damn about phytoplankton. All these tiny mechanisms provide the preconditions of our planetary life. To say we do not care is to say in the most literal sense that we choose death.
Barbara Ward

A camel is a horse designed by a committee.
Unknown

example, both GAA and GAG code for glutamic acid. These simple chains can contain a staggering amount of information. Each cell could store the entire 30-volume *Encyclopedia Brittanica* three or four times over, and each salamander sperm or lily seed can store some 60 complete sets. Some amoebas can store up to 1,000 complete sets.

New dewdrops don't emerge from existing dewdrops, but new DNA emerges from existing DNA. Computer *ROM* (Read Only Memory) can only be written to once, after which the information may be accessed (read) but not altered or removed (written). The same is true of DNA, except for the inevitable copying errors. A mutated DNA strand that confers a benefit gets "written" in the form of better instructions when the beneficial error gets passed on, spreading that single evolutionary change into many subsequent DNA copies. Changes in genetic content ultimately drive evolution.

It is tempting to think of genes as a master blueprint for building a body, and that one need only follow the instructions to the letter to pop out a fully formed organism. This is not quite the case: DNA acts more like a recipe that is followed by each individual cell. There is no "central control" telling all of the cells what to do, simply a collection of up to trillions of cells acting alone, while unwittingly depending on their peers for their own survival. A liver cell has no idea that the chemical reaction it is performing (such as detoxifying alcohol after a night of partying) is helping a larger organism, a heart cell has no idea that its rhythmic contractions are keeping the entire body alive, and a kidney cell has no idea that it is a cog in a wastewater treatment plant. Each cell is playing the game for its own benefit, which coincidentally helps the organism.

Medieval scientists thought that each human sperm contained a *homunculus* (miniature human) that would then grow inside the mother. This idea has obvious problems, such the infinite regression that occurs if one realizes that there is no hard limit to how long humans can remain a viable species, and that countless thousands of generations may be born before our eventual demise. Children get some of their traits from their mothers as well, and thus look like intermediaries between the parents because of the summed differences in their genes. (Of course, children have their own mutations and are therefore not clones of their parents.) This leaves open the question of

---

A common mistake that people make when trying to design something completely foolproof is to underestimate the ingenuity of complete fools.
Douglas Adams

What more fiendish proof of cosmic irresponsibility than a Nature which, having invented sex as a way to mix genes, then permits to arise, amid all its perfumed and hypnotic inducements to mate, a tireless tribe of spirochetes and viruses that torture and kill us for following orders?
John Updike

just how an embryo develops into a viable offspring with the proper shape and functionality needed for a normal life. Research is ongoing, but here again, we can say that this process more resembles a recipe than a blueprint. As always, small differences can have drastic results, just like swapping yeast for baking soda can ruin a cake, which explains both birth defects and congenital disorders.

Genes get passed on, but not bodies. Children are randomly different from their parents, but those who survive to reproduce are not random, because they have managed to mature fully. A body that survives brings its successful genes with it, which benefits those genes. No conscious processing is necessary for this to happen, except where mate selection is involved—and even this may be far less conscious than anyone might want to admit. As I outlined in *The Natural Savage*, and will discuss in detail in *The Romantic Savage*, there is plenty of reason to suspect that we can subconsciously appraise each other's genomes, and are attracted to people whose genes are as different as possible from our own. This provides the greatest chance for genetic mixing and evolution to occur. Even more importantly, the single greatest selector is life itself; death is the strongest possible rejection that nature has.

What happens with children of different genders? A man with an exceptionally long penis who has a daughter may then have similarly well-endowed grandchildren. The daughter in this example carries the gene, but does not express it for obvious reasons. The man's own ancestors probably had genes for enhanced genital endowment as well. In the case of competing needs, assume that females of a flying species like long tails that are also aerodynamically inefficient. If 2 inches is the best length for flying but females prefer four inches, then males will probably end up with tails averaging 3 inches long, assuming that all else is equal. How did gazelles end up eating grass with tigers, lions, and cheetahs eating gazelles? Things could have turned out the other way around; however, the familiar pattern continued under its own momentum once it got started.

Deleting a file from a computer only deletes the pointers to that file, while leaving the file itself on the disk. The space is available for reuse, but that does not mean that it is blank, hence the success of "undelete" programs and forensic inves-

---

The precise form of an individual's activity is determined, of course, by the equipment with which he came into the world. In other words, it is determined by his heredity.
Henry Louis

Heredity is nothing, but stored environment.
Luther Burbank

tigations of computer storage devices. Unlike your filing cabinets, a given file is not all stored in the same location on the disk; bits and pieces are scattered all over the disk with a master index that tells the computer where to find the pieces and how to reassemble them. It's as if you took all the books off your shelves, removed their pages, and then randomly placed the pages back on the shelf while noting their locations for future reference. We see apparent orderliness when we open, say, a file containing a book chapter or spreadsheet, but that is misleading. Also, saving the same file repeatedly does not place the pieces back in their original locations, but scatters them around the disk once again and updates the master index. Computer storage systems thus have several copies of the exact same data, with only one of those copies labeled as active. Genes work the same way, with snippets of instructions scattered around and pointers that indicate where the rest of any given instruction set is located. The "coding" portions of the DNA molecule are randomly surrounded by "non-coding" DNA that was until recently deemed to be useless "junk," as we learned in Chapter 19.

> *We do not know, in most cases, how far social failure and success are due to heredity, and how far to environment. But environment is the easier of the two to improve.*
> J.B.S. Haldane

This seems nice and neat, except that DNA and RNA do not self-replicate but require enzymes to do it for them. Neither can function without the other. Which came first and how?

Mutation and Selection

> *There is nothing exempt from the peril of mutation; the earth, heavens, and whole world is thereunto subject.*
> Walter Raleigh

Natural selection appears to have done everything from sculpting our bodies to shaping our beliefs and morals. We do not yet know how natural selection got started, but we do know how it works. We also know that the rise from bacteria to humans was easier than the rise from amino acids to bacteria, because the former already had the fundamental building blocks and processes in place. As I noted in *The Enlightened Savage*, the MYH16 gene helps develop jaw muscles and may have gotten shut off in humans. A weaker jaw places less stress on the skull; this could have been what allowed our brains to grow. We may literally be sick monkeys.

In 2005, two teams at the University of Chicago discovered two genes called ASPM and microcephalin. Disabling these genes causes *microcephaly*, a condition in which the cerebral cortex is much smaller than normal. These could be reminders of our own genetic legacy. Modern humans only emerged fewer than 100,000 years ago, and these two genes have begun evolving faster over the last 37,000 years, which proves that the human brain is very much a work in progress. We can also identify about 700 areas of the human genome where natural selection has been active as recently as 5,000 to 15,000 years ago, such as in the development of racial characteristics that represent nothing more than very superficial adaptations to different climates. (See *The Natural Savage*.)

Earth itself had to have properties that allowed self-replicating genes to evolve. The normal laws of physics themselves must allow this phenomenon, along with a slow but steady error rate and the ability of each generation to have at least a little control over its own fate and reproductive potential. This also had to allow for ever more replication and diversity over time. For example, bodies evolved brains that in turn evolved the ability to communicate, which in turn fostered cultural development that allowed ideas to replicate and spread through non-genetic means and spur further cultural development. In Chapter 9, we saw that ideas with the capability for self-replication are called memes. Memes are driving cultural evolution far more quickly than biological evolution, which explains how we managed to find ourselves blessed with supercomputers and saddled with nuclear weapons while trying to process all of this with brains that would much rather be avoiding capture on the Serengeti. Jokes that spread like wildfire are examples of memes, as are other persistent ideas, such as religious dogma, some taboos, fashion, and many others. The study of how ideas self-replicate within and across cultures is called *memetics*, a fascinating field of study that has attracted its share of both researchers and advertisers seeking to inject their products into the cultural zeitgeist.

If genes are the real stakeholders in the evolution game with humans as mere mixing bowls for different genetic combinations, then we could see memes in a similar light. Blackmore points out that we humans like to think that we are the ones building computers, telecommunications links, and other means of spreading ideas; however, from the meme's point of view, we are simply their copying machines being used to build a global idea distribution system. I personally find this idea both humorous and little scary at the same time. After all, if all ideas are memes, then so are each of our own thoughts. It

If Nature denies eternity to beings, it follows that their destruction is one of her laws. Now, once we observe that destruction is so useful to her that she absolutely cannot dispense with it from this moment onward the idea of annihilation which we attach to death ceases to be real what we call the end of the living animal is no longer a true finish, but a simple transformation, a transmutation of matter. According to these irrefutable principles, death is hence no more than a change of form, an imperceptible passage from one existence into another.
Marquis De Sade

becomes impossible to distinguish meme from non-meme, which potentially undermines the validity of all human knowledge. Thankfully, no one has proven that memes exist, unlike genes.

> *Mutation. It is the key to our evolution. It has enabled us to evolve from a single-celled organism into the dominant species on the planet. This process is slow and normally taking thousands and thousands of years. But every few hundred millennia, evolution leaps forward.*
> Jean Grey

Most random molecular changes have no discernible effect and are therefore benign, although such a change in one generation could have dramatic consequences for later generations. Such small steps are how evolution gradually moves toward new species, with no end goal in mind beyond having each step be more effective than the last. Survival and reproduction are the only metrics that matter. Most mutations are random and can be caused by copying error, as we have already discussed, or by external factors such as toxins, radiation, and other *mutagens*. Solar radiation, the Earth's own background radiation, and the cosmic rays constantly bombarding our planet are examples of natural mutagens. The rise of technology and modern chemistry has unleashed new waves of mutagens with very little knowledge of their long-term effects. We know that mutation drives evolution; however, just because a certain mutation might benefit a species does not mean that said mutation actually occurs.

DNA does contain proofing mechanisms to repair and/or dispose of bad copies, which reduces the error rate in *E. coli* bacteria from 1 in 10^6 to 1 in 10^9, and is therefore 99.9% effective. This is the rough equivalent of copying the Bible about 285 times with only one error. By contrast, a good secretary makes approximately one error per page. If one secretary manually copied a document and then passed it to another secretary to make another copy and so on, the text would quickly degrade to 99% accuracy, and less than one percent of the original would remain by the 10,000th generation. This astonishing degree of accuracy in DNA is further curtailed by natural selection, which weeds out the majority of the changes. If natural selection were to suddenly cease, we would expect the actual rate of evolution to increase to the maximum mutation rate. Mutation is the foot pressing the gas, while natural selection is the foot riding the brake by forcing members of a species to compete with each other for survival and reproduction. This lateral distribution and competition yields vertical results.

> *Natural selection, as it has operated in human history, favors not only the clever but the murderous.*
> Barbara Ehrenreich

Physics allows us to date rocks, because radioactive compounds decay at predictable rates; all we need do is measure the decay. Radiocarbon dating works for dating objects that are thousands of years old, but other means are needed to accurately measure million- or billion-year histories.

Thankfully, molecules change at constant rates. All we need to do is measure the differences to determine how long it has been since two species were the same. The story of evolution is that of a single ancestor species that branched out again and again in a tree-like manner into many millions of individual species. The study of how the "tree of life" branched out and where each species belongs on that tree is called *taxonomy*. Our tree of life happens to be *cladistic*, which means that separate branches can never come back together again, as far we know. Taxonomy is not as easy as it seems. For example, three separate niches in three different continents have been filled by different mammals, thanks to an evolutionary process that created species that look very similar to each other while having completely different evolutionary histories. Any given ecosystem fills its various niches (burrowers, grazers, predators, etc.) with different specialists based on the genetic material available to that ecosystem.

The overwhelming majority of species that have emerged over the last 3.5 billion years have gone extinct, with new species normally emerging at a rate that balances extinction. A very lucky few (like the coelacanth) have found themselves in very stable environments, and have been able to simply coast right along. One can presume that coelacanths have a mutation rate just like any other animal, and that natural selection has consistently cleaned up the messes and preserved the species unchanged for hundreds of millions of years.

Don't underestimate the power of genetic programming. Soldier ants are constantly prepared to fight and die to defend the queen, the colony's sole source of DNA, because that is what they are programmed to do. We can speculate that they feel neither love for the queen nor hate for the enemy to justify their heroics, but we cannot be certain of this. We can be certain that any such emotions they may feel are also thanks to evolution, just like maternal love is a direct result of evolution, because mothers who love their children will take better care of them and help ensure successful passing on of the genetic

Change the changeable, accept the unchangeable, and remove yourself from the unacceptable.
Denis Waitley

Every body continues in its state of rest, or of uniform motion in a right line, unless it is compelled to change that state by forces impressed upon it.
Isaac Newton

> *When you are content to be simply yourself and don't compare or compete, everybody will respect you.*
> Lao Tzu

football. Fish can do basic math, because that ability confers a survival benefit. The very air we breathe is laden with programming in the form of pollen, spores, baby insects, and more, just as if it was raining USB thumb drives.

Life is too beautiful, complex, and exquisitely tuned to have emerged by pure chance. Darwinian evolution solves this problem by calling for small changes over time, with each step being both simple relative to its predecessor and directed by non-random natural selection. Waves are non-random, because their patterns are predictable. Holes are non-random, because anything smaller can fall in. Dawkins uses the example 28-character sequence, "Methinks it is like a weasel." (Period added.) There are 27 ways to fill each slot (26 letters and one space), meaning that there are 2728 (1.197×10^{40}) random ways to write a sequence of 28 alphabetical characters. Cumulative selection that keeps correctly placed characters and discards the others reduces this to as little as 43 generations, which literally shaves millions of years off the process.

This example demonstrates the power of cumulative natural selection, and uses the goal of the desired target phrase to simulate the non-random nature of natural selection; it is not intended to imply that evolution has any goals or outcomes in mind, or even that evolution has any sort of mind at all.

Competition

> *A competent and self-confident person is incapable of jealousy in anything. Jealousy is invariably a symptom of neurotic insecurity.*
> Lazarus Long

Evolution is a real bitch, the epitome of cold, ruthless efficiency. There is nothing soft and cuddly about it. Plant and animal species are locked in a state of perpetual warfare that is driving a relentless arms race. Predator versus prey, parasite versus host, and male versus female are just three examples of the trench warfare known as natural selection. We can only imagine what evolution might have looked like if the weather and nonliving factors were the only things life had to worry about. Long-term evolution has made plants and animals able to handle normal weather patterns. When it comes to competition with other living species, the term "enemy" refers to anything that interferes with a species' life. Everything that eats or is eaten is an enemy of what it eats or gets eaten by, in a merciless spiral that creates ongoing improvements in fits and starts, and sometimes even backpedals over time with no net gain for anyone. Gazelles that evolve faster speeds to outrun lions are quickly countered by faster lions. Humans that evolve defenses to a particular virus are quickly targeted afresh by mutated versions of the same virus. Sooner or later, everyone loses.

Trees are tall, because taller trees get more light and air at the price of expending resources and lowering efficiency. All trees would be more efficient if they were all shorter, but somewhere one tree got a mutation that made it grow taller than those around it. This tree survived better than its shorter peers, and the trend toward increasing height began, culminating so far in giant redwoods that are well over 300 feet tall; the tallest, named Hyperion, stands at 397.1 feet. Natural selection does not care about long-term implications, only on what works best at the time. The sole goal at every stage of this particular competition has been to be taller than other trees. Most animals are at almost constant risk of being killed and eaten, and their traits make perfect sense when viewed against the traits of their enemies.

Parasites and hosts are locked in an endless arms race, which explains why the males and females of most species place a premium on the appearance of health such as shiny eyes, glossy coats, smooth unblemished skin, etc. Only truly healthy specimens can display all of the signs of health, which is why evolution favors them. There are various species of cicadas with 13- and 17-year recurrences. These relatively rare events are true plagues that overwhelm and starve out all opponents. Why 13 and 17 years? Why not 14? The simple answer is that a 14-year lifespan would make them vulnerable to predators with 7-year life spans. Something is special about 13 and 17 years. If the underlying cause ever changes, the cicada life spans will soon change to match.

Mammals have various encephalization quotients. Rabbits have an EQ of 0.40, as do rats. Dogs have an EQ of 1.17, chimpanzees 2.48, bottlenose dolphins 5.31 and humans 7.44. Monkeys, apes, and humans are above average in this regard, and we can speculate that arboreal movement requires extra processing power for safe navigation. Monkeys and apes that eat insects and fruits have higher EQs than their leaf-eating cousins, because finding fruit and insects is harder than finding leaves. The overall trend is toward higher EQ over time, because smarter animals can outwit dimmer opponents. In

I don't compete with other discus throwers. I compete with my own history.
Al Oerter

Men often compete with one another until the day they die; comradeship consists of rubbing shoulders jocularly with a competitor.
Edward Hoagland

general, this trend leapfrogs back and forth from prey to predator. A prey animal needs more brainpower for better defenses, while the predator needs it to improve its strategies. The prey animal is at more of a disadvantage, because a rabbit runs for its life, while a fox only runs for its dinner. Genes are selected for their ability to interact with the environment, and other genes are certainly part of the environment. This can also lead to symbiotic relationships, such as those between flowers and bees or between termites and certain fungus.

Genes can also work in concert, because they can never evolve in a vacuum. For example, a gene that makes teeth better shaped for eating meat is of limited effect without a gene that helps the animal digest meat.

Transitional Species

If we consider the superiority of the human species, the size of its brain, its powers of thinking, language and organization, we can say this: were there the slightest possibility that another rival or superior species might appear, on earth or elsewhere, man would use every means at his disposal to destroy it.
Jean Baudrillard

Darwin himself said it best: If any complex organ could not have formed by steps, then his theory breaks down. Thus, we should expect to find *transitional species*, or species that existed between a given ancestor species and its modern counterparts. In other words, we should expect to find evidence of animals changing over time, including humans. Most species on Earth remain single-celled to this day. We already know that single-celled organisms can evolve to multicellular organisms, because that happens to every single person at conception. Creationists constantly demand to see transitional species, and the fossil record is supplying evidence of just such changes. Failure to find these changes would falsify Darwin by his own admission, but the exact opposite is happening. Darwin is being validated in a big way, because we are finding these species in well-layered deposits that reveal how and when the evolution occurred. Some of the many examples of evolutionary change we have discovered include jaw bones migrating to the inner ear and *Tiktaalik*.

It is not the strongest of the species that survive, nor the most intelligent, but the one most responsive to change.
Charles Darwin(?)

We already know that evolution works in small steps over time. If Animal A lived 25 million years before Animal B and B has a tail that is 50 inches longer than A's, then we know that the tail grew at an average rate of 2 inches per million years. It is preposterous to think that this happened at a constant 2 millionths of an inch per year; rather, we would expect this growth to occur in fits and starts over time, with periods of rapid growth punctuated by periods of little to no progress. As we saw in Chapter 13, Eldredge and Gould proposed that maybe the fossil record is more complete than we think, and that we should not be looking for a complete series of gradual transitions. To them, evolution could have periods of stasis followed by rapid bursts. We have already seen that the coelacanth has remained unchanged for millions of years. We have also seen that human evolution has sped up very recently to (among other things) produce different ethnicities. We also have hundreds of dog breeds that have been created from wolves over only a few thousand years. Punctuated equilibrium seems to be the way of the world. Fossils from the Cambrian explosion display many highly evolved species that have no apparent history before then. This may be a huge gap in the fossil record; then again, it may be evidence that previous species had either no bones or soft bones that decayed.

The uneven pace of evolution, the possibility of evolutionary leaps thanks to quantum waveforms and/or biocentric evolution (see Chapters 23 and 29), the fact that we can only find a few of the many fossils in the world, and the fact that all surviving fossils represent only a tiny fraction of all the plants and animals that lived and died at any point in time all mean that gaps are inevitable and must be expected *a priori*. In fairness, all it would take to falsify punctuated equilibrium would be one fossil that is demonstrably out of place in the geological record. So far, none has been found. This situation can be likened to a very nuanced shade of gray that is constantly shimmering either lighter or darker with each succeeding discovery; however, creationists and intelligent design advocates want black and white answers, because they see the question of life's origins as black and white and have their entire belief set invested in white.

This absolute viewpoint cannot grasp that imperfection is acceptable, that (for example) some vision beats total blindness. Their ideas about irreducible complexity are nothing more than grasping at intellectual straws in the face of overwhelming evidence. As Dawkins points out, plenty of people need glasses, yet not one of them would prefer blindness or thinks their eyes don't work just fine and provide invaluable service, albeit with a little help from the local optician. Evolutionary biology had falsified Behe, Dembski, and others long before they put fingers to keyboard.

Our wretched species is so made that those who walk on the well-trodden path always throw stones at those who are showing a new road.
Voltaire

I believe we are still so innocent. The species are still so innocent that a person who is apt to be murdered believes that the murderer, just before he puts the final wrench on his throat, will have enough compassion to give him one sweet cup of water.
Maya Angelou

> *It is a pleasant thing to reflect upon, and furnishes a complete answer to those who contend for the gradual degeneration of the human species, that every baby born into the world is a finer one than the last.*
> Roland Barthes

It is possible for genes to become inactive only to be switched back on later, provided they have not been removed by natural selection. It is also possible for a gene to take on entirely new functions over time. These simple truths utterly refute *saltationism*, a mechanism proposed by creationists to try to demonstrate how God could repeatedly direct the course of evolution by causing new species to appear fully formed and ready to go, like the planet receiving a new piece of furniture with no assembly required and all setup included at no charge. Small changes make positive changes much easier compared to wild swings in the same way that tiny turns of the steering wheel are the best way to keep your car in its lane on the road. Stretching an airplane is a lot less complex than it looks because it mostly involves using more of the same components that are already being used. Snakes grew or shrank by increments of one or more vertebra at a time, annelid worms grew or shrank by one segment at a time, etc.

Punctuated equilibrium is not the same thing as saltationism. All members of the same species can breed with other members of the same species. A population that migrates away or somehow gets cut off from the rest can still breed with the original group in theory. Changes come gradually until two distinct species have emerged that can no longer interbreed. Both groups may evolve at different rates, depending on each group's circumstances.

Arguments For Design

> *Myth is neither a lie nor a confession: it is an inflexion.*
> Charles Dickens

Some people see evolution as a product of random chance and natural selection, others as design. Either way, it is a fallacy to assume that evolution is a single peak, or that some forms of life are more or less evolved than others. Evolution resembles a mountain range with many peaks, some of which represent failed attempts that have gone extinct, others representing a species that has successfully evolved to fits its niche and is still alive. This much both sides can agree on, but the disputes begin almost immediately thereafter. Does the evidence I presented above and in other chapters in this book really point to random pointlessness?

Let's begin by looking at the much-trumpeted claims of similarity between humans and apes. Humans and chimpanzees are close genetic cousins, but that is not the crux of the question. Apes are not good analogs for human thinking, and our observations are very easily warped by observer bias. Furthermore, humans and chimps rarely share anything resembling a close emotional or intellectual bond. Fantastic claims of ape mental abilities are overblown. Maimonides wrote that, "In the time of Adam there were other animals that looked like humans but that lacked the spark."

We already know about Paley's watchmaker argument, and that mainstream evolutionary science sees that watchmaker as utterly blind, but... What if Paley's watchmaker was not blind but divine? What if all life is the result of intelligent planning and design instead of quasi-random evolution? At least some of the life on Earth is far too complex to be the result of chance or anything but active divine influence. The traditional creationist view of God creating the world fully formed without evolution is an example of downward causation with God up and us down. This is not satisfying, because it paints an anthropomorphic picture of a king figure; however, nature is vastly more complex than any watch, so life must be a product of God's design. Evolution is a creative process geared toward producing ever-higher levels of intelligence and other qualities that religion defines as godly. Believers need not fear that science can ever usurp or displace God.

The evidence for design is overwhelming, and evolution proves the truth of intelligent design theory. The close relationship between chemistry and physics (the foundations of evolution) is clear evidence of design by God, who is both an objective organizing principle and a creative and intelligent designer. The relationship between evolution and intelligent design is similar to that of arches in a church ceiling: Intelligent design builds the arches, and evolution decorates them.

Materialism

A mechanical, Godless view swept over science in the 19th century and has persisted ever since. Some of this may well be an ongoing backlash against religious treatment of anything that disagreed with dogma over the previous centuries. Materialists see blind mechanical determinism, in which the future and past states of a system can be extrapolated from the present state. Quantum mechanics has thrown a huge wrench in

Statistically, the probability of any of us being here is so small that you'd think the mere fact of existing would keep us all in contented dazzlement of surprise.
Lewis Thomas

Once is an instance. Twice may be an accident. But three or more times makes a pattern.
Diane Ackerman

this idea, as we will discover in Chapter 23, but this has not stopped some physicists from trying to generate theories that include determinism in quantum systems.

Recently, a small but growing number of scientists have begun leaving materialism in search of new explanations, in light of ongoing observations that are undermining materialism—a trend that seems similar to the gradual abandonment of medieval ideas about our solar system. Observations disagreed with the Ptolemaic model of planetary motion. In response, more and more *epicycles* (orbits within orbits) were added to the model. Planets using this model would literally be doing a jitterbug through the heavens. The new understanding of the solar system completely upended the old model and revealed a very simple and elegant system of elliptical orbits. A similar revolution may be starting to happen in physics and biology. One of the main problems with Darwinism, some say, is that it is so general that it can justify almost anything.

When science from creation's face enchantment's veil withdraws, what lovely visions yield their place to cold material laws.
Roland Barthes

Probability

Chance does not preclude necessity.
Christian de Duve

Again, complex objects and mechanisms are very unlikely to come together by chance (the whirlwind making an airplane in a junk yard argument). Biological beings are far more complex than even the most advanced jumbo jet. Any collection of assorted parts is unique and just as improbable as any other collection of parts, just as every possible bridge hand is equally improbable. One can spin the wheels on a combination lock and point out the probability of getting any combination as being 1 in x, but the uniqueness of the combination that opens the lock has been decided in advance, just like the rules of bridge that make a hand of all spades an excellent hand. Fine, complex, viable living structures are extremely unique. It is this uniqueness that places the rise of life far beyond the limit of what we can expect from random evolution. This can only imply intentional design.

The materialist position is that no "supernatural" entity is needed. Darwinism is a mix of random mutations and natural selection; however, we can see evolution as the expression of how God decided to create life in the universe, which requires one to accept the idea of a clockwork universe with no room for free will. Thankfully, this view has been debunked by quantum mechanics to allow free will, but God must still step in many times to guarantee our eventual appearance. Theologians are trying to figure out how God could act in such a universe.

Sickle cell anemia is a serious disease caused by the body forming red blood cells that are not round, but sickle-shaped. These cells carry less oxygen and are prone to getting stuck in capillaries and breaking apart inside blood vessels. Malaria is another debilitating disease usually transmitted by mosquitoes. It just so happens that a single sickle cell gene protects against malaria; a mother whose children were thus immune to malaria would be blessed by seeing her children survive and multiply—that is, until the percentage of people with this single gene grew to the point where people started having two copies of the sickle cell gene and coming down with sickle cell anemia. So far, this is a classic example of evolution doing the best it can with the tools it has while being unable to predict future consequences.

Fortunately, a workaround is available in the form of C-Harlem hemoglobin, which confers all the malaria prevention benefits of sickle cells without any of the drawbacks. The odds of this happening by chance are 1 in 10^{16}, which should require longer than the current age of the universe to happen by chance. (Of course, just because something is so improbable that we can only expect it to happen once in however long a time period does not mean that it cannot happen at any moment. It is therefore misleading to say that we "should" wait any length of time for anything to happen.) Animals that have been injected with random synthetic chemicals almost always form antibodies for them, which is a phenomenon that we should not necessarily expect to see.

Are the chances of life appearing on Earth too low for the available time frame? The existence of a common ancestor to all subsequent species argues against random chance. For example, the odds of randomly duplicating two identical protein chains with 100 amino acids are 1 in 10^{130}, which is far less than the roughly 4.3×10^{17} seconds that have elapsed since the universe. Obtaining a single protein by chance would require 10^{110} trials each second since the beginning of time. Running these trials would require 10^{90} grams of carbon; however, the Earth's total mass is only 6×10^{27} grams, and 10^{90} grams is far more than the estimated mass of the entire universe. Even

Random events cannot account for the origin of life, at least not in the time available.
Harold Morowitz

On Earth, a long sequence of improbable events transpired in just the right way to bring forth our existence, as if we had won a million-dollar lottery a million times in a row. Contrary to the prevailing belief, maybe we are special.
Robert Naeye

resorting to extraterrestrial origins does not solve this problem.

The best evidence we have says that life on Earth began just as soon as conditions were favorable. So where did this first cell come from, and how did it evolve? Are we really expected to believe that all of the enzymes, hundreds of genes, etc. just popped into existence by sheer dumb luck? There are either 20^{32} or 10^{42} ways to string 32 amino acids into peptides. If random primordial chemistry synthesized one molecule of each peptide, then we would need 10^{18} tons of organic material, which far outstrips the available supply. We could help things along by allowing multiple planets to solve part of the problem, but we still run into the problems described in the previous paragraph. We also have to account for how this material could have been transported effectively across vast tracts of space.

There is a way to sidestep the probability problem by postulating a "megaverse" that is much larger than the known universe, which could give life so many more chances for success as to make it practically inevitable. This line of thinking is not given much credence.

Irreducible Complexity

Reductionism has done a great job of explaining life at a certain level; however, attempts to dig deeper lead one in a completely unexpected direction. A digital life simulation called Tierra developed by ecologist Thomas S. Ray (1941-) tended to evolve toward simplicity instead of complexity. Common descent is a fact, and we just saw that something nonrandom had to account for its beginning. If evolution is unavoidable once life begins, then how much effort does God need to keep investing in the process? What is the minimum input needed? Materialist science says zero; God can be infinitely aloof, infinitely lazy, or both.

Physicists do not take matter for granted, and have peeled atoms apart to reveal particles and small-scale forces. Their problem is the ultimate origin of the universe we know and live in, and the natural laws that made it possible. Biologists face the opposite problem of explaining how simple things became more complex. Their work is complete when they

---

Faced with the enormous sum of lucky draws behind the success of the evolutionary game, one may legitimately wonder to what extent the success is actually written into the fabric of the universe.
Christian de Duve

Most mutations that built the great structures of life must have been nonrandom.
Michael Behe

find entities that are simple enough to pass on to the physicists. Meanwhile, all complex life on Earth (and possibly beyond) requires explanation.

The ongoing process of avoiding death prevents the body from returning to a state of equilibrium with its environment, which means that the quantity of electrical, biological, and chemical activity in a body is markedly different from its surroundings. All living things would die and blend into their environments without ongoing active effort. Each body is complex, with many parts working on concert, and one must apply physics to the parts and not just the whole. The body's behavior emerges from the behavior of its parts.

As a means to explaining life, the unrelenting reductionist approach is doomed to failure.
Johnjoe McFadden

Flagella and eyes are two examples of organs that are too complex for randomness. Some eyes can detect a single photon, while even the most sensitive film can only detect 25. Eyes, wings, etc. are only useful as functional wholes. Fractionally developed organs are useless. A giraffe's long neck requires a different head, musculature, circulation, and other factors working together to function. It is not credibly possible to believe that this happens by chance. Most mutations must therefore be non-random, because random mutations can only produce jumbled results. If random evolution cannot explain the emergence and subsequent progression of life, then we must assume that evolution is nonrandom. This assumption is validated by the fact that DNA has changed in many beneficial ways that defy random odds. Intelligent design is the best way to explain this.

Beyond Competition

The golden rule is universal, but why? According to Lewis, friendship itself has no survival value. On the contrary, friendship is something that makes survival valuable. Which begs the question: Do we even deserve to survive? One must look to God for the answer.

Loneliness is perhaps the most terrible condition, a state which we fear from childhood to the grave.
Louise B. Young

Traditional evolutionists see only war and arms races between species, such as antelopes that run faster to evade cheetahs who then become faster themselves; however, there are other ways for prey animals to defend themselves that don't require constantly pushing the evolutionary envelope. There are also too many examples to list of cooperation and even friendship

Transitional Species

> *The simplest living cell could not have arisen by chance.*
> Johnjoe McFadden

The earliest known vertebrate is a two-inch creature that looks like a lamprey or eel called *Pikaia gracilens*. We are here because (among a great many other things) a small vertebrate evolved, survived, and thrived through the Burgess Decimation, a major die-off that occurred around the same time as the Cambrian explosion and may have wiped out more species than are currently alive. Before and since that time, we have been led to believe that evolution has worked through many small steps. Darwin told us that we should find many fossils belonging to transitional species, but the fossil record still contains plenty of gaps, despite ongoing discoveries that are helping narrow some of those gaps. According to Eldredge, this pattern of punctuated equilibrium is not what Darwinian evolution tells us to expect.

Living things are too beautifully complex and designed to have come into being by random chance. Darwin says that evolutionary change consists of tiny simple steps relative to the earlier species, but these steps cannot be random, because they are driven by non-random survival. Cumulative selection is inherently non-random. Dawkins' "Methinks it is like a weasel" example only works if evolution has "Methinks it is like a weasel" as the desired outcome. God uses His prior creations as templates to build on towards His intermediate and ultimate goals, which explains the great similarity of body plans across both vertebrates and invertebrates.

An Apology

> *It is through cooperation, rather than conflict, that your greatest successes will be derived.*
> Ralph Charell

In this section, I am not using the word *apology* to signify an expression of regret, but as a term used to describe an explanation or defense.

My research included reading *Thank God for Evolution* by Michael Dowd (1958-), a former fundamentalist preacher who embraced evolutionary theory. He offers a refreshing way to bridge the divide between believers and nonbelievers that is built on the conviction that both science and religion must learn to compromise. What if we saw God, not as some transcendent super being, but as a grand organizing principle that is also capable of downward causation (to distinguish it from the materialist idea of upward causation)? The six days of creation could thus be a metaphor for periods of rapid, punctuated evolution. God's creativity could come in sudden instantaneous burst that explains the gaps in the fossil record, where God does not create from scratch but improves on His previous work. Creative evolution may be a way to vindicate all sides of the creation-versus-evolution debate. Quantum mechanics speaks only of possibilities contained within a waveform that the observer collapses to create reality. This observer effect is discontinuous, which fits perfectly with the overall idea of creative evolution.

Thou canst not stir a flower without troubling a star.
Francis Thompson

Monotheism can therefore be seen as the realization that reality is a single unified whole. All theologies assume the existence of complexity, and we have valid theories that demonstrate how complexity can emerge from simplicity. Life is an incredible statistical improbability, and non-random survival is also chancy. Cumulative selection is the mechanism that makes the system work; evolution can break down seemingly impossible odds and explain apparent miracles.

Many conservative believers reject evolution. They must be commended, because we need to find more sacred ways to communicate this to them; however, knowledge of the universe is not incompatible with belief in God. As one of Mr. Dowd's parishioners said, "You know, the more I learn about this universe, the more awesome my God becomes!" Part of the challenge is that people like having preset meanings, which is why some of them cling to the literal words in the Bible; however, this too is problematic, because the words and their meanings have changed over the centuries. (These changes have sometimes been significant, as we learned in Chapter 8.) This is in keeping with the natural order of things, because the universe itself is moving in a direction of increased diversity, complexity, awareness, and self-understanding.

Earth and life look just like they can be expected to look if there is no designer God.
Victor Stenger

The universe began as homogenous, undifferentiated energy that became more specific and self-aware. The universe is learning to see itself and experience itself more intimately. Despite this, millions of years of tides on beaches will not create a sand castle, nor will a tornado ever create a working bicycle (let alone a jumbo jet) in any junk yard. Also, as Gould has

> *Our part in the universe may possibly in some distant way be analogous to that of the cells in the body, and our personalities may be the transient but essential elements of an immortal mind.*
> Francis Galton

said, rewinding the tape of life and starting over from any arbitrary point would probably yield very different results. Problems and breakdowns are both normal and healthy. Every advance owes itself to challenges, and the greatest progress has come on the heels of the greatest challenges. The Yucatan meteor that fell 65 million years ago portended a really bad day for the dinosaurs, but small mammals in burrows were able to emerge and evolve to become humans.

We should therefore be thankful for all of life, both good and bad. After all, the universe can get along just fine without us, but the reverse is not true. We must give service to both sides, alpha and omega, realizing that both are part of the ultimate reality that creates—and destroys—all things. We must also have enough faith to interpret new discoveries in ways that both affirm life and give glory to God. Ancient texts become backward and limiting when we insist on interpreting them literally instead of focusing on facts, which constantly reveal God's process. We need not know what God looks like from the outside; all we need to know is what God looks like in our own hearts.

Dawkins is a leading voice among atheist materialist scientists who attest—with good reason—that blind faith is a bad thing; however, he makes assertions that border on logical fallacies. First, he says that evolution as a natural process accounts for all life and removes the need for God; however, this does not disprove God. Second, he says that religion is anti-rational, which it is; however, Dawkins defines faith as existing in absence of evidence or even in the face of contradicting evidence, which is a caricature of real faith. Third, he cites the great harm done in the name of religion; however, it is the people who must be implicated, not the faith itself.

The Idealist View

> *A felicitous mix of law and chance might be generalized to cosmology, producing directional evolution from simple states through complex, to life and mind.*
> Paul Davies

Neurophysiologist John Eccles (1903-1997) asked why consciousness is necessary when we could do everything necessary for life without an "I" experience, or even any true awareness at all. Evolution is purposeful, because the universe is becoming aware of itself through humans and other conscious species on Earth and possibly elsewhere, a theme I will keep returning to. Physical objects obey natural laws. Biological objects obey both natural laws and programming. One of the most sublime and fascinating properties of consciousness is that materialists are utterly unable to explain how the subjective experience arises. If matter can neither process meaning nor give rise to an organized state of consciousness, then how can it both create meaning and give rise to consciousness as an adaptation open for natural selection? The inability of matter to process meaning sinks materialist speculation that meaning may itself be an adaptation open for natural selection. The traditional view is that mind is a product of brain; however, the idealist view is that brain is a product of mind. A cell's physical workings explains neither our sense of self nor our feelings. Biologists looking to evolutionary adaptation for an explanation are employing the same kind of knee-jerk reaction as religious people who ascribe all to God.

The watchmaker is blind. But does this mean, as Dawkins claims, that there is no watchmaker? Not all orthodox evolutionists, including Darwin himself, have seen this as a logically compelling inference.
Christian de Duve

Traditional science defines *nature* and *natural* as comprising the material universe, while dismissing everything that does not fit this mold as "supernatural." Slapping a dismissive epithet on something one does not understand or believe in is prejudice and bigotry, plain and simple. Materialism arises from the idea that everything can be understood in terms of matter. This assertion that is anything but proven, except as a matter of faith. Can one unproven assertion be used to prove another? Materialism sounds disturbingly similar to the religious faith that is so despised by scientists. These scientists claim that materialism is a proven truth, and that it is absurd to even think of believing something else. Those who swim against the materialist tide tend to be marginalized and ridiculed by their peers, just like religious people who are expelled from the flock for not toeing the ideological line. Resorting to nonmaterial organizational principles is dualism, which is a religious view that both material and spiritual worlds exist. As we have seen, this view is attractive, but fails when one realizes that there is no way for spirit to interact with matter according to the laws of physics as they exist in our universe.

Any organism created by natural selection is, by default, under the illusion that it is special.
Robert Bowker

Even the most hard-core materialists and Darwinists accept the reality of intelligence as a tool that aids survival; they simply deny the idea of an intelligent cause or design for the universe or life. Similarly, it is extremely difficult to deny that humans and other animals possess creativity, which means that biological beings are not deterministic. If we were, then we could not be creative. Saying that quantum uncertainty governs biology means that the uncertainty was there to begin with, without the need to invoke evolutionary adaptation. We cannot have it both ways.

> *Just the facts, ma'am, just the facts.*
> Sergeant Friday
> Dragnet

Idealism holds that life was not designed by a transcendent God, but rather evolved by a living and immanent God. This view does not conflict with mainstream science in the slightest, nor does it validate the intelligent design crowd. The God of Genesis creates matter and then life, but the quantum God of idealism creates everything at once. The universe remains in a quantum superposition of all possible possibilities, until the emergence of the first conscious life form collapses the wave into an actuality. This collapse includes the entire history required to get to that actuality, which completes a self-referencing meaning circuit in the same way that flipping a light switch completes a circuit that allows a bulb to illuminate.

Chemist Robert Shapiro calculated that the maximum possible rate of random events is 2.5×10^{51} per billion (10^9) years, or 2.5×10^{42} random events per year. In 2003, astrophysicist Arne Wyller used some conservative assumptions to arrive at the conclusion that creating multicellular organisms with 1 billion cells requires more than $10^{1,000,000,000,000}$ random events, which is many orders of magnitude greater than the maximum possible number of random events that can occur. It is also far greater than the total number of Planck time intervals that have transpired since the beginning of the universe. Each second consists of approximately 10^{43} Planck time intervals. Multiplying this by the total number of seconds that have passed since the universe began (4.3×10^{17}) gives us a grand total of 4.3×10^{60} Planck intervals that have occurred to date. If either of these calculations is even remotely correct, then we are forced to concede that Darwinian evolution is not the only game in town. For example, idealism would explain punctuated equilibrium and apparent gaps in the fossil record.

> *A quasi-mystical response to nature and the universe is common among scientists and rationalists. It has no connection with "supernatural" beliefs.*
> Richard Dawkins

The basic tenets of idealism are that:

- Consciousness is the foundation of the entire universe.

- Consciousness is unified, non-local, and objective.

- Collapsing a quantum waveform makes consciousness self-referent by giving the sensation of a physical object and a subjective experience of a self that experiences the object as separate from itself, when no such distinction actually applies.

Why does pure consciousness need to create evolution and the illusion of separate and distinct selves as both subjects and objects? Because daydreaming and actually doing something are two different things. Separating consciousness from what is possible is the only way to undertake real experiences. Polymath G. Spencer Brown (1923-) put it best when he said that, "The world we know is constructed as to see itself but must cut itself up into at least one state that sees and other that is seen."

The idea that consciousness collapses a quantum wave smacks of dualism at first; however, the brain also consists of quantum particles and possibilities in their own superposition. Two probability waves interacting with each other simply create a bigger wave, according to mathematician John von Neumann (1903-1957). We cannot have a collapse without an observer, but there is also no observer without a collapse. This seeming paradox remains unresolved if we stay in the material world, but vanishes if we postulate idealism. What happens to a wave that remains uncollapsed? Idealism lets us at least partially redefine entropy as objects moving toward maximum probability, which is arguably the same as maximum entropy.

The delayed choice experiments (see Chapter 23) demonstrate that light is also a probability wave that stretches back through time, until an observation collapses the wave and manifests the path the light took to arrive at the observer from any distance or time in the past. This requires us to completely redefine our ideas about time, which we will do in Chapter 28. A similar experiment in 1993 tracked radioactive decay and sealed the results away from all observation. Psychics tried to influence the results away from their expected values, after which the seals were broken and the results examined. They all demonstrated small but noticeable shifts in the desired direction every time.

All of this validates the Copenhagen interpretation of quantum mechanics (see Chapters 15 and 29), which says that nothing can be deemed to exist unless and until a measurement (observation) occurs. Books are meaningless without conscious readers, music is meaningless without conscious lis-

When shall we come to realize that health is just as contagious as disease, virtue as contagious as vice, cheerfulness as contagious as moroseness?
Anonymous

We humans are not separate creatures on earth in the universe. We are a mode of being of earth, an expression of the universe.
Michael Dowd

teners, the scent of flowers is meaningless without conscious smelling, and so on. A measurement without anyone to view or interpret is also meaningless. Thus, a measurement by any means remains meaningless until a conscious observation is made. The entire universe was a superposition of all possible universes until sentient life evolved to collapse the possibilities all the way back to the first moment of creation and see itself as separate from the environment. Quantum collapse creates both observer and observed. As Jeans said, "The universe begins to look more like a great thought than a great machine." Descartes' famous, "I think therefore I am" becomes, "I choose, therefore I am and my world is."

Shared Reality

Wheeler pondered the problem of many observers in a participatory universe and how we manage to avoid chaos caused by multiple disparate viewpoints. The answer is that all observers participate to make the universe orderly. The Wigner's friend thought experiment has a friend of his perform the Shrödinger's cat experiment (see Chapters 23 and 29) while Wigner himself is not in the room. Wigner does not know the result until he comes back into the room and asks his friend what happened. The question is, who collapsed the wave? Was it the friend who peered into the box, or Wigner, who asked his friend? To Wheeler, observers choose from the nonlocal perspective of a cosmic consciousness that maintains order and ensures mutual consistency across all observers. The choice therefore does not occur from the ego-based "I" consciousness we are all familiar with, but from the quantum consciousness that we tend to regard as God, which gives rise to downward causation. (Shrödinger probably selected a cat because he believed that cats are not conscious enough to collapse their own waves; however, this view is not compatible with quantum mechanics.)

The consciousness looking at the cat and making the choice of life or death is the quantum consciousness that lies far beyond any individual ego. We could say that it is the Godhead. This view of God is the kernel on which all religions are built. Quantum mechanics allows modern scientists to rediscover truths that mystics have known for many thousands of years, if not longer. There is nothing unscientific about this God,

---

I absolutely love how imperfect and clumsy we are as humans. I mean, almost any situation can be laughed off if we only step back and take a look at why we feel the way we do. I love it! I wouldn't want it any other way.

Halsey Barlow

Man has learned to fly like a bird in the sky; man has also learned to swim like a fish in the ocean; alas, man has yet to learn to walk like a human being on this Earth.

Unknown

and the hypothesis of God formed in Chapters 13 and 14 remains valid.

Quantum Evolution

If life consists of chemical reactions, then we need to explain how these reactions can cause life, which requires us to go beyond standard chemistry and into quantum mechanics. All cellular activity involves orchestrating and directing the movements of subatomic particles in long sequences of complex orderly movements that bear little resemblance to purely dissipative constructs such as steam engines. It is quantum mechanics and not mere thermodynamics that is driving what is happening inside our bodies. Movements of particles in DNA molecules drives evolution as quantum events cause mutations that affect evolution. Quantum evolution theory does not replace Darwinian natural selection; it simply allows life to use quantum measurements/collapse to direct its own evolution in gradual or punctuated steps.

Peptides by themselves are not life, because they cannot self-replicate. The earliest cells were unable to protect themselves against quantum events for long. The environment took notice and measured the system, thus solving the problem. According to the Copenhagen interpretation, this happened some 3.5 billion years after the first life appeared, which means that the first observation by a sentient life form created 3.5 billion years of appropriate history in a single stroke. A slightly different viewpoint says that growing peptides drifted back and forth between quantum and macro (classical) states, and increased the probability wave until it collapsed into a given state for good when the peptide learned how to replicate itself. There is little practical difference between these viewpoints; you either believe that the Copenhagen interpretation allows a single cell to make a measurement/collapse or that conscious (and presumably multicellular) life evolved first. Either way, the evolution of life was driven by quantum mechanics. Each cell performs quantum measurements on its own particles. This does not mean that cells can choose their own actions; however, the ability to make quantum decisions is the foundation of consciousness. A cell dies when it loses the ability to make these internal measurements.

I never lose sight of the fact that just being is fun.
Katherine Hepburn

Flesh is merely a lesson. We learn it, and pass it on.
Erica Jong

Poems are made by fools like me but only God can make a tree.
Joyce Kilmer

Genetic determinism is only part of the quantum evolution picture. Darwinian selection by itself cannot collapse the probability wave, because it is not conscious. Seeing consciousness as the agency making choices makes things much easier, and opens the door to unifying biology and quantum mechanics. New species often come from conscious acts of creation, which explains punctuated equilibrium. Possibilities expand, interact, and accumulate until the moment of collapse. Mainstream biology assumes that this happens without consciousness, but consciousness and downward causation is necessary to prevent logical paradoxes during the collapse.

Shroeder and zoologist James Brough have noted that no new phyla have appeared on Earth since the Cambrian explosion. (Examples of phyla include chordates and arthropods.) *Classes* within phyla stopped appearing about 400 million years ago. (Mammals are classes of chordates, and insects are classes of arthropods.) New *orders* stopped appearing approximately 60 million years ago. (Primates are an example of an order.) Overall, macro evolution seems to be slowing down.

In developmental biology, the term *morphogenetic field* refers to a group of cells that can respond to discrete local biochemical signals that lead to developing specific structure of organs. Embryonic fields are very dynamic, and contain collections of cells that form various structures. For example, the cells in a limb field will form a leg, while cells in a cardiac field will form a heart. These fields allow routine maintenance and repair functions to take place without conscious input. Meanwhile, gene mutations can accumulate in waves that can remain unselected and uncollapsed until needed. Consciousness chooses when enough micro-level possibilities accumulate to form macro-level possibilities. According to Sheldrake, collapse occurs when the macro-level possibilities align with the form's morphogenetic blueprint. A quantum leap happens at this point, and consciousness creates a suitable physical manifestation of the evolved form. There are no fossils of intermediate stages because no such stages exist.

Nothing has ever been more insupportable for a man and a human society than freedom.
Feodor Dostoevsky

A living organism is a measurement device that operates in a tangled hierarchy and is capable of self-reference, because the collapse of quantum waves is a product of consciousness. On a different but closely related note, mental creativity is a non-linear insight that also requires a tangled hierarchy. Computers can write excellent poetry, but they can't tell us what it means; creativity seems to be another function of consciousness.

As previously mentioned, the ultimate creator is the quantum God working through a tangled hierarchy of quantum and manifest (ego) consciousness, which may explain why creation is not always to our liking. The manifest consciousness in each of us prepares and manifests gains made real by quantum leaps, and alternates back and forth with the quantum consciousness in an going cycle of do, be, do, be, do, be.

Understanding that a single consciousness divides into life and environment and living versus nonliving at the same time reveals the symbiotic relationship between each of us and God. On a more mundane level, it explains the symbiotic relationships between bees and flowers, or between cleaner fish and eels, to give just two examples. Consciousness simply collapses the wave for both bee and flower, fish and eels.

It does not matter that in a million years nothing we do now will matter.
Thomas Nagel

Chakras

The preceding theory of creative quantum evolution may also imply the existence of noncorporeal subtle bodies that contain the morphogenetic fields for both individuals and species. *Chakras* are points on the body where consciousness collapses the movements of both morphogenetic fields and the organs that represent those chakras. The term *chakra* means circle, which reminds us that quantum collapse occurs in a tangled hierarchy that simultaneously collapses the movements of morphogenetic fields along with the organs that represent them, while also ensuring the rise of self reference at the chakra points. All of this is highly speculative, but cannot simply be dismissed or wished out of due consideration.

Life Goes On

To paraphrase the Celine Dion song from the movie *Titanic*, "Life goes on and on and on." We have looked at arguments for and against design, as well as a theory (idealism) that could resolve the entire debate; however, the question of how life arose remains open for now. So what do we know, and where does that knowledge lead us?

Life is either a reproducible, almost commonplace manifestation of matter, given certain conditions, or a miracle. Too may steps are involved to allow for something in between.
Christian de Duve

One of the great mysteries of evolution is why no new phyla have appeared since that Cambrian explosion of life.
Gerald L. Schroeder

- By some estimates, our galaxy contains 10^{68} protons, or 1 per cubic centimeter. The time it would take to make a star by chance is $10^{1,504}$ years, but we see many billions of stars in the sky because there are nonrandom processes that allow stars to form far more rapidly.

- A human being contains about 10^{23} bits of information and has approximately 10^{360} possible DNA configurations, of which only a very few are viable for life.

- The original Urey-Miller experiments consistently produce most of the 20 possible amino acids, each of which can be used for multiple purposes within an organism. The presence of oxygen in the primordial atmosphere helped encourage amino acid evolution. Amino acids can be found throughout the galaxy.

- Our planet's surface has undergone dramatic changes, but the inside has remained relatively unchanged. The total underground biomass may rival that of surface dwellers.

- Minor changes to the laws of physics could easily result in a universe that is either too unstable or too stable for life to form. This would have drastic consequences, because large molecules copy themselves using a quantum procedure. The addition of chaotic quantum effects requires a non-deterministic approach to nature that gives evolution plenty of room to operate.

- The Standard Model of quantum mechanics cannot explain particle masses, but we do know that minor increases in mass would create a universe devoid of everything but black holes.

Western science has labored under the bias that the best way to understand a physical phenomenon, whether a frog or an atom, is to dissect it and study its respective parts.
Michael Talbot

- The universe is almost completely flat to within 1 part in 10^{60}. Excess space-time curvature could result in a universe that fails to expand enough for life to form, or one that expands too rapidly.

- Without inflation (see Chapter 26), the universe would need extremely precise starting conditions.

- Matter can decay into nonexistence in a time period that averages 10^{33} years, which is much longer than the current age of the universe, which should be able to remain habitable for 1,000 times its current age, or 13.7 trillion years.

- Empty galactic regions between stars are one million times as densely packed with matter as intergalactic space, which has an average density of only 1 proton per cubic meter.

- Galaxies formed about 1 billion years after the Big Bang and will continue to exist for another 100 trillion years. Most stars are red dwarfs, and the majority of stellar evolution is yet to occur. As our Sun dies, it will provide material for new solar systems in the ongoing cycle of birth and rebirth.

- The longest-lived radioactive isotopes have half lives of about 14 billion years.

We will continue exploring what all this means in the next several chapters. Meanwhile, look up at the stars some evening and ponder that what you are experiencing is the Milky Way, and the entire universe learning about itself. No matter which version of evolutionary theory proves correct, this is still a heady thing to contemplate.

> *When a story becomes scripture, it ceases to evolve.*
> Michael Dowd

Well?

Materialism says that we did not exist before birth and will never exist again after we die, that there will be nothing of us left to even know that we once existed. Religions offer up the promise of life after death, packaged as either a one-way trip (JCI) or virtually endless round trips (Buddhism). So which is it? Faith may be the only way to answer this question for now, because we may have no way to answer the question for certain. Either way, Hoyle was absolutely correct when he said that each round of scientific inquiry consistently yields finer concepts and structures than we previously imagined. Something is going on. But what? In his *Pensées*, Pascal wrote God's answer to a researcher as, "Be consoled; you would not look for me if you had not already found me."

Wilson laments that so many people continue to turn to religion despite its demonstrable falseness, because people prefer to believe instead of know. Ancient creeds are still alive, well,

> *Man has a rope stretched between the animal and the superman—a rope over an abyss.*
> Friedrich Nietzsche

> *The components of living cells, stripped of context, seem fundamentally no different to inanimate chemical systems. Life seems to emerge only at higher levels.*
> Johnjoe McFadden

and spreading. Dowd urges us to, "Surrender to what will be with utter trust and faith that what comes beyond, if anything, is perfect for us and our loved ones."

Intelligent design as presented by Behe and others is a lame attempt to validate religious dogma by disguising it in scientific-sounding terms. The intelligent design movement has all the hallmarks of being a conclusion in search of supporting facts, as opposed to a conclusion drawn from the best available evidence. The arguments made by Dawkins and other materialists against this theory are damning—if not outright fatal—to intelligent design. So where does this leave us?

> *Creation is a process of evolution of which man is not merely a witness but a participant and a partner as well.*
> Theodore Dobzhansky

Intelligent design theory is based on validating the JCI god, which we thoroughly debunked in Chapters 3 through 12; however, as we saw above and in Chapters 13 and 14, there are other ways to look at God that seem to align far more closely with both Eastern philosophy and mystical traditions from around the world that long ago mutated to form organized religions. In other words, intelligent design theory as presented is laughable, but that does not leave God out in the cold. In fact, idealism gives us an insight into a possible model of reality that could help us answer the central question of this book: What happens when we die?

Darwinian evolution is a valid theory that has withstood every experiment and attempt at disproof to date. This life may indeed be all we get. This life may not exist as anything but a stubbornly persistent illusion, to paraphrase Einstein. Either way, one truth remains: We are the universe made self-aware. We are the universe standing in awe of itself and struggling to comprehend. We are the universe discovering ever deeper truths and revealing more wonders at every turn. As Professor Elizabeth Johnson said, "We are the cantors of the universe." We did not come into the world; we grew out of it like a fruit on a tree.

A materialist universe leaves no room for any conscious observer. An idealist universe knew we were coming all along. Can these two views be reconciled? Keep this question in mind as we leave evolution and biology behind and dive into physics from classical to quantum and beyond. This far in our journey, almost all things remain possible.

594 | *The Divine Savage*
Revealing the Miracle of Being

PART SIX

The Physics

Chapter 21

Ancient Physics

Truth is stranger than fiction; fiction has to make sense.
Leo Rosten

Let no one untrained in geometry enter here.
Nicolaus Copernicus

The word *physics* comes from the Greek word *physis*, which means, "The true nature of things." Unlike people today, the ancient Greeks did not separate science, religion, and philosophy. To them, all was the study of the universe and how it works, and every discipline had a role to play in ascertaining the truth—an approach I have tried to emulate in writing this book. Further, many ancient sages saw the universe as one giant living organism. In the fifth century BC, Parmenides thought of the entire universe and everything in it as part of a single, eternal reality. The idea of dualism, the idea that matter moves at the behest of external spiritual forces, was also born in the 5th century BC, as was the concept of the atom. Aristotle noted that living beings seem to be motivated by some larger purpose. These ideas had profound effects on the Jewish and later Christian religions as Greeks pushed into Israel and then Egypt, because they provided the framework around which the JCI concepts of heaven and hell were built.

This was also an era of no small scientific importance. Eratosthenes (276-195BC) calculated the circumference of the Earth to within 15%, at 21,300 miles. (The actual figure is 24,900 miles.) The Phoenicians are believed to have reached South America, based on possible artifacts found on both sides of the continent. It is easier to reach the West side of South America by going around Africa, which leads to speculation that the Phoenicians may have been the first to circle the globe. Based on his observation of planetary movements across the sky, Ptolemy created a model of the universe with the Earth at the center and the Sun and planets orbiting in circles with epicycles. The Catholics adopted the Ptolemaic system despite known problems with the model identified by Ptolemy himself. Over 1,000 years later, the Copernican discoveries would contradict the Bible and threaten the church's vision of salvation. Occam's razor demands simplicity, which inevitably spelled the end of the terracentric model. Further discoveries continue to knock humans further and further from the center of the universe.

Let us take a very brief look at just a few of the key figures in the history of physics.

Aristotle

Aristotle knew that the Earth is round, based on seeing different constellations in the sky as he traveled north or south and the circular shadows cast during lunar eclipses. Contrary to some contemporaries who believed in atoms, he held that the universe is composed of five elements that can be subdivided to infinity:

- **Fire:** hot and dry
- **Earth:** cold and dry
- **Air:** hot and wet.
- **Water:** cold and wet.
- **Ether:** divine stuff that stars, planets, etc. are made of

These elements are naturally arranged in order from inner to outer, with earth at the center followed by water, air, and then fire. Elements that are out of their natural place have natural motion toward that place, which requires no external cause. This is why solids tend to sink in water, bubbles tend to rise, rain falls, and flames rise. The ether is always moving in circles. As we will see, the concept of the ether endured for well over 1,000 years and provided one of the springboards that led to the quantum revolution early in the 20th century.

To fatigue the minds with notions that cannot be grasped by them and for the grasp of which they have on instrument, is a defect in one's inborn disposition or some sort of temptation.
Maimonides

What is everything made of?
Thales

Aristarchus

> *He who has not contemplated the mind of nature which is said to exist in the stars is not able to give a reason of such things as have a reason.*
> Plato

Aristarchus saw the universe as a series of spheres with first the Earth and then the Sun in the center, the planets revolving around the Sun with the Moon orbiting the Earth, and the stars fixed in their positions. His *On The Sizes and Distances of the Sun and Moon* is the only surviving work attributed to him. Archimedes' book *The Sand Reckoner* references a different book by Aristarchus that describes a heliocentric model of the solar system as an alternate hypothesis:

"You (King Gelon) are aware that 'universe' is the name given by most astronomers to the sphere the center of which is the center of the Earth, while its radius is equal to the straight line between the center of the Sun and the center of the Earth. This is the common account as you have heard from astronomers. But Aristarchus has brought out a book consisting of certain hypotheses, wherein it appears, as a consequence of the assumptions made, that the universe is many times greater than the 'universe' just mentioned. His hypotheses are that the fixed stars and the Sun remain unmoved, that the Earth revolves about the Sun on the circumference of a circle, the Sun lying in the middle of the Floor, and that the sphere of the fixed stars, situated about the same center as the Sun, is so great that the circle in which he supposes the Earth to revolve bears such a proportion to the distance of the fixed stars as the center of the sphere bears to its surface."

Aristarchus believed that the great distance of the stars from Earth accounts for the lack of visible *parallax* (motion of stars relative to each other as the Earth orbits the Sun). The stars are much farther away than the ancients realized. They interpreted the lack of parallax as suggesting an Earth-centered universe and tended to reject the idea of the Sun being at the center. This rejection was so strong that Cleanthes, the head of the Stoics during this time, called for Aristarchus to be prosecuted for impiety for daring to suggest that the sky is standing still while the Earth moves.

> *All works that display both means and an end presuppose a workman; therefore, this universe, made up of moving forces, of means each of which has its end, presupposes a highly powerful and intelligent workman.*
> Cicero

Hipparchus

The astronomer Hipparchus (200-126 BC) is presumed to have been born in what is now northwestern Turkey, and to have traveled to Rhodes to conduct his studies. Ptolemy's work superseded Hipparchus some three centuries later, but even this later work relied heavily on the latter's ideas. Hipparchus is credited with advancing the idea of the Earth's *precession* (the change in the orientation of a rotating body's rotational axis over time), and for compiling the first comprehensive catalog of about 850 stars and a corresponding map with over 1,000 stars arranged into six orders based on brightness, a system that is still used today. Sirius is the brightest star in the night sky, and is therefore a first-magnitude star. He also estimated the distance of the Moon from Earth. His model of the universe placed the Earth in the center, but still allowed accurate predictions of planetary positions. Much of his work was based on amazingly precise naked-eye observation, but he also examined Middle Eastern records, especially those of the ancient Babylonians. His astronomical calculations were aided by the invention of *trigonometry* (mathematics that deals with the relations between the sides and angles of plane or spherical triangles)

He himself said it, and this 'himself' was Pythagoras.
Cicero

Claudius Ptolemy

Claudius Ptolemy (90- 168AD) built the first truly mature model of the universe. His *Almagest* drew on centuries of Greek, Near Eastern, and Middle Eastern work, and was accepted as correct by European and Middle Eastern astronomers for over 1,000 years. Ptolemy placed the Earth in the center of the universe, because half the stars are above and below the horizon at any given moment, and because of the assumption that the stars were far away from the center of the universe; moving the Earth from the center would create an uneven distribution of stars.

Each planet (1) in the Ptolemaic system moves on at least two circles, the *deferent* (2) and *epicycle* (3). The deferent is centered on a point (5) halfway between the *equant* (point directly opposite Earth from the center of the deferent, (4) and the Earth (6). The epicycle circle is centered on the deferent, and the planet moves along the epicycle.

The earth is found to move and is no longer the center of the universe.
William Drummond of Hawthornden

All things are numbers. The cosmos is based on melody.
Pythagoras

The epicycle moves at the same speed relative to the equant. The deferent moves around the Earth, while the epicycle rotates around the deferent, which makes the planet move closer to and away from the Earth at different points in its orbit. Their motion also seems to slow down, stop, and reverse in the observed loop-the-loop *retrograde* motion (see diagram below). The epicycles of Venus and Mercury are always centered along the line between Earth and the Sun, with Mercury being closer to the Earth. From Earth, the spheres were arranged as follows:

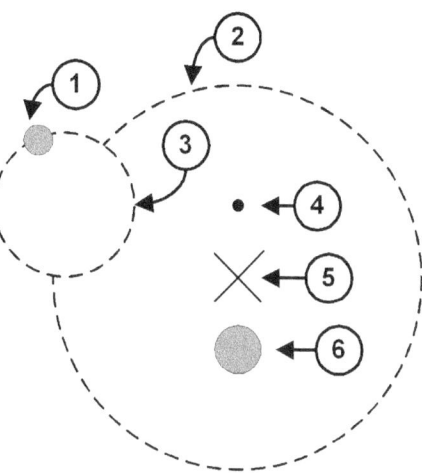

- Moon
- Mercury
- Venus
- Sun
- Mars
- Jupiter
- Saturn
- Fixed Stars
- Sphere of the Prime Mover (God)

We find nothing in the mathematical sciences that would lead us to believe that eccentrics and epicycles exist.
Averroes

The Greeks used this basic model for centuries before Ptolemy, with and without an *eccentric* (deferent off-center from the Earth, as shown in the illustration above). This system did not match observations precisely despite being a great improvement over Aristotle's system, chiefly in that a planet's retrograde loop would appear either smaller or larger than predicted. The equant solved this problem; an observer standing at this point would see the center of the planet's epicycle always moving at the same speed. The planet itself therefore moved at various speeds, depending on where the epicycle was along the deferent. The eventual system did enable reasonably accurate predictions but was an unwieldy affair, with each planet having its own deferent, epicycle, and equant. There is some debate as to whether Ptolemy saw this arrangement as a conceptual model for performing observations or as representing the actual motions of the planets.

Nicolaus Copernicus

Copernicus is considered the father of the modern model of the solar system. His major contribution was advocating a heliocentric model instead of the terracentric Ptolemaic model, a switch that had huge implications beyond astronomy because it demoted Earth to one planet among many, humans from the center of God's universe, and stars into other suns that made the universe seem infinite. This revolution was just one of many that knocked humanity off its high horse and forced us to admit that we are little more than cosmic also-rans.

Copernicus went to great lengths to be discreet, and to present his work as purely hypothetical in order to avoid religious wrath and persecution. His famous book, *De revolutionibus orbium coelestium* (*On the Revolutions of the Celestial Spheres*) was first published the year he died. This book quite literally touched off the shift away from the idea of the anthropomorphic universe. To Copernicus, Earth orbited the fixed Sun once per year, while revolving on its axis once per day. His system used perfectly circular orbits because circles were deemed to have more aesthetic value, which meant that he still had to use epicycles, though not nearly as many as Ptolemy.

The Copernican theory can be summarized as follows:

- The motions of celestial bodies are uniform and either circular or made up of several circles (epicycles).
- The Sun is located near the center of the universe.
- In order, the planets orbiting the Sun are Mercury, Venus, Earth and Moon, Mars, Jupiter, Saturn, and the fixed stars.
- The Earth orbits the Sun annually, rotates daily, and tilts on its axis annually to create the seasons.

For my part, I believe that gravity is nothing but a certain natural desire, which the divine providence of the creator of all things has implanted in parts, to gather as a unity and a whole by combining in the form of a globe.
Nicolaus Copernicus

What indeed is more beautiful than heaven, which of course contains all things of beauty?
Nicolaus Copernicus

- The fact that the Earth is moving explains retrograde motion.
- The Earth is much closer to the Sun than to the stars.

Tycho Brahe

The body of the earth, large, sluggish and inapt for motion is not to be disturbed by movement any more than the ethereal lights (stars) are to be shifted, so that such ideas are opposed to physical principles and also the authority of holy writ.
Tycho Brahe

Tycho Brahe (1546-1601AD) created a large number of observations of planetary and stellar locations that were far more accurate than anyone before or during his time. He wanted to have his observations always be within one arc minute of the actual positions, but fell short of that goal; his average error was about 2 arc minutes. (An arc minute is 1/60 of a degree.) Incorrect transcription by hired scribes resulted in much larger published errors. Still, his records allowed his assistant, Johannes Kepler (1571-1630AD), to discover both elliptical orbits and the equal area law of planetary motion, which in very general terms states that a planet nearer the Sun will move faster than a planet further from the Sun. Kepler used these two laws to achieve a new level of accuracy that supported his heliocentric model of the solar system. He was unable to convince Brahe to adopt the heliocentric model, because Brahe believed the Earth to be too sluggish to be moving. He believed that one should be able to observe stellar parallax, an observation based on his assumption that the stars are much closer than they actually are. This parallax does in fact exist, but can only be detected with telescopes.

Brahe's own system had the Sun and Moon orbiting the Earth while the rest of the planets orbited the Sun, a safe medium that helped astronomers who were dissatisfied with older models but not ready to accept the Earth not being in the center. This model also managed to avoid arousing the church, which had officially decreed that placing the Earth anywhere but at the center of things was both philosophically and scientifically inaccurate.

Yea, the first morning of creation wrote what the last dawn of reckoning shall read.
Omar Khayyam

Brahe also devised corrections for atmospheric refraction that alters the apparent positions of objects near the horizon. He also developed an algorithm for approximating the results of long series of multiplications based on trigonometry, the forerunners of logarithms.

Johannes Kepler

Kepler published *Mysterium Cosmographicum* (*The Cosmographic Mystery*), the first defense of Copernicus's system. His Eureka moment came in 1595 while teaching in Graz, Austria with the realization that regular polygons bind one inscribed and one circumscribed circle at definite ratios that could form the geometrical foundations of the universe. He tried to find such an arrangement of polygons that fit with current astronomical observations, but gave up and started experimenting with 3D *polyhedra* (many-faceted figures). This was the breakthrough: Kepler discovered that each of the five Platonic solids (tetrahedron, cube, octahedron, dodecahedron, and isocahedron) could be inscribed and circumscribed by unique spherical orbs. Nesting these encased solids inside each other like Russian dolls yielded six layers corresponding to Mercury, Venus, Earth, Mars, Jupiter, and Saturn. (Uranus, Neptune, and Pluto were not yet discovered.) Placing these solids in the other octahedron, isocahedron, dodecahedron, tetrahedron, and cube placed the spheres that accorded within the relative sizes of each planet's orbital path, given the limits of observational accuracy of the day, provided all of the planets orbited the Sun.

This led Kepler to believe that he had discovered God's geometrical plan. He felt that the universe was itself an image of God, with the Sun (father) at the center, stars (son) at the outside, and the holy spirit in between them, and devoted a complete early chapter of the *Mysterium* to reconciling the heliocentric model with the Bible. The Tübingen university elders allowed him to publish this book with that chapter removed, and with a simpler and clearer explanation of both the Copernican system (which Kepler admired) and his own ideas. Kepler modified and refined his system over time, but never abandoned it completely.

Galileo Galilei

Galileo Galilei (1564-1642AD) is heralded as the father of modern science, and the man who took on the church and eventually won. He was famous for creating working telescopes that varied from about 3x to about 30x magnification

> *I used to measure the skies, now I shall measure the shadows of the earth. Skybound was the mind, earthbound the body resides.*
> Johannes Kepler

> *Myth is neither a lie nor a confession: it is an inflexion.*
> Roland Barthes

> *The fool [Galileo] wants to turn the whole art of astronomy upside down.*
> Martin Luther

based on earlier, incomplete information. Building and selling telescopes to merchants and sailors provided him with a profitable sideline to his astronomical work.

On January 7th, 1610, Galileo used his telescope to spot "three fixed stars, all invisible by their smallness" close to Jupiter, and lying on a line that extended through the planet. Subsequent observations showed that these objects were moving, which was impossible for fixed stars. On the 10th, he noted that one of them had disappeared behind Jupiter. Three days later, he discovered a fourth such object. It did not take him long to understand that these objects were orbiting Jupiter; Galileo had discovered the Jovian moons of Io, Europa, Ganymede, and Callisto. This discovery completely defied the Aristotelian view that all heavenly bodies must orbit the Earth, and was therefore widely disbelieved until confirmed by independent observations. By the middle of 1611, Galileo had managed to estimate the moons' orbital periods with a high degree of accuracy.

Galileo also observed that Venus displayed a full set of phases similar to the lunar phases. The Copernican system predicted that all phases would be visible, because Venus's orbit around the Sun would cause its illuminated hemisphere to face the Earth when the planets were on opposite sides of the Sun, and away from Earth when the planets were on the same side of the Sun. By contrast, the terracentric Ptolemaic model made it impossible for any planet's orbit to intercept the Sun's supposed orbit. Galileo's observations invalidated Ptolemy. Many astronomers adopted hybrid models, such as Tycho's, which explained the phases without throwing out the terracentric system altogether. His other observations included Saturn's rings and sunspots. The latter threw yet another wrench into the idea of eternal celestial perfection, while also demonstrating that the Sun itself rotates. Annual variations in sunspot models provided yet another argument against both Ptolemy and Brahe. Moreover, his observations of the Moon and knowledge of light and shadow revealed that the Moon is not a perfect and translucent sphere, but a rocky world with peaks and valleys. His observations of the Milky Way proved that it was not a ring of cosmic gas but a huge number of stars packed together so tightly as to be indistinguishable to the

---

It is believed by most that time passes; in actual fact, it stays where it is. The idea of passing may be called time, but this is an incorrect idea, for since one sees it only passing, one cannot understand that it just stays where it is.
Dogen

For the wise all 'things' are wiped away.
Buddha

naked eye. He further revealed that stars are spheres of varying sizes and distances from the Earth.

Galileo is also famous for his belief that science should only deal with what can be observed or demonstrated, with no room for intuition or authority. This of course was a direct affront to the Catholic orthodoxy. This plus the great number of observations Galileo was producing and the devastating effect said observations were having on the religiously sanctioned Ptolemaic system, finally forced the church to respond. Galileo was arrested by the Inquisition, forced to recant, and placed under house arrest. The Catholics finally forgave Galileo in 1992, a full 350 years after his death. By then, their monopoly on thought had been broken, and they had been forced to recognize the validity of science. In desperation, the church that once freely silenced science and stifled thought hired scientists of its own to advise it, in an ongoing attempt to fit the increasing pace and scope of scientific discovery with nearly two thousand years of orthodox dogma. It is well worth noting that no scientist has theologians for advisers.

All science is either physics or stamp collecting.
Ernest Rutherford

Chapter 22

Sir Isaac Newton

O world invisible, we view thee. O world intangible, we touch thee. O world unknowable, we know thee. Inapprehensible, we clutch thee!.
Francis Thompson

My God, it's full of stars.
Keir Dullea as David Bowman

Newton is considered the father of modern physics; indeed, when we refer to physics, we are often referring to Newtonian or "classical" physics (as opposed to quantum or cosmological physics). The theories Newton constructed back in the 1600s remain in use today. As we will see in Chapter 23, developments in physics since that time have not invalidated Newton's work. On the contrary, his equations are alive and well today, wrapped within layers of more recent discoveries. We can therefore think of classical physics as an approximation—albeit an excellent approximation—for objects that are larger than quantum particles and atoms moving at speeds where time dilation is negligible. (See Chapter 24.) Newtonian physics allows us to build the tallest buildings, design safer vehicles, launch probes to the furthest reaches of the solar system, and more.

One interesting aspect of Newtonian physics is its portrayal of the universe as a deterministic mechanism, where one can extrapolate a future or past state based on knowing the present state. This "clockwork" universe is one of cause and effect, and contains discrete objects moving in a single, objective, and absolute reality that leaves no room for mind, soul, or God. Nevertheless, Newton did believe in a "prime mover" God as the ultimate cause for the universe. To the best of our knowledge today, he never intended to remove God, mind, and free will from the universe, but that is exactly what he managed to do by describing all that was known of nature at the time in terms of simple mathematical laws. Objects in the Newtonian universe interact by means of physically real forces imparted by other objects. Beyond this, objects could be considered in isolation. The astute reader will note that this position is about as far as one can get from mystical ideas about cosmic oneness and the illusion of separation. Problems such as free will remained interesting but largely unimportant and benign paradoxes for centuries, until the discovery of quantum mechanics placed these issues front and center, as we will discover in the coming chapters.

Newton's *Philosophiæ Naturalis Principia Mathematica* (published in 1687 and often referred to simply as the *Principia*) is the straw that broke the spell of religious dogma and introduced the Age of Reason. This epic book is still used by mathematicians today. Newton's equations worked well, described observed experience perfectly, and seemed to explain just about everything. No longer would physical science be bound together with spirituality.

The revolution begun by Copernicus culminated in the discovery that the entire universe obeys the same natural laws—laws that geologists would later use to prove that the Earth is far older than the 6,000 years believed by the JCI religions. The once-mysterious ways of nature could now be understood as parts in a giant machine whose workings rivaled the finest timepieces. The universe suddenly seemed both rational and predictable. For example, in the *Principia*, Newton showed that his three laws of motion combined with his law of universal gravitation explained Kepler's laws of planetary motion. Astronomer Edmond Halley (1656-1742) accurately predicted the return of the comet that now bears his name using Newtonian physics, an astonishing feat made even more impressive because Halley's comet only returns every 76 years. This proof that comets are just cogs in the cosmic machine finally removed the old superstitions surrounding their appearances.

Just as importantly, the religious stranglehold on thought and inquiry described in Chapter 8 was over for good. Atheism and agnosticism became acceptable, and even commendable. Author Thomas Paine (1737-1809) published *The Age of Reason*

I do not know what I may appear to the world, but to myself I seem to have been only a boy playing on the seashore and diverting myself in now and then finding a smoother pebble or a prettier shell than ordinary, whilst the great ocean of truth lay all undiscovered before me.
Isaac Newton

about 100 years after the *Principia*, and Darwin was less than 200 years away.

Newton's Three Laws of Motion

Isaac Newton invented physics, and all science depends on physics.
John Gribbin

Newton's three laws of motion form the foundation of modern physics by describing how forces act on bodies, the relationships between those forces, and how bodies move because of those forces. These three laws are:

The First Law

Bodies at rest or moving at a constant speed remain in those states, because the sum of the forces acting on those bodies is zero. The state of rest or constant motion will only change if the sum of forces acting on the body changes from zero by altering one or more forces and/or by introducing a new and unbalanced force. A spacecraft remains in orbit because its forward motion precisely matches the gravitational pull to create a rounded path around the Earth. This law is often summarized as, "A body at rest tends to remain at rest while a body in motion tends to remain in motion."

The Second Law

Acceleration is directly proportional to the force applied to a body, and inversely proportional to the mass of that body. If the mass of the body is m and the force is F, then the acceleration a is determined using the formula $F=ma$.

The Third Law

This law is often summarized as, "Every action has an equal and opposite reaction." If a body A exerts a force F on a body B, then body B exerts a force equal to $-F$ on body A, where F and $-F$ have equal magnitudes and opposing directions. Both action and reaction occur simultaneously.

In every age there is a turning point, a new way of seeing and asserting the coherence of the world.
Jacob Bronowski

Reference Frames

In Newtonian calculations, bodies (objects) are considered to be particles in the sense that the size, deformation, and rotation of the body do not matter. For example, when calculating a planet's orbit, one can treat the planet as a particle. There is also the matter of defining the variables that form the components of the three laws: The first interpretation assumes that mass, acceleration, momentum, and force are externally defined quantities. The second (and less common) interpretation is that the laws themselves can be seen as defining these quantities. These may not seem like important considerations but are in fact very important, especially at extremes of size and speed.

Movement, stillness, and forces are great, but what are we using for comparison? How do we know that an object is at rest or in motion? How do we know which way a force is acting or how strong the force is? The answer depends on how you look at it, as Einstein would later demonstrate in his theories of relativity. (See Chapter 24.) Newton's laws are only valid with a set of reference frames called either *Newtonian* or *inertial reference frames*. Some believe that the first law of motion defines the inertial reference frame, which means that the second law is only valid when measured from an inertial reference frame, which in turn means that the first law cannot be proven to be a special case of the second law. If this sounds confusing, it is; the concept of reference frames was not developed until many years after Newton's death.

Nature and nature's laws lay hid in sight. God sad, let Newton be! And all was light!
Alexander Pope

Limitations

Newton's laws have been verified by repeated experiments over the centuries since he first discovered them. As I said above, they are excellent approximations for everyday use on ordinary scales of size and speed. These laws combined with the universal law of gravitation provided the first-ever quantitative explanation for many natural phenomena. Again, they also revealed a universe that seemed supremely ordered and rational, a concept that persists to this day.

What hinders the fix'd stars from falling upon one another?
Isaac Newton

We saw that Newton's laws of motion as originally written deal with idealized objects that behave like single particles; however, these laws are inadequate for dealing with moving rigid and deformable bodies. Mathematician Leonhard Euler (1707-1783) generalized these laws in what came to be known as Euler's two laws of motion, which apply to both rigid and deformable extended objects, that is, objects that consist of more than one theoretical particle. This one limitation was fairly easy to circumvent; the others pose far more serious challenges.

The universe would be a very different place if the properties of the matter and force particles were even moderately changed.
Brian Greene

Newton's laws break down at very small scales and/or very high speeds. They also don't contain the Lorentz transformations required to account for time and space dilation (named after physicist Hendrik Lorentz, 1853-1928; see Chapter 24), which means they cannot be used to explain common phenomena such as superconductivity, extreme gravity, or optical properties of different materials that require the more sophisticated mathematics of quantum mechanics and relativity. These limitations became significant as more and more scientists began exploring the tiniest and largest extremes of our universe. Having said this, Newton's laws remain valid when dealing with non-quantum objects and speeds that are very slow compared to the speed of light. Modern physics has adopted the laws of conservation of momentum, energy, and angular momentum, which basically state that none of these three things can be created or destroyed. These laws are more useful than Newton's original three laws, because they apply to light and matter, and in both classical and non-classical (quantum or relativistic) physics. The law of conservation of energy was not discovered until long after Newton because of the complexity of understanding certain types of energy, such as heat and non-visible light.

Newtonian Nonlocality

Innumerable suns, and an infinite number of earths revolve around those suns.
Giordano Bruno

Newton's third law assumes instantaneous action between distant particles, or nonlocality. This raised no small amount of philosophical criticism, to which Newton admitted that he had no answer when he said, "I feign no hypotheses." Nonlocality has since been confirmed (see Chapter 23), and various interpretations offered to try to explain that and other quantum strangeness. Interestingly, all of these interpretations involve consciousness, even those that try to avoid it. We will learn more about this in Chapter 29.

Other Contributions

Newton's contributions extended beyond physics to mathematics and optics.

Mathematics

Newton is credited with advancing every type of mathematics known during his era. He worked with calculus, infinite series, and argued with Leibniz about priority in developing infinitesimal calculus. Today it is thought that Newton and Leibniz developed this branch of mathematics independently using completely different notational systems. This dispute escalated and continued until Leibniz died in 1716. Newton is also credited with creating the generalized binomial theorem that is valid for any exponent. He also served as the Lucasian Professor of Mathematics at Cambridge University, a seat also held by Hawking. In those days, fellows of Cambridge and Oxford had to be ordained Anglican priests. Newton argued for—and received—an exemption to this requirement from King Charles II.

Optics

Newton lectured on optics and examined refraction. He demonstrated that a prism separates white light into different colors, and that adding a lens and a second prism can reassemble the spectrum into white light. He also showed that the properties of the colored light remained constant no matter how the light was manipulated by lenses and additional prisms, and observed that color results from objects interacting with the light, as opposed to creating the color themselves. This allowed him to conclude that any refractive lens would have at least some prismatic effect, called *chromatic aberration*, that affects all refracting telescopes. Newton bypassed this problem by building reflecting telescopes that relied on mirrors instead of lenses to perform the magnification.

As for light itself, Newton believed that it consists of particles that refracted when accelerated into a denser medium, such as water or glass. Some of his descriptions appeared to be discussing waves, an early forerunner of the wave/particle problem we will explore in Chapter 23. Later physicists believed

> *Consciousness is the one phenomenon that we know of which time needs to 'flow' at all! The way in which time is treated in modern physics is not essentially different from the way in which space is treated. The time of physical description does not really 'flow' at all. The temporal ordering that we 'appear' to perceive is something that we impose upon our perceptions in order to make sense of them.*
> Roger Penrose

> *When people deny the existence of God today, they are often rejecting the God of Newton, the origin and sustainer of the universe whom scientists can no longer accommodate.*
> Karen Armstrong

that light functions as a wave until the advent of quantum physics once again brought this problem into stark focus. Newton also proposed the ether, a medium undetectable to any of our senses, as a mechanism for transmitting forces between light particles. The quest to detect the ether in the 1800s would help touch off the quantum revolution.

Newton's *Opticks* of 1704 presented his particle or "corpuscular" theory of light, in which he basically said that light is made up of small particles, while matter is made up of larger particles. This belief has been proven correct because photons can act as particles and are massless, unlike the particles that make up normal matter. He also postulated that light could be converted to matter and vice-versa, asking, "Are not gross Bodies and Light convertible into one another, and may not Bodies receive much of their Activity from the Particles of Light which enter their Composition?" Two hundred years later, a young patent clerk named Albert Einstein would prove Newton's prescience when he discovered what I believe to be the single most profound equation ever discovered: $e=mc^2$.

614 | *The Divine Savage*
Revealing the Miracle of Being

Chapter 23

Quantum Mechanics

If you were to measure the distance from Los Angeles to New York to this accuracy, it would be exact to the thickness of a human hair. That's how delicately quantum electrodynamics has, in the past 50 years, been checked-experimentally and theoretically.
Richard Feynman

Those who are not shocked when they first come across quantum theory cannot possibly have understood it.
Niels Bohr

Potential and kinetic energy are the foundations of physics. The sum of both types of energy always remains constant. Thanks to $e=mc^2$, we know that matter is simply "frozen" energy, in much the same way that ice is frozen steam. As we saw in Chapter 22, Newton's laws work when one is discussing the "middle" level of reality between the extremely small and the extremely large. The question is, how much certainty do we need for any given purpose? For example, quantum fluctuations and time dilation are not factors one must consider when designing an airplane. Describing the behavior of subatomic particles with any accuracy requires a much higher level of certainty, which is obtainable... to a point.

Physicist Michael Faraday (1791-1867) invented the concept of a *field*, or a type of invisible force. Magnetic fields are one example.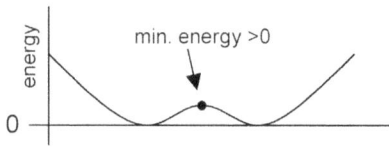
Electrical fields are another example: It turns out that lightning and static electricity are very closely related cousins. We know a great deal about how many types of fields operate, and we all feel their effects every day. What we don't know is what a field actually is. All we can do is describe its behavior and try to explain it using various subatomic particles that "carry" the field, possibly through being created and annihilated or chang-

ing into different particles. Not absurd enough? Common sense says that the lowest possible energy state of anything—including a field—is zero; you either have energy or you don't. It turns out, however, that the minimum possible energy state in a field is not necessarily zero. It's like a bowl with a W-shaped bottom, where the peak in the middle represents the lowest possible amount of energy for that field. This is a classic example of how not knowing has not slowed us down; today, well over half of the Earth's economic output depends on the electromagnetic force.

The Quantum Revolution

Science progressed extremely rapidly after Galileo and Newton. By the 1800s, scientists were delving into the mysteries of electromagnetism and evolution. Everything seemed explainable. In fact, many luminaries of the day were forecasting an end to science, because humanity was on the verge of knowing everything there was to know about the universe and our place in it. In 1900, physicist Lord Kelvin (1824-1907, after whom the Kelvin temperature scale is named in honor of his discovery of absolute zero, the lowest possible temperature) famously declared that only "two small clouds" remained on the horizon: explaining the propagation of light and black body radiation. Kelvin could not have been more wrong. To this day, the science taught in many high schools remains an approximation of what happens at extreme scales and speeds.

> *There is nothing new to be discovered in physics now. all that remains is more and more precise measurement.*
> Lord Kelvin

Black Body Radiation

Black body radiation seems simple: Shine light at any object, and that object will radiate heat, because the energy from the light causes the molecules and atoms in the solid to vibrate. A black body is a theoretical object that absorbs all electromagnetic energy that strikes it. (No real object will absorb 100% of any energy that strikes it.)

Experiments had confirmed physicist James Maxwell's (1831-1879) prediction that oscillating charges would emit electromagnetic radiation, which is the principle behind radio and other wireless communications and technologies. Maxwell had also demonstrated that this radiation traveled at the speed of

> *The universe is not only queerer than we suppose, but queerer than we can suppose.*
> John Haldane

Chapter 23
Quantum Mechanics | 617

light, which led to the conclusion that both light and heat are electromagnetic waves. Heating a body thus produced vibrations on the atomic and molecular level, which in turn caused charge oscillations that would be detected as light and heat. This seems simple, even obvious. The catch is that if any given frequency carries a certain amount of energy, then higher frequencies will carry higher amounts of energy, which is both obvious and correct. Following this logic using classical physics leads to the inevitable conclusion that raising the frequency would keep raising the energy level without any hard upper bound. In other words, this logic leads to infinite energy, which goes against both common sense and direct experience. It is obvious that the universe does not have infinite energy.

We are forced to confront the fact that something hidden in the void is controlling not just the subtle properties of matter but the destiny of the universe.
Sean Odenwald

What actually happens is that the intensity increases and the wavelength decreases as energy is applied to the black body. Eventually, it begins giving off visible red light. Keep heating it, and the light will turn blue. Keep going, and you will start seeing ultraviolet, X-rays, and eventually gamma rays; however, the intensity of the radiation will actually begin to decrease, avoiding the infinity problem altogether.

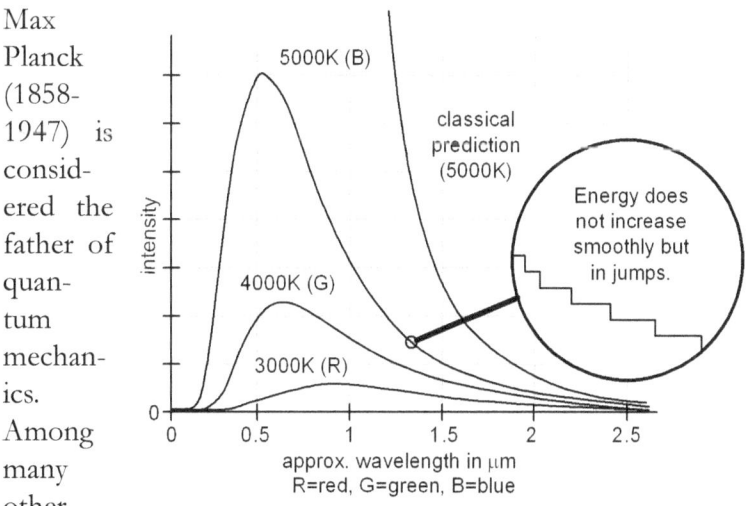

One Sun by day, by night ten thousand shine. And light us deep into the deity. How boundless in magnificence and might.
Edward Young

Max Planck (1858-1947) is considered the father of quantum mechanics. Among many other contributions, he discovered that energy emissions do not rise smoothly, but in discrete jumps. The smooth curve predicted by classical physics is actually more like a staircase. Light comes in *quanta* (packets or bundles) of energy. Changing the frequency of the light gives each photon more or less energy. Increasing the intensity of the light simply puts out more photons without changing the amount of energy in each photon.

These jumps and associated quanta are very small, but they do exist. Incidentally, Planck initially thought that the outside world was absolute and separate from consciousness, as did Newton before him. Planck later reversed his position.

Light Propagation

Maxwell's experiments and what was known about the properties of light such as absorption, reflection, refraction, and interference led to the conclusion that light is a wave. But how does light propagate? Waves were thought to require a medium to carry them based on our experience with sound, which does not travel in a vacuum. This is why Newton proposed the ether as a medium for carrying light. Assuming that light travels at a constant speed through the ether, it should seem to be going faster or slower depending on the relative motion of the light source and the observer. There was just one problem: Wherever science looked, the speed of light seemed to remain constant, no matter what. This problem would have to wait until Einstein's theories of relativity forever altered our understanding of space and time.

The unwearied Sun from day to day does his creator's powers display, and publishes to every land the work of an almighty hand.
Roland Barthes

The End of Determinism

As I mentioned in Chapter 22, classical physics is deterministic, meaning that one can predict future and past states if one knows the current state. Quantum mechanics changes all that. Inherent uncertainties in measurement (see below) make it impossible to know everything about a current state, which means that we can only predict the probability of future and past events. The clockwork universe we think we see is actually a giant game of chance, in which things are only definite when suitably observed, the act of which collapses all quantum possibilities into a single unpredictable outcome. Even then, we can only measure within certain limits and can never know everything with precision. For example, we cannot measure a particle's precise velocity if we measure its position precisely, nor can we measure its position precisely if we measure its velocity precisely. We can measure both properties within limits, meaning that the best we can do is a good approximation. Also, knowing the outcome of an experiment does not mean we can know what happened during the experiment.

You say you've got a real solution, well, you know, we'd all love to see the plan.
John Lennon and
Paul McCartney

> *Chaos is not disorder; it is a higher form of order.*
> Alan Garfinkel

I have already mentioned that quantum equations have the same general structure as classical equations, and that quantum mechanics incorporates classical physics; Newton's physics remain valid for just about everything we do in our normal daily lives. We can therefore say that the heart of quantum mechanics is little more than a set of rules for predicting the possibility of different observed outcomes. In fact, quantum mechanics essentially says that reality is simply a collection of all possible realities that consist of everything we could expect to find when we perform a measurement. But what exactly is a measurement? It may shock you to learn that the only kind of measurement that can collapse a quantum probability seems to involve consciousness. In short, consciousness may well influence—if not outright dictate—the outcome! More on this in Chapters 29 through 32.

Classical physics sees a whole as the sum of its parts, but quantum mechanics turns that on its head as well. $ab=ba$ is a staple of classical mathematics. For example, we all know that 3x2=2x3=6; however, this rule does not necessarily hold true when dealing with the quantum world. Classical physics also implies a single, unique past that can be extrapolated from the current state, whereas quantum mechanics only implies a past that is consistent with the current state—and the two need not bear any resemblance to each other. To put this into context, you may not have lived the life you think you have from birth to this moment. The events that brought you to this point may not form the nice, smooth continuum you think they do. Your past could be riddled with starts, stops, and jumps that combine with your current state to resemble a smooth flow. In quantum mechanics, the past is an average of all possibilities that are compatible with the present time. This seems counterintuitive if one is thinking of time as a flowing river; however, that commonsense view is also wrong, as we will learn in Chapter 28.

> *It was not possible to formulate the law of quantum mechanics in a fully consistent way without reference to the consciousness.*
> Eugene Wigner

We seem to live in a material world, and Madonna claims to be the Material Girl, but this too may not be quite accurate. The unparalleled success of quantum mechanics, including the Shrödinger wave equation, Heisenberg uncertainty, and more add weight to the idea that the universe itself is a giant waveform. Wave/particle duality adds even more credence to this idea, because our observations of light and other phenomena reveal photons to be either particles or waves, depending on what we are looking for. Quantum objects have been seen to tunnel, change energy states, influence each other across space and time, and more. Quantum electrodynamics (QED) goes so far as to say that matter is nothing but a denser field. Everything we call "real" is thus literally made of unreal things. This would be the stuff of science fiction, except that it isn't fiction.

Repeated experiments have shown that our commonsense view of the world is utterly wrong. Either the experiments are correct, or there is something fundamentally wrong with quantum mechanics. Either is possible; however, if quantum mechanics is fundamentally wrong, we should know this by now. What we do know is that quantum mechanics works perfectly. It has never been found to be in error, and it forms the foundation of all other sciences.

Quantum mechanics poses scientific and philosophical problems that are almost as intractable as that of consciousness. The math works, but how are we to make sense of it? Many think the mysteries of quantum mechanics and consciousness are intertwined, not least because of the observer effect that will discuss in more detail below. Quantum theories of consciousness have also failed. There is nothing in quantum physics that says that consciousness must exist or why it exists. In quantum mechanics, we may have come full circle from the magical age of mysticism and ancient religions with dualist concepts of soul to the Age of Reason that saw mind as a by-product of the brain, and finally to the modern era where more and more scientists are questioning what role quantum mechanics plays in consciousness, if any. Each point on this journey has been marked by some huge assumptions about what is correct, what is wrong, what is heretical, and even what is "natural" or "supernatural."

Quantum mechanics is typically invoked to deal with the probabilistic actions of extremely tiny objects. The term *decoherence* refers to a loss of information to the environment, such as when the building blocks of DNA molecules in a dead body break down and disperse to become part of something else. This works for things inside the universe, but the universe itself has no environment; it is a closed system, which means that information is always retained. This suggests that the universe is a quantum system that remains coherent. If this is cor-

Only an arbitrary distinction in thought divides form of substance from form of energy.
Sri Aurobindo

And if you gaze for long into an abyss, the abyss also gazes into you.
Friedrich Nietzsche

rect and quantum mechanics applies to everything in the universe, then we need to start thinking in terms of a quantum cosmology. As we will see in Chapter 25, loop quantum cosmology theory holds out the promise of uniting quantum mechanics and relativity, and plays nicely with holography theories. Both of these are perfectly compatible with the concept of time that we will examine in Chapter 28.

Cosmologists deal with the very largest structures in the universe. They tend to avoid quantum mechanics because of the different scales involved; however, superstring, brane, and loop quantum theories are bridging the gap, and rightly so, because the universe is riddled with quantum effects that lie at the foundations of existence. For example, we cannot rule out the possibility that a tiny region of the universe might form a bubble where the Standard Model of quantum mechanics no longer applies. This bubble could theoretically spread across the entire universe at the speed of light. We would never see it coming or know what had hit us when it arrived. In fact, we would not even know we had been hit. The odds of this happening are extremely low, but we cannot rule it out entirely. We can't definitively rule out much of anything; all we can say is that some things are far more likely than others. In my opinion, a deterministic universe is a very comforting thing, while a probabilistic universe is a much more liberating thing.

Humanity has perhaps never faced a greater challenge; for by his admission [that humanity is not the center of the universe], how much else did not collapse in dust and smoke: a second paradise, a world of innocence, poetry and piety, the witness of the senses, the conviction of a religious and poetic faith.
Goethe

Grainy Reality

It was an act of desperation.
Max Planck

As I mentioned above, common sense indicates that our universe is a smooth place that is governed by natural laws full of nicely rounded curves, but this is only true because we are seeing these curves from a distance. Zoom in close enough, and the smooth curves become jagged upward and downward leaps like staircases gone mad. One of Planck's biggest contributions to science was the discovery of fundamental units that form certain boundaries, beyond which nature dares not tread, or that at least mark the point at which our theories break down. These units are:

- **Planck length:** $1.616\ 199(97) \times 10^{-35}$ meter. This is the smallest possible distance. If we enlarged a region of this size to the size of an electron, the electron would be enlarged in turn to the size of a football. The formula used to derive this value contains both Planck's constant (critical for quantum mechanics) and Newton's constant (important for gravity). We can therefore expect that this distance will be very important in any theory of quantum gravity that unifies quantum mechanics with gravity.

- **Planck mass:** $2.176\ 51(13) \times 10^{-8}$ kilogram. This is the smallest possible mass.

- **Planck time**: $5.391\ 06(32) \times 10^{-44}$ second. This is the smallest possible time, defined as the length of time it takes light to travel the distance of one Planck length.

- **Planck charge:** $1.875\ 545\ 956(41) \times 10^{-18}$ Coulomb. This is the smallest possible electrical charge.

- **Planck temperature:** $1.416\ 833(85) \times 10^{32}$ Kelvin. This is the temperature at which our theories break down for lack of a quantum theory of gravity.

I journeyed through all the elements and came back. I saw at midnight, the Sun, sparking in white light.
Lucius Apuleius

This set of base units lets us derive a number of additional units including the Planck density, which corresponds to a Planck mass compressed into a cube whose sides all equal one Planck length, and is equal to compressing one trillion of our suns into a space the size of a proton. Scientists believe that the universe had a mass density of roughly one Planck unit at one Planck time interval after the Big Bang.

All Planck values are derived from Planck's constant. This value determines the size of the grains of reality, from changes in energy levels to the size of the very mesh of space-time.

Layers of Science

As I mentioned above, quantum physics contains all of classical Newtonian physics within it.

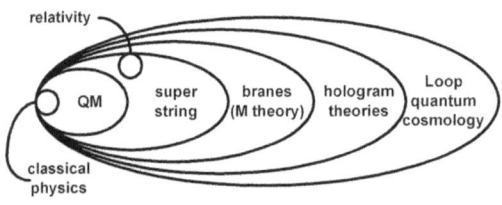

All our science, measured against reality, is primitive and childlike.
Albert Einstein

Quantum mechanics blends into classical physics when objects become large enough to effectively cancel out quantum effects and collapse probabilities to either 0% or 100%. Still, it does not contain relativity, which has stymied the search for a Grand Unified Theory (or Theory of Everything) that unifies the four fundamental forces (strong force, weak force, electromagnetism, and gravity). Three of these forces have already been unified; gravity remains the holdout. Superstring theory was developed to resolve this problem, but created some problems of its own. Brane theory (also called M theory) resolves some of the problems, as does holography. Most recently, the field of loop quantum cosmology (LQC) seems to be gaining traction as a possible candidate for a GUT. LQC allows for (and even possibly demands) a cyclic universe that continually comes into and goes out of existence—a very Buddhist concept.

> *Calculate what man knows and it cannot compare to what he does not know.*
> Chuang Tzu

Particles

The ancient Greeks were the first to speculate that matter is not infinitely divisible into ever-smaller pieces, but that it can be broken down into atoms and no further. The concepts of air, water, earth, and fire as elements was the first step toward discovering molecules, because different substances could be made out of different proportions of the fundamental elements. The Greeks who subscribed to these theories were more right than they knew, even if their numbers were off a bit. (There are 94 natural elements ranging from hydrogen to plutonium, and 24 synthetic elements ranging from americium to ununoctium.) Physicist Richard Feynman (1918-1988) said that the discovery that matter is indeed made of atoms is the most important scientific discovery ever. He may well be right.

> *Everything—everything at all—is at the same time particle and field.*
> Erwin Shrödinger

One of the most astounding things about atoms and the electrons that orbit their nuclei is how empty so-called solid matter is. As I mentioned in Chapter 14, the perception of matter as solid is an illusion; expanding an atom to the size of the Empire State Building (1,250 feet) would enlarge the nucleus to only the size of a grain of sand. Solid matter is nothing more than tiny specks in a huge void. Kick a ball, and the atoms of your feet and the ball will never meet. This and other discoveries revealed the limits of Newtonian physics. Even the idealized model of an electron orbiting a nucleus like a miniature solar system is flawed; actual electrons and even atoms are more like smeared out waves of probability until observed.

> *But how can a particle without a property of mass be a building block?*
> Drasco Jovanovic

The basic structure of the atom was figured out in 1912. Protons and electrons were discovered in 1920. The atomic mass of hydrogen was double what researchers expected, which led to speculation about the neutron. This particle was discovered in 1932. In 1935, it was discovered that *mesons* (a type of subatomic particle) carry the strong force, but the meson originally thought to do this turned out to be the wrong one. The weak force also has particles associated with it. Physicists have now identified over 400 subatomic particles, most of which serve to carry forces to or from parts of other particles; every layer contains still more layers within it. The fundamental particles of matter are called *quarks*, of which there are only six varieties. Pairs of quarks form *baryons*, and trios form mesons. In addition to all this, what we think of as empty space is in fact full of virtual particles that are just as real as their "real" counterparts.

The atoms or elementary particles themselves are not real; they form a world of potentialities or possibilities rather than one of things or facts.
Werner von Heisenberg

Just how precise is quantum mechanics? According to Feynman, physicist Paul Dirac (1902-1984) predicted that the quantum measurement of a proton's size would yield 1.00115965246. The actual measurement turns out to be 1.00115965221, which is the equivalent precision of measuring the distance from New York to Los Angeles to within the thickness of a human hair. A few physicists such as Stenger scoff at this by saying that we can get any level of precision we want if we create the right units; however, the fact remains that fact validated theory. One of Dirac's other predictions included a positively charged electron. (Electrons usually have negative electrical charges.) Just such a particle, the *positron*, was discovered in 1932. According to Dirac's negative energy solutions, electrons can travel backward in time, and these particles are what we experience as forward-moving positrons.

A Slight Surplus

Anyone who has ever seen *Star Trek* knows that matter and antimatter annihilate each other on contact. So why are we here? The simple answer is that particles decay a little more slowly than antiparticles, which is lucky for us, since we could not be here otherwise. (We could be here if the situation had been reversed, and there been more antiparticles than particles.)

To say 'it is' is to grasp for permanence. To say 'it is not' is to adopt the view of nihilism. Therefore a wise person does not say 'exist' or 'does not exist.
Siddha Nagarjuna

It is tempting to think of particles as solid objects, but this is not an accurate image, as I indicated above. In fact, the Copenhagen interpretation of quantum mechanics (see Chapter 29) implies that quantum potentials only form particles when we are watching them. When we are not looking, particles can theoretically wink in and out of existence—assuming they exist independently of any observation, of course.

It is a primitive form of thought that things either exist or they don't exist.
Arthur Eddington

Physicists use particle colliders to smash different types of particles together at extremely high energies, attempting to recreate the conditions present at the Big Bang. These experiments have shown matter to be completely mutable. Any particle can (among other things):

- turn into any other particle
- be created from energy
- vanish into energy

Classical physics simply does not apply to these situations. The extent to which it approximates everyday circumstances becomes apparent when one grasps the idea that the universe itself may consist of nothing more than exquisitely linked patterns of energy or geometry. The tiny surplus of positive energy remaining once the opposing forces of matter and antimatter wiped each other out just after the Big Bang is responsible for our being here. It is also illusory, because gravity provides negative energy that makes the universe add up to a giant nothingness. In other words, adding up all of the positive energy in the universe and subtracting gravity leaves 0 energy. Einstein said that $e=mc^2$, and it turns out that $e=0$.

If I could remember the names of all the particles, I would have become a botanist.
Enrico Fermi

All of this begs a question: How do we know this? How can we possibly study particles that are smaller than the atoms of any lens we could use to look at them? The short answer is that this is accomplished using indirect observation that studies the relationships between particles. Physicists select different combinations of particles, slam them into each other in colliders, and see what happens. Which begs another question: If the Copenhagen interpretation of quantum mechanics is correct, what are physicists seeing when they look at subatomic particles? Are they seeing some form of even slightly objective reality or merely a projection of their own minds? The latter possibility could easily explain why quantum physics is by far the most successful science ever attempted.

The "God" Particle

The search is currently on—at facilities such as the Tevatron at the Fermi lab near Batavia, Illinois and the Large Hadron Collider at CERN near Geneva, Switzerland—for the Higgs boson, a large elementary particle whose existence is predicted by the Standard Model of quantum mechanics (although some dissenters say the Standard Model is fundamentally flawed). Detecting the Higgs boson would be a tremendous achievement because it would validate a theory that has been used for a long time without proof. A relatively simple explanation of why the Higgs boson is so important is that it represents space-in-itself. Space, of course, constitutes the whole of reality, while at the same time constituting no part of reality.

Some modern-day mystics believe that light is the foundation of reality. Light is indeed significant; there is no anti-photon, and photons themselves can be shown to be either particles or waves. Photons travel at a constant velocity regardless of the relative movements of both emitter and viewer, and thus do not experience time as a flow. Light is pure, eternal energy. The mystics may have a point.

> *Everything has its 'that' everything has its 'this.' a state in which 'this' and 'that' no longer find their opposites is called the things of the way.*
> Chuang Tzu

Exclusion and Stability

It is a good thing that classical physics is only an approximation, because matter that obeyed classical laws would cease to exist almost instantly. From a classical perspective, an electron would behave just like any other electrical charge by emitting all its energy and falling into the atomic nucleus; however, an electron is not a classical object, but a wavelike quantum object. It is therefore restricted to gaining and losing energy in the discrete jumps described above. This allows atoms to exist in different energy states that cannot become arbitrarily small; as we saw, the lowest possible energy level (called the *ground state*) is not zero. An excited atom can emit energy as photons until it eventually reaches its ground state.

Physicist Wolfgang Pauli (1900-1958) discovered what came to be known as the Pauli Exclusion Principle: that two similar particles cannot exist in the same state within the limits of Heisenberg uncertainty. But for this exclusion, quarks could not form separate, distinct, and well-defined protons and neutrons, nor could separate atoms form; the universe would

> *Physicists now believe that entanglement between particles exists everywhere, all the time, and have recently found shocking evidence that it affects the wider macroscopic world we inhabit.*
> Michael Brooks

essentially consist of soup (though not any soup we would recognize). We will see in a few moments that particles seem to take all possible paths to their destinations at once when we discuss the dual slit experiment; meanwhile, electrons seem to test all possible orbits at the same time, which is the literal equivalent of your knocking on every door on your street at the same time. The orbits they settle into are determined by their waveforms and their associated wavelength. The orbiting electrons are therefore "excluded" from simply collapsing into the nucleus, and matter as we know it can go on existing.

Pauli also introduced the concept of indistinguishability, which says that two particles with identical mass, spin, charge, and other properties have no separate identities, and there is thus no way to distinguish one particle of any given type from another particle of the same type. This accords with Leibniz's thoughts on the identity of indiscernibles, which says that two things that are identical in all respects must be one and the same. This principle helps explain how materials behave when heated, why metal feels cold or warm to the touch, and more.

As bizarre as it seems, a gram of rose petals and a gram of uranium contain identical amounts of energy.
Gerald L. Schroeder

My own vain hopes of overthrowing quantum mechanics were overthrown by the data.
John Clauser

Nonlocality implies transcendence.
Amit Goswami

Entanglement

Put two classical objects together, do anything you want to them, and then send them on their way confident that neither object will retain any memory of the encounter. The same is not true of quantum objects, which not only "remember" the encounter, but also "know" what their partner objects are doing and respond in exactly the same way, no matter how far apart they are. This is similar to dropping objects in water; the shape of the spreading wave will retain an imprint of the original object. This should come as no surprise, given that particles demonstrate both particle-like and wave-like qualities. This phenomenon is called *entanglement*, following Schrödinger, and has been observed in the dual slit and delayed choice experiments we will discuss in a few moments.

Entanglement is an instant connection that extends across space and time to connect objects that, by all classical definitions, should be completely isolated. The separateness we see between objects is nothing more than an illusion caused by our limited perception. The entire universe and everything in it is connected in ways we are only beginning to grasp.

Close relationships create *coherence* or a deep connection, as anyone who has ever been in a serious relationship can attest. Physicist John Bell (1928-1990) is the man behind Bell's Theorem, also sometimes called Bell's Inequality, which has been called one of the most significant scientific discoveries of all time. According to Bell, if quantum mechanics is correct, then entanglement between particles should take place as if they are not separated in space. All particles were entangled at the Big Bang and remain in a massive superposition until measured; no interpretation that uses local hidden variables (see Chapter 29) can be compatible with quantum mechanics. Nonlocality seems to be a fact, and Einstein seems to have erred when he opposed the ideas of "spooky action at a distance."

It is interesting, even comforting that the laws that determine atomic interactions in cosmic interstellar dust are the same laws that determine the interactions of molecules on the surface of brain cells.
W. Ross Adey

The entire universe seems to be a single entangled object, and thus still "one," which is what the mystics have been telling us for thousands of years. The discrete objects we see may be mostly illusory, or even completely so. Some speculate that the degree of coherence seen in all living things may be caused by entanglement, which could be an evolutionary tool for coordinating activities across cells, or even members of the same species. Awareness may be either caused by or related to entangled relationships between cells or even particles in our own brain. Macro objects are too big for us to see their constituent particles, which gives the illusion of separation.

Nonlocality

"I dwell in possibility," said poet Emily Dickinson (1830-1886), a statement that may well be similar to how a subatomic particle might describe itself. Nonlocality implies instantaneous action at any distance without any signals or other communications or information transfer, in seeming violation of the cosmic speed limit of 186,000 miles per second (the speed of light). Particles can run but not hide and may also share properties with other particles (such as spin). It can also refer to a particle's ability to tunnel itself instantly to anywhere in the universe. As one writer said, "Prepare your oven and then sit back and wait; there is a nonzero probability that a turkey from a nearby grocery store will materialize in your oven." This is the universe we live in: a mesh of forces, fields, and matter that can come and go at will. The question is, are these jumps and entanglements really random, or do they just seem

Biology is the study of the larger organisms; whereas physics is the study of the smaller organisms.
Alfred North Whitehead

that way? The Heisenberg Uncertainty Principle (see below) indicates that randomness prevails. Hidden variable interpretations say that there is a deeper, undetectable level of order. My personal belief is that the hidden variable is consciousness. More on this later.

One is constantly reminded of the infinite lavishness of nature. No particle of her material is wasted or worn out. It is eternally flowing form use to use, beauty to yet higher beauty.
John Muir

Nonlocality also means that a quantum object, such as a particle, can be in multiple places at one time (superposition) thanks to the Schrödinger wave function; it does not exist in any particular position or state until an observation collapses the wave and forces the particle to jump into one state or another. If the particle is entangled with another particle, then observing either particle immediately affects the other particle regardless of distance. The waveform spreads over time, meaning that the probability of finding any particle we observe anywhere near its previous location shrinks over time.

The Aspect experiments in the 1980s proved that entangled objects lose their autonomy and behave like single entities. This is not a normal cause and effect relationship, because nothing travels between the different locations, and the waveform must exist beyond space-time. As counterintuitive as it is, the only viable conclusion we can arrive at is that the universe is nonlocal. As physicist Henry Stapp said, "The fundamental process of nature is outside space-time but creates events that can be found in space-time." Bell's Theorem prevents any interpretation that does not recognize nonlocality.

If the statistical predictions of quantum theory are true, an objective universe is incompatible with the law of local causes.
Henry Stapp

As we saw in Chapter 15, Bohm believed this phenomenon must depend on the state of the whole system, and rejected any idea of an independent objective reality. To him, entanglement implies a common cause that is part of all physical events. This cause spreads infinitely fast without exchanging information, and thus does not violate relativity. All systems are part of larger systems and subject to common influences. This much makes sense, except that Bohm went one step further by saying that the whole is implied in the parts as well, which is compatible with holographic theories, as we will discover in Chapter 27. Bohm also said that location ceases to exist at sub-quantum scales; all points in space become equal to one another, and it is therefore pointless to think of anything as separate from anything else.

Like Wigner's friend, the phenomenon of entanglement has led a small army of modern spiritualists, self-help experts, "gurus," and others to declare that everything is connected to everything else in the universe, and that we can literally create our own reality, complete to "manifesting" great amounts of material wealth using the "Law of Attraction." Well, on a purely technical level they may be correct, if for no other reason than millions of people are snapping up products like *The Secret* and greatly enriching their authors. In reality, entanglement is a very delicate thing; as a particle's wave smears out between observations, the entanglement becomes impossible to detect, and is thus erased for all practical purposes. This is not to say that the entanglement ceases to exist, merely that it becomes negligible. This is also not to say that quantum mechanics does not have powerful implications for self-help; it's just not quite where people have been looking. (See Appendix B for a complete explanation.)

This phenomenon also has another interesting feature: No classical system can truly simulate nonlocality. Any nonlocal information in our brains is a program that simply cannot be simulated, which may at least partially explain why our ongoing search for the seat of consciousness has turned up empty.

Delayed Choice

Imagine firing two entangled photons with identical polarization in opposite directions, changing the polarization of the detector before the photons have traveled the distance, thus changing the outcome of the measurement on the other side—in short, making a decision extending backward in time. Alternatively, consider switching a mirror's position after a photon has passed, but before it has hit a target, and having the photon hit the target it would have hit had the mirror been moved before it passed. Now imagine being able to do this with light that left a distant star or quasar millions or even billions of years ago, and have your decision extend across vast reaches of space and time. It may shock you to realize that these experiments have already occurred (with the exception of the quasar, but there is no reason to think the results will be any different). This happens because of the entanglement and nonlocality described above, with no signals traveling between the correlated photons and/or the experimental apparatus.

[Aspect's findings] indicate that we must be prepared to consider radically new views of reality.
Basil Hiley

These experiments prove that the quantum world transcends space and time. It's as if the particles knew what would happen beforehand. It also seems to confirm spiritual claims that consciousness made it happen.

It is a remarkable fact that shining a light through a hole can reveal truly deep mysteries about the world we live in.
Johnjoe McFadden

The success of Aspect's experiments, as well as others performed by Zeilinger and other researchers, force us to reject local hidden variables because of Bell's Theorem, and to accept either the Copenhagen or many-worlds interpretation. (See Chapter 29.) Further, accepting this startling outcome means ditching our common view of time as an arrow with us forever riding on the very tip, and replacing it with a transactional view that has the future affecting the past in just the same way that the past and present can affect the future. You could think of the starlight reaching out to your eyes from the past, and of waves from your eyes reaching out to the star from the future relative to when the light set off on its journey. This exchange gets the photon to go along with the experiment being conducted, and also affects any other entangled photons. No paradox occurs, so long as we exclude the idea of an objective reality that exists independently of observation—a seemingly overwhelming confirmation of the Copenhagen interpretation.

Dual Slit

The experimental results we report and our explanation of them with quantum theory are completely undisputed. it is the mystery these results imply beyond physics that is hotly disputed.
Bruce Rosenblum

Quantum mechanics allows us to conclusively prove that light is composed of particles called photons. Quantum mechanics also allows us to conclusively prove that light is composed of waves. This *complementarity*, or wave-particle duality, is yet another of the mind-boggling realities of our universe. Even more amazing, the answer we get to the question of whether something is a particle or wave depends on how we set up the experiment, the detectors we use, etc. All particles are waves and vice-versa.

Around the year 1800, polymath Thomas Young (1773 -1829) performed the first dual slit experiment and concluded that light is a wave

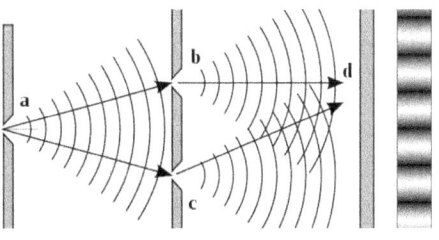

This experiment is very simple: aim a beam of light from point

a in the above diagram toward a barrier with two narrow slits *b* and *c*. The peaks and valleys of the waves interfere with each other as shown in the diagram, with the wave peaks *d* creating the interference pattern shown on the right.

The diagram on the right shows how waves either reinforce or cancel each other out. Adding two waves with opposing peaks cancels both waves (dark spot on the interference pattern), while adding two waves with synchronized peaks complements both waves (light spot on the interference pattern). Intermediate situations create partial brightness, resulting in smooth gradients. All of this can be seen in the interference pattern on the previous page.

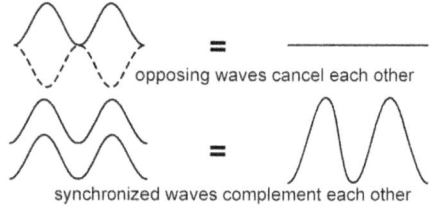

Truth cannot be cut up into pieces and arranged in a system. The words can only be used as a figure of speech.
Buddha

So far so good—or at least it would be, if that was all there is to it; however, observing an atom at a given place causes it to be there, as we will see later in this chapter when we discuss wave forms. Looking for a subatomic particle in a given location also causes it to be there; otherwise, it is in a superposition, and is at both places at once. This is a demonstration of both nonlocality and seemingly backward causality. Delayed choice even applies here; looking at the slit after the particle goes through lets you determine which slit it passed through, and destroys the interference pattern. Stop looking, and the interference pattern returns. This brings us to what the dual slit experiment should look like if we are dealing with particles. The diagram in this paragraph shows two streams of particles flowing through the slits and creating two piles on the detector. Try this setup with sand or any grainy material, and this is exactly what you will get. The size of the particles does not matter.

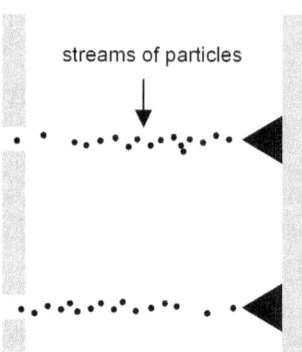

There can hardly be a sharper contrast between the everlasting atoms of classical physics and the vanishing 'particles' of modern physics.
Melic Capek

In 1924, de Broglie wondered why electrons could not behave like waves, since photons can act as particles. The answer is, they can. Photons are not the only type of particle to behave this way; the New York telephone company experienced a problem in some of its vacuum tube equipment that turned out to be caused by an interference pattern of electrons. Later dual slit experiments with electrons showed that they too display wave/particle duality, even when the beam is so weak that only one electron at a time is passing through the slits. This means that single electrons can interfere with themselves (and so can photons). Zoom in for a closer look, and the interference pattern vanishes. It's as if the electrons know all about the experiment, including whether one or both slits are open and whether we are looking or not, and are determined to adjust their behavior to confound the experimenter. Feynman proposed a radical idea: Perhaps the electrons are traveling through both slits at once and take all possible paths to reach them! This does not conflict with natural laws, but then again the laws don't tell us why this occurs.

We can't avoid some anthropic component in our science, which is interesting, because after three hundred years we finally realize that we do matter.
Paul Davies

Wave/particle duality does not stop with particles. Any particle or object can display complementarity. The physical nature of an object depends on how you decide to look at it. The double slit experiment has been performed with different particles and molecules, and it has worked the same every time. As I mentioned in Chapter 15, it has even worked with buckyballs, proving that quantum effects extend into the macro world. We see solid objects instead of wavy smears in the normal world because the waves involved are extremely small—far too small for us to see or otherwise perceive directly. This allows us to ignore the wave-like properties of matter in almost all normal situations. The length of the wave depends on the mass of the object. An electron in a home appliance has a wave about 1×10^{-6} cm in length, while the average bacterium has a wave that is smaller than the radius of an atom. Each can tunnel through a barrier whose thickness corresponds to the overall wave. People and other large objects have waves so short that we can safely ignore them. As I have said, there is a nonzero probability that you will suddenly find yourself orbiting some distant star, or even sitting across the room from your current location. Just don't hold your breath waiting for it to happen.

Mass and energy are not just interconvertible, in way that I can convert dollars into gold or vice-versa. They are the same thing: mass-energy.
Nigel Calder

Stenger disagrees, saying that, "Light is a particle, not a wave, period." He claims that photons are particles that act like waves; place a detector in the impact areas, and each photon will show up as a dot, with succeeding dots gradually filling in to look like an interference pattern. This seems like a compelling argument, until one looks at recent "quantum eraser" experiments that have reaffirmed complementarity, wave/particle duality, and delayed choice in a single experimental setup. Stenger's argument also fails to account for the disappearance of the interference pattern when you are observing the path of the objects. A full discussion of these experiments is beyond the scope of this book, but there are plenty of online resources if you are interested.

The light from quasars set off on its journey to Earth billions of years ago. It should be theoretically possible to use a galaxy as a sort of double-slit using gravitational lensing to affect the light even though billions of years have elapsed. Such an experiment would provide dramatic proof of nonlocality that extends through time as well, as once again confirm other dual slit findings.

Particles and buckyballs cannot collapse their own paths. Both paths and particles alike exist only as possibilities, until actualized through observation. This means that turning possibility into reality happens retroactively, in a backwardly causative manner. We can see the dual slit experiment as another variation on the delayed choice experiment described above. Dual-slit experiments have also been conducted with specific delayed choice components added to them. According to the Copenhagen interpretation, no observation means no matter. Scale does not matter, because the entire universe is subject to the same laws. The implications of this are truly staggering. For example, it is possible that the universe evolved in a giant superposition until the first entity capable of measuring its environment became possible, and created both itself and the history of the universe all the way back to the Big Bang by interfering with itself and collapsing the cosmic waveform. In this scenario, the first sentient being created both itself and the history of the cosmos leading up to itself in one instant, bringing space, matter, energy, and time into being in the same moment. Abandoning our limited notion of time as a one-way arrow makes this scenario even more plausible, as we will learn in Chapter 28.

It is possible to force an object to behave like either a particle or a wave, as we saw above with the dual-slit experiment. Observing the path destroys the wave-like interference. The

Let man then contemplate the whole of nature in her full and great majesty, and turn his vision from the low objects which surround him.
Blaise Pascal

From the moving flux of subatomic particles that make up physical reality- including our own bodies-we somehow mentally 'create' the sold, stable world of experience.
David Fontana

nature that seems so orderly on macro scales does some decidedly odd things at small scales, but always avoids logical paradoxes. How interesting that this is so.

Weirdness

The once believed ultimacy of the line of division between the 'self' and the 'not self,' the subjective and the objective, is rejected as untrue.
K. Venkata Ramanan

Those who dismiss quantum effects in the macro world forget one very important fact: that quantum mechanics is the foundation on which all physics rests. Physics is the foundation on which all other sciences are built, from chemistry to the "soft" sciences like psychology and sociology. For example, people think, feel, and act in accordance with psychological concepts that are built on neurology, which is built on biology, which is built on organic chemistry, which is built on chemistry, which is built on quantum mechanics. I am writing this book on a computer that obeys—and is even made possible by—the laws of quantum mechanics. All scientific and technological roads lead to the same place. Our current technology limits our study of quantum effects to very small things, but that does not change the foundation at all. As I have mentioned before, quantum mechanics is astonishingly successful; not one prediction has ever been wrong.

According to quantum mechanics we cannot exclude the possibility that free will is a part of the process by which the future is created.
Edward Teller

With that said, quantum mechanics only describes potentials and probabilities in a way that is both fuzzy and more than a little imaginary. People have been able to debate philosophical speculations on commonsense grounds for millennia, but that is no longer an option. The entire universe rests on quantum mechanics. Our inability to make sense of it means that we don't comprehend the universe as we once thought we did. Nature has no reason to care whether we can tell two particles apart or not, or whether we understand what we see or not. Still, quantum mechanics is of more immediate relevance to us than other scientific revelations—such as those by Galileo, Newton, Copernicus, Darwin, Freud, and others—because it deals with who and what we are at every moment. Discoveries such as the Heisenberg Uncertainty Principle and Shrödinger's Cat have revealed a universe that is far more complex, nuanced, and rich than we had ever dared to imagine. What we know now hints at discoveries to come that will only reveal an even richer natural tapestry.

Heisenberg

Position and speed are measured visually by bouncing electromagnetic energy, such as visible light or radar waves, off the object being measured, and using the travel time of that energy to calculate distance. Successive measurements allow us to measure speed, because we can measure the changes in the travel time and the rate at which those changes occur. This method works well enough in the macro world, but wreaks havoc on quantum scales.

Heisenberg demonstrated that only one variable can be measured with absolute precision. For example, we can measure the position of a particle with exact precision, but cannot also measure position, and vice-versa. The uncertainty equation is: $\Delta x \Delta p \geq (\hbar/2)$ where Δx is an expression of velocity uncertainty and Δp is an expression of position uncertainty. Reducing one of these variables to zero raises the other to infinity. The total amount of uncertainty must always be at least equal to half the reduced Planck's constant $\hbar/2$. There is no such thing as total precision.

It is this uncertainty that allows quantum tunneling, and that causes the minimum energy state of an atom or particle to always be above 0, as we saw at the beginning of this chapter. Uncertainty extends all the way to the macro world as well, which forces us to ask just how much we can ever really know about reality. All we can know is the result of our observations, and those are limited. Even using a single photon with a very short wavelength to make the measurement can never be more precise than the wavelength of that photon. Also, the more we try to pin down the measurement using higher- and higher-energy photons, the more those same photons disturb the measurements. If we cannot know a particle's path, then can we really know it is taking a path from Point A to point B at all? If we cannot observe with absolute precision, then can we say that absolute precision exists? The Heisenberg uncertainty equation refers to Planck's constant, which indicates that the limits to what we can know are intimately connected to the graininess of the universe itself. It also seems to prevent quantum phenomena from being their own cause.

Heisenberg worked with Bohr, first as his student then as a collaborator. Bohr reviewed and rejected Heisenberg's first

If we think we can picture what is going on in the quantum domain, that is one indication that we've got it wrong.
Werner von Heisenberg

How come the quantum?
John Archibald Wheeler

attempt at describing uncertainty, because it contained a fundamental flaw. Embarrassed, Heisenberg tried again, and succeeded with the equation described above. In his own words, "I had the feeling that, through the surface of atomic phenomena, I was looking at a strangely beautiful interior, and felt almost giddy at the thought that I now had to probe this wealth of mathematical structures that nature had so generously spread out before me."

But where is the rationale—the raison d'être—for our universe having these features?
Brian Greene

Bohr and physicists since then agreed fully with Heisenberg's revised equations. Bohr himself saw reality as governed by uncertainty that created probabilistic descriptions of motion. Even so, as we appreciate the profound implications of uncertainty for things like quantum tunneling and ground state energy, we must remind ourselves that the quantum world is not a complete free-for-all. As Heisenberg said, "It was a strange experience to find that many of the results of Newtonian mechanics like conservation of energy could be derived also in the new scheme."

A serious challenge to uncertainty arose in what came to be known as the *EPR* (Einstein-Podolsky-Rosen) paradox. This thought experiment involves entangling two particles and then separating them. It should be possible to measure one property of one of the particles and the other property of the other particle to obtain an exact measurement without any uncertainty. Einstein was extremely uncomfortable with the idea of uncertainty, saying that, "God does not play dice with the universe." He also rejected nonlocality, calling it "spooky action at a distance." Experiments after Einstein's death demonstrated that the paradox does not arise, because nonlocality is indeed real; measuring one property of one particle affects the other particle, which upholds uncertainty and avoids paradox.

Waveform

There is no absolute truth at the quantum level.
Werner von Heisenberg

Set up as many identical experiments and starting conditions as you like, and you would be forgiven for expecting the same results every time. Without this consistency, science itself would grind to a halt. Students could not be taught, and discoveries could not be validated through replication. This works well in the macro world; however, the quantum world is not nearly so compliant. Repeating the same experiment with the same starting conditions can indeed yield different results.

All objects have a wave function (also called a waveform) that measures the probability of finding that object at any given location in space-time. We cannot know this location until we observe it, and the particle exists in a superposition until the observation occurs. What we think of as solid bits of matter are waves that extend throughout the universe, unless and until we measure them. Before a measurement, all we can do is predict possibilities. If we had all of the information about a particle, we could predict the exact results, but uncertainty prevents us from ever having all of the information or the resources to crunch all of the numbers. We are forever limited to educated guesses.

If we are still going to put up with these damned quantum jumps, then I am sorry I ever had anything to do with quantum theory.
Erwin Shrödinger

The waveform itself represents the probability of finding an object in any location within the wave. The height of the wave at any point represents the probability of finding the object at that exact location. This is not the same as saying that, for example, my son has a 99% chance of being at school during school hours on any given day. My son is either at school during that time or he isn't; however, for a quantum object, the waveform is the same as the object itself. The waveform of an atom is the atom itself. Looking for my son at school does not cause him to be at school; he is either there or not, independently of my observation. Or is he? In the quantum world, observing a particle at a given location causes it to be there, and observing a particle not at a given location causes it to not be there. Finding the object instantly collapses the waveform to 1 (100%) in that location and 0 (0%) in all other locations. The pre-observation waveform only tells us the probability of what we will observe. We can only know where something is when we are staring right at it. Having to take the observer into account turns classical reality on its head.

My son should be at school right now. The GPS tracking feature on his cell phone indicates that he is within 40 yards of his classroom, which is located at a corner of the school. He therefore has a waveform that is 80 yards in diameter whose highest peak is centered on his desk in his classroom. Going to his school and looking in the classroom window is the only way for me to collapse his waveform, which will then spread right back out as soon as my back is turned. His waveform could also be collapsed by someone else observing him in his classroom, which collapses the waveform for everyone else

All matter is just a mass of stable light.
Sri Aurobindo

> *Relativity and quantum theory have shown that it has no meaning to divide the observing apparatus from what is observed.*
> David Bohm

who looks through the window immediately afterward. Reporting that information to me (such as by calling me) would collapse my view of his waveform. My son's teacher is normally the first person to see him in his classroom. We can conclude that he is not in his classroom before his teacher observes him. This example applies equally to the quantum world: There are no actual atoms in any actual place unless and until we look for them, a fact that seems hard to swallow, since each of us is looking at collections of countless atoms that form the objects we are used to seeing, and that don't seem to be going anywhere.

Physicist Max Born (1882-1970) saw the wave, not as a particle smeared across the universe, but simply as a probability function that tells us the likelihood of finding a particle in one location or another. This seems simple enough, except that the Copenhagen interpretation (see Chapter 29) makes it meaningless to ask where the object is before an observation is made. Can we then distinguish between wave and object? Does the object have a position when we are not looking at it? Does the Moon exist when nobody is looking at it? The short answer is, "Quite probably, but not definitely." The more precise answer is that the question itself is meaningless.

> *An independent reality in the ordinary physical sense can neither be ascribed to the phenomena nor to the agencies of observation.*
> Niels Bohr

A probability wave is large in a given area, but quickly drops to near zero with distance. My son's 80-yard waveform peaks above his desk in his classroom, shrinks rapidly to near zero by the time it reaches the edges of his classroom, and finally drops to zero 40 yards out. I know that it drops to zero, because I made an indirect observation using the GPS locator. Absent this observation, my son's waveform would extend across the entire universe. The same thing happens with quantum objects, meaning that the odds of one particle tunneling any noticeable distance is extremely small. Multiply this by the number of particles involved, and you will soon discover why you need not worry about suddenly finding yourself somewhere else. It also explains why the "Law of Attraction" that stars in more "get-rich," "empowerment," and "mastery" schemes than I can count is both technically real (after a fashion) and absolutely wrong, as you can learn in Appendix B.

There is just one major problem: Nothing in quantum mechanics describes the collapse or what exactly causes it to occur; it is simply postulated. We cannot even say for sure that any collapse actually happens after 70 years of trying. The collapse itself is inserted by hand, because there is no defined mathematical or experimentally validated way to do this. The most extreme view says that objects don't exist at all when they are not being measured. Also, what makes measurements special? Is consciousness required, or can some detector making observations not seen by any conscious entity cause collapse? Is the collapse caused by said detector or by the person who finally checks the results on the machine? Can animals cause collapse? Plants? Extraterrestrials? Ghosts or spirits? Are humans special or not? Given how science has continually thrust us ever further from the cosmic center, starting with heliocentric theories of the solar system, the answer is, "probably not."

Then again, Bohr believed that phenomena do not exist until the waveform collapses, which makes phenomena the fundamental units of existence. The world is therefore defined in probability only, with humans (and/or other conscious species) acting as arbiters and referees of reality. Particles, stars, and everything in between may not be able to exist without us, or may be able to exist only in possibility until observed/collapsed. This viewpoint forces us to abandon the concept of a single unique reality.

Let's push the waveform concept further by replacing the word "particle" with "universe." Physics does imply that the universe is a giant whole, but can say little about that whole. If we substitute our terms, though, we can see that a waveform can be used to describe the set of all possible universes. The starting point for our own universe must therefore be an infinite number of possible universes. The resulting waveform is very high near our own universe, but tapers off rapidly for all other universes. This substitution works in theory, but has yet to be proven right or wrong; however, it does allow us to expand our definition of *universe* from "all that exists," to "all that can exist."

As we have just seen, it is possible that micro-scale collapse causes macro-scale collapse, because the two scales are inseparable. We must either accept this or reject the idea that the waveform represents reality, which is possible if properties have determinate, not probable values that render collapse unnecessary. If this is the case, then we need a new theory to

Light travels in a wave but departs and arrives in a particle.
Ralph Baierlein

When a scientist states that something is possible, he is almost certainly right; when he states that something is impossible, he is very probably wrong.
Arthur C. Clarke

explain why the waveform and resulting collapse seem to work as well as they do. One possibility is that the waveform collapse is nothing more than mathematical shorthand, because some interpretations of quantum mechanics (many worlds and hidden variables) don't use waveforms. We will examine this more in Chapter 29.

Nothing outruns photons.
Brian Greene

If we accept the waveform, then we must also accept that possibilities are not any less real than actualities. In fact, reality may actually be the other way around, because possibilities are timeless, while actualities are fleeting instants. Classical physics tells us that objects can only be in one state at a time, while quantum physics tells us they must exist in infinite states at the same time. Quantum determinism replaces classical determinism; knowing the waveform at any instant allows us to predict its evolution over time using well-established mathematical rules. Knowing all of the waveforms in the universe would allow some super intellect to know the entire past, present, and future of the universe and all in it. Knowing the waveform in a human brain would allow us to predict what that person would do at any arbitrary time. Knowing the waveform of an object determines the probability that a given event will take place at some future point.

If we knew the waveform of the entire universe, then we would truly know the mind of God; however, we cannot ever know this waveform, only that of individual objects on quantum scales. Our own waveform interacts with the object's waveform to produce the outcome. We cannot know both waveforms, and must therefore conclude that the universe is probabilistic; however, we can also see that the universe is probabilistic in a determinate way, which brings us dangerously close to the idea of a divine plan. More on this in a moment.

Shrödinger's Cat

The entire system would contain equal parts of living and dead cat.
Erwin Shrödinger

We have already discussed Shrödinger's cat in Chapters 15 and 20, but it bears examination in further depth. Is the cat in the box alive or dead? Classical physics says that the cat is either alive or dead with 50/50 odds. Quantum mechanics purists say that the cat is in a superposition of both states that is neither dead nor alive—nor really existing at all—until the box is opened by an observer, who collapses the waveform by doing so. Before observing the outcome, all we can say is that the cat in the box is only there because someone observed the cat going in. If Observer A places the cat in the box and Observer B opens the box to reveal a dead cat, does this mean that Observer B killed the cat? No, because s/he did not set up the experiment, and because the waveform collapse is random. Anything that provides information is considered an observation that collapses the waveform. The cat itself has all possible histories; arguing one or the other uses classical views. Both histories can coexist. The observation thus creates not only the reality but the history that is appropriate to that reality.

Shrödinger was aware of the formidable challenge of isolating something as large as a cat, but the logic of his thought experiment stands, because there is no hard boundary between small and large; any object can therefore be in a superposition. We saw that Zeilinger has already placed buckyballs in superposition and is increasing the size of the objects he is working with, budget being his only limitation so far.

The Shrödinger waveform is the complete representation for any object, because it describes both exactly what the possible outcomes and the probability of each possible outcome. Place 100 cats in 100 boxes, and the waveform will tell us how many live and dead cats to expect after a certain time. This aspect of the waveform is deterministic. What the waveform would not do is tell us which cats would be alive and which would be dead. Assuming a 50/50 chance that any cat will emerge alive or dead, we should expect to find 50 live and 50 dead cats; however, the actual number we find would fluctuate around 50 in the same way that tossing a coin 100 times would give us about 50 heads. Repeated trials would bring us closer and closer to the idealized average of 50/50, and would also needlessly kill a lot of cats. Thus, the waveform is deterministic in that it tells us what is possible. It is also probabilistic, because it does not tell us which of those possibilities will transpire.

Shrödinger's equation predicts what we will see, but his goal of getting rid of what he called nonsense failed. As author John Gribben (1946-) wrote, "What quantum mechanics says is that nothing is real and that we cannot say anything about what things are doing when we are not looking at them." Shrödinger thought he had discovered an oddity. What he actually discovered goes to the heart of reality. Still, there is

When I hear about Shrödinger's cat I reach for my gun.
Stephen Hawking

Brother, can you spare a paradigm?
Unknown

something missing from his equations, because we end up with two equations and two complete sets of solutions. History has demonstrated that extra solutions are often clues to some fundamental aspect of reality, because they are answering a question we have not thought to ask. This is how Dirac found positrons, and how we were introduced to the concept of antimatter. What is the question waiting to be asked that has already been answered?

The Zero Point Field

We and all we see are frozen energy.
Gerald L. Schroeder

Aristotle argued that empty space is not really empty. As we saw above, Faraday introduced the concept of a field. Fields that don't necessarily obey classical laws form the most fundamental layers of reality. We already saw that no particle can ever be truly at rest, that its ground energy state must always be above zero. Quantum mechanics also allows matter and energy to spontaneously appear from nowhere, provided it disappears again just as quickly. Empty space is not truly empty, and the universe can be thought of as a ripple in this energy.

Modern physics says that each part of space has fields with waves of varying wavelengths. If we calculate the minimum amount of energy a wave can have, we arrive at the startling conclusion that every cubic centimeter of space has an incredible amount of energy that exceeds the amount of energy in matter by 10^{40}. According to Feynman, every cubic centimeter of space has enough energy to boil all of Earth's oceans. This Zero Point Field (ZPF) gets routinely subtracted out of quantum calculations. Leaving it in reveals a universe that is a vast sea of interconnected energy. For example, mass could be an illusion caused by the ZPF resisting the movement of objects. The Van der Waals effect, named for physicist Johannes Van der Waals (1837-1923), sums up attractive and repellent forces between molecules—or parts of the same molecule—that exist because of localized imbalances in the ZPF. This allows matter to change states, such as water turning into steam. Physicist Andrei Sakharov (1921-1989) speculated that gravity could be an aftereffect of the ZPF caused by matter-induced alterations to the ZPF.

We are monkeys just out of the trees. And for us to be so arrogant as to imagine we're close to understanding the universe is just insane.
Dean Radin

Biologist Alexander Gurwitsch (1874-1954) believed that weak radiation of ultraviolet photons from tissues (what he called *biophotons*) stimulates growth in nearby tissues of the same organism. All living things do emit photons, although there is debate about whether the strength of these emissions is enough to cause change in other cells, and whether healthy emissions are coherent and periodic, while cancer is disordered and multiple sclerosis is too ordered.

Mystics see the ZPF that physicists cancel out of their equations as proof that we live in a sea of light. To them, the ZPF could provide a scientific explanation of spirituality, because it can be seen as the be all and end all of everything in the universe. Some see it as the Akashic record. Some say that the imprint of missing limbs on the ZPF could be the cause of phantom limb sensations in amputees. Some even think that the ZPF is what holds everything together, and that we may need to rethink just what life is. Life could be made of quantum energy exchanging information with the ZPF. This conclusion is highly debatable, but may not by as farfetched as it seems, because our senses and thoughts depend on quantum processes to some extent.

Every particle, every field of force, even the space-time continuum itself, derives its function, its meaning and its very existence directly or indirectly from apparatus-elicited answers to yes or no questions.
John Archibald Wheeler

Implications

The Royal Society of London was founded in 1600 with the motto *nullis in verba*, which literally means, "Take no one's word for it." This is not to say that theory implies uncertainty; all valid scientific theories must have many predictions validated without a single contradiction. One such contradiction or incorrect answer requires modifying or even abandoning the theory, which means that no theory is ever totally reliable.

Both quantum mechanics and Einstein's relativity (which we will discuss in Chapter 24) make it clear that space, time, matter, and energy depend on the observer. The classical theory of objective reality and the realist theory that a world exists independent of observation have been thoroughly falsified. Quantum uncertainty, wave/particle duality, and superpositions are all staple ingredients of quantum mechanics that are not paradoxical, if one sees them as implying that the universe is not deterministic, and that reality is wavelike at its most basic level. Materialism is nothing more than a hypothesis that continues to be rammed down our collective throats as if it is

Matter has reached the point of beginning to know itself. [Man is] a star's way of knowing about stars.
George Wald

real; all we can say is that it is easy to see where they are coming from without validating their beliefs.

We may not have all the answers yet, but that did not stop us from producing 100 billion transistors every second in 2006—all of which depend on quantum mechanics—or from reaching the point where half of our economy depends on products that are based on quantum mechanics. If nothing else, quantum mechanics is forcing us to reassess existence itself.

The existence of existence is amazing, awesome.
Gerald L. Schroeder

According to Bohr, "The discovery of the quantum of action shows us, in fact, not only the natural limitation of classical physics, but by throwing a new light upon the old philosophical problem of the objective existence of phenomena independently of our observations, confronts us with a situation hitherto unknown in natural science."

I mentioned earlier that quantum effects are being seen in larger and larger objects. Even the huge metal bars being used to detect gravity waves must be measured as quantum objects. Cosmologists are writing waveform equations for the entire universe to study the Big Bang. Weinberg believes that quantum mechanics will remain essentially unchanged in any final Theory of Everything.

Is the universe deterministic or probabilistic? The answer depends to some extent on how we frame the question. Professor Timothy Ferris (1944-) puts it this way: A close-up of a hammer hitting a wire seems deterministic. Pulling back to see 88 sets of hammers and wires seems voluntary. Pulling back to see a player piano seems deterministic. The composer who wrote the original melody seems voluntary. On and on we go.

I postulate that we are witnessing—and indeed participating in—a creative act that is taking place throughout time.
Louise B. Young

I must, in fairness, confess that the inadequacy of current materialist explanations does not make any non-materialist system true; however, reality does seem to be both transcendent and immanent. All living things have survival as their top priority. The evolution of senses good enough to aid in the quest for survival by detecting externalities has turned inward in at least one case, which of itself is no more astounding than aiming a camcorder at the TV screen it is hooked up to; however, if everything in the universe depends on observation for its existence, then everything is self-referential. We have good reason to suspect this may be the case, because mathematical patterns always have a reason.

Quantum mechanics may or may not point the way to eternal life, but we must always remember that eternal life itself is meaningless, because time has no effect on meaning and is not the place to look. In fact, as we will see in Chapter 28, time may be totally irrelevant.

Physics is nothing but the ABC's. Nature is an equation with an unknown, a Hebrew word which is written only with consonants to which reason has to add the dots.
Johann G. Hamann

Chapter 24

It's All Relative

> *The current recognition of the theory of relativity is that geometry is a construct of the intellect. Only when this discovery is accepted can the mind feel free to tamper with the time-honored notions of space and time, to survey the range of possibilities available for defining them, and to select that formulation that agrees with observation.*
> Henry Margenau

Long before Einstein, Descartes postulated that an object only moves in relation to other objects, being careful to present his belief in terms that would not arouse the suspicions of the Inquisition. Newton believed in an immovable space within which all motion occurs, which helped introduce the concept of absolute time and motion. Leibniz argued against this idea, lest God be led into a conundrum. Later, physicist Ernst Mach (1838-1916) removed the power of space-time and vested it in the actual contents of the universe, saying that time is an abstraction that we think we see based on how things change. This caught Einstein's attention. The resulting theories of relativity show time being influenced by matter. According to general relativity, this can mean that time itself stops under extreme conditions. Einstein's general and special theories of relativity showed that Newton's ideas were flawed, that space and time are not separate, but are simply different aspects of each other—a conclusion that defies common sense.

As the name implies, relativity is indeed relative. Motion of one body can only be measured relative to motion of another body; however, this only becomes apparent under extreme conditions, such as high speed or very strong gravity. Newton's classical physics remains a fantastic approximation under ordinary circumstances. There is a key difference at all speeds, which may seem largely semantic in our daily lives, but is actually very important: We do not sense motion, but rather changes in motion. It is impossible to measure absolute velocity because anyone traveling at a constant speed thinks s/he is at rest. Relative velocities are the only meaningful speeds.

Newton believed that space is space, even when empty. This space provides a place in which material objects can move. Mach disagreed, saying that a person spinning in an empty universe would feel nothing, while a person spinning in a universe that is less dense than ours would feel reduced outward pressure, because forces are based on the combined effects of all matter in the universe. Thus, all motion is relative. Einstein's theory of relativity has contributed enormously to both physics and astronomy over the 20th and 21st centuries, because it supersedes Newton's theories of mechanics. Relativity revolutionized science. It was later demonstrated that Einstein's equations contain Mach's principles, albeit so well hidden that Einstein didn't see them.

Einstein came up with his theory by imagining what it would be like to travel at light speed. Maxwell's work did not include any standard of rest against which to measure light speed; it therefore seemed that light moves at 186,000 miles per second relative to everything else. In other words, light moves at an absolute speed relative to all observers, while everything moving slower than light speed moves relative to everything else. This astonishing idea only works if one abandons the idea of fixed space and time that are treated separately. Time is now understood to be a fourth dimension that unites with three-dimensional space to form a curved space-time. The perception of time now depends on velocity, and this contraction can become a significant factor at high enough speeds. The impact of relativity extends across all of the physical sciences, and has left changed theories and methods in its wake.

Einstein published two versions of his theory. The first, special relativity, appeared in 1905, and applies to particles and interactions between them. It was largely accepted in physics by 1920, and became in indispensable tool in atomic, nuclear, and quantum physics. The second, general relativity, appeared in 1916, and works in cosmology, astrophysics, and astronomy. It didn't seem to be of much use, since it was perceived as only

God does not play dice with the universe.
Albert Einstein

Einstein, don't tell God what to do.
Niels Bohr

making minor corrections to Newtonian gravitational theory. Its full import did not become apparent until the 1930s.

There initially appeared to be little applicability for experimenters, as most applications were for astronomical scales. It also seemed limited to only making minor corrections to predictions of Newton's gravitation theory and contained incredibly dense mathematics. New mathematical techniques made calculations easier, and revealed visible concepts. This coincided with the discovery of astronomical phenomena that made general relativity very relevant, such as quasars, the cosmic background radiation, pulsars, and black holes.

Special Relativity

The little formula [e=mc²] sums up all action and creation in the universe.
Nigel Calder

Special relativity deals with the structure of space-time by replacing Galileo's classical mathematical transformation with those by physicist Hendrik Lorentz (1853-1928) that describe how two observers' individual measurements of space-time can be converted to each other's viewpoint. In very simple terms, different observers moving at different speeds may measure different distances, elapsed times, and even different sequences of events, thanks to two postulates that are contradictory in classical physics but complementary in relativity:

- All observers share the same laws of physics when in uniform motion relative to each other.
- The speed of light in a vacuum is the same for everyone, regardless of relative motion or the motion of the light source.

Let's take a look at some of the many surprising consequences of special relativity.

All Together Now

This is the authentic content of the doctrine of eternal recurrence: that eternity is the now, that the moment is not just the futile now, which is only for the onlooker, but the clash of a past and a future.
Martin Heidegger

Observer A sees two events occurring simultaneously; however, these events may not appear to occur simultaneously to Observer B if both observers are moving relative to each other because each observer "slices" time differently. (See Chapter 28.)

Unification of Space and Time

I have already mentioned that relativity unifies space and time and eliminates the possibility of a single, universal time in favor of personal times that are unique to each observer. Length, width, and height are not independent; they can be transmuted into one another. For example, we need only turn in space to make an object's height its width. We are all familiar with how changing our viewing angle changes spatial references. What we probably do not know is that our velocity transforms some spatial distances to time, and vice-versa.

Space and time are both changeable and dynamic, meaning that the form of space-time depends on the matter it contains. Space-time is not a simple, straight-edged, four-dimensional hypercube; it curves under its own built-in tensions that are caused by gravity. Put it this way: If space-time is a novel, then the book is one of the characters. Each of us sees space and time differently depending on our motion, meaning that the subjective experience of space and time is in the eye of the beholder. Each of us experiences the passing of time differently.

There was a young lady named Bright who traveled much faster than light. She set out one day, in a relative way, and returned on the previous night.
A. H. R. Buller

Time Dilation

Moving clocks have been proven to tick more slowly compared to slower-moving reference clocks. For example, two atomic clocks were synchronized. One was then placed aboard a U.S. Navy aircraft and flown above Chesapeake Bay for several hours. Upon landing, the clock in the plane was found to have ticked more slowly than the one on the ground.

It is important to note that the discrepancy only occurs when comparing different reference points. As far as each clock is concerned, it is ticking at the same speed, no matter what happens. The clock in the airplane literally logged less time than the clock on the ground. This difference was extremely tiny, but very real.

Imagine two identical twins who are born at the exact same moment and who will lead identical life spans, with both of them dying precisely at midnight on their 80th birthdays. Twin A remains on Earth, while Twin B is instantly accelerated to 80% of light speed for his entire life. Twin A experiences 80 years of life, and nothing seems amiss. Twin B experiences the

So little do we understand the time that perhaps we ought to compare the whole of time to the act of creation.
James Jeans

Space is not an entity.
Siddha Nagarjuna

same thing. One minute before his death, Twin B returns from his voyage, having lived 79 years, 364 days, 23 hours, and 59 minutes that seem to him to have taken no more or no less than the same time experienced by his twin. Imagine his surprise upon stepping out of the spacecraft and realizing that 136 years have passed on Earth! Twin B has been dead for 56 years. Both twins have lived 80 full years, but one has taken 136 Earth years to do so, while the other has only seen 80 of those years.

This is not mere speculation or mathematical sleight of hand. Relativistic time dilation makes some particles seem to live much longer than they should, which gives us more time to study them. If the particles could see and think, they would see themselves living their normal spans and see us moving extremely slowly.

Length Contraction

Be it clearly understood that space is nothing but a mode of particularization and that is has no real existence of its own. Space exists only in relation to our particularizing consciousness.
Ashvaghosha

This discrepancy in time must be accounted for, because Nature must always balance her books. The key to doing this lies in the unification of space and time, where distance cannot be treated as a quality that is distinct from time. Distance is simply the time light takes to travel from Point A to Point B in a vacuum. The answer is both simple and fantastic: Objects appear shortened in the direction they are moving relative to the observer. A speeding object converts some of its motion though time into motion through space. Increasing speed slows down the flow of time (as viewed from an outside perspective), while the combined total of space and time remains the same.

An object traveling at light speed experiences neither distance nor a flow of time. Moving at light speed leaves no room for moving through time, which is why c is the ultimate speed limit. A simple visualization can help make sense of this: A loaf of bread remains the same no matter how we slice it. Absolute space and time don't exist, in the sense that we can cut the loaf any way we want it; however, absolute space-time does exist, because the loaf remains the same. The following diagram shows the path through space-time for an object moving in a straight line at a constant velocity, moving in circles at a constant velocity, or accelerating at constant velocity.

Einstein saw past, present, and future as existing at once, even though time travel to the past seems to be impossible. Time, like everything else, turns out to be just another illusion. Do we each go on living inside our own personal time references? Does time have any role to play at all? We will look at these questions more in Chapter 28.

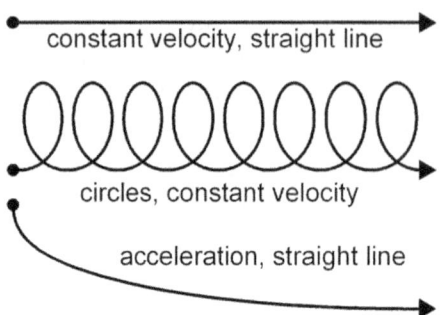

Mass–Energy Equivalence

Having erased the distinction between space and time, it seems only fitting to take the next step and erase the distinction between mass and energy. As we have already seen, the famous equation $e = mc^2$ literally means that energy and mass are both equivalent and transmutable. Even the most solid-looking matter is nothing more than insubstantial energy patterns. This one equation sums up both action and creation in the universe.

The four-dimensional space-time manifold is only a fabrication, only a theory.
John Wheeler

Mass and energy are also relative. For example, the amount of energy being emitted by a star depends on who is measuring it. To different observers, the star is either losing or gaining mass. From our perspective on Earth, our own Sun is losing 4 million tons of mass per second as light, which carries that mass converted into energy, and thus keeps Nature's books balanced.

Ultimate Speed Limit

No physical object, message, or anything can travel faster than light, period. The only catch is that light speed is not relative; all observers see light as moving at light speed. For example, a person traveling at 75% (or any fraction) of light speed shining a light out the front of the spacecraft would see the light s/he is emitting as moving away from the spacecraft at light speed.

I believe that in every instant we experience creation directly.
Julian Barbour

A person on Earth would see the spacecraft moving at 75% of light speed, with the emitted light moving away from the ship at 25% of light speed, for a total of 100%. The difference is that the person in the spacecraft would see the light shifted toward the blue end of the spectrum, because s/he is moving toward it, for the same general reason that the siren of an approaching ambulance seems to have a higher pitch. If the light were shining out the back of the spacecraft, then the person in the ship would see the light moving off behind them at light speed. This light would be shifted toward the red end of the spectrum, for the same general reason that the siren of a departing ambulance seems to have a lower pitch.

I am led to believe that everything we have seen... exists within us.
Denis Diderot

Red and blue shifting occurs because light cannot travel any slower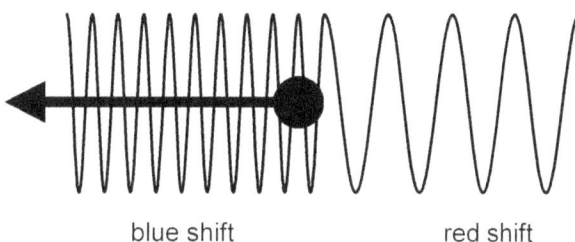
or faster than light speed. Blue light has a shorter wavelength than red light, which is why moving toward a light source will cause blue shifting. Red light has a longer wavelength than blue light, which is why moving away from a light source will cause red shifting. The discovery that starlight is red-shifted was one of the clues that led to the Big Bang theory, as we will discover in Chapter 26.

General Relativity

This requirement of general relativity takes away from space and time the last remnant of physical objectivity.
Albert Einstein

Einstein developed general relativity during the years 1907 to 1915. He began with the *equivalence principle*, which equates being under accelerated movement and being at rest in a gravity field. According to this principle, free fall is inertial movement. The object is falling because objects move that way when no force is acting on them—an important distinction from both the classical and special relativistic definition that the fall is occurring because of gravity. These theories do not allow inertially moving objects to accelerate relative to each other, but free-falling objects can do just that. Einstein solved this problem by proposing that space-time is curved, and created the field equations that relate space-time curvature with the mass, energy, and momentum contained therein.

Einstein realized that gravity is instantaneous, which contradicts special relativity and the upper speed limit. He resolved this by seeing gravity and accelerated motion as two sides of the same coin. Gravitational forces and acceleration forces feel the exact same, thanks to the equivalence principle. Someone subject to gravity is always accelerating. Only people who do not feel any forces can say they are not accelerating, and theirs are the only valid references for discussing motion. Thus, Newton's head came up to meet the apple instead of the other way around.

Curves in Space

Another of Einstein's many Eureka moments came when he realized that 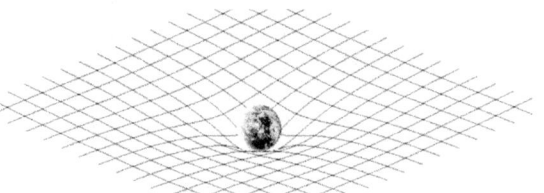 matter distorts space. He was then able to use general relativity and astronomical information about star density to study cosmology and determine the shape of the whole universe. Gravity can be explained as curves in space-time. For example, a marble on a warped floor will follow a path based on the curves in the floor. Space-time without mass is flat, but the addition of mass adds curvature, as shown here. It is because of this curve that you feel heavy; you are trying your level best to fall to the center of the Earth. Your muscles are resisting this force, and the surface of the Earth is keeping you from going too far. Avoiding falling is what gives us our sensation of weight.

There may not be any locations in the universe that are more subtly or delicately organized than human brains.
Julian Barbour

Curved space-time does not work well with traditional Euclidean geometry that deals with flat surfaces. Mathematician Bernhard Riemann (1826-1866) developed a theory of higher dimensions and a geometry capable of dealing with curved surfaces. These equations were a huge help to Einstein.

The Expanding Universe

Before Einstein, the universe had been thought of as existing in a steady state, with no firm date of creation or beginning. Einstein was shocked to discover that the universe is actually expanding. Evidence of a moment of natural or divine creation was the last thing materialist science expected to dis-

Ye gods! Annihilate but space and time, and make two lovers happy.
Alexander Pope

cover, but there it was. Einstein was so opposed to this idea that he introduced a number called the *cosmological constant* to zero out the expansion, a decision he would later call the biggest blunder of his life.

Linking Space and Gravity

So far, we have united matter with energy, space with time, and mass with gravity. We must now unify space and gravity. In simple terms, space-time is the incarnation and embodiment of gravity. Gravity is not a field within space-time; it is space-time itself, and should be treated as warping of space-time. If gravity created matter as some suspect, then space-time itself created matter—which begs the question of how space came into existence.

> *One feature of our physical world is so obvious that most people are not even puzzled by it—the fact that space is three-dimensional.*
> Heinz Pagels

Implications of General Relativity

Some of the implications of general relativity include:

- Clocks run more slowly in areas with lower gravitational potential, i.e. when in areas with lesser amounts of curvature. Here again, the clock is not running any slower or faster as far it is concerned; the difference is relative to observers. This is called *gravitational time dilation*.

- Orbits precess (wobble) in ways that Newton's theory of gravity cannot explain; general relativity explains this.

- Gravity bends light.

- Rotating masses "drag" nearby space-time with them; this is called *frame dragging*.

- The universe is expanding; the most distant parts are moving away from us faster than the speed of light, which means we cannot see into those regions. (Nothing is moving through this space faster than light; it is space itself that is expanding.)

- The red shift of a galaxy is directly proportional to its distance from Earth.

- General relativity predicts the point at which it breaks down. We will see more of this in Chapter 26.

> *The now is what is real.*
> Janet Baker

The Divine Savage
Revealing the Miracle of Being

Chapter 25

Strings, Branes, and Loops

Anyone who is not shocked by quantum theory has not understood it.
Niels Bohr

As I indicated in Chapter 24, our knowledge of the universe is based on the two mutually exclusive and contradictory sciences of quantum mechanics and relativity. Most of the universe can be described in either set of terms, but not all. Quantum mechanics describes the matter in the universe, but does not cover gravity, space, or time. These subjects are squarely in the realm of general relativity, which is almost completely independent of the quantum. Understanding the origins of the universe will require us to resolve these two disparate sciences in a way that incorporates everything we already know. What we need is a quantum theory of gravity.

One might think that simply combining the two sciences would give us the information we need. The problem is that doing this to calculate a probability returns an infinite result. This is a problem, because anything over 100% is meaningless, which proves that the attempted marriage of the sciences is not working. The good news is that this conflict in the laws of physics probably means that there is a deeper truth awaiting our discovery. The search is on, but meanwhile the theories we have today do not give a full description of the universe.

The only laws of matter are those which our minds must fabricate, and the laws of mind are fabricated for it by matter.
James Clark Maxwell

One key problem is that we have to make the transition from a set of wild and chaotic conditions that we presume existed at the moment of the Big Bang to an almost perfectly homogenous period of inflation in order to yield the universe we see today, presumably without having to invoke a creator. We need a theory that can handle extreme conditions. Three candidates are currently vying for the lead:

- Superstring theory
- M (or brane) theory
- Loop quantum cosmology

Let's take a quick look at each of these theories.

Superstrings

Quantum mechanics includes an assortment of particles that come together to create protons, neutrons, and others. Superstring theory says that these particles are not tiny grains of matter, but bundles of energy that are far too tiny for us to probe directly. Vibrating these bits of string-like energy in different ways produces different particles; a string vibrating in a certain manner becomes an electron, while another becomes a neutron, and so forth. Reducing particle physics to a single object opens the door to unifying the strong, weak, electromagnetic, and gravitational forces, which must be done if we are to create a viable Theory of Everything. Space-time would no longer be seen as carrying gravity, nor the electromagnetic field as carrying electricity. Everything would be a single object whose vibrations create all of the forces and bits of matter on which they act. One easy way to see how a string can become different particles is to imagine a guitar string that can produce a wide range of notes by vibrating differently.

We already know that mass can be produced from energy, and vice-versa. In superstring theory, mass is simply the energy of the vibrating string, meaning that heavier particles result from stronger vibrations. If this speculation is correct, we could explain the apparent fine tuning of the universe as coming from the resonant vibrational patterns that the strings can reproduce—which begs the question of just why they can vibrate in these special ways and makes the anthropic principle even more interesting, at least in my never-humble opinion.

The real revolution that came with Einstein's idea was the abandonment of the idea that the space-time coordinate system has objective significance as a separate physical entity. Instead of this idea, relativity theory implies that the space and time coordinates are only the elements of a language that is used by an observer to describe his environment.
Mendel Sachs

In our knowledge of physical nature we have penetrated so far that... here is neither suffering nor evil nor deficiency but perfection only.
Herman Weyl

Grand Unification?

> *For some reason, the 10- or 26-dimensional string worlds of scientific theory are acceptable, but the "supernatural" realms of mysticism are judged to be mere superstition.*
> Bernard Haisch

String theory automatically includes all preceding theories; in fact, quantum mechanics and relativity require each other. At first glance, this seems like a marriage made in heaven, since we can combine the laws of both the very small and very large in a single framework. It gets even better, because the Standard Model of quantum physics has 19 parameters that can be adjusted to fit results, while string theory has none. The question of why we keep finding so many subatomic particles as we probe is also answerable, because a string is about 100 billion billion times smaller than a proton. If we could magnify a particle enough, we would find that it is simply a bit of vibrating string, which eliminates the seeming chaos of having so many particles. The four fundamental forces also fit into this framework, because they too are based on vibrating strings. The entire cosmos becomes a grand symphony.

In Chapter 23, we learned that true emptiness is not compatible with quantum mechanics; the ground state can flutter around 0, but can never be exactly 0. This is what we call the *vacuum fluctuation*, which proves that there is no such thing as nothing in the classical sense. What seems like empty space is full of activity. We can only wonder whether quantum mechanics applies beyond the universe to give us the bubbling cauldron that theorists believed set off the Big Bang.

> *Common sense has been tested as a guide to quantum mechanics and been found wanting.*
> John Gribbin

Space at lengths smaller than the Planck length is full of activity. Relativity theory calls for curved space-time but runs into Heisenberg's uncertainty, which predicts turbulence at these very small scales, almost like a fractal that keeps zooming in to reveal ever more detail. This potential problem with two of the cornerstones of quantum mechanics is avoided, because the fluctuations end at the string length, which happens to be the Planck length that limits just how small something can be. This limit is just enough to both avoid conflict between quantum mechanics and relativity and to join the two theories, which could be evidence of even more cosmic tuning. Superstring is the first theory of gravity to contain finite corrections; all prior theories failed this key criterion. Applying the constraints that strings place on space-time reveals Einstein's equations, which means that string theory allows one to derive relativity.

All of this hints that space-time itself may be only an approximation of something even more fundamental, and points the way to new concepts awaiting our discovery. The universe just keeps getting grander.

Higher Dimensions

Abbott wrote a fascinating tale about an attempt to get a two-dimensional creature living in the imaginary Flatland to comprehend the three-dimensional Spaceland, and what it means to go "up" and "down." This journey also takes us to one-dimensional (Lineland) and zero-dimensional (Pointland) space. As he wrote, "In Flatland thou hast lived; of Lineland thou hast received a vision; thou hast soared with me to the heights of Spaceland; now, in order to complete the range of thy experience, I conduct thee downward to the lowest level of existence, even to the realm of Pointland, the abyss of no dimensions."

To expect microscale system to conform to our macroscale intuitions is one of worst forms of bigotry and political incorrectness; it is scalsim.
Stephen D. Unwin

Our own situation is similar to the Flatlanders: We can imagine lower dimensions easily enough, but have a much harder time visualizing higher dimensions. For example, we tend to think of time as a one-way river that exists independently of space, when in fact there is no distinction between space and time. More on this in Chapter 28; meanwhile, one of the most interesting features of superstring theory is that it only works in a universe with 10 dimensions of space-time. (Some variations require 11 or even 26 dimensions.) How do higher dimensions work? The most common example used is a tightrope whose surface has large and small dimensions. This rope seems one-dimensional from afar, but an insect walking on its surface would see three dimensions.

If this is correct, then it would seem that life exists in three spatial and one temporal dimension, with the other 6 dimensions curled up tight. This flies in the face of thousands of years of mathematics that assumed the fourth dimension did not exist because we cannot see it, at least not as directly as we can see the lower three dimensions. This may also seem like a novel concept, since sting theory originated in 1969, but the idea of higher dimensions has been around for quite a while. Some 500 years ago, Kabbalists theorized that the universe expanded as God contracted after creating 10 dimensions, with only 4 of those dimensions being measurable, and the

To build matter itself from geometry—that in a sense is what string theory does.
David Gross

others being shrunk to sub-microscopic levels. This theory is based on the fact that Genesis contains the phrase "and God said" 10 times. The similarities between this Kabbalistic theory and modern science are too many to ignore, and reignite the questions about just how much our ancestors knew about the origins of the universe, as we saw in Chapter 3.

Mystics have no monopoly on theories of higher dimensions. Riemann said that forces have no lives of their own. They result from geometry, specifically wrinkles in the fourth dimension. Thus, forces are simply higher-dimensional geometry. In 1919, mathematician Theodor Kaluza (1885-1954) suggested that there may be more than three spatial dimensions. He rewrote Einstein's equations in five (not four) dimensions, and showed that adding this extra dimension retained all of Einstein's work while also adding Maxwell's theory of light to create a theory built on pure geometry. Kaluza and physicist Oskar Klein (1894-1977) theorized that light may consist of vibrations in the 5th dimension, and that its ability to cross the vacuum of space means that the vacuum is vibrating as well.

Star wars and string theory... neither ambition can be accomplished with existing technology, and neither may achieve its stated objectives. Both adventures are costly in terms of scarce human resources. And the Russians are trying desperately to catch up.
Sheldon Glashow

As I mentioned above, strings vibrate consistently in either 10 or 26 dimensions, depending on which flavor of superstring theory you prefer. A more robust version, M-theory, requires 11 dimensions. We will look at M-theory in a moment; meanwhile, one thing we can be sure of is that superstring theory is giving us unmistakable hints that we have only scratched the surface of whatever reality our universe is made of. New facilities, such as the Large Hadron Collider, may allow us to test some aspects of string theory. If proven correct, it will mean that what we see is just a thin layer over a very rich reality, a continuation of the ongoing discoveries revealing a far grander universe than we could have imagined beforehand.

Mass grips space by telling it how to curve; space grips matter by telling it how to move.
John Archibald Wheeler

Again, this is a purely geometric theory. Space at the Planck length may be a grid containing extra dimensions, which means that time itself could also be grainy (as evidenced by the Planck time). Trying to get any smaller is meaningless; either space and time stop at the Planck scale, or they turn into something even more fundamental. Another possible consequence of a purely geometric theory is that the universe may be a giant hologram, a concept we will explore more in Chapter 27.

Kaluza-Klein theory seemed arbitrary until string theory later corroborated their work. The equations of string theory also determined the shape of the added dimensions, which are called Calabi-Yau shapes after Eugenio Calabi (1923-) and Sung-Tung Yao (1949-). Our movement through three dimensions is actually movement through all of the dimensions, except that we don't notice this other movement. The number of dimensions constrains the types of vibrations a string can perform, a constraint without which superstring theory would fall apart. The limits imposed by 10-dimensional space-time were circumvented by adding an 11th dimension in M-theory.

There is open conjecture about how the universe split into 4D and 6D spaces. One idea is that the universe before the Big Bang was a region of unstable 10-dimensional space that cracked apart to form the universe we know today. If this is correct, then the Big Bang was a mere aftershock of a much larger event, in which space-time literally broke apart from the rest. The energy driving the expansion of the universe is found within 10-dimensional space itself, and the red shift is thus caused by this energy. This theory predicts that our 4-dimensional universe has a tiny 6-dimensional cousin that is too small to be observed. This universe could be our salvation; we can speculate about our future descendants being able to tunnel into these higher dimensions to watch the rest of the universe crunch back into a singularity and explode outward again. (Of course, who says it hasn't happened already?)

Why did some dimensions curl up but not others? The overall goal of superstring theory is to explain all four fundamental forces in one theory, in the same way that Einstein described gravity as being a disturbance caused by matter. In superstring theory, the vibrating strings can curl space into very tiny loops, to where we cannot observe the 6 higher dimensions except through the effects they have on particles; however, the strings themselves are nothing but vibrating space waves. Which begs the question: Is this all there is?

Limitations of String Theory

String theory has a lot of promise, but it's way too early to declare victory. For starters, there are millions of ways for a 10-dimensional universe to break and curl up. Why did the universe settle on this state? This seems like a fundamental

The underlying unity of the natural world is, we have seen, evidenced in its universal embedded rationality, which the sciences assume and continue to verify.
Arthur Peacocke

String theory is twenty-first century physics that fell accidentally into the twentieth century.
Edward Witten

question that exists because nobody has yet solved string theory. We find ourselves in the frustrating position of staring a TOE straight in the eye while being unable to solve it.

Supersymmetry is the ultimate proposal for a complete unification of all particles.
Abdus Salam

The concept of *supersymmetry* in particle physics refers to the process of relating elementary particles with one spin to other particles whose spin differs by half a unit, and are called super partners. A theory with perfect supersymmetry means that every type of boson must have a corresponding fermion, with the same mass and internal quantum numbers. The reverse is also true. A *fermion* is a particle that obeys Fermi-Dirac statistics, while a *boson* obeys Bose-Einstein statistics. Only one fermion can occupy the same quantum state at a time, while several bosons can do this. This feature of bosons often makes them force carrier particles, while fermions are usually associated with matter. (The distinction between matter and energy is tenuous at best, but still.)

There are five ways to incorporate supersymmetry into string theory called Open Type I, Closed Type 1, Open II-A, Open II-B, and heterotic. Heterotic theory has two sub-types called SO(32) and $E_8 \times E_8$. All of these are simply different ways of describing and analyzing the same theory. Physicist Edward Witten (1951-) discovered M-theory as a way to unify the different string theories and provide a way to move between the theories as a sort of Rosetta stone. This is not a complete unification, but is a huge advancement nonetheless. Even so, there are more solutions to superstring theory than there are protons in the entire universe, which renders string theory both an amazingly beautiful and powerful theory and utterly useless for practical purposes because it cannot make any kind of testable concrete predictions or restrict unknown phenomena.

Insane in the M-Brane

The universe must be viewed as a quantum system not merely in its early stages but at all scales and times.
Menas Kafatos

Witten drew on the work of leading string theorists to devise M-theory, and used it to explain many previously observed dualities. This sparked renewed interest in superstring theory. What Witten showed is that the various types of superstring theory are related by dualities that allow physicists to relate an object's description in one string theory to the description of a different object in a different theory. This indicates that the superstring theories are different aspects of a single underlying theory that Witten called M-theory. The "M" originally stood for "membrane" (or "brane" for short), because branes are constructs used to generalize the strings from the various string theories. Witten was noncommittal about the use of the word "membrane," and said that one can interpret the letter "M" to signify most anything s/he wants.

M-theory extends superstring by adding an 11th dimension to supersede and unite all five string theories. String theory can also work with membranes of 2, 3, or up to 11 dimensions. The universe itself may be a giant membrane, which would be compatible with holographic theories. The three spatial dimensions we know and love may be a 3-brane, which would make space-time just as real as a particle. Higher dimensions may be large, but we don't see them because of our limited viewpoint. Adding time means we are stuck in a 4-brane, but not totally cut off from the other dimensions.

Gravity moves across all dimensions, and could potentially be used as a means for detecting the higher dimensions. For example, any deviation of gravity from the inverse square rule originally set forth by Newton would be a telltale sign. We have not yet detected these waves, because the distances involved are so small that quantum effects pose significant interference; however, the extra dimensions could be as large as 0.1mm, and we would never know it. If the higher dimensions were larger, then gravity could be much stronger and the strings much larger, because gravity is spread across all dimensions.

There is a cyclic M-theory model, where branes collide and restart the universe in a grand cosmic rebooting that sounds awfully similar to Buddhist concepts of a cyclic universe; however, this leaves too much unexplained, such as how the cycles got started, since we don't want to get into a case of infinite regression.

Thrown for a Loop

The gravitational force governs the universe across large distances. Once stars, and galaxies, form, the resulting gravita-

Myth is neither a lie nor a confession: it is an inflexion.
Roland Barthes

The views of space and time which I wish to lay before you have sprung from the soil of experimental physics, and therein lies their strength. They are radical. Henceforth space, by itself, and time, by itself, are doomed to fade away into mere shadows, and only a kind of union of the two will preserve as an independent reality.
Hermann Minkowski

tional interactions hop into the driver's seat. We do have a solid gravitational theory in the form of relativity, which is in perfect accord with all observations to date. We also know that this theory is not complete, because we have not yet unified relativity with quantum mechanics. This much is well established. Of course, any successful combination would represent a tremendous leap forward in our understanding, particularly regarding what happened during the Big Bang. This Theory of Everything is the holy grail of science.

The grand objective of physics is to understand the universe in terms as simple as possible.
Nigel Calder

There is an emerging field of study around quantum gravity, many areas of which remain speculative and not yet in the realm of theory, despite plenty of successful mathematical testing. Of these, loop quantum gravity (also called loop quantum cosmology, or LQC for short) provides a framework in which we can say that a universe existed before the Big Bang. It also lets us take an educated guess about what that universe might have looked like, based on its influence on our current cycle's expansion. Such a theory should even be testable, given observations of enough sensitivity. This sounds promising, so how do we go about building such a theory?

There are always two solutions to Dirac's wave equation for every solution to the Schrödinger wave equation that have different energy signs (and possibly charges as well), but that agree on other properties, such as mass. One of the most cherished core beliefs of theoretical physicists is that every solution to a theory must either be consistent with a mathematical principle, or correspond to something real. Dirac's combination of quantum mechanics and special relativity predicted a whole new type of matter, specifically that every known particle must have an anti-particle with the same mass and opposite charge. These particles have since been found, all from a mathematical exercise that combined two theories.

The Infinity Problem

As they are currently formulated, general relativity and quantum mechanics cannot both be right.
Brian Greene

Will the universe expand ad infinitum or eventually reverse course and collapse? If the amount of matter in the universe falls below a critical density, then gravity will not be strong enough to stop the expansion. If it is above the critical density, then gravity will at some point be able to stop the expansion and make the universe collapse back on itself into the Big Crunch. It just so happens that the density of the universe is very near this critical point, and we are unable to know for sure what the ultimate fate of our universe will be.

Matter cannot have infinite density, because there is only a finite amount of matter that can only be packed so tightly; however, our equations indicate that space-time curves by an infinite amount to form a singularity, an infinitely dense, zero-dimensional point. The mathematical possibility of singularities therefore poses a serious problem that will either force us to abandon general relativity or face the challenge head-on by figuring out just how a singularity works and what—if anything—lies on the other side.

For practical purposes, our current mathematics need only be compatible with the universe of today. Virtually all of these theories include at least one singularity where space and time end, and where general relativity breaks down. This happens because general relativity does not contain any mechanism for reversing the attractive nature of gravity. General relativity allows matter to influence both space and time by curving it, as we saw in Chapter 24. This literally turns gravity loose on space and time, which must now go along with every whim of matter without any means of stopping a full collapse. This is why the search for a valid TOE remains ongoing. Previous theories have focused on the quantum nature of matter, while ignoring the quantum nature of gravity, space, and time. This may seem a little hard to believe, since matter and energy are transmutable, as are time and space—but this is the current theoretical landscape.

As we saw in Chapter 23, quantum mechanics describes matter as smeared out waves with no definite boundaries, instead of particles. The Schrödinger equation describes how a waveform evolves over time, but does not include Einstein's realization that space and time are complementary. The two theories are therefore fundamentally incompatible. Furthermore, the Planck length is very small but not zero, which demonstrates that certain conditions such as the Big Bang can only be properly understood within a quantum theory of gravity.

If energy at small wavelengths and/or distances does not behave as predicted by general relativity (a non-quantum theory), then we should expect a modified form of space-time as determined by matter, where the curvature and implied gravi-

---

In the beginning was the great cosmic egg. Inside the egg was chaos, and floating in chaos was P'an Ku, the divine embryo.
P'an Ku myth

A remarkable feature of 'random' numbers-numbers that pass all the tests is that two such numbers may be related to one another in a nonrandom way.
Heinz Pagels

tational force do not behave classically. This is of special importance in cases such as the Big Bang or black holes. The stability of atoms and finite nature of black body radiation indicate that this may in fact be the case.

Unifying Space, Time, and Matter

LQC gives space, time, and matter equal—but not entirely unified—treatment. Applying quantum concepts directly to space-time yields a structure that is similar to that of matter. This geometric approach is important, because geometry is considered even more fundamental than what that geometry contains. In LQC, everything (space, time, and matter) consists of a fluctuating mesh with internal relationships that we perceive as change. This is accomplished by simply applying quantum concepts to the space-time defined by general relativity.

We know that there is not one time and one space only, but that there are as many spaces and times as objects.
Jacob von Uexkull

Matter is made of atoms. A region of space-time without atoms is in a vacuum state that LQC treats as the foundation on which all configurations of matter rest. LQC treats space as being atomic; the only difference is that spatial atoms are abstractions, some properties of which can be described by giving them a one-dimensional loop shape instead of a spherical shape, hence the name "loop quantum cosmology." In this model, any volume of space can only grow in discrete steps by either adding or exciting spatial atoms. The total volume of the universe can expand without the need for an external source, meaning that the universe can grow by itself and within itself. The interactions between space, time, and (possibly) matter as the universe expands excites spatial atoms that subdivide once they get big enough to form new spatial atoms.

All descriptions of reality are temporary hypotheses.
Buddha

Space does not exist independently of spatial atoms; however, even the lack of space is a non-zero physical state that is distinct from both empty space and the material vacuum. Remove all the spatial atoms, and neither space nor volume remains. This state is almost mind-bendingly empty, with no lights, no sound, no matter, and no space, and with only time offering a potential escape. This state, as distinct from a matter vacuum, would be a state of infinite temperature. Physicist Klaus Fredenhagen (1947-) called this the "State of hell." LQC must deal with this state somehow, and it turns out that this state is closely linked to the problem of singularities where space has completely collapsed. Even regions with many spatial atoms retain an imprint from their earlier hellish state.

Loops combine to form a framework that contains area, volume, and distance in a lattice structure that both is space and that supports space. Surface area is determined by the loops that intersect that surface. Many intersections of loops create the space we are familiar with. The rubber sheet used to illustrate gravitational bending of space-time becomes an intricate network of rubber bands. All of space is full of loops; even small volumes contain many billions, trillions, or more loops. This mesh is always shifting, which creates time. This mesh is also subject to quantum laws, meaning that space is very active at very small scales. We generally think of space as a medium for matter and energy; however, everything resides in this mesh of loops. Nothing moves from loop to loop, because there is nothing in between these loops. Rather, it is the ever-shifting connections between loops that move matter and energy, spatial atom by spatial atom.

Absurdity of absurdity of absurdity is equivalent to absurdity.
Luitzen Brouwer

Infinity Avoided

The volume of the universe has the same quantum uncertainty as matter. Thus, it seems to expand smoothly when seen from afar, but a closer inspection would reveal small irregularities as the universe grows in jumps, loop by loop, with each loop being roughly the size of the Planck length. Growth seems to be continuous, because the Planck length is extremely small.

The process of adding space loop by loop must begin at some initial state; removing loops is one way to get us there, which eventually leaves us in Fredenhagen's hell. The loop structure of space requires us to add quantum corrections to Einstein's equations. The continuity of relativity creates the relativity problem; quantum gravity is discontinuous. Standard quantum equations also break down at this extreme. The Big Bang remains a mystery. Quantum gravity examines the actual size of each step of spatial growth and offers some key advancements by only allowing time and space to change in multiples of the smallest possible step. This change occurs where loops meet, but not in between loops. The universe thus has less time, because all time that slips between the loop mesh is eliminated.

When by patient inquiry we learn the answer to any problem, we always find, both as a whole and in detail, that the answer thus revealed is finer in concept and design than anything we could ever have arrived at by a random guess.
Fred Hoyle

The eternal silence of these infinite spaces frightens me.
Blaise Pascal

General relativity allows energy densities to rise to infinity as the universe shrinks, which goes against the idea of a lattice with limited storage space. This also goes against common sense, since the universe is not infinite. It is a cosmic version of the black body problem we saw in Chapter 23. Filling the lattice beyond its storage capacity forces it to act repulsively, just like a sponge that has absorbed too much water.

Similarly, a grid of time can only permit oscillations that are larger than the size of the grid because no smaller amounts of time exist. Smaller waves would vanish, because they would fall between the mesh. This prevents both overly rapid oscillations and overly large energies. Each time step is small (the Planck time), so the lattice can absorb a huge—but not infinite—amount of energy.

But wait, there's more: This structure does not simply prevent a singularity; it also implies a time before the beginning of our universe. This means that time does not begin at the Big Bang, and that it becomes meaningful to ask what happened before our universe came along. In other words, we now have a prehistory with a form of space and time that can only be explored using LQC, because quantum mechanics and relativity break down long before we get back this far. The universe before our Big Bang was shrinking and collapsing into ever-smaller sizes under its own weight, and eventually reaching the level of heat and density of the Big Bang, which was not a singularity. (The concept of the Big Bang as a point of infinite density arises from a limitation in our theoretical language that does not necessarily reflect reality, or even possibility.)

Having worked out a non-infinite solution to our own beginnings, we can now ask a series of questions such as:

How extraordinary that anything should exist.
Ludwig Wittgenstein

- Would a conscious entity in the previous shrinking universe perceive time backward, by remembering the future and looking back into the past?

- Was life possible in the previous universe, and/or did the conditions in the previous universe help make this universe fit for life?

- What caused the previous universe? Did it also arise from a collapsing ancestor? Was there an original Big Bang, and if so what caused it? Is there an endless string of universes stretching back to infinity?

- What will happen to our own universe? Will it go on expanding forever or will it eventually collapse and form the seeds of a new Big Bang for the next universe?

Universal Evolution

Quantum gravity acting repulsively prevents a singularity, saves space-time from annihilation, and gives us a peek into a universe before our own. As we saw in Chapter 23, knowing a waveform at any given time allows us to extrapolate forward or backward in time. Thus, knowing the state of the universe today would allow us to extrapolate back before the Big Bang using the equations permitted by LQC. Different equations give different results, but we can still get a decent idea of what the universe might have looked like before our Big Bang. The prior collapse and repulsive action of quantum gravity during the transition should leave an imprint on our expanding universe. The overall picture is of a balloon that deflates to a certain point and then reinflates.

The general picture we get is one of complete reversal, as if we were looking into a giant mirror. For example, a right-handed person who made it through the transition would find herself left-handed. Space itself would be turned inside out like a balloon that keeps on collapsing through itself and ends up expanding again. If this happens by sheer inertia, then inside is now outside as spatial atoms move through each other and completely invert space. The sizes of quantum fluctuations before the Big Bang are not necessarily the same afterward. The old value still influences our universe, but only to a very small degree, as if the universe has forgotten what went before. This "quantum forgetfulness" gives us a mixture of linear and cyclic views that makes the end of every cycle a new beginning. Each new universe is born in an extremely dense foam that breathes new life into it, just as we can recycle metal endlessly with each new form starting afresh. For example, the metal in a toaster can be used to make a car, which can then be used to make an airplane, and so on. The old value is still there, but is so small as to be irrelevant. Thus, some properties are born and die with one universal cycle, while some other properties may live on.

We cannot know for sure how many cycles have gone before us or whether there was ever a true beginning or simply infi-

Why sometimes I've believed six impossible things before breakfast.
Lewis Carroll

The universe is an evolving product of an evolutionary process. It is not an accident; it is an enterprise.
Theodosius Dobzhansky

nite cycles, but this is not terribly important, thanks to cosmic forgetfulness. That is not to say that such forgetfulness is unimportant; as it happens, it is very important.

No matter which logic one adopts, one has to come to terms with the fact that we are living in a very peculiar cosmos.
Robert Lanza

A universe that is experiencing expansion, collapse, and rebound in finite cycles allows the tiny influence of preceding cycles to help set up the very special conditions in our own universe that have allowed life to evolve. It is easy to see how many—if not most—of the preceding cycles could have been downright inhospitable, while others could have been perfectly friendly. Assuming infinite cycles, universes such as ours are inevitable; however, appeal to infinite regression is hardly a satisfactory explanation. Ah, but there is an alternative: The very earliest universes could have been both stable and very simple, with the subtle influences of each cycle adding more and more complexity until conditions suitable for inflation arose and made us possible, if not inevitable. (See Chapter 26.) Plenty of theorists have demonstrated how complexity can arise from simplicity in areas as diverse as chemistry, biology, evolution, and cosmology. Why not apply the same concept to a string of evolving universes?

What About Time?

Space and time... are names.
Siddha Nagarjuna

The idea of time travel could be based on a misconception. We experience motion. All motion occurs in space-time, which means that we cannot move only in space or only in time. We therefore cannot talk about moving through space independently of time, or of moving through time independently of space. All motion entails a chance of spatial position during a time interval. Given that space and time are inextricably linked, the only alternative to directed time would make all forms of motion impossible. We will examine this more in Chapter 28.

Time marches on as the universe expands, but what about when it stops expanding and starts to contract? The volume of space would decrease and develop backward. Time might do the same thing—an idea originally proposed long before the development of LQC by astrophysicist Thomas Gold (1920-2004). Then again, maybe not. Maybe time keeps marching forward as the universe grows ever warmer and denser.

Qumans

In *Once Before Time*, physicist Martin Bojowald (1973-) concludes by pondering the possibility of *qumans*, quantum beings who can live from universe to universe. This of course is speculation, a mere flight of fancy... or is it?

Being here is the supreme gift.
Julian Barbour

Chapter 26

Universal Origins

> *Why is there something rather than nothing?*
> Lucretius

Look up into the night sky some evening, and take a few moments to ponder the fact that some of the light you see set off on its journey hundreds, thousands, millions, or even billions of years ago. Some of the light you are seeing originated when humans were just leaving the trees for the African grasslands. Other bits of the light you see were in transit before the Earth was even formed. Stare up into the night sky, and you are staring back toward the dawn of time. How did we get here? Where did this all come from? Why is there something instead of nothing? Questions like these have absorbed countless mystics, theologians, and scientists for thousands of years.

We still don't know what happened at zero time, but we have gotten very close, to within incredibly small fractions of a second. We don't know why the universe exists in the precise form that it does, but we do know much about how. We don't know why the structure of the universe seems so complicated, or why it seems full of random ingredients in random quantities. We can't say for sure whether the entire universe—and our presence therein—is accidental or part of some hidden plan. We aren't quite sure why the universe has areas of high energy instead of a uniform distribution, though we do know that the darkness of the night sky is a powerful argument against an infinite universe. Some feel there are huge gaps in our knowledge that we cannot explain. Others feel these ques-

> *Nothing comes from nothing; nothing ever could.*
> Julie Andrews in
> The Sound of Music

tions have already been asked and answered one way or another. Let's take a detailed look about what we know—and don't know—about the origins of the universe.

Timeline

The generally accepted timeline of events in our universe's history is as follows:

- **Planck epoch:** This period extends from zero time until about 10^{-43} seconds after the Big Bang. At these extremely high energy levels, the four fundamental forces could have all had the same strength and been unified in a single fundamental force. Different theories propose different conditions during this era. For example, we saw in Chapter 24 that general relativity predicts a singularity, while LQC avoids this problem. More work is needed to generate a fully robust theory of what happened during this first time interval.

- **Grand unification epoch:** This period extends from 10^{-43} to 10^{-36} seconds after the Big Bang. Gravitation begins to separate from the other four forces.

- **Electroweak epoch:** This period extends from 10^{-36} to 10^{-12} seconds after the Big Bang. The grand unification of the four forces dissolves when the strong and weak forces separate from each other, which occurs when the universe cools to 10^{28} degrees Kelvin. This initiates inflation.

- **Inflationary epoch:** This period extends from 10^{-36} to 10^{-32} seconds after the Big Bang. We don't know the exact temperature or time for the inflationary epoch. What we do know is that the spatial curvature of the universe flattened, and the universe began a homogenous, even phase of extremely rapid expansion that sowed the seeds for future structures through tiny irregularities. Some particles formed, but decayed rapidly. Energies were still high enough at the end of the inflationary epoch to produce W, Z, and Higgs bosons. As this period ended, the leftover energy reheated the universe.

Looking up at the stars, I know quite well that for all they care I can go to hell.
W. H. Auden

The highest wisdom has but one science, the science of the whole the science explaining the creation and man's place in it.
Leo Tolstoy

We are stardust, for every carbon atoms in our bodies, every iron atom in our blood's hemoglobin was made in stars and scattered by supernovae explosions before the earth existed as a planet.
Arthur Peacocke

- **Quark epoch:** This period extends from 10^{-12} to 10^{-6} seconds after the Big Bang. If our theories are correct, particles obtain mass from Higgs bosons. The four forces (gravity, strong, weak, electromagnetic) were now operating as they do today; however, the universe was too hot to allow quarks to bind together.

- **Hadron epoch:** This period extends from 10^{-6} to 1 second after the Big Bang. The plasma of quarks and gluons forming the universe cooled until baryons, such as protons and neutrons, could form. Neutrinos decoupled about 1 second after the Big Bang and started traveling freely through space, forming a cosmic neutrino background that will probably never be observed in detail, but which is analogous to the cosmic background radiation that was emitted far later.

- **Lepton epoch:** This period extends from 1 to 10 seconds after the Big Bang. Most of the hadrons and anti-hadrons annihilated each other at the end of the hadron epoch, leaving leptons and anti-leptons as the dominant mass in the universe. The universe cooled to where no more lepton/anti-lepton pairs were being formed. Most of these pairs annihilated each other, leaving behind a small excess of leptons.

- **Photon epoch:** This period extends from 10 seconds to 380,000 years after the Big Bang. With most of the lepton/anti-lepton pairs gone, photons held most of the universe's energy. These photons continued to interact with charged protons, electrons, and atomic nuclei for the next 300,000 years.

- **Nucleosynthesis:** This period extends from 3 to 20 minutes after the Big Bang. The universe cooled to where atomic nuclei could form. Protons (hydrogen atoms) and neutrons began combining into other elements via nuclear fusion. This process continued for about 17 minutes, until the universe had cooled to where nuclear fusion could no longer continue. The universe at this point contained about three times more hydrogen than helium by mass, and trace quantities of other elements.

To sum up, if the earth did not lie in the middle of the universe, the whole order of things which we observe in the increase and decrease of daylight would be fundamentally upset.
Michael J. Crowe

- **Matter domination:** This period begins 70,000 years after the Big Bang. The densities of atomic nuclei and photons are equal. The Jeans length determines the smallest possible structures that can form during the competition between gravity and pressure. This length shortened, and minor irregularities in the universe started to grow. One theory has dark matter dominating this era and setting the stage for gravitational collapse to enlarge the tiny irregularities left over by inflation, which makes dense regions more dense and empty regions more empty; however, we do not yet have a theory about how dark matter formed to begin with.

- **Recombination:** This occurred about 377,000 years after the Big Bang. Hydrogen and helium atoms that were initially *ionized* (lacking electrons) captured electrons as the universe continued to cool, and the nuclei became electrically neutral. This process of recombination was fairly rapid. Photons were free to travel once the atoms had become neutral, and the universe was now transparent. These photons are what we see as cosmic background radiation, which means that this radiation is a snapshot of the universe as it existed at the end of this epoch.

- **Dark Ages:** This period lasted between 150 million and 800 million years after the Big Bang. The universe was opaque before recombination, but was now transparent, and neutral hydrogen was emitting very weak radiation. We are devising ways to detect this radiation, which could tell us even more than the cosmic background radiation, as it is theoretically an even more powerful tool for studying the early universe. The Dark Ages are currently thought to have lasted between 150 million to 800 million years after the Big Bang. Structure formation so far can be thought of as coming from the tiny irregularities that existed during the Big Bang, and that were not completely smoothed out during inflation. This is a purely linear model until this point, at which nonlinear structures begin to form.

- **Reionization:** This period occurred between 150 million and 1 billion years after the Big Bang. Structure formation since the Big Bang was hierarchical, and

It is this love of the contemplation of the eternal and the unchanging which we constantly strive to increase, by studying those parts of these sciences which have already been mastered by those who approached them in a genuine spirit of enquiry, and by ourselves attempting to contribute as much advancement as has been made possible by the additional time between those people and ourselves.
Michael J. Crowe

> *Since the time of Copernicus, almost 500 years ago, humankind has had to learn and relearn the salutary lesson that there is nothing special or privileged about the Earth.*
> Paul Davies

proceeded from small to large. The first large structures to form from gravitational collapse were quasars, which are believed to be early galaxies. These structures emit intense radiation, which reionized a universe that was primarily composed of plasma. The first stars formed and began turning the light elements left over from the Big Bang (hydrogen, helium, and lithium) into heavier elements. We have not yet observed any of these stars.

- **Galaxy formation and evolution:** Matter collapsed to form galaxies. The oldest known quasar is CFHQS 1641+3755, which is 12.7 billion light years away. In October of 2010, the galaxy UDFy-38135539 was measured as being 13.1 or even 13.2 billion light years from Earth. Our own Milky Way is estimated to be between 6.5 and 10.1 billion years old. Gravitational attraction pulled galaxies into groups, clusters, and superclusters. Our solar system formed around 5 billion years ago around a late-generation star with the debris from earlier generations of stars that initially formed the heavier elements.

- **Today:** Our best estimate is that the universe is 13.7 billion old, give or take 170 million years. Expansion is accelerating, which means that superclusters are probably the largest objects that will ever form. The current rate of expansion makes it impossible for any more inflationary structures to become visible, and prevents new gravitationally-bound objects from forming.

Our Expanding Universe

> *Twinkle, twinkle little star, I don't wonder what you are, for by spectroscopic ken, I know that you are hydrogen.*
> Ian D. Bush

One advantage of a steady-state universe is that we need not explain what has gone before, or even much about how the universe came to exist. This would be a very convenient thing. Unfortunately, that's not quite how it happened, as we saw above. For starters, we know that the universe is not both infinite and static, because the sky over our heads would be infinitely bright, and we would quite literally be fried. Astronomer Edwin Hubble (1889-1953) discovered that the universe is expanding, which strongly hinted at a starting point. As I mentioned in Chapter 24, Einstein had gotten indications that the universe is expanding during his work on relativity but added a cosmological constant to zero it out, because he did not like the implications of a starting point. He later referred to this as the biggest blunder of his life. I must pause to note that an expanding universe does not preclude a creator; it simply limits when that creator did the job.

It is important to understand that galaxies are not traveling through space. Rather, space itself is expanding. The light we see from these galaxies is shifted toward the red end of the spectrum, because it has to traverse this stretching space. The faster the galaxy is moving, the redder the light appears, until it eventually becomes invisible. This point of invisibility is called the *horizon*. To see an example of this, drop two stones into some water one after the other. The waves from the leading stone will pick up speed and move away from the waves created by the second stone.

The effort to understand the universe is one of the very few things that lifts human life a little above the level of farce, and gives it some of the grace of tragedy.
Steven Weinberg

The Big Bang

The universe has no hard edge, meaning that it is much like a bubble with no outer membrane. This is not to say that the universe is infinite, because one could conceptually reach the end of the universe and keep right on going. The universe also has no special time, no special position, nor even a center to speak of. This concept may seem obvious, but it took a long time for us to wrap our minds around it, possibly because of religious insistence that we are smack in the middle of God's creation.

The standard Big Bang model does not describe the origin of the universe, just the events that unfolded an instant afterward and from there onward. It also does not say anything about time itself. Adding inflation to the theory solves some problems with "pure" Big Bang theories, such as how the universe managed to begin with such low entropy. If the universe was a dot, then time and space began with the universe. If the universe is expanding into infinite space, then the Big Bang happened everywhere at once and caused space to expand without increasing in overall size, since it's already infinite. The standard interpretation says that the Big Bang created time, space, and matter, and that there is neither an "outside" to the universe nor a "before." Where did the Big Bang happen? Everywhere. The Big Bang did not move into space, it created space, and continues doing so today. What happened before

To say that the universe is infinitely old is to say that it had no beginning—not a beginning that was infinitely long ago.
Keith Parsons

Chapter 26
Universal Origins
679

the Big Bang? That is a meaningless question. Loop quantum cosmology (see Chapter 25) solves the infinity problem and gives us hints about what our ancestor universe might have looked like, but from our perspective time began with the beginning of our own universe.

Plain vanilla Big Bang theory explains Hubble's findings and the cosmic background radiation but has some problems, chiefly the flatness of space-time. There is no reason to expect space-time to be so flat, unless the Big Bang had precisely the correct density tuned to 1 part in 10^{55} per Guth's calculations; space should either be curled up in a little ball or have such extremely negative curvature that all matter would have been blown away from all other matter far too quickly to make any kind of structure. The universe would either collapse back on itself or expand far too quickly. The horizon problem presents more challenges, because it seems that the Big Bang exploded all at once, while normal explosions happen progressively. The universe is so homogenous that we must assume the entire explosion happened at once, which means that about 10^{83} "detonators" would have to fire all at once.

We also have the problem of how the universe could have appeared from nothing at all. As I mentioned in Chapter 13, the universe has 0 total energy, because gravity acts as negative energy. This means that the creation of the universe did not violate the law of conservation of energy. Some scientists believe the universe came from random quantum fluctuations, which begs the question of whether those fluctuations obey the quantum laws we are accustomed to. If not, what laws did they follow, and how did our quantum laws come into being? If so, where did the laws come from?

Again, loop quantum cosmology gives us a way out of this conundrum, because it gives us a workable model of how our universe might be oscillating or cyclical, In this model, the crunch of one universe leads to the expansion of a new one; however, this model runs up against the Second Law of Thermodynamics, which says that entropy must always increase. It may be preferable to avoid this problem by saying that the universe had a definite beginning and will have a definite end; however, it is also possible that the Second Law works in reverse as the universe contracts. It is even possible that the "cosmic forgetfulness" I mentioned in Chapter 25 could reset

We have assigned the wrong words for the very beginning... it could not have been a bang of any sort... it was something else, occurring in the most absolute silence we can imagine. It was the great light.
Lewis Thomas

If Galileo had only known how to retain the favor of the Jesuits, he would have stood in renown before the world and he could have written about what he pleased about anything, even about the motion of the Earth.
Christopher Grienberger

the entropy for the new universe. Then again, the idea of cosmic forgetfulness could be wishful thinking.

Hawking and Penrose also understood that there was no singularity at the beginning. Relativity does not imply such a thing, because quantum mechanics does not apply at distances less than the Planck distance or to times less than the Planck time. These are the smallest and shortest things possible, which means we could never have had a singularity. This is another argument for a prior universe and maybe even for ongoing cycles of universes.

Black Holes

The Big Bang is somewhat similar to a huge exploding black hole. Understanding black holes may help us shed a little more light on the Big Bang. A black hole is a region of infinitely curved space-time and infinite density, a singularity that ends time for anything unlucky enough to fall into it. A photon stuck at the edge of a black hole is literally frozen in space and time. Stars that are above the *Chandrasekhar limit* (named after astrophysicist Subrahmanyan Chandrasekhar, 1910-1995) of about 1.5 times our Sun's mass will not be able to support themselves under their own gravity once their fuel runs out, and will collapse in on themselves. Stars smaller than this limit could end up as white dwarves. Physicist Lev Landau (1908-1968) discovered that a star between 2 and 3 times our Sun's mass could collapse during a supernova and end up as a *neutron star*. As the name implies, such a star consists almost entirely of neutrons and is prevented from collapsing any further by the Pauli exclusion principle we discussed in Chapter 23. A neutron star could have a radius of about 10 miles and a density of hundreds of millions of tons per cubic inch.

A star destined to become a black hole runs out of fuel and collapses in on itself. Quantum mechanics takes over at both the singularity and the event horizon, in that a black hole can emit Hawking radiation that would eventually cause it to dissipate entirely. This process is not rapid; it could take far longer than the current age of the universe for a larger black hole to dissipate completely, but dissipate it will, because quantum mechanics demands the preservation (conservation) of information. Hawking radiation must be non-random in order to preserve and remove the information that fell into the hole.

Chaos is but unperceived order; it is a word indicating the limitations of the human mind and the paucity of observational facts. The words 'chaos,' 'accidental,' 'chance,' 'unpredictable,' are conveniences behind which we hide our ignorance.
Harlow Shapley

The secret, if one may paraphrase a savage vocabulary, lies in the egg of night.
Loren Eiseley

For example, if you were to toss a book into a fire, you could—at least in theory—reconstruct every letter based on studying the ash and smoke.

It was once thought that a black hole represents the maximum amount of entropy possible for the volume of space it occupies; however, later discoveries show that the entropy is actually proportional to the area of the event horizon, the point of no return beyond which not even light can escape the gravitic maw. In other words, the amount of entropy a black hole can contain is a lot less than one might think.

Cosmic Background Radiation

We saw in Chapter 23 that a black body is an ideal object which absorbs all of the electromagnetic radiation that hits it. This quality also makes it the best possible emitter of incandescent heat in a spectrum that depends on the temperature.

Is it not shocking to know that... all the heavens including all the luminaries whose lights are measured to reach this Earth after millions of years are said to be mere bubbles in the ocean of eternal emptiness?
D. T. Suzuki

The space between stars seems empty and completely dark, but is in fact suffused with a pale microwave glow that is almost the same in all directions, and that is not associated with any celestial body. Observations indicate that this radiation is amazingly uniform across the sky. It took several generations of increasingly sophisticated probes to reveal the tiny wrinkles in this radiation, remnants from the irregularities that were present when the universe formed. The universe was—and remains—almost perfectly uniform. The tiny wrinkles in the background radiation are providing a wealth of data about what the early universe looked like and how it worked. The galaxies we see today formed from the density variations in the very early universe that we can still see today as tiny anomalies in the background radiation. Einstein's equations tell how matter behaves in an expanding universe; we can use current values to extrapolate backward to see matter distribution at very early points in the universe's history.

It is often said that there is no such thing as a free lunch; the universe, however, is the ultimate free lunch.
Alan Guth

The wrinkles we see began as quantum fluctuations of matter in a tiny space that have expanded across the entire universe we see today. Study is ongoing, but we already know that the actual radiation measurements are perfectly in line with theoretical predictions, a huge validation for Big Bang theory. Any number of phenomena could create black body radiation, but only the Big Bang model can explain the fluctuations.

Inflation

The Big Bang, be it a singularity or not, is accepted as the beginning of our universe, which appears to become more and more homogenous the further back we look. This is taken as good news, because it indicates that our very complex universe started out very simply; however, such a state creates problems for cosmology. The inflation theory is attractive, because it gets around all this. Also, we can understand how it could have arisen by random quantum fluctuations, based on the assumption that empty space consists of particles coming into and out of existence. During the inflationary epoch, the universe expanded by a factor of about 10^{80} (see below) from—10^{-26}cm to 10cm—while the temperature plummeted from 10^{27} degrees Kelvin to nearly 0. All matter in the universe is thought to have been created during this period, and the overall density of the expanding universe is thought to have remained the same. The newly created matter precipitated out into the forms we know today at the end of the inflationary epoch. Creating matter in this way requires adding gravity to the mix to avoid violating the law of energy conservation. Energy-carrying particles must draw this energy from somewhere. Gravity provides this energy.

As I previously mentioned, the expansion of space is not an explosion into something, but space itself expanding. The farther galaxies are from each other, the more space they have between them, and the faster they are moving apart; however, clocks in each galaxy stay synchronized because the galaxies are moving with space, not through space. Relativity restricts the speed of something moving through space to light speed; it places no restriction on how fast space itself can expand.

The Big Bang itself happened at the beginning of time, but inflation only happened when the time was right, which need not be the instant of creation; however, we must still ask why the energy of the early universe was shaped correctly for inflation to occur, why space and time were contained within it, and why—as Leibniz asked—there is something rather than nothing. Again, the total energy required for inflation to occur is zero. All we need is a bubble of space with the right configuration, and the rest can happen pretty much on its own. So how did the universe arrive at that one correct state? One pos-

> *In the quantum world, what you see is what you get, and nothing is real; the best you can hope for is a set of delusions that agree with one another.*
> John Gribbin

> *The paradox is only a conflict between reality and your feeling of what really ought to be.*
> Richard Feynman

sibility is that there were many inflation bubbles before the right one came along and got the universe off the ground.

Guth theorized that a Higgs field (see below) infused space with energy while adding uniform negative pressure in a manner very similar to Einstein's cosmological constant. A supercooled Higgs field affects expansion by exerting a negative gravitational force that causes expansion. This field need not be constant; it must only have had a plateau strength high enough to exert a strong enough push. Model rockets are a good example: They contain small motors that fire for a brief period, after which the rocket continues ascending on its own. The Higgs field value would have to be about 10^{100} times larger than Einstein's constant, and the total amount of expansion could have been anywhere from 10^{30} on the conservative side to 10^{100} at the upper end. To put this into perspective, imagine a DNA molecule expanding to the size of the Milky Way in an extremely tiny fraction of a second. One of the many benefits of inflation is allowing a universe that is much larger than we might have had otherwise. In fact, we may only be able to see an infinitesimal amount of the total universe. How much bigger might the universe be? One model says that if the universe were the size of Earth, the space we could see would be the size of a grain of sand.

At the end of the inflationary epoch, the energy that propelled the expansion turned into matter. Again, the energy for the mass and energy in the universe was essentially borrowed from gravity. The matter and radiation in the expanding universe is running down because of gravity, which provided the energy for the initial expansion in the first place. Nature always balances her books. Quantum uncertainties present before and during the inflationary epoch stretched out and provided the seeds for galaxies. Inflation theory demonstrates that galaxies are quantum mechanics writ large across the sky. We may not be able to unify quantum mechanics and relativity yet, but that has not stopped the universe from doing so. Adding quantum mechanics to the very early stages of the universe seems to avoid some of the need for fine tuning. Tiny irregularities lead to regions that don't expand quite as rapidly as others, which leads to regions with denser matter, which collapse under gravity to form galaxies.

We cannot know as a principle the present in all its details.
Werner von Heisenberg

I have seen nothing in the world that is ultimately real.
Yeshe Tsogyel

Inflation also helps explain the very low entropy present at the beginning of the universe and the smoothness of spatial curvature. Inflation was so rapid that the wrinkles were literally pulled out of the cosmos, in the same way that removing a frog from a puddle and placing it in a lake reduces the density of frog to water. Inflation happened everywhere at once, which means that matter was distributed perfectly evenly, within quantum uncertainties. The 10^{80} atoms and associated radiation we know of as the universe formed at the end of the inflationary epoch. This is an enormous amount of entropy; however, the initial uniformity of the inflation left a huge gap between the potential and actual entropy in the universe. We could not be here without entropy, which helped form the clumps of matter that led to galaxies, stars, planets, and ourselves. We also could not be here had there been too much entropy.

The concept of substance has disappeared from fundamental physics.
Arthur Eddington

Inflation can account for the mass and energy we see in the universe. The energy density remained the same, which means that the total amount of energy grew by the same factor as the space itself, or by anywhere from 10^{30} to 10^{100}, as we saw above. Triggering the inflation therefore did not require much energy at all A nugget of matter measuring about 10^{-26}cm and weighing 20 pounds with an inflationary field would be energetic enough to account for everything we see in the universe today. This is in stark contrast to "plain" Big Bang theory, in which the singularity needs an immense amount of energy from the very beginning. Inflation theory does not answer Leibniz, but it does make the answer a lot more believable.

We all know that life is anything but smooth sailing, that there are conflicts and decisions at every step. It is almost impossible to remain in a completely balanced state for very long. Symmetry is also virtually impossible to maintain in the universe, because it is a very unstable state. This instability led to the fractionating of the 10-11 dimensions in the universe (see Chapter 25) into the four dimensions we are familiar with plus the 6-7 higher dimensions.

The proportion of matter to space is even less than a grain of sand in the Albert hall.
John Gribbin

The inflationary model was put forth in an attempt to solve difficulties with the "pure" Big Bang theory and avoid having to resort to fine-tuning and/or purposeful creation. Inflation goes a long way toward doing this, but does not resolve all of the problems. First, the line between an open and closed uni-

> *Those scientists who claim that science tells them nothing about ultimate origins are not being quite honest, be they atheists or creationists.*
> Fred Heeren

verse is very fine; small shifts to one side could yield a universe expanding too rapidly or not rapidly enough. It also has implications for the ultimate fate of the universe, as we discussed above. Second, negative pressure must be established and maintained for a long enough time. The simplest model of inflation shows negative pressure working long enough to provide enough energy for the universe to undergo the kind of expansion theorized. Without this, matter could not possibly be as homogeneously distributed as shown by the cosmic background radiation. Third, inflation explains how the universe got going in a chain reaction once the triggering event (Big Bang) occurred; it does nothing to define the empty space into which the universe emerged, nor does it explain how that space came to have quantum laws that would allow the Big Bang to begin with.

Even with inflation, we are still left attempting to explain creation *ex nihilo*. We can imagine a universe with different laws governing different regions, much as individual cities, states, and nations have different laws that are not always compatible with one another. For example, it is perfectly legal to smoke marijuana in the Netherlands and illegal to do so in the United States; however, even this model has limits, because where did quantum fluctuations come from? Loop quantum cosmology solves some of this by postulating an ancestor universe and making inflation a result of the universe's basic properties, such as the grainy nature of space-time.

We have seen that inflation can explain a great many things, such as the homogenous distribution of matter, how matter was created from nothingness, courtesy of the ultimate loan shark called gravity, and how we need not rely on a perfectly tuned singularity; however, this model still leaves many questions open in its own right.

The Higgs Field

> *It is simpler to postulate creation ex nihilo—divine will constituting nature from nothing.*
> Edmund Whittaker

The Higgs field and associated Higgs boson are named after physicist Peter Higgs (1929-). The Higgs field is thought to be behind the phenomenon of inertial and gravitational mass. As the universe cooled, this field's value became equivalent to the lowest possible vacuum energy, which we already know must be greater than 0.

Swing your arm, and you are moving all of the quarks that make up the particles that make up the molecules of your arm through the Higgs fields. The resistance you feel is the Higgs field, which resists all accelerated motion, but not constant motion. Quarks are bound together by gluons, which carry the strong nuclear force and add mass provided by the Higgs field. Different particles interact differently with this field, which explains their varying masses. Most physicists believe that matter would not have mass, but for the Higgs field. I should digress for a moment to explain that this field is not the ether of old theories. Protons are massless, and therefore do not interact with the field. The Higgs field could also be the lattice of space-time called for in loop quantum cosmology.

If this field is real, then there should be a Higgs boson, which is sometimes called the God particle. Experiments at the Large Hadron Collider and other facilities are underway to try to find this particle. If it does not exist, then physics for the last several decades will need to be re-thought. If confirmed, then empty space will not be truly empty. Theologian Henry More (1614-1687) said that space is not empty, because it is filled with divine spirit. He may be right, albeit not quite in the way he imagined.

There is no doubt that a parallel exists between the Big Bang and the Christian notion of creation from nothing.
George Smoot

A Zero Sum Game

The Big Bang and subsequent inflation created some 10^{80} atoms from energy borrowed from gravity. The universe as a whole has 0 total energy. The debt owed to gravity must be repaid at the end of the universe, theoretically in the Big Crunch. We have already seen that a closed universe could have come from quantum fluctuations. Many systems within the universe can be functionally isolated from each other on different levels, such as stars, galaxies, and galactic clusters. Each of these systems can have non-zero energy and angular momentum. Adding these up and subtracting gravity should equal 0. This would be a complete accident in a Newtonian universe, but becomes mandatory in a Machian universe. Mach's law predicts that effectively isolated systems will behave as Newton expected, which means that Newton's laws may apply on local levels throughout the universe but not to the universe as a whole.

One may find it easier to believe in an infinite array of universes than in an infinite deity, but such a belief must rest on faith and not on observation.
Paul Davies

Dark Energy and Matter

In the language of its nihilist proponents, a pointless universe has no conceivable outcome except the grim maximization of entropy.
Bernard Haisch

There is increasing evidence for a mysterious form of matter and energy called *dark matter* and *dark energy*, respectively. We call them "dark," because we cannot yet see or otherwise detect them, except through indirect observation and theoretical necessity. For example, rotating galaxies should spit stars out into space because of centrifugal force, in a manner similar to spinning around while holding an object at arm's length and then letting go. Also, the rate of expansion in the universe has been speeding up. This repulsion of mass seems to be at direct odds with attractive gravity, and the mechanism causing it remains a matter of much debate. Calculations indicate that there may be 5 times more dark matter than normal matter in the universe, which would either make us little more than cosmic flotsam or part of an even more finely tuned universe than previously imagined, depending on how one looks at it.

One view of this accelerated expansion is that the initial energy of inflation dissipated and "froze" into ordinary matter and energy. Gravity slowed down the expansion, but the thinning universe allowed the expansion to pick up steam. One theory says that all but the closest galaxies will move away from us at more than light speed over the next 100 billion years, meaning that we will see less and less of the universe.

The light created at the very beginning is not the same as the light emitted by the Sun, the Moon, and the stars, which appeared only on the fourth day.
Haggadah

Space itself need not be flat. It could be positively curved (like a sphere), planar (flat), or negatively curved like a saddle. According to general relativity, curvature causes gravity, but space is almost perfectly flat on large scales, with curvature being caused by mass. Only completely empty space-time could be perfectly flat. The amount of ordinary matter seems to be far too little than expected for flat space. The only ready solution to this problem seems to be to postulate an unseen form of energy that calculations indicate makes up about 70% of the energy in the universe. This is where the concept of dark energy comes in. This presents another problem: The amount of mass not attributed to dark energy is much larger than matter seen in stars, which leads to the concept of dark matter. Unlike dark energy, dark matter does not cause accelerated expansion. Together, dark energy and dark matter form one model to explain accelerated inflation.

Multiverse?

The multiverse theory (not to be confused with the "many worlds" interpretation of quantum mechanics) postulates a large number of universes to answer the question of just how likely it is to get the correct conditions for inflation. Gambling with fair odds will eventually win a jackpot, and some area of any primordial space will eventually get the correct conditions. In other words, we got lucky, and our universe may be one bubble in a space full of bubbles. There could be many inflationary universes. Then again, the idea of a multiverse could be a metaphor for a deeper undiscovered reality, because we can translate whatever conclusions we can come to into any system we want. From a purely scientific standpoint, if what happens in another universe stays in another universe, then we should focus on our own universe and leave the others alone. If the multiverse is really just different parts of our own universe, then we have to determine whether one set of laws applies everywhere herein, or whether our home is a patchwork of local jurisdictions.

> *There could have been millions and millions of different universes created each with different dial settings of the fundamental ratios and constants, so many in fact that the right set was bound to turn up by sheer chance. We just happened to be the lucky ones.*
> Clifford Longley

Revisiting the Anthropic Principle

If the Big Bang is correct back to the very beginning of time, then the initial conditions of the universe require an extremely fine set of adjustments. Our universe is the way it is because of the energy and matter values it has. As we saw in Chapter 11, even tiny differences would create an uninhabitable universe. Why does the universe have so many properties set so perfectly to allow life to evolve and look back to the beginning of time to ask these questions? The more we try to evade the need for fine tuning, the more the true level of tuning becomes apparent.

We have already discussed expansion. Any slower, and the universe would have collapsed in on itself. Any faster would stretch space too much for stars and galaxies to form. At the Planck time, the mathematical coefficient of expansion to gravity was within 1 part in 10^{60}, and this problem does not necessarily go away by adding inflation. Furthermore, the density of matter and energy in the universe is almost precisely the critical density to create mostly flat space, a feat that required

> *In the beginning there were only probabilities. The universe could only come into existence if someone observed it. It does not matter that the observers turned up several billion years later. The universe exists because we are aware of it.*
> Martin Rees

> *The universe in some sense must have known we were coming.*
> Freeman Dyson

> *The odds against a universe like ours emerging out of something like the Big Bang are enormous. I think there are clearly religious implications.*
> Stephen Hawking

the density to be within 1 part in 10^{12} at just one second after the Big Bang, or well after the inflationary epoch ended. The strength of the electromagnetic force is less than 1% of the strong nuclear force, the weak nuclear force is 1,000 times weaker still, and gravity is about 10^{-35} times as strong as the weak force. If the strength of gravity varied by 1 part in 10^{40}, then stars and planets could not exist. Gravity only works because the electrical forces are usually in balance.

The absence of antimatter is another puzzle. We know from $e=mc^2$ that matter is simply dense energy, and that the distinction between matter and antimatter is one of electrical charges. Matter and antimatter are governed by the same laws of physics, so how did we end up with a surplus of matter? It turns out that matter and antimatter decay slightly differently, with antimatter decaying more quickly. For every 1 billion antiparticles that existed in the split second before quarks and electrons formed, there were 1 billion and 1 particles.

Any way one looks at it, our very presence here requires a long laundry list of cosmic coincidences. It is very difficult to explain this without invoking a god or other super intellect who intended for us to evolve. Perhaps the initial conditions were such that the universe could only unfold as it did, which only punts the question back in time without resolving it. Random starting conditions should allow only 1 chance in 10^{120} for life to evolved. Penrose placed the exponent at 10^{123}, or 1,000 times smaller still. Either way, these odds are far lower than the odds of plucking a particular atom from the 10^{80} the universe has to choose from.

> *There is no reason why the laws of physics cannot have come from the universe itself.*
> Victor Stenger

The anthropic principle can imply just about anything from our presence being totally random in one universe out of many to a universe hand-built by a creator god. Some interpretations say that some regions of a universe with infinite space and time will be suitable for finite creatures like ourselves, which is a bit like a wealthy person who does not see poverty. Others essentially say that the universe must be the way it is, or we would not be here.

The very fact that we are here guarantees that our universe is good for life. Fish swim in a warm, homogenous medium full of nutrients. It has to be that way, or fish would not exist. It would be possible to have a fishtropic principle—at least until the fish got caught in a net. The anthropic principle can lead to more than a little narcissism, which is one reason why some people don't like it one bit. Then again, the mere fact that our science is as successful as it is argues against random beginnings.

Self-Service Evolution

After the Big Bang, several generations of stars were required to create the heavier elements needed for life. We are recycled star dust; there is no other way to make heavier elements. The list of coincidences from the Big Bang to the formation of the Earth, the evolution of life, the meteor that wiped out the dinosaurs, mitochondrial Adam and Eve (Chapter 9), and even the survival of our own parents is extremely long, and we would not be here but for getting everything just right.

> *The anthropic principle is incredibly vague. You can use the anthropic principle, if you want, to explain almost anything. And it never gives precise predictions; it only explains after the fact that what you saw was, in some sense, acceptable.*
> Alan Guth

This book keeps returning to the idea of biocentrism, and it is appropriate to do so again. The biocentric theory says that we created the universe. Quantum mechanics has properties created by measurement, according to the Copenhagen interpretation. Physicists agree that  quantum mechanics is universal, to the best of our knowledge. This universe is hospitable to life, because we could not have created it any other way. Do our eyes looking back at the Big Bang make what happened back then real? Wheeler believed that every particle, field, and even space-time gets its function, meaning—even its very existence—from answers to yes/no questions. If this is correct, then we must assume that the universe we live in is the one with the best odds of producing conscious life on Earth, and quite probably elsewhere.

God in the Machine

The "ghost in the machine" was originally a response to Descartes' dualist philosophy in which mental action parallels, but does not equal physical action. As we have seen, the biggest problem with dualism lies in the fact that no known mechanism exists to mediate between the two worlds.

> *A commonsense interpretation of the facts suggests that a superintellect has monkeyed with physics as well as chemistry and biology.*
> Fred Hoyle

We have seen how Big Bang theory and inflation have been used to argue against a creator, and understand that the laws of physics could predate our universe. But where did these

> *I find it as difficult to understand a scientist who does not acknowledge the presence of a superior rationality behind the existence of the universe as it is to comprehend a theologian who would deny the advances of science.*
> Werner von Braun

laws come from? Our theories get us to within one Planck time interval after the Big Bang, and possibly back even before the beginning. We know that the universe began as an intense state of pure quantum potential in an extremely small amount of space. But what is a quantum state? According to some of the implications of the Copenhagen interpretation of quantum mechanics, all such states are states of mind, and fluctuations are the stuff of consciousness and will.

Mind could well have been the beginning of our universe, and could be the first cause we seek. This mind would exist independently of time and be purely non-local. If this seems far-fetched, consider once again that $e=mc^2$, and that the sum total of energy e in the universe is 0. If e is 0, then mass m must be 0 as well. What did any theoretical Creator create? Nothing! Nothing, that is, except a persistent illusion of matter and energy flowing along a river of equally illusory time. If this seems unbelievable, consider that you do the same thing in your dreams every night, whether you are aware of it or not.

Assuming this God is real, did s/he have a choice about how to create the universe, or were the divine hands tied? In a sense, yes; if God wanted to create this result, then this what had to happen. Even an OOO being must be bound by logic, even if that logic is whatever it decides to create. If this is true, then the laws of science are nothing less than the expression of God's will. As the Gospel of Thomas says, "I am he who exists from the undivided. It is I who am the light. Everything came from me and everything extends unto me. Split a piece of wood and I am there. Part the stone and you will find me." No wonder the Gnostics thought matter evil!

The Dying Universe

> *A great while ago the world begun, with hey, ho, the wind and the rain, but that's all one, our play is done.*
> Shakespeare
> (Twelfth Night)

We already know that entropy is the measure of disorder in a system and/or lack of knowledge about said state. The Second Law of Thermodynamics states that entropy always increases with time. The universe is running down. The First Law of Thermodynamics says that energy can neither be created nor destroyed, and $e=mc^2$ says that matter and energy are equal. The universe faces heat death, in which all energy will become frozen into matter that will cool down further to its ground energy state at a temperature just above absolute zero. Only quantum uncertainty will keep the energy from dropping all the way to zero. Before this happens, one theory says that protons will decay into positrons after 10^{30} years. If this is correct, the universe will simply evaporate away. Electrons and protons will eventually meet and annihilate each other. The outlook is grim, no matter how we look at it. We cannot win. We cannot break even. We cannot get out of the game.

Entropy measures the number of states a given system can have. There are only a few orderly states a system can have, compared to a great many chaotic or disordered states. For example, there are far more ways for a cup to be broken than whole. A book whose pages are in order has only one way for the pages to be in order, but many ways for the pages to be out of order. For example, this book is approximately 1,000 pages long. These 1,000 pages have only one way to be in correct order, and 1,000! (1,000*999*998... *3*2*1) ways (approximately $4.02 \times 10^{2,594}$) to be out of order. Even a small change in the number of pages makes a huge difference. A 10-page book has 10! (3,628,800) ways to be out of order, while an 11-page book has 11! (39,916,800) such ways, and so on. There are also a lot more ways to be dead than alive. Death is the final surrender to the Second Law. The overall increase of entropy is what gives time its apparent unidirectional flow.

For a parallel to the lesson of atomic theory, [we must turn] to those kinds of epistemological problems with which thinkers like the Buddha and Lao Tzu have been confronted, when trying to harmonize our position as spectators and actors in the great drama of existence.
Niels Bohr

Localized decreases in entropy are permitted, so long as the total amount of entropy increases. As we saw in Chapter 19, the Earth can evolve more complex life that seemingly violates the Second Law, because the Sun's entropy is increasing by a vastly larger percentage. The Earth is thus an open system whose survival depends on being able to use the Sun's entropy to decrease its own. Life is therefore an elaborate antientropy system. Life grows and evolves by becoming more systematic and reducing its overall entropy.

A black hole represents the maximum possible entropy for the area of the event horizon, which is far less entropy than one might guess. The mathematics may defy common sense, but they

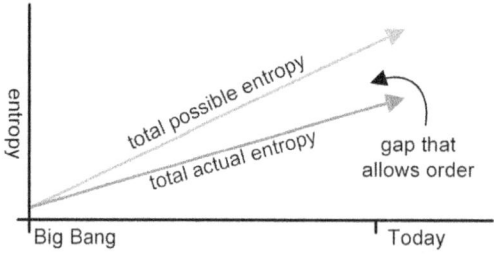

Bohr did not propose an answer, not philosophize, not go an inch beyond the fullest statement of the inescapable lessons of quantum mechanics.
John Archibald Wheeler

Myth is neither a lie nor a confession: it is an inflexion.
Roland Barthes

never lie. High entropy is the norm; it is low entropy that requires explanation. The cosmic background radiation is proof that the universe began in an extremely ordered state. Entropy has been increasing ever since, at less than the maximum possible rate. The clumps of matter in the universe are caused by gravity, which means that clumps are the rule; gravity is causing entropy that is masquerading as order. The gap between the total amount of entropy in the universe and the maximum possible amount of entropy possible provides room for order. Entropy was as high as possible at the beginning, before the expansion of the universe created the gap that in turn allowed structure.

Here is where things get really interesting because the increase in entropy over time, while very probable, is not certain. The Second Law of Thermodynamics is probabilistic, with no built-in sense of time. It could therefore work just as well when applied to the past as it does to the future. For all the overwhelming evidence that entropy will be higher in the future, that same evidence works just as well to say that entropy was higher in the past. No law of physics says that a flat tire cannot or will not spontaneously reinflate; however, the odds of something like this happening on a macro level are extremely remote. See some partially melted ice, and you can assume it will be more melted in the future; you just need to assume that it could have been more melted in the past. Any assurance we can possibly have about the future applies just as well to the past.

It is easy to see that only a restricted range of laws of nature are consistent with galaxies and stars, planets, life and intelligence.
Carl Sagan

Boltzmann pointed out that the universe could be chaotic, but isn't. Where did the order come from? To him, it is far more possible that the universe came into being from an unusual transition from high to low entropy than from popping into being one fine day. Toss as many coins in the air as you like enough times, and they will eventually land all heads. This is far more unlikely than any combination with at least one tail in it, however one could argue that a sudden drop from high to low entropy is a far better explanation for our presence on Earth than the incredibly long string of successes demanded of evolution—and that is just one small example. Any way one spins it, the universe is an incredible statistical anomaly that cries out for a definitive explanation.

The idea that entropy increase is equally likely going either forward or backward in time is a clear indication that something is missing in our examination. We can explain evolution from low to high entropy, assuming some very special starting conditions, but maybe there is an alternate explanation. Why should the Big Bang be the one explosion to ever produce order instead of chaos? We either popped into existence (not terribly likely) or the universe started in an extremely low-entropy state (also not terribly likely). Faced with these two distinctly unsavory explanations, common sense forces us to choose the latter possibility because that at least allows us to trust our memories and records, and provides a nice linear explanation. But it ain't necessarily so.

De Temporum Fine Comedia

The way in which our universe was born will ultimately determine its demise, because there is an unavoidable causal link between the shape and mass of the universe and its ultimate fate. We do not yet know whether the universe is open and will expand forever, or closed and will eventually stop expanding and fall back in on itself. The inflationary model is by far the leading theory, and is compatible with both standard and loop quantum cosmologies. It is a very good model, but inflation cannot make an open universe closed or vice-versa. We do know that we are very close to the dividing line, and have some reason to think that the universe is closed.

If this is true, then the universe will eventually stop expanding and start to collapse. In the final second, particles will turn to energy, and the universe will simply wink out of existence. We already know that particles can appear and disappear spontaneously. Is the entire universe just another particle? Or is the loop quantum cosmology view correct in predicting that the end of our universe will trigger a new expansion, like a phoenix rising from its own ashes? Either way, the Big Crunch, if it happens, will leave behind no space, no time, and no matter. As far as we are concerned, it will truly be the end of all things.

If the universe is open, then it will continue to expand forever. Entropy will increase until all energy has frozen into matter and that matter has cooled to its ground energy state. Stars will go dark, and the universe will become a giant tomb.

> *The great debate over whether the universe is open or closed comes down to the question of whether everything will fall back on itself only to repeat the cycle or whether the last bits of matter and radiation will disappear into a darkness that expands forever. This is, in a sense, the last, the ultimate, question of science. The cosmic switch has already been thrown; the answer, though unknown, is already ordained, and man cannot affect the outcome.*
> James Trefil

Chapter 27

Holography

Listen, there's a hell of a universe next door. Let's go!
E. E. Cummings

Atoms are not things.
Werner von Heisenberg

The word *holography* comes from the Greek words *hólos* (whole) and *grafē* (writing or drawing). In modern usage, holography is a method of recording light scattered by an object, and later reconstructing it to make it appear as though the object is in the same position relative to the recording medium that it was when initially recorded. The holographic image changes based on a viewer's position and orientation changes in exactly the same way as the original object would, which gives the hologram the appearance of being three-dimensional.

Holography is not the same as 3D photography, such as is used in a growing number of movies. Rather, it is the visual equivalent of recording audio in such as way that the original sound field can be recreated during playback. Modern audio systems with 5 or 7 audio channels plus an added bass channel can provide extremely realistic three-dimensional audio. My first experience with this was in high school, when I dated a girl whose father owned a very high-end audio visual system. I remember watching a tank battle during the movie *Red Dawn* and hearing the sound of artillery shells flying back and forth over my head.

Making a Hologram

Creating a hologram begins with placing the recording medium in a suitable location. A beam of coherent light is then shined on a beam splitter. Half of the beam illuminates the object, which scatters the light. Some of this scattered light strikes a recording medium. The other half of the beam is shined directly onto the recording medium. A seemingly random pattern appears on the recording medium. Afterward, illuminating the recording medium with the original reference beam causes the hologram to diffract that beam and produce a light field that is identical to the one scattered by the recorded object(s). Someone looking at the hologram sees the object(s) as if they were actually present. The hologram is thus a three-dimensional illusion in two dimensions.

There is neither seer nor seeing nor seen. There is but one reality—changeless, formless and absolute. How can it be divided?
Shankara

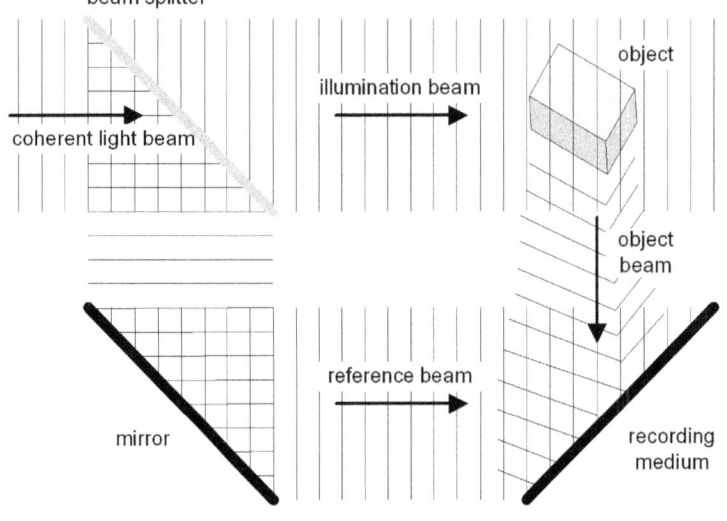

Cut a hologram in pieces, and you would expect each piece to contain only the appropriate portion of the original image. Amazingly, each piece of the hologram contains the entire image,

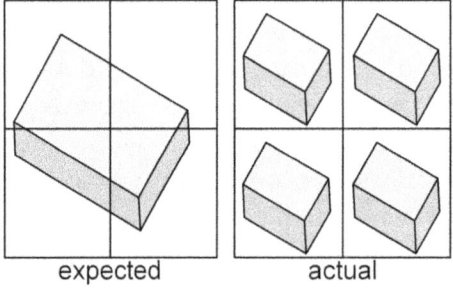

In a holographic universe, even time and space could no longer be regarded as fundamentals.
Michael Talbot

no matter how you cut it. The recorded object loses clarity as the hologram is cut ever finer, but still carries the entire image. But that's not all. In fact, we're just getting started. You see, it just so happens that the entire universe may be nothing but a giant hologram.

Grainy Holographic Reality

Holography might be a clue to the ultimate theory of reality.
Jacob D. Bekenstein

The GEO600 facility near Hanover, Germany is a 600m long detector that has been seeking gravitational waves from extremely dense objects such as neutron stars and black holes. This ongoing experiment has come up empty in that search, but in the process may have accidentally made one of the most important discoveries in the history of physics. GEO600 had been beset with noise that the experimenters could not explain or correct for, until physicist Craig Hogan stunned them by saying that he had predicted the noise before learning that they were detecting it.

If Hogan is correct, then GEO600 has encountered the limits of space-time, where smooth continuity breaks into grains. It also means that we could be living inside a giant hologram. This is not mere speculation; we can describe our physics from the perspective of taking place at the edge of the universe. Superstring theory contains a hypothetical realization of holography inside it. This idea is also compatible with loop quantum cosmology. Here we have yet another clue that space-time as we know it is not the fundamental reality, but simply a decoration. It could certainly help explain how the entire universe seems to be humming along on zero total energy.

Black Holes and Information

The very concept of an atom with a definite location and motion is meaningless. What right have we to say that an atom is a thing if it isn't located somewhere, or else has no meaningful motion?
Paul Davies

Physicists Leonard Susskind (1940-) and Gerard 't Hooft (1946-) developed their ideas about a holographic universe based on pioneering work conducted by other physicists on the nature of black holes. Hawking showed that black holes are not entirely black, but emit weak radiation that eventually causes them to evaporate and disappear entirely. This radiation does not carry information, which means that all of the information about the star that formed the black hole is destroyed along with the black hole, in violation of the law of conservation of information. This is the *black hole information paradox*. Physicist Jacob Bekenstein (1947-) discovered that the entropy of a black hole is proportional to the area of its event horizon, the theoretical surface that marks the point of no return for incoming light and matter.

Later theorists showed that microscopic quantum ripples at the event horizon encode the black hole's information, thus preventing information loss as it evaporates. (A black hole's entropy is roughly analogous to the amount of information it can contain.) This in turn allowed physicists to discover the absolute limit on how much information a given volume of space or quantity of matter can hold. The universe might be "recorded" on a two-dimensional surface, like a hologram that we see as three-dimensional. Susskind and 't Hooft took this example and extrapolated it to cover the entire universe on the basis that the universe itself has a horizon, which is the boundary beyond which light has not yet reached us after 13.7 billion years. Other physicists have confirmed that this research is on the right track.

Structuring the Universal Hologram

Hogan believes that seeing the universe as a giant hologram forces us to radically change our view of space-time. The conventional belief is that space-time at Planck scales is full of activity. Reality on this scale is made up of individual grains of space-time. Loop quantum cosmology sees the universe as a mesh of Planck-length loops. In a holographic universe, the surface of the sphere bounding our universe would be covered in Planck-length squares that each contain one bit of information. The amount of information on the edge of the universe must equal the number of bits contained within the volume of the universe.

There is just one problem: The volume of the universe is far greater than the area of its outer surface. Hogan solved this problem by saying that the grains inside the universe must be larger than the Planck length. In other words, the inside of the holographic universe needs to be blurry. The Planck length itself is far too small to ever detect; however, the holographic projection of Planck-length graininess could result in areas that are about 10^{-16} meters long, which is much larger than the Planck length. Assuming we are in a holographic universe, we should be able to measure the blurring. Enter GEO600.

We must be clear that, when it comes to atoms, language can be used only as in poetry.
Niels Bohr

When one knows that the great void is full of chi'i, one realizes that there is no such thing as nothingness.
Chang Tsai

Finding the Evidence

Standard big-bang theory, for example, essentially explains the propitious universe in this way: 'Well, we got lucky.'
U.S. News and World Report

The GEO600 is an extremely accurate ruler. A gravitational wave passing through the detector will stretch space in one direction while squeezing it in the other. To measure this, a single laser is fired through a beam splitter that splits the light into two beams. These beams pass down the perpendicular 600-meter arms and bounce back again. The two beams merge at the beam splitter and create an interference pattern. Any shift in the pattern is a dead giveaway that the relative lengths of the arms have changed, which is exactly the sort of thing that a gravity wave should do. This system is sensitive enough to detect length changes that are much smaller than the diameter of a proton.

There are five gravity wave detectors on Earth. Hogan knew that GEO600 should be the most sensitive to what he was looking for. His theory was that there should be random noise in the experiment if space-time fluctuations were hitting the beam splitter. He surprised the GEO600 researchers by sending them his predictions, and they returned the favor by telling him that they had indeed been detecting unexpected noise in frequencies between 300 and 1,500 Hertz. They sent Hogan a plot of the noise, which matched his predictions precisely. Nobody is claiming that GEO600 has found the smoking gun indicating that our universe is holographic; extensive verification remains ongoing to eliminate or account for other potential noise sources.

Suppose that... we find an unmovable finger obstinately pointing outside the subject, to the mind of the observer, to God or even only gravitation? Would that not be very, very interesting?
John Bell

This is no mean feat. The breathtaking sensitivity of gravity wave detectors makes noise a perennial problem. Everything from distant traffic, seismic activity, and even passing clouds can cause noise. No clear earthly candidate has been identified for this particular batch of noise, but the need for an abundance of caution should be self-evident. So far, the GEO600 team has eliminated temperature changes across the beam splitter and are planning upgrades to the system to make it even more sensitive, while also eliminating some possible sources of noise. If this particular noise still shows up after those upgrades, well... GEO600 will be useless for detecting gravity waves, but will have ended up discovering something far more significant. Plans are also in the works to build detectors that are both optimized and dedicated to examining holographic noise.

Implications

The holograms on your credit cards are small two-dimensional images that give the illusion of being 3D images when light hits them. We just saw that this principle could well apply to the entire universe. If GEO600 ends up discovering holography, then the small loss of not finding gravitational waves will be more than offset by the magnitude of that discovery. This would not be the first time such unexpected discoveries have been made. The noise we are picking up right now could be a modern analogy to the noise originally detected by the Bell Labs antenna in 1964 that turned out to be the cosmic background radiation. The universe we perceive could be a four-dimensional system with an alternate set of laws working on the three-dimensional boundary of space-time.

If the universe really is a giant hologram, then everything we think of as stable is nothing more than a will-o-the-wisp, all permanence is illusory, and consciousness might be the only eternal thing left. A holographic universe is compatible with both every law of physics we know and with the view of time we will examine in Chapter 28. It is also very compatible with Buddhist philosophy, for it would prove that everything we think we see, do, and are is nothing more than maya. Then again, maybe the materialists are right, but we can never know. All we can do is ask others, and in so doing return ourselves to a reality created by mind.

I mentioned above that cutting a hologram makes increasingly blurry copies of the entire hologram. Holography may be at work in our own brains. Experiments with mice removed various pieces and quantities of brain tissue, but the mice retained their memories, as evidenced by successfully navigating mazes despite having crippling brain injuries. If this is true, then we may think we are viewing images sequentially in a flow of time, but all of the images may be there all of the time. If so, they can neither be created nor destroyed. This has some free-will implications that I will discuss further in Chapters 29 and 32. Meanwhile, if Bohm is correct about our consciousness being located in the implicate order, then our minds and holo-

The particular form of the laws of physics that reigns in our universe allows and perhaps even requires the development of complex structures and life.
Fred Adams

Our everyday experience even down to the smallest details seems to be so closely integrated to the grand-scale features of the universe that it is well-nigh impossible to contemplate the two being separated.
Fred Hoyle

graphic records of our own past are in the same place as everyone else's. We may need only shift our attention to access the past—and possibly the future as well.

Physical objects may ultimately be holographic. Our physical bodies may simply be regions of higher density in the hologram. This would allow us to identify parts as wholes because of the commonality that binds our consciousness to our experiences. We will see in Chapter 30 that consciousness seems to be holographic, at least on a functional level. Consciousness does not seem to have any fixed structure.

It may seem silly to think that we live in a hologram, but this is nothing more than an extension of our best knowledge about black holes and has very sound theoretical foundations. Holographic theory has also been of help to physicists struggling to develop new theories of reality at its most fundamental levels. It would help researchers complete the unification of quantum mechanics and relativity. It would also eliminate all incompatible interpretations of quantum mechanics and take a large and significant step past field-based theories. How would your philosophy and religion change if holographic theory was affirmed, and if everything you do and are owes itself to things happening at the very edge of the universe?

In plain language, things are the way they are in our universe because if they weren't, we wouldn't be here to notice.
Brian Greene

Being under illusion means perceiving objective appearances and mental appearances as having independent reality.
Bokar Rinpoche

702 | *The Divine Savage*
Revealing the Miracle of Being

Chapter 28

About Time

> *Time is what happens when nothing else does.*
> Richard Feynman

Time is a mystery. We seem to be locked into a one-way flow of time, where we begin existing at a certain time, continue existing for some time, and then cease to exist. We cannot avoid this flow, cannot change it, and cannot skip ahead or behind.

Astronomers use *sidereal time t*o help them track celestial bodies. In general, a body that appears in a certain *azimuth* (direction) and *declination* (height above the horizon) in the sky at a given time of day will generally be at or very near the same spot at the same time on the following day. The positions of stars in the sky seem to be fixed relative to each other and sidereal time is very accurate; a trained person can tell time to within 15 minutes very easily, and more detailed observations can tell time to within about one minute.

> *Time is just a parallelogram.*
> Jerry Lewis

A globular cluster may contain one million stars. This cluster is not rigid. All stars move independently because of gravity; however, one can find the cluster's angular momentum by choosing three mutually perpendicular axes and calculating the net spin around each axis. It is always possible to find three such axes where the spin around all three equals zero. These axes are like arrows, and the net spin remains constant over time, an astonishing feat of natural bookkeeping because each star is moving on its own. This is a strong clue that some profound underlying principle is at work. It also indicates that time may not be what it appears to be.

Newton could not fathom the idea of zero space and time, and theorized that both are absolute. Einstein changed all this by proving that each of us has our own personal time flow that need not match anyone else's, and that space and time are one and the same, merely different aspects of one another. By saying that entropy will always increase, the Second Law of Thermodynamics seems to give time an arrow. Loop quantum cosmology (Chapter 25) goes one step beyond Einstein, adding matter to what could be called space-time-matter. Holography (Chapter 27) indicates that everything that was, is, and shall be already exists, like frames on a movie reel that seem sequential until one sees the entire film reel at once. We already know that space and time are grainy approximations of a smooth flow, and that we have good reason to suspect that our universe may be holographic. Let's take a closer look at this thing we call time.

There is nothing more precious than time, for this is the price of eternity.
Louis Bourdaloue

What is Time?

On a practical level, we use time to plan and sequence events, track and compare event durations and intervals between them, and quantify rates of change (such as saying that a dropped object on Earth accelerates at a rate of 32 feet per second per second). Each of us is intimately familiar with time. It is a critical feature of religion, philosophy, and science, but defining time has proven to be almost as difficult as the problem of consciousness. The challenge is multiplied by the need to define time in a way that is non-controversial to all of the affected fields of study. This definition has proved very elusive.

Never believe anything anyone tells you without checking again and again.
Julian Barbour

We use time to define other quantities, such as velocity and acceleration. It is one of the few fundamental physical quantities in the International System of Units. We cannot define time by any of the units that are wholly or partially defined by time, because we would end up with a useless, circular definition. We have had great success defining time as a given number of standard cyclic events, such as the transition between two ground state levels of cesium 133, sand in an hourglass, or a swinging pendulum. This is great for practical purposes, but completely sidesteps the question of whether there really is such a thing as time that has an actual measurable flow. The unification of space and time achieved by Einstein complicates things, because questions about one are automatically questions about the other.

People like us, who believe in physics, know that the distinction between past, present, and future is only a stubborn, persistent illusion.
Albert Einstein

From a philosophical standpoint, there are two primary opinions about time. The first is the standard realist view that time is part of the structure of the universe, a special dimension where events take place sequentially. Newton supported this viewpoint, which is sometimes called *Newtonian time*. All past, present, and future times exist on a filmstrip that we can either play normally or rewind/fast forward to travel back and forth in time. Time can be viewed as a container for everything that did, is, and will happen, just like a filmstrip is a container for everything that can possibly happen in that particular movie. The second view holds that time is not a container, and that nothing moves through or flows with time. In this view, time is neither an event nor a thing, but a mental construct that helps us sequence and compare events along with space and number. As such, time can be neither measured nor travelled. Leibniz and Kant held this latter view.

Moving Through Time

The old men ask for more time while the young waste it. And the philosopher smiles, knowing there is none there.
R. S. Thomas

Time seems to have a nice smooth flow from future to past, with the present forming a sort of sliding window that we are powerless to change. We are born, live a little while, then die. It all seems that starkly simple. But is this really the case?

Time's Arrow

We have memories of the past, but not the future. Objects roll downhill. Hot things cool down. Entropy always increases. Or at least so it seems; in Chapter 26 we learned that entropy can decrease, tea and milk can separate. Any event that can happen can also unhappen. Of course, there are many ways to have high entropy and only a few ways to have low entropy, as we saw with the example of the 10, 11, and 1,000-page books.

We just saw that the two basic ways to see time are Newtonian and non-Newtonian. Let's peek under the hood.

Newtonian time is a special state that has been degrading from an initial or previous state of higher order. Place a bunch of atoms in a box, wait long enough, and they will eventually experience all states, even highly ordered ones. There are lots of possibilities, of which order and structure are a distinct minority. Religion calls these states miracles. Science either calls them statistical possibilities or invokes the anthropic principle. (See Chapter 26.) None of these explanations really work, because each of us is aware of a distinct entropy gradient, with time flowing "downhill" toward ever-higher entropy. But what if entropy suddenly decreased? Boltzmann opined that two beings on either side of a point of lowest entropy would experience their time lines flowing in opposite directions.

It moves. It moves not. It is far, and it is near. It is within all this, and it outside of all this.
Upanishads

There is also gravity. Leave a cloud of homogeneous gas alone long enough, and it will clump under its own gravity, thus giving an arrow to time. Black holes represent the ultimate clumps that seem to give time an irreversible arrow that is identical from all viewpoints, regardless of entropy. The single direction of time is also implied by the chain of cause and effect, in which the present separates past from future and cause from effect. The general consensus is that the effect cannot precede the cause. This too would seem to be the case regardless of the entropy gradient, and Boltzmann's assertion seems to be skating on thin ice. Just because quantum mechanics works regardless of which way time flows does not mean that time in fact flows in either direction—or in any direction at all.

Leibniz proposed looking at the universe not as space, time, and matter but as a fusion of them all, where instants of time are called "Nows." This is perfectly compatible with Einstein's view of unified space-time, where one can swap one for the other, and with both loop quantum cosmology and holography, which are themselves compatible with both quantum mechanics and relativity. A Now would therefore contain space, time, matter, and energy, all of which are simply different aspects of each other—a powerful validation of pure-geometry theories that have been extensively tested for over a century and found valid. Any of these things could be seen as redundant compared to the other, and a Now could be seen as

Time forks perpetually toward innumerable futures.
Jorge Luis Borges

both a container for these items and the thing-of-itself, in roughly the same way that a house is its own container.

Relative Time

Both Newton and the general and special theories of relativity see space and time as the most fundamental underpinnings of the universe, with objects being derived from those fundamentals. It is possible that this view is backward, that things exist while space and time are the derivatives. As if this wasn't mind bending enough, Einstein even elevates the simple concept of "now" to a true challenge. Simultaneity is nothing more than an abstraction we use to describe events in our world; it is not a property of the world itself.

We cannot understand relativity if we think of time as passing independently of the world. The common time measure we use that is defined by the Earth's rotation is inaccurate; tidal forces shift the Earth's mass and affect its rotation. Moon-based *ephemeral time* is much more accurate. Today, we use atomic clocks to measure time with extreme precision. Ephemeral time only exists because of the laws governing the solar system. It is therefore a property of the entire solar system, not just the Moon, and works as well as it does, because the solar system is functionally isolated from the rest of the universe; however, it is not perfectly isolated. No system within the universe is. This means that the universe itself is ultimately its own clock. Which begs a question: If there are multiple universes, is each universe perfectly isolated from all others, or do we need to take those other universes into account? If so, how should we do this? How can we do this?

Reference time describes how you see yourself relative to the world and is measured within your own personal inertial frame. Your answers remain consistent compared to your own frame, but measure across frames and the answers start to vary, such as the example I gave in Chapter 24 about time dilation. The one thing everyone across all reference frames will agree on is that light always moves at 186,000 miles per second, no matter what. Thanks to relativity, a single event can occur at multiple times for different observers, and the sequence of events seen by each observer may vary. Some of what someone sees as the past will be seen as the future by others. Time is therefore not a series of instants, but of smears

---

The past the future, are nothing but names, forms of thought, words of common usage, merely superficial realities.
T. R. V. Murti

All this took a long time, or a short time: for, strictly speaking, no time on earth exists for such things.
Friederich Nietzsche

that conceptually sound very similar to quantum waveforms. The idea of "now" has been reduced to a purely subjective judgment call. As Einstein said, "People like us, who believe in physics, know that the distinction between past, present, and future is only a stubbornly persistent illusion." Given the fundamental role that time seems to play in our lives, it is not much of a leap at all to understand Einstein's related comment that, "reality is merely an illusion, albeit a very persistent one."

Bidirectional Time

At its most basic, time is simply change; however, as I have said repeatedly, it seems to only work in one direction because the past is bounded, while the future is either not bounded or far less so. Hand someone a pile of individual movie frames, and the entire film could be pieced back together in perfect order easily enough, even if doing so would be a time-consuming process (There are 129,600 frames in a 90-minute movie running at the standard 24 frames per second.) This job is feasible because the illusion of time's direction is so strong.

All things flow, nothing stays still: nothing endures the change.
Heraclitus

Entropy suggests that time can reverse, but that it is not terribly likely to do so. Symmetry says that time must be able to flow in both directions without reversing on itself. Either way, very few things in physics care which way time flows. As far as a photon is concerned, it is present at the Big Bang, today, the end of the universe, and all times in between. Time does not flow for a photon. Everything in the universe is connected by both photons and other radiation that in essence "sees" everything, which could be seen as validating the mystics.

An extension of special relativity allows for faster-than-light travel; such particles are called *tachyons*. By this definition, all matter consists of sub-light *tardyons*. Conservation of energy, linear momentum, and angular momentum automatically appear in models that do not single out special moments in time, positions in space, or directions in space. Mathematician Emmy Noether (1882-1935) pointed out that Einstein's relativity follows, if we do not single out a direction in time.

Time flies like an arrow but fruit flies like a banana.
Groucho Marx

This brings us to an interesting conundrum: Space-time as a whole is real, but moment-by-moment time is not quite so real. Every moment is fundamentally the same as every other moment. Reality includes past, present, and future, and the flow of time truly is an illusion. Why is this illusion so stubborn? It is possible that conditions during the Big Bang imparted a direction to time; after all, breaking an egg—as opposed to unbreaking it—is the correct analogy for describing the beginning of the universe!

Rewriting History

Nobody has ever noticed a place except at a time, or a time except at a place.
Hermann Minkowski

If both quantum mechanics and relativity allow time travel and if later theories don't disallow it, then why can't we travel through time just as easily as we travel through space? Why are our motions through one of the space-time dimensions like being glued to our own personal conveyer belt, when we can presumably go anywhere we like in the other dimensions? And yet time travel seems to be impossible for all practical purposes because—to the best of our knowledge—no tourists from the future have ever visited us. It is of course possible that they have indeed visited and that history has been altered once or many times; in each such case, each of our histories would be altered, our memories and histories and current associations would change to match, and we would have no way to know. If this sounds like science fiction, then consider the delayed choice experiments we discussed in Chapter 23, in which the future literally affected the past.

The quantum concept of "past" is a little different than the classical definition, because all points in time are nothing but possibilities until an observation occurs. This observation bridges the gap between the quantum and classical worlds, and makes certain possibilities more likely than others. Collections of observations eventually combine to create a strand that may—or may not—restrict future possibilities. So what effect do future measurements have on this moment? Those measurements do not change our current measurements, but they do restrict how we can describe those current measurements in the future This is yet more evidence that everything we think is in our past is an illusion that could change at any time, with us none the wiser that anything was different.

We perceive duration as a stream against which we cannot go. It is the foundation of our being, and, as we feel, the very substance of the world in which we live.
Henri Bergson

History seems to be made of a mix of classical and quantum physics, of which we only have records. Those records—in various forms, such as memories—are all currently existing phenomena. For example, your thinking about and remembering your own childhood is happening in the present. The actual past makes no difference. There is no need to link successive world configurations into a single path.

Beyond Time

We may never know why we experience time, but we may be able to explain the structure found in each instant. The first thing we need to get our minds around is that the idea of an ever-moving present simply doesn't work. If the present leaves an impression on the past, then the future must likewise leave an impression on the present. Breaking time up into components labeled future, present, and past does not do nearly enough to explain time itself. Each component is part of a whole, and all components are linked. Future events are both contingent on the present and limited by the past. What we think of as flowing time is nothing more than a movie-like, frame by frame projection of individual experiences. If we view events as transactions, then we can see signals traveling forward and backward in time. For example, your choice to go to medical school could be seen as signaling the future that you will be a doctor. Similarly, the delayed choice experiments demonstrate that signals travel backward in time. Also, your memories of the past and the beliefs you formed because of them (see *The Enlightened Savage*) are constantly sending signals to your present. Everything is intertwined.

Physicist Fred Alan Wolf (1934-) says that creativity is nonlocal in time. What is creative today is pedestrian tomorrow. Could we be borrowing ideas from the future? Are we really sending signals back and forth? Maybe not. A famous equation by Wheeler and fellow physicist Bryce DeWitt (1923-2004) called the Wheeler-DeWitt equation says that the universe is static. If this is true, then nothing is ever truly happening—another lesson mystics have been trying to tell us for millennia. So if nothing ever happens, and if future, present, and past are not real, then where does the illusion of motion come from? Our brains have a lot of computing and processing power, and could be the source of this illusion. This idea is compatible with the idea of a timeless God wanting to have experiences within a static framework. A flying bird is another illusion, albeit one built around the very real phenomenon of flight.

Time, space, and causation are like the glass through which the absolute is seen. In the absolute there is neither time, space, nor causation.
Swami Vivekananda

There is a circle in all things that have natural movement and coming into being and passing away. Even time is thought to be a circle.
Aristotle

> *People normally cut reality into compartments, and so are unable to see the interdependence of all phenomena. To see one in all and all in one is to break through the great barrier which narrows one's perception of reality.*
> Aristotle

To paraphrase Barbour in *The End of Time*, all change on a quantum level comes from interference in stationary states with different energy levels; however, no change takes place in a system described by a stationary state. Machian dynamics meet quantum mechanics here... and if Mach was correct, then the universe has zero total energy, time does not exist, quantum cosmology has no dynamics (change), and the entire universe is both timeless and frameless. Here again, the implication is that time does not exist. Many stories and fables have the universe existing forever in stasis. The Big Bang beginning is today's scientific orthodoxy, but what if the universe is literally recreated from instant to instant? No two instants are exactly alike, so how can we say something was created in the past and continues to exist today?

Mathematician Hermann Minkowski (1864-1909) beat Einstein to the punch when he said that viewing space and time separately is incorrect. He predicted that all future natural laws would agree with relativity, and that finding those laws would require treating time as though it were space. Mach's view of the universe is a constant switching of configurations that creates a path, which in turn gives the appearance of history, In the end, the path is nothing but the universe moving from state to state in the same way a movie moves from frame to frame. Either way, there is nothing beyond the universe to time it as it changes states; history is simply a curving path through the universe's *configuration space*, or pool of possible configurations. To quote Mach, "It is utterly beyond our power to measure the changes of things by time. Quite the contrary, time is an abstraction, at which we arrive by means of the changes of things."

This sounds absolutely nonsensical—and it is, in a materialist world, or any belief system that interprets what we perceive as what we actually get. We will soon see that this idea makes perfect sense when viewed from a slightly different perspective.

Endless Nows

> *From wild, weird clime that lieth, sublime, out of space, out of time.*
> Edgar Allen Poe

Time seems to come in linear instants that are themselves created not by time, but by things. Each instant may be a three-dimensional image that is built from comparing many two-dimensional images. One three-dimensional image forms a

Now. Everything we were, are, and will be is all encoded in timeless structures. If our brain state at any one moment is a movie frame, then the flow of time comes from seeing one frame after another, but all possible frames always exist. We could say that God handed out all possible cards and then let us sort out which ones we would see and the order we would see them in to create the illusion of flowing time. Pierce the illusion, and every conceivable state can be present at once with nothing remotely resembling a flow of time.

The smooth arc of my cat jumping from my desk to the floor seems like one motion, but in reality there are a great many versions of my cat in all possible states. I somehow abstract one cat from that collection. This is why I think I see her jump. This has limits: Give someone a pile of snapshots of a ball rolling toward the edge of a table and ask her to predict where the ball will land. She will tell you that this cannot be done without a sense of time, because Heisenberg long ago showed that we can calculate position or momentum, but not both. Speed is not distance divided by time, but distance over actual state changes. Being able to see all instants of time at once would let us see the effects of laws of motion.

The Greek philosopher Zeno (ca. 490-ca. 430 BC) posed a paradox in which a bow fires an arrow that travels halfway to its target, then half of the remaining distance, half of that remaining distance, and so on. Such an arrow would never actually reach its target, and this is not a realistic scenario according to our sense of time; however, in the timeless universe, the arrow that leaves the bow never hits its target, because the arrow that finally hits home is not the same arrow.

Are there limits to the Nows we can select? Does Now A imply Now B, or can it be simply random? Boltzmann correctly pointed out that there are far more unstructured states than structured states, which means that structured Nows must be a rarity. Selecting only structured Nows must therefore be an extremely discriminating process. We could think of the universe as being a collection of all possible structures, a few of which have Nows, and a subset of those which have mutual consistency. We are exposed only to that tiny latter fraction. The ability to select only structured, mutually consistent Nows must by definition be extremely rational, even if

If even our concept of time has to change, then it seems we can be sure of very little.
David Fontana

Eternity is time. Time, eternity; to see the two as opposites is man's perversity.
Book of Angelus Silesius

> *We are all just visitors to this time, this place. We are just passing through. Our purpose here is to observe, to learn, to grow, to love... and then we return home.*
> Aborigine philosophy

you are not aware of it; everything you perceive has already happened, and is no longer a Now.

Returning to the playing card example, if all possibilities are contained in a full deck, then each living thing (and possibly non-living things as well) has its own Nows and slices its own space-time differently. This explains why each of us has different ideas about what happens at every moment and different ideas about just what reality is. If your reality consists of all the cards in your Now list, and if all Now lists are equally valid, then reality consists of all events in space-time. Again, every frame in a movie exists whether or not it is playing, and one can navigate the available frames any way one wants. Your brain makes selections and gives you the illusion of flowing time. Each leap to a new Now is caused by consciousness collapsing the waveform at each card, which determines which way(s) we can go. In this example, each card represents one Planck time interval.

Each Now also contains instants of consciousness and everything else needed to produce a complete universe in that instant, just like every movie frame contains the entire picture and sound track for that 1/24th of a second. The probability that we will select a given card for our next Now is the sum of the probabilities for all Nows that contain our individual experiences. All we experience is therefore manifested into existence by the experience embedded in each Now. We will return to this concept later.

Sensing the Nonexistent Flow

> *In any attempt to bridge the domains of experience belonging to the spiritual and philosophical sides of our nature, time occupies a key position.*
> Arthur Stanley Eddington

Explaining our sense of time requires us to explain consciousness. (See Chapter 30.) Meanwhile, it is important to understand that we do not see things as they truly are, but however our brain interprets those things. Any book of visual illusions will tell you just how easy it is to fool our senses. As for time, it is possible that our brain contains several Nows at once that it buffers and plays for us one at a time. Each Now may contain information about possible successor states, and the brain whips up the illusion of motion. Assuming our brains are Newtonian, how can we sense time? It does not make much sense to say that we can directly perceive an invisible time flow when we cannot directly sense anything else. Where does the sense of time come from? One possible answer is that Nows have an ultimate cause that is beyond time. Any individual Now that contains consistent records of processes that occurred in "the past," according to some yet-to-be-discovered natural laws, could trigger the sense of flowing time, especially when paired with the extreme organization of the average human brain.

Current records store everything we think we know about our "past," which seems more real with increasing consistency. Still, the past itself is never more than we can at least infer using present information. We could thus emphasize the definition of a Now as including all information needed to create the illusion of a consistent story.

The pioneers of astronomy and physics we met in Chapter 23 taught us to see motion where we sense none. Embracing the concept of Nows may help us see the stillness that is behind the supposed motion.

All we know about the past is actually contained in present records.
Julian Barbour

Choosing Nows

If time does not exist—and if the universe is a network of Nows that each contain information about consistent histories and where one can go from here—then we have to once again ask whether free will exists, or whether everything is deterministic. A holographic universe seems to argue for the latter, since it contains everything that is possible. In fairness, we can define Nows as either one universe at all possible times, or many universes that each form a separate Now. It seems to make more sense to go with the former as a model, because it eliminates the need for any real flow.

The flow that occurs is one of possibility. The nearer Card B is to Card A, the higher the possibility that Card B will come up instead of some other card. (This is not physical distance but probabilistic distance.) Imagining a deck of cards laid out on a table with a waveform as a blob overlaying the cards is a good visual. This opens the door for individual consciousness and choice to act within the constraints of what is possible. Thus, free will is real, and so is a small degree of determinism. We move across the grid of cards from A to B to C to D based on consciously collapsing the waveform at each card that influences the direction(s) we can take. Everything that can happen is waiting for us like a buffet that has been prepared in

I spit into the face of time that has transfigured me.
Yeats

advance; how we move through this flow (the specific items we eat) is up to us. The Plinko game on the television game show *The Price is Right* is another good example of this.

Potential arises from both direction and purpose. Without direction, possibilities could not become actualities, even if the possibility is staring us in the face.

Young Earth Creationism

The hardest thing of all is to find a black cat in a dark room, especially if there is no cat.
Confucius

Many people are uncomfortable with the idea of time having a beginning, because it seems to indicate divine intervention. It is possible that the universe has a finite but unbounded time, which could be seen as eliminating the need for a creator. Either way, we have several methods available for measuring the age of the universe, and all of these methods are giving us a consistent 13.7 billion years. Not everyone accepts this result, however. The Young Earth Creationism movement asserts that the JCI god Jehovah created the universe and everything in it sometime between 5,700 and 10,000 years ago. They also believe that God literally created the universe in six 24-hour days, based on a literal reading of Genesis. The people who buy into this idea are far from a fringe group. Various polls place the number of American adults who believe this as being somewhere between 10% and 45% of the total adult population. They even claim that scientific evidence supports their beliefs, and that any evidence that contradicts them is being misinterpreted. The Bible gives God 6 days to get the job done. Science says the process took 13.7 billion years. Who is correct?

You cannot step twice into the same river for fresh waters are ever flowing in upon you.
Heraclitus

In Chapter 38, I will explain how both groups can be literally correct. Meanwhile, we saw in Chapter 3 that Genesis 1 paints an uncannily accurate picture of the evolution of the universe, along with other creation myths. Perhaps we need to invoke science to reconcile the seeming difference, specifically relativistic time dilation and the Doppler effect we discussed in Chapter 24. A completely literal reading requires a patently dishonest God who embedded fossils in the ground, decayed elements, and placed photons in transit to make the universe appear to be much older. How such a God can then forbid bearing false witness in His own Ten Commandments is conveniently not explained. Nevertheless, we cannot say that any measure of time is the same today as it was at the Big Bang and thereafter. Thus, 6 days in God's frame could equal 13.7 billion years in ours.

I am not suggesting that this is in fact the case. I am simply opening the door toward a potential reconciliation of two seemingly divergent viewpoints. First, though, we must conclude our tour of the sciences and their implications by finally looking into our own heads to see what we can learn about the nature of reality, and about the fate that awaits us after die.

The implications are as profound as they can be. Time does not exist.
Julian Barbour

PART SEVEN

Mind and Matter

The Divine Savage
Revealing the Miracle of Being

Chapter 29

Interpretations

> *Physics is like sex: sure, it may give some practical results, but that's not why we do it.*
> Richard Feynman

I had a hard time deciding where to place this chapter. On one hand, it is about how different people interpret the results of quantum mechanics and related theories, and therefore belongs with the chapters about the science. On the other hand, our interpretations are all about belief, and belief comes from our minds. Also, we will see that every one of the interpretations we are about to look at inevitably leads back to consciousness, no matter how strongly its adherents claim otherwise. The efforts to remove the human mind (and any other sentient mind) from the equation have failed, some more spectacularly than others. It therefore seems appropriate to place this chapter here as part of our exploration of brain, mind, and consciousness.

> *Nothing is accidental in the universe—this is one of my Laws of Physics—except the entire universe itself, which is Pure Accident, pure divinity.*
> Joyce Carol Oates

Quantum mechanics is widely accepted, along with the experimental results and its possible explanations about the nature of reality. How to interpret those results and explanations is where the dispute lies. There are two basic schools of thought. The first says that all observations need an observer, which means the universe must have consciousness. Wigner believed that quantum mechanics proves the existence of a universal consciousness. The second basically ignores the problem. Most scientists are in this latter camp. There is a concerted effort underway to remove probability and any interpretation that separates the system and the observer. These efforts have not been successful yet, but we will soon see that it makes very little difference. Still, we owe it to ourselves to take all interpretations seriously, while keeping in mind that every interpretation has its own challenges and limitations.

The Schrödinger equation is the heart of quantum mechanics. Different interpretations have different workarounds, but the simplest way to see it is that it holds. Thus, all physical systems are fully described by a waveform that evolves according to Schrödinger's equation. Anything else is simply an elaborate attempt to circumvent the very simple truth that the universe exists in a superposition that nevertheless feels discrete. Everett took the superposition to its logical extreme by assigning a different observer to each position. Each will see the world pretty much as we expect it, and will be unaware that any sort of superposition has occurred. This seems like an overly complex interpretation, far more complex than the others, and there seems to be little reason to accept it.

On a subatomic level, things seem to just happen without any apparent cause at least some of the time, such as the waveform that simply collapses to an outcome, particles that simply wink in and out of existence, or the "random quantum fluctuations" that so many physicists love to talk about. All of these examples and more seem to violate causality. We must either assume that cause and effect is not universal, or that there is something very wrong with any model that asserts the existence of an objective and universal reality. This leaves us with the observer-created reality that quantum mechanics seems to need in order to be a viable theory. But just what is an observer?

Some interpretations seem to require a conscious observer to collapse the waveform. This puts consciousness in the mix, but does nothing to explain consciousness beyond simply assuming that it exists and that it plays a role; however, these theories are anything but reductionist. Quantum mechanics may have something to do with consciousness. It further seems that any theory of consciousness that is solely based in matter—where consciousness is seen as a function of our brains—is doomed to failure.

This creates a problem: What is the difference between the consciousness of the observer and that of the observed? This

---

It should be possible to explain the laws of physics to a barmaid.
Albert Einstein

I am somehow less interested in the weight and convolutions of Einstein's brain than in the near certainty that people of equal talent have lived and died in cotton fields and sweatshops.
Stephen Jay Gould

presents a paradox, unless we accept an incomplete solution that does not give any privilege to any observer. The easiest way to do that is to simply leave consciousness out of the equation, which avoids the problem altogether.

Traditional science rests on three fundamental assumptions:

- Realism asserts the existence of a separate, independent, and objective reality. Our individual experiences are subjective, but they all occur within the context of this objective reality.
- We therefore believe that science describes this objective reality that is completely independent of consciousness or observation.
- We further believe that the results of Experiment A should not of themselves affect, or be affected by, an Experiment B that is conducted elsewhere. In other words, we believe in locality. As we have seen, quantum mechanics conflicts with this belief.

The many possible interpretations of quantum mechanics basically boil down to these three primary options:

- Copenhagen
- Many worlds
- Hidden variables

We have talked about all of these interpretations before, but let us now take the time to examine them in depth.

Copenhagen

The Copenhagen interpretation is so named, because it was created in Copenhagen by Bohr and Heisenberg circa 1927 as an extension of the probabilistic waveform created by Born. In this interpretation, questions about what happened before an observation are meaningless; an observation selects one of the possibilities allowed in the wave function in accordance with the probabilities assigned to each possibility. This can occur randomly. Thus:

There are children playing in the street who could solve some of my top problems in physics, because they have modes of sensory perception that I lost long ago.
Robert Oppenheimer

I tell you the solemn truth, that the doctrine of the Trinity is not so difficult to accept for a working proposition as any one of the axioms of physics.
Henry Brooks Adams

- Each system is completely described by a waveform that represents an observer's knowledge about that system, per Heisenberg.

- Nature is probabilistic. The probability of any event is related to the square of the height (amplitude) of the waveform related to that event. This is known as the Born rule.

- The Heisenberg Uncertainty Principle says that one cannot know the values of all properties of a system. The unknown properties are described using probabilities.

- Matter has both particle-like and wave-like properties. Experiments can show either set of properties, but cannot show both at the same time, per Bohr.

- Measuring devices are classical and measure classical properties, such as position and momentum.

- The quantum description of a large system should be a close approximation of the classical description for that system, according to Bohr and Heisenberg's correspondence principle.

Nevertheless, all of us who work in quantum physics believe in the reality of a quantum world, and the reality of quantum entities like protons and electrons.
John Polkinghorne

Observer-created Reality

The contents of your desk drawer do not exist until you open it and look inside, thus collapsing the waveform. Position and momentum are meaningless without observation. Unless and until an observation is made, everything exists only in possibility. This is the standard Copenhagen interpretation. If it meaningless to ask what happens before an observation because possibility is all that exists, then the only logical explanation is that the observer creates the reality by "collapsing" the waveform into a single reality. It is the measurement that creates and/or modifies reality.

Measurements are never benign. This includes "null" measurements, such as the detector in a dual slit experiment that simply watches a particle sail by. Measurements collapse the waveform, period. Some believe the observer can be a device, such as a Geiger counter; however, such measurements are meaningless unless and until seen by a conscious observer. Until that occurs, the measuring device simply becomes part of the experiment and adds another set of probabilities to the

Even when I was studying mathematics, physics, and computer science, it always seemed that the problem of consciousness was about the most interesting problem out there for science to come to grips with.
David Chalmers

> *The observer, when he seems to himself to be observing a stone, is really, if physics is to be believed, observing the effects of the stone upon himself.*
> Bertrand Russell

waveform. Conventional wisdom says that we should use quantum mechanics to describe things on very small scales and classical physics for the regular world, ignoring consciousness. This sounds good, except that the dual slit experiment has demonstrated that quantum weirdness is not confined to individual particles. A buckyball is much larger than an electron or photon, yet it still exhibits wave/particle duality. Conventional wisdom therefore seems harder and harder to justify.

There's more. The Zeno effect occurs when one places an unstable particle under constant observation. This "suicide watch" prevents the particle from decaying, even if the observation carries on much longer than one could reasonably expect the particle to survive on its own. An inverse Zeno effect can also occur, because repeated observations can force a system to evolve in a measured direction. Either way, it is impossible to separate the observer and the observed. The same thing happens with each of us: Focus on a single idea, and we can keep it fresh in our minds. Lose focus and the idea soon decays and vanishes. Extensive focus can also cause us to remember things we would otherwise forget. For example, repeating a phone number several times helps us remember it far better than simply hearing it and moving on to other things.

Von Neumann believed that particles do exist in definite states before observation, and that the state can change when the observation occurs. This avoids the "existing only in possibility until observed" problem at first glance, except that we cannot say for sure which state the particle is in before making the observation. Note that I am using "observation" and "measurement" interchangeably, because a measurement requires an observation, and an observation is automatically a measurement. For example, simply observing an object in your vicinity collapses that object's waveform, even though you are not trying to measure anything.

So what—or who—is doing the measuring?

The Role of Consciousness

> *I am now convinced that theoretical physics is actually philosophy.*
> Max Born

Let's return to Schrödinger's cat. We already know that the Copenhagen interpretation says that the cat is in a superposition that is both alive and dead at once until we open the box and look inside, which collapses the waveform to one outcome or the other. This alone implies that mind is superior to matter, because the observation and associated observer seem to play such a central role. The fact that observation impinges on the world is well-known because—among many examples—each of us has seen and felt the effects of staring and being stared at from time to time. This impingement was thought to be pretty minor, even negligible in most cases, but it turns out that observations are part of the bedrock of reality and cannot be ignored or otherwise compensated for. The Zeno effect and dual slit experiment place the observer front and center because, what s/he chooses to measure directly impacts the outcome.

Indubitably, magic is one of the subtlest and most difficult of the sciences and arts. There is more opportunity for errors of comprehension, judgment and practice than in any other branch of physics.
Aleister Crowley

What is a measurement? A measurement occurs whenever someone observes or interacts with a system. As I mentioned above, it is easy to say that this happens when the measuring device enters the system, but the fact is that the system simply grows larger and more complex, because the device itself is covered by the Schrödinger waveform. The system now has two components, the objects being measured and the device, but we only see one outcome. We know that the actual collapse happened either earlier, when the device entered the system, or when we looked at it. Either way, we only know the outcome when we look at it. To us, the entire system, measuring device included, remains in superposition, unless and until we look at the device. This directly implies that consciousness is involved, and that consciousness causes the collapse to occur. Nothing else seems to be able to cause collapse, only to expand the waveform. We are forced to conclude that consciousness influences—if not outright creates—matter.

Why does adding devices and other components only make the system more complex, with successively wider collections of possible states? What makes conscious interaction different and why? It would seem that conscious observation completes a kind of circuit, just as turning on a light switch completes an electrical circuit. But how? What is being completed? Why is consciousness seemingly able to do this, but nothing else? There seems to be some hidden layer to what is going on. Physicists tend to reject this idea, because the hidden layer is, well, hidden, and it is impossible to find something "real" that can make this happen. Wigner came very close to throwing in

The new formula in physics describes humans as paradoxical beings who have two complementary aspects: They can show properties of Newtonian objects and also infinite fields of consciousness.
Stanislav Grof

the towel when he said that the content of consciousness is the ultimate reality. He stopped short of going all the way and calling consciousness a new entity, but the implication could not be any clearer.

Physics is experience, arranged in economical order.
Ernst Mach

Is physics tied to human awareness? Can animals cause collapse? If so, does this extend to all animals or only a subset? Can plants cause collapse? If only some life can cause collapse, what distinguishes between those species that can and those that can't? Throwing consciousness into the mix adds a whole new set of questions that may have no easy or fixed answers. This is why many physicists are so eager to avoid the problem by any means necessary. The Ghirardi, Rimini, Weber (GRW) theory tries to avoid the problem by saying that waveforms are unstable and eventually collapse on their own anyway. An extreme version of the Copenhagen interpretation says that the quantum world does not exist. This is how far some scientists will go to avoid tackling the problem head-on.

Physicists are not gods, and quantum mechanics is imperfect, but neither can hide what their theories and interpretations omit, which is an explanation of just how consciousness fits into the picture. I personally cannot imagine a more intriguing topic, hence this book.

Biocentrism Revisited

We must be physicists in order to be creative since so far codes of values and ideals have been constructed in ignorance of physics or even in contradiction to physics.
Friedrich Nietzsche

Quantum mechanics tells us that the universe seems to have been around for billions of years. Our Sun is a third-generation star, and all of the material on Earth is made of third-generation materials. Carbon, the element on which all life is based, has a quirk: The steps from hydrogen to carbon involve unstable nuclei. Three helium nuclei must collide at the same time to create carbon, which is more than a little unlikely even inside stars. Something seems to be happening inside stars that is helping the process along. Carbon itself seems to have a resonance that helps stars make lots of it, despite the instability problem. This resonance depends on the strong force that keeps atomic nuclei together. What an interesting coincidence.

Would the universe exist if it lacked life, perception, and consciousness? Creation myths say that life is the universe's fundamental creative power and often express the beginning of the universe as a kind of birth, as we learned in Chapter 3. We have glanced at the subject of biocentrism before, and it bears closer examination now. The Copenhagen interpretation implies biocentrism, that the universe existed only in potential until the first observation created it at that moment along with the compatible history all the way to the Big Bang. From there, the inverse Zeno effect described above helped guide the universe along and possibly even guided the evolution of life on Earth, which would certainly explain the punctuated equilibrium we saw in Chapter 19.

The extremely fine-tuned structure of the universe can be explained using biocentrism, which makes sense because of both the Copenhagen interpretation and the blindingly obvious fact that the universe is hospitable to life. Biocentrism holds that space, time, energy, and matter exist because of—and are expressions of—consciousness. This becomes especially plausible in a timeless universe that consists of holographic Nows. (See Chapters 27 and 28.) Materialist science can offer no comfort to those who are afraid to die, but biocentrism implies—if not flat out promises—something more than simply winking out of existence. Space, time, energy, and mass are equal. The sum total of energy in the universe is 0, which means that space, time, and mass are also 0. The direct implication is that consciousness created the universe and continues to create it. If this is true, then how can consciousness be extinguished?

Let's revisit entropy for a moment. Imagine a box with two compartments: one full of oxygen, the other full of nitrogen. Open the door between the two compartments, and come back a while later. You will find that the gases have mixed thoroughly. We know from Chapter 26 that there are many more ways for the gases to be mixed than to remain separate; it is this inequality that we call entropy. This seems to be irreversible, because of the long odds against the gases separating spontaneously. We can see the movement of the atoms, not as time, but simply as movements through Nows, which makes it just another natural phenomenon. When we look at it this way, it becomes apparent that what we call entropy may simply refer to our mind's inability to see order, which does not mean that order does not exist. Our sensation of being on the leading edge of constantly moving time, where tomorrow has not happened and yesterday is gone, is an illusion whose purpose

---

Could Hamlet have been written by a committee, or the Mona Lisa painted by a club? Could the New Testament have been composed as a conference report? Creative ideas do not spring from groups. They spring from individuals. The divine spark leaps from the finger of God to the finger of Adam, whether it takes ultimate shape in a law of physics or a law of the land, a poem or a policy, a sonata or a mechanical computer.
Alfred Whitney Griswold

is to create an organized pattern. A biocentric universe does not have—and cannot have—have a linear time flow. Seeing all Nows at once would be more than a little overwhelming.

It should be obvious that the Copenhagen interpretation is perfectly compatible with a biocentric universe; otherwise, it makes no sense whatsoever. You cannot have it both ways by saying that the observer creates the measurements that create quantum results that only exist as possibilities beforehand, only to turn around and say that the universe itself can exist independent of observation. This is why Wheeler believed in a participatory anthropic universe that requires observers to come into existence; the pre-life universe came into being retroactively, the largest-ever delayed choice experiment.

We can think of the universe as a closed system with nothing beyond it. Even if there are other universes, we need to presume that each universe is completely cut off from all others, in the absence of any evidence to the contrary. Seeing the universe as a closed system helps explain why nothing exists independent of an observation/measurement. An open universe would be subject to outside influences, and we would be unable to validate the Copenhagen interpretations.

The only explainable, non-arbitrary criterion for waveform collapse is that it happens when the system affects some being's consciousness. This implies duality, or the separation of brain and mind. We already know that dualism is untenable, because there is no mechanism for transferring information (energy) between the physical and spiritual worlds. We are left with idealism as the only viable option. Conscious collapse becomes a fundamental property of the universe, which means that the Schrödinger equation governs all physical systems. Again, no observer means no measurement, which means that the physical world exists only in superposition. The smoothness we perceive in the macro world means only that our perception is too limited to see the grainy quantum world, just like a slightly blurry photo that will not show the finest details. The buckyball dual slit experiment shows us that the separation between the quantum and macro worlds is literally one of perceived degree and not one of either kind or actual degree.

It seems to the ordinary observer that nothing can be more remotely and widely separated than some so-called act of 'consciousness' and a material object.
Hugh Elliott

The reason why our sentient, percipient, and thinking ego is met nowhere within our scientific world picture can easily be summed up in seven words: because it is itself that world picture.
Erwin Shrödinger

The Greek philosopher Anaxagoras (ca. 500BC-ca. 428BC) believed that the spirit is the reason for everything that has total power over all motion. He seems to have been more right that he could ever have imagined. This image bears an uncanny resemblance to the formation of structures during the inflationary epoch that arose from an initial structure-less state that itself arose from quantum fluctuations—the "spirit hovering on the waters" from Genesis 1:2. The random irregularities in the inflating universe evolved in all possible forms, until the first creature with sufficient consciousness perceived itself and caused everything to snap into place. Sadly, only a few physicists hold this view, presumably because of the strong, almost visceral urge science has to distance itself from anything that smacks of religion. After the way religion treated science for 2,000 years and continues to treat it, I can't say I blame them in the slightest.

Subjectivity and objectivity are only two sides of one consciousness.
Sri Aurobindo

Given that cosmic background radiation was only discovered in the 1960s by physicist Arno Penzias (1933-) and astronomer Robert Wilson (1936-), we could say that the waveform for that radiation remained in superposition until the 1960s. Further measurements, such as COBE and WMAP, collapsed the waveform even more. The cosmic background radiation reflects the quantum uncertainties that existed during the inflationary epoch, which are ultimately responsible for all of the structure we see in the universe. We therefore cannot rule out that Penzias and Wilson collapsed the waveform of the entire universe right up to and including their own births when they first measured that radiation. Alternatively, we could see earlier observations as collapsing the universe waveform, which in turn restricted the number and type of superpositions available to the cosmic background radiation.

How does this work? We can look to Everett (see the discussion about the Many Worlds interpretation, below) for an explanation. Everett believes that the universe splits into multiple copies whenever a quantum event occurs with multiple possible outcomes. We could therefore see the waveform of the universe splitting off into nearly infinite branches like a tree run amok, with each split (and split of a split) representing different possible histories. The moment the first conscious observer evolved on one of those branches and completed the proverbial meaning circuit by making the first

Matter is not that which produces consciousness, but that which limits it.
Ferdinand Schiller

observation, all of the branches collapsed into a single line that represents an actual history. Thus was an immanent reality chosen from among many transcendent possibilities. This is similar to a film editor who may have to go through many hours of film to extract a few minutes of usable footage.

Problems with Copenhagen

The single largest problem with the Copenhagen interpretation is that it forces us to take consciousness seriously, and possibly even view it as being more fundamental than matter—a view that discounts materialism. Since the vast majority of scientists subscribe to materialism with the same fervor normally reserved for the devoutly religious, it is no wonder that alternative viewpoints such as this are having such a hard time reaching the mainstream. This does not mean that the mainstream is correct or that biocentrism represents magical or wishful thinking. After all, Semmelweiss was essentially committed for suggesting what we take for granted today: that disinfection could dramatically lower hospital mortality.

Hidden Variables

The hidden variables interpretation holds that objective reality does in fact exist—that objects have definite position and momentum—but that we cannot detect or determine what those values are. What we call random becomes a function of our own limitations, and thus a sort of pseudo-randomness, because order underlies everything. The Copenhagen interpretation is probabilistic, meaning that it predicts the probability of various outcomes rather than the outcomes themselves. This is how measuring the same property in two seemingly identical systems can yield different results. The obvious question is whether some deeper layer of reality exists beneath quantum mechanics that requires its own theory. This new theory would then be able to predict the outcome of all experiments with absolute certainty. The seemingly probabilistic nature of quantum mechanics could then be replaced with a deterministic model. In other words, the Copenhagen interpretation could be just another approximation.

I am acutely aware of the fact that the marriage between mathematics and physics which was so enormously fruitful in past centuries, has recently ended in divorce.
Freeman Dyson

I study myself more than any other subject. That is my metaphysics, that is my physics.
Michel de Montaigne

This is not mere speculation. Bohm demonstrated that the Schrödinger equation can be written in classical terms using a quantum potential that is similar to—but much smaller than—the waveform. The objective fundamental property underlying the seemingly random universe is the hidden variable. Where Copenhagen implies the primacy of consciousness in creating reality, the hidden variables interpretation says that all properties do exist, even though they may not be measurable; Copenhagen is incomplete because it does not represent all possible realities.

David Bohm

Bohm felt that what we perceive as tangible realities in life are illusions, that deeper layers of existence underlie what we perceive. These layers are the source of what we see and experience in the physical world, which functions like a hologram. Bohm called this the *implicate* (connected or enfolded) order. There may be infinite degrees of order. What we call random may in fact be a level of order that is too high for us to detect. For example, picture a tub of glycerin with a handle that can turn the glycerin. Put a drop of ink in the glycerin, and turn the handle. Notice that the ink disappears. Now turn the handle backwards, and watch the ink reappear. What you are seeing is varying degrees of either manifest (explicit) or hidden (implicit) order.

Bohm saw that quantum mechanics is not consistent with actual positions and velocities, a view that inspired Bell to explore the matter more deeply, as we saw in Chapter 23 and will soon see again. To Bohm, quantum mechanics does nothing more than reveal the limits of our ability to know. Where Einstein linked space and time, Bohm linked everything. Look around you, and imagine everything you see as being part of the same thing. This is not to say that the universe is simply a blob; the analogy in this case would be the many ripples, waves, and eddies in a river that each have their own characteristics while also being part of the same river. In this example, it is impossible to tell exactly where each feature in the water begins or ends, and it is therefore meaningless to make the distinction.

Dividing the universe into living and nonliving is similarly meaningless, according to Bohm. Everything in the universe

If you miss one day in physics, that's it.
Robert Iler

The pace of science forces the pace of technique.
Erich Fromm

has life and intelligence, even a rock. Past, present, and future are all present everywhere. Bohm thus gave spiritualists more reason to believe that the universe is a single whole. Thus, the nonlocal hidden variables governing a particle's trajectory may need to take other galaxies into account, because everything in the universe helps compose the global waveform that governs particles throughout the universe. This implies a lot of complexity behind something that seems simple enough, if counterintuitive. We should therefore be suspicious of this view, while also keeping in mind that the overall pattern of scientific discovery has been to reveal an ever more complex and intricate universe.

> *Quantum mechanics is very impressive. But an inner voice tells me that it is not yet the real thing. The theory yields a lot, but it hardly brings us any closer to the secret of the Old One.*
> Albert Einstein

Bohm also believed that particles are real even when not observed. Most physicists might be inclined to disagree, but at the same time, they either avoid considering the meaning of this or simply deny it altogether. Nevertheless, Bohm felt that this hidden world can be understood without any recourse to consciousness or potential states. But in the end, all he did was rename parts of the Schrödinger equation, albeit in a way that allows us to expand on the idea that more is happening than meets the eye. So what has changed when we repeat an experiment and get different answers? Are we seeing God playing dice? This question becomes even more relevant when one considers that validating Bohm would require some mechanism for guiding the particles to determine the outcome... and right back to consciousness we go.

Bell's Theorem

> *Nothing can be more incorrect than the assumption one sometimes meets with, that physics has one method, chemistry another, and biology a third.*
> Thomas Huxley

Bell is best remembered for Bell's Theorem, which shows that interpretations of quantum mechanics that rely on local hidden variables (as opposed to nonlocal) require certain conditions that are violated by measurements performed on entangled quantum systems. If an objective reality exists, then things must behave in definite, testable ways. Any interpretation of quantum mechanics can give us a world that looks random and that can even work with the Heisenberg Uncertainty Principle; however, the fact that one particle acquires definite properties when its entangled twin is measured means that quantum mechanics must disagree with any possible theory of hidden variables.

Tests that did not violate Bell's inequality would mean that objective reality had been proven true; however, results that went the other way would make it impossible to presume that any objective reality exists. Imagine two correlated photons headed toward two polarized filters. The chance that the second photon will go through its filter if the first one goes through depends on the angle between the two filters. Quantum mechanics says that each photon is both individual and probabilistic. If the photon has a different angle then it chooses whether or not to go through the filter, as if it had a plan; however, both photons cooperate so that both make it through the filter if one does. Quantum mechanics says that going through the filter must depend on the angle between the two filters, no matter what angle the other photon's filter is set to, even though the photons have no way to communicate.

Bell's theorem says that there are limits to what correlated photons can do even if they plan ahead. So is quantum mechanics wrong? Tests have proven that quantum mechanics is correct; the photons have no advance plan, but seem to communicate telepathically. Nonlocality is therefore proven, and objective reality is disproved, even when the filter angle is changed so quickly that the photons could not "know" about it. The photons have no plan, there are no local hidden variables, the classical worldview is nothing but an approximation, and all objects are correlated to some degree, because everything in the universe interacted with everything else in the universe at the beginning. (In fairness, these correlations are too weak to detect in the vast majority of cases.)

Bell's theorem has been subjected to an exhaustive battery of tests that have both ruled out local hidden variables and affirmed nonlocal "spooky action at a distance." Some physicists have proposed two loopholes in Bell's arguments, the detection and communication loopholes. Each proposition prompted fresh rounds of testing that reaffirmed the original results. Bell's theorem has been affirmed by an enormous amount of supporting evidence, and is therefore considered one of the cornerstones of modern quantum mechanics. Some have even called Bell's theorem the most important discovery of the second half of the 1900s.

Universal connectedness is real. Objects that once interacted continue to influence each other across time and space.

Quantum mechanics brought an unexpected fuzziness into physics because of quantum uncertainty, the Heisenberg uncertainty principle.
Edward Witten

It is impossible to trap modern physics into predicting anything with perfect determinism because it deals with probabilities from the outset.
Arthur Eddington

Events happening anywhere in the universe affect each of us, however slightly and undetectably. Physicist David Mermin (1935-) once said that, "Anyone who is not bothered by Bell's experiment has to have rocks in his head." For good reason: Bell's theorem puts a powerful torpedo directly into the heart of materialism.

Problems with Hidden Variables

We have already learned that local hidden variables do not represent reality. D'Espagnat goes one step further by invalidating hidden variables altogether for three reasons: First, eliminating hidden variables simplifies the mathematics. Second, the present Copenhagen model works plenty well to predict outcomes. Third, adding more variables does not explain anything that quantum mechanics does not already explain, and therefore violates Occam's razor. Further, ongoing experiments have revealed that an increasing number of hidden variable theories are incompatible with observations. For example, an experiment involving a system of trapped ions shows that quantum mechanics conflicts with hidden variable theories, no matter what quantum state the system is in.

Relativity must replace absolutism in the realm of morals as well as in the spheres of physics and biology.
Thomas Cochrane

Many Worlds

Many worlds theory (the theory that the universe splits at every quantum event) should not be confused with multiverse theory (that our universe is one of many, and thus the apparent tuning for life is nothing but a statistical probability). I should also point out that these two theories are not mutually exclusive.

Bohr did not take his theory to the logical conclusion, choosing to view the experimenters and any machinery involved in the experiment as separate from particles. He only cared about the measurement itself. The standard Copenhagen interpretation is based on a clear distinction between a physical system that follows quantum rules and a measurement that follows classical rules. The particle has a wave function of "here or there" that collapses to "here not there" when measured. We cannot treat the system and the observer differently. If we believe that the universe runs on quantum mechanics, then the

It is hardly to be believed how spiritual reflections when mixed with a little physics can hold people's attention and give them a livelier idea of God than do the often ill-applied examples of his wrath.
G. C. Lichtenberg

wave function must include the measurement as well. If this is true, if the system evolves in a deterministic fashion, then why are we even bothering to talk about probabilities? We saw how Bohm answered this by allowing both definite position and speed, while also cloaking them in a hidden layer.

Everett did not like any of this, and devised his "many worlds" theory to solve the sticky philosophical problems raised with the Copenhagen and implicate order interpretations. His approach takes quantum mechanics literally, in that anything that can happen does happen, and the universe simply splits up every time, thus avoiding waveform collapse. Schrödinger's cat is resolved by saying the cat is alive in one universe and dead in another. The observer also splits into copies. Neither copy is aware of the other or the fact that a split has taken place. The universe splits into as many copies as needed to cover all of the possibilities contained within a waveform, and these copies also split up. Splitting occurs at every quantum opportunity when some A or B possibility opens up. One can see that the number of universes quickly becomes practically infinite. This is a purely deterministic interpretation, of course.

Sidestepping the Observer

On one hand, Everett's solution is extremely attractive because it is laissez faire—in that everything that can happen will happen—while also completely cutting consciousness out of the loop. The fate of each universe is deterministic, but the collective possibilities are probabilistic, as each instant brings forth countless more minds that each have just as much right to call themselves Anthony Hernandez (or you, or whoever) as any other copies. According to Everett, each of us is an army. On the other hand, this theory is both messy and seems to completely violate Occam's razor. Then again, all interpretations of quantum mechanics are a bit crazy.

The three leading interpretations of quantum mechanics seem to be:

- **Wigner:** Consciousness causes waveform collapse
- **Bohm:** Nonlocal hidden variables; the trajectory of any particle depends on the trajectories of all particles in a giant tangled hierarchy

---

To the pure geometer the radius of curvature is an incidental characteristic - like the grin of the Cheshire cat. To the physicist it is an indispensable characteristic. It would be going too far to say that to the physicist the cat is merely incidental to the grin. Physics is concerned with interrelatedness such as the interrelatedness of cats and grins. In this case the cat without a grin and the grin without a cat are equally set aside as purely mathematical fantasies.
Arthur Eddington

- **Everett:** The universe splits whenever a quantum probability exists. This is by far the simplest theory. It is local, needs no new mathematics, and is completely compatible with what we perceive as reality. One offshoot of this is that each of us will be dead in infinite copies of the universe and alive in infinite copies. At least half of our infinite copies are guaranteed immortality. Whether the copy of you reading this book or the copy of me writing this book is one of the lucky ones is anybody's guess, but at least 50/50 odds are better than nothing.

Problems with Many Worlds

It would be better for the true physics if there were no mathematicians on Earth.
Daniel Bernoulli

The chief problem with the many worlds interpretation is that it is anything but an efficient answer, because it quickly leads to infinite regression. Science seeks simple and efficient answers that avoid infinite regressions, instead of welcoming and even mandating them. Even Bohm's theory lacks economy. We cannot rule out the possibility that the universe does indeed split in two at every quantum opportunity, but then neither can we conclusively rule out a teapot orbiting between Mars and Jupiter (or elsewhere for that matter). Communications between split universes cannot happen, which means we cannot test Everett's theory. Science mocks religion for presenting untestable theories—and must hold itself to the same standard—but nobody is mocking Everett.

Einstein used a cosmological constant to avoid the implication he discovered within his own work that indicated that the universe is not static. He abandoned the idea after Hubble observed the red shift of receding cosmic objects that showed that the universe is expanding. The recent discovery that the expansion of the universe is accelerating has renewed interest in a cosmological constant. If Everett's theory is correct, the only tangible effect of billions of other universes would be to keep the cosmological constant at 0. Of course, it should be noted that the discovery of a cosmological constant that equals 0 would not by itself be evidence that Everett is correct.

We live in a Newtonian world of Einsteinian physics ruled by Frankenstein logic.
David Russell

The most severe problem with the many worlds theory is that it offers no way for an "I" as a unified experience to emerge from incoherent waveforms within a brain. On a similar note, yesterday's "I" is today's "we." How does any version get a sense of "me," and what are anyone's swarms of copies doing/thinking/feeling/experiencing at any given moment? Philosopher Derek Parfit (1942-) argues that personal identity consists of nothing more than facts, such as the physical continuity of one's body, memory, etc. Each copy is therefore just as much "me" as I am. This is a rather unsettling conclusion, because it reduces personal identity to an illusion; however, that does not make it wrong.

Quantum physics thus reveals a basic oneness of the universe.
Erwin Shrödinger

Chapter 30

The Universe in our Heads

[The brain] is the barrier that stands between us and the limitless potential of the universe.
David Darling

We just saw in Chapter 29 that every interpretation of quantum mechanics involves consciousness, whether we like it or not. Follow-on theories such as superstring, M theory, and holography build on quantum mechanics, and must therefore incorporate the same interpretations. This begs a very fundamental question, one that is central to the main topic of this book: What is consciousness?

To answer this question, we must distinguish between the following three components:

- **Brain:** physical organ in our heads
- **Mind:** the functioning of the physical organ
- **Consciousness:** our subjective experience of mind.

This is a fuzzy distinction at best, and I will do my best to observe these boundaries while acknowledging that they are more than a little subjective.

The Human Brain

In our infinite ignorance, we are all equal.
Karl Popper

The brain is a very special organ. It is the only organ completely encased in bone and separated from the rest of the body in its own appendage. Its primary connection to the rest of the body is also encased in protective bone. Both brain and spine are also surrounded by a unique cushion of cerebrospinal fluid. Filters restrict the chemicals that can access the central nervous system from the bloodstream. Medical science has created prosthetic limbs, artificial hearts, intravenous feeding, external dialysis, iron lungs, and more. Doctors can also transplant just about any organ from one person to another. Despite all these advancements, the brain is the one organ that is both absolutely beyond reach and utterly irreplaceable.

If the brain was so simple we could understand it, we would be so simple that we couldn't.
Lyall Watson

The brain remains a mystery after thousands of years of surgery, dissection, vivisection, psychology, psychiatry, imaging, and other studies and probes. This begs a simple yet profound question: Can we ever fully understand what's happening inside our own heads? Gödel's theorem stipulates that a computer of a given size and capacity can only model another computer that is smaller/simpler than itself. By this logic, human brains will never be able to understand the human brain, whereas a sperm whale—whose brain is six times larger than our own—should have no problem doing so. In plainer terms, if the human brain were simple enough for us to understand, then we would be too simple to understand it. Then again, we have managed to get some very intriguing insights.

Anatomy

We saw in Chapter 9 that the brain generally consists of the cerebrum, cerebellum, and medulla. This diagram provides a more detailed look at the brain's internal components. We will explore these areas and more below.

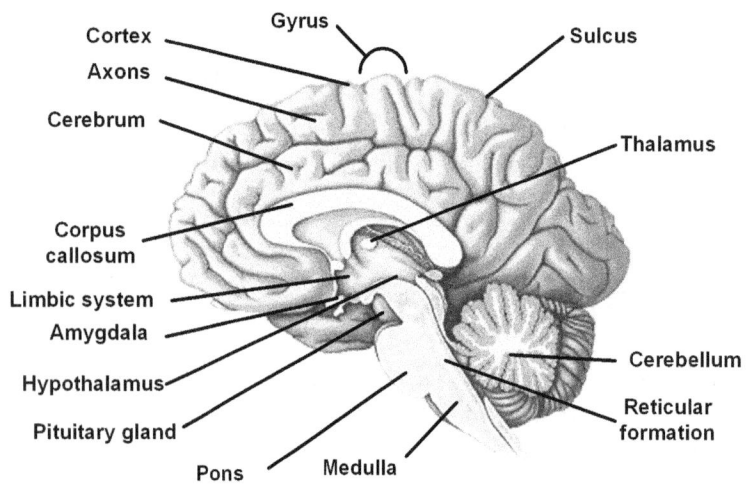

[The brain] isn't a deterministic machine with an invariant code; it's statistical, probabilistic.
Alan Gevins

> *Neurologically, truth and fiction are subjective values created by the brain.*
> Andrew Newberg

The cerebrum is the largest part of the brain. It is bisected into left and right hemispheres. This is where higher functions occur, such as interpreting sensory data, speaking, reason, emotions, learning, and fine motor control. The cerebellum is located under the rear portion of the cerebrum, and coordinates muscle movements to maintain posture and balance. The brainstem at the bottom of the brain consists of the midbrain, pons, and medulla. It connects the cerebrum and cerebellum to the spinal cord, and from there to the rest of the body. It also controls many automatic functions, such as breathing, heartbeat, body temperature, *circadian* (wake/sleep) cycles, clearing the lungs (sneezing/coughing), and digestion (swallowing/vomiting/etc.).

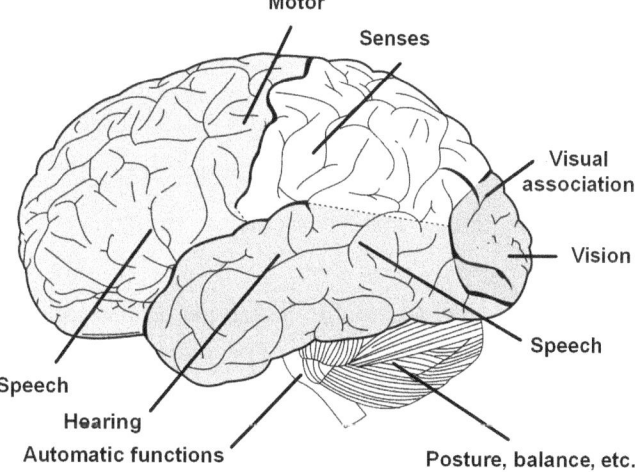

The surface of the cerebrum contains many folds that increase the overall surface area. This area is called the *cortex*. The cortex contains about 70% of the brain's 100 billion neurons, whose bodies give it a grayish color. This is commonly called *gray matter*. The underlying *white matter* contains the axons that connect neurons to each other. The folds in the cortex allow many more neurons to fit inside the skull, which in turn allows more higher functions. Each fold is called a *gyrus*, while each groove between folds is called a *sulcus*. Different folds and grooves have different names, which we need not concern ourselves with here.

> *One of the best definitions of anxiety is, that emotion that 5mg of Valium makes better.*
> Philip Berger

The two hemispheres are divided along a line that extends from roughly the bridge of the nose straight up and over to near the base of the skull. These hemispheres are connected by a fibrous bundle called the *corpus callosum*, which passes messages back and forth. Each hemisphere controls the opposite side of the body, meaning that the left hemisphere controls the right side of the body and vice-versa. The left hemisphere generally controls speech, understanding, mathematics, and writing. The right hemisphere generally controls creativity, spatial ability, and artistic/musical skills. Roughly 92% of people have left-dominated hand use and language, which is why most people are right-handed.

We looked at the lobes of the brain in Chapter 9, but a brief recap is in order here. Each hemisphere is divided into the frontal (forehead), temporal (sides), parietal (top), and occipital (rear) lobes. Each lobe contains areas for very specific functions, but these areas do not operate in isolation. There is extensive cooperation between different areas of the brain. Many different messages pass back and forth between lobes and hemispheres. These messages can travel from gyrus to gyrus, lobe to lobe, across the hemispheres, and/or to and from deep-brain structures, such as the thalamus.

The taste of the soup is the average of all of the ingredients influencing other ingredients.
Arnold Mandell

In very general terms, the lobes contain the following functions:

- **Frontal lobe:** Personality, behavior, emotions, reason, planning, problem-solving, speaking, writing, movement, intelligence, concentration, and self-awareness.

- **Parietal lobe:** Language interpretation, sensory input, spatial and visual perception.

- **Occipital lobe:** Light, color, and movement.

- **Temporal lobe:** Language comprehension, memory, hearing, sequencing/organization.

The innermost portion of the brain contains the following structures:

- **Thalamus:** Waypoint for most information going to and from the cortex that helps with pain sensation, attention, alertness, and memory.

- **Hypothalamus:** Helps control and/or regulate hunger, thirst, sleep, sexual arousal, body temperature, blood pressure, emotions, and hormones.

As long as the brain is a mystery, the universe, the reflection of the structure of the brain, will also be a mystery.
Santiago Ramón y Cajal

- **Limbic system:** Center for emotions, learning, and memory. This system includes the amygdala, which regulates emotions related to safety and wellbeing.

- **Pituitary gland:** Controls the body's endocrine system and glands, and also secretes hormones that control sexual development, bone and muscle growth, stress, and immune response.

- **Pineal gland:** Helps regulate the body clock and sexual development.

Neurons

We often lose sight of the fact that the brains we carry in our heads are not the last word in nervous systems.
Daniel Robinson

I have already mentioned that the typical human brain contains about 100 billion neurons, which happens to be the approximate number of stars in the Milky Way galaxy. Each neuron is connected to roughly 7,000 other neurons through its synapses. The gaps between neurons use ions to communicate. These ions behave in a quantum manner, which seems to indicate that our brains function according to the laws of quantum mechanics, and not classical physics, because the "decision" to fire a synapse or not may completely alter our overall response. It would therefore seem that we can forget about determinism in the human brain, because we saw in Chapter 29 that everything seems to remain in possibility until collapsed. What used to seem like a complex clockwork mechanism now seems fuzzy at best. Neuroscience is therefore becoming more and more complex.

There is no reason at all for believing that our brain is the supreme ne plus ultra of an organ of thought in which the world is reflected.
Erwin Shrödinger

Some neurons map to areas outside the brain, such as those for sensation and movement. You may think you are feeling your fingers holding this book, but the actual sensation is happening inside your head and being perceived as happening in your fingers. This helps explain *phantom limb syndrome*, where amputees often report feeling sensations from the area of the missing limb. This occurs because the brain maps have not changed. Maps can change over time to accept input from other sources, which is why people with one or more deficient senses may have heightened abilities in their other senses. More neurons mapped to a specific sensation or area equals greater sensitivity to that sensation and/or area.

Groups of neurons also band together to form circuits that perform different functions, such as facial recognition or pleasure/reward. Ramachandran believes he may even have found a "God circuit" that may predispose us to religious belief. This circuit could have evolved to maintain social order. Religious belief corresponds to how enhanced these circuits are.

Neurons are full of water (about 80%), hormones, and neurotransmitters that transmit information back and forth. The molecules in each neuron are replaced about 10,000 times during the average lifetime, and yet each of us retains a continuous sense of being an "I," a unique self moving through time. As Radin has said, all of the material initially used to create and express the pattern called "I" has come and gone many times over, but yet the pattern remains. Can we really conclude that our sense of self arises from nothing more than biochemical interplay among neurons?

Inside our heads is the most complex and sophisticated device in creation.
Greg Peterson

This question becomes more interesting, because one species of earthworm contains only 200 neurons, but displays behavior that is far too complex for the small amount of impulses that tiny brain can produce. *Aplysia* sea slugs remember conditioned messages for weeks. Further study revealed that this learning occurred in the synapses, in the form of altered numbers of chemical quanta (or "packets") sent from specific neurons. All of this takes place in what can only be described as an extremely simple brain—and the human brain is orders of magnitude more complex. It may be the most complex structure to evolve to date by some measures, although we are learning that animals are not nearly so different from us as we used to imagine. (This of course neither implies nor refutes the idea of any sort of divine plan or intelligent design.)

Our physical brain anatomy provides a road map where certain traits are more likely to be emphasized than others, while other traits are more likely to be de-emphasized. This is the "nature" side that gives us bare-bone personality traits, such as emotional propensities, intellectual capacity, and sexual orientation. Experience, the "nurture" side, fills in the map throughout our lives. Learning entails forming new connections in as little as 1 to 15 seconds, which validates the expression "making the connection." But how do the neurons know where to connect?

The surgeon knows all the parts of the brain but he does not know his patient's dreams.
Richard Selzer

Quantum Processing

> *We are descended from robots, and composed of robots, and all the intentionality we enjoy is derived from the more fundamental intentionality of these billions of crude intentional systems.*
> Daniel Dennett

Computer scientists around the world are racing to create the world's first quantum computer that directly uses quantum phenomena, such as superposition and entanglement, to perform calculations. Such a machine could literally solve in seconds what would take years for a classical computer to solve. It is true that a brain is not like any current or projected future computer, because the latter is linear, while the former is non-linear; however, we cannot discount the idea that our brains are in fact advanced quantum computers that have been in development for about 54 million times longer than the current technological explosion that began with the first transistor in 1947. In fact, the idea that our brains use quantum mechanics is both not farfetched and very well supported by the available evidence.

We are led to believe that our brains use electricity to communicate and possibly to store information in a manner that may be analogous to computer memory banks; however, the brain does not create electricity. Further, electricity is said to flow through nerves, which should generate heat, which has not been detected. It would also seem that we don't really perceive objects at all, but rather quantum information, based on the Copenhagen and other interpretations we saw in Chapter 29.

The brain does obey quantum laws, as does everything in the universe, which only deepens the mystery of how it can act in a coherent manner. This problem exists whether the brain's actions are quantum or not. How does this incredibly complex organ give rise to a unified—and singular—sense of "I" ness? Quantum mechanics could provide a framework for an explanation; for example, any waveform present at the synapses could indicate that wave/particle duality could be part of the foundation of human consciousness, although this does not begin to explain the high degree of order in what should be an extremely chaotic environment. It is possible that particles are acting in concert to produce the unified experience, but this possibility would require explanation in its own right.

> *The organism is a theory of its environment.*
> Walter Weimer

It is possible that subatomic particles influence—or even cause—emotions, thoughts, and actions. The trouble is that following particles around won't tell us a thing about behavior, while watching people will tell us about behavior without telling us anything about the quantum phenomena influencing that behavior—a textbook "chicken or egg?" conundrum.

Anesthesiologist Stuart Hameroff (1947-) studied neurophysiology in detail and identified *microtubules* as the possible source of quantum brain activity. Microtubules are tiny tubes made of the protein tubulin that make up cellular skeletons including neurons. Tubulin can take either extended or contracted forms, and could in theory exist in a superposition of the two. Electrons within these tubes could also be in superpositions that collapse. It is conceivable that a collapse of one could influence others, in a cascade that leads to a conscious or subconscious decision. Hameroff and Penrose showed how this type of quantum system could act like a large quantum computer that could be the source of consciousness. This seems hard to believe, except that modern computers have been around for only 63 years, while life and intelligence have been evolving for 3.5 billion years.

This is a strong-sounding argument, but it is still not enough for us to declare that the brain is a quantum mechanism that evolved in a quantum universe. For example, Tegmark calculated that the quantum model cannot apply to the human brain, and decided that it must be a classical organ working according to Newtonian physics. This could help explain why the brain tends to act in predictable ways that are still subject to perturbation by quantum events from time to time. The debate rages on. Meanwhile, repeated experiments have proven that we prepare decisions in our subconscious minds before we become consciously aware of them, which leads us to ask whether consciousness can be explained by mere electrochemical activity in our brains, or whether there is more to it than that. More on this in a moment.

If our brains are indeed quantum computers, then they may not be storage and processing devices, but transceivers. Physicist Walter Schempp (1938-) believes that our memories and brain processes may not be handled in the brain at all, but may reside in the Zero Point Field (see Chapter 23), a possibility whose side effect would be blurring the line between individual and group consciousness. This may sound like the stuff of science fiction, but we cannot dismiss it after some of the things we have learned about quantum mechanics.

If the universe is indeed the expression of an idea, the brain may be the sole antenna with circuitry tuned to pick up the signal of that idea.
Gerald L. Schroeder

Anything can be reduced to simple, obvious mechanical interactions. The cell is a machine. The animal is a machine. Man is a machine.
Jacques Monod

The Holographic Brain

Maybe the world is a hologram!
Karl Pribram

Brains convert sensory data into the frequencies that make up that data using Fourier transformations, named after mathematician Jean Fourier (1768-1830). Fourier showed that any pattern of arbitrary complexity can be transformed into simple waves that can then be mathematically reassembled. Modern digital cameras use a similar process, whereby each pixel records the relative brightness of the red, green, and blue components of the light being received and later reassembles that information to generate a viewable image on a screen or printer. A Fourier transformation of a musical chord would be a mathematical representation of the amplitudes of the frequencies included in that chord. Researchers Russell and Karen De Valois showed that brains respond to Fourier transformations of visual patterns; our brains are using the mathematics of holograms, which is very interesting in light of what we saw in Chapter 27.

If holography is correct, then we would seem to be living in two worlds, one within the hologram where objects appear solid and have discrete locations in space-time, the other as energy without any definite existence or location. Where does this leave mind? If our minds and brains are holograms, then just what is the mind? Researchers Benjamin Libet (1916-2007) and Bertram Feinstein at the University of California at San Francisco (UCSF) showed that subjects are not consciously aware of pinpricks for up to half a second after they happen. They also pushed buttons before being aware of doing so. This reignites the free will question. Is free will an illusion? Our brains have decided before we are aware of it, so just what is making the decision?

Our uncanny ability to quickly retrieve whatever information we need from the enormous store of our memories becomes more understandable if the brain functions according to holographic principles.
Michael Talbot

Our brains may also use Fourier transforms to coordinate bodily movements. An experiment with people photographed while moving dressed in black leotards painted with white dots revealed Fourier transformations that allowed the researcher to predict future movements with extraordinary precision. If a remote observer can do this based on photographs, then it stands to reason that we should be able to do the same thing for our own movements with both far greater precision and much less effort.

Pribram pointed out that this could explain how we can quickly learn complex tasks, such as riding a bicycle. We learn holistically instead of piece by piece, which makes sense, because it would be awfully hard to learn part of how to ride a bicycle. This holistic and fluid approach can be explained using Fourier transformations.

Taking holography to its logical extreme implies that objective reality is illusory; there may be nothing out there but a mass of wave patterns that our brains convert into this thing we call the world and everything within it. We learned in Chapter 27 that holography is wave interference. Our quantum brains could be making observations/measurements that transform a world lacking both space and time into the "real world" based on nothing but those interfering waves. Our world of solid matter may be nothing more than a virtual world, with little more in the way of reality than a game, such as the popular *World of Warcraft* by Blizzard Entertainment, Inc. that creates an immersive virtual fantasy world. Of course, our brains are only sensitive to certain waves and limit what they take in.

Psychologist Karl Lashey (1890-1958) conducted experiments where he trained rats to navigate a maze and then destroyed various portions of their brains in an attempt to locate the seat of memory. He subjected hundreds of rats to this experiment and sliced and diced in all different ways, but the rats never forgot the path. They may have been barely able to walk, but they still knew which way to go. Other experiments with salamanders (see below) yielded similar results. This led to the theory that memories are not stored in any special locations, but are stored throughout the entire cortex, possibly holographically.

Our larger, more complex brains have blessed us with a limited degree of freedom from instinct, albeit far less than many people know or care to admit. Nevertheless, it appears that our brains can use images—or rather wave patterns representing images—to control our bodies and thoughts, which extends to creating more images and associated wave patterns ad infinitum, another perfect example of holography in action. Even removing large parts of the visual cortex or severing most of the optic nerve still allows us to perform complex tasks. Our brain's neural activity behaves like an image being projected on our cortex.

Memory seems to be both everywhere and nowhere in particular in the brain.
Brian Boycott

We are each rather like a prisoner in a tower permitted to look out through five slits in the wall at the landscape outside. It is presumptuous to suppose that we can perceive the whole of the landscape through these slits-although I think there is good evidence that the prisoner can sometimes have a glimpse out the top!
Raynor Johnson

Psychologist Stanislav Grof (1931-) concluded that our existing neurological models of the brain are inadequate, and that only holographic models can explain our archetypal experiences, the collective unconscious, and altered states of consciousness. Wolf says that the holographic model explains lucid dreams as visits to parallel realities. Physicist F. David Peat (1938-) says that the holographic model explains *synchronicities*, experiences of two or more non-causally related events that are unlikely to occur together by chance, but that occur together in a manner that conveys meaning. Some aspects of psychosis may also be attributable to holography, in that psychotics may be able to see more of the deeper reality than most, and may find themselves overwhelmed and unable to make sense of it all. Going further, the sayings of many mystics have an eerie resemblance to Bohm's implicate order (Chapter 29). Further still, Ring speculated that a holographic universe could also explain NDEs, and that death may be nothing more than shifting one's consciousness from one hologram to another.

> *If the concreteness of reality is only a holographic illusion, it would no longer be true to say the brain produces consciousness. Rather, it is consciousness that creates the appearance of a brain.*
> Keith Floyd

Memory

I mentioned Lashey's experiments with rats who had ever-larger amounts of their brains removed but could still run mazes despite showing levels of impairment roughly analogous to the amount of cortex material removed. Biologist Paul Pietsch (1929-2009) conducted over 700 experiments with salamanders in which he sliced, diced, and mixed their brains in every way imaginable, but they still fed normally. These experiments pretty much debunked the idea that memories inhabit specific areas within the brain. Furthermore, human patients with parts of their brains removed lost no specific memories. Some of their recollections may have become somewhat hazy, but there were no gaps, such as forgetting half of their friends or discrete portions of a previously read book.

> *The essence of life, of all life, is the storage, organization, and processing of information.*
> Gerald L. Schroeder

I mentioned Penfield's electrically stimulating patients' brains during surgery in Chapter 15. The patients relived their past experiences in detail, but remained aware of their current location on the operating table and what was going on around them. The electrodes triggered one stream of consciousness without disturbing the currently running stream. These results led Penfield to conclude that "mind can only be something quite apart from the neuronal reflex action," or that the mind is not in the brain. Critics were quick to point out several flaws in this line of thinking, chiefly that the "real" stream of consciousness was not being stimulated by Penfield's electrodes, and that both streams probably depend on brain activity. Also, the fact that the recalled experiences are both random and routine makes it a safe bet that the brain stores it all. Third, stimulating the same part of the brain triggered the same recall, while stimulating a different area triggered a different memory.

Memory ought to be impossible, yet it happens.
Karl Lashley

How are we to explain these contradictory results? Our brains may store multiple redundant copies of memories, or may store memories as a series of processes all over the brain that can be triggered by stimulating one spot. Maybe Penfield's results were somehow flawed. Maybe there is some other explanation, such as patients recalling bits of various experiences or tapping into an alternate stream of reality and interpreting it as memories. Any way you choose to look at it, the fact is that memory is anything but reliable, as any trial attorney will tell you. The list of things we forget starts with our own infancies and grows from there.

As for whether the self resides in the brain, we cannot deny that events such as comas, deep sleep, anesthesia, or injury can cause us to either leave our brains or simply wink out of existence during those times. Still, it does seem that the self, whatever that is, resides everywhere in the brain—if it is in the brain at all.

To Sleep, Perchance

"To sleep, perchance to dream. Aye, there's the rub." Hamlet was absolutely correct. If the cortex functions as our main switchboard, then we may be little more than glorified dream machines. Our neurons make images that are the only world we can ever know. We cannot break into the machine, because we are the machine. This does not mean that dreams are not real, at least as far as lucid dreams are concerned. Psychophysiologist Stephen LaBerge (1947-) trained subjects to send signals when they were having lucid dreams, and found those signals on the EEG. The lucid dreamers proved capable of alerting the waking world that they were dreaming while they were dreaming. It turns out that people who experience lucid

Things are not what they seem to be, nor are they otherwise.
Lankavatara Sutra

dreams experience less psychosis and depression, and have both higher self-esteem and better emotional balance than those who do not have these kinds of dreams. Is the waking world simply a higher-order dream, as mystics have been trying to tell us? The single largest difference may be that the waking world is a collective dream, while the world of lucid dreams is totally private. But how can we know?

Dream-state sleep is often characterized by rapid eye movements, and is therefore referred to as *Rapid Eye Movement* or REM sleep. This state is a complete mystery. Dreaming is not only an escape from this world but a health necessity, without which we can suffer severe consequences—up to and including death. In general, sleep deprivation can have side effects that include muscle aches, hallucinations, tremors in the hands, headaches, higher blood pressure, increased risk of diabetes and/or fibromyalgia, irritable temperament, memory problems, rapid involuntary movement called *nystagmus*, obesity, tantrums in children, and psychiatric symptoms that resemble ADHD and/or psychosis.

Fish and reptiles don't dream, because their brains don't have cortexes; however, all animals with cortexes dream, including newborn humans. Birds have a formation called the *wulst* that is similar to the cortex, and also have dreams. The pons in the medulla stops motor commands while dreaming; cats with this feature disabled act out their dreams in their sleep, as would we if our pons were disabled. The common nightmare of being chased may occur because of the disconnect between parts of the brain commanding muscles to move, while other parts of the brain belay those same commands by suppressing certain neurotransmitters. This "kill switch" feature called *REM atonia* protects the body from making potentially dangerous movements during sleep.

Dreams themselves consist of images, sounds, and/or emotions that play in the mind during REM sleep. We do not fully understand dream content or purpose, and these topics have been the subject of spirited discussion and debate for thousands of years. People have several dream periods during the night. The first typically lasts for about 10-12 minutes, with successive periods getting progressively longer, to a peak of around 15-20 minutes. The last dream period may be remembered as several dreams because of momentary arousals inter-

---

Are you going to believe me or your lying eyes?
Groucho Marx

After more than a century of looking, brain researchers have long since concluded that there is no conceivable place for such a self to be located in the physical brain, and that it simply doesn't exist.
Michael Lemonick

rupting sleep at the end of the night that eventually lead to full wakefulness. Humans and other animals have complex dreams, and are often able to remember and recall long, complex sequences of events while sleeping.

Dreams only occur when activity in certain parts of the brain rises to the point where consciousness can occur, while the rest of the brain remains essentially dormant. We can therefore think of dreams as partial waking states that may be controlled by the *Reticular Activation System* (RAS). The idea of a master circuit that sends reveille messages to the brain makes sense, so long as we do not confuse waking with consciousness. The RAS may get the waking process started and may control dreams, but other circuits are presumably involved in the waking cycle.

A Delicate Electrochemical Balance

Immune cells throughout the body have receptors for *neuropeptides*, chemicals the brain uses to communicate. Other cells have these as well, which has led neurologist Candace Pert (1946-) to conclude that we cannot say for sure where the brain begins and ends. We saw the role this plays in creating patterns of belief and action using the ETEAR process in Chapter 9.

At first blush, it seems like molecular biology may be capable of explaining consciousness and our individual subjective "I" experiences in purely materialist terms. For example, schizophrenia may be caused by having too much dopamine. On the other hand, dopamine deficiency is responsible for Parkinson's disease. A group of patients who were rendered practically moribund were given the drug L-dopa during a series of trials conducted by Dr. Oliver Sacks (1933-), with the astonishing results chronicled in his book *Awakenings* and the movie of the same name. Patients came out of their states and either picked up where they left off or reported having felt trapped inside their own bodies, unable to reach the outside world for years or even decades. This miracle was short-lived, as side effects from withdrawal to psychotic episodes took their toll, and patients began slipping back into themselves. Sacks felt these experiments betray the complete inability of the mechanical worldview, because he had seen firsthand that the brain is far

The brain is not an organ that generates consciousness.
Cyril Burt

Gracious one play, your head is an empty shell, wherein your mind frolics infinitely.
Sanskrit proverb

> *What is the point of holding out hope of being able to think and remain conscious when the brain is dead, if we can't even do it in the depths of sleep?*
> David Darling

more than a simple machine that can be repaired with the judicious application of a bit of one chemical or another.

Our bodies secrete endorphins, which are potent pain killers that may help us tolerate things like acupuncture and other severe pain, and may also play a role in the placebo effect. Different chemical cocktails work for different emotions in humans and animals. The reviled cockroach has neurotransmitters that prepare it to have sex, fight, or retreat. Even bacteria demonstrate likes and dislikes. This stands to reason: Making things that aid survival pleasurable provides incentive for the animal to do those things, while applying punishment for things that hinder survival provides more incentive to remain on the straight and narrow.

In their material form, our brains employ a complex programming language that consists of electromagnetic waves of different amplitudes, frequencies, shapes, and distributions, which also explains why electrodes work as well as they do. Applying minute currents to subjects' temporal lobes causes hallucinations, déjà vu, and other effects. Epileptics experience confused electrical discharges in their temporal lobes that may cause euphoria, dream states, and seizures. Probing this area spurred "total recalls" of past events in full living color, as we saw above. Penfield also concluded that we retain the full event, even though we may not be aware of it or able to consciously recall that information. Epileptics are also more prone than most to feel that they alone possess the absolute truth, sans the normal checks and balances normally provided by the cortex. They can also experience sudden memory recalls, believe they are receiving messages from God, or even believe they are conversing with God during their seizures. We saw in Chapter 9 that a good number of religious figures may have been epileptics.

> *The greatest manufacturer and user of drugs is the human brain.*
> Judith Hooper and Dick Teresi

Alzheimer's patients experience a gradual brain shutdown, where memories typically erode from most recent to oldest. It is localized in its early stages, but greatly affects brain metabolism as it progresses and gradually causes the patient to lose her or his grasp on the world. It can even be fatal.

Environmental conditioning also helps shape our brains. For example, put on a pair of glasses that inverts the light entering your eyes, and you will adapt within a few days and begin seeing everything right side up. Remove the glasses, and your brain will adapt back to the old orientation again. Another example concerns a series of strange phone calls to the New York City police department reporting that the callers just didn't feel right, or that they were hearing strange noises. It turned out that the El train line running through that area had recently gone out of service. The residents were calling at the times the trains used to run, because they were so used to hearing the trains that the sudden quiet spooked them.

All of these are examples of chemical and/or environmental conditioning causing marked changes in our brains, which does suggest a purely biological basis for consciousness and self. But then how can we explain homeopathic medicine? This branch of medicine dilutes drugs to the point where one can expect to be ingesting 0 molecules of the active ingredient, and yet people experience the same effect. It has even been reported that the "signals" in the homeopathic medicine has been emailed around the world to "imprint" water with high levels of success. It is easy to dismiss this as quackery foisted on a gullible public; however, I have taken homeopathic medicine myself, knowing full well that it was, and experienced the effects. My foreknowledge of what I was taking rules out the placebo effect. I have also seen the placebo effect at work, most notably when I gave my wide-awake son a completely harmless fish oil capsule that I said was a fast-acting sleeping pill, and watched him fall fast asleep within seconds.

The brain secretes thought as the liver secretes bile.
Pierre Jean Georges Cabanis

Are We In Our Heads?

Under normal circumstances, everyone perceives the entire universe in her or his own head. No matter how we try, we can only ever know other people and our environments indirectly, from within the confines of our skulls. It is tempting to wonder what our evolutionary ancestors thought about besides brute survival, because those thoughts helped shape who we are today.

We are aware of a certain level of reality that seems both objective and to operate by classical rules. We are enlightened in some ways, because we can understand the universe on this and other levels, and because the approximations from which we build our daily lives serve us very well. On the other hand, we live in near complete ignorance of the reality of the

I've stopped seeing the brain as the end of the line. The brain is just a receiver, an amplifier, a little wet mini receiver for collective reality. We make maps, but we should never confuse the map with the territory.
Candace Pert

> *The light of memory, or rather the light that memory lends to things, is the palest light of all. I am not quite sure whether I am dreaming or remembering, whether I have lived my life or dreamed it.*
> Eugene Ionesco

Ground/Godhead or whatever we want to call it, because we filter raw perceptual information several times as it passes through different mental levels on its way to entering our awareness. Interestingly, it seems that the healthiest people have the best filters, and that mental stability is directly related to the brain's ability to censor perceptual data. Creative people tend to have fewer filters, which can work well when combined with high intelligence. Weak filters combined with low intelligence can lead to psychosis.

If we are searching for the seat of consciousness in our brains, then looking between input and output is a good place to begin. It is interesting to speculate what our gods, religions, philosophies, instincts, and customs might look like if we inhabited different bodies. For example, if we navigated by sonar like bats and dolphins, then would our sciences and arts be any different? What if we could sense the Earth's magnetic field like birds, or see ultraviolet light like many birds and insects and fish, or sense infrared heat like some snakes and other animals? What if we walked on all fours, or swam, slithered, or flew? What if our sense of smell was more or less developed, or our visual acuity higher or lower? We can only ponder and get some very general ideas by observing other animals, but trying to see things from a fundamentally different perspective can often give us insights into our own makeup.

> *We have these complex human emotions, which we have always believed were of the soul.*
> Philip Berger

The illusion of a classical universe obeying classical laws makes it seem painfully obvious that some things are internal while others are external, but how can we tell? Even if we assume that the illusion is real, we cannot escape the subjective nature of our personal experiences. The problem deepens when we take the illusion into account and realize that there is no such thing as an objective reality, and that classical laws are only approximations. Our brain shows us heavily filtered views of infinity by assigning different sensations to the same general patterns of information, based on the affected receptors (tongue, eyes, ears, etc.). Each of us also displays unique handwriting and EEG patterns that distinguish every person from everyone else. Where exactly should any line be drawn?

The problem becomes more complicated when we consider the case of people with multiple personalities. Each personality within a body can have its own EEG pattern, vision (different personalities can require different glasses prescriptions), different voices, menstrual cycles, and more. Spectral analysis of the different voices shows distinct differences that are not the result of acting. Vocabularies, accents, body language, clothing styles, hair styles, and fears are also markedly different. Even scars, cysts, birthmarks, and other physical features may vary between personalities. There is a strong correlation between this disorder and child abuse, with 85-90% of cases involving people who were abused as children. The personalities may be the result of self hypnosis designed to help the person cope with the stress and trauma of the abuse. It is very easy to dismiss this as a mental disorder caused by a single brain that deludes itself into thinking it is more than one person; however, the degree of change between personalities makes such a dismissal difficult.

It is also possible that normal people have two personalities, one in each side of the brain, that communicate back and forth and give off the outward appearance of unity. This becomes apparent in patients who have had the corpus callosum cut, and who then develop split personalities. Both multiple personality disorder and split brain cases raise the question of just what a self is. Also, how does it feel to share a single body among multiple minds? Are all of the personalities responsible for the actions of one of them? If one personality commits an atrocious crime, while the others are perfectly law abiding, should we place the body harboring all of the personalities in jail? If so, why? If not, how do we punish the one personality without punishing what could be innocent bystanders who are just as horrified as anyone else by what has occurred?

A peasant in India once saw a locomotive and asked what made it move. He received a detailed explanation about all of the machinery and how it works, to which he replied, "Yes,. but what makes it move?" When it comes to looking for self in the brain, we are looking for the ghost in the machine, and that may be exactly the wrong place to look. An individual brain may play host to one, two, or multiple personalities, but that is a far cry from saying that the brain is the self or selves. We know that we only see and experience the world from our own perspectives that seem locked inside our heads, and that we receive a very filtered and censored view of reality. This seems to argue that we are indeed in our heads and that our

Now our skulls are not empty. but what we find there, in spite of the keen interest it arouses, is truly nothing compared when held against the life and the emotions of the soul.
Erwin Shrödinger

The brain is amazing. The mind even more so.
Gerald L. Schroeder

When I say the mind, I do not necessarily mean the brain.
Jim B. Tucker

If it is for mind that we are searching the brain, then we are supposing the brain to be much more than a telephone exchange. We are supposing it to be a telephone exchange along with subscribers as well.
Charles Sherrington

The mind creates the abyss; the heart crosses it.
Unknown

brains are the source of consciousness; however, the presence of multiple personalities and other anomalies render this conclusion nowhere near as cut and dried as it may seem.

Mind

The search for the human mind has been compared to fish trying to understand the nature of water. This may be true — but on the other hand, we already know much, and may be able to infer much more from what we know.

Where Is the Mind?

Where are our minds when we are in a coma, under anesthetic, or when another personality is in control? Does someone with total amnesia have a mind? Are our minds localized inside our heads or are they nonlocal in nature? Psychologist Theodore Barber (1927-2005) posed the following question: If we return to the egg and sperm, when does the mind enter into the picture? When does the mind switch on, and at what point during the development of the fetus? Does there need to be a critical mass of brain or other cells for this to happen? Do the molecules that inhabit the sperm and egg have purpose and planning? Are they somehow also mind, or are we to believe that the whole process of gestation and metamorphosis really is blind? Throughout this book, we have seen intriguing hints that our minds are not just in our brains, and that we may well be something greater than brains with a lot of ancillary support machinery.

Any psychotherapist will tell you that the mind is the best possible agent of change for the brain. Altering one's beliefs or resolving old issues can literally rewire the brain's connections between neurons. Is the brain changing itself, or is the change coming from somewhere beyond our heads? The psychoneural translation hypothesis says that brain and mind can interact because they are different aspects of the same transcendent reality, in which the mind works in the first person while brain works in the third person. Mental states initiated inside our brains cannot be reduced to electrical or chemical incidents because they are nonlocal in nature.

The simple truth is that we cannot understand our brains (physical organ) separately from our minds (that which makes us individuals). The term *qualia* refers to packets of experience, such as the experience of the color red, or the sweetness of milk chocolate, or of light as particle or wave. Sounds are waves, but we don't experience the waves themselves, only their effects on our eardrums. Pains, emotions, and thoughts don't feel electrochemical in nature. Maybe they aren't.

We know how color works. We know the chemical makeup and properties of chocolate. We know that light presents itself either way depending on how we look at it. What we do not know is how any of this translates into actual experiences of seeing red or tasting sweetness. Our brains could perform this analysis just fine in a zombie state with no consciousness or awareness, just like a computer can process extremely complicated equations without needing any awareness at all. But this awareness and subjective experience does exist, which forces us to try to determine what role the brain has in generating and supporting the mind. Identical twins are perfect genetic copies of each other; however, while they may display the same tendencies and traits, we cannot say that their minds are exact duplicates. It is common for identical twins to have different personalities and completely separate belief systems. Those differences are large enough to call into question any theory that says that brain gives rise to mind.

Freud was Newtonian to his core. He tried to explain psychological states as being determined by the states of particles in the brain. This may well be, but not in the way Freud thought. One can only introduce free will into a classical deterministic system by leaving mind out of physics, lest we postulate a bottom-up causality that has particles running the show—and doing so deterministically to boot. This brings us back to quantum ideas about the mind. These ideas are fine, but they demand that we impose order on particles that seem to act in manners that are anything but orderly.

Lucretius felt that mind is a phenomenon of brain for the simple fact that hitting someone in the head can have significant effects on both brain and mind. Spinoza thought mind and body were made of matter that embodies God within it. Leibniz saw people as monads that move according to God's directions. This invites speculation about whether philosophy

There can be no demonstrative argument to prove that those instances in which we have no experience resemble those of which we have had the experience.
David Hume

If you work on your mind with your mind, how can you avoid an immense confusion?
Sang Ts'an

included God as a genuine field of inquiry or to avoid rousing the suspicions of the church.

What about RSMEs? If mystics are to be believed, maybe Huxley and others were correct when they suggested that the brain may be a reducing valve for consciousness. Persinger felt that magnetic fields can explain RSMEs and OBEs, as we learned in Chapter 9; however, his experiments have not been replicated as far as I know. As for the RSMEs themselves, mystics don't use them for practical matters but as scientific explorations designed to understand the ultimate reality that lies beyond our normal perceptions. People who claim to hear voices, or to be in communication with spirits, or even God, are not necessarily crazy. For example, Joan of Arc led a demanding and intensely scrutinized life. She claimed to hear voices giving her guidance and instructions, and the information she passed on was generally sound, which discredits the idea that she suffered from some disorder.

Most canines are inveterate, optimistic believers in the goodwill of their masters.
Andrew Newberg

The way in which mystics go about their work deep in meditation and come back to communicate their findings to the world demonstrates that they are not in it for personal gain, but for the benefit of all humanity. It is hard for materialists to take these experiences seriously, but I for one feel that the uniformity of the reports coming from mystics over thousands of years needs to be taken seriously. We have devoted many resources looking into the head for answers; maybe it's time we look outward as well. In other words, maybe our minds are not in our heads. Maybe they are nonlocal. Maybe there are greater minds than ours on Earth and/or elsewhere. Regarding the central question of this book, a nonlocal mind can theoretically accommodate some form of life after physical death.

A World of Ideas

The brain does for the mind what a radio does for music.
Gerald L. Schroeder

The human brain can only contain or process a certain number of patterns, no matter how one alters it with drugs or by other means. All human brains are similar, and all have the same limitations. We also know that our brains filter our sensory input, possibly as a way of limiting the flow to a rate and type we can process. We further know that each of our individual worlds consists of these subjectively filtered patterns, which means that we live in a world of conscious and subconscious ideas. Our ability to experience ideas and extrapolate them in the abstract adds richness to our lives, and can even be a survival tool. Researchers at the National Institute of Mental Health were asked to stop sacrificing rats with a guillotine in front of the other rats, because the surviving rats were traumatized by the experience, to the point where it was impacting other experiments. Rats clearly live in a world of ideas as well, and are just as capable of extrapolating those ideas as we are.

With that said, there does not seem to be any rational connection between an experience and the emotions it can create. We can understand fear and horror at the sight of slaughter, but this reaction is not mandatory; plenty of people have been known to enjoy carnage. Humans are emotional animals. We know from Chapter 9 that experience causes emotions; however, we cannot predict the emotions that will arise with any degree of certainty.

We can also perceive the same data in different ways, such as the two examples shown below. These simple drawings should be proof enough that our eyes and brains are not cameras but active participants in altering sensory data, and, in a sense, creating the world we see before we are conscious of it. Only about 50% of what we perceive may be "out there," with the rest filled in and/or made up in our own heads. This percentage may vary widely by individual, perhaps even within the same person, and is itself a made-up number to some extent. We can all agree that some people are more "reality" oriented while others tend more toward "fantasy," but which is which?

I think you can manufacture a reality for yourself that is indistinguishable from true reality.
Elizabeth Loftus

Old woman?

Young lady?

Goblet?

Two faces?

The miracle is that the universe created a part of itself to study the rest of it, and that this part, in studying itself, finds the rest of the universe in its own natural inner realities.
John C. Lilly

Each of us can store memories by encoding our experiences for later recall, thus creating a *datum* that we can reference later. This datum can include information from all five senses plus our logical and/or emotional responses to the thing that triggered us to remember that experience. We can also create and store original ideas, ideas about experiences, and so on. Each of these represents additional abstractions of the original sense data. Are these memories real? Yes, because we have them—or at least think we do. No, because they do not necessarily have any basis in fact. For example, the same event can trigger many different levels of many different emotions and thoughts in different people. For example, falling could literally scare someone to death, while a skydiving aficionado may not feel totally alive unless s/he is plummeting toward the ground.

We each live in a world of our own ideas, and that world is surprisingly limited. There are literally tons of ideas floating in the air surrounding us that we will never perceive or experience for ourselves, including light beyond the visible spectrum, pollen, pheromones, and much more.

The Power of Suggestion

To be alone in your own version of reality, a reality created by your every decision, is terrifying in the extreme.
Anthony Peake

An experiment was performed in which a subject was hypnotized. A person then stood in front of the subject, and a pocket watch was held against that person's back. The subject was told there was nothing in front of him, and was asked to read the watch. The subject did so perfectly, as if the intervening standee had not been there. Psychologist Charles Tart (1931-) had two students hypnotize each other. Both students saw and experienced being on the same beach together, in an interesting case of shared hallucination or dream. The EEG of a hypnotized person is very close to that of a normal waking state, which leads one to wonder whether said normal waking state is itself a form of hypnosis. What we call objective reality could be as simple as a sort of a collective, shared hypnosis. Everything you have read thus far in this book should indicate that this idea is far more than idle speculation.

Intuition is reason in a hurry.
Holbrook Jackson

Dr. Jeanne Archterberg discovered that people who do not understand cancer mortality tend to have lower rates of cancer than the general population. Her research has led her to the conclusion that a sick person should think health as much as possible. Some disease—or even all of it—may originate in the mind, which does not mean that it is psychosomatic or unreal. The implication is that our ability to create sickness should be matched by our ability to create health. Cognitive therapy can change how patients see themselves. Doctor attitudes may also influence the outcome.

The placebo effect is caused by belief. Many experiments involving giving people one type of medication (such as a powerful stimulant) and telling them that it is a different medication (such as a powerful sedative) often cause the believed effects to occur instead of those expected, given the type of medicine delivered. Drugs are tested against placebos, because placebos do work. A drug must perform statistically better than a placebo to be licensed for use, a restriction that makes no sense if we believe that mind is either nonexistent or impotent. This works with physical interventions as well: One patient received superficial incisions and sutures on his knee but went on to recover as if he had received the full operation.

We have to get rid of emotions... in order to reach feeling.
Metropolitan Anthony

Despite this overwhelming evidence, mainstream medicine continues to push physical and/or pharmacological intervention, even though centers that employ a "whole person" approach to treatment that includes mental, emotional, and spiritual wellbeing are springing up. Medical knowledge, technology, and methods have come a long way over the past century. Treating people like machines has been enormously successful; however, this mechanical approach also has its limitations. Materialists tend to see placebos as hoaxes, because to do otherwise means that an immaterial mind can change a physical brain, an assumption that upends materialism by default. The emerging field of *psychoneuroimmunology* takes a patient's brain, mind, and immune system into account by treating what Pert calls the *psychosomatic network*.

Cultural beliefs also play a huge role in shaping outcomes. For example, a tribe living on the Kiriwina Islands near New Guinea approves of premarital sex, but not premarital pregnancy. Unmarried people have unprotected sex as a matter of course, yet pregnancies and abortions are rare, occurring at rates far lower than in other nations.

Thought constitutes the greatness of man.
Blaise Pascal

The holographic model implies that the mind cannot tell the difference between internal and external holograms. (Whether there is really any difference between the two is open to debate.) We can substitute the word "idea" for "hologram," and the same logic applies. An idea is an idea, and ideas can be extremely powerful things.

The Power of Belief

Flout 'em and scout 'em-and scout 'em and flout 'em; thought is free.
Shakespeare

The standard materialist orthodoxy is that praying for someone cannot cause any form of remote healing, because mind comes from brain. If this is in fact true, then A cannot affect B; however, we know from Chapter 23 that entanglement and nonlocality are very real. Quantum experiments prove that the answer depends on the question, and that the beliefs of the experimenter can—and do—impact the outcome, if not outright create it. Belief is powerful and has real impacts. We accept our beliefs on faith, which is nothing less than uncritical acceptance—something I show people how to avoid in *The Enlightened Savage*. Faith is fine, so long as it is based on conscious observation and backed by reason. Blind faith of any sort is where problems begin. People die every day because of their own, or someone else's, beliefs. Other people die because they did not double-check their beliefs. We know this, and yet we cling to our beliefs like babies to blankets.

This seems both simple and obvious, as do the maxims that wise people seek supporting evidence instead of simply asserting truths, that science allows one to distinguish between belief and knowledge, that statements are either true or false, or that knowledge is connected to a selfless reality that is distinguishable from fantasy. The difference between fact and belief is that the latter is asserted without verification. We come into this world ignorant of reality and build a reality for ourselves based on beliefs that separate us from that reality. Again, this seems painfully obvious, except that we know that things are not necessarily as we believe them to be.

What is mind? No matter. What is matter? Never mind.
Thomas Hewitt Key

Egocentric people have many wants that can provide some illusion of security—a statement that really offends these same egocentric people, who eschew investigation and understanding in favor of protecting their beliefs. Taking offense at something is a great way to hide, which is why it is such a safe bet that the best way to offend someone is to reveal some truth that they might not want to face. It is entirely possible to dislike something that someone else loves, which demonstrates that there is nothing intrinsically good or bad about the object in question. Thus, the cause of the offense lies totally within the offended person—just don't try too hard to get them to admit this. One of the best ways to distinguish a true mystic from one of the "born again" rabble is that the former will probably be far more imperturbable than the latter.

Here is where religion gets ugly, because it forces people to surrender the minds that God supposedly gave them in favor of beliefs that are both unverifiable and demonstrably false. These beliefs run the gamut from thinking that bread and wine turns to Jesus tartare in the form of actual flesh and blood after being swallowed to thinking that flying a plane into a building is a good way to meet women. The idea that such a system of forced suspension of disbelief can possibly be conducive to morality, physics, metaphysics, philosophy, or even basic logic is ludicrous on its face. The idea that giving up whatever certainty one can derive from science and questioning in favor of assertions that fly in face of grade-school education is baffling. The idea that giving up one's own humanity by adopting the belief that one was born unworthy is a grievous insult to the god that supposedly made us all. Responding violently to any and all criticism is beyond the pale.

And yet religious beliefs can be beneficial, as remote prayer experiments have demonstrated. The actual religion matters far less than the nature of one's beliefs; positive beliefs can be incredibly affirming despite the problems outlined above and the legacy described in Chapters 3 through 11. This does not replace medical treatments, nor does it imply that mind alone can cure everything; all it affirms is that mental states and choosing one's area of focus are important for maintaining or restoring health. It is no accident that patients want their doctors to be aware of their spiritual beliefs, and that doctors are more likely to have spiritual beliefs than other scientists. Some spiritual practices aid recovery; others impede it.

Most healing prayer studies have people praying to some god to intervene on behalf of a specific patient, making statistical and other analysis difficult at best. How should we define prayer? Which prayers should be counted in the study? Does it matter which god one prays to? Does the patient even want to survive and recover? We may never know for sure. What we do know is that beliefs are extremely powerful, and that reli-

---

Man is made by his belief. As he believes, so he is.
Bhagavad Gita

The mind will be explained as an epiphenomenon of the neural machinery of the brain.
Edward O. Wilson

Consciousness

The human mind is a computer made out of meat.
Marvin Minsky

The physical world seems both separate and isolated from individual minds, because the sensory information we are aware of is not real, and yet we act on this quasi-reality. Our minds make inferences about the world based on these imperfect experiences, with our left brain usually handing *deductive* thinking (building conclusions based on premises) and our right brain usually handing *inductive* thinking (building conclusions that do not necessarily follow from the premises). This of course requires awareness, which is not necessarily the same as consciousness. We can be conscious without being fully aware of our surroundings, such as when we are focused on a single task, distracted, or daydreaming. We can also be aware without being conscious, like a sleeping person who will wake up if s/he senses a disturbance. Seen this way, we can loosely define awareness as the passive foundation for active consciousness. Some spiritual traditions define these terms the other way around, with consciousness being the passive observer and awareness being active.

Are humans the only species with consciousness? Our fancy brains seem to be blessed with an abundance of self-perception and consciousness, while some animals such as insects seem to have none, with other animals fitting somewhere in between. This may seem logical until we rename consciousness as "essence" and see the folly of trying to figure out which species gets how much of it. The question may therefore be moot, especially because it should be easy to see that each animal probably considers its own life just as real and just as precious as we do our own lives. I am not suggesting that animals experience the world in the same way we do, but that is not the same thing as saying that a mosquito does not truly enjoy its life.

Consciousness defines our existence and our reality. But how does the brain generate thoughts and feelings?
Stuart Hameroff

Which animals and/or other entities have consciousness, and how much of it do they have? Does the brain have all of the consciousness, or is it distributed throughout the body, as Pert suggests? Does consciousness attach to a single being, and, if so, how? Is one consciousness different from another, and if so, how? These are all excellent questions. Let's do our best to answer them.

The Illusion of Self

Is the appearance of you and me as separate entities merely a convincing illusion? We explored this idea in Chapter 5 and will revisit it in Chapter 32, because this concept is central to the theme of this book, namely, what happens when we die. Meanwhile, a quick recap is in order.

Philosopher Ludwig Wittgenstein (1889-1951) saw no place in objective language for a subjective "I," because otherwise we would be able to measure consciousness and place it inside the brain. Jung said that the collective unconscious has no means of existing according to modern physics; however, holography offers a suitable mechanism. We have to come to terms with the fact that, like quantum waveforms, we may have no firm edges. We may be *personnes sans frontiers*, persons without borders. This would certainly explain our current inability to map consciousness to specific areas of the brain and processes within the brain, and would also resolve both the problem of shared reality and conscious collapse of quantum possibilities specified in the Copenhagen interpretation. (See Chapter 29.)

Peel away the layers, and there may well be only one consciousness, a non-thing that both is and contains everything at once. We affirm consciousness as a something, even though we cannot categorize or even locate it; however, we saw in Chapters 21 through 28 that everything in the universe may well be non-things as well. Zen Buddhists say that Zen is not a state of mind, but pure consciousness that exists first before everything else, and yet is none of those things. Hofstadter points out that Zen masters define enlightenment as a state where the borders between self and universe vanish, and one loses the sense of subject/object split, along with the desire to perceive, a definition he equates to death; however, a Zen master who asks, "When there is neither I nor you, who is the one that wants to see it?" is not saying there is no one. He is referring to reality seeking to know itself, and to the consciousness that is in everything—a definition that is very far removed from the idea of death as annihilation!

> *I've dated guys who couldn't pass a Turing test!*
> Anonymous

> *What or where is the unified center of sentience that comes into and goes out of existence, that changes over time but remains the same entity, and that has a supreme sense of moral worth?*
> Steven Pinker

In Chapter 17, we learned that our bodies are constantly changing. Our skins renew themselves every 27 days, our bodies replace 500,000 cells per day, and all of the molecules in our bodies get replaced many times over. From a purely physical standpoint, we are constantly becoming entirely new beings, and yet the pattern that we call "I" persists. We are therefore little more than intricate mirages. The term "I" can be used as a kind of shorthand that encompasses a great many things, including our perceived oneness and distinction from everything and everyone else in the universe. In most cases, this shorthand becomes mistaken for reality. But if "I" is an illusion, then brain cannot be the source of consciousness, and we are still confronted with the problem of trying to define and understand consciousness.

What is Consciousness?

Strictly speaking, at present there is no scientific evidence even for the existence of consciousness!
B. Alan Wallace

Any talk of consciousness quickly turns to talking about the brain, which brings on the problems of distinguishing the brain from the external world, and needing to remember that the external world we are aware of perceiving is inside our heads. Ultimately, we see our own consciousness and ourselves, and the results of the images created for us. Our being conscious is about the only direct experience we have. If awareness and consciousness are not synonymous, then they are entangled, which is not the same as saying that they are different but intermingled. We have both consciousness and the things we are aware of and experience. The problem is where to draw the line, because awareness itself is not necessarily conscious, such as in the example of the sleeping person smelling smoke.

Introduce a gap between synapses, a junction with room for uncertainty and the brain becomes less certain, more probabilistic.
Steven Rose

Some mental states can be correlated to activity in different parts of the brain. Knowing the state of the brain at any moment could tell us the person's conscious state at that instant. Our Standard Model of reality says that there is an external world "out there" that functions according to a set of natural laws, and that there is a strong correlation between that world and our experience. This seems simple enough, so why is consciousness such a problem for science?

The key problem is that defining consciousness is no easy task. Second, there are many ways to interpret what we do know, and each of these interpretations exists for a good reason. Third, the increasing amount of pseudoscience being carried out by people ranging from self-proclaimed experts claiming that, "our brains can tap into the intelligence of the universe," to those who misinterpret spiritual teachings—and even those who mean well but lack understanding—is making it harder and harder to take on the dilemma of consciousness, much less resolve it—a problem I am trying hard to avoid.

Author and researcher B. Alan Wallace (1950-) pointed out that if consciousness can be reduced to an emergent phenomenon in our brain, then the relationship between consciousness and the brain is unlike any other emergent property in the world. Standard neurology says that consciousness resides within our bodies and nowhere else, but this is a huge leap of faith, because we cannot say what gives rise to consciousness, nor can we detect consciousness in anything living or dead. In short, the standard materialist line is an assertion that wants for evidence. Physicist Fritz-Albert Popp (1938-) theorized that consciousness at its most fundamental level may be coherent light.

Materialists argue that understanding the brain's electrochemical activity will reveal the secrets of consciousness. Others insist that consciousness is irreducible, that experience is primary. Both sides use quantum mechanics to back their positions. It should be noted that the Copenhagen interpretation relies on measurements to create results; however, the link between measurement and consciousness was not universally embraced. Bohr never completely bought into the idea advocated by Pauli and Heisenberg that consciousness plays an active role in measurement and waveform collapse.

Hameroff observes that anesthesiologists turn consciousness on and off at will, which may explain how it works in the brain. Francis Crick (1916-2004), one of the co-discoverers of the double-helix structure of DNA molecules, said that, "You, your joys and your sorrows, your memories and your ambitions, your sense of personal identity and free will, are in fact no more than the behavior of a vast assembly of nerve cells and their associated molecules." If Crick is right, then any sense we have of being more than a complicated series of chemical reactions is an illusion.

Nobody understands how decisions are made or how imagination is set free. What consciousness consists of, or how it should be defined, is equally puzzling.
Roland Barthes

Self may be a sleight of the brain, but that doesn't make it any less important from a subjective viewpoint.
David Darling

> *The analysis of mental activities in the context of brain physiology indicates that our own self, our ego, is not so unique or even independent as Freud pointed out many years ago.*
> Jose Delgado

On the other hand, philosopher David Chalmers (1966-) believes that mere physical process cannot tell how consciousness arises, because our experiences go beyond what any purely physical theory can tell us. For example, we know why water is wet and can even define wetness, but we cannot explain the feeling of wetness. We saw in Chapter 29 that Everett saw consciousness as being associated with a particular universe and not with the entire process, and that each version of ourselves is only aware of its own version both before and after splitting off, which must be occurring countless times per second. What Everett did not explain is which elements of his theory correspond to conscious experience.

Some believe that Chalmers's theory of consciousness predicts Everett's, even if the universe is only a superposition until subjectively experienced; however, all this does is "explain" consciousness by attempting to sweep it under the rug of infinite regression. Why are there different observers, and one with an increasingly confused mental state? How does the incoherence of splitting into virtually infinite copies lead to mind at all? Everett's theory does not account for the relationship between mind and body, and simply assumes that a brain state in superposition will have numerous, distinct experiential subjects that will somehow cause the entire universe to split into copies of itself. How can this assumption be made without a theory of consciousness? Everett's eagerness to avoid the problem of consciousness seems to be a bridge too far. Penrose agreed, saying that we need a theory of consciousness before we can reconcile Everett's theory with observations.

Researchers Robert Jahn and Brenda Dunne believe that consciousness may be probabilistic, assuming that reality consists of an interplay between consciousness and the environment, and may display wavelike and particle-like qualities, where the former implies unity and the latter implies separation. Bohm might agree, because he saw consciousness and matter as different aspects of a fundamental implicate order. (See Chapter 29.)

> *The common root from which scientific and other knowledge must arise is the content of my consciousness.*
> Arthur Eddington

At its purest level, consciousness is not thinking, nor is it thinking about consciousness. It is not self-reflection, and needs neither word nor deed nor anything but itself; deaf-mutes are just as conscious as anyone. Remove all thought, and consciousness is still there like an image on a screen, where our bodies are the set and our brains are the circuitry. The image is neither the tube nor the screen nor even the light, but may possibly arise from the coherence of all three. The image is ephemeral and yet is the entire point of the television. It exists whether or not the television is active, and whether or not the television even exists, because the station is still transmitting it in wave form. The TV is simply the device that filters the waves into images and sound, just like the human brain filters incoming sensory information using Fourier transformations, as we saw above.

If we see consciousness in this way, then asking questions such as whether a melting ice cube feels pain are not as silly as they may sound. If we see consciousness in this way, then it soon becomes apparent that we have never really seen, felt, heard, smelled, or tasted anything, because there is no objective world. The glass you hold in your hand and the keyboard I am typing on are not external. You are the glass, and I am the keyboard. Our very bodies are thus part of consciousness and depend on consciousness, not the other way around. Space and objects exist in our mind more as illusions than anything else. We don't experience consciousness; we are consciousness. If you can accept this information as a wonderful revelation, then you have reached what Japanese Buddhists call *satori*, the state of understanding. Mind is thus eternal. At least this is one way to look at it. Only the mystics know for sure, because they alone have seen this firsthand.

But where does consciousness come from, and how does it work? Basic identity theory says that consciousness equals the brain's neurophysiological state, and that our experience of consciousness is an ongoing sequence of these events, making the "problem" of consciousness simply one of semantics. This functional view is elegant in its simplicity but fails, because it does not tell us anything about how or why some events within the brain lead to consciousness, while others don't involve any consciousness at all. Saying that all the processes that we know lead to consciousness have consciousness themselves is both easy and a cop out. We are left with a catalog saying, "This leads to consciousness and this doesn't," without any idea how or why the catalog has been assembled that way.

If we look at consciousness as a phenomenon separate from physical matter, then we must tentatively accept the idea of

The mind that searches for contact with the Milky Way is the very mind of the Milky Way galaxy in search of its inner depths.
Brian Swimme &
Thomas Berry

Nevertheless all these experiments and descriptions of brain activation processes do not explain how neural activity is the cause for consciousness.
Stephen Patt

dualism. Is consciousness linked to electrons and/or quarks? Reducing consciousness to interactions of a limited group of particles sounds like a promising idea. The problem is that we cannot tie our own consciousness to any single particle, because doing so makes that particle both conscious and probably isolated from other particles and events. Trying to link consciousness to any single thing does not seem to account for consciousness involving many areas of the brain. How about space-time? We cannot pick one or the other, and even picking both may not explain the immediate conscious contact that sometime occurs between separate individuals. No force fields govern or give rise to consciousness. We cannot look to gravity or the strong or weak nuclear forces; the exchange of particles involved in the nuclear forces cannot account for consciousness.

The 'home' of the mind, as of all things, is the implicate order. At this level, which is the fundamental plenum for the entire manifest universe, there is no linear time. The implicate domain is atemporal; moments are not strung together serially like beads on a string.
Larry Dossey

This brings us to electromagnetism, which most people see as the only possible candidate. Consciousness slips away during sleep, but our brain waves are still present. If consciousness depended on those brain waves, then we could expect our experience of consciousness to change during sleep, but not for it to disappear altogether. We would also expect exposure to electromagnetic fields to alter or disrupt consciousness, but we are subjected to powerful fields every day with no discernible effects. Even standing next to a radio or TV antenna pumping out electromagnetic waves far more powerful than anything our brains can produce has no effect, and many of us do that every day when we use cellular telephones. We need to look elsewhere for consciousness. We can also rule out electromagnetism as a bridge or carrier for consciousness, because we would be unable to tease apart the conscious processes from the unconscious maintenance processes.

The private kingdom of our memories gives us continuity of self.
John Eccles

Furthermore, each of the 100 billion neurons in our brains has its own electromagnetic field, which means that our brains are full of overlapping fields. Consciousness could be a product of those overlapping fields that could also influence neuronal firing, but such a consciousness field should be both detectable and subject to outside interference. Further, many people have had large parts of their brains destroyed while remaining fully conscious.

For example, railroad worker Phineas Gage (1823-1860) survived having his left frontal lobe destroyed by a 1.25" tamping rod blown through his skull, as shown here. He recovered and was fully conscious, although people did report significant personality changes in him. There are other examples of people born with skulls full of fluid and practically no brain beyond a thin layer, who have led perfectly normal lives and experienced normal consciousness and intelligence. Each of these examples and many others involve significant alterations to the brain's electromagnetic field.

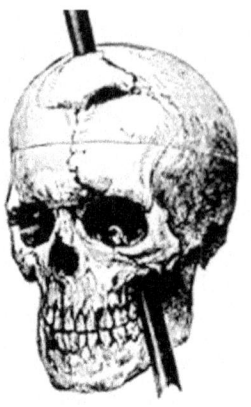

This field does seem to have some correspondence to consciousness (hence the EEG differences between waking and sleeping), but does not actually make up our thoughts and feelings; however, the self-feedback loop created by neurons influencing each other may be needed for consciousness in the same way that a television needs electricity that does not compose the actual image. The presence of these fields in our brains is a form of dualism that is firmly grounded in the brain's physical matter, which needs neither soul nor deity. We could see our unconscious mind as matter and our conscious mind as energy, with both coming from the same place. Electromagnetism plays a role in consciousness and may even be required for consciousness to exist, but is not conscious itself.

Cognitive science and neurology are making great progress and interesting and insightful discoveries. We have many detailed theories of cognition; however, consciousness remains a mystery. We think it comes from our heads, but have no idea how the brain produces it We don't know why we have egos, experiences, and feelings. The theories we do have about consciousness either deny it exists or try to define it in ways that will make the problem go away. Each of us knows consciousness better than anything else, but we understand just about everything else better than consciousness. Presented with everything we know about the brain, we would not guess that consciousness exists if we did not experience it

---

The psyche's attachment to the brain, i.e. its space-time limitation, is no longer seen as self-evident and incontrovertible as we've hitherto been led to believe. It is not only permissible to doubt the absolute validity of space-time perception; it is, in view of the available facts, even imperative to do so.
Carl Jung

I cannot transcend experience, and experience is my experience. From this it follows that nothing besides myself exists; for what is experience is its (the self's) states.
F. H. Bradley

---

for ourselves. We would not even need to postulate consciousness if we knew all of the brain's chemistry and physics.

Consciousness does not exist in a vacuum; it is tied to cognitive processes, and may arise from those processes in some way. This has led some researchers to develop cognitive models, such as David Rosenthal, who said that consciousness is a mental state that involves higher-order processing. Dennett believes that many thought processes compete for attention, with the loudest one winning and having a role in later processing. Consciousness has no central control in this model; it is simply an assembly of different thought channels with varying amount of influence. This model does provide some good ideas about how information may influence actions, but breaks no ground in explaining how consciousness emerges from brain processes. Nevertheless, Dennett claims his model solves all problems, because explaining functional phenomena explains the whole.

Models built on the brain's biology have much going for them, such as cataloging different types of awareness and discovering brain processes that are associated with consciousness; however, no theory leaps the gap to explain how or why these processes should—and do—lead to consciousness, which is treated as a brute fact. It may indeed be a brute fact, but not for the reasons materialist scientists might think.

These theories sound good, except that they offer no specific definitions of "higher order" processes, nor do they provide any kind of mechanism by which higher order processing begins or how it leads to consciousness. They don't describe the phenomenon of consciousness. These and other reductive models work well for describing cognition, but do nothing to solve the larger problem of consciousness, or speak to how—or even why—it arises. The central question is left both unanswered and untouched.

Consciousness does not *supervene* (follow or result as an additional, adventitious, or unlooked-for development) on the physical world, and defies all attempts at physical explanation. The relationship between consciousness and physical facts is different in kind from relationships between facts that don't involve this phenomenon. Starting with our own conscious experience, we can infer facts about macro-level objects fol-

---

Science is based on personal experience or on the experience of others, reliably reported.
Werner von Heisenberg

The mind does not understand its own reason for being.
Rene Magritte

lowed by quantum objects. The regularity of these objects and their behavior allows us to infer physical laws and facts. We can also see the relationship between consciousness and experience, which allows us to infer psychological laws and, by extension, the laws of how others experience consciousness; however, the problems of what consciousness is and where it comes from remain unanswered. We are left knowing all about the locomotive without having the slightest idea why it moves. Just about everything in the universe can be explained from physical facts, but consciousness has resisted all attempts at purely physical description.

Dr. John Lilly (1915-2001) conducted experiments with isolation tanks in which he experienced dreams, reveries, trances, mystical states, and even OBEs. He came to the conclusion that, "A human being is a biorobot with a biocomputer in it, the brain. But we are not that brain, and we are not the body. A soul essence inhabits us, and under acid, under anaesthesia, in a coma, you'll find that the essence isn't tied to brain activity at all. Brain activity can be virtually flat, and you can be conscious, off somewhere in another realm."

The mere fact that we can conceive of something does not make that something possible. For example, we can imagine accelerating to "Warp 9" and covering interstellar distances in hours, but nothing even close to the fictional starship *Enterprise* has ever been built or even seriously proposed; however, conceivability must be part of any explanation. If all of the processes that humans perform can occur without consciousness, then no reductive explanation of consciousness can be viable. Consider a zombie that is physically and functionally identical to a human, only without conscious experience. From an external viewpoint, consciousness is only active under certain conditions, such as a waking state. It does not necessarily appear as a result of brain activity or thinking, because our brains are very active while dreaming. Consciousness seems to switch the brain on and off—or maybe the alternative occurs.

We need not seriously speculate about zombies; all we need to do is infer a world where the facts about conscious experience are different from our own world. If any fact about conscious experience in our world does not hold in any alternate world, then consciousness is not logically supervenient. The zombie

To a very large extent men and women are a product of how they define themselves.
Jeremy W. Hayward

The mind creates the brain.
Jeffrey M. Schwartz and Sharon Begley

> *In other animals ignorance of the self is nature; in man, it is vice.*
> Boethius

world suffices for this, because it is very easy to imagine our world working in much the same way, without any need for consciousness at all. We could even discover an intelligent species that claims to be conscious and conclude that they are only sophisticated machines. The fact that we can so easily conceive of such a scenario means that consciousness is a surprise added feature of our universe. We could know all there is to know about the universe, and we would not have to postulate the existence of consciousness—as evidenced by the many theories that fit observed facts perfectly while ignoring or sidestepping consciousness—except that we are each both conscious and conscious of being conscious.

So what if we do figure out everything there is to know about the brain? What if we plumb the deepest secrets of the visual system and discern the neurology of color and image processing. Will that tell a color-blind person what redness is? Will that tell a deaf person what music is? No amount of theorizing or explaining can substitute for the actual experience—and experience could be the whole point of living. Any materialist theory of consciousness must overcome these hurdles. None of them manages to do this. Consciousness is not logically supervenient, and here again, any reductive explanation is doomed to failure. Consciousness may play many roles, but it is not defined by those roles.

Conscious roles are distinctive, because they are phenomena we experience that cannot be adequately defined by defining the function. The physical facts of most objects and phenomena entail their existence. Fact implies function, and vice-versa. Consciousness does not fit this model, because we must explain consciousness itself beyond the objects and functions that may or may not contribute to it. There is no bridge connecting consciousness to lower-level facts, because consciousness is not logically supervenient on those facts, and yet it begs description; the first-person nature of being conscious leads us all to believe that there is something here that needs explaining. Such a problem might not seem nearly so pressing, were it something that could only happen in the third person.

> *Blind, deaf, dumb! Infinitely beyond the reach of imaginative contrivances!*
> Seccho

What kind(s) of organisms can have consciousness? How simple can an organism get while maintaining consciousness? Is it possible to predict the essence of an experience using purely physical models? A complete set of physical and psychological laws may give us an understanding of the entire universe, but will such a path truly explain everything? Is consciousness tied to knowledge? If so, then we can speculate that the thermostat on the wall is conscious, and that consciousness is everywhere. Every causal relationship involves information, and experience exists everywhere there is information. Even rocks and subatomic particles contain information. We would be hard pressed to say that a rock has experiences or consciousness, but that does not mean it doesn't. We could say that the thermostat has some glimmer of consciousness, and that the rock could contain conscious systems. Such an approach is counterintuitive but also satisfying, because it integrates consciousness into the natural order of things instead of treating it as somehow different. Consciousness would thus be seen as a property of the entire universe.

Materialists remain undaunted by problems like this, and keep promising that a viable reductive explanation will come along; however, the problems with such approaches involve fundamental components of the approach itself. No reductive attempt will ever fully explain the mystery of consciousness. Doubtless, we will obtain new insights and valuable discoveries along the way; however, this is ultimately a dead end.

At some point, we have to stop explaining and simply accept some things as given. For example, we could choose to take consciousness for granted, but even this simple step forces us beyond reductive explanations.

All awareness is awareness of something.
Husserl

Quantum Consciousness

Hofstadter says that consciousness is inevitable in a suitably complex brain, and that a "strange loop" of self-awareness arises naturally when a sufficient number of categories is available, which gives rise to "I." How exactly this happens is not explained; however, this proposition reminds us that we can view consciousness as either being or not being somehow different than simply an outcome of physical laws. Neither path is easy, but seeing consciousness as magically popping into existence in a sufficiently complex brain is much easier than the alternative, which forces us to consider dualism or even idealism.

Science has provided a massive amount of evidence suggesting we need not postulate the existence of an entity such as a soul or mind in order to account for life and consciousness.
Nancey Murphy

> *Can you really believe that all your thoughts and meanings, your feelings and struggles with values, and indeed your consciousness itself, are the results of a random dance of elementary particles or genetic determinism?*
> Amit Goswami

Some views of the quantum waveform say that it only represents information about possible measurement outcomes and should not be confused with the entire system, or even thought of as describing the system. This attempt to limit the role of consciousness misses the mark, because it forgets that the outcome of the entire experiment depends on the entire system. This must include the observer, as we learned in Chapter 23. The idea that consciousness creates reality seems new but goes all the way back to the *Vedas*, if not before. Wigner was correct when he said that, "It is not possible to formulate the laws of quantum mechanics in a fully consistent way without reference to the consciousness."

Would it be possible for some creature to experience the superposition states themselves? Such a creature would see atoms in both boxes, a cat that is both alive and dead, and would consider all of this perfectly natural. They would have no measurement problem, because they could see all possible outcomes at once.

Quantum mechanics gives us a mechanism for allowing free will, because it is inherently probabilistic and strange, especially if consciousness does in fact collapse the waveform. In this sense, it does not matter whether biocentrism is true, whether Bohm's implicate order that says that the entire universe is entangled is true, or whether our brains are quantum processors; the outcome is the same any way one looks at it. This replaces the traditional religious view of an OOO God that decided everything from His transcendental perch outside space and time, which does not allow much room for free will, and which precludes anyone from being responsible for her or his actions. A human brain has approximately 10^{26} particles. We know the fundamental questions, but even solving one equation per particle per second would take us far longer than the current age of the universe (4.3×10^{17} seconds). In fact, we would need 2.31×10^9 (2,314,815,000) times 13.7 billion years, or 31,712,965,500,000,000,000 years. Suffice it to say that we will never be able to measure the complete state of a human brain at any one instant, let alone know for sure what the person will say or do next.

> *If every particle interacts with every other particle, the brain itself must be viewed as infinitely interconnected with the rest of the universe.*
> Michael Talbot

The quantum world we explored in Chapter 23 may be the ultimate source of consciousness. In fact, a close look at the quantum world may erase any distinction between "physical" and "mental," and reveal that the foundation of reality is built of consciousness parsing a huge amount of information. This is, again, in stark contrast to the classical Newtonian system that has no need of an observer. Consciousness is the universe looking back at itself; if reality is built on consciousness, then the whole universe manifests this self-reflective order. Human consciousness then becomes self-reflective awareness built on the overall sense of order. In this way, the entire universe is conscious, and we must accept that both consciousness and order exist. After all, the universe could not exist sans consciousness, and physics could not exist sans order.

Adopting this view means that we must see cognition as connected to the whole, because human consciousness both exists within and participates in a conscious universe. The presence of consciousness, whether as the foundation for the universe or as emerging from the universe, means that the universe itself is conscious. There is just no other way to look at it. It is of course possible that superpositions only exist in our imagination and that, for example, the cat really is alive or dead in the box even before we open the lid to get a peek. It is true that the macro environment tends to smooth out quantum effects by making them so small that we can safely ignore them most of the time. Then again, adding the environment to the mix reveals that consciousness really isn't all that special, because it is part of that same environment.

Quantum mechanics places observer and observed on equal footing. Macro-level smoothing still requires actualization, and we must still wonder just how a particular outcome is selected. We can use classical laws to approximate the quantum world at macro scales, but that's about as far as that goes. Consciousness seems to contain all of reality and vice-versa. There is no distinction between mental and physical, but there can still be individual features within the whole, such as the eddies in a river. Quantum mechanics bridges the gap between the external world that we perceive as objective and our private internal worlds. Advancements in our understanding of quantum mechanics can therefore help us better understand consciousness, and vice-versa. Pribram noted that descriptions of spiritual experiences were often similar to descriptions of quantum mechanics.

The analysis of the physical world, pursued to sufficient depth, will lead back in some hidden way to man himself, to conscious mind, tied unexpectedly through the very acts of observation and participation to partnership in the foundation of the universe.
John Archibald Wheeler

Many aspects of the natural world are in fact counterintuitive to our biological expectations.
Pascal Boyer

Chapter 30
The Universe in our Heads

Our worldview has shifted from myths and religion to the idea of objective reality and back to the subjective through quantum mechanics, relativity, and other theories. The four elements of the ancients have been replaced by the strong, weak, electromagnetic, and gravitational forces. Revealing the subjective nature of reality underscores the fact that the more we look at our universe, the more how, where, and why becomes important. In short, science itself is an experiment.

The physicists who created the Copenhagen interpretation found themselves unable to confront what they were seeing. They left their own work incomplete by not explaining what an observer is or the nature of the consciousness responsible for causing waveform collapse. Physicists are in many ways the high priests of science, because everything from chemistry to biochemistry, neurology, and psychology ultimately depends on the laws of physics; however, none of these disciplines has yet cracked the mystery of consciousness, because they are trying to include it on their terms instead of on its own terms—a sharp contrast from Newton, who sought and deciphered gravity on its own terms. Newton asked, "How does gravity work?" instead of, "How can I describe gravity in terms of this particular theory or branch of science?" Examining consciousness on its own terms may finally force mainstream science to conclude what mystics have known for thousands of years: that conscious observers are part of a reality that is far richer than imagined by any scientific theory.

We saw in Chapter 29 that the hidden variable and many worlds interpretations do not avoid consciousness. We may therefore have to break down and finally admit that consciousness itself is primary and irreducible. We would then be left looking for a connection. Maybe consciousness connects to the brain and the world through a waveform. Quantum tunneling in synapses may play a role in triggering neurons, making synapses a great place to start looking for the mind/brain link where mind meets the physical world. One could say that consciousness collapses waveforms in the brain, which then affects the outside world. Evan Walker Harris calls quantum mechanics the mechanism of the mind where, "Mind is the forge of creation and the echoes are the universe."

Taking the Copenhagen interpretation to its logical conclusion entails quantum consciousness. Proving the primacy of con-

---

We are spiritual beings with souls in a spiritual world, as well as material beings with bodies and brains existing in a material world.
John Eccles

If what you mean by 'soul' is something immaterial and immortal, something that exists independently of the brain, then souls do not exist. This is old hat for most psychologists and philosophers, the stuff of introductory lectures.
Paul Bloom

sciousness requires results that are simultaneously meaningful, testable, predictable, and falsifiable for us to be able to measure consciousness. We can do this for emotions in a relative sense, which is fine, because everything is relative. How can we speak of numbers associated with consciousness, while at the same time denying that we can measure consciousness itself? Granted, this is a rather more subjective undertaking than measuring physical systems, because we don't have any device for measuring someone else's consciousness. We only know of consciousness from our own experience and associating that with other data that we experience at the same time.

In Chapters 15 through 17, we saw that many experiments have indicated that mental processes can affect the physical world by such means as altering machine outputs even across time. We know that consciousness plays a decisive role in quantum experiments; however, these results remain hotly debated by scientists of different disciplines, many of whom seem to be out to validate their particular worldviews rather than judging solely on the merits. Still, we must allow that the brain is indeed involved with quantum processing, and that physical bodies and systems are possibilities that manifest as local, finite structures. The idea of "soul" can therefore be seen as referring to nonlocal potential that may move from one body in one time at one location to a different body in another time and location.

Can any theory of physics entail consciousness, and if so how? The physics of the universe itself must—and does seem to—entail consciousness as influencing—if not outright creating—the particles, fields, waves, and other objects described in the mathematics. Allowing physics, the bedrock science on which all other sciences rest, to explain consciousness prevents any attempt at reduction. Most methods of trying to link consciousness with physics use quantum mechanics, which suffers from the problem of conflicting with relativity, as we saw in Chapter 24. Penrose felt that a theory linking quantum mechanics and relativity may be the key to understanding consciousness, and he was right. Superstring, M-theory, and holography build on quantum mechanics without conflicting with it, which means that they all have room for consciousness. I would argue that the links forged between these different theories can only strengthen the argument for the primacy

And how can a machine have a soul.
David Darling

Mental formations with regard to objects and mind agitate the fundamental consciousness like waves on water.
Kongtrul Lodrö Tayé

of consciousness, for all of the reasons we have explored throughout this book.

I could be wrong. Some unknown wave or force may collapse the waveform with no consciousness needed. Waveform collapse inside microtubules deep in our neurons may drive cognition. Nonlocality may play a role. Still, we have no reason to assume that any of these processes can or must lead to consciousness. One thing we can be sure of, is that it's damned difficult not to think of consciousness as being somehow different than any other process.

A Theory of Quantum Consciousness

> *So with all due acknowledgment to the fact that physical theory is at all times relative, in that it depends on certain basic assumptions, we may, or so I believe, assert that physical theory in its present stage strongly suggests the indestructibility of mind by time.*
> Erwin Shrödinger

Let's take a closer look at quantum consciousness by focusing on the synapse junction between neurons. In this section, we will look at the theory of consciousness presented by Harris in his book, *The Physics of Consciousness*. I am not presenting the numbers Harris included in his theory, because they are not terribly important for this discussion. What is important is the philosophy and the mechanism Harris describes.

The Conscious Electron

A synapse releases neurotransmitters in small packets called *vesicles*. A large synapse may release hundreds of vesicles when it fires. There is no guarantee that a synapse will fire. The behavior of synapses and their associated neurons depends on smaller-scale factors that may operate according to quantum laws (instead of classical) in order to cause waveform collapse.

> *We create our own reality.*
> Fred Alan Wolf

A synapse can make the binary decision to either fire (1) or not fire (0). This is where quantum mechanics may have its best opportunity to affect brain behavior, and may be the window through which the observer interacts with the quantum world. The problem is order, or rather the lack thereof. Information processing requires order, which seems at odds with apparently random quantum noise. Each synapse must therefore depend on some interconnection and communication between pre- and post-synaptic neurons, and make some sort of quantum decision about which particular synapses will fire and when. This is the only way in which we can tie the observer effect to what we call consciousness. The only problem is that we cannot rely on quantum mechanics to play a starring role, because we cannot expect atoms to jump across synapses at precise enough intervals or in significant enough quantities to have any effect.

Electrons are always in motion. The time it takes a synapse to fire could give an electron billions of chances to tunnel across the synapses, a feat that is possible at electrically polarized synapses and that allows quantum mechanics to enter our brains. The post-synaptic (receiving) end provides the tunneling electrons that also provide the energy needed to open the vesicle gates. Electrons themselves do not have enough energy to jump directly to a sodium (Na), potassium (P), or calcium (Ca) ion. Instead, they jump to a macromolecule in the vesicle gate. The electron that is already there must be neutralized, which the calcium ion does when it attaches to the gate molecule. The post-synaptic electron then jumps the gap. This model has been tested and validated by researchers B. Katz and Ricardo Miledi, and the theoretical vs. actual electron tunneling curves matched, which would be impossible if the synapse did not function as just described.

Does this tunneling give rise to consciousness, and if so how? We have not yet discussed how or why synapses firing are part of conscious experience, or why a synapse firing in someone else's head does not affect our own consciousness. (Then again, maybe it does!) Still, the mere presence of a complex neural network is not in and of itself any reason for consciousness, any more than the mere presence of the Internet causes any information to flow across it. But back to the electrons, which can move from synapse to synapse by hopping from molecule to molecule, and across synapses on large molecules (messenger RNA) like stepping stones. (As an aside, metals whose outer 1-2 electrons are free to move are perfect for conducting electrical currents. The easy mobility of electrons is one reason why the best conductors are very malleable compared to other metals.)

An electron can travel across the entire brain in the time it takes a synapse to fire. This means that the electron has plenty of time to either hop to the nearest neighbor across a single synapse or to interact with any other synapses in the brain. Each synapse has many electrons available for potential journeys to other synapses. A strong enough continuous stream of

Let man then learn the revelation of all nature and all thought to his heart; this, namely; that the highest dwells within him; that the sources of nature are in his own mind.
Ralph Waldo Emerson

Man, the start of the analysis, man the end of the analysis - because the physical word is, in some deep sense, tied to the human being.
John Archibald Wheeler

> *The explanations of different phenomena most likely to survive are those that can be connected and proved consistent with one another.*
> Edward. O. Wilson

electrons moving around the brain and interacting with synapses may be necessary for consciousness to switch on. This may be the functional mechanism by which quantum mechanics allows coherent mental functioning and a conscious "I" experience to exist. Weakening this stream below a certain threshold may shut off consciousness and result in sleep.

To paraphrase Crick, we, our joys and our sorrows, our memories and our ambitions, our sense of personal identity and free will, may in fact be a single electron setting off across the brain. As Harris said, "The Great Oz is but one electron." What, exactly, is consciousness? It is that electron, roughly speaking. Both the electron and the synapses it crosses are part of consciousness, and yet consciousness itself is neither of these things, but the collection of possibilities that develop as the electrons in the brain interact. Consciousness brings about the possibilities inside our heads, and all of the interwoven collections that weave these possibilities together. The act of creating a possible experience and selecting which synapses to fire is what allows the mind to turn our thoughts, experiences, and actions into realities.

We can therefore surmise that consciousness may be part of all natural quantum processes. The difference between them and us is that our consciousness is part of a sophisticated logic machine that turns out to be the brain of a living organism. Life, thought (data processing), and consciousness are three separate things. An organism needs neither consciousness nor thought to live, such as a plant (even though this is open to debate). A brain does not need consciousness in order to think. Any of these three things can exist on its own or together with any combination of the others. A nonliving machine may thus be perfectly capable of thought and, someday, consciousness. Given that an ant has roughly the processing power of a Macintosh LCII computer (see *The Natural Savage*), our machines may already be there by some definitions.

> *Everyone is a mystic by virtue of their existence, but not everyone know is, or accepts it.*
> Wayne Robert Teasdale

There is just one more thing: If we accept that life, thinking, and consciousness can exist on their own, then we must accept that consciousness can exist without being tied to either life or data processing. Everything that exists ultimately owes that existence to quantum events; the universe may therefore be inhabited by an incredibly large number of quasi-discrete, non-thinking entities, most of which do not process data.

A Quantum Theory of Sleep and Identity

Harris's theory also includes a fascinating idea about sleep that also ties in with identity. Ultraviolet rays stimulate *melanin* production, a mixture of large organic molecules that absorbs a wide range of electromagnetic radiation, including ultraviolet light from the Sun and scattered light in our eyeballs that might otherwise blur our vision. Melanin protects the irises of the eyes, and also gives the cells in the brain cortex their distinctly gray color. The brain's need for a radiation shield may shed light on our circadian rhythms and why we sleep. At first blush, sleep makes little evolutionary sense, since it leaves us extremely vulnerable to predators. One could be forgiven for wondering why natural selection has not selected insomniacs. The body uses the down time to heal and restore itself for the next day, but this can't be the whole story. Sleep must be absolutely indispensable for life, and possibly even consciousness.

The electron theory we just discussed may explain this. Simply allowing electrons free rein inside our heads could cause consciousness to become a blur, until a huge surplus of electrons chokes it off altogether. Melanin could prevent this blurring by absorbing some of the electrons and excess energy. This would be a delicate balancing act, because too many electrons blur consciousness, while too few cease being able to support consciousness. The rate at which melanin absorbs electrons combined with the synapse firing rate could provide a *consciousness interval*, which is roughly analogous to the number of frames per second we can perceive. Harris puts this figure at a maximum of about 0.04 (1/25) second, which he says explains the speed at which movies need to run in order to appear smooth without any flickering. Our brains *interpolate* (fill in between) frames to provide the illusion of smooth movement.

Not all electrons go on to synapses. Some stop on a messenger RNA stepping stone, effectively preventing that molecule from being able to assist with consciousness. Keep this up long enough, and the RNA will get used up to where consciousness can no longer be sustained. Sleep ensues and continues until the excess electrons clogging the works dissipate, and the RNA returns to its ground state, a process that takes

Maybe reality isn't what we see with our eyes. If we did not have that lens—the mathematics performed by your brain—maybe we would know a world organized in the frequency domain. No space, no time, just events. Can reality be read out of that domain?
Karl Pribram

We are, each of us, a part of the universe seeking itself.
Gerald L. Schroeder

> *As long as we refuse to admit into the debate the forever private awareness each person has of himself, his thoughts and his feelings, his judgments and rationality, and as long as we insist on public and purely behavioral signs of these, radical materialism can remain in the debate.*
> John Eccles and
> Daniel N. Robinson

hours. Harris estimates that we can remain awake for 16-20 hours without too much problem. The brain keeps functioning during sleep and forming thoughts, with synapses firing at a slower rate. Activity in some areas of the brain occasionally exceeds the local consciousness requirement, and we begin dreaming. The dreams themselves are thought bubbles that rise up into consciousness for limited periods, during which we act out our inner thoughts, ideas, fantasies, fears, and hopes.

In general, the idea that consciousness is a quantum phenomenon is very pertinent to the idea of individual identities. Western religions believe that the soul maintains its personal identity after death, while Eastern religions think some identity carries forward as karma, but that personal identity is an illusion. (See Chapter 5.) We have seen that materialist science equates identity with the physical body and brain. The quantum theory we have been discussing equates identity with a sort of software that is independent of the brain, and that could theoretically run on any human brain—a concept that sounds similar to the idea of a soul living in a body. A classical machine cannot have consciousness, but a brain working by quantum principles transcends mere computer abilities and acquires an identity that exists both as and within a conscious state. We can see identity as the continuity of this quantum process.

Sleep interrupts this process, which begs the question of whether we retain the same identity when we awaken, or if we begin a new one. Should we assume that we remain the same because of memory, or should we assume that quantum processes create something new each evening? Alternatively, should we conclude that our identity persists? Sleep does not stop the brain, but simply dims it, which seems to argue for continuity of identity. We therefore have three possibilities:

> *Your mind is bigger than the entire universe.*
> Gary E. Schwartz

- The brain slows, but not enough to break continuity. This is the commonsense solution, but does not explain how personal identity is recovered following brain flatline during clinical death; however, this interpretation could work under normal circumstances, with the second interpretation filling in during those rare instances when someone is brought back.

- We could awaken as duplicates of our previous identities, which are absolutely indistinguishable, and therefore functionally identical. Tipler argues for this interpretation in *The Physics of Immortality*, in which he says that an exact replica of a person down to the quantum states is the original person. We die and cease to exist until a future time when our exact quantum states can be reproduced, which causes us to be "born again." Since the quantum states are identical, the copy itself is new, but carries with it everything that makes a person an individual. For example, when I die, my body will decompose, and I will cease to exist until my exact clone is created, at which point I will switch back on. This reawakening may take billions or even trillions of years to occur, but as far as I am concerned, I will close my eyes in one instant and open them again the next, blissfully unaware of having been completely annihilated in the interim. There are many philosophical problems with this interpretation that I won't go into here, but this scenario could work in cases where a person experiences clinical death but is saved. In this case, the brain has not decayed, and the quantum state at the time of recovery is "close enough" to the before-incident state to allow identity to continue on.

- We could be someone new every time we awaken, having died the previous evening and been reborn in the morning. This interpretation requires no soul, no eternity, nothing. When we die for real, the light goes out for good. This interpretation does not make much sense to me, because, again, the brain dims but does not shut off during sleep.

This part of Harris's theory has several problems, such as:

- Why does melanin shield us from internal electromagnetic radiation but not external fields that can easily reach many orders of magnitude higher than our own? Furthermore, how can internal fields create and maintain personality while remaining immune to much larger external fields? It follows from this theory that things like radio antennas, cellular telephones, MRI scanners, and nearby power lines should have grave consequences for us, but they don't. The electromagnetic field around

What we perceive comes as much from inside our heads as from outside.
William James

In a universe in which individual brains are actually indivisible portions of the greater hologram and everything is interconnected, telepathy may merely be accessing of the holographic level.
Michael Talbot

If all this seems dehumanizing, you haven't seen anything yet.
V. S. Ramachandran

some power lines is strong enough to light fluorescent bulbs many feet away, as evidenced by the many photos one can find of people standing under power lines holding those bulbs. Why are the same fields not affecting those people?

- The consciousness interval Harris uses corresponds nicely to the 24fps rate of movies shown in darkened rooms; however, the human eye can detect far faster frame rates, such as the visible flickering of 60Hz (and sometimes even faster computer monitors) that are notorious for causing eyestrain, headaches, etc. Also, some online video formats can reduce the frame rate to 15 frames per second while still giving the appearance of smooth movement.

- Doctors, elite soldiers, and others are trained to remain awake for 48 to 72 hours, often under extremely demanding circumstances. The numbers Harris uses in his theory do not seem to allow such cases to occur.

- Why do only parts of the brain wake up during sleep, but not the whole? Assuming that this occurs randomly, how does this lead to organized dream states?

- What keeps the brain from "hunting," that is, sleeping until just enough messenger molecules have reset to allow consciousness, waking for a short while until the molecules are saturated again, and going back to sleep? What keeps the brain asleep long enough to completely reset the molecules and awake long enough to use them all up? What role, if any, does physical exertion play in determining how long we sleep?

No matter how old you are, don't stop learning. For this is a process, I gather, that goes on for eternity.
Anonymous

- What principle or mechanism causes the electrons to behave in a manner that is coherent enough to create consciousness in the first place?

These are serious challenges; however, the beauty of Harris's theory is that it does offer a potential bridge between the quantum and the classical, and in so doing goes far beyond other theories of consciousness.

The Consciousness Data Stream

Contrary to what some scientists believe, Harris held that our brains don't push too much data around too quickly. In fact, he believed that some household appliances can easily outmatch the brain in sheer data processing ability. Information can be moved and processed at different rates. For example, a modern household Internet connection may be capable of moving 2 million bits of information per second (and some go substantially faster), while some of the earliest data connections could only handle 300 bits per second. Harris set out to determine just how much processing power the human brain has, and how fast it can push data around.

All the contradictions and conflicts we experience in the world are born deep within our consciousness.
Irwin Kula

We saw in Chapter 28 that time may consist of movement across Nows, in which all that can possibly happen already exists just like a movie exists in its entirety. Moving from Now to Now works similarly to viewing sequential frames in a movie, which gives rise to our sensation of time. This is of course different than the classical view, which says that time is a one-way street. Each Now has a duration equal to one Planck time, each of which is far too rapid for us to detect. This super-fine mesh of time allows each species to have its own smooth, non-flickering consciousness interval. We can thus speculate that a snail thinks of itself as moving at a good clip, while we see it as moving almost interminably slowly. By contrast, the life span of some insects is measured in days, but this may seem just as long to them as our lives seem to us. The fact that a fly can easily outmaneuver a swatting hand indicates that is has a much shorter consciousness interval than we do. Thus, a consciousness interval consists of a bundle containing a higher or lower number of Nows.

It may seem silly to speak of internal body clocks and measuring the length of consciousness intervals when the flow of time is an illusion, and when consciousness itself is beyond space and time; still, Harris maintained that this is essential for understanding consciousness. One consciousness interval is thus the briefest chunk of conscious awareness that makes up the stream of consciousness that we experience. Understanding this helps us understand the connection between mind and brain. A consciousness interval of an hour would result in our seeing flashes of blurry events like a stop motion film. By contrast, a much shorter interval would make things seem to hap-

Conscious experience is part of the natural world, and like other natural phenomena it cries out for explanation.
David J. Chalmers

> *This consciousness that is myself of selves, that is everything yet nothing at all—what is it? And where did it come from? And why?.*
> Edwin T. Jaynes

pen awfully slowly. This makes sense if we are talking about different intervals based on the same life span of somewhere around 70-100 years; however, a species with a shorter or longer life span could benefit from a different consciousness interval, as we saw above.

Harris calculated our consciousness interval based on a standard movie frame rate of 24fps, which works perfectly well, given that professor Rahul Serpeshkar provides refresh rates of 12Hz for rods and 55Hz for cones. Rods are very sensitive to light but only detect grays, while cones are more sensitive to color. Running a movie that is half dark at 24 frames per second works perfectly (24 is half of the quotient of the faster refresh rate divided by the slower one), while a television that only changes color without going dark must run much more quickly at 60Hz. But even this is not necessarily fast enough; turn on a television in a darkened room, look off to one side while keeping the TV in your peripheral vision, and you will notice a distinct flicker.

Similar flickering in fluorescent light bulbs and computer monitors is both a leading source of eye strain and a strong argument against Harris's proposed consciousness interval of 0.04 second; however, the idea of a consciousness interval with interpolation between frames does remain valid. Harris also postulated a data rate within the brain of between 45 and 200Mbits per second as the conscious data rate. (Again, the numbers are not nearly as important as the ideas behind them.)

> *Consciousness is more than an accident.*
> Paul Davies

Neurons send impulses along paths resembling a network of electrical wires and junction boxes, although at much slower speeds than normal electrical circuits. The average speed of impulses is about 100 meters per second, which means that there is a lag between perception and registering that perception in the brain, plus another lag before we become consciously aware of the perception. This means that we are always running a bit behind, although our brains are doing a great job of convincing us that we are in the present. One estimate says that someone driving down the freeway at 60 miles per hour is 11 feet in front of where s/he thinks s/he is—a convincing argument for leaving plenty of room between yourself and the vehicle in front of you! There is no guarantee that an impulse will reach its destination or cause any conscious or unconscious thought, emotion, or action, because each impulse must run a gauntlet of synapses that can either pass it along or suppress it. The synapses must be where mind and brain connect, and quantum tunneling must play a role.

Sharing a Consensus Reality

We noted above that consciousness can exist without being tied to life as we know it or to data processing. Everything that exists is ultimately the result of quantum events, which allows us to speculate that the universe contains a vast number of conscious entities in the form of particles and larger objects, most of which are non-thinking. These entities determine the outcome of each quantum event within the probabilities provided by the Schrödinger equation that describes the limits of any collective freedom of action. In other words, each entity can act of its own accord within the realm of what is possible under any given circumstance, and we can predict the possible outcomes using the Schrödinger waveform. We can therefore say that the particle itself chooses to collapse the waveform.

The universe is creative; you and I in our creativity are the living proof.
Amit Goswami

There are rules that govern how information must be configured for transfer across space and/or time. Quantum mechanics also allows two distinct ways of handling information. The first is information in potential states that can be actualized by interacting with material, such as a particle that is in several potential locations until observed. Second, something eventually has to give, where one of the potential states becomes the actual state and something happens, such as a synapse firing.

The selection of which potential state will become an actual state is a distinct type of information that is enmeshed with state preparation, which leads to consciousness. This the information that gets applied to the waveform collapse, which is the result of the potential possibilities. Consciousness thus observes what can happen in the brain and then selects a state from among the available choices. This is the will portion of consciousness. The mind thus consists of both consciousness and will, and gives us three distinct processes: unconscious computing, quantum sampling of that computing to collect possibilities, and the selection of a potential state that becomes an actuality.

A total oneness is as close as the several trillion neural connections in our brains can come in our quest to discern the infinite.
Gerald L. Schroeder

In this context, will works as follows:

> *Every man's world picture is and always remains a construct of his mind and cannot be proved to have any other existence.*
> Erwin Shrödinger

- A state of mind associated with conscious experience.

- The mind must be able to affect events and to control the body as an active participant, and not merely as a passive observer. Physical laws must allow a range of possibilities for the body, at which point will makes the selection. The process that makes the selection must be unconstrained in its ability to do so.

- Will interacting with mind and body leads to one of the actions permitted by physics for the situation at hand. In other words, the interaction leads to brain states and actions. One synapse or group of synapses among all those that could fire at that instant does fire, which sets brain and body in motion, which in turn creates new possibilities for what can happen next.

It stands to reason that the will stream is a subset of the consciousness stream. Given that much of our consciousness is taken up by perception and other routine functions, this does not leave much room for will. Will can thus be said to use only a tiny fraction of the bandwidth available to consciousness.

> *If a man loses consciousness as soon as his brain is injured, it is clearly as good an explanation to say the injury to the brain destroyed the mechanism by which the manifestation of the consciousness was rendered possible, as to say that it destroyed the seat of consciousness.*
> Ferdinand Schiller

Objects are in a state of superposition before waveform collapse. After observation, the objects are in a single state. When a synapse fires, everything in the universe connected to that firing must collapse to a single state. Will is a small percentage of consciousness, but it is not restricted to the brain, nor to any particular location or time. Will causes things to happen in the brain, and every other affected object in the universe goes along with it. One mind can affect other minds and brains, and they can affect us in turn. Thanks to nonlocality and superposition, we can all be said to be telepathic and psychokinetic in a way, Einstein's worst fears come true. Each observer is thus connected to all observers, and all observers collectively select reality. A hypothetical "perfect" observer would always get what s/he wants, because her or his will would be uncluttered by the noise of ordinary consciousness. Just because we consciously want something does not make that desire part of our will, and therefore does not mean we will get what we want. This is yet another reason why the so-called "Law of Attraction" that so many self-help experts talk about is worse than useless at best.

The amount of bandwidth taken up by will is so small as to be nearly impossible to distinguish from consciousness. Both are part of mind, and both arise from quantum events in synapses. Will reflects our ability to affect things both directly and indirectly anywhere in the universe, while consciousness deals with information about the here and now and is essentially noise as far as will is concerned. Harris placed the ratio of will to consciousness at 0.00124. (Again, the actual number may vary, and we can argue about both the formulas he used and the values he plugged into those formulas, but the basic idea remains fascinating.) Our wills are limited by the laws of physics, which explains why jumping off a cliff is a terrible idea, no matter how convinced you may be of your ability to fly.

That said, time is an illusion, and nonlocality is a staple of quantum mechanics. Tunneling is another staple. We surmise from the dual slit experiment that particles take all possible paths to their destinations at once. Two particles that collide bounce off each other into a superposition of all possible outcomes and jump into a definite state upon observation. The Schrödinger equation allows us to predict how a waveform evolves between observations that reveal the particles jumping from location to location. These jumps are so small that they appear smooth and classical on the macro scale. The Heisenberg uncertainty principle limits what we can observe with precision. For example, we cannot predict the outcome of a die toss after the first bounce, as the die goes into a linear superposition that only collapses after it comes to rest and an observer gets a look at it. All of this we know from Chapters 15 and 23. But what if we could see the possibilities before they happen?

Humans and animals have a certain degree of foresight that can help us predict outcomes. If we could predict the outcomes with anything approaching perfect reliably, then we could get just about anything we want, like the Biff character in *Back to the Future*, who gets his hands on a book listing the outcome of every major sporting event for the next 50 years and uses that information to make a fortune. Our ability to make predictions is not nearly that good, but we must assume that some correlation exists between the observation and the state of the observer; the latter is part of the system, and even looking at something becomes part of the physics of what is

---

We find that we can best understand the course of events in terms of waves of knowledge.
Sir James Jeans

This world faces us with the impossibility of knowing it directly. It is a world whose nature cannot be comprehended by our human powers of conception.
Max Planck

happening. Consciousness and will could therefore be the hidden variables that select the state that in turn selects the moment from among all possible moments.

How does the will get what it wants? It's not like we select a result like pushing buttons on a vending machine. The will selects what is being willed. Both will and that being willed can be seen as a tangled hierarchy that cause each other, and thus as almost the same thing from a practical perspective. Thus, the state that occurs is the state that was willed, and the state that was willed is what occurs. A will equal to consciousness would get its way all the time, but this does not actually happen in actuality, because the will is very small. We therefore only get a very small amount of what we will for.

The ultimate creative principle is consciousness.
Dalai Lama

All observers must be treated equally. We cannot always say for sure who saw what, since there is no absolute Newtonian frame of mind. In Chapter 24, we saw that relativity means that different observers may see things happening in a different sequence, because everyone has her or his own frame of reference. Everyone involved collapses the wave nonlocally and independently of time and space. Each observer is part of the system, and each observer contributes to collapsing the waveform. The collective wills of the observers become one of the hidden variables, because all of their minds are entangled as if they were one as far as a given observation is concerned. It is therefore possible that a person observing many things at once will be involved in many such links with different sets of observers, one set for each event.

Our consciousness determines the kind of space in which we live. The way in which we experience space, or in which we are aware of space, is characteristic of the dimension of our consciousness.
Lama Anagarika Govinda

This conclusion is both reasonable and almost forced on us by quantum mechanics and relativity. Nonlocality has been confirmed by tests of Bell's theorem. Experiments have confirmed Einstein's relativity predictions about time. Purely geometric theories link and equate space, time, energy, and matter as simply different aspects of each other. Further, the fact that everything in the universe is entangled to one degree or another means that all conscious observers play a role in every observation anywhere in the universe. The strength of the entanglement varies;. For example, one would expect the people watching a sporting event to have a much stronger effect on the outcome of that event than people who are not watching it but who are on Earth. This second group would have far more influence than conscious beings on a distant planet. Everyone's hands are in every cookie jar to one extent or another. This collection of wills could be what Goswami refers to as "God consciousness."

During an observation, consciousness impinges on the material world, causes collapse, and retreats until the next observation. This is a safe assumption, because we cannot say that conscious observation is omnipresent, when personal experience confirms that this is not the case. Consciousness therefore remains hidden and undetectable. It forms one of the terms present in the system to close a measurement loop, in the same way that a light switch completes an electrical circuit. This term vanishes along with the waveform as the measurement loop closes. Consciousness is part of the loop and one side of the reality that combines with the opposite side (superposition) to cause collapse. The entire loop is time-independent and nonlocal, which means that all observations affecting the loop happen in a single moment, and the loop itself is a single Now that includes a piece of consciousness/will experience shared by all observers.

We saw in Chapter 28 that time and space are inherently grainy Nows. Each observation selects a new Now from among all available Nows for everything touched by the measurement loop. Loops interact with each other billions upon billions of times per second, tying brains with both themselves and the outside world to form patterns. Each loop contains the consciousness and will of its observers. Loops and the Nows they select can feed back to the brain(s) involved in ways that affect neural circuits. These circuits can affect our sense of time, emotions, thoughts, and beliefs. They can also influence the will, which in turn influences the next Now that is selected. This is how patterns of belief and behavior become entrenched in humans, in everything from learning to walk, talk, and use a fork to self-esteem, drug addictions, and more. This also explains how pleasant experiences can seem to fly by, while painful ones seem to last forever. The physiology of the brain and body are also affected, as we saw in Chapter 9, and as I discuss in detail in *The Enlightened Savage*.

Once made, an observation cannot be unmade, because the next Now has already been selected. Nature enforces a strict no returns, no refunds policy. This is how we experience a flow of time that is both unidirectional and highly personal.

I believe that human consciousness is a conjuring trick, designed to fool us into thinking we are in the presence of an inexplicable mystery.
Nicholas Humphrey

There's a democratic assumption that we're all at the same level of consciousness, and that's wrong.
Theodore X. Barber

Chapter 30
The Universe in our Heads

We get a glimpse of the Nows available from the current Now, make our selection, and move on to the next Now like hopping across stepping stones. The following diagram demonstrates how this occurs: The universe contains all possible Nows, everything that could ever possibly happen. Selecting a Now limits the Nows one can select, in the same way that selecting a Thai restaurant limits your available food choices to Thai food. Will selects a Now, which has its own set of available choices, and the process continues.

Whether this immaterial 'mind' persists beyond ultimate biological death is purely speculative.
Michael Sabom

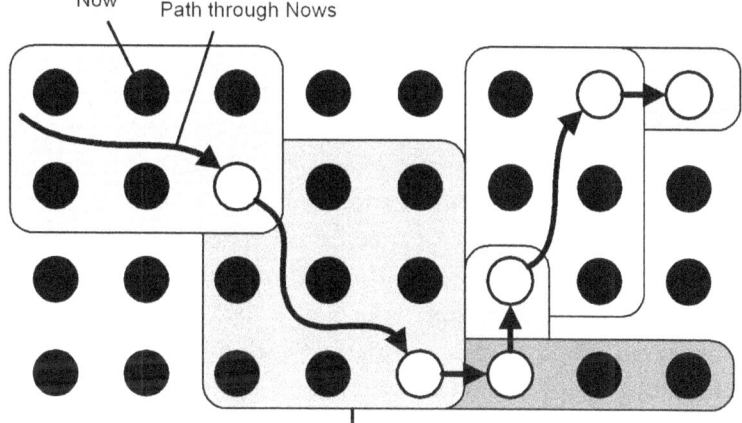

In this diagram, each Now is represented by a circle. The Nows available from a selected Now (white circle) are represented by shaded boxes. The arrows represent one possible flow from Now to Now. The flow itself can be arbitrary, and is only limited by where it can go next on a Now by Now basis; however, it is will that makes the selection and creates the flow. The range of possibilities from any given Now is itself a waveform that extends across all Nows in the universe, which is how you could theoretically find yourself orbiting Pluto at any given moment; however, the waveform beyond a limited number of Nows from any given location is extremely small.

To see what is in front of one's nose requires a constant struggle.
George Orwell

Our interpretation of a Now gives that Now a location in the space-time we perceive. Each Now receives a unique space-time location, and we thus experience a flow of time. Space-time serves as a filing system that allows us to organize our experiences; however, the delayed choice experiment reveals that future events can affect past ones, which opens the door to the biocentric theory we have looked at in Chapter 29.

The standard Schrödinger equation is incompatible with relativity. We must therefore use Dirac's equation to combine Schrödinger with relativity. Interestingly, we can replace the mass term with an information term and see how raw information becomes the source of particle mass in physics. Writing the resulting equation with space-time and information terms tied to consciousness gives us back the Dirac equation, and gives us the clear insight that space, time, energy, and matter do not exist, per se; there is only the conscious observer carrying out observations, weaving the illusion of space-time and matter from the measurement loops.

Harris's theory does a brilliant job of tying consciousness into the universe we experience and resolving the seeming paradoxes involved with multiple observers, to present a view of reality that combines dualism and idealism in a manner consistent with what we know of quantum mechanics. Superstring, M theory, holography, and loop quantum cosmology are all compatible with quantum mechanics by necessity, which means that this theory is compatible with all of the above. It also explains how time can be both illusory and very real at once, and how we navigate between Nows. It even goes so far as to speak to the hypothesis of God we discussed in Chapter 14. Observer and observed are two separated halves of the same ultimate reality coming together. We perceive a tremendous gulf between the two, but everything we have learned so far indicates that there is neither space to live in nor time to die in, because they are products of our own creation and not the other way around. As Harris said, "Observation is the stuff of space that reaches across the universe."

We have not yet explained what consciousness is, nor have we plumbed that which connects consciousness to the material universe. We may not even agree with Harris's figures about the number of neurons, synapses, bit rates, and consciousness intervals. We must, however, agree with his logic and his conclusions as a viable theory for all of the reasons we have explored thus far in this book, and continue trying to find a workable definition of consciousness. We will take another look at mysticism and religion in light of what we just learned, and then determine whether we should adopt a materialist, dualist, or idealist interpretation of reality. We will then be ready to examine the evidence and arrive at our conclusion.

The real puzzle, perhaps, is not that pathological/extraordinary consciousness should on occasion resemble one of the chimera—the sphinx, the manticore, the centaur—that combine the torso of one species with the head or hindquarters of another but that 'ordinary' consciousness should be unitary at all.
Wilder Penfield

Everything is essentially consciousness, purity, and joy.
Shankara

Chapter 31

Models of Reality

> *In effect, science has usurped our religions and scientists have become our new high priests.*
> David Darling

What is the ultimate nature of reality, and what model best describes it? Descartes' goal was to have science deal with matter and religion with mind, a paradigm that worked for some 200 years. His philosophy split people into separate components of mind and body, which seemed to make sense, until modern science refuted it for want of a means of communication between the physical and spiritual worlds. Circumventing this problem requires one to suppose either that mind and body do not interact (which contradicts everyday experience), or that they are one and the same (which medicine seems to bear out), which gives rise to materialism. On the other hand, one can postulate that mind is the stuff of reality with the physical world being simply a by-product, literally a figment of imagination. Bishop George Berkeley (1685-1753) famously said that a tree remains where it is overnight, because God is watching—a clear example of idealism.

> *The more closely matter is examined from within its own premise that finiteness is fundamental, the more complex and contradictory it becomes.*
> Daniel Cowan

Materialism believes just what the name implies: that everything in the universe is physical, period. This worldview is correct if the positive facts about the universe logically supervene on physical facts throughout the universe. The great successes of science demonstrate that this is indeed true in virtually all situations, with the possible exception of consciousness. In this scenario, all God had to do was set the universe up and let it run on its own—if indeed God exists. A strange blend of materialism and dualism pervades Western cultures; for example, scientists used to try to weigh the soul at the moment of death. Today's science supports monism since, again, there is no way for soul and body to communicate. Such communication would require energy transfer, and the amount of energy in the physical world remains constant. Faced with the great success of reductive approaches and centuries of religious mistreatment, science took the side of materialism.

The emerging "new" science is only now beginning to reexamine materialist tenets, and increasing numbers of scientists are making the intuitive leap between experimental results and mystical/spiritual concepts. The mere fact that a zombie universe going through the exact same motions without a flicker of consciousness could theoretically exist compels this examination. New science is not mainstream yet by any stretch of the imagination, but it's getting there by fits and starts. One problem is that people tend to find what they are looking for; a materialist will see evidence supporting materialism everywhere, while a religious person will see that same evidence as supporting her or his particular beliefs. Our universe seems disposed to show people exactly what they seek.

Our physical bodies are just as much a part of nature as any other plant or animal. We are no more outside the circle of life than our lungs are outside our bodies; however, our bodies are composed of trillions of quasi-independent cells, and play host to many billion more microscopic creatures, over whom we have little—if any—direct control. We are the ultimate puppet leaders in our own physical manifestations. Looking at a human body cell by cell gives us no reason to suspect the existence of consciousness in the host, and yet we feel that we are in the body and speak of ourselves in terms of a knower who is different than the known, a phenomenon for which science offers no explanation. The Aspect experiments and Bell's Theorem (see Chapter 23) compel us to keep searching for deeper levels of meaning. Idealism—the opposite of materialism—holds that what we call empirical, objective reality is nothing more than a product of consciousness, in stark contrast to those who believe in objective reality.

I should note that nonmaterial worldviews such as idealism are not technically anti-material, because idealism easily encompasses materialist science, just as quantum mechanics

There is a reality even prior to heaven and Earth, indeed, it has no form, much less a name.
Dai O Kokushi

We are in the midst of a revolution; physics are yielding slowly to metaphysics; mortal mind rebels at its own boundaries; weary of matter, it would catch the meaning of Spirit.
Mary Baker Eddy

encompasses classical Newtonian physics. I should also note that non-material views also have plenty of room for OBEs, NDEs, psi, ghosts, and paranormality in general.

We have already discussed materialism, dualism, and idealism from the point of view of God and religion in Chapters 13 and 14; however, we must reexamine these concepts in more detail in light of what we have since we learned about physics and the problem of brain, mind, and consciousness.

Materialism

What is a man in the infinite?
Blaise Pascal

As we have already discussed, materialists take mind to be a product of brain and dismiss any ideas of a soul, spirit, afterlife, or purpose. The universe is a ball of mass and energy created by natural forces, with no deity or other "supernatural" entity present. This bleak definition is partly a result of the Old Testament God and a reading of Genesis 1 that refutes modern scientific views. To a materialist, the interpretation of Genesis 1 presented in Chapter 4 does not hold water. Another huge factor has been the spectacular success of science that had painted a fairly complete picture of the material universe by the beginning of the 20th century. This progress owes much to Newton, whose idea of inert matter helped drive the Industrial Revolution, which was at least partially focused on owning and controlling matter. The universe seemed like a clockwork deterministic place. This view has persisted to modern scientists, who think that life is utterly devoid of any higher purpose.

The ancient covenant is in pieces: man at last knows that he is alone in the unfeeling emptiness of the universe, out of which he has emerged only by chance. Neither his destiny nor his duty have been written down.
Jacques Monod

We saw how materialists argue against so-called intelligent design in Chapter 20. Let us now take a high-level look from a physics standpoint. For example, Stenger built a cottage industry writing books such as *God: The Failed Hypothesis* and *Quantum Gods* that seek to disprove the JCI God and the idea that quantum mechanics has revealed anything extraordinary about the universe. (We have seen some of his work in Chapters 13, 23, and elsewhere.) Stenger's arguments sound airtight, until we come across statements such as, "Photons are particles, period" and the assertion that Paul Davies is a materialist despite Davies himself saying that "materialism is dead." Stenger also claims that his work requires no peer review.

I mention Stenger, because he represents a perfect example of what I call fundamentalist materialism. It is instructive to compare the arguments of fundamentalists like Stenger and Dawkins with those of religious fundamentalists. It is very true that both parties are arguing for completely different—and seemingly incompatible—models of reality; however, the logic and types of arguments used are a little more alike than either side might like to admit. Both are willing to misquote sources and exclude evidence that does not fit their respective paradigms. I say this not to excoriate Stenger or anyone else, merely to point out that our attempt to answer the central question of this book (what happens when we die) cannot possibly be complete without doing our utmost to examine the available evidence from all possible perspectives to try to determine how to create the best fit.

Denying the Anthropic Principle

We discussed the anthropic principle in Chapter 11, and must ask ourselves whether or not it is a valid concept. It is entirely possible that what we call "fine tuning" comes from misunderstanding just how arbitrary our own measurements are. All numbers involving dimensions use arbitrary units based on such things as the distance of our planet from the Sun, or the Earth's rotational and orbital period. These units may have significant meaning for us, but would be utterly meaningless to anyone living outside our solar system. Planck's constant, the speed of light, and Newton's gravitational concepts are examples of dimensional numbers. One could say that dimensionless numbers are the only meaningful numbers. We can even imagine different laws of nature based on different constants that could also support life in this universe or elsewhere, which seems to torpedo the anthropic principle beyond repair.

If this is correct, then concepts such as "precision" are meaningless, because one's choice of units must be taken into account. For example, if the mass of neutrinos was 5×10^{-34}kg instead of 5×10^{-35}kg, then the universe would probably have collapsed on itself—a fact that sounds like fine tuning to one part in 10^{35}, but that is nothing of the sort. Moving on, the elements needed for life just happen to be the most common elements in the universe, so it's no surprise that life consists of those elements. Carbon forms 0.0007% of the mass of the

---

I think we follow the basic law of nature, which is that we're a bunch of chemical reactions running around in a bag.
Dean Hamer

It is really quite amazing that an unguided, purely material process can produce the fantastic complexity of living organisms. Yet it did.
Victor Stenger

I have come to believe that the whole world is an enigma, a harmless enigma that is made terrible by our own mad attempt to interpret it as though it had an underlying truth.
Umberto Eco

universe, and 95% of the matter in the universe has nothing to do with life. Our Sun is a colossal waste of resources, since only about 2 photons per billion heat the Earth. Improbable events, such as the spontaneous evolution of life, are to be expected in such a vast universe. Complex behavior need not require complex laws, because complexity can come from simplicity, as we saw in the fractal example in Chapter 18. Those who argue for fine tuning have the burden of proving that no other types of life are available.

The anthropic principle can be seen as arguing against the JCI God. If this God created the universe with humans in mind, then it should be more hospitable to humans, in that it should be easier for us to roam across space. Furthermore, the universe is an awfully big place, where the furthest visible galaxies are about 13.2 billion light years away. We can see about 13.7 billion light years, because light from further away requires longer than the current age of the universe to reach us. For example, light left the Abell galaxy 13.2 billion years ago, meaning that the galaxy is about 40 billion light years from Earth. Assuming that inflation theory is correct (see Chapter 26), then the size of the visible universe is only about 10^{61} larger than the Planck distance. A God who created the universe for us thus wasted a lot of space. This God also wasted a lot of time, because almost 9 billion years elapsed before the Earth formed, and another 4.5 billion years passed before humans evolved. Humans have been on Earth for less than 1/100th of 1% of the Earth's history. Besides, why would an OOO god need six days and not one instant?

It is therefore difficult to believe that the universe exists for our benefit, or that of any other civilizations that are separated by plenty of wasted space. This is especially true, because the universe has 0 total energy. We could expect a miracle if this total energy was anything other than 0, but that is not the case. In the end, the laws of physics come not from structure, but from a decided lack of structure.

Against Quantum Spirituality

Materialism of one sort or another is now a received opinion, approaching unanimity.
Daniel C. Dennett

The wave function of the universe sets no boundary and entails no need for a God. As for the laws of physics, they serve to restrain physicists, and not matter itself. Quantum mechanics can only be used to support mystical claims by those who do not understand it. It is true that some experiments seem to show future events affecting past events, but this does not provide any evidence of backward causality on a practical level, since these effects only occur at quantum scales. It does not matter whether subatomic particles are "real" or not, because the outcome is the same either way. Far from proving Newton wrong, classical physics remains valid today, with many practical applications, and is in fact contained within quantum mechanics. Mass is a form of energy and the two can be interchanged, thanks to $e=mc^2$.

Some chaotic systems can self-organize to form complexity from simplicity without outside assistance. This does not mean that these systems lack determinism, merely that we may not be able to predict them, because we cannot measure with great enough precision. We do not have the mathematics to get us from the initial state of a chaotic system to the final state, which may make it seem indeterminate; however, bodies on a macro scale obey deterministic Newtonian physics. If we cannot determine the end state, it is only because we do not know the starting state.

The universe may contain next to no information. Our limited, subjective perspective makes it seem like the universe contains a huge amount of information, but that does not make it so. Quantum mechanics implies random superposition of all realities. Spiritualists use this phenomenon to claim that we are all part of some inseparable whole, and that we can create whatever reality we choose; however, the people making those claims are aging just as rapidly as everyone else. A sufficiently large, random universe can be expected to develop some pockets of complexity. This is what we see, because we happen to be in the middle of such a pocket. Nothing "supernatural" is needed to account for this.

No God Need Apply

Philosophically speaking, materialism denies that the spirit is independent of the body, which is not the same as saying it does not exist at all; however, a spirit that is dependent on the body is just as mortal as the body itself. Both theory and observation prove that there is no spirit world, and dualism fails as a theory.

The wave function collapse is not an actual physical event, but represents the change that occurs in our knowledge when we become aware of the result of a measurement.
Nick Herbert

His situation, insofar as he was a machine, was complex, tragic, and laughable. But the sacred part of him, his awareness, remained an unwavering band of light. And this book is being written by a meat machine in cooperation with a machine made of metal and plastic... And at the core of the writing meat machine is an unwavering band of light. At the core of each person who reads this book is an unwavering band of light.
Kurt Vonnegut

Hawking provided a theory in which our universe could come into existence by tunneling through a barrier from another universe, which is creation *ex nihilo* by purely natural means. Hawking's zero bound scenario proposes that the odds of having something rather than nothing are 60/40, which means that the natural state of things is to have something over nothing. God would thus have to maintain nothingness versus somethingness. The very fact that we have more than nothing is precisely what we should expect if there is no God.

Matter seems to be capable of handling everything without the need for spirits or God. Even theologians are being forced to admit that the concept of a separate soul is not viable given modern science, but they must still try to find room for God in order to remain employed. Sadly, their attempts at nonreductive theories do little more than rehash existing natural laws that already allow for complexity to emerge from simplicity, as the fractal example in Chapter 18 demonstrates.

The "proofs" offered by Aquinas became church dogma for centuries, but Galileo showed us that perhaps no Prime Mover is necessary to explain the universe. Matter either predates the universe or was created with the universe. Either way, the idea of the Logos is disproved.

- The cosmological "proof" is disproved, because the conservation of mass and energy is self-sufficient without recourse to any outside agency.

- The teleological proof is disproved, because evolution can create complex forms from simple ones.

Dawkins challenges us to explain just what we might be, if not complex biological robots. Edward Wilson seems to agree by suggesting that sociobiology could eventually replace the "soft" sciences, such as psychology, sociology, or anthropology. Some try to say that life is irreducible because it cannot be explained by mere physical facts, but modern science has demonstrated how physical processes can in fact account for life. Materialism cannot completely refute the existence of a soul, but it does not look promising.

At its most fundamental level, life itself is an ongoing massacre, which totally contradicts the JCI picture of a loving God. The Buddhist teaching that all life is suffering is much closer to the truth. If God indeed created humans in His own image,

---

All the labors of the ages, all the inspiration, all the noonday brightness of human genius, are destined to extinction in the vast death of the solar system; and the whole temple of man's achievement must inevitably be buried beneath the debris of a universe in ruins.
Bertrand Russell

Praying to an otherworldly god is like kissing through glass.
Brian Patrick

then God is flawed to the point of hilarity or despair, given the sorry state of His works. The God of the JCI religions is too good to be true—by the religions' own admissions, even though none of them will ever admit to it—and the religions surrounding this concept of God are too reassuring to have any credibility whatsoever. The promises of religion are all deliverable after death, meaning that one can go one's whole life with nothing more than a promise whose veracity cannot be tested objectively. The delivery of this promise is open to all who believe and obey, no matter what deeds they may perform—just so long as they do not deny the Holy Spirit, that is.

God (or at least the JCI god) matches our desires too perfectly to be real. Mistaking faith for knowledge and attempting to impose that faith by force is fanaticism, a desperate attempt to shield ourselves from our own ignorance. We cannot know the absolute reality, and defining that reality as God is the pinnacle of hubris in the absence of any corroborating evidence. God could settle the question any time simply by appearing. The religious assertion that He hides to preserve our freedom is ridiculous; but what better way to end all faith than by proving it? Ignorance may seem to be freedom, since we can believe anything we want, but knowledge is far more liberating. Any human father who hid from his children could be labeled insane, abusive, or both. Why does God get a free pass?

Believing in God means adding an extra layer of explanation over and above physical facts, which means we are forced to explain something we cannot possibly understand—a distinctly unsatisfying intellectual exercise. Any thought of God as a supreme being smacks of anthropomorphism. If we can say nothing more substantive about God, then how can we really say He exists? This takes care of the JCI God; however, we can extend the same logic to cover other types/concepts of what role a deity might play. It is impossible to disprove all possible types of god; however, there is no reason to assume that they exist, because materialism does a robust job of explaining nature without the need to appeal to any god.

Not the Final Word

These arguments may seem convincing at face value, but they are anything but definitive.

I really like the idea of an evolutionary revival. Revivals are American. Revivals evolved to meet a spiritual need. Why let the anti-revolutionists have all the fun?
Russ Genet

I have loved my fellow men, and lived to learn that they are neither fellow nor men but machine robots.
D. H. Laurence

> *Mankind is nothing but a bundle or collection of perceptions, which succeed each other with an inconceivable rapidity and are in a perpetual flux and movement.*
> David Hume

It is true that many of our mathematical and other units are arbitrary and based on values that are only relevant to our own experience. We should expect that life on other planets would have very different units of space, time, mass, etc. What matters is not the units, but the proportions. If you have 5 containers of equal size where 3 are filled with red paint and 2 are filled with blue paint, then 60% of your paint is red, no matter what units you choose to apply. Changing our units of mass, time, and space may make the relationships and amount of tuning seem more or less extraordinary, but nothing we can do can change the relationships among the values being measured. Gerrymandering one unit of measurement to eliminate the evidence of tuning would only alter the relationship to other units; it's a zero-sum game that preserves the anthropic principle any way one looks at it.

Classical physics is based on objects moving in three-dimensional space. Relativity adds a fourth dimension, while string and brane theories push the total number of dimensions to 10 or 11. The most recent theories involve pure geometry that inextricably links space, time, matter, and energy. It is therefore impossible to avoid dimensional numbers, and absurd to say that only dimensionless numbers can be meaningful.

It is one thing to say that the laws of nature favor having something instead of nothing, and that we would need God to maintain nothing far more than we would need God to maintain something. It is something else entirely to say that the universe has zero total energy, which is what we should expect if there is no God. It is contradictory to present both arguments in defense of the same theory, especially in light of $e=mc^2$, which equates energy with matter. If e is zero then m must be zero as well. So which is it? Something is more probable than nothing, or nothing is what we should expect if God does not exist? Materialism cannot have it both ways.

> *We may begin to see reality differently simply because the computer provides a different angle on reality.*
> Heinz Pagels

Saying that the universe contains a lot of wasted space and energy sounds a little arrogant coming from someone who cannot create a single particle *ex nihilo*, let alone create a universe that operates according to natural laws. Could God have set up the cosmic equivalent of a terrarium, in which the universe is just big enough for a planet full of humans with perhaps an exercise wheel or three and a Sun whose light shines directly and exclusively on Earth? Sure. Could this universe exist as a closed system unto itself? Possibly. But anyone knowledgeable about aquariums, terrariums, and other animal enclosures can tell you that those that mimic nature most closely—especially those that are at least somewhat self-sustaining—are among the most highly prized by true connoisseurs of the subject. Just because a child may appreciate a bowl with a goldfish and a plastic plant does not mean God must be so limited. On the contrary, a god wanting to create a self-sustaining system might well have to create a universe as big as ours to allow life to emerge from the natural processes s/he set in motion.

It is a mistake to compare the laws of physics to laws such as traffic laws. Traffic laws were created by people and enjoin—but do not constrain—our behavior. The mere fact that the speed limit on a certain street is 25 miles per hour does not in and of itself prevent someone from driving faster than that. The mere fact that a red light is illuminated does not prevent someone from driving right past. The mere presence of painted crosswalk lines does not ensure that cars will yield to pedestrians. In short, traffic laws only constrain those who choose to obey them.

The laws of physics do not work like this at all, because they were discovered, not invented. It would be one thing if some physicist decided that nothing could travel faster than light, because then s/he would be prevented from generating higher velocities in her or his experiments. It is quite another to have discovered that nothing travels faster than light and to proceed to discern why this is the case. Thus, natural laws do not constrain physicists at all—on the contrary, they are expressions of constraints that existed long before the physicists who discovered them had begun to evolve from the proverbial primordial soup.

One similarly cannot say that current and/or future events affect past events only on a quantum scale, any more than one can say that something happening to the foundation of a building does not affect the entire building. Quantum mechanics is the foundation on which matter is built. Just because macro-level events may or may not be discernible does not mean they don't occur. For example, a building may have rollers, dampers, or other mechanisms to shield itself against the foundation's moving during an earthquake. This

The World cannot be a giant machine, ruled by any pre-established continuum of physical law.
John Archibald Wheeler

But there is that which does not belong to materialism and which is not reached by the knowledge of the philosophers who cling to false discriminations and erroneous reasoning because they fail to see that, fundamentally, there is no reality in external object.
Buddha

may make it seem that only the foundation is being affected, but no one would seriously claim that the earthquake is not affecting the entire building.

Materialism is monistic, meaning that it rests on one leg and can be destroyed by knocking that leg out from under it. This is not to say that materialism is evil; the materialist worldview has done an incredible amount of good and advanced human knowledge by leaps and bounds; however, it has limits that are similar to the limits of quantum mechanics and relativity. Subsequent theories (such as superstring, brane, and LQC) incorporate all previous theories. Any non-material model must incorporate materialism within itself, while also going beyond materialism. That is a tall order to accomplish. Nevertheless, we have explored reasons to doubt materialism throughout this book, and will do so again later in this chapter. The universe is not deterministic, and life is a mystery.

Materialism is a metaphysical philosophy. It is impossible to prove that our brain generates mind and consciousness. In fact, we have reason to believe that the opposite may be true. For example, spiritual sages all around the world have consistently told of reaching a place where they can see that all is grounded in limitless, united consciousness. Their reports have been corroborated by countless other similar reports and by experiments, such as those discussed in Chapters 15 to 17 and 23, that—at minimum— cast serious doubt on materialist philosophies.

Why do people persist in materialism despite the mountains of evidence to the contrary? For the same reason that people cling to religious beliefs that have been utterly debunked, and why all of us cling to self-destructive beliefs despite whatever desire we have to change. (See *The Enlightened Savage* and *The Natural Savage*.) People catch monkeys by putting chickpeas in a jar. The monkey grabs the peas in its fist, which then becomes stuck in the bottle. The monkey could free itself instantly just by letting go, but it refuses to let go of the food. Materialism could be the philosophical equivalent of chickpeas, and releasing it as the dominant scientific paradigm could yield tremendous benefits to all humankind.

Maslow's hierarchy of needs lists self-actualization (which includes spirituality) as the highest need. Materialism effec-

---

Why, for example, should a group of simple, stable compounds of carbon, hydrogen, oxygen, and nitrogen struggle for billions of years to organize themselves into a professor of chemistry? What's the motive?
Robert M. Pirsig

The ultimate aim of the modern movement in biology is in fact to explain all biology in terms of physics and chemistry.
Francis Crick

tively blocks people from reaching those highest levels despite all claims to the contrary. Materialist dogma is reflected in our materialist, consumption-based society that keeps buying the latest and greatest gadgets in search of a happiness that never comes. I remember the day I bought my dream motorcycle, a 1993 Yamaha FJ1200, from a location about 150 miles from home. I rode the bus to go get it, paid the seller, and was soon headed home on the freeway. I loved the motorcycle—still do—but realized as I drove home that I was still me, with all of my strengths, weaknesses, opportunities, risks, successes, and problems. Nothing had really changed.

You can only retain your faith in materialism by assuming—on faith—that any contrary evidence you read about must be wrong.
Mario Beauregard

Materialism persists despite mounting evidence to the contrary. Skinner's behavioral and conditioning research sought to remove the problem of mind altogether, presumably to avoid having to face it head-on. Computer analogies of mind and brain are valid for practical purposes as long as we don't take them too far, as I demonstrate in *The Enlightened Savage*; however, when it comes time to move beyond the useful analogy and explain what is really going on, materialists are utterly unable to explain consciousness. Even the books I recently picked up—whose covers include such testimonials as, "Ground breaking!" or "A completely new theory!"—simply rehash the same old ideas, albeit slightly recombined. Nevertheless, materialists keep telling us not to worry, that consciousness will eventually be explained in purely material terms, a promise that has become known as "promissory materialism" that demands faith to accept—the same type of faith required to sustain the religions that the materialists are so fond of decrying.

Dualism

One of the central ideas of dualism is that a so-called zombie world may not be identical to ours. Dualists thus believe just what John 3:6 says, that "Flesh is born from flesh, and spirit is born from spirit." Matter influences mind, and mind influences matter. This of course leads to the mind-body problem: Given that mind and matter are separate and distinct, how does mind receive data (such as sensory input) from the body and vice versa? Descartes (philosophy) and Eccles (neuroscience) are two examples of people who believed in some form

Materialism is dead.
Paul Davies

of dualistic interaction between the physical world and the "supernatural" soul.

Some see dualism as conceding to religious faith. This viewpoint can lead to materialism and the perceived erasure of any perceived or actual differences between brain and mind. As we just saw, materialism entails just as much a leap of faith as any religion despite all claims to the contrary.

The Attraction of Dualism

The materialist must decide for himself whether the ghostly remains of matter should be labeled as matter or something else; it is mainly a question of terminology. What remains is in any case very different from the full-blooded matter and forbidding materialism of the Victorian scientist..
James Jeans

It is possible that mind evolved from matter in an emergent fashion where matter eventually becomes mind. This is a form of "naturalistic dualism" that may or may not imply the existence of a soul depending on how one looks at it, because it need not involve anything transcendental—a sort of expanded version of materialism, where mind is still dependent on brain. On the other hand, if one cannot be reduced to the other, then we are forced to accept the validity of dualism as proposed by the major religions. According to Eccles, consciousness cannot be ascribed to any natural laws, physical or otherwise; humans consist of both a spiritual and a material side, and both must be nurtured. We can think of mind and brain like a television commercial and the television circuitry, respectively; the former cannot replay itself simply by tinkering with the latter. This is a perfect example of dualism.

We can leave aside the search for energy transfer or other mechanisms of dualistic interaction for now and simply assume a practical dualism without attempting to go beyond that, to say the following:

- Consciousness is real, nonphysical, and irreducible to monads or anything else.

Here was an educated, intelligent man telling me that he will not give up materialism, no matter what.
Neal Grossman

- The physical world connects to consciousness via a single fundamental quality.

These two assumptions can help us get a grip on the mind-body problem and begin searching for methods to resolve it. Assuming dualist properties is the only viable option by which we can see consciousness in a coherent and naturalistic way with all of the mystery stripped out.

Problems with Dualism

Even if we assume that no energy transfer takes place between mind and body, we are still left with the problem of determining the mechanism by which the two worlds interact, and how they manage to do so coherently. We are also left to speculate about just what makes a brain—particularly a human brain—so special, compared to plants or inanimate objects.

Thought seems to take some material form, or to at least be connected to material, given the experiments we discussed in Chapter 30, where stimulating parts of the brain with a probe triggered powerful recalls of past events. It also seems to be affected when the brain is injured, such as during a stroke or trauma. If mind and brain are separate, then the question remains: How does one affect the other? We must either ignore the problem, or say that mind and spirit somehow control each other. Neither path leads to the dualism envisioned by Descartes, Eccles, and others.

The best mainstream evidence we have indicates that the physical world is a closed system with no need—or room—for a ghost in the machine. Quantum indeterminacy is not nearly a big enough loophole to allow the full and coherent functioning of a nonmaterial mind. Thus, the dualism we are talking about seems to be limited to property dualism, where consciousness involves nonphysical traits that are not entailed by physical traits, but which may depend on these physical traits. This would place consciousness above—but not apart from—the physical brain. Again, postulating some sort of physical/spiritual interaction would require gaps and/or loopholes in physics that just don't seem to exist.

Dualism that postulates separate realms fails for lack of a viable mechanism of operation. Emergent dualism that has mind emerging from—and dependent on—brain fails, because it is just another form of materialism. Materialism also seems to have failed. This leaves us with only one other option...

Idealism

There is a proverb of two monks arguing about a flag flapping in the wind. The first monk insists that the flag is moving. The

---

This is a material world, and I am a material girl.
Madonna

Madonna may be a material girl, but this is not a material world!
Evan Walker Harris

second insists that the wind is moving. A third monk walks by and says, "Your minds are moving!" This is the essence of idealism. Mainstream science is dominated by Western thought and therefore tends to sneer at the possibility that consciousness creates reality, because it smacks of Eastern mysticism. In general, it is amazing the lengths to which science will go in order to avoid anything perceived as religious or spiritual, especially when doing so entails a form of religious thought. Nevertheless, there do seem to be some signs of a subtle paradigm shift that is beginning to crack the foundations of monism in favor of a more holistic worldview.

Can you be "explained" in terms of mechanics, electromagnetism, physics, or chemistry?
Judith Hooper and Dick Teresi

We already know that idealist thought holds that the physical world is not the real world, and that our minds are capable of understanding the real world. To an idealist, consciousness is central in shaping the world. Reality (the Ground, Godhead, Source, Ein Sof, Tao, Brahman, or whatver you want to call it) is the unitive, transcendent foundation of the entire universe. Reality takes on the illusion of immanence and separateness in each one of us. We are all formed of consciousness, and the entire universe—including matter—is simply a manifestation of the transcendent forms of consciousness. With idealism, Descartes' *cogito ergo sum* becomes *eligo ergo sum*.

Wilson once pointed out that the human brain has been probed to the point where no place remains to host a nonmaterial mind—an argument against dualism but, ironically, in favor of idealism, because consciousness is hosting the brain, not the other way around. A brain that does not seem to have any seat of consciousness and no means of interacting with a nonmaterial spirit is just what we would expect to see in an idealist universe, where consciousness is not a phenomenon, but where everything else are phenomena within consciousness. If we look at reality in this manner, we begin to understand that we are not anywhere near the geographic center of the universe (a meaningless concept), but that we are central to the universe, because we are that which gives it meaning.

One benefit of switching humanity to a correct perception of the world is the resulting joy of discovering the mental nature of the universe. We have no idea what this mental nature implies, but-the great thing is—it is true.
Richard Conn Henry

It is funny that biologists are moving toward the 19th-century brand of hard-core materialism, while physics is starting to show some signs of moving away from it. Biologists would be well advised to heed developments in physics, because biology depends on physics, and not the other way around. One can only imagine how neuroscience might progress but for the *a priori* expectation of materialism. We can trace Western idealism to Plato, who spoke of reality as people in a cave facing the walls with sunshine casting shadows that the people think are real. Consciousness is the light in this example. Most physicists don't delve into the philosophical implications of their mathematics. Most are content with the knowledge that the equations work and allow them to continue their studies. Something may be going on that we don't—and possibly can't—understand (hidden variables), or nothing may be happening. The math works either way.

Idealism gives us a world in which perceptions are all that exist. The cat is both alive and dead until the box is opened. Independent reality does not exist, and most of us cannot see or understand that. Those few of us who can are in both heaven and hell at once.

> *Matter is derived from mind, and not mind from matter.*
> The Tibetan Book of the Great Liberation

We Are All One

We have touched on the possibly illusory nature of self before, but this important topic deserves another look. Abolishing the self is not the same as ending consciousness; it is simply the erasure of the hard lines that make us separate, isolated, and finite. Tagore called Nirvana not blowing out a candle, but extinguishing the flame because day has come. Don Juan Matus (a possibly fictional character in books by author Carlos Castaneda, 1925-1998) is quoted as saying that we are perceivers and awareness. We are not objects, we are not solid, we have no boundaries, and the world of physical objects is just a way to help make our lives on Earth convenient—an illusion designed to aid us. The trouble is that we mistake this illusion as being real and get ourselves into a vicious rut that can often last an entire lifetime.

Our own perceptions, thoughts, and emotions are all each of us can ever know about the world. It is impossible to ever prove anything real beyond one's own self. Taking this fact to its logical extreme can lead one to solipsism. Idealism does not support this brand of solipsism, because it is based on the premise that we do have individual thoughts and feelings, but that we don't "have" consciousness, nor are we ultimately composed of separate, conscious entities. Instead, idealism takes the stance that we are consciousness. This resolves the Copenhagen interpretation of quantum mechanics, because

> *Carried to its logical extreme, the bootstrap conjecture implies that the existence of all consciousness, along with all other aspects of nature, is necessary for self-consistency of the whole.*
> Geoffrey Chow

everything in the universe is the product of a single consciousness. (See Chapter 29.) This pure consciousness is both immanent in every sentient being and transcendent in its unity, which blurs—if not outright erases—the definition of "I" and the distinction between subject and object. As Meister Eckhart (1260-1327) said, "The eye with which I see God is the same eye with which God sees me."

All of this seems very hard to fathom for most people, because we experience ourselves as separate and distinct entities interacting with other separate and distinct entities and objects. We do this in several ways, such as:

- Being aware of sensory data that seems to come from beyond ourselves.

- Building patterns of sensory data and memories that reinforce self-identification.

- Confronting entities and objects that seem to be outside of ourselves, which reinforce the idea that we are separate from our environment.

- Considering ourselves to consist of our bodies by associating mind with body because of the apparent interactions between the two.

- Relating our personal interpretations of our experiences to reality.

All of the above and more requires an immanent world, where things and entities can be made manifest in order to create a framework in which souls can exist as selves, but where each self think it is separate. Constructing states that see and states that are seen is the world's way of seeing itself. The subject/object split occurs in our minds where self-reference happens.

Initial stimuli create responses that affect our minds and that predispose us to respond to future stimuli. Each such learned experience increases the likelihood of similar responses in the future, following the ETEAR model described in Chapter 9 and *The Enlightened Savage*. This has the effect of making quantum measurements seem almost classical in many cases, because the older someone gets, the more predictable they tend to become. This also reinforces the illusion of separation and individual self or *ego*, which my dictionary defines as, "The 'I' or self of any person; a person as thinking, feeling, and will-

---

The recognition that physical objects and spiritual laws have a very similar kind of reality has contributed in some measure to my mental peace... the principle argument against materialism is that thought processes and consciousness are the primary concepts, that out knowledge of the external world is the content of our consciousness, therefore, cannot be denied.
Eugene P. Wigner

Because of the infinite regression of cause and effect, the whole universe may owe its 'real' existence to the fact that it is observed by intelligent beings.
John Gribbin

ing, and distinguishing itself from the selves of others and from objects of its thought." The ego and associated feelings of identity permeate us until we forget all about the nonlocality of our consciousness. The ego serves a valuable purpose as the instrument for acting within the material universe, but nonlocality and unity never completely vanish. Free will exists but becomes increasingly biased toward conditioned responses—a valuable survival tool that can also go awry in many situations.

Imagine a situation where a prisoner is confined inside an arrow-shaped room. A friend arrives to help the inmate tunnel out. Assuming neither party can communicate with the other, how can 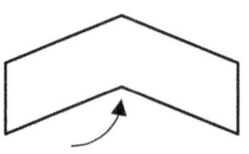 they both know where to dig? Chances are that they will start digging at the corner indicated by the arrow, because it is the only concave corner when viewed from the outside and the only convex corner when viewed from the inside. This is an example of a waveform with 36 possibilities, because each person has 6 corners to choose from and 6x6=36; however, the waveform will be much higher than 1:36 for the special corner, and much lower than 1:36 for any other set of possibilities. We can essentially treat this experiment as two loaded dice. It is also an example of a tangled hierarchy.

Where does the illusion of "I" come from? Each of us makes conscious decisions all the time, while remaining blissfully ignorant of the underlying unconscious processes driving those waking thoughts, feelings, and actions. This helps explain how we begin to perceive of ourselves as separate "I" beings, instead of part of a "we" collective consciousness. This illusion of separation also involves a tangled hierarchy; however, the reality is that the consciousness of subject and object are one and the same. Goswami explains it thusly in *The Self Aware Universe*: There is only one consciousness (the atman, holy spirit, inner light, etc.), and we are each just that. The Buddhist reference to the consciousness beyond this level as no-self causes a lot of confusion, because people think of it as nothingness, and thus as annihilation; however, what this ultimate level really contains is the unborn, unoriginated, uncreated, and unformed Ground/Godhead. Without this, the born/originated/created/formed would not be possible.

Pure in its own nature and free from the category of finite and infinite, universal mind is the undefiled Buddha-womb, which is wrongly apprehended by sentient beings.
Lankavatura Sutra

It is difficult for the matter of fact physicist to accept that the substratum of everything is of mental character. But no one can deny that mind is the first and most direct thing in our experience, and all else is remote inference.
Arthur Eddington

> *In fact, what is called the world is only a thought.*
> Sri Ramana Maharshi

"I" is thus not a thing, but a relationship between conscious experience and the immediate environment in a world that has been divided into subject and object and combined with memory to create ego. Breaking the unified consciousness spreads pieces of the one single intelligence among each of the parts. Each part can fulfill its purpose by becoming part of a greater whole beyond what it was designed for, like pieces of a puzzle that come back together to form a whole that both contains all of the constituent pieces and is more than the mere sum of those pieces. Each part contains part of the entire picture. The cutting implies both design and place in the overall scheme. A puzzle piece is designed to come back together and to contribute to the whole; the same is true of every individual piece of consciousness in the universe.

As Bohm put it, "Information about the whole system can be folded into a part that reflects the whole. Each human is not an independent actuality that interacts with other humans and nature. All are projections of single totality. In the implicate order, mind enfolds matter and body, body enfolds mind and also universe."

Surfing the Cosmic Quantum Waves

> *Although quantum theory is an abstruse and formidable field, its philosophical and theological implications reduce to one shattering effect: the overthrow of matter.*
> George Gilder

We cannot expect an entangled particle to know what to do to achieve the observed correlation on its own; however, observer and observed are part of the same system, which allows us to implicate consciousness in the process. This is what Wheeler meant when he spoke of a participatory universe, where experiments such as delayed choice and dual slit depend on the question we ask and how we choose to ask that question. What we call reality is a product of our own making, whose assembly occurs in each thought, emotion, action, and belief we have.

I must digress to explain that a conscious universe does not imply anthropomorphism, nor does it necessarily imply or refute religion, God, or design. It also does not imply any form of survival after death. With that said, the ancient and medieval scientists made one crucial error: The Earth is not the center of the universe; each of us is. Only consciousness and possibility exist before a measurement, the act of which creates and separates life and environment. Sight and choice collapse the transcendent quantum possibilities embedded in the waveform into an immanent actuality; consciousness manifests a result from among all possible results, according to the laws of physics. As poet Robert Pack said, "If I had been, in the beginning, God brooding upon the womb of absence, I might well have pondered: matter plucked from emptiness."

Again, this is a far cry from "manifesting any reality we want," the "Law of Attraction," or any of the quasi-scientific buzzwords used by the self-help industry. What we are talking about here is at once far more subtle, limited, and yet profound than that. Consciousness obeys the laws of quantum mechanics. The choice of outcome is constrained by the wave function, which is why the "Law of Attraction" is useless for all practical purposes. (See Appendix B.)

Materialist scientists—and even some dualists—reject the idea that consciousness collapses the quantum waveform, because they see no mechanism to mediate how consciousness interacts with the environment within the laws of physics. Even machines have such mechanisms, such as the Hardware Abstraction Layer (HAL) in computers that connects software to hardware. Quantum mechanics is a form of calculus that allows one to calculate the probabilities of all possibilities within a wave function. We saw in Chapter 23 that this calculus upends the traditional materialist view of upward causation from particle to atom to molecule, cell, neuron, brain, and consciousness, and replaces it with downward causation that begins with consciousness. The latter view removes the apparent paradoxes from quantum mechanics.

Subatomic particles can be transcendentally explained by the nuclear structure, which can be transcendentally explained by the atomic structure. The same applies to molecules, macromolecules, cell structure, tissue structure, organ structure, physical activity of life form, awareness of life-form, and finally consciousness. In this model, consciousness itself is self-explanatory and explains all else.

We also saw that Bell's Theorem forces us to reject the idea that actual physical states (location, charge, momentum, etc.) and corresponding values exist before a measurement. The subatomic behavior of all particles and objects is unavoidably connected to an observer, without whom they can only exist as undetermined superpositions described by probability

All phenomena in the world are nothing but the illusory manifestation of the mind and have no reality of their own.
Ashvaghosha

Nature tools along, not knowing that it's unified.
Allan Sandage

> *It is my contention... that ideas created by a spiritual consciousness are the cause and basis of the physical world.*
> Bernard Haisch

waves. This implies biocentrism, as we saw in Chapter 29. It also implies that time does not exist independent of perceiving changes in the universe, which is perfectly compatible with a holographic universe made of Nows, as we saw in Chapter 28. Nonlocal particles either violate relativity or are not nearly as separate as we might like to think. Everything in the universe was connected to everything else, and that entanglement persists today.

The Schrödinger equation works both "forward" and "backward" in time, which is perfectly compatible with the concept of Nows. Idealists see time as a transcendent two-way street that only seems one-way in the immanent world. Whatever arrow time has occurs at the moment of waveform collapse, which is irreversible. The string of irreversible collapses, and the constant progression from potential to actuality, is what gives time its apparent arrow and what makes that arrow seem one way; however, this hopping from Now to Now is not the classical flow of time from fixed past to open future that we think we experience; it is jumping from stepping stone to stepping stone, where the stones always exist, and where the sum total of all of the stones represents everything that is possible. There is nothing preventing us from returning to a previously visited Now, although we would do so by collapsing a string of Nows to circle back, and thus would not experience it as going back in time.

> *Hence this life of yours which you are now living is not merely a piece of the entire existence, but is, in a certain sense, the whole; only this whole is not so constituted that it can be surveyed in one single glance.*
> Erwin Shrödinger

Observations are discrete events, and it is impossible to fill the gaps in between them. Collapse involves recognition and choice by a conscious observer—and ultimately, there is only one observer. So what would happen if we placed a human inside Schrödinger's box? Would he not experience himself as being alive the whole time, thus removing the 50/50 chance of death within an hour? No. Consider that most of us are not aware of our own bodies most of the time. The person in the box collapses his wave function during his moments of awareness but returns to superposition at all other times, which makes him just as susceptible to the 50/50 odds as the cat.

This points to the need to distinguish between consciousness and awareness, because the former does not necessarily imply the latter. Saying that a transcendent consciousness causes collapse smacks of the JCI god, and the concept of an OOO god causing the collapse runs up against the problem of free will.

Measurement cannot occur without an immanent awareness, such as a human mind. The collapse of the transcendent waveform occurs when an immanent mind becomes aware of that waveform. Mere detection of a waveform by an unconscious detector does not cause collapse, because that detector obeys quantum laws and is itself part of the system.

That which we call our unconscious is consciousness that lacks awareness. We can perceive events that we are not aware of perceiving, and can think and feel without being aware of those thoughts. Phenomena such as *blindsight*, where people see things without being aware of seeing them, are proof positive that perception and awareness are not necessarily connected. These unconscious—or unaware—processes affect our awareness. Choice and collapse involve awareness.

Where materialism and dualism strip away meaning and freedom of will, idealism infuses life with both freedom and meaning. Consciousness must be nonlocal in order to collapse a nonlocal waveform. That has tremendous implications for paranormality, because nonlocality implies psi, and also opens the door to explaining NDEs, OBEs, past lives, and more. We can think of consciousness as having two aspects: the local material aspect we are familiar with, and the transcendent unity. The former is person, while the latter includes and contains all local personal experiences plus everything else.

The unity of nonlocal consciousness also allows us to assume that objective reality exists for practical purposes. Under normal circumstances, most people will agree about enough aspects of reality to allow that to be called objective. This may seem paradoxical; if multiple observers see the same waveform at the same time, who gets to collapse the wave and why? Here again, a single consciousness solves the problem, because there is only one thing causing the collapse. The variances in perception that exist among the conscious fragments (individuals) making the observation are part of the waveform. Paradox is thus avoided without resorting to solipsism.

This is of course an approximation, but so is Newtonian physics. It will also have to make do for the majority of us who are not mystics, because our own mind experiences are the only things we can know. Nonlocal consciousness collapses the waveform from beyond space-time and the subject/object

Millennia passed before humankind discovered that energy is the basis of matter. It may take a few more years before we prove that wisdom and knowledge are the basis of and can actually create-energy, which in turn creates matter.
Gerald L. Schroeder

Failing to recognize me, you objectify me as an external entity. But when you finally discover me, the one naked mind arises from within, absolute awareness permeates the universe.
Yeshe Tsogyel

> *The material universe cannot be the thing that always existed because matter had a beginning. This means that whatever existed is non-material. The only non-material reality seems to be the mind.*
> Robert M. Augros and George N. Stanciu

split. Combined, the self of self reference and universal consciousness equals self consciousness. Mind moves, and matter moves to follow in accordance with quantum laws. As I mentioned above, ascribing all of reality to consciousness solves the observer problem. Consciousness conceives of existence, and experience also flows from consciousness. Space itself is consciousness, which is absolute, changeless intelligence that only functions in an eternal present and provides the ultimate source of experience, which is an externalization and reductive function of consciousness. Matter is the stuff of collapsed waveforms, which means that the waveform contains all possible forms or designs that collapse into a single manifestation. We could therefore say that the waveform itself is the design.

Contradictory items often imply each other. For example, nothingness can imply somethingness, and vice-versa. This may seem counterintuitive, but in mathematical terms, a *set* is an empty construct that can accept contents in the form of individual members and/or subsets. All sets are empty unless and until filled, which means that the concept of emptiness is an inherent part of all sets. We can therefore define space as both nothingness and part of reality without any logical contradiction. We cannot have existence (individual fragments or pieces) without nonexistence (the whole). The form of any given unit is a spatial aspect of that unit's nothingness. Nothingness explains itself. It is self-explanatory consciousness.

The Quantum Godhead

> *A blanket statement that anything that conflicts with a materialist view of the universe must be false risks becoming considered one day as shortsighted as the past rejections of mainstream science of phenomena such as meteorites are now.*
> Jim B. Tucker

It follows from everything we have discussed so far that consciousness is the supreme intelligence that bears ultimate responsibility for the universe and all in it. This God is not OOO, because it is limited in how it creates experiences. This is no mere speculation: The proven existence of nonlocality across space and time that we explored in Chapter 23 implies the existence of an unbroken wholeness in the universe, and the evidence from countless quantum experiments as seen by the Copenhagen interpretation, implies—if not outright demands—that observation literally creates space and time. This causal thread is what connects each conscious individual with the universe, and with that which can only be called the Ground, or Godhead, or Source.

This all sounds very nice, but what does it mean? Ongoing experiments in laboratories and probes dispatched to the depths of the oceans and into space are working to discover just who we are, why we are here, and the purpose of life. Particle colliders and telescopes have looked back almost to the beginning of time, searching for the ultimate answers to the ultimate questions. Does God exist? If so, does this god care about anything or anyone? Is there any purpose or meaning to life that transcends the six core functions of life (predator avoidance, social status, food, shelter, reproduction, and death)? Does the soul exist? Do dead people go on living somehow somewhere? Is God a personal, caring God? How many of us fear we know the answer, and how many of us have succumbed to the skepticism we fear is true? Lucretius asked these words over 2,000 years ago. Those questions remain just as poignant today as ever, if not more so in light of the recent disasters in Haiti, Chile, and Japan.

Promissory materialism is simply a religious belief held by dogmatic materialists who often confuse their religion with their science.
John Eccles and
Daniel N. Robinson

Materialists paint a bleak picture of birth as the beginning and death as the end, in a cold heartless universe with no god and no miracles, where all that's left of deceased loved ones is memories that eventually wink out as descendants die in their own times. I have already noted that this reductive view freed us from the bonds and tyranny of religion and improved the quality of life for billions of people; however the cost of all these benefits was quite literally the human soul. Materialism starts to look more and more like the ultimate deal with the devil.

By contrast, idealism understands that the image of a big bag of space filled with clouds of particles is not all there is. Consciousness created quantum potential in accordance with the laws of physics that it also created. The universe is an ongoing act of pure will. Our individual consciousness, minds, and the will of God are one and the same. Each path, be it quantum mechanics, relativity, or something else, leads back to this supreme consciousness from which all springs and that both unifies and constrains us as seemingly separate beings. The Kingdom of God that mystics speak of—and that religion insists comes as a reward for a proper life—is right here, right now. We are fragments of this consciousness temporarily attached to the illusions of bodies, but this is not our real existence.

There is no god but reality. To seek him elsewhere is the action of the fall.
Unknown

Chapter 31
Models of Reality | 819

Logical and natural supervenience are different concepts. If B is logically supervenient on A, then B is a given once we have A. The Godhead/Ground created physical laws, and life evolved as a consequence of those laws. Life adheres to—and arises from—physical laws, but we need not extend those laws to achieve life; life entails no new laws of its own. If it did, then mere physical laws would not suffice to evolve life, and logical supervenience would not hold. Natural supervenience is different because that would allow A to exist without needing B. B would require more work to exist. Pain is an example of natural supervenience. We can create the sensation of pain simply by stimulating a nerve without causing any harm or "real" pain whatsoever. The interpretation of certain nerve impulses as pain—versus any other type of data—is not logically supervenient on a nervous system, because a nerve can serve for many purposes; it is naturally supervenient, because pain is one possible type of data that nerves can carry.

Reality is that which, when you stop believing in it, doesn't go away.
Philip K. Dick

Strip away the trappings and differences of religion to expose its mystical roots, and we find that people all over the world share an extremely similar view of God. For example:

- "My being is God." - Catherine Adorna
- "Our very self nature is Buddha and apart from this nature there is no other Buddha." - Hui Neng
- "Thou art neither ceasing to exist nor existing. Thou art He without one of those limitations." - Ibn al-Arabi
- "When God launched creation it was called He. When it unfolds in love bliss and being it is called you. But the supreme manifestation of God is called 'I.'"- Moses de Leon
- "I receive that God and I are one." - Meister Eckhart
- "I am the truth!" - Monsoor al-Halaj
- "I know that I am all." - Shankara.
- "My father and I are one." - Jesus Christ

Reality: what a concept!
Robin Williams

The act of creation that causes the Godhead to lose its wholeness and become immanent limits its ability to act in the universe, in a manner similar to passengers on a ship, who are constrained by the ship's limits of space and amenities. We could say that the act of creation destroys God. Furthermore, we know from Chapter 20 that the sum total of energy in the universe is 0, because the positives of matter and energy is perfectly balanced by the negative of gravity. We also know that everything exists in complementarity, that particles can be waves and vice-versa, and that particles have anti-particles. This stands to reason, because the only way to create something from nothing *ex nihilo* is to create opposites. God is also subject to this complementarity, which means that God is also the Devil or Antichrist. This explains suffering and evil; it also points the way to happiness and bliss, because all things balance out in the end.

Beyond Birth and Death

It is tempting to see what we have been talking about in this chapter as validating either Eastern beliefs over Western beliefs or one religion or set of religions over another, but this is not the case at all. We just saw that mystics from all around the world share common visions of the Godhead. Religion buries this universal insight under layers of dogma and hierarchy that expose and highlight differences. These differences inspire and fuel the worst atrocities the world has ever known, and continue to present the most exigent threat to the ongoing survival of our species. I will say it again: Religion is the ultimate insult to the God it claims to revere—especially because snippets of the original mystical teachings can still be found in religious texts, including the Bible. For example:

- Luke 3:6 says that everyone will see God's salvation, which means that nobody will not be saved—a direct contradiction of "fire and brimstone" preachers. This echoes Isaiah 40:5, which says that all will see the Lord.

- Luke 17:21 explains that people cannot see the kingdom of God, because it is in them—a rather Gnostic-sounding idea.

- Luke 17:33 says that those who try to save their lives will lose them, but that those who die will live again in another existence. Man cannot save his own life, because salvation requires death, which ushers in a new phase of existence.

- Galatians 3:3 says that we are all descended from the original spirit.

I believe the reason nobody can find the soul or the spirit inside the brain or the body is because it is not there. We are looking in the wrong place. We are not in the body, the body is in us; we are not in the mind, the mind is in us; we are not in the world, the world is in us.
Deepak Chopra

> *Physics, the most empirical science, seems ultimately based on consciousness.*
> ruce Rosenblum

- John 14:6 is the famous passage that says, "I am the way, the truth, and the light. No one comes to the Father except through me." Christians have taken this passage as commanding belief in Christ as the superhuman son of God who must be worshipped and obeyed; however, those who have been reading closely will see the mystical meaning embedded within these words and realize that abject obedience is not what this passage is calling for at all. Heaven is within each of our minds, and we can refer to the original mind as Adam, part of whom dwells within each of us.

- A student once read passages from the Christian Bible to a Zen master, who said that the author was enlightened and not far from Buddhahood.

It seems for all practical purposes that our bodies and brains retain their unique identities. If idealism is correct, then mind is immortal and survives the death of our bodies and brains. This survival can take the form of passing down from parents (a view compatible with materialism), or reincarnation; either scenario satisfies the "survival" definition, although the former is not clear about what happens to the mind of someone who, like me, does not have biological children. If the latter scenario is correct, then our minds may have experienced pre-existing identities and can take on a new identity without remembering details from the previous identities.

> *The external world is only a manifestation of the activities of the mind itself and the mind grasps it as an external world simply because of its habit of discrimination and false-reasoning. The disciple must get into the habit of looking at things truthfully.*
> Buddha

In an afterlife/reincarnation model, we can speculate that the body ceases to function at death, as we saw in Chapter 1. Our mind also ceases to function in a reductive/deductive manner, and is freed from the illusory confines of the physical universe. Mind as a fragment of the Source/Godhead is separate from the body that is a manifestation of mind. The extent to which mind believes that body is real is the extent to which mind is confined to the body's point of view and tendency toward reductive/deductive thinking. In this limited sense, we can see that dualism is a valid concept, so long as we do not confuse separation of degrees (mind and manifestations of mind from the Godhead) with kinds. We can speculate that freeing the mind from its association with the body frees it from the limitations of body and allows it to understand its true identity as a monad and part of the Godhead. This is the model I discussed briefly in Chapter 15.

If this model is true, then why do we have such a persistent belief in—and fear of—death? The short answers are that people tell us we will die, we see people seeming to die and decompose never to return, and we associate ourselves with our own bodies. Materialism offers no hope to those afraid to die, and religious dualism stokes those fears with threats of eternal torture that make the oblivion of nonexistence sound like paradise by comparison. The idealist dualism described here explains both our attachment to this life and the freedom that awaits. Again, it is also perfectly compatible with biocentrism (see Chapter 29) and with the science we examined in Chapters 21 through 28.

Mach once said that, "In wishing to preserve our personal memories beyond death, we are behaving like the astute Eskimo who refused immortality without his seals and walruses." He had a valid point. "I" is at once nothing and everything, zero and infinity, nothing because there is no personal pigeonhole to contain us, and everything because we are nothing less than the universe and Godhead seeing itself through a unique vantage point. Immortality is staring us right in the face, if we can only open our eyes to see it. This view seems to disagree with the Aboriginal belief that we are visitors to this lifetime who then return home; however, the quantum monad concept preserves this concept if we expand it slightly to see each lifetime as a play in which we take on different roles.

The fear of death is itself paradoxical. Most of us in today's world think death is bad and actively deny that it can or will happen to us. Primitive people, by contrast, are not afraid of dying, but do tend to have a healthy fear of the dead. Death—like anything else—can only affect us if we believe in it, which makes disbelief a potent weapon. Materialism may have removed the fear of eternal hellfire, but in so doing reinvigorated the fear of death. To our materialist eyes, the death of another person seems like the end, but from our own perspective, as reported by NDE survivors, it is liberating.

Consciousness identifies with the body on various levels, including cells and organs, such as the brain. Death removes those identifications, as consciousness stops collapsing waveforms associated with the physical body. This process begins in the brain and proceeds to our organs and cells, although death is virtually complete once consciousness has left the

Everything material is also mental and everything mental is also material.
David Bohm

The universe can be best pictured as consisting of pure thought.
James Jeans

brain and returned to the monad. If we adopt a Buddhist model, then memories of a lifetime and associating with a specific identity (such as Anthony Hernandez) may fade, but the long-term memory and changes wrought by each passage persist, and that is what forms karma.

Science based on irreducible consciousness helps us define life while resolving some of the problems with materialism. It also defines death as the moment when consciousness removes its transcendent, self-referencing supervention from matter—or the illusion of matter, to be more accurate. It is difficult to accept the idea that material life gives way to the immaterial after death, particularly because events like drugs and trauma can alter the brain and change what and how measurements occur. On the other hand, it may be difficult to dismiss this illusory world as superfluous or deserving of anything less than marvel and respect, because it is incredibly beautiful and an integral part of the wonder of creation. Disrespect for this world is also disrespect for that which created this world, which is no less marvelous, be it God or mindless natural forces. Heaven is right here right now, and we are all partaking.

Refuting Materialism

Materialist physicists and other scientists such as Dawkins, Stenger, and Weinberg claim that the universe and life are pointless; however, as I point out in *The Enlightened Savage*, meaning is up to each of us to decide for ourselves. Reading materialist literature demonstrates that it is chiefly concerned with protecting its own existence over examining the evidence for what it is. It impairs our ability to understand our minds and brains, and is out of step with many of its own experiments; for example, we saw in Chapter 23 that quantum mechanics refutes classical ideas of causality.

Materialism dismisses religion as the desire for a parent figure who can help us avoid—or otherwise circumvent—death. It attempts to explain away mysticism through magnetism, epilepsy, or other physical states. Religion has treated science abysmally for thousands of years, and few scientists condone behavior accepted as normal by religious people. Materialism is a very natural backlash against religion that has enjoyed great success; however, the pendulum seems to have swung to

---

The entire universe and everything in it is conceptually designated. The sutras state that all these phenomena are designated by thoughts.
Gen Lamrimpa

We regard promissory materialism as superstition without a rational foundation.
John Eccles and Daniel N. Robinson

the point where materialism seems to be a religion in its own right that does not fit its own data.

Both dualism and idealism can easily account for changing brain states—as shown by various imaging studies, the placebo effect, and paranormal phenomena—by acknowledging that brain and mind are different things altogether. Swimming requires a body of water, but not the other way around. The effort to reduce mental states to purely physical processes is doomed for reasons we have covered in detail throughout this book. Even if we could know what it is in our head right now, that is no guarantee that we can tell what will be there at any time in the future. Humanity is itself irreducible; it is impossible to know what it feels like to experience something without having that experience for ourselves. We may be able to guess, but that is never the same as knowing.

Physics cannot explain consciousness, yet its biggest mysteries all involve consciousness. We cannot fully understand life by looking at its smallest components; however, we cannot separate physical existence from the structures that control our perceptions, because we only see what we perceive.

What we believe is reality is only the endpoint of a process that involves consciousness. Animals also perceive the world in ways that are doubtless just as real, and just as rich and fraught with meaning to them, as ours are to us. An ant's life must be just as precious to it as ours is to us. We believe in a world "out there" beyond our senses, and yet everything we perceive, think, and feel is inside our heads. Experiments by Libet and others have shown that unconscious brain activity occurs up to 1/2 second before people take conscious action, and that one can predict which hand a subject will raise up to 10 seconds in advance. Our brains make choices on a subconscious level, while giving us the feeling that we took action. This does not repudiate idealism; it merely highlights the difference between consciousness and awareness.

Space and time are not discrete objects or phenomena. Neither forms any sort of absolute matrix for anything within the universe. More on this in Chapter 32. Meanwhile, the lack of amazement shown by scientists who say the universe simply popped into being is itself amazing, because we cannot turn to space or time to explain consciousness.

The belief that we are frail biochemical machines controlled by genes is giving way to an understanding that we are powerful creators of our own lives and the world in which we live.
Bruce Lipton

A reality completely independent of the spirit that conceives it, sees it, or feels it, is an impossibility. A world so external as that, even if it existed, would be forever inaccessible to us.
Henri Poincaré

> *Myth is neither a lie nor a confession: it is an inflexion.*
> Roland Barthes

Atoms and proteins do not explain consciousness either. In fact, the quantum behavior of particles makes it all the more difficult to imagine that mere matter can sustain the level of organization required to generate a coherent picture of consciousness. Nerve impulses carry information, but do not entail any more consciousness than a hard drive that also contains information. We would not mistake the raw information and processing of a hard drive and its associated computer for consciousness, so why make the mistake simply because the circuits are biological and not purely electrical?

Materialism is natural, because we normally only see matter; indeed, our minds are so good at creating the physical universe that we rarely stop to ask questions, and even then usually in a theoretical context; however, we can get past this if we look hard enough, a process some call enlightenment. It is natural to look for mind and God by looking for genes, structures in the brain, or even the so-called God helmet that can sometimes induce mystical experiences using magnetism. It is natural to conduct experiments, such as delayed-choice or dual slit, that demonstrate nonlocality; it is not natural to refuse to accept the implications of this material, and to respond to valid critiques by simply shouting louder. There is no materialist theory that allows for spirituality and its effects without distorting the evidence, such as that seen in Chapters 16 and 23. It also cannot explain life's origins, which becomes easy if we see the universe as a product of consciousness.

As I have mentioned, we can imagine a parallel universe where consciousness does not exist; consciousness is thus over and above physical facts, because it is not logically supervenient from matter but a separate phenomenon. Consciousness that is only naturally supervenient should not have much of any capacity to make a difference in the physical world.

> *Materialist ontology draws no support for contemporary physics and is in fact contradicted by it.*
> Mario Beauregard

Upholding idealism requires the following core assumptions:

- Conscious experiences are real. If we deny this, then we can say nothing about consciousness, and must conclude that consciousness as a phenomenon does not exist.

- Conscious experience is not logically supervenient on the physical world. Denying this is the only way in which the so-called zombie world could exist, and in which materialism could hold. If a phenomenon is not logically supervenient on the physical world, then materialism is false.

- The physical world is causally closed. Denying this is the only way to allow dualism.

Fractured Unity

We all know that a pie can exist as a whole that retains its wholeness function until cut. We cannot have slices of pie without a whole pie. A pie cut into slices that have not been removed yet is a whole pie without its wholeness function. This function is entailed in all whole things, because there must be a whole of any whole thing. Each slice is a unit of the whole pie that is distinct while also retaining its pie-ness. The only difference between the whole pie and its constituent slices is functional, and we can ascribe pie-ness to everything from a tiny drop of filling to the whole thing. This difference exists between the universe as a whole and its constituent parts. The fractionating of space into functional units creates thing-ness, which is the stuff of experience. Matter, space, time, and energy are the "what" of experience.

Someone who has learned to accept that nothing exists but observations is far ahead of peers who stumble through physics hoping to find out 'how things are.'
Richard Conn Henry

Nothing and Everything

This all sounds good, but how can we prove the existence of a higher-level intellect? Any explanation must include how this intellect interacts with the physical world. We have made some great strides throughout this book, and will start tying everything together in Chapter 33. Meantime, I must concede that assertions and proof are two very different things.

The fundamental assertion of idealism is that the physical world is a manifestation of the single consciousness that we can call the Godhead/Ground. The term "consciousness" refers to all of reality, but not to any part. Consciousness projects and derives meaning from experience. Experience cannot exist without consciousness. There can be no minds without consciousness. It is possible to have consciousness without awareness; for example, I am wide awake right now, but my awareness is focused on my computer and the words I am typing. There are other people in the house right now doing

One nature, perfect and pervading, circulates in all natures. One reality, all-comprehensive, contains within itself all realities.
Lankavatura Sutra

things I am not aware of. We cannot, however, have awareness without consciousness.

Animals and humans have awareness, which cannot be explained by matter. We experience things and are aware of experiencing them. Experience is real; indeed our lives consist of nothing but experience. Awareness is a component of consciousness that transcends the body. Consciousness is the core component and source of experience. Without consciousness, there could be no existence and no things to experience. We cannot rely on space, time, matter, or energy as givens, as we will see in Chapter 36. The one thing that we can rely on is consciousness. Consciousness must be a given, and must be both self-explanatory and the explanation for everything else.

The plurality that we perceive is only an appearance; it is not real.
Erwin Shrödinger

Physics may be the bedrock science on which all others are built; however, physics itself is built on mathematics, which is explained by set theory, which is explainable by logic. In short, logic is the ultimate foundation of all knowledge. Based on all the evidence we have examined thus far in this book, the only logical conclusion we can come to is that idealism is either the correct model or very close to it. If we accept this, then we can also accept that we transcend this lifetime. As Kübler-Ross said, we can look in three directions to understand existence: at matter, outward, or inward. We will have our answer when we can look anywhere and understand the same thing. That has been the central aim of this book: to look in as many directions as possible to see how the seemingly disparate pieces of evidence fit together, if at all.

The physical world is entirely abstract and without 'actuality' apart from its link to consciousness.
Arthur Eddington

The final chapters of this book will be devoted to the attempt to construct a model of reality that best fits the available evidence, and to demonstrate why this model is valid for answering the question of what happens when we die. Ultimately, there is only one way to find out; however, knowing the true nature of a thing is just as much about inference as it is about direct experience, if not more so.

828 | *The Divine Savage*
Revealing the Miracle of Being

Chapter 32

Philosophy

There are two ways to live your life. One is as though nothing is a miracle. The other as though everything is a miracle.
Albert Einstein

In the great drama of existence we ourselves are both actors and spectators.
Neils Bohr

Light travels more slowly when traversing a medium than it does in a vacuum. Mathematician Pierre de Fermat (1607-1665) is famous for many things including the Least Time Principle, which states that refracting light passing from a rarer medium to a thicker one will take the path of least overall transit time. In

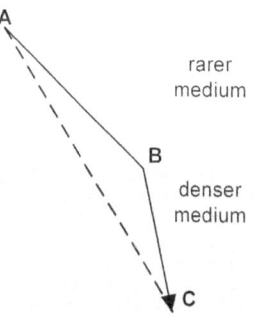

this example, following the straight line would result in a longer overall transit time, because the light would spend longer in the denser (slower) medium. We can see this same principle at work whenever we deliberately choose to take a longer route to avoid traffic or other delays that would make the shorter route less efficient. We are also conscious beings with the ability to foresee consequences. Light has neither of these properties, if materialism is to be believed.

Fermat's principle fascinated Leibniz because of the clear indication that there may be some optimization principle at work in the universe. It is possible that the universe contains as much variety as possible while also containing as much order as possible, which would make it as perfect as possible. Too much order would stifle evolution, growth, and experience.

Too much variety would not allow life to take hold, and would not allow relatively stable evolution.

Dirac observed that classical physics lets us build mental images of the grand scheme of things in space and time, but that natural law does not work in quite the way it appears to work. For example, we might not necessarily think of refraction where light passes through Point B on its way to C until we observe it by reaching to grab something under water and realizing that the object is not where it appears to be. Tribal fishermen learned to account for refraction when spearing fish while wading in shallow water, and aimed for the correct spot. Any casual observer would be forgiven for thinking the fisherman would miss. We could then build a mental image of refraction that would allow us to hit a fish underwater, but would not necessarily suspect that efficiency is part of that model.

Nature has been kinder to us than we have any right to expect.
Freeman Dyson

If you have been paying attention while reading Chapters 21 through 29, you have already realized that Fermat's principle both skewers materialism and provides yet another example of nonlocality and entanglement that is the stuff of Copenhagen and idealism. Let's take a closer look at how this might work in an idealist universe, and take a look at how the mechanics of reincarnation might work while we're at it.

Metaphysics

Metaphysics is not as widely accepted as it once was, but there are answers out there. Truth does exist, and the searcher honest enough to accept that truth no matter what will find the answers s/he seeks and will be forever changed as a result. I humbly offer myself as an example.

I have put duality away. I have seen that the two worlds are one. One, I see, One I know, One I see, One I call.
Jalal-uddin Rumi

It takes many years of training to become a physicist. It also takes years to learn how to have mystical experiences. Like scientists, mystics seek to repeat their experiences. It is safe to say that mystics are scientists, because they have both the same level of training and the same goal of explaining the universe we live in. Mystics describe their experiences as removing all barriers between self and universe. It is equally incorrect to label this state as nothing or everything; it is a cosmic consciousness without subject or object, no cause or effect, just pure being. Achieving these states is crucial for exploring the fundamental reality that cannot be explored by any means other than direct experience. This is the same process of experiment and discovery used by scientists, albeit with the challenge of being unable to present empirical data.

I was thrown out of NYU my freshman year for cheating on my metaphysics final. I looked within the soul of the boy sitting next to me.
Woody Allen

Crank up a debate between scientists and mystics, and my money is on the mystics to win. There are far too many parallels between physics and metaphysics to ignore, which is fascinating, because physics set off in the exact opposite direction only to come full circle. I confess to a similar fate, having started off to prove materialism and the lack of any God or afterlife. I fully intended to tell you, dear reader, that you face annihilation and to just suck it up and enjoy life before it all goes black. It should be blindingly obvious by now that my research has led me down a much different path.

Walker had a very valid point when he said that, "Physics is the tool we must use to learn about reality. But it has its hazards. If we are not careful, she will ensnare us. If we are not careful, we may begin to believe there is nothing else but this (material) reality." This sentiment was echoed by author Sir Arthur Conan Doyle (1859-1930) when his character Sherlock Holmes said that, "When you have eliminated the impossible, what remains, however improbable, must be the truth."

Nothing and Everything

It is not how things are in the world that is mystical, but that it exists.
Ludwig Wittgenstein

Eastern philosophy uses terms like "empty" and "void," but these are misleading, because they are talking about infinite creative potential and consciousness in a manner that is similar to descriptions of quantum superpositions, multiple paths, or complementarity. The word "nothing" is really "no-thing," which is the opposite of "thing." Nothingness and thingness are therefore complements. If truth speaks of things, then falseness speaks of no-things, which is why mystics have such a difficulty translating their experiences into words that others can hope to understand. Anything we can say about the nothingness is wrong, and yet we must say something in any attempt to describe it.

We will explore this concept in much more detail in Chapter 39; meanwhile, it may help to think of 0 and infinity as complements. We already know that complements, such as waves and particles, are equal and opposites. Zero can therefore be said to equal infinity. Wrap your mind around that, and you will start to understand.

Exposing the Illusion

Patterns can be semipermanent, which can give them the aura of depth and reality and helps explain why so many people think that manifest reality is also the ultimate reality. Author Daisetsu Teitaro Suzuki (1870-1966) wrote that words destroy Zen and logic kills mind—a possible reference to the fact that humans are emotional beings first and foremost, as I said in both Chapter 9 and *The Enlightened Savage*. I also said that logic is the root of understanding; however, that layer of understanding also serves as a barrier to the Godhead/Ground. There is no substitute for direct experience. Hofstadter agrees, saying that we need to step outside logic to be enlightened. Buddha went one step further by calling the truth incomplete and enlightenment only one step on the way. It seems safe to say that neither science nor our quest for understanding will be over any time soon.

For if he once, by chance, uttered the most perfect thought, he would not know so himself. For only delusion is given to all.
Xenophanes

Our quest to find reality has led us from mythology to shamanism to religion and eventually to science that forced us to look in the mirror to try to see what we really are. Consciousness is real, and it is our task to discern what that means. Capra tells us that the goal of Eastern mysticism is to experience all of the phenomena in the world as manifestations of Brahman/Dharmakaya/Tao/Ground/Godhead/Source. Physics is closer than any other science to achieving this goal.

There is a growing recognition of creativity in physical processes that is blurring the line between living and nonliving. This is not to say that no such distinction exists. Easterners recognize differences while also embracing them within unity, which is far different than saying that all things are equal. Buddhists see all objects as fluctuating processes and deny material substance by adopting monistic idealism. All agree that it is difficult to impossible to be constantly aware of the underlying unity. Physics understands the universe as a whole that includes observer and observed. Any way you slice them, our age-old beliefs about space, time, energy, matter, locality, and causality just don't stand up to the facts.

Nature tends to hide herself.
Heraclitus

The illusion itself exists to allow us to try to understand the myriad possibilities and potential of consciousness. This understanding is called *ananda*, or spiritual joy. If you think that life is suffering, then you will be freed from the cycle of birth and death. If you think that life is joy, then you will have the choice to be born again or not. The practice of *death yoga* (part of tantra) aims give people that very choice. Whether it achieves this purpose or not is anyone's guess; however, at the very least it claims to be able to help someone stay conscious during the dying process, to give them a real choice about their next life. There is only one way to find out for sure.

Knowledge is information that accurately represents and corresponds to what can be known of reality, while beliefs are not necessarily connected to anything real. We can see beliefs as shortcuts that can aid survival (see Chapter 9, *The Enlightened Savage*, and *The Natural Savage*) but they can—and do—impede knowledge. Learning requires dedication and experience, and human nature makes it difficult to truly learn anything—hence the ancient saying, "Have your ears heard what your mouth has spoken?" Pragmatism is not necessarily the same as materialism, because the former asks what the practical implications or values are without any preconceptions—or at least with fewer preconceptions than the latter. Mystics see value in consciousness. Quantum mechanics has an incredible amount of value for understanding reality and consciousness, although that was not the original intention of the researchers who pioneered it. Pragmatism seeks knowledge based on evidence, no matter where that evidence leads.

Skepticism is healthy, because it keeps us looking. A religious person who peddles faith may sincerely believe that s/he is leading people to God and toward salvation, when in fact s/he is leading them away from the most fundamental truths and away from the Godhead/Ground. The simple fact is that we cannot know God unless and until we have a detailed understanding of the world God created. I have spent decades studying the big questions, seeking answers, and researching and writing this book. I can hold a conversation about any of the material in any of my work; however, I am neither naive nor arrogant enough to think that my level of knowledge is anywhere near enough to give me anything more than a glimmer of knowledge about the true nature of God.

Apparently the master wants to get across the idea that an enlightened state is one where the borderlines between the self sand the universe are dissolved. This would truly be the end of dualism, for as he says, there is no system left which has any desire for perception. But what is that state, if not death? How can a live human being dissolve the borderlines between himself and the outside world?
Douglas Hofstadter

Spiritual claims contain at least a kernel of truth, albeit non-empirical truth in most cases; however, they are extremely valuable because we can compare them against empirical evidence to see how well they stand up. The implication thus far is that they stand up perfectly, but we have yet to tie up the final loose ends.

A Model of Reincarnation

There is a story of a dying sage whose disciples begged him not to go, to which he replied, "But where would I go?" We explored the amazing degree of accuracy found in creation myths from around the world in Chapter 3. Does the same hold true for death? Integrating the concepts of purgatory (Christianity), reincarnation, and eventual liberation from reincarnation demonstrates how the chain of lives is functionally very similar to a purgatory, if not identical. Furthermore, Paul's claim that resurrection is not physical but spiritual is easy to accept, if we accept monads. The Cherokee believe that each animal has a certain lifespan; an animal that dies prematurely receives a new body. Once the animal lives out its full lifespan, the body dissolves, and its spirit becomes free to join its relatives in the sky.

You are mourning for those who are not to be mourned for, and you speak wise words in vain. The truly wise mourn for neither the living nor the dead.
Krsna

We have known about lightning from the first time one of our evolutionary ancestors saw it arcing across the sky, but only recently have we started to understand its many subtleties. The same is true of death. Where we used to see it as black or white, alive or dead, we now have different types of death, such as clinical death or brain death, and understand that death is a process and not an event.

We saw just a few of the countless examples of seemingly verified reincarnation stories in Chapter 16. In an idealist universe, it is not unreasonable to postulate the reality of reincarnation. The trick is to determine whether these stories are in fact examples of reincarnation, and how it may work. In building a possible model of what reincarnation is and how it works, we must heed Occam's razor by keeping that model as simple as possible, while also remembering Einstein's admonition not to get any simpler than necessary lest we lose something important.

When you close your doors, and make darkness within, remember never to say that you are alone, for you are not alone; god is within, your genius is within.
Epictetus

Pondering the Soul

The successive existences in a series of rebirths are not like the pearls in a pearl necklace, held together by a string, the 'soul' which passes through all the pearls; rather, they are like dice, one piled on top of the other. Each die is separate, but it supports the one above it, with which it is functionally connected. Between the dice, there is no identity but conditionality.
Dalai Lama

We saw in Chapter 17 that the Hayflick limit essentially consists of planned cell death that gives humans an average top lifespan of about 100 years. It is true that survival is critical to evolution, but it makes sense to have finite lives as part of a finite system. There is no way that any physical body could last forever, because the universe will eventually either freeze to near absolute zero or come back together again in the Big Crunch. Reincarnation also allows monads to experience many different identities, which thus allows the Godhead to explore all possible permutations of life and possibility. Physical existence is not a problem, because this assumes that matter is a product of consciousness that can—theoretically at least—start over again whenever it wants to.

We also saw that all human cultures have at least some belief in a spirit world, which seems to indicate a genetic predisposition for such beliefs. Chapter 9 identified a candidate gene. But why? If religion is genetic and/or arises from fear of death, then natural selection must have played some role. Even if this is not the ultimate reason for religious and spiritual beliefs, we can rest assured that their presence serves some measurable survival purpose.

It is natural and beneficial to fear death and to seek to avoid it as long as possible. Our life or death does not matter to the atoms that make up our bodies at all; however, their dissipation after death seems to argue against anything of our identities or selves continuing on. If we are simply bags of meat, then our lives are valuable to the extent that we successfully contribute to the gene pool, period. Our dreams, aspirations, joys, frustrations, and very identities are meaningless beyond this reproductive function.

Being is eternal; for laws there are to conserve the treasures of life on which the universe draws for beauty.
Goethe

An idealist will tell you that the real question is not whether some forms of consciousness or self persist beyond the grave but how they do so, because consciousness and all associated with it are nonlocal and entangled at many levels. This one fact seems to lie at the heart of all religions and spiritual traditions in one form or another.

I need to make clear what I mean by immortality. Mere physical immortality is impossible for the reasons I described above. This form of immortality remains within time and space, and is therefore not true immortality, because even this will end. True immortality, if it exists, must lie in a realm beyond time and beyond space, energy, and matter as well, for reasons we will see in Chapter 36.

Paranormal evidence, such as that described in Chapters 15 through 17, indicates that belief in an afterlife of some sort may be warranted. The possibility of higher dimensions that we saw in Chapters 24 through 28 means that death could be nothing but the metaphorical equivalent of shedding a snakeskin, or of an actor who wraps up playing a certain role and moves on to the next. We can also look at child prodigies. Genetics and training alone don't seem like quite enough to explain geniuses like Mozart, who could play the piano at age four, or Ramadan, who learned mathematics without any training. These are examples of skills that could possibly have come from a previous life. Also, as I said above, the sheer number of verifiable stories for which reincarnation is the best explanation seems to rule out hoaxes, mistakes, and delusions.

Magic is the best theology for in it true faith is grounded.
Jacob Boehme

There are many different ways to look at the soul, such as:

- The classic dualist view of a spirit that goes to a heaven or hell, depending on the life lived while on Earth.

- A phenomenon of brain that ceases to exist when the brain dies.

- Identical to consciousness, which means the soul can never die.

- A carrier of attributes, such as mind, themes, contexts, and quantum memories.

- Quantum monad

- Karma, in which no identity is passed, only certain traits and tendencies, such as transferring a configuration file from one computer to another.

Any of the above—or any number of other imaginable scenarios—are possible either singly, or in many combinations. The important thing is to remember physics, so as not to fall into the trap of dualism.

I have seen God's house. It is very nice. You do not know everything that is there.
Sunita Kandelwal

Quantum Processing

> *I myself believe that the evidence for God lies primarily in inner personal experiences.*
> William James

All objects obey quantum laws and spread into increasing possibilities according to the Schrödinger equation, but the waveform is not encoded in those objects. Goswami speaks of both classical and quantum memory, where the former is recorded in the brain, and the latter is more akin to the non-material Akashic record. Computers can process symbols, but only our brains can give meaning to those symbols. For example, your brain is engaged in converting symbols (the letters and spaces on this page) into meaning right now.

Our entire lives revolve around meaning, which needs symbols, which also need symbols, and so it goes. Non-living things behave lawfully, while living things behave programmatically. Goswami believes that consciousness uses the physical body as a measuring device to make representations of the material world, in the same way in which we use the characters in our dreams to interact with worlds of our own creation. These representations affect different subtle bodies, which we saw in Chapter 15, and which we will examine again in just a few moments.

The idea of reincarnation is bothersome if we postulate that our individual egos survive death. Our egos identify with brain and body to the point that we are almost forced to say that ego and memory perish when our bodies perish. Adding a monad that preserves a higher level of identity resolves this problem. The only drawback is that it may not fully account for people who remember past lives, because the monad may be unchangeable and unable to hold ego content from life to life. If this model is correct, we can say that the soul continues on but leaves the mind behind; however, someone who is dying consciously can become aware of the fetus getting ready for the next life, and nonlocally collapse the waveform of both bodies and minds. This sharing with the child allows it to gain memories from the dying person that it may act on later. Of course, most children never actively recall a past life; fewer still do anything about it, but the memories remain.

> *There is a central order to the universe, an order that can be directly apprehended by the soul in mystical union.*
> Albert Einstein

This image reminds me of the Vulcan mind meld from *Star Trek*, where Mr. Spock can transfer his identity to a different body, such as near the end of *The Wrath of Khan*. It also makes me wonder what happens to someone who does not die consciously. Not all of us are Tibetan monks, and not all Tibetan monks have the opportunity to see death approaching. For example, I am sure that more than one seemingly healthy monk has died in his sleep.

A monad that carries only karma and stored quantum tendencies is virtually indistinguishable from the personal annihilation postulated by materialism. Also, an unchanging monad that is incapable of carrying classical memories seems like it should not be able to carry karma and quantum memories, since these are continually updated in between lives. NDE reports have indicated that people become aware of information they could not know in this lifetime about the past and future, which may indicate that something of ourselves does indeed survive. This seems to fit holographic theories of the universe, and is also compatible with both nonlocal quantum consciousness and conservation of information. More on this in Chapter 37.

If a man never contradicts himself, the reason must be that he virtually never says anything at all.
Miguel de Unamuno

From our perspective, reincarnation seems to happen in fits and starts where a person dies, leaves this world, and returns as a different person; however, we learned that everything is connected by quantum nonlocality and possibly quantum consciousness as well. This consciousness collapses waveforms to form the events of individual lives. We are all connected by correlated nonlocal experiences.

We know that experiments have noted a 0.5-second delay between stimulus and response. We can explain this lag using quantum consciousness, because there are many possible responses in the waveform. Collapsing this waveform leads to subject/object split awareness, because the subject has the freedom to choose how to collapse the waveform. This quantum processing also causes the individual mind and intellect. (The higher quantum consciousness and processing cannot normally map to the brain, which is why we only rarely experience being ourselves, except under certain conditions, such as RSMEs or NDEs.)

Forget distinctions. Leap into the boundless and make it your home!
Chuang Tzu

Individual events also affect the subject's own waveform, which impacts how the subject collapses future waveforms. Over time, this builds increasingly entrenched patterns that become increasingly difficult to get out of, because the waveforms affecting the subject become more and more skewed in favor of continuing the pattern. This is the quantum explanation of the ETEAR process (Chapter 9). The half-second delay is the time in which the stimulus waveform interacts with us on a preconscious level before it hits our belief filters. A person who exercises her or his freedom to collapse the wave at this instant accesses her or his quantum self. These instances are often associated with insight, serendipity, creativity, or luck. The stimulus waveform continues interacting with the subject's waveform, which alters the probability distribution of the possible outcomes. It is this conditioning that identifies with ego, and which creates personality. Only quantum consciousness can interact with the "pure" waveform.

Revisiting the Quantum Monad

We seem to have plenty of information to be able to guess that monads are real. The question is, how does the monad work? Does it convey information between lives? If so, which information? Someone who claims to be resurrected or reincarnated bases that claim on the supposed existence of a mind that can function away from the physical world, which implies that the mind survives death and carries at least some classical memories from lifetime to lifetime. If this is true, then mind may well be immortal, because we can assume that it existed before this lifetime. Even a mind that has not been resurrected or reincarnated could have been passed down from parents, presumably in the form of genetic predispositions and an upbringing that often resembles how the parents were brought up—a clear form of karma.

Death ends the physical body and may take the ego and ego-related memories with it; however, the structureless monad (soul) does not die. The monad contains quantum memories and conditioned responses, and forms the bridge between incarnations by preserving the character, thought patterns, emotional patterns, beliefs, joys, and fears that are collectively known as karma, the sum of good and bad habits. Each monad is a unique individual that carries memories and learns over time. The monad brings karma to each new incarnation, and then presses on after death with the karma gained during the current lifetime. For example, the memory of discovering romance and of falling in love with a particular person may not survive death, but the knowledge about love stays on.

But if a man becomes attached to the literal meanings of words and holds fast to the illusion that words and meaning are in agreement, especially in such things as Nirvana which is unborn and undying, the he will fail to understand the true meaning and will become entangled in assertions and refutations.
Buddha

We are the cantors of the universe.
Elizabeth Johnson

Each monad undergoes numerous cycles of birth and death, because it is impossible to learn everything in a single lifetime. As musician Elton John (1947-) famously sang in *The Circle of Life*, "There's more to see than can ever be seen, more to do than can ever be done. There's far too much to take in here, more to find than can ever be found." The concept of a monad that survives death is perfectly compatible with nonlocal quantum consciousness that can also transfer content from lifetime to lifetime nonlocally.

Death marks the end of the public domain where we can see and interact with the people around us and vice-versa; however, it need not mark the end of the private domain. Monads that are between incarnations may not have subject/object split awareness and may not be able to grow and change in between lifetimes. They may only carry the conditioning picked up during the lifetime that gets added to the conditioning from all lifetimes while being unable to alter that conditioning on their own, because they lack a sufficiently tangled hierarchy that includes the material world. If this is the case, then the monad does not exist separately from the Godhead. It carries a limited form of identification that splits off from the Godhead to experience a lifetime and/or pursue opportunities before returning home. The God-consciousness can collapse waves in both monads and sentient creatures at once. God and the monad are thus very similar, although it is appropriate to speak of the monad when looking up from the bottom and of God when looking down from the top.

A person in a state of nonlocal consciousness during the dying process may have the choice to either reincarnate or remain a monad free of reincarnations. The fears, tendencies, etc. of the current lifetime pass to the new through the monad. As we saw above, this mechanism, if it exists, could easily explain geniuses and child prodigies, to name just two examples. Einstein, Edison, Heisenberg, and others may have leveraged experiences gained over many prior lifetimes.

There is no reason to suspect that humans are the only species to have monads. All conscious species should have them, although there is some disagreement about whether individual animals have them, or whether all animals in a given species share a species-wide monad. I believe the former, because of the ongoing processes of evolution and hybridization, plus the

---

When mind assumes the form of a sentient being, it has suffered no decrease; when it has become a Buddha, it has added nothing to itself.
Haung Po

See all things not in process of becoming but in being, and see themselves in the other. Each being contains in itself the whole intelligible world. Each is all, and all is each.
Plotinius

> *The universe produced phenomenally in me, is pervaded by me. From me the world is born, in me it exists, in me it dissolves.*
> Ashtavakra Gita

fact that different members of the same species can have dramatically different habits.

Goswami explains ghosts, angels, and other nonwordly creatures as *Sambhogakāya*, people who no longer identify themselves with physical bodies and who also no longer need a monad to transfer information from life to life. Their monads are available to everyone; people who receive their services become fulfilled monads themselves. This is one reason why some speculate that the monad retains an individual identity over and above that mentioned above.

A monad or quantum self is universal, transcends personality, and is a unifying force, unlike the ego that operates in a divisive manner. Some believe that someone who is conscious during the stage of dying where the ego identity fades can see themselves as a creator of worlds—as God. A monad that has earned its freedom exits the cycle of birth and death, and remains an immortal spirit that experiences all Nows at once, instead of as a stream of individual Nows.

Do We Need Subtle Bodies?

> *Your duty is to be and not to be this or that. I am that i am' sums up the whole truth. The method is summarized in 'be still.'*
> Sri Ramana Bhagavan Maharshi

In Chapter 15, we touched on various ideas about subtle bodies, and the topic bears another look. Are there layers of nonmaterial bodies such as the astral plane, ethereal plane, and others? Metaphysics based on Tibetan beliefs believes in monistic idealism where consciousness has five levels:

- **Physical:** The material body.
- **Vital body:** Provides the physical body with energy and keeps the physical body alive as long at they remain together.
- **Mental:** Mind and senses/self.
- **Supramental:** Intellect, etc.
- **Causal:** reflection of atman and bliss.

All bodies are created by consciousness and represent quantum waveforms within consciousness that handles the interactions and maintains synchronization. Tradition says that death ends the physical bodies, while the subtle bodies live on. (This seems like dualist thinking, until we remember that all bodies, material and non, are products of consciousness and thus avoid the issue altogether.)

According to Goswami, only the physical body can collapse waveforms. The subtle bodies live in potential, and reality only happens when a collapse occurs and a Now is selected. Consciousness chooses the reality from among all available options, and experiences the selected choice. Consciousness is the ultimate child in the ultimate candy store that consists of a universe full of Nows. This collapse occurs across all bodies, which retains the gist of dualism with none of the problems. The idea of subtle bodies may open the door to God's mind.

In this model, the physical body acts out the collapse like an actor on a stage—or perhaps closer to a marionette, including classical memories obtained during the current lifetime. A body that receives a stimulus passes that stimulus to the quantum consciousness, which collapses the waveform. Repeated stimuli build quantum memory.

We seem to experience our physical bodies as being both external to our selves and as part of a shared reality that includes the body. Our thoughts and feelings seem internal by contrast, as does our feeling of being alive. (This can change during paranormal encounters, such as via telepathy.) According to this model, the death of the human body releases us to become supramental beings that are both ourselves and yet so much more, without really being ourselves at all anymore.

The question we now have to ask ourselves is whether the vital bodies are real, or if they are conceptual constructs designed to facilitate comprehension of reincarnation and self-identity. We will return to this very important question in Chapter 37.

Karma

Buddha once said that we are everything we have ever thought. This much seems obvious in the context of our own lives, but karma (if real) adds another layer by carrying forward habits, behaviors, and tendencies from previous lives. We identify with our egos during each individual lifetime; however, our monads provide an ongoing context to our karmic evolution by carrying karma with them, and giving that context to each new incarnation that becomes a new ego. This

Behold but one in all things; it is the second that leads you astray.
Kabir

To err is human, to forgive, canine.
Unknown

helps give us our character and predisposition. Karma can go up or down. Someone who is having a bad life now could be said to be working off bad karma from a previous life, while someone who is leading a good life could be said to be enjoying good karma from a previous life.

Bad karma can be used up through one or more lifetimes of penance/doing right, and good karma can be used up as well. Getting closure for ended relationships is important, both for relationships in this life and relationships we have had in past lives. Bad things happen to good people because of karma, which is difficult to acknowledge, because we tend to see events as only occurring within the context of a single lifetime, and not that of multiple lifetimes. The movie *Groundhog Day* illustrates this concept perfectly, and I highly recommend that you watch it if you have not done so already.

It is theoretically possible for people to remain karmically connected over more than one lifetime, which may help explain otherwise unexplainable connections/love between people. There are plenty of biological explanations for this, which I explain in *The Natural Savage*, and which will be explained in more detail in *The Romantic Savage*; however, we cannot be sure whether mere biology is responsible. If we assume an idealist universe where consciousness creates matter, we must at least allow for the possibility that consciousness alters our genes to help carry karma forward—which may also explain odd birthmarks that resemble old injuries.

I mentioned above that good karma can get used up just as well as bad karma. One who is caught up in this cycle will keep riding the wave up and down; however, a person who is clear about both how karma works and their dharma and who lives their life with pure intentions does not use up her or his karma, because s/he is doing right. The key difference is that such a person is not simply *laissant les bon temps rouler* (letting the good times roll), which often involves a degree of exploitation and selfishness. Such a person is living a life of service to the world. This does not necessarily imply austerity or poverty, because the material things themselves are not important; it is what we do with them that counts.

The way to achieve good karma (or at least to avoid building bad karma) is to follow the right actions listed in Chapter 5. In

Men are cruel but Man is kind.
Rabindranath Tagore

The ultimate work of civilization the unfolding of ever-deeper spiritual understanding.
Arnold Toynbee

essence, these actions boil down to non-attachment and conscious choice. There is nothing inherently good or bad in life, and we have the power to choose how to perceive events, if only we would use it.

The concepts of karma, dharma, and neutrality can help explain suffering and evil. If the Godhead is seeking to have all possible experiences, then that must include the bad along with the good. It all comes out in the wash over the long run, because the bad that someone suffers will be made up for over the course of multiple lifetimes, and the good someone enjoys in this lifetime will eventually come back to haunt them, unless they take steps to avoid that. If someone is not following her or his true calling in life, then that could explain whatever strife they are having. Above all, remaining neutral is the key. It is OK to be happy or sad about anything at all—just be aware that it is coming from you and not from the event itself. The massive earthquakes in Haiti, New Zealand, and Japan are perfect examples. Natural disasters facilitate suffering, but the actual suffering only occurs in those people who choose—consciously or not—to suffer. I am not saying that any of these people deserved what happened or that disasters are good; I am saying that life is what we make of it no matter what.

We are the music makers and we are the dreamers of dreams wandering by lone sea-breakers and sitting by desolate streams; world losers and world forsakers, on whom the pale Moon gleams; yet we are the movers and shakers of the world forever, it seems.
Arthur O'Shaughnessy

To me, karma only makes sense in the context of multiple lifetimes. The alternative is to simply shrug one's shoulders and say, "Shit happens." This may well be true, but then again maybe not.

The Case for Reincarnation

It is nice and comforting to think that the "I" we each know and identify with will somehow return after death; however, the idealist position is that death is an illusion, because only consciousness is real. Everything else is the illusion of maya. The thing to remember is that death is just not the big deal that religion and materialism would have us believe—a truth that is extremely difficult to truly accept.

And what in fluctuating appearance hovers, ye shall fix by lasting thoughts.
Goethe

We have explored plenty of reasons why it is both perfectly reasonable and scientifically valid to assume that reincarnation does occur. We have plenty of evidence that our mind is not located in our physical brains, and that at least some of it survives physical death. Seeing our lives in the context of a bigger picture can help our path through this or any other life, by showing us that death is part of a creative process and not a destructive descent into oblivion. Understanding this one fact alone can have significant therapeutic value.

It is only in the face of death that a man's self is born.
Augustine

I mentioned above that genius may well be the result of many lifetimes of experience and training that we mistake for a gift or talent that just got handed to someone. Some souls may be older than others, and it stands to reason that these souls would have more experience than their younger peers. A single lifetime is not long enough to create and nurture the kind of receptive disposition that opens the door to the full power of creativity.

Death is the inevitable result of the struggle between entropy and creativity. Entropy always wins, because there are so many more ways for disorder to exist than order. In the reincarnation model, we come back again and again and put countless physical bodies through their paces in the process, while our monads carry the karma and other baggage. The amount of time one spends between incarnations, and the amount of time one spends in a given lifetime can vary. Also, there is no reason why the Godhead cannot split off and reunite monads at will, which easily explains population shifts over time, including the presently rising human population.

If death is not the end of all existence, then it follows that we carry our joys and sadness, our triumphs and turmoil, and our opportunities and limitations with us. Suicide is therefore not the answer, because the problems that drive a person to this final desperate act remain with her or him. Suicide solves nothing and provides anything but the relief the person is so desperately seeking.

And now it's up to you to walk into the world of spiritual awareness and spiritual knowledge, and to explore and experience your truth.
Gloria Chadwick

The whole purpose of life may be to manifest possibilities and explore potentials, just like our dreams do for each one of us. If idealism is correct—and there is every reason to think it is—then consciousness does not die with death. Realizing this within one's self is the key to unlocking the mystery of death. The primacy of consciousness from which all flows means that death neither eradicates an individual nor her or his mind. How this works within the context of a monad and Goswami's assertions that classical memories perish with death is something I will explore in Chapter 37.

Knowledge Limits

Examining mysteries is a fascinating pursuit. In seeking to unravel them, we inevitably come face to face with ourselves and the strengths and limitations we bring to bear. If we could step outside of the universe and look back, we would see those strengths and weaknesses in stark relief; however, doing so from within that which we are trying to solve is far more difficult because of the classic problem of being unable to see the forest through the trees. Still, we can infer a great many things about a forest from its individual trees, and the same is true when it comes to figuring out what we can and cannot solve.

Into this mysterious world we are plunged at birth with no set of instructions and no maps or guideposts.
Louise B. Young

The Drive to Know

The human animal is an extremely curious creature. We have an insatiable desire to learn, explore, and ask questions. This desire can last a lifetime, provided that it is not stifled by educational methods that inhibit creativity and different modes of learning, or by religious or other indoctrination. Even the Bible acknowledges this fundamental drive in passages, such as John 5:31, where Christ tells people not to take his word for things too seriously, or 1 Thessalonians 5:21 that tells the faithful to test and prove all things and only hold on to what is good. It is a shame that orthodox religions have forgotten these passages, the consequence of which can only be stagnation at best—or tyranny at worst. The historical record contains plenty of examples of both.

Knowledge is a function of being. When there is a change in the being of the knower, there is a corresponding change in the nature and amount of knowing.
Aldous Huxley

Science fiction author David Brin reminds us of the crucial role learning plays in the following online exchange between two characters from his novel *Earth* about the very first command in the Bible:

Query by TM: Monseigneur, according to the Bible, what was the very first injunction laid by the Lord upon our first ancestor?

> *It's never enough.*
> Allison Klupar

Reply by Monsignor Bruhuni (MB): By first ancestor, I assume you mean Adam. Do you refer to the charge to be fruitful and multiply?

TM: That's the first command mentioned, in Genesis 1. But Genesis 1 is just a summary of the more detailed story in Genesis 2. Anyway, to "multiply" can't have been first chronologically. That could only happen after Eve appeared, after sex was discovered through sin, and after mankind lost immortality of the flesh!

MB: I see your point. In that case, I'd say the command not to eat of the Tree of Knowledge. It was by breaking that injunction that Adam fell.

TM: But that's still only a negative commandment... "don't do that." Wasn't there something else? Something Adam was asked actively to do? Consider: Every heavenly intervention mentioned in the Bible, from Genesis onward, can be seen as a palliative measure, to help mend a fallen race of obdurate sinners. But what of the original mission for which we were made? Have we no clue what our purpose was to have been if we hadn't sinned at all? Why we were created in the first place?

MB: Our purpose was to glorify the Lord.

TM: As a good Catholic, I agree. But how was Adam to glorify? By singing praises? The Heavenly hosts were already doing that, and even a parrot can make unctuous noises. No, the evidence is right there in Genesis. Adam was told to do something very specific, something before the fall, before Eve, before even being told not to eat the fruit!

MB: Let me scan and refresh my... ah. I think I see what you refer to. The paragraph in which the Lord has Adam name all the beasts. Is that it? But that's a minor thing. Nobody considers it important.

> *The pain of the mind is worse than the pain of the body.*
> Publius Syrus

TM: Not important? The very first request by the Creator of His creation? The only request that has nothing to do with the repair work of mortality, or rescue from sin? Would such a thing have been mentioned so prominently if the Lord were merely idly curious?

MB: Please, I see others queued for questions. Your point is?

TM: Only this: our original purpose clearly was to glorify God by going forth, comprehending, and naming the Creator's works. Therefore, aren't zoologists crawling through the jungle, struggling to name endangered species before they go extinct, doing holy labor? Or take even those camera-bearing probes we have sent to other planets... What is the first thing we do when awe-inspiring vistas of some faraway moon are transmitted back by our little robot envoys? Why, we reverently name the craters, valleys, and other strange beasts discovered out there. So you see it's impossible for the End of Days to come, as your group predicts, until we succeed in our mission or utterly fail. Either we'll complete the preservation and description of this Earth, and go forth to name everything else in God's Universe, or we'll prove ourselves unworthy by spoiling what we started with: this, our first garden. Either way, the verdict's not in yet!

Everyone now knows how to find the meaning of life within himself. But less than a century ago men and women did not have access to the puzzle boxes within them. They could not name even one of the 53 portals to the soul.
Kurt Vonnegut

MB: I... really don't know how to answer this. Not in real time. At minimum you've drawn an intriguing sophistry to delight your fellow Franciscans. And those neo-Gaian Jesuits, if they haven't thought of it already. Perhaps you'll allow me time to send out my own ferrets and contemplate? I'll get back to you next week, same time, same access code.

The preceding is a work of fiction about a conversation taking place over a computer network chat program, but Brin's message is absolutely on point. Religious people of all stripes would be well advised to heed its message. The human quest for knowledge ranks just below survival and reproduction.

Inherent Imprecision

We have to face the fact that there may be hard limits to what we can observe no matter how good our observational machines get. A quantum universe may not be completely describable via mathematics because of Heisenberg uncertainty, and because it is impossible to separate the universe from an observer within that universe. Given that the universe seems unable to predict future behavior for even a tiny part of itself, it would seem that the universe is its own best simulator.

The real meaning of the dharma... must be directly experienced.
Siddha Nagarjuna

We must also account for the fact that we are both finite and somewhat less than perfect, which places serious limits on our ability to conceptualize. Even our ideas of the infinite carry some arbitrary upper limit, whether we want to admit it to ourselves or not. The sad truth is that we may not be able to find a viable Theory of Everything no matter how hard we try.

Gödel's Incompleteness Theorem

> *The things causing insecurity are the very things which, if understood, would lead to real security.*
> Jerry D. Wheatley

In this theorem, Gödel proved that any computationally self-evident system that is both consistent and powerful enough to describe the arithmetic of natural numbers cannot be complete, nor can it prove the consistency of its own axioms. Gödel arrived at this theorem after decades of attempts by mathematicians to find a complete set of axioms for all branches of mathematics. We will never find them all, nor can we ever hope to compute all mathematical questions.

To put this in plain language, no system can be both complete and logically consistent. Any logically complete system is logically incomplete. Some mathematics will therefore remain unprovable by logic, because any system that can process enough logic to handle all branches of mathematics can never prove its own consistency. All logical systems contain assertions that can never be proven. This theorem is a bit like the Heisenberg Uncertainty Principle for mathematics, except that the limits described here are computational and not observational.

The question arises, if we can neither observe nor compute with total precision, then how can we possibly hope to find a TOE? Observational limits blur the input, and mathematical limits blur the output.

No Proof of Meaning

> *It is a riddle wrapped in a mystery inside of an enigma.*
> Winston Churchill

Let's say we can overcome all obstacles and create a TOE. Gödel's theorem would apply, meaning that this theory, as impressive at it would be, would not be able to answer all questions asked of it. We can speculate that it would describe space, time, energy, and matter, but could it ever give us any insight into the purpose of life, if any? Could it prove or disprove God? Would it prove or refute materialism? Even a so-called TOE could never truly claim to be truly complete, nor could it even be complete. Weinberg is a diehard materialist, but even he had to concede that the search for a TOE seems pointless. At some point, it becomes easier to believe that we are puppets being guided by an unseen puppeteer. Weinberg sees this is a limitation. I think it may be an opening to proving or at least conclusively demonstrating the value and plausibility of idealism.

Intellectual Limits

The single greatest obstacle to learning is that we see the world as we want it to see it, which may or may not resemble how it really is. All of us build mental models that reflect our individual beliefs. The limitations our beliefs impose on us were designed to help us survive predation and other hazards. (See *The Enlightened Savage*.) Our brains simply cannot tell the difference between survival instructions and intellectual curiosity when it comes to applying the filter of belief. We smother ourselves with our own security blankets.

A man can do what he wills to do but he cannot determine what he wills.
Arthur Schopenhauer

We have seen many examples of people clinging to beliefs like a drowning person clinging to a piece of floating wreckage. This is perfectly understandable, predictable for any species that must quickly absorb survival instructions, and very limiting at the same time. The quest for knowledge can be both frustrating and painful, but that does not excuse anyone. To paraphrase Professor David Barash (1946-), all people have at least a little unspoken hypocrisy, in that we assume that determinism governs our intellectual pursuits and professional lives while experiencing our subjective selves as free will. Meanwhile, we speak without defining ourselves and listen without question, forgetting that not taking action is a decision in its own right, whether we are aware of it or not.

As social and hierarchical creatures, we do all we can to encourage conformity. We identify ourselves with friends, family, tribes, nations, causes, politics, religions, philosophies, employers, and competitors. Each of these examples and more can be wellsprings of knowledge and inspiration. They can also be sources of control. Going along with the crowd and fitting in is a powerful survival strategy because humans are prey animals, and outcasts must face the dangers of predation alone. Group status is one of the six core life functions. (See *The Natural Savage*.) Its importance cannot be overstated because it has helped keep us alive and off the menu for millions of years.

Chance favors only the prepared mind.
Louis Pasteur

Belief is powerful, because we subconsciously interpret beliefs as survival instructions. We therefore suppress anything that runs counter to our beliefs and stifle our own natural curiosity. This is doubly true when the belief itself encourages self-repression, such as religion. This keeps us from asking too many questions, asking the wrong questions, violating some taboos, and/or sacrificing creativity to at least some degree. It happens to all of us; there is no escaping it. Some of us even go so far as to exclude the possibility of cognition in the same natural world that we are both part of and trying to understand using cognition.

If a man begins with certainties, he shall end in doubts; but he will be content to begin with doubts, he shall end in certainties.
Francis Bacon

We also have the capacity to compartmentalize our personalities and show different aspects of them under different circumstances. We assume different roles under different circumstances and play them faithfully, even when they contradict each other. I am aware of this tendency in myself. Examine yourself carefully enough, and you will discover it in yourself as well. It is universal. People lacking healthy amounts of skepticism even go so far as to surrender their own individuality. The pejorative term "drone" is often used to describe such a person. Each of us can think of at least one drone.

Religious and scientific people love to contradict one another, but the truth is that they are all in the same boat. Both make claims without proof, and both exclude evidence that runs counter to their expectations. Some say that agnostics are honest, because they can at least admit not knowing; however, even this can become a mantra. Maimonides said that those armed with both scientific and religious knowledge are sitting in a room with a king, while everyone else is outside trying to get in. I like that analogy. Science today follows a decidedly Western paradigm that mostly ignores—and even dismisses—Eastern contributions, unless those contributions come from Western-style thinking. How much scientific insight and discovery is humanity losing as a result?

The plural of anecdote is not data.
Frank Kotsonis

We can also compare science to theology, because both are searching for the answers to the same questions. As Capra said, science asks how while religion asks why.

Perceptual Limits

Our drive to learn is constrained by uncertainty, incompleteness, beliefs, and the fact that our brains can only process so much information. These constraints are further limited, in that we can only perceive a tiny amount of all there is to perceive. We can only perceive a subset of the electromagnetic spectrum, hear only a subset of available frequencies, or smell only a tiny fraction of the many chemicals in the air. We can only survive within narrow ranges of temperature, pressure, atmosphere, gravity, etc. On top of all this, the ETEAR process described in Chapter 9 limits what we can do with what little we do perceive. Given these limits, we have every reason to suspect that we may not be capable of comprehending the true nature of nature.

It is data which are the final arbiter of hypotheses.
Harald Wallach and Stefan Schmidt

This is not to say that the search is a lost cause. The concepts we can generate do correspond to both observable and non-observable facts, which do correspond to reality; however, we often only experience fragments. For example, we experience space and time as separate phenomena, despite their being one and the same. Which poses another problem: If sensory data is only a representation of the real world, then we cannot empirically prove that said world exists. This does not falsify the external world, because both the laws of nature and the fact that others seem to corroborate most of our own perceptions strongly suggest that there indeed is something "out there."

The Theory of Everything

We have already examined contenders for the title of Theory of Everything/Grand Unified Theory in theories such as superstring, branes, loop quantum cosmology, and holography in Chapters 25 through 28. We have also seen how each new theory is grander than the last. The ultimate goal is to create a single theory that describes the entire universe. We may have such a theory and not know it yet for reasons we saw above and will see more of in a moment. This theory, if we ever find it, will be a key first step, because it will tell us how and perhaps even why. Even so, uncertainty will prevent it from making absolute predictions beyond a certain level of precision;

We're not sure who discovered water, but we're pretty sure it wasn't the fish.
Marshall McLuhan

however, this precision should exist in all predictions. Such a theory would be a tremendous scientific achievement. It would also answer many philosophical and possibly theological questions while also opening up many more.

Accounting for Consciousness

A valid Theory of Everything should ideally be able to address the role of consciousness, if consciousness does indeed have a role. Some of the questions we will need to answer include whether consciousness is somehow part of—or associated with—some or all of the following:

- Everything covered by the Schrödinger equation
- Any single point or subatomic particle
- Space
- Time
- Energy
- Matter
- One or more of the four fundamental forces (strong, weak, electromagnetic, or gravity)

Everything in the universe involves one or more of the above items. If quantum mechanics is correct and observers do play a role in a measurement—and especially if we play the starring role—then it stands to reason that the interface between consciousness and the material universe must involve one or more of the channels described here. We can safely say that the interface does happen through things like atoms and molecules. Any other channel would require one or more undiscovered forces, which implies dualism.

This is a very tall order for any theory to attempt, because the assumption is that the theory should be able to answer the ultimate questions about the nature of reality, who we are, what consciousness is, and what it all means.

Requirements for a Final Theory

A complete Theory of Everything should be able to answer some basic questions, such as whether the logical structure of the universe is inevitable or whether it is a special feature of

To see a world in a grain of sand and heaven in a wild flower, hold infinity in the palm of your hand, and eternity in an hour.
William Blake

Science is not a dogmatic belief system or an ideology; it is a method of inquiry.
Rupert Sheldrake

our own universe that could—for example—set it apart from other universes. This theory should be falsifiable, in the sense that it is only necessary to find one white crow to falsify the theory that all crows are black.

A key feature of any TOE we may discover must be an explanation for why our universe is the way it is, along with some constraints or limits that allow these parameters to be as they are and/or that prevents them from having different values.

The TOE must answer all of our basic questions and smoothly include all facts already discovered to date in a complete vision of reality. If everything in the universe came from random events, then we need to have a framework that explains both the events and their framework; however, I doubt that will happen. The universe follows a long series of natural laws; but where did those laws come from?

Ultimately, a TOE must also be the simplest possible theory, certainly far simpler than the concepts it represents and explains. Occam's razor demands no less. It stands to reason that there is a very simple explanation for the universe and everything in it, and we already know that great complexity can come from utter simplicity.

Can We Find a Final Theory?

What if all of the laws of nature can be distilled down to a single universal equation that describes everything? What might this look like? Would this theory say anything definitive, or would it be only statistical? Would it prove that there is only one universe or the existence of others? Would it say anything at all about God, or would it disprove that concept once and for all? Would it affirm or refute mysticism, spirituality, and paranormality? We can imagine a large number of universes with different properties arranged along a bell curve, which removes the problem of how ours became so hospitable, but which does nothing to explain ultimate origins.

We may never know. Philosophers do not have the mathematics to tell us this. They may even be struggling to catch up with ongoing scientific discoveries, and thus may not be much help in this matter. There is also the ongoing problem of people using science to further agendas, such as intelligent design or

With all your science, can you tell how it is, and whence it is, that light comes into the soul?.
Henry Thoreau

Science, in clearing away the fog of myth and mysticism that shrouded the world in the dark ages, has exposed not only the sharply delineated islands of knowledge but also boundless seas of ignorance.
Hans Christian von Bayer

materialism, instead of simply following the data wherever it may take them.

Whether we achieve this holy grail of knowledge or not, the fact remains that science is continuing to reveal a universe that is far richer and more magnificent than anything we could have imagined before each new discovery; however, we must also be aware of the fact that theories may or may not reflect reality. For example, the Ptolemaic model of the solar system that we examined in Chapter 21 does a good job of predicting planetary motions, while not remotely resembling how the solar system actually works.

Science must be open to following data wherever it leads without ideology; however, many scientists still harbor classical, Newtonian beliefs that we can loosely describe as, "I'll believe it when I see it." Such an outlook could mean dismissing data that contradicts with a certain hypothesis or that disagrees with expected results. One can only wonder where we would be if we truly let the data speak for itself.

From strings and loops, matter clumps together in ever-larger forms starting with particles, atomic nuclei, atoms, molecules, and more. More than once we have placed an upper or lower bound on this continuum, only to have to move the bar. Even galaxies have membranous structures shrouding their interiors as if they were gigantic cells; the very structure of the universe resembles both its origins and the structures in its smallest constituents. This all seems very complex; however, a humble chessboard with 32 pieces and 64 squares has more possible combinations (at least 10^{123}) than the number of atoms in the universe (about 10^{80}).

On one hand, we have reason to hope that a TOE is possible. Nobody has ever—or could ever—return from a black hole to recount her or his experiences, but that does not prevent us from coming up with a viable theory of how black holes work. Quantum mechanics seems like a likely candidate from which to build such a theory. The biggest questions may well lie beyond quantum mechanics (which is why superstring and other theories exist), but it's a good place from which to continue the search.

There are three possibilities:

Good better best, never let it rest, until the good is better and the better is best.
Anonymous

The map is not the territory.
Alfred Korzybski

- There is a TOE that we can and may eventually discover.
- There is no final TOE beyond a chain of successive theories.
- There can be no TOE, because we cannot predict outcomes beyond a certain level of precision, which would limit us to simple solutions for simple situations.

What If We Succeed?

In Chapter 39, we will discover that zero is a less-than-enumerable value, while infinity is a greater-than enumerable value. We think of zero as nothingness (which may be different than no-thing-ness), but this may be only an approximation; even a line of zero length has just as many points as line of infinite length.

The surprise is not that our theories are flawed, but that they work at all.
John Gribbin

We saw in Chapter 23 that the vacuum of space contains a theoretically infinite amount of energy. This vacuum is space, which is mathematically represented by 0, while infinity represents parts contained within space. Zero is indivisible, a whole but not a part, and there are unlimited zeros. Time exists as part of space, not separately. Without space, there could be no time. Matter and time are also linked; increasing speed also increases mass. Relativity links space and time, energy and matter. It also links matter to space-time, which ends up meaning that space, time, energy, and matter are linked. Everything is nothing, and yet nothing is everything—a rather startling fact with amazing implications that we will explore in Chapter 39.

The important thing is to not stop questioning.
Albert Einstein

We can wonder, ponder, research, and analyze all we want. We may even find a viable TOE someday. But will this convince anyone that life is worth living? In many ways, that is the ultimate question. Stay tuned, because the end of this chapter marks the end of our examination of the various sciences. The final eight chapters of this book will bring everything we have looked at together and lead us to our conclusion.

PART EIGHT

The Thesis

The Divine Savage
Revealing the Miracle of Being

Chapter 33

Evidence from Religion

> *Infidelity does not consist in believing, or in disbelieving. It consists in professing to believe what he does not believe.*
> Thomas Paine

> *Most Christians gerrymander the gospels and carve an idealized self portrait out of the texts.*
> Walter Kaufmann

You may recall that I leveled some strong charges against religion in Chapters 7 through 11, and I can understand any confusion you may feel as to why I might look to religion for any part of the answer to the question of what happens when we die. The short answer is that the mysticism that lies at the very core of all religions is very real, as we have seen throughout this book. Myth represents our earliest and most enduring form of science. It is tempting to think of mysticism as quackery and myth as old wives' tales, but the truth is that they form the foundation for all human knowledge, religious or otherwise. Religion began as a quest for truth with the promise of elevating people to the divine, but degenerated into a quest to seize and maintain power. We have seen how all religions faced this crucial choice point and how the vast majority of them opted for the trappings of earthly power, forgetting their original purposes. Nevertheless, the core idea and promise remains: that humanity can know the divine and be transformed in the process. It is to that core that I turn.

The fundamental premise of all religions is that something more is happening than meets the eye. We have seen throughout this book that this premise is entirely valid. This is not necessarily anything to do with spirituality or even with God, although I am limiting my discussions to religions that do deal with the divine (as opposed to secular religions, such as the

North Korean personality cult around Kim Il Sung, Kim Jong Il, and now Kim Jong Un). It is about acknowledging that metaphysics is valid, even if no religious dogma can ever hope to prove this.

Religious Shortcomings

I stand behind the idea that a partnership between science and religion that is dedicated to discovering the truth regardless of dogma or texts could accomplish much, because it would help drive mainstream science beyond the materialism that has plagued it beginning with Galileo and especially with Newton. Instead, religion chose to defend its dogma against anything that might contradict it because of its insistence that it holds a monopoly on moral and cosmological truths. The blowback has been inevitable: Science today rejects religion and actively attacks it, both sides forgetting that they used to be one and the same. We forget the example of Islam preserving and advancing science, while the West remained mired in the Dark Ages. Religion has responded by digging in its heels and refusing to see how science is uncovering the original truths that religion once focused on.

Religion spoke its last intelligible or noble or inspiring words a long time ago.
Christopher Hitchens

Dogma, rituals, and texts are fine to the extent that their adherents see them as reminders and metaphors; for example, we saw in Chapter 3 how creation myths from around the world are metaphors for the scientifically accepted history of the universe that we saw in more detail in Chapter 26; however, religion and science have instead chosen to engage in petty territoriality and turf wars to where today's religious leaders from the Pope to most pastors have forgotten the teachings of the person they claim to be following. No amount of dogma or ritual can replace the reality of a mystical experience, and it is ludicrous to think that God actually cares what we eat or how we dress. People like Origen, Tertullian, and Irenaeus represent the very worst that religion has to offer.

If religion addresses a genuine sphere of understanding then it should be susceptible to progress.
Sam Harris

In Chapter 7, we learned that polytheism is the natural human state that was only replaced by monotheism after more than a thousand years of struggle. (Even today's monotheism is only thinly veiled polytheism, thanks to the throngs of saints, holy people, and relics.) At its best, monotheism acknowledges that the universe is all one and thus has the right idea; however, we can see polytheism as an expression of God as a fractured whole, which is in line with Eastern thought.

The core message of religion has nothing to do with belief, faith, sin, or any of that nonsense. Belief in God is not—and should not be—limiting, if we understand God as being an immanent part of and within all that is. Seeing God as a single transcendent overlord creates a hierarchy where none exists; the pagans who celebrated harvest and seasons had a lot more going for them than any orthodoxy.

False Atheism

The Bible is not my book, nor Christianity my profession.
Abraham Lincoln

The JCI God does not exist, period. That much just about any non-religious person can agree with. What they fail to realize is that disproving one model of God does not disprove the existence of another type of god, such as the one we have discussed in this book. Saying that no god exists because the JCI God does not exist is just as big a leap of faith as the religious beliefs that atheists decry. These same atheists demonstrate their own religiosity by clinging to materialism, despite the mountains of evidence (much of it gathered by materialists) that disproves their positions. This knee-jerk atheism is just as intellectually dishonest—and just as religious—as religion itself. As I said in Chapter 2, one does not imply the other.

No Sacred Texts

The New Testament compared with the old, is like a farce of one act, in which there is not room for very numerous violations of the unities. There are, however, some glaring contradictions, which, exclusive of the fallacy of the pretended prophesies, are sufficient to show the story of Jesus Christ to be false.
Thomas Paine

The Torah, Bible, Koran, hadith, Vedas, and other "sacred" texts have all been edited beyond virtually all recognition. As we saw in Chapter 8, not even the original teachings remain, having been altered accidentally and deliberately to where modern religious beliefs often bear no resemblance to the original lessons. Even if these books had been translated without change or error throughout the thousands of years since their creation, we would still not be able to agree on all points, because we each interpret things differently. I am writing this book knowing full well that everyone who reads it will interpret it differently, and am proceeding in the hopes that most readers will get out of it pretty much what I want them to.

There is just no way that any book can possibly claim to be the inerrant word of God and expect to be taken seriously. Similarly, there is no intelligent way to believe in the truth of any such claim. Those religious people who believe that their books contain anything but human words written by humans for human needs are fooling no one but themselves, and blinding themselves to the true glory of the universe—a tremendous, deliberate, and ongoing insult to the God they claim to revere, and in whose name they claim to be acting.

Religion is what keeps the poor from murdering the rich.
Napoleon Bonaparte

With that said, all of the "sacred" texts contain germs of truth because all spring from the common fountain of myth and mysticism. Here again, seeing these books as human works written and edited by humans for humans could shed light on both history and the never ending human quest to answer the biggest questions ever asked. It is these nuggets of commonality and hints of a deeper truth buried beneath the many layers of dogma and obfuscation that hold the keys to the treasure. By this measure, all books that contain knowledge can be said to be sacred—including books on science, philosophy, business, and even fiction. There is nothing particularly sacred about religious texts simply because they are religious.

Limitations

Religion has done much to hamstring itself. It has no legitimate claim as the source of any morality or goodness. The extent to which it focuses on self-preservation and demands abject belief and blind credulity at the expense of knowledge and reason is the extent to which it deserves no credible role in civilization. It is utterly ridiculous to have clergy acting as consultants in matters of governance.

Just what a meme is and how it is distinguishable from beliefs, I find difficult.
Lewis Wolpert

Religion could be a source of knowledge and inspiration simply by returning to its roots and giving to God what is God's and leaving to Caesar (secular government) what is Caesar's, to paraphrase Christ. It could exalt the unknown divine. It could demand evidence for the divine of the same caliber as that demanded of any scientist. It could, as Sagan once said, look up into the sky and say, "Wow, this is bigger than any of us!" It could revere the fact that, as Einstein said, "The only incomprehensible thing about the universe is that it is comprehensible." It could offer evidence and proofs of its claims to science and welcome testing and falsification. Even better, it could assume its rightful obligation to work to prove its claims instead of making scientists do their work for it. It could give scientific evidence the same level of respect it demands for itself. It could be open to growth and change and enrichment to evolution. Religion could give us so much—if only it would get over itself even a little.

Universal Truths

> *I have come to this place more noble than any place.*
> Pyramid Texts

Tear away the layers of denominations, sects, cults, dogmas, creeds, rituals, ranks, trappings, texts, and everything that separates the different religions, and you would see an almost alarming degree of similarity among them, East and West alike. The astonishing similarities of worldwide creation myths attest to this, as we learned in Chapter 3. Christianity has no monopoly on the truth, and neither do Judaism, Islam, Hinduism, Buddhism, or anything else (although I personally believe that Buddhism comes closest to whatever truth may be out there). This common foundation stretches across history all around the world. Something is going on. There is a deeper truth. How else could so many groups of people with no technology or other means of disseminating messages come up with essentially the same story?

The Afterlife

> *We that acquaint ourselves with every zone and pass both tropics and behold the poles, when we come home, are to ourselves unknown, and unacquainted still with our own souls.*
> John Davies

All religions believe in some form of existence after this life is over. Western religions believe that a person carries on much as she or he is in life, that Anthony Hernandez will remain Anthony Hernandez once I die, and that you will remain whoever you are once you die. This sounds wonderful, except that we are all born deficient and guilty of Eve's original sin of eating the wrong apple at the wrong time. Thus, unless we repent and live by the religion's rules, all we have to look forward to is an eternity of misery.

Buddhism essentially says that personal identity and life-specific memories cease to exist at death, but that karma is passed along, as are knowledge of different emotions and states, such as love. This feels like cold comfort to those hoping that something of "us" persists after death. It also smacks of original sin, because the new baby being born will be saddled with the results of a cosmic lottery that it played no role in creating. The comfort one can take from this model is that the baby at least stands a chance of inheriting good karma and a correspondingly good life. This stands in stark opposition to the automatic eternal damnation awaiting JCI children barring conversion and devotion.

Still other models see each individual lifetime as a role being played by an actor. The role perishes, but the actor lives on and comes back time and time again to play new roles, until it finally reunites with the Godhead and becomes everyone and everything at once.

These may seem like irreconcilable differences, but here again, a closer look reveals far more similarities than differences. We are born with a sort of "original sin" that consists of separation from and forgetfulness of the Godhead who is both one and all things, and forget that we ourselves are at once fragments of God and yet completely God, just like pieces of a hologram that carry the entire picture. From the perspective of this lifetime, each of us faces a virtual eternity in the hell of endless death and rebirth or the heaven of enlightenment and reunion with the Godhead like a drop of rainwater rejoining the ocean from where it came. The entire universe, including ourselves, is little more than an illusion; it stands to reason that whoever we are in this lifetime will fade away when we die; however, we also have reason to suspect that we are much more than we think we are.

It does not require much in the way of interpreting metaphors to see that all religions agree both that death is not the end and on many of the aspects of what awaits us, even if their descriptions are mired in cultural language. This is where science comes in, providing the universal language of physics and mathematics as a sort of Rosetta Stone that bridges and reconciles all seeming differences.

Universal form, I see you without limit, infinite of eyes, arms, mouths, and bellies. See, and find no end, midst, or beginning.
Arjuna

Young Earth

Assuming that a second is a second, or a year is a year—in other words, that our perception of time accurately represents the flow of time—then so-called Young Earth Creationism has been utterly debunked, because the universe shows too many signs of being 13.7 billion years old for any claim that it is really only 10,000 years to be taken seriously; however, we learned in Chapters 24 and 28 that time is relative, that each of us experiences our own personal flow of time. We also know

How is it that so many philosophers and cognitive scientists can say so many things that, to me at least, are obviously false?
John Searle

> *If you're going to ham it up, go the whole hog.*
> G. I. Gurdjieff

that our units are based on our own local conditions, such as the Earth's orbit around the Sun (year) and the Moon's orbit around the Earth (month); a being on another planet might have a completely different idea of how quickly time flows, and this may also extend to species on our own planet. For example, an insect whose life span is measured in days may perceive its own life as lasting just as long as ours, or even longer. A tortoise that can live 150 or more years may not perceive of itself as particularly long lived. In short, the definition and perceived duration of a day (or any other unit of time) is purely subjective, even though it is based on stable natural phenomena.

We also know that time is a rather fungible thing, thanks to relativity and dilation. We even know that time itself is an illusion, as we saw in Chapter 28. It is entirely possible to start at Now A and take two paths that lead to Now B, where one path takes 10,000 years, while the other takes 13.7 billion years. If the biocentric theory is correct, then billions of years of history could have been created in a single instant, where life evolving at Now A created its own backstory (a form of myth, if you will) extending back to a Big Bang that is both absolutely real and absolutely fictitious.

The reasons Young Earth Creationists use to justify the idea of the universe being only 10,000 years old (spurious calculations based on the Bible, plus the belief that God gave the universe a faux patina of age) are utterly laughable; however, the fact remains that just because the universe appears to be 13.7 billion years old does not necessarily make that true in the literal sense of roughly 5 trillion 24-hour days having elapsed since the Big Bang.

All Have Sinned

> *Most institutions demand unqualified faith; but the institution of science makes skepticism a virtue.*
> Robert K. Merton

All religions speak of sin. The JCI religions speak of sin as originating when Eve ate the wrong apple from the wrong tree in direct violation of God's orders and gave humanity the knowledge of good and evil. This seems to be a direct attack against logic and reason in favor of blind faith and obedience—the only tools that have a prayer of making people believe some of the preposterous claims made by these religions (pun fully intended). Church writings and dogma have reinforced the idea that one must check one's brain at the door and swallow the literal or metaphorical Kool Aid without question, on pain of eternal damnation. I have said it before and will say it again: Forbidding the use of one of the organs God supposedly gave us is one of the gravest insults imaginable against that God. Combine that with rules requiring genital mutilation, and God's perfect creation seems to be fundamentally flawed as originally designed and built.

The Gnostics and Eastern religions also have a concept of sin, wherein people forget their divine nature and see themselves as temporary beings that are separate from their environments and each other. This is a clear parallel to the Christian idea that, "all have sinned and come short of the glory of God." This does not mean that we have displeased some otherworldly father figure who will then bend us over his divine knees and spank us forever; it means we have forgotten our own divine natures and forgotten that God resides within each one of us—that we are each God in both fact and deed, and in both separation and wholeness.

Any Christian can tell you that the wage of sin is death. What they probably do not know is that Buddhists and Hindus would agree with them. The price we pay for the sin of not realizing our God nature and true selves is the suffering of death, again and again, until we get it right and reach enlightenment and beyond. The former see this as punishment, while the latter see it as infinite patience and mercy. A key difference between seeing God as transcendent versus immanent is that the former is placed in a position of authority and enforcement, while the latter cannot punish or enforce, because all that exists is God. To punish us would be to punish God. In fact, we can see sin as good and essential in a way, because every cycle of birth and death is a new opportunity for God to experience all that is possible—which is not incompatible with the fundamental goal of life being to reunite with God.

When we look at it this way, we start to realize that the only real sin lies in not using every intellectual, logical, intuitive, emotional, and other means at our disposal to answer life's biggest and deepest questions for ourselves. The answers themselves are in a way less important than the questions, in the same way that winning a marathon is by itself less important than being fit enough to run it in the first place.

Nothing that is contrary to, and inconsistent with, the clear and self-evident dictates of reason, has a right to be urged or assented to as a matter of faith, wherein reason hath nothing to do.
Thomas Locke

We can easily forgive a child who is afraid of the dark; the real tragedy is when men are afraid of the light.
Plato

None of this is intended to excuse the ongoing misery and suffering that religion continues to inflict on the world, nor does it change the fact that religion poses the single gravest threat to the ongoing survival of the Earth. Religion began with the whisper of possibility to help us understand the universe and our place it, and may well go out with a bang.

In Conclusion

The concepts which now prove to be fundamental to our understanding of nature seem to my mind to be structures of pure thought. The universe begins to look more like a great thought than a great machine.
James Jeans

Strip away all of the crime, dogma, ritual, stigma, and stultification of religion to arrive at its deepest core, and we find the simple message first expressed by mystics and validated by an ever-growing body of scientific knowledge that continues to reveal a universe far grander than anything previously imagined: that there is more to us than our bodies, that our true natures are far more profound than most of us will even comprehend during this lifetime, and that each of us can touch the divine.

This is a very encouraging start. We need to keep digging.

868 | *The Divine Savage*
Revealing the Miracle of Being

Chapter 34

Evidence from Evolution

If God created the universe as a special place for humanity, he seems to have wasted an awfully large amount of space where humanity will never make an appearance.
Victor Stenger

I said at the beginning of this book that I am a firm believer in evolution. There simply is no rational alternative, especially when the evidence is all around us. I also said I believe in materialism, and stand behind that statement as well. One day in 2003, I fell ill with both vomiting and diarrhea—a set of symptoms that kills many people each year. Today, most of these deaths are confined to impoverished nations that lack even basic medical care, but that was not the case before modern medicine came along. At the hospital, I received a dose of antacid laced with a common anesthetic to calm my stomach, and intravenous fluids to rehydrate my body. This simple treatment transformed a potentially life-threatening situation into a minor inconvenience, and I left the hospital feeling like a new man within just a few hours. My case, and many millions more each year, prove that materialism works.

We saw in Chapter 23 that classical Newtonian physics works, only to discover in Chapters 24 and 25 that it breaks down and needs replacing at extreme scales. These replacements (quantum mechanics and relativity) supersede classical physics, but do not invalidate it; on the contrary, Newton's equations remain buried within these newer sciences. Can we say the same for materialism? Is it an approximation that works brilliantly under ordinary circumstances, only to break down and

A theory for the origin of life remains under heavy construction.
Fred Adams

need replacing by a larger, all-encompassing worldview under extreme conditions?

We know how helium, a colorless, odorless, and almost completely inert gas transmuted into every known element over billions of years. We also know how single-celled life evolved into multicellular life that eventually begat animals capable of writing messages for others to read later. What we don't know is how the latter evolved from the former, how nonlife came alive. Even solving this vexing problem would not explain where consciousness and self-awareness come from. We know that matter is not inherently conscious, and can imagine an identical universe completely devoid of consciousness in which my fingers tapping my keyboard are doing so robotically with no intention behind them, just as my laser printer blindly spits out pages for me to review when asked.

Edit and interpret the conclusions of modern science as tenderly as we like, it is still quite impossible for us to regard man as the child of God for whom the earth was created as a temporary habitation.
Carl Becker

Darwin demonstrated how evolution works through mindless natural selection that picks and chooses the results of random genetic mutations like an artist selecting from a random color palette. This process is ruthlessly efficient and quick to eliminate both individuals and entire species that prove themselves unable to cope with current or changing conditions. The vast majority of the species that have evolved on Earth over the past 3.5 billion years have gone extinct and been replaced by new species at a rate that tends to cancel extinctions. Notable exceptions include the Cambrian explosion of new species and periodic die-offs, such as the one that sealed the dinosaurs' fate and allowed mammals to emerge.

This much we know, but is there more to the story? In Chapter 33, we saw that religion says yes. Many see religion and evolution as opponents, but neither by itself has anything to say about the validity of the other.

The Blind Watchmaker

The single defining characteristic of life versus non-life is that the former actively resists entropy, while the latter passively accepts whatever comes. Either way, there is no free lunch; nonliving things simply decay, while living things inflict entropy on other living things through eating for self-preservation. On a purely mechanical level, we understand that

So it is that, instead of living, we hope to live.
Blaise Pascal

Let me not pray to be sheltered from dangers but to be fearless in facing them.
Rabindranath Tagore

everything we experience consists of tiny particles suspended in a rich web of various forces and fields that somehow translate into individual experiences unique to each experiencer. The process of sensing and analyzing one's surroundings is understandable; what is not understandable is the subjective experience that arises from these activities.

The same natural selection that could not possibly care any less about us seems to be responsible for everything from our physical shape to our beliefs, morals, likes, dislikes, and more. Virtually all thoughts and feelings are visible as brain activity during scanning (such as an MRI), and neurologists have identified many brain "circuits" involved with everything from facial recognition to religion. Many animals show traces of their ancestral species, such as marine mammals that retain vestigial legs that may or may not be used as fins or flippers.

Life took hold on Earth as soon as it became habitable, and living organisms have been found everywhere from the upper reaches of the atmosphere to miles underground. Life seems to be ubiquitous, a predictable outcome once conditions are even remotely tolerable. Does this mean that life is deterministic, bound to arise and evolve along Darwinian lines? Or has life on Earth taken one of many possible roads? Did we play any role in creating ourselves? Did the first living thing collapse the waveform of the entire universe back to the Big Bang (and possibly beforehand)? Biocentrism is perfectly compatible with the idea of a blind watchmaker; for example, we should expect to see less-than-optimal designs, because nothing can create something more complex than itself. Even the most complex machines built by humans are laughably primitive compared to even the simplest living thing.

We have seen no evidence against biocentrism thus far in our explorations, nor did I find any such evidence when researching this book. That alone does not mean that biocentrism is the way of the universe, but it is intriguing nonetheless.

If God created the world, where was He before creation? Know that the world is uncreated, as time itself is, without beginning or end.
Mahapurana

Mind and Brain

All of this talk of idealism, biocentrism, and any form of afterlife sounds wonderful until we remember that things like anesthetic, trauma, sleep, or coma can all alter and/or seemingly extinguish consciousness. On a related note, where does each personality in a person with multiple personality disorder go when another personality takes over? This begs a crucial question: Is my mind inside my head and your mind inside your head? If so, does that have any implications for any sort of life after death? If not, then how can physical events affect a nonphysical mind? Basic anatomy seems to argue for the former. Most brains occupy their own appendages, and are completely encased in bone with fluid cushions.

Electromagnetism?

In mid-2011, I purchased a new computer with the latest and greatest processor then available. This chip is many thousands of times more powerful than the 16MHz processor used in the Macintosh LCII computer, which means it has the combined brain power of many thousands of ants, or of a single more advanced animal. Simple extrapolation posits that all living things have some form of subjective experience of themselves as "I," and we have plenty of evidence to conclude that they enjoy their lives just as much as we do in their own unique ways. This means that ants have some form of consciousness and some form of subjective experience, as do countless other species. My computer runs rings around many species in the brain power department, but nobody I know would seriously suggest that it has any sort of consciousness or self-awareness whatsoever.

Most people see electromagnetism as the only viable source of consciousness; however, there is no necessary causal connection between electromagnetic information processing and consciousness, because we can presume that animals with less brain power than my computer do experience some form of consciousness. If mere electromagnetism can cause a non-designed animal to become conscious to even a small degree, then a very carefully designed computer should have no problem achieving that same state—or at least of giving a convincing impression thereof—but no such phenomenon is forthcoming.

We know that brain scans highlight brain activity during different states of consciousness, such as emotions or performing various tasks such as movement or facial recognition. We

Scientists are in the strange position of being confronted daily by the indisputable fact of their own consciousness, yet with no way of explaining it.
Christian de Duve

Consciousness poses the most baffling problems in the science of the mind. There is nothing that we know more intimately than conscious experience, but there is nothing that is harder to explain.
David J. Chalmers

can therefore say with certainty that different location in the brain correspond to these functions and map those locations with a good deal of precision; however, they cannot account for the experience and awareness of those activities.

We like to think that consciousness vanishes during times like sleep or coma, but how can we know for sure? We saw in Chapter 24 that Einstein proved that time is a purely subjective thing. Chapter 28 went one step further and revealed time as an illusion caused by moving from Now to Now. For example, nothing prevents Person 1 from going to sleep at Now A and moving directly to Now Z (solid line in the diagram above), while Person 2 experiences a long series of Nows between A and Z (dashed line) and concludes that the sleeping person has been unconscious the entire time. All that has to happen is for the paths of Persons 1 and 2 to diverge at Now A and reunite at Now Z, provided that nothing inconsistent happens along the way. The brain waves that sleeping people emit are perfectly consistent with this scenario, because we would expect something to always be happening in a living body. The same concept applies to coma, anesthetic, and other similar situations.

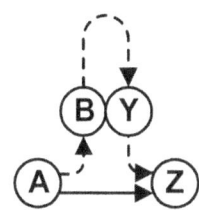

What possible evolutionary use could there be for a pathetic eternal life wish for an animal doomed to live for less than a century?
Fred Heeren

The existence of consciousness does not seem to be derivable from physical laws.
David J. Chalmers

The Elusive Mind

If mind equals brain, then the worm with only 200 neurons defies explanation, because its behavior is far too complex for that tiny brain to handle on its own. Furthermore, that brain—like all brains—obeys quantum laws, which only deepens the mystery, because brains are inherently coherent, while quantum mechanics is anything but. Lashey's and Pietch's experimental scrambling of rat and salamander brains failed to destroy memories or core behaviors, such as eating. Libet and others have demonstrated that humans brains act up to 1/2 second before we take conscious action, and that one can predict which hand a subject will raise up to 10 seconds in advance.

The common explanation is that our brains are acting on autopilot while giving us the illusion of being in the driver's seat taking immediate actions. I offer a different theory: Assuming that the holographic theory we saw in Chapter 27 is correct, then all the Nows that can ever exist do exist. What we experience as motion through time is simply seeing knitting individual Nows into a smooth flow just as we do at the movies. Assuming that each Now contains one Planck time interval (the smallest possible duration), we can presume that the delay is caused by the brain choosing which Now to jump to from among all the available choices. We can then presume that the brain can process an absolute maximum of one Now per Planck time, because each Now contains the entire universe within it. There are approximately 1.85×10^{43} Planck time intervals in a second; half of that is approximately 9.28×10^{42}. This latter figure could represent the upper bound of free will, within which the available Nows available at any given Now are predetermined, while the actual Now one selects is probabilistic. The universe could be both deterministic and probabilistic with no paradox whatsoever.

For example, think of a buffet where everyone gets a plate immediately upon entering, which represents a Now. From there, one can choose to eat anything one wants from the available choices. The menu is predetermined, while the individual decisions about what to eat are entirely probabilistic. Each diner weaves her or his individual path through the available dishes without fear of paradox, because the available Nows are so constructed as to ensure mutual compatibility. We could go one step further and postulate that the buffet is a potluck where everyone contributes a dish. In this example, everyone collaborates to define what is possible and contributes a piece of the puzzle. An outsider looking in would see one party, while the participants would see themselves as individuals within a larger context. Next time you visit a buffet or attend a dinner party, take a moment to consider that reality may well be a buffet potluck writ large.

We can see the entire party as a Mind that is at once transcendent (because the combined dishes form more than the mere sum of their parts) and immanent within the minds of the individual participants, where the latter are expressions of the former that has both existence (in the form of the party) and non-existence. This example is limited, because the party itself has a beginning and an end, but the basic idea is correct. In the case of the universe, the Mind (God) broke itself into pieces (individual plants, animals, particles, etc.) and is similarly transcendent and immanent at once. Mind created all that

If consciousness is non-local then it is infinite in space and time. If something is infinite in space, it is omnipresent. If it's infinite in time, it's eternal or immortal. So you can see that from the get-go there's theological dynamite that's hooked up with this idea of the non-local mind.
Larry Dossey

Somehow, the dust spewed into space by the nuclear furnace of a bygone supernova has become a human brain that learned to make nuclear reactors here on Earth.
Gerald L. Schroeder

Chapter 34
Evidence from Evolution

is possible, and the resulting minds choose from among those possibilities.

Is your mind inside your head? Yes, because altering a brain seems to alter the corresponding mind. No, because your mind created your brain as a convenient way to navigate the maze of possibilities before it.

Idealism in our Heads

> *Real simplicity, so far from being foolish, is almost sublime.*
> Fénelon

Consciousness seems to be rooted in biology, because each of us associates ourselves with a single body that grows, matures, ages, and eventually dies. We are prisoners in our bodies, peering out at the universe through a series of sensors and filters whose only purpose seems to be to restrict both what we can sense and what we can make of those sensations. Our senses are almost painfully limited; we can only see light across a small portion of the spectrum, and our other senses are unable to detect the masses of information literally being carried on the wind in the form of pheromones, pollen, etc.

We are also unable to detect the countless thousands of radio, television, telephone, Internet, and other information that is literally passing through our bodies at every moment of every day, until we switch on the appropriate device to translate that information into forms we can detect. The device mediates and translates the signal but is not the signal itself, nor is the information we receive inside the device. Television broadcasts 24 hours a day, whether our TV is on or not; the only role we play is choosing whether or not to tap into those signals and, if so, which ones. Tuning different TVs to the same signal displays the same show on all sets. The next time you are visiting your local electronics emporium, take a moment to look at the wall of TVs all showing the same thing, and ponder the fact that you are looking at an excellent example of nonlocality. No TV produces the signal; similarly, our brains may not produce consciousness.

> *Liberation cannot be achieved except by the perception of the identity of the individual spirit with the universal spirit.*
> Shankara

The body's electromagnetic activity weakens and slows down during death, until it eventually ceases altogether. These extremely low levels of activity must be described with a quantum waveform instead of the classic descriptions normally used. The diminishing and eventual cessation of electromagnetic activity in the body or any activity associated with consciousness does not eliminate the waveform. During death, the body's classic activities cease, but the quantum potential remains undiminished. These waves can interact with whatever activity is taking place in the brain, which can account for NDEs. We experience ourselves as classical beings during ordinary life with all of the limitations of body, time, and space that we are accustomed to; however, we can see how this is but a small aspect of a quantum level of consciousness that is nonlocal and has neither beginning nor end.

Short- and long-term memories may form and be retained by altering the brains' waveforms, which explains why Lashey and Pietch were unable to destroy memories and instincts by destroying brains in every way they could imagine. This does not conflict with the idea of specialized brain circuits that perform narrowly defined functions, which explains how Libet was able to trigger specific memories by stimulating specific parts of his subjects' brains. The brain loses thousands of neurons and their synapses every day; however, each such neuron and synapse has already left its own indelible mark on the brain's overall waveform. The ETEAR process shapes the waveform over time, which explains how habits and patterns form, and why they are so hard to shake.

The only existence we can prove is that each of us lives in our own little world created for us by our own minds. We see through the windows of our own making into a universe of our own making. The same thing occurs when we dream; the only difference is that we acknowledge dreams as existing inside our own heads, while insisting that the waking world really does exist somewhere "out there." This seems silly; dreams are often more liberating than waking life, because the laws of physics don't necessarily apply. I have had many dreams of flying while either hanging onto or sitting on a model airplane wing, only to land and be unable to take off again, because I've realized that the wing is way too small to lift me. In this example, the laws of physics themselves are constructs of a sort—just please don't succumb to any urge you may feel to test this idea in any potentially harmful situation. These minor variances aside, our dreams all conform to the laws of physics we are familiar with, in that each dream universe is hospitable to our own existence; we can postulate—but cannot create—a universe in which we cannot exist.

The universe and the observer exist as a pair. I cannot imagine a consistent theory of the universe that ignores consciousness. I do not know any sense in which I could claim that the universe is here in the absence of observers.
Andrei Linde

Research into consciousness is in its infancy, so what actually conjures the sentient presence is anyone's guess.
Robert Hercz

> *The one and only thing of paramount importance to us is that we feel and think and perceive.*
> Erwin Shrödinger

Our bodies play the same roles in both dreams and waking life; however, we acknowledge the former as phantasms created by our imagination, while clinging to the notion that the latter are somehow more real. How can we possibly tell? Our dream bodies seem just as real to us when we are dreaming as our waking bodies seem to us while we are awake. In fact, the only difference we can demonstrate between dreams and waking is that the latter seem to fall into a continuously ongoing pattern, while the former are individual random episodes with no necessary connection between them.

Our minds, and the minds of the characters in our dreams, do a completely convincing job of seeming to occupy brains, but we know that is not the case. It is very telling that science with all of its tools has not been able to locate the seat of consciousness or even generate a coherent theory of what consciousness is, despite probing the brain down to the molecular level and beyond. Our dreams and every person, animal, plant, and subatomic particle therein depend on consciousness for their very existence. It is impossible to dismiss the possibility that the same is true of everything and everyone in our waking lives—especially when neuroscience tells us that the brain has no built-in switch to tell dreaming from waking.

This smacks of solipsism, the philosophy that only the self exists or can be proved to exist. It is very easy to dismiss solipsism with a laugh. We see and interact with a world full of people and other things every day; however, none of the characters in our dreams—including our own dream selves—have any real independent existence. All depend on the unseen mind of the dreamer. The same can be said of the waking world, with the sole difference that we see ourselves as locked into one fixed role per person for our entire lives, and thus may not be the ultimate dreamer; we can directly postulate the existence of a single Mind that is greater than ourselves and the universe, and can thus see ourselves as figments of that mind. Solipsism is therefore both viable and possibly correct, albeit on a much grander scale than originally imagined.

> *Creation of new information is habitually associated with conscious activity.*
> Henry Quastler

Our brains are built to offer glimpses of this larger reality and the Godhead/Ground. Mystics retain crystal-clear memories of their experiences and their reality to a far greater extent than any dream, hallucination, or delusion. Those of us who have experimented with drugs may be able to recall our "trips." Most everyone remembers some dreams, but nobody considers those experiences real. Mystics see their experiences as being every bit as real as any normal experience is to any of us. In short, mystical experiences are on a completely different level than more mundane forms of altered consciousness states.

Materialist models of the brain have lots going for them, not least of which is the increasing ability to repair damage and injury to the brain, in the same way that a repairman might fix a television set without affecting the signal itself one bit. Through materialism, we have learned about different forms of awareness, and learned about brain processes and neuronal circuits that handle all manner of functions; however, the fact remains that consciousness does not supervene on any of this.

Models built on the brain's biology have much going for them, such as cataloging different types of awareness and discovering brain processes that are associated with consciousness; however, none of them can explain consciousness. That which we perceive as reality is simply the tail end of a process involving consciousness. Philosopher Ludwig Wittgenstein (1889-1951) saw no place in objective language for a subjective "I," because otherwise we would be able to measure consciousness and place it inside the brain.

The subatomic particles in our dreams can be explained on a transcendental level by their nuclear structure, which can in turn be explained by their atomic structure, and so on through molecules, cells, tissues, organs, organisms, actions, awareness, and ultimately the consciousness of the dreamer which is—as far as the dream world is concerned—where the buck finally stops. The dreamer's consciousness is utterly transcendent, because it created that world, utterly immanent because it manifests in every aspect of that world, and utterly irreducible because it is both self-explanatory and required for the dream world to exist. The same example may apply to the waking universe: According to an idealist, consciousness in the form of the Godhead/Ground is self-explanatory and irreducible.

To an idealist, evolution is the process of becoming more and more aware of the true nature of the universe. We can see this progression in the relentless march toward increased complexity and mental/emotional capability found across all species,

While the traditional model of psychiatry and psychoanalysis is strictly personalistic and biographical, modern consciousness research has added new levels, realms, and dimensions and shows the human psyche as being essentially commensurate with the whole universe and all of existence
Stanislav Grof

and can see hints of a reality far more profound than we are accustomed to. We can see evolution itself as arguing for idealism, despite all mundane material appearances to the contrary.

Waste Not

Adaptation is the gold standard against which rationality must be judged along with all other forms of thought.
David Sloan Wilson

We know that only about 5% of DNA codes for proteins. The rest was seen as junk and trotted out as evidence against any explanation of evolution by other-than-random means. We also saw that this position is being revisited in light of a growing body of evidence that this so-called "junk" DNA is actually crucial for life. It is just one more example of the maxim that no trait exists in any species without a beneficial purpose or use; everything is there for a good reason. We may not be able to see or understand that reason, but our limitations are no reason to deny that assertion. Consciousness exists to carry out a function—or functions—that cannot be carried out any other way. Furthermore, our genetic predisposition to spirituality and religion must also have some use. It is very easy to dismiss this as a need for hierarchy gone awry, except that humans have plenty of hierarchical structures and leaders that need no "supernatural" traits to work just fine.

No one organism can possibly yield to its owner the whole body of truth.
William James

Spirituality (and by extension religion) exist because they are beneficial to our species. Only since the advent of agriculture have societies been able to support a caste of full-time priest, who have time and time again been shown to be acting in their own self-interests above and beyond those of their fellow human beings. I submit that spirituality is just one more area where agriculture has upset the natural order of things, which does not for a moment discredit spirituality. Even the priests' self-indulgence is perfectly explainable; faced with a free ride and the power to tell people what to think and do without all the bother of becoming a king, who would not jump at the chance? Peel away the influence of agriculture, and we are left with beneficial spirituality that seems to have no practical use for daily life. After all, many of us can get by just fine without it.

Quantum DNA

There are some who believe that DNA is not a carrier for heredity, but that it acts as a nonlocal mediator for both heredity and morphogenesis. We know through experimentation that cells communicate nonlocally, making nonlocal DNA information storage, retrieval, and exchange plausible, if not likely. This would also explain why some donor organ recipients report feelings, ideas, and/or knowledge that later research reveals as matching the dead donor. It could also explain the intricate symbiotic relationships that exist between many species, such as termites and fungus, bees and flowers, and many others.

Schrödinger opined that DNA might be a sort of quantum antenna for nonlocal communications, which means that it could mediate between an organism and nonlocal space. Popp and scholar Marco Bischof (1947-) believe that the DNA molecule may create fields of coherent *biophotons* (photons created by biological sources). Neurobiologist Herms Romijn believed that electromagnetic fields, possibly based on "virtual" photons, could either carry or be produced by consciousness. This is in keeping with what we discussed in Chapter 29, where all interpretations of quantum mechanics involve consciousness in some way.

Remove a live ant queen from her colony, and life goes on as normal. Kill the removed queen, and the colony erupts into chaos as productive work shuts down. This is a clear example of nonlocality, in which the queen may be creating and/or maintaining a collective unconscious among her subjects. The queen's DNA, and that of the workers and soldiers, may play a role in coordinating all of this. It could also explain how termites manage to build intricate mounds up to seven feet or even more in height that create air currents to maintain a constant interior temperature of 87 degrees, while the outside air temperature may fluctuate between 35 and over 100 degrees.

DNA preserves heredity in the form of physical possibility and behavioral predispositions that go a long way to distinguishing individuals. It may also help coordinate cells, organs, and systems across the organism, and even beyond. The DNA in each cell includes an interface function that may help exchange heredity information between local and nonlocal

---

I accept the universe.
Margaret Fuller

Deep in man, at the core of his being, there exists the need for experiences of truth. Around this need everything else in him is arranged like planets around the Sun.
Jacob Needleman

space and consciousness, which could help explain how our overall body patterns remain identical throughout our entire lives. This is only a hypothesis, but it fits perfectly with everything we learned about evolution and physics throughout this book. If this hypothesis is correct, then the most valid conclusion we can make is that life is a result of the expression of conscious will.

What Evolution Tells Us

The true delight is in the finding out, rather than in the knowing.
Isaac Asimov

At face value, birth and death are biological processes like any other. It is certainly possible that mind evolved from—and emerges from—matter in a "functional" dualist sense that requires no separate spirit world. This scenario is functionally identical to pure materialism, in that we did not exist before this lifetime and will have no existence after death. Still, we have seen some compelling evidence that death may not be the end, that it may be only an illusion. If death is an illusion, then what exactly survives death?

In Chapter 26, we saw that the universe began as almost perfectly homogenized energy that became more discrete, varied, and self-aware over time. The universe is quite literally becoming aware of itself, and is looking back on itself through all conscious species on Earth (and quite probably elsewhere). This fact alone can be seen as implying purpose, as if the universe is actively seeking to explore, learn, and grow.

No microscopic property is a property until it is an observed property.
John Archibald Wheeler

Assuming that idealism is correct, why should the Godhead/Ground create evolution and the illusion of separate and distinct individuals that can be subject, object, or both? The simplest answer is that daydreaming and actually doing something are two completely separate propositions. You can sit back and daydream about skydiving to your heart's content, and I think you'll agree that is nowhere close to actually standing in the doorway of a perfectly good airplane preparing to throw yourself out into thin air and trust your life to a backpack full of fabric. You will probably also agree that it helps to see an airplane, the ground below, and yourself as completely separate objects in order to complete the experience. Separating your consciousness from the possibilities (exhilarating ride or death plunge) and facing that actual choice is the only way to actually have an experience; anything less just doesn't cut it. By breaking itself into pieces with little to no awareness of their true oneness, the Godhead can act out all that is possible and bring the Nows to life.

This is all speculation, of course. Nothing can alter the fact that the best available mainstream evidence seems pretty damning in that humans may well be nothing more than cosmic flotsam adrift in the cosmos, with each of us destined to live a few short decades before winking out like a blown light bulb and returning to the oblivion from which we emerged. The Second Law of Thermodynamics is both clear and inescapable: Each of us is destined to lose the battle between entropy and creation, because there are so many more ways to be disordered than ordered, and we can only duck the odds for so long. The fact that anything from drugs to trauma can impede or destroy a person's ability to think, feel, or even "be themselves" may tell us all we need to know about the material nature of our minds.

Religion tells us that there may be something more to life than the mundane existence we perceive on a daily basis. Evolution shows us either that religion may be on to something, or that religion is completely false. We must keep looking.

It would be strange if a single ear of corn grew in a large plain or if there were only one world in the infinite.
Metrodorus of Chios

Chapter 35

God and Intelligent Design

> *If God did not exist he would have to be invented.*
> Voltaire

If religion hints at the presence of a god and a life that encompasses far more than our brief sojourn on Earth, then it is appropriate to return to the topic of God and whether ID theory has any validity to it, keeping in mind Lisa Baker's poignantly expressed desire to know God and reality from Chapter 11. Primitive people used anthropomorphism to describe their world, and built myths to answer life's big questions. Did the concept of God arise from these early attempts to understand the universe? Or is there something deeper happening, as religion suggests?

Do We Need God?

> *All men need the gods.*
> Homer

Hawking devised a theory that explains the origins of our universe as material tunneling through a barrier from another universe. At face value, this satisfies the definition of creation *ex nihilo* without the need to resort to a god; however, it does not solve the problem of infinite regression, because we must explain the existence of a potentially infinite number of universes across an infinite span of time. Hawking's theory is turtles all the way down. He also postulated the zero bound idea, where the Big Bang happens, only it doesn't really happen, and the universe essentially pops into existence. Then, in *The Grand Design*, Hawking offered a theory whereby the universe could have come into existence thanks to quantum fluctuations. This also sounds good, except that we must then account for the quantum fluctuations. The changing theories also give the impression that Hawking may be grasping at straws, trying to eliminate consciousness from the discussion.

Dawkins has said that something capable of creating something as complex as the universe would have to be even more complex and would require a yet more complex creator, and so on—a case of birds all the way up, if you will. Dawkins then expounds at great length about how simple starting conditions can create complexity through a combination of random and non-random (but not directed) events, such as random mutation and non-random natural selection. Thus, God does not need to be complex; on the contrary God can be almost absurdly simple. Dawkins answers his own question without realizing it.

The cosmological argument for God says that every effect must have a cause; at a certain point, we run out of natural explanations for the universe, and must therefore turn to God. In short, one cannot get something for nothing. The First Law of Thermodynamics agrees, saying that energy can be converted to matter and vice-versa, but that nothing can simply wink into or out of existence. Physicists will point out that the sum total of energy in the universe is 0, with all matter and positive energy coming on loan from gravity, which balances the books and provides a literal example of nothing coming from nothing. The zero-energy problem forces us to conclude either that the universe simply popped into existence out of nothing at all, or that idealists are correct when they say that matter and energy are all illusions produced by consciousness.

Which choice is the right one, and how can we know it is right? The correct answer must square with everything we know across all branches of science. For example, an answer that satisfies physics while contradicting biology is probably not the right answer. This is why I have endeavored to include as many sciences as possible in this book—to have enough information from enough sources to be able to arrive at a meaningful decision.

So far things seem to be looking good for God. It is true that scientific explanations increasingly do not require God, and

The yearning for forever is an essential part of being human.
Irwin Kula

There is no reason for believing that any sort of gods exist and quite good reason for believing that they do not exist and never have. it has all been a gigantic waste of time and a waste of life. It would be a joke of cosmic proportions if it weren't so tragic.
Richard Dawkins

> *Today people need proof in order to believe, and they deserve that proof.*
> Evan Harris Walker

Stenger is correct when he says that any observations that demonstrate a non-zero sum total energy in the universe are what we should expect if God is real. The only catch is that the god in question happens to be the JCI God, which is not the only type of deity we can imagine or that fits what we know of the universe. As an aside, physics does state that the lowest possible energy of any matter or field need not be zero; however, gravity (negative energy) should account for any such imbalances.

Newton's clockwork universe with its single objective and absolute reality left no room for minds, souls, or God, despite his personal belief in God as a prime mover or first cause for the universe. Later science revealed an astonishing amount of apparent fine-tuning in the universe that moved Hawking to ask, "What breathes fire into the equations and makes a universe for them to describe?" We just saw that his answer turns out to be, "Nothing." This stands to reason, for any assertion that God needs no creator can imply that the universe does not need God to create it. Again, this is perfectly consistent with the entire universe being a no-thing with zero sum energy.

If God exists, did s/he have any say in how the universe was created, or is this the only universe that could exist? The answer may well be both yes and no. Superstring theory predicts a maximum of 10^{500} universes, not all of which will be hospitable to life as we know it, or even any form of life. Within each universe lies an extremely diverse collection of possibilities encapsulated in Nows. All that remains is to navigate those Nows to create a history in the form of an illusory flow of time. We can take this literally to mean separate universes, or figuratively to mean a tremendously abundant and divergent set of possibilities in a single universe where the laws oh physics are how they are.

Unintelligent Design

> *There are no coincidences. They are miracles for which God doesn't want to take credit.*
> Unknown

In Chapter 19, we discussed the tendency of ID adherents to use the eye as an example of something irreducibly complex, that either works perfectly or not at all. We also debunked that idea by showing that eyes exist in all possible configurations, from patches of light-sensitive cells to optical marvels that put the finest cameras to shame, and how eyes evolved on several independent paths. We have also seen how jawbones eventually became the bones that help us process sound and balance information. The list of gradual adaptations based on existing material could—and does—fill a great many books.

Just because ID is false on its face, because it attempts to find room for God and religious dogma instead of studying the problem at hand, does not say anything about whether God exists or not. That is how poor a job ID does. If an object includes features that a competent engineer would include for a clear purpose, then we can presume that the object is designed. Every single trait of every single species exists for a purpose that benefits that species, whether we can identify that purpose or not. Thus, all life on Earth is indeed designed. The only question is whether that designer is intelligent or merely the deaf, dumb, and blind forces of natural selection. Paley's watchmaker is very real; but is it/she/he divine?

Descartes thought it impossible to see design or deduce first principles by studying nature, but thought we could find evidence for God by studying human consciousness. In this, he was absolutely correct, as we saw in Chapters 33 and 34. The JCI God has been totally debunked. Any theory that attempts to preserve the JCI God as a viable answer to the biggest questions in the universe is doomed to fail. Having said that, biocentrism remains a viable possibility that is perfectly compatible with a God designing as s/he goes. It stands to reason that a god that broke itself apart into objects with the illusion of separation would start simple and see where the path leads. This would account for both the amazing fine-tuning of the universe and the many real and perceived "design flaws" we see in nature. God could be both intelligent and blind, just as we are when we create dream worlds in our sleep.

> *A man who says, 'if God is dead, nothing matters,' is a spoilt child who has never looked at his fellow man with compassion.*
> Kai Nielsen

What God Is

Western religions have made the supreme mistake of forgetting the lessons of the mystics and adopting a view of a God that is absolute in power, knowledge, presence, and transcendence. This supreme lawmaker makes the rules, and anyone who flouts them does so on pain of eternal hellfire and damnation. This tyrant, like so many of His terrestrial counter-

> *The wise man is the one who understands that the essence of Brahman and of Atman is pure consciousness, and who realizes their true identity.*
> Shankara

> *Oh god! If I worship thee in fear of hell, burn me in hell; and if I worship thee in hope of paradise, exclude me from paradise; but if I worship thee for thine own sake, withhold not thine everlasting beauty!*
> Rabiah

parts, requires constant genuflection and praise in the form of elaborate rituals and sacrifices. Such a concept of God is clearly modeled after human royal courts, where even something as simple as getting dressed in the morning can involve a long and tedious ritual that more than one monarch has found tiresome in the extreme. One can only suppose what God might think of the many synagogues, temples, churches, mosques, etc. dotting the landscape and elaborate costumes, songs, dietary restrictions, genital mutilations, and other mind-boggling activities that take place in His name. Even my young son found the whole thing nonsensical when some friends of his invited him to church with them. The God that religion has packaged and made comprehensible is a God that is very much created in our image, not the other way around.

Monotheism began as an attempt to consolidate rule under a single king, helping ensure the people's loyalty to that god and the divinely chosen monarch. Again, a rigid and strong-armed God justifies rigid and strong-armed rule. Despite these inauspicious beginnings and its decidedly inauspicious history ever since, monotheism nevertheless indicates that the core of reality may be a single consciousness that is both everything and no-thing at once. This is a classic case of being right for all the wrong reasons.

Zen Buddhists will say that Zen is not a state of mind, but pure consciousness that exists first before everything else, and yet is none of those things. The question, "When there is neither I nor you, who is the one that wants to see it?" is not saying there is no one; it refers both to reality trying to learn about itself and to the consciousness that is, and is within in everything—a definition completely at odds with any concept of death as the end of existence. Consciousness exists when thought is absent, like the image on a TV screen that is not the set itself. The TV simply translates the incoming data into a form we can perceive, just as the brain filters raw sensory data. This is perfectly consistent with the idea of a fractionated God seeking to reunite with its eternal Self.

Constrained Omnipotence

> *I knew a mathematician who said, 'I don't know as much as God, but I know as much as God did at my age.'*
> Milton Shulman

We return to the problem of God's omnipotence. An all-powerful god has no constraints, and yet a god must be constrained in order for free will to exist. Furthermore, we know that everything we perceive is a product of our own internal ETEAR cycles, which begs the question of how we can have shared experiences. How can we resolve these seemingly intractable problems?

First, we must recognize that the Godhead/Ground's power is limited, either by its own choice or by its very nature. If there is indeed a Godhead that created the universe, then the Godhead is indeed limited; the self-sacrificial act of creation that fractionated the Godhead and resulted in seemingly discrete and separate objects, in turn results in an immanent Godhead, whose ability to act within its own creation is constrained. This is similar to passengers on a ship, whose freedoms are constrained by the ship's physical construction, layout, amenities, and itinerary. In a sense, creating the universe destroyed God. This is how materialists and other atheists can look at their data to proclaim that the universe was created from nothing, and that there is no need for God; however, just like the cosmic background radiation that helps explain the origin and fate of the universe, mystics and scientists alike are finding evidence that there is more to existence than materialism. Reductionism works, because the universe itself is reductive, like a colored slide that subtracts from the white light passing through it to create a coherent image.

What is the difference between gods and humans? That many waves before each from an eternal stream; the waves lift us up; the waves overcome us, and we are swept away.
Goethe

We could say that the Godhead is unlimited until it fractionates into an immanent creation, yielding both a designed and undesigned universe and the evolution taking place within it.

Free Will

We discussed the problem of free will in the previous chapter. The JCI religions speak of God knowing everything that will ever happen before it ever happens, while at the same time giving us the freedom to choose. This paradox has formed the topic of countless books, letters, articles, and sermons, as scholar after religious scholar attempts to grapple with a problem that is no mere academic exercise. Remember that the JCI religions see life on Earth as a one-way ticket to either eternal hellfire or heavenly bliss, depending on our deeds and thoughts. If God knows all that I will do before He creates me, then I have no room to choose what kind of life to live, and am thus rewarded or punished for a life that I am not truly responsible for living.

They are on the way to the truth who apprehend God by means of the divine, light by the light.
Philo

In the buffet model we explored above, God can be both omniscient—in the sense that someone putting on a buffet knows every possible dish being offered—and can calculate every possible combination of dishes a diner might select. Free will also exists in the sense that each diner makes her or his own choices from among the available items. Similarly, each of us is free to choose from the possibilities the Godhead sets before us. Thus, God knows everything we can do but not necessarily everything we will do, although the Schrödinger waveform makes it possible to see which choices are more likely than others. God can guess and suspect, but no more.

Suffering and Evil

> *Most men worship the gods because they want success in their worldly undertakings. This kind of material success can be gained very quickly by such worship here on Earth.*
> Bhagavad Gita

If we postulate a supremely loving God, then we must grapple with the existence of suffering and evil. Here again, the JCI religions are rife with endless tracts attempting to resolve this problem. Here again, the solution is obvious: Evil and suffering are only a problem if we believe that each person only lives once. If we allow for reincarnation or some other mechanism by which an individual consciousness or monad lives beyond a single lifetime, then we can see that everything balances out in the end. The Godhead ceases to be a paradox or a tyrant, and instead becomes an infinitely patient and loving thing that gives us as many chances as we need to proverbially fall in mud puddles and scrape our knees before finally getting it right. This, of course, does not justify evil or allow us to turn our backs on suffering, but it does set these things against a far larger context than the JCI religions imagine. It also demonstrates that everything from bliss to despair is inherently neutral. The difference between good and bad, just and evil, bliss and suffering, or night and day is one of definition.

How God Acts

> *Myth is neither a lie nor a confession: it is an inflexion.*
> Roland Barthes

I cannot stress enough that the uncertainties within quantum mechanics are not there because we lack knowledge or understanding; they are part and parcel of reality. They may also give the Godhead room to act in the universe, making the Godhead a type of hidden variable in a tangled hierarchy.

None of us is a god in the quantum mechanical sense, and the "Law of Attraction" and "manifesting" peddled by legions of self-help experts are just so much snake oil from a practical standpoint; however, idealism opens the door to being god-like. This seeming paradox may manifest itself in inspiration and creativity that often seem to arise out of nowhere. Consciousness must be nonlocal to collapse a nonlocal waveform. All waveforms are inherently nonlocal, because they extend across the entire universe. This, of course, has tremendous implications for paranormality, as we will see in Chapter 37.

Wheeler's participatory anthropic universe (see Chapters 11 and 15) solves the problem of shared reality by postulating that everyone participates in creating the order we see in the universe. The consciousness that observes the cat in the box and chooses life or death is the quantum consciousness, which is both beyond and greater than any individual ego. God is both one and many. It is only by grasping the concept of one consciousness that fractionates into life, environment, living, and nonliving at once that we can resolve the problem with shared reality and see the deeply symbiotic relationship between all things in the universe.

We can take this to its logical conclusion by saying that the act of collapsing the probability wave leaves room for the Godhead to act and even requires the Godhead, because everything we perceive exists in actuality and not in possibility.

> *The person who is certain, and who claims divine warrant for his certainty, now belongs to the infancy of our species.*
> Christopher Hitchens

Praising God

I have said it before and will say it again: A God such as the one described in this book requires no praise, no submission, no obedience, no rituals, donations, churches, holidays, fancy dress, genital mutilation, abandonment of reason, sacrifice, nothing. We are each pieces of the Godhead interacting with other pieces of the Godhead. We are literally the Godhead playing with him/her/itself in both the prurient and mundane senses of the term.

> *I am so made that I cannot believe.*
> Blaise Pascal

Finding God

If there is a God, how can we find him/her/it? Mystics claim to have been doing this for thousands of years, and those claims are becoming increasingly corroborated by other sciences, as we have seen throughout this book. Eastern religions

> *Those who seek the truth seek God, whether they realize it or not.*
> Edith Stein

> *Great is the cosmic order, for it has not changed since the time of Osiris, who put it there.*
> Ptahotep

say that any individual can find God on her or his own without the need for any outside assistance, and that salvation is not necessary. By contrast, the JCI religions emphasize that people are separated from God because of Eve's original sin. The latter seeks to shut off intellect in favor of blind faith, while the former seeks to use all faculties at our disposal. One might expect that the Eastern religions would be the closest to proving that God exists.

Of course, proving whether or not God exists is not quite as easy it may sound, especially since any hypothesis about the nature of God must take feelings and other subjective, non-scientific data into account. No such hypothesis can be subjected to precisely controlled experiments, nor can it be falsified, which by definition makes a hypothesis of God unscientific. Materialists have pounced on this as evidence—if not proof—that God does not exist, yet another example of conflation; there is no necessary connection between replication, falsifiability, and veracity.

Religion is free to ask about natural selection and any other scientific concept, and science is free to ask about any religious concept, at least in theory. The former would do well to remember that their roots are grounded in science (see Chapter 3), while the latter should acknowledge that materialism is not the end all, be all of existence. I blame religion for this rift, since it is religion that has persecuted science in the name of blind faith and devotion.

At heart, there should be no incompatibility between differing viewpoints. Evolution does not preclude either the belief in or the possibility of God, not because it is wrong, but because it is right. Similarly, belief in God does not mean that one cannot believe in the reality of evolution and natural laws as God's mechanisms for acting in the universe.

Debunking Religion

> *When you have eliminated the impossible, what remains, however improbable, must be the truth.*
> Sherlock Holmes

Again, we have thoroughly debunked the JCI religions in Chapters 7 through 11. There is no way that any deity who created the entire universe with hundreds of billions of galaxies, stars, and planets could possibly have chosen to reveal Himself to a tiny backward tribe cursed with the misfortune of living on the world's doorstep with conquering armies tramping back and forth over their land. As for the Eastern religions, we have found much to admire about them throughout this book. Having said that, the fact remains that the Godhead must be both transcendent and immanent, while all religions are purely immanent in nature, because humans are immanent animals. Thus, the best that any religion can hope to do is to devise an approximate model of God that fits with the culture doing the creating. If horses had gods, then those gods would probably resemble horses. Asian people have Asian gods. African people have African gods. Everywhere one looks, humankind has created models of God in their own images. Not one major religion worships a god that looks nothing like the people worshipping it; if there was such a religion, it would deserve serious consideration for just that reason.

A single consciousness, an all-encompassing wisdom, pervades the universe.
Gerald L. Schroeder

As we learned in Chapter 33, religion gives us clues; it does not give us any final answers. Any priest, rabbi, imam, or other theologian who says otherwise is deluded, fraudulent, or both. Religion is therefore of dubious use—if it has any use at all.

On Immortality

Our brains are Now-selection machines, which severely limits our ability to directly experience the timeless Ground/Godhead. If brain is a product of mind/spirit, then immortality consists of spiritually participating in the eternal Now that is the Ground—in short, complete removal from the concept of time. The illusion of Nows serves a valuable purpose, since it allows the Godhead to explore countless different avenues, just as the imaginary flow of time in our dreams allows us to have those experiences; however, beyond experience lies the utter stillness and tranquility of simply being.

Any flea as it is in God is nobler than the highest of the angels in himself.
Meister Eckhart

Physics offers enticing hints that time is an illusion, as do holography and relativity. Physics is also time-independent, which means that it works with time flowing in either direction. If time was a real flow, then we should expect that the equations should only work if we assume that time is a one-way street. So-called time travel is perfectly possible in theory but impossible in practice, because the odds of either returning to a previous Now or of selecting a specific Now by jumping ahead are incredibly tiny. If we do somehow manage to revisit a previous Now, then the only way we can do it is by consequence of selecting it from a different Now where the selection will seem not like stepping outside the perceived space-time continuum but of remaining within it. For example, if I were to wake up in the year 1920 tomorrow morning, I would not experience it as going back in time, but as being a continuation of what I perceive to be my normal timeline.

Who the gods love dies young.
Menandes

If we accept that the first moment of the universe's existence exists without a cause and that a moment of consciousness requires a prior moment of consciousness, then we must conclude that consciousness is eternal. It does not travel through time. It did not come before time. It will not endure beyond time. It simply is. True immortality does not mean living through an endless series of Nows; it means literally stepping outside the space-time continuum. It means losing the immanent parts of ourselves to become our true transcendent Selves. Anything else is mere survival, which beats oblivion, but not by much. The "immortal" elves in Tolkien's *Lord of the Rings* saga are not truly immortal, because they can be killed in battle; they are survivors, because they are born and are subject to death. True immortals are not born and do not die.

Is There a God?

The so-called scientific argument is sustained simply by a bald assertion that nature did it and not by evidence that god could not have done it.
Gregory Kouki

In Chapter 3, we saw that the Jain myth berates fools for saying that a creator made the world, for that would require God to exist before creation, and the immaterial to make material. Einstein seemingly concurred when he called reality a stubbornly persistent illusion. Mystics tell us that dreams and waking life are both unreal. Despite this, science is built on the belief that the universe and everything in it has a rational explanation and that the rules of logic, mathematics, and logic apply everywhere. To the best of our knowledge, we live in a supremely rational universe that is learning to see and experience itself more intimately. That, in fact, may be the whole point of life: to explore all that is possible to explore, do all that there is to do, and know all that there is to know.

If the Godhead is a fractionated whole, then it represents the collective will of all consciousness in the universe and has the potential to fulfill its drive to do and learn, free of the limits of any individual mind. Davies said it best when he said, "Through conscious beings, the universe has generated self-awareness. This can be no trivial detail, no minor by-product of mindless, purposeless forces. We are truly meant to be here."

Does God exist? Is the picture of God that we painted in Chapters 13 and 14 real? Based on everything I have researched for this book, my answer is an unequivocal yes.

If scientists and nonscientists fail to communicate with each other over religious questions, it is because they are talking past each other, referring to entirely different gods.
Michio Kaku

Chapter 36

Evidence from Physics

> *The answer is yes. But what is the question?*
> Woody Allen

Physics is the bedrock on which all other sciences are built. Does physics vindicate my assertion that God is absolutely real? We shall see. Meanwhile, I urge you to find a peaceful location away from all city lights some clear evening, and spend some time just looking up into the night sky. Wrap your mind around the fact that you are peering back in time, because the light you are seeing set off on its journey anywhere from seconds to billions of years before you saw it. Ponder that there are more stars in the Milky Way than there are neurons in your head, and that there are also more galaxies than there are neurons in your head.

> *Truth is what stands the test of experience.*
> Albert Einstein

Relax your eyes, and let your focus wander until the hard separation between you and the magnificence before you blurs and fades away. Try to comprehend the sheer magnitude of the smallness of your problems, your joys, your opportunities, your challenges, your loves, your family, everything, and you yourself. You are many orders of magnitude smaller compared to the universe than a grain of sand is compared to the Earth. Now take a deep breath and try to fully understand the fact that you are capable of understanding all of these things. You are a child of the universe born from the stuff of stars, and if the conclusions this book has been drawing us to are valid, then the stars are also your children in a sense, because your looking at them may be what is making them real.

Where We Came From

The Big Bang model picks up the story of our origins from a fraction of a second after the universe began, but without describing the beginning itself. The universe is finite, and we can safely conclude that no infinitely dense singularity existed. We can, however, conclude that something very dense did in fact exist, there being a difference between "very" and "infinitely" dense. This model also says nothing about time itself.

Any way we look at it, the Big Bang is widely accepted as the starting point of our universe, which appears more and more homogenous the further back we look. Inflation theory accounts for the mass, energy, and uniformity we see in the universe. Loop quantum cosmology shows us how a cyclic universe could exist. LQC does provide a true origin for the universe as we know it; however, no theory explains our ultimate origins. Science uses the infinite regression problem to debunk the JCI God by saying that such a god would Himself need a God, and so on ad infinitum. It does this while at the same time either sweeping its own infinite regression problem under the carpet or wearing it as a badge of honor by offering theories that require infinite universes to hold up.

Fighting fire with fire does work under certain limited conditions; however, accusing someone of arson while lighting your own forest fire is more than a little disingenuous, because all we are doing is swapping one problem for another.

Assessing existence while failing to embrace the insights of modern physics would be like wrestling in the dark with an unknown opponent. By deepening our understanding of the true nature of physical reality, we profoundly reconfigure our sense of ourselves and our experience of the universe.
Brian Greene

The Lawful Universe

In Chapter 21, we learned that Copernicus fired the mortal shot into the religiously enforced belief that the Earth is at the center of the universe. This launched the scientific awakening continued by Galileo, Newton, and others who discovered that the entire universe obeys a single set of natural laws—laws that were used to prove that the Earth is far older than 6,000 years, among other things. In Chapter 23, we learned that physicists in the 1800s and 1900s discovered that Newton's laws are approximations that work brilliantly at ordinary scales and speeds, but that break down at extremely small and extremely large scales. The blackbody and speed of light problems ushered in the era of quantum mechanics, and revealed

Myth is neither a lie nor a confession: it is an inflexion.
Roland Barthes

that the universe is a far more bizarre place than initially imagined. More importantly, Newton's clockwork universe was replaced by a statistical one, in which almost anything is possible. This, of course, does not mean that anything is likely; on the contrary, most things are spectacularly unlikely.

> *A new scientific truth does not triumph by convincing its opponents and making them see the light, but rather because its opponents eventually die.*
> Max Planck

I must again emphasize that natural laws have nothing to do with humans. All science has ever done is discover these laws, which the universe has been obeying since it first came into existence. Nature cannot violate these laws, unlike a traffic law that people can—and do—break all the time. In natural laws, we are seeing nature's self-imposed limits and not attempting to impose our own will on nature. The laws of quantum mechanics pose some serious challenges. The math works, but we don't know quite how or why; the best theories we have seem to involve consciousness. We are learning throughout this book that other sciences seem to be corroborating the findings from quantum mechanics, which makes sense, because quantum mechanics is the foundation for all other sciences. Despite this, Newton's equations remain valid and cocooned inside every latter form of physics.

Quantum mechanics is the most precise science ever discovered and has passed the many thousands of tests it has been subjected to, not just well, but perfectly. No other science in human history can claim this distinction.

The Nature of Reality

> *The world does not happen, it simply is.*
> Hermann Weyl

Quantum mechanics, relativity, string, brane, loop, holographic, and other theories have revealed some astonishing things about our universe. Here is a brief recap of what we have learned and some of the conclusions we can draw from that knowledge.

The Discontinuous Universe

The smoothness we see in the universe is an illusion caused our own perceptual limitations. At the smallest (Planck) scales, the universe is a very grainy and discontinuous place. Time, space, energy, matter, and density are all subject to absolute limits, just like a movie can be broken down into individual frames but no further. In Chapter 28, we saw how a complete frame of space, energy, matter, time can be said to encompass the entire universe into a single Now that spans a single Planck time interval.

Out of Time

There are two leading philosophical schools of thought about time: The standard view says that time is a separate dimension within the universe where events happen in sequence. The second view sees time solely as a mental construct that helps us with sequencing and comparisons, and that nothing moves through or travels with time. The latter view is the one we saw in Chapter 28. We sense a flow of time, because we make choices by selecting Nows, and these choices cannot be undone; nature has the ultimate "no return" policy, which is why we remember the "past" but not the "future."

Relativity proves that each of us has our own reference time, which describes how we see ourselves relative to the universe. We each select Nows completely independently from everyone else. A single event can literally occur at multiple times for different observers, who may each see a different sequence of events; despite this, the universe always manages to avoid paradox—a level of built-in rationality that I personally find very telling. Space-time is real, but moments of time are not quite real, because past, present, and future are all illusions. This is consistent with holographic theories, as we saw in Chapter 27.

The bottom line is that time as we think we know it does not exist. All that can possibly exist is already out there waiting to be selected, like a girl waiting to be asked to dance. The hologram always exists; we simply choose how to view it. On a certain level, I am not me; I am a vast collection of copies, each of which is in a slightly different position. There is no single entity typing these words; there is a collection of entities, each at every stage of the writing process. I can even speculate that there are copies of me that I will never select, but that exist in potential. To be even more specific, my consciousness is navigating a path of Nows and seeing multiple copies of me, one for every Now, and weaving all of this into a sense of oneness moving through a river, just like we weave individual movie frames into a smoothly flowing sequence. As we saw in Chapter 35, this view of time is very compatible with both determinism and free will.

The experimental method is the most powerful tool we have, that's how we find truth and non-truth.
Michael Persinger

Science is wonderful at explaining what science is wonderful at explaining, but beyond that it tends to look for its car keys where the light is good.
Jonah Goldberg

Out of Space

Rejection of evidence that cannot yet be measured with instruments in a laboratory is contrary to the scientific spirit of inquiry.
Bernard Haisch

We learned in Chapter 27 that cutting a hologram into pieces results in each piece containing the entire hologram; the image may get fuzzy, but the entire picture remains. This has tremendous implications, because a holographic universe is a giant illusion where consciousness is the only "real" thing left. After all, if everything that can ever happen in the universe is contained in a giant hologram that resides on the edges of the universe and projects inward to create the illusion of three-dimensional space, then where could all of these possible states have come from but from consciousness? A holographic universe is perfectly compatible with all of the laws of physics we reviewed in Chapters 21 through 28. It also resolves the infinity problem created by simply throwing quantum physics and relativity together.

Out of Matter

He who pursues learning will increase every day. He who pursues the Tao will decrease every day.
Lao Tzu

The entire universe seems to be a single entangled object, and is thus still "one," a lesson mystics have been trying to impart for millennia. We saw in Chapter 23 that matter is not composed of specks of stuff, but of probability waves that collapse into matter when observed, according to the Copenhagen interpretation (as well as the hidden variable and many worlds interpretations, as we saw in Chapter 29). Thanks to the dual slit experiment, we know that matter is composed of particles; we also know it is composed of waves. This complementarity manifests itself based on how the experiment is set up—in other words, on what question we ask and how. Quantum effects exist on scales, as the dual-slit buckyball experiments attest; however, we cannot see them under ordinary circumstances, because the effects are far too small to notice.

For all of its seeming flawlessness, quantum mechanics does not describe actualities; rather, it describes potentials and their probabilities in ways that defy common sense and easy comprehension. This is what allows tunneling to occur, and why the ground state is never 0. Probabilities are expressed as a waveform that depicts the possibility of finding any given object at any point within that wave.

Relativity, superstring, brane, and loop theories all treat space-time as pure geometry. Time is not a river through which space flows; it is simply another geometric dimension (that almost by definition cannot have any flow). This has been proven using Einstein's equations that equate space (S) and time (T).

Out of Energy

Relativity also proves that energy (E) and mass (M) are identical thanks to $e=mc^2$. It just so happens that the universe has zero-sum energy, which means that on the largest scale $e=0$. This has stunning implications, which we will revisit in Chapter 39.

Very Much in Mind

We saw in Chapter 29 that all of the leading interpretations of quantum mechanics involve—and require—consciousness, no matter how hard their adherents try to deny it. Quantum mechanics may be the bedrock of all science, but it owes its existence to consciousness, while not being able to explain consciousness in the slightest.

Whether facts be moral or physical, it makes no matter. They always have their causes.
M. Taine

If idealism is correct—and we have every reason to think it is—then consciousness created quantum potential (Nows) according to the laws of physics that the same consciousness also created. In other words, the Godhead created the laws of physics and then created the Nows in accordance with those laws. Quantum entanglement nonlocally connects particles and minds alike across space and time. Each observer is free to select her or his own Nows in a manner that is consistent with other Nows being selected by others. The coherence among all entangled minds on Earth and possibly beyond maintains the order that we all experience, which explains why Lorentz transformations work so well to convert one observer's viewpoint to another's. Each observer is created equal as far as observations are concerned; waveform collapse occurs when consciousness assesses the Nows that are available from the current Now and makes its choice by selecting one of those Nows. A measurement occurs whenever consciousness interacts with a waveform and selects one of the possibilities.

I am proposing that spiritual truth, in analogy to poetic, musical, and artistic truths, is something other than the brute force of logic.
Stephen D. Unwin

> *How shall I grasp it? Do not grasp it. That which remains when there is no more grasping is the self.*
> Panchadasi

It is impossible to replace a conscious observer with instrumentation, because the instruments themselves become part of the waveform and affect the results of the experiment (thanks to Heisenberg uncertainty). It is only when consciousness reads the instruments and makes a measurement that the waveform collapses according to the appropriate waveform, in keeping with the laws of physics. The need to take measurements to select Nows is what creates the illusion of separation between subject and object, in the same way that you see yourself as a separate and distinct entity in your own dreams.

The worlds in your head are not deemed to be real, but your dream self and all of the other characters in your dreams receive sensory information and act on it, which creates the illusion of separation and isolation. The same thing happens in the waking world on a far larger scale.

If the many experiments performed on quantum scales have taught us anything, it is that the view of the world that we experience every day and take utterly for granted is utterly wrong. Bell's Theorem forces us to abandon the idea that any "real" state exists before a measurement. There simply is no such thing as a definite location, momentum, charge, size, shape, or other property without a measurement. The waveform is such that the probability of seeing the same objects in pretty much the same places as I last saw them is extremely high, but that does not alter the fact that a nonzero possibility exists for me to see something quite unexpected. Quite simply, the contents of my desk do not exist until I open the drawer to take a look. Measurement is never inconsequential, because every measurement selects a Now. This is the mechanism through which free will operates. This is perfectly compatible with an omnipotent, omniscient Godhead that chooses to limit itself. We can posit that the Godhead sees all possible states and experiences the superposition states themselves. Accepting the waveform means accepting that possibilities are as real as actualities—or more so—because possibilities can exist in perpetuity, while a Now comes and goes in an instant.

> *The exact relationship between the elements of scientific models and whatever true reality lies out there is not of major concern.*
> Victor Stenger

Both materialists and some dualists reject the idea that consciousness plays a role in collapsing a waveform, because they don't see any mechanism through which this can occur; however, as we have seen, the mechanism is the exact same one that allows both you and the characters insider your head to act inside your dreams every night. As the old children's song says, life is but a dream. This sounds like a nice pat answer, except for one glaring problem: Quantum mechanics neither describes the collapse nor what causes it; collapse is postulated, and that's that. The mechanism that causes collapse does not exist within space, time, energy, or material; if it did, we should have found it by now. We must therefore conclude that this mechanism lies beyond space, time, energy, and matter. In Chapter 39, we will see that space, time, energy, and matter are all reducible, convertible, and subject to waveforms and measurement. The measuring mechanism must therefore be irreducible and inconvertible. Consciousness is the only possible candidate, especially since it does not supervene either naturally or logically on that upon which it acts. An ancient saying from India goes, "Coming and going is all pure delusion; the soul never comes nor goes. Where is the place to which it shall go when all space is in the soul?"

In *Consciousness Beyond Life: The Science of the Near-Death Experience*, Van Lommel offers the following calculation:

- Number of neurons in a single cubic centimeter of cerebral cortex tissues: approximately 100,000,000 (10^8).

- Number of synapses in that cubic centimeter: approximately 100,000,000,000 (10^{11})

- Total number of synapses in the brain: about 100,000,000,000,000 (10^{14})

- The brain performs about 10^{24} actions per second when awake. Each action may involve 1 or more bits. (A *bit* is a single 1 or 0 of information; a *byte* is typically a string of 8 bits. For example, the capital letter A can be translated to 01000001 using 8-bit ASCII coding.).

- Including memory, the brain needs a storage capacity of 3×10^{17} bits per cubic centimeter.

According to Van Lommel, the brain by itself is unable to store or process this much information, which indicates that it may not be the seat of consciousness. It is possible that future discoveries in neuroscience will explain this biologically; however, that will not address the problems posed by quantum mechanics, where the best solution is to concede that consciousness plays the leading role in creating the universe.

Faith in reason is the trust that ultimate natures of things lie together in a harmony which excludes mere arbitrariness. It is the faith that at the base of things we shall not find mere arbitrary mystery.
Alfred North Whitehead

In this model, consciousness creates the brain as an intermediary or proxy to interact with the universe. This explains how mind can alter matter, in the sense that learning literally alters brain circuitry, as do activities like meditation or psychotherapy. For example, simply talking to a psychotherapist can gradually alter the brain's wiring, thus "curing" the problem that caused the patient to seek the therapy.

On the Nature of Consciousness

A Bose-Einstein condensate is a special condition in which supercooled atoms become indistinguishable from each other, and essentially become a single entity where the identities of the original atoms are lost. According to physicist Herbert Fröhlich (1905-1991), living cells could emulate a Bose-Einstein condensate at normal temperatures, thus forming coherent wholes, like individual musicians forming a single band. Living systems are self-organizing; patterns and coherence emerge from interactions with each other and the world, without being caused by those interactions. Romijn believed that the electromagnetic fields of neurons could be an example of quantum coherence that could either carry or be produced by consciousness, a view that fits nicely with Pribram's holographic theories—where memories are stored in the coherent patterns formed by neural networks, which is inherently nonlocal. We seem to nonlocally select holographic Nows. The nonlocal aspect may help explain perception during NDEs, as well as the life review.

Zeilinger pointed out that one cannot have a half-thought, half-feeling, half-yes, or half-no; there are only complete thoughts and feelings, and definite yeses and nos—a binary system. Consciousness thus consists of quanta, or packets; the consciousness interval posited by Harris (see Chapter 30) appears to be very real, and each such interval is a Now. We interpolate these quanta into smooth-seeming movement in the same way we interpolate movie frames into a smooth flow.

Stapp combined aspects of quantum mechanics, Newtonian physics, neuroscience, and other disciplines into a theory of consciousness with some interesting results. As he said, "The connection between consciousness and the brain is primarily a problem in physics and addressable by physics—but only the correct physics. The causal irrelevance of our thoughts within

---

The reasonable man adapts himself to the world; the unreasonable one persists in trying to adapt the world to himself. Therefore, all progress depends on the unreasonable man.

George Bernard Shaw

the classical physics constitutes a serious deficiency of that theory." In Stapp's theory, a decision made by mind (consciousness) affects the brain, which alters the brain's waveform, which alters the waveform of the experiment to create the results. Measurement is an active function of consciousness.

No Place for Snake Oil

I have said this several time before, and will say it again: The concepts I am describing in this book have helped line the pockets of far too many self-help gurus to count. Hardly a week goes by without someone releasing a product claiming that quantum entanglement means we can create our own reality out of thin air, if we can only believe in our own abilities. The "process of manifestation" is real enough in a sense, and one's beliefs play a critical role in how any person carries out her or his life. The so-called "Law of Attraction" does not exist—or to be more specific, it does exist, but the odds of using it for any practical outcome are so low that you are truly better off acting as if it didn't exist. The concept we should be talking about is, "The Law of Realization" or, to put it more precisely, "The Law of Collapse."

As to the ultimate things we can know nothing, and only when we admit this do we return to equilibrium.
Carl Jung

Chapter 37

Evidence from Paranormality

All that belongs to human understanding, in this deep ignorance and obscurity, is to be skeptical, or at least cautious; and not to admit of any hypothesis, whatsoever; much less, of any which is supported by no appearance of probability.
David Hume

Consciousness, the subjective experience of an inner self, could be a phenomenon forever beyond the reach of neuroscience.
David J. Chalmers

"You're flying today, aren't you?"

Those were Sarah's first words to me when I called her a couple weeks before writing this chapter while on my way to the airport. I have made a habit of giving Sarah a call to ask her what she thinks of my prospects for a safe landing when needed. Her track record with me is such that I would flat-out refuse to board a plane if she expressed any serious concerns. Still, it's not like I only call her when catching the train to San Francisco International Airport; nobody else would have any reason to suspect that I was calling for flight recommendations. The only plausible explanation is that Sarah knew I was going to be getting on an airplane that day, which is just one of the many paranormal anecdotes I have about her. (See the Introduction for more.)

Beyond Normal Explanation

We discussed paranormality in Chapters 15 through 17, and I provided several examples of paranormal encounters that have no "rational" materialist explanation at face value. These are but a small handful of the tales one can find without any trouble at all. Ask around, and it's a safe bet that a significant percentage of respondents will report paranormal experiences ranging from déjà vu to meeting/speaking with dead relatives, NDEs, and more. It is entirely possible that some of these experiences are not what they seemed, that the person is mistaken or lying; however, a person who is convinced of what s/he saw will describe the experience with unshakable certainty.

My experiences with Sarah are little more than anecdotal, because I did not set up any controlled experiments or take records of predictions and results that would satisfy any scientific criteria. Still, I know what I have experienced. There is no way she could have known what my property looked like when she located my chain saw blade, there is no reason for her to randomly zero in on my ex-wife's right front tire, predict the date of a major life event months in advance, or any of it. Peel away the hype and outright fraud surrounding paranormality, and the conclusion becomes inescapable: Something is going on; but what?

The first principle is that you must not fool yourself, and you are the easiest person to fool.
Richard Feynman

Revisiting NDEs

This book focuses on the question of what happens when we die, if anything. It is therefore fitting that we take one last look at NDEs.

Universal Experiences

Read enough NDE accounts, and your eyes will eventually glaze over from sheer boredom as you read essentially the same story over and over and over again, as if every NDE survivor had been issued a copy of the same script. Virtually all NDEs include some or all components of the following sequence:

1. Unpleasant noise.
2. Awareness of being dead.
3. Feeling of peace, wellness, and detachment from the world.
4. Leaving one's body, which may include looking back at one's body, and possibly seeing medical personnel trying to revive it.
5. Moving through a dark tunnel toward a bright light.

I believe that consciousness and its contents are all that exist. Space-time, matter and fields never were the fundamental denizens of the universe but have always been, from their beginning, among the humbler contents of consciousness, dependent on it for their very being.
Donald Hoffman

6. Feeling unconditionally loved.

7. Meeting spiritual entities that may include deceased relatives, friends, or other loved ones.

8. Detailed life review that passes in an instant, the proverbial "life passing before one's eyes."

9. Receiving additional information about life and/or the universe.

10. Approaching a final border, gateway, or other point of no return.

11. Making the decision (with or without support) to return to the body, despite any reluctance to do so.

Beyond Normality?

This sounds very compelling, but skeptics all point to one inescapable fact: Every NDE survivor has come back to tell the tale. To the best of our knowledge, no one has returned from beyond the grave to corroborate these stories... or have they? At least one NDE report includes meeting a dead person whose family did not know was dead until hearing the survivor's story. At least one NDE report includes seeing objects far removed from the hospital room that the survivor would have no way of seeing beforehand, even under normal circumstances. Also, most mediums and people who claim to see ghosts may be mistaken and/or fraudulent, but we cannot with any confidence say this is true for all situations.

Beyond Skepticism

Skepticism is good, because it forces us to examine new evidence and outlandish-sounding claims in the cold light of reason; however, skepticism for its own sake becomes just as faith-based and dogmatic as any religion. Only someone whose mind is truly open to following the evidence wherever it leads can examine it on its own merits and come to a correct conclusion. When examining evidence, there is no right or wrong and no good or bad, only the possibility of a framework in which all of the available evidence can fit. Evidence that does not fit this framework is either specious due to mis-

---

As the reigning paradigm in evolutionary psychology has produced questionable results, the evolutionary study of human psychology is still in need of a guiding paradigm.
David Buller

Through conscious beings the universe has generated self-awareness. This can be no trivial detail, no minor by-product of mindless, purposeless forces. We are truly meant to be here.
Paul Davies

take, wrong, or an indicator that the current framework may need alteration or expansion.

This should be true for all evidence; however, many scientists accept extremely long odds for "natural" phenomena while applying a double-standard to any evidence that seems to indicate the reality of anything paranormal or "supernatural." To the extent that anything paranormal is real, paranormality ceases to be "supernatural" and is instead part of the natural order of things. Opening one's mind to paranormality may smack of religious faith but conflating the two is a huge mistake as we saw in Chapter 2.

When the heart weeps for what it has lost, the spirit laughs for what it has found.
Sufi aphorism

Science and religion would work together to follow the data wherever it leads in my ideal world, but this can never happen. Religion has trampled science for thousands of years, and scientists have had enough of it. The need for power trumped the need for knowledge, and for that I blame religious leaders past and present. For shame! That said, I also have little love for scientists who consistently ignore the growing body of evidence that what we see is by no means all that we get.

Facing Death

In Chapter 1, I asked you to imagine looking into the eyes of your own child as s/he lies on her deathbed in the final moments of life. What would you say? What would you tell your son or daughter? Would you regale them with religious nonsense, promise them an afterlife, or tell them to be brave? What will you want people to say and do for you as you approach your own inevitable end of your sojourn on this planet? It is very easy, very tempting, and very wrong to sweep death under the carpet and to pretend it does not exist, or that we have a "while" or "long time" before having to face death.

Perhaps the only limits to the human mind are those we believe in.
Willis Harman

I have done my best to accept the fact that my wonderful partner Jennifer could die at any moment; every time she walks out our front door could be her last. Every time Logan puts on his helmet, adjusts his backpack, and hops on his bike to pedal off to school could be the last time I will see him alive. There are absolutely no guarantees that I will live out this day, much less the week, month, year, or the remainder of a "normal" lifespan. My situation is not unique; every single human on the planet is in the same boat... including you, dear reader.

I am not saying this to be morbid or to depress you. On the contrary, I see death as a powerful motivator to live and to love. Every day I tell the people I cherish how I feel about them, and every day I do my utmost to show them how I feel by treating them as such. My philosophy is simple: If every encounter with someone I love could be my last, then I owe it to them and to myself to act from love. I also owe it to everyone to choose whether or not I want to be with them every single day, and to mentally refresh my relationship with them every morning when I wake up. In this respect, I am like the king who kept putting off Sheherazade's execution for just one more night in *The Thousand and One Nights*. That king reevaluated his relationship with Sheherazade every day and made the conscious decision to continue it. Likewise, I make the conscious and deliberate choice to continue my relationships with Jennifer, Logan, and others every day.

> *Take me from the unreal to the real. Take me from darkness to the light. Take me from death to immortality.*
> Upanishads

> *Neither a man nor a nation can live without a higher idea, and there is only one such idea on earth, that of an immortal human soul; all the other higher ideas by which men live follow from that.*
> Fyodor Dostoyevsky

What Paranormality Tells Us

Paranormality is perfectly compatible with the laws of physics and relativity. It is perfectly compatible with timeless Nows in a holographic universe, and with just about any other imaginable construct of reality that is based on today's science. Quantum waveforms extend throughout the entire universe. Consciousness entails waveforms; thus, Sheldrake's belief that consciousness is a field and that the human brain is a glorified TV set is affirmed. There are indeed levels of reality beyond normal perception and awareness, and that may well include an afterlife. The sciences we have explored throughout this book indicate that an afterlife is possible, and provide all of the framework needed to allow it. Still, simply being open to something does not make it so. Paranormality gives us direct evidence that everything we have learned from physics and neuroscience is in fact real.

910 | *The Divine Savage*
Revealing the Miracle of Being

Chapter 38

The Multivariate Monoverse

> *The objective world arises from the mind itself.*
> Buddha

Any attempt to answer the question of what happens when we die must be based on understanding how the universe works and what the applicable sciences have to say, which is why we have spent so much time looking at the various sciences. Any explanation of what happens when we die must occur within the framework of an explanation about why the universe we live in is the way it is and what limits the parameters of this universe to their extremely hospitable values. In other words, an explanation of what happens when we die must include at least a rudimentary TOE that answers basic questions and incorporates all known science.

I set out to find just such a thing.

Methodology

> *Physics is nothing but the ABC's. Nature is an equation with an unknown, a Hebrew word which is written only with consonants to which reason has to add the dots.*
> Johann G. Hamann

The overwhelming majority of the books, articles, Web sites, and other resources I used to research this book focused on a single science—such as physics, evolution, physiology, neurology, or astronomy—because they were written by specialists in those fields. Some authors arrived at very compelling conclusions based on the evidence from their own fields of study that resulted from countless hours of education, experimentation, and examining of evidence; however, a comprehensive theory of reality that can explain both life and death in order to give us a window to anything that may lie beyond must embrace all of the key sciences. I therefore set out to read as much as I could across as many sciences as I could, including the so-called sacred texts; my disdain for religion is no excuse for not researching it thoroughly.

My body of knowledge and information grew, but I eventually found myself running up against the law of diminishing returns, in that I started reading the same basic arguments over and over again. Even books hailed as ground-breaking and new contained the same old stuff, sometimes with new packaging. I pressed on until eventually the well ran dry, and all I was reading were repeats. Then I sat back to take a look at what I had found and where it all might lead me. It soon dawned on me that the authors I read fell into two distinct camps, materialist and otherwise. Most of them back up their arguments with excellent evidence. Many of them even use the same evidence to arrive at contradictory conclusions. The only way this could happen is if the conclusions are ultimately based on faith—the same kind of faith that most scientists lambaste. Any decision I made about who was right and who was wrong, no matter how carefully I made that decision, would also be based on faith. This conundrum puzzled me for many months. How could I avoid the trap of making the same type of faith-based conclusion that I object to?

One day, it hit me: The only other possible conclusion is that everyone is right, despite their seeming contradictions. This seems outlandish, and yet I had no choice but to examine that possibility very seriously while doing my best to avoid conflating concepts that have no necessary causal or other relationship. I also had to examine whether the evidence from the various sciences could be unified into a cohesive theory. The focused specialization of individual sciences and professions has yielded tremendous advances and advantages for all of humanity; it has also made it impossible to see the forest for the trees. Each science can contribute to a comprehensive picture of reality, but no science—not even quantum mechanics—can claim otherwise. Gödels incompleteness theorem makes sure of that. Not even combining all of the sciences can give us a final answer, but we can get tantalizingly close.

The only reality is mind and observations, but observations are not of things. To see the universe as it really is, we must abandon our tendency to conceptualize observations as things.
Richard Conn Henry

We have ignored a critical component of the cosmos, shunted it out of the way because we didn't know what to do with it. This component is consciousness.
Robert Lanza and Bob Berman

The Inescapable Conclusion

> *For when all is said and done, we are in the end absolutely dependent on the universe.*
> William James

Looking for evidence to back a pre-selected conclusion is not scientific. I am guilty of that, because I initially set out to prove the correctness of materialism and call this book something depressingly inspiring like *The Mortal Savage* or *The Finite Savage*. The same materialism that has yielded most of the technology we take for granted today is utterly unable to explain why intelligent life exists and why it exists only for short bursts of time. A universe that arises from random events must exist within a framework that explains both the events and the framework itself, which means it must have a strong nonrandom component. In short, the universe must be supremely rational, even if that rationality only exists because of our limited perceptual and cognitive abilities. A rational universe implies an organizing principle or force, which implies will, which implies consciousness. As we have seen, consciousness is neither logically nor naturally supervenient on nature; according to the laws of nature, it is perfectly possible to have an identical universe without consciousness.

Biocentrism is the best explanation for how the universe evolved as it did, given all of the evidence we have examined. The universe evolved in potentiality (which is just as real as actuality, as we learned in Chapter 37) until the first organism was able to measure its environment and create both the past and future history of the universe by selecting an appropriate Now. We could say that the Godhead created all possibilities and looked at all of them until finding the Now that began everything we are familiar with. This explains both the amazing fine-tuning in the universe and blind design at a single stroke.

> *Sometimes it almost appears that the theories are not a description of a nearly inaccessible reality, but that so-called reality is a result of the theory.*
> Hendrick Casimir

Which interpretation of quantum mechanics is correct? The Godhead has every option and yet no other option open to it at once. The Copenhagen interpretation is therefore correct. The Godhead is the hidden organizing principle, which means that the hidden variables/implicate order interpretation is also correct. Superstring theory allows up to 10^{500} universes to exist; everything seems to be compatible with string theory, rendering it untestable, unfalsifiable, and therefore unscientific. Multiverse and many-worlds theories also postulate many different universes, most of which are probably incapable of sustaining life. The concept of Nows means that we can see these universes as either entirely separate entities, or as unused components of one and the same universe, in which life has chosen a path steering away from incompatible possibilities. Furthermore, each organism can choose different Nows and take different paths, which we can say exist within this universe or within personal universes as we please. Thus, superstring and other multiple-universe theories are correct, as are theories that postulate a single universe. (The difference is roughly akin to seeing different organs as either separate entities or as parts of the same organism.) There is one shared universe or many universes, depending on how you choose to look at it, just like space and time are either separate or different aspects of the same thing depending on how you look at it.

Materialism is correct in the same way that Newtonian physics is both correct and contained within later discoveries. Dualism is correct to the extent that it acknowledges the existence of something more than mere materialism. Idealism encompassed and supersedes both materialism and dualism without necessarily contradicting them.

The objective reality that exists in the universe is the sum total of every subjective reality in the universe, plus the wholeness function that allows us to share so much of our subjective realities and that is greater than just the sum of its parts. This objective reality is what we refer to when we refer to the Godhead/Ground/Source.

I refer to this grand ensemble of many universes within one universe that all consist of the same nonlocal holographic buffet prepared by the original consciousness as the Multivariate Monoverse—a place where just about anything goes, so long as the things that go on in any one place remain consistent at all times.

Can the Multivariate Monoverse theory be tested? No. Can it be falsified? No. Thus, I must be the first to confess that it is by definition an unscientific theory; however, it is also the theory that seems to best fit the available evidence.

This is as strange a maze as e'er men trod, and there is in this business more than nature was ever conduct of.
Alonso

There are two ways to be fooled. One is to believe what isn't true; the other is to refuse to believe what is true.
Søren Kierkegaard

Chapter 39

Zero and Infinity

> *The opposite of a fact is a falsehood but the opposite of a profound truth is very often another profound truth.*
> Niels Bohr

I have mentioned several times that the universe consists of nothing, or more specifically "no thing." I have also said that this nothingness contains everything there is and everything that could ever be, and that references to "nothing" are anything but nihilistic. In Chapters 31 and 32, I referred to the concept of zero and infinity being equal. Let's explore this concept in a little more depth.

Nothing to See Here

> *Unconditional love is forever on the horizon because it is infinite and we are finite.*
> Irwin Kula

In Chapter 30, I told the story of the peasant in India who asked what makes a locomotive move, only to repeat the question after receiving a thorough explanation of all the moving parts and how they operate—an example of a straightforward explanation that turns out to be not quite so straightforward after all. Similarly, Eastern philosophy and religion use words like "empty" and "void" in a manner that seems misleading, because we are so used to those terms being used to describe the total absence of anything whatsoever, when in fact they are describing infinite potential and consciousness. These descriptions are similar to descriptions of quantum effects, such as superposition, multiple paths, and complementarity. The word "nothing" can thus be seen as the opposite of "thing," which does not necessarily mean a complete absence.

Nothing and everything are complements in the same way that waves and particles are complements. Zero and infinity are also complements, which means we can say that zero actually equals infinity. If you can wrap your mind around that, then you can begin to understand the deeper truths of the Multivariate Monoverse.

Materialism believes that only material in the form of space (S), time (T), energy (E), and/or matter (M) exist, with all else being epiphenomena of these four components. STEM (the combination of all four components) exists, and a complete description of STEM would yield the elusive TOE or GUT.

We know from relativity that S and T are not separate and distinct; rather, they are intimately linked geometric dimensions, which are simply different aspects of each other. For comparative purposes, we can say that S=T. This assertion is validated by the fact that movement through S impacts our personal T through time dilation, which becomes noticeable at very high speeds and infinite at c, the speed of light; a photon literally exists everywhere and everywhen in the universe at once.

Relativity also tells us that E and M are also different aspects of each other, thanks to $e=mc^2$. Matter is literally frozen energy. Matter and time are also linked, because time dilates and mass increases with higher velocities. This indicates that matter and energy are also geometric, an idea that is validated by superstring, M theory, and loop quantum cosmology. Further, based on what we just saw above, we can see that E and T are linked. We can therefore say that S=T=E=M. Here is where things get extremely interesting, because quantum mechanics and cosmology tell us that the total sum of energy in the universe is 0. Adding this to the equation we just saw gives us S=T=E=M=0, which literally means that space, time, energy, and matter are nothing!

We know that a strict materialist sees consciousness as an epiphenomenon that can be explained by thoroughly describing STEM. We have also just seen that such a complete description might just add up to 0, a concept that the late science fiction author Isaac Asimov (1920-1992) had some fun with. This does not necessarily mean that anything exists besides STEM; it does mean that any description of STEM must be able to describe the spontaneous generation of STEM from

As rivers flow into the sea and in so doing lose name and form, even so the wise man, freed from name and form, attains the supreme being, the self-luminous, the infinite.
Mundaka Upanishad

If you don't break your ropes while you're alive, do you think ghosts will do it better?.
Robert Bly

nothing at all. Some theories postulate that the universe could have begun with as little as 20 pounds of matter. Where did those 20 pounds come from? Any appeal to a different or prior universe risks the same kind of infinite regression that science sees in religion.

Everything to See Here

Life is an epileptic fit between two nothings.
Edmond Goncourt

We all know that a pie can exist as a whole that retains its wholeness function until someone cuts it. We also know that it is impossible to have slices of pie without a whole pie, and that a pie cut into slices that have not yet been removed is a whole pie that has lost its wholeness function. A slice of pie implies an entire pie, and an entire pie implies slices of pie. We can extend this logic to see that nothingness can imply somethingness and vice-versa, in the same way that sadness implies happiness or that black implies white. In Chapter 31, we saw that a set is simply an empty shell that can contain either individual members and/or subsets. All sets are empty until filled; emptiness is an inherent property of all sets. This means we can define space as both nothingness and an integral part of reality without any contradiction or paradox. It is impossible to have fragments (existence) without the whole (nonexistence).

Nothingness is self-explanatory. It is pure consciousness, from which comes infinite potential manifested as the buffet of Nows, from which consciousness can pick and choose. In this idealist universe, the cat really is both dead and alive until the box is opened. Independent reality does not exist, because independence does not exist; all things in the universe are entangled to one degree or another. Most people fail to grasp that in a way that they can truly appreciate. Those of us who do understand this are literally in heaven and hell at once, because we can see the wholeness function and unity in the universe while at the same time being painfully aware that most people are unaware of this. We are ignorant of our own evolutionary history and how our own brains tick, which causes no small amount of suffering. We are equally ignorant of reality beyond narrow worldviews that see only separation and differences, instead of similarity and unity.

Life is nothing between two epileptic fits.
Anthony Peake

Materialism is dead as a description of how the universe works; however, it serves an excellent pragmatic purpose, because the maya of STEM gives us a good framework from which to interact with each other and with the entire universe. None of us really exists; and yet here we are.

Space, time, energy, and matter are mutually interchangeable and complementary aspects of each other, separate dimensions that exist as pure geometry. STEM is also reducible to 0. Consciousness (C) cannot be so reduced. It is the source, the Ground and Godhead, the alpha and omega, the be all and end all. It exists outside of time and yet becomes immanent within time. That which is beyond time need have no beginning as we understand it, and can have no ending. Infinite regression is avoided, and we need not invoke multiple universes and/or past universes to explain our own existence.

Consciousness itself is extremely simple. The argument that a complex universe would require a complex creator is demonstrably false; a mirror is simple. Shattering that mirror creates thousands of shards that are each more complex than the whole from which they came. Each shard is part of the original mirror, but can yet look at and reflect every other shard as if all were truly separate entities. In this example, the mirror is the Godhead, and the shards represent conscious individuals.

The ultimate truth for all of us is that we live in infinite love and oneness. We are expressions of the divine.
Roger Teel

Now what is history? It is the centuries of systematic explorations of the riddle of death, with a view toward overcoming death. That's why people discover mathematical infinity and electromagnetic waves, that's why they write symphonies.
Boris Pasternak

Chapter 40

The Divine Savage

> *Universal mind is like a great ocean, its surface ruffled by waves and surges but its depths remaining forever unmoved.*
> Buddha

So here we are, dear reader. We have come full circle and can finally answer the question of what will happen to us when we die with a high degree of certainty. I hope that you have had all of your core beliefs about the nature of reality both challenged and affirmed in ways you did not expect when you first picked up this book. If you are a materialist, then I hope this book has opened your eyes to a new way of seeing the universe we live in. If you are religious, then I hope you have been inspired to see your religion as a metaphor for the truly divine, and not something to take literally. In short, I hope our long journey together has changed you in some way that will comfort and benefit you going forward. As I said in the Introduction, if I have succeeded however minutely in this quest, then this book will have been an abject success.

> *To be conscious that we are perceiving is to be conscious of our own existence.*
> Aristotle

Our quest to find reality has led us from mythology to shamanism to religion, and eventually to science that forced us to look in the mirror to try to see what we really are. Having examined all of the evidence and drawn our conclusions, we can now finally answer the big questions.

Everything Stands

I cannot emphasize strongly enough that the model contained in this book does not invalidate anything I say in the other *Savage* books; on the contrary, this model contains and expands upon my previous model within itself, just as quantum mechanics contains and expands upon classical Newtonian physics. For example, the model and methods contained in *The Enlightened Savage* and explored more fully in *The Natural Savage* are designed to help you in life, and are well suited for that job.

The body dies but the spirit that transcends it cannot be touched by death. That means that I am the deathless spirit.
Sri Ramana Bhagavan Maharshi

Who Are You?

The short answer is that you are not who you think you are. The Beatles had it right when they sang that, "I am he as you are he and we are all together." You are not an individual with whatever name you received at birth. The ego that makes you "you" is not a thing, and has no separation or distinction from me, anyone or anything else you interact with on as daily basis, or the Godhead. You are not a label, and yet you aren't not a label; you are; it is that simple. At the most simple level, you and your ego are nothing but descriptors of a relationship between conscious experience and the local environment that provides the illusion of subject-object split, which in turn combines with memory to create the ego.

You are the crown of creation.
The Jefferson Airplane

When we say "I," we are referring to our egos. Our unconscious is simply consciousness without awareness. It is perfectly possible to perceive events without being aware of those perceptions, and we are constantly thinking and feeling, whether or not we are aware of it. Collapsing the waveform requires awareness, because will can only come from awareness. It is nice to think that our individual egos will survive death and somehow live on to fight another day; however, we have learned that death itself is an illusion, which means that birth must also be an illusion. Our egos will not survive death, because they have never existed after all. Only consciousness is real. All else is maya. Wanting to stay in this lifetime and retain our current egos is like a caterpillar not wanting to become a butterfly—a process that can involve seeming death and decay before the final form appears. We cannot even say for sure that we are not entirely new people when we wake up in the morning; our memories could come from a single Now.

We are not our thoughts. We are not our feelings. We are not our perceptions. We are not separate and distinct from anything or anyone in the universe. We are to the Godhead what shards are to a mirror. Drop a simple mirror, and it will fragment into thousands of complex shards, each of which can reflect the lights from every other shard. Each of us is a shard of God... and thus, each one of us is God. God is the mirror seeking to rebuild itself.

Why Are We Here?

One of the most misleading representational techniques in our language is the use of the word 'I.'
Ludwig Wittgenstein

Aristotle held that all we can know and discover comes through reason, and that truth can be found through discussion and logical arguments—a philosophy I have done my utmost to uphold throughout this book. Experience is a valid learning tool, but only certain types of experience contribute to learning. Truth must be experienced directly and not merely sought through reason. For example, the only way to truly know why you should not touch a hot stove is to touch it. This simple truism is the ultimate meaning of life: There is only one way to find out.

The human mind cannot be absolutely destroyed with the human body, but there is some part of it which remains eternal.
Baruch Spinoza

Breaking the unified Godhead unto shards spreads the one intelligence into separate-seeming bits that can then experience all of the joys and tribulations of interacting with each other in every imaginable way during their quest to come back together again. A puzzle piece is made to contribute to the entire image; the same holds true for every shard of the Godhead in the universe—in other words, for the entire universe. This painstaking process of reassembling the puzzle culminates in the unimaginable and unlimited bliss of rediscovering our true natures—of rediscovering the God within each of us.

In the meantime, our journey resembles a long and winding road that at times seems to lead anywhere but to this liberation, but that nevertheless draws us inexorably toward our goal. Our monads, like actors, take on roles and play those roles for a lifetime before moving on to take on other roles and continue the process. The period between death and rebirth gives us a glimpse of what awaits us, just as the period between roles gives an actor a semblance of a normal life free of rehearsals and a long series of performances. Before long, the actor takes on a new role; our monads take on a new identity, and we throw ourselves headlong into the new lifetime, forgetting what—and who—we truly are. If this model is correct, then death is simply a rite of passage.

The whole purpose of life is to make possibilities real and to explore potential in the same way that our dreams do for us at night when we sleep. Consciousness does not die with our bodies. Realizing this on our deepest levels is the key to unlocking the true meaning of death, just as the Essenes and others have done for millennia in rituals such as lying in caves for three days to bring on a death-like state. Death does not extinguish an individual monad any more than wrapping a movie extinguishes the actors or crew.

What Happens When We Die?

As I just said, the end of the self is not the same as the end of consciousness. The question is not if consciousness survives death, but how. We already know that the Western (JCI) religions believe that the ego continues in a unique soul, that Anthony Hernandez will remain Anthony Hernandez after I die and will be sent to heaven for eternal bliss or—far more likely—to hell to endure eternal punishment. We also know that the Eastern religions say that the self/personality does not pass on. The Eastern religions are much closer to the mark, because the only way "I" can survive physical death is in a dualist universe, which has been conclusively disproved.

We can see the individual ego/personality we are familiar with as a unique combination of soul and body, just like the characters in your dream have the same "soul" while each inhabiting different ersatz bodies. This combination makes it impossible for the characters in your dreams to live when the dream ends and winks their universe out of existence. We have seen that the dream-within-a-dream model is viable and compatible with all of the science we have studied throughout this book. Thus, Anthony Hernandez will indeed cease to exist in a way, as will you and everyone you know and love; however, our monads will continue, and will carry on both memories and karma. We may not all be able to remember past lives, but

---

The boundaries between life and death are at best shadowy and vague. Who shall say where one ends, and the other begins?
Edgar Allen Poe

Separated from the mind there are not objects of senses.
Ashvaghosha

some of us can, and there is every reason to suspect that we will have full knowledge of our true identities.

Some Buddhists see death as the end, with nothing surviving beyond the flame of karma that passes to a newborn who will then reap the benefits or suffer the consequences. This view combines the worst aspects of personal annihilation and original sin, because a monad that carries only karma and stored quantum tendencies is functionally identical to the personal annihilation that materialists seem to crave so desperately; however, we can postulate a monad that carries identity with it until it merges with the Godhead and becomes everything and everyone at once. With such a monad, death becomes the annihilation of the caterpillar in exchange for the butterfly that is both the caterpillar and yet so much more.

Mysticism affirms the unity of all life and all things in the universe. Evolution opens the door to biocentrism. Physics allows—if not demands—the primacy of consciousness. Neuroscience demonstrates that our brains are not the source of consciousness, but merely transceivers. Paranormality indicates that death is not the end, and that something of who we are carries on. Together, the evidence from these sources affirms the existence of the monad that loses itself in the role of each lifetime, only to experience its true nature once the play is over.

Each of us is a consummate actor, totally convinced of the reality of this lifetime. On one hand, this may seem to argue against the model I am presenting; however, on the other hand, can we really expect any less convincing a performance from a shard of the Godhead?

A person who says s/he has lived one or more past lives must base that claim on the idea that mind is both deathless and able to function out of the material world. The only way past lives can be possible is if the mind carries at least some classical memories with it from lifetime to lifetime. If this happens, then mind is indeed immortal, and the monad model holds. Postulating an idealist universe implies that consciousness does not die with the body; after all, the body does not exist, and neither does STEM as we know it. The loss of the selves we have grown accustomed to in no way precludes an expansion of consciousness. No mystic has ever reported being any-

---

There is no one in the world who cannot arrive without difficulty at the most eminent perfection by fulfilling with love obscure and common duties.
J. P. de Caussade

When perfect enlightenment shines, it is neither bondage nor deliverance.
Prunabuddha Sutra

thing less than her or his self during an RSME. Buddha once said that we are everything we have ever thought—a clear reference to some type of continuous existence.

Other Opinions

Goswami believes that monads are both structureless and unchanging. The former makes perfect sense in an idealist universe that lacks structure by definition. The latter is a little more difficult, because Goswami distinguishes between classical memories (such as what you ate for lunch yesterday and who you were married to in this lifetime) and what he calls *quantum memory* (the memory of food in general, or the knowledge of what love is). According to Goswami, the monad carries quantum memories but not classical memories. There is no way to parse this without concluding that the monad undergoes some sort of change, which directly contradicts his stated belief in a changeless monad. If the monad can carry quantum memories, then it ought to be able to carry classical memories as well. I respect Goswami very much in general; however, I must disagree with him here.

Tipler believes that death is the end, that body and mind will decay and cease to exist; however, he sees technology expanding across the entire universe, and eventually gaining the capacity to recreate bodies and their associated brains/identities by replicating their quantum states. All of this activity takes place during the last few seconds of the universe's existence as it collapses into the Big Crunch and allows us an eternity of paradise thanks to time dilation. According to Tipler, recreating the quantum state would recreate the actual person, and not merely a fully loaded copy. To the resurrected person, untold billions or even trillions of years would have passed in the blink of an eye, because they were sleeping the deepest sleep of all for the entire time. I like Tipler's thinking and admire his imagination; however, his model does not seem to fit the evidence we have examined.

A Note on Suicide

I have been on the verge of suicide twice in my life, both times as a young adult who had no end of problems fitting into society. Both times, I sat on a chair in my living room with a pistol

Perpetual inspiration is as necessary to the life of goodliness, holiness and happiness as perpetual respiration is to animal life.
William Law

There is no creature who perceives all of what is and what happens.
Judith and Herbert Kohl

in my mouth and my finger on the trigger. Both times I was saved by remembering a seemingly trivial event from high school in which Mr. Keily, my probability and statistics teacher, tossed a coin during a lesson. He then asked whether the result of that toss could affect the outcome of a subsequent toss. (The answer is no.) That ridiculously simple lesson saved my life, because I realized that my life to that moment was like the first coin, which did not preclude the rest of my life from turning out much better. The only thing I would accomplish by blowing my brains out the back of my head would be depriving myself of the chance to find out.

If death does not end our existence, then it stands to reason that our joys, sadness, triumphs, turmoil, opportunities, and limitations go with us. In other words, suicide is not the answer.

The Joy of Death

Realization is nothing to be granted afresh; it is already there. All that is necessary is to get rid of the thought, 'I have not realized.
Sri Ramana Maharshi

I used to be afraid of dying, to the point of near panic or extreme sadness. Thanks to writing this book, I have come to see death as joyful. Don't get me wrong; I don't have a death wish, but I am no longer afraid. If the model in this book is completely wrong and death is the end, then it is pointless to worry, since I will be the last one to ever know I've died.

Many hospital and hospice workers report patients having joyful visions just before dying. This may happen because the dying person is letting go of her or his attachment to the body and starting to remember her or his true, eternal nature. Schizophrenics and psychotics often have periods of complete lucidity shortly before dying. This makes zero sense in a materialist universe, where death must by definition be something to fight at all cost. In such a universe, there is no reason for death to be the least bit happy or pleasant—quite the contrary.

Self realization is an exalted state of inner attainment which transcends all illustrations.
Buddha

The only way these reports make sense is if the "I" who died never existed, if there is no death, and if each of us is part of an eternal process that has neither beginning nor end. Death is therefore liberation from the profound isolation of this life, the final throwing open of the prison doors into the bright, clear light of possibility and union. It is the emergence of self into Self. Can you imagine anything more beautiful or joyful? I can't.

Where Did We Come From?

We came from consciousness that has no beginning, no middle, and no end. We can say that this consciousness created the Big Bang, that evolving life retroactively collapsed the universal waveform, that the universe came into being *ex nihilo*, that it arose from an earlier universe, that it is the only universe, that it is one of many in a collection of universes that resemble gigantic soda bubbles, or pretty much anything. Each of these models is perfectly admissible and perfectly compatible with the Multivariate Monoverse.

> *A mathematical truth is timeless, it does not come into being when we discover it. Yet its discovery is a very real event.*
> Erwin Shrödinger

Where Are We Going?

This universe will eventually fade until all energy is frozen into matter and cools to the ground state. (See Chapter 23.) The Sun will eventually swell to consume the Earth. Later, galaxies will move away from each other, until an observer on any galaxy will only see that galaxy, with no hint of the cosmic background radiation or the Big Bang. Future observers will therefore believe in a static universe. Trillions of years later, molecules, atoms, and particles will be torn asunder, and the pace of universal expansion will accelerate unchecked, as the amount of gravity available to slow it down decreases. Through it all, the Godhead will remain, imperturbable, steadfast, and eternal.

> *We shall not cease from exploration and the end of all our exploring will be to arrive where we started and know the place for the first time.*
> T. S. Eliot

The Divine Savage

We humans are slightly glorified chimpanzees, apes who evolved from mammals that evolved from reptiles and so on, right back to the first life form on Earth. Again. the model presented in *The Enlightened Savage* and the other *Savage* books is perfectly valid; we are indeed savages. We are also divine. We live our lives as savages who are almost completely unaware of our divine nature, but we are evolved savages who can gain glimpses of that very same divine nature.

As we rise to this new awareness, we will see ourselves and everything in this universe and beyond for what we truly are: fragments of God sent forth to experience all that can be

> *Believe nothing just because a so-called wise person said it. Believe nothing just because a belief is generally held. Believe nothing just because it is said in ancient books. Believe nothing just because it is said to be of divine origin. Believe nothing just because someone else believes it. Believe only what you yourself test and judge to be true.*
> Buddha

experienced across lifetime after lifetime before once again returning to the whole in a joyous reunion. That is our destiny, period...

... or not.

There is my truth; now tell me yours.
Friederich Nietzsche

928 | *The Divine Savage*
Revealing the Miracle of Being

Appendix A:

Words to Live By

It seems only fitting to finish this book with some small selections taken from various texts from around the world. We may debate their origins and may argue over their meanings and precise translations. We may or may not agree on their divinity; however, we can all agree that there is much comfort and wisdom to be found in these words, especially in light of the long journey we have taken together throughout this book. These words inspire me, and I hope they will inspire you too.

Excerpts from the Gospel of Thomas

Whoever discovers the interpretation of these sayings will not taste death.

- Those who seek should not stop seeking until they find.
- The person old in days won't hesitate to ask a little child seven days old about the place of life, and that person will live. Know what is in front of your face, and what is hidden from you will be disclosed to you. For there is nothing hidden that won't be revealed.
- Don't lie and don't do what you hate because all things are disclosed before heaven. There is nothing hidden

that won't be revealed, and there is nothing covered up that will remain undisclosed.

- The human one is like a wise fisherman who cast his net into the sea and drew it up from the sea full of little fish. Among them the wise fisherman discovered a fine large fish. He threw all the little fish back into the sea, and easily chose the large fish.

- This heaven will pass away, and the one above it will pass away. The dead are not alive, and the living will not die. During the days when you ate what is dead, you made it come alive.

- I am not your teacher. Because you have drunk, you have become intoxicated from the bubbling spring that I have tended.

- If you fast, you will bring sin upon yourselves, and if you pray, you will be condemned, and if you give to charity, you will harm your spirits. When you go into any region and walk about in the countryside, when people take you in, eat what they serve you and heal the sick among them. After all, what goes into your mouth won't defile you; what comes out of your mouth will.

- Have you found the beginning, then, that you are looking for the end? You see, the end will be where the beginning is. Congratulations to the one who stands at the beginning: that one will know the end and will not taste death.

- These nursing babies are like those who enter the kingdom. When you make the two into one, and when you make the inner like the outer and the outer like the inner, and the upper like the lower, and when you make male and female into a single one, so that the male will not be male nor the female be female, when you make eyes in place of an eye, a hand in place of a hand, a foot in place of a foot, an image in place of an image, then you will enter the kingdom.

- There is light within a person of light, and it shines on the whole world. If it does not shine, it is dark.

- If the flesh came into being because of spirit, that is a marvel, but if spirit came into being because of the body, that is a marvel of marvels. Yet I marvel at how this great wealth has come to dwell in this poverty.

- Whoever blasphemes against the Father will be forgiven, and whoever blasphemes against the son will be forgiven, but whoever blasphemes against the holy spirit will not be forgiven, either on earth or in heaven.

- Congratulations to those who are alone and chosen, for you will find the kingdom. You have come from it and you will return there again.

- We have come from the light, from the place where the light came into being by itself, established itself, and appeared in their image.

- When will the rest for the dead take place, and when will the new world come? What you are looking forward to has come, but you don't know it.

- If circumcision were useful, the Father would produce children already circumcised from their mother. Rather, the true circumcision in spirit has become profitable in every respect.

- Whoever has come to know the world has discovered a carcass and whoever has discovered a carcass, of that person the world is not worthy.

- Congratulations to the person who has toiled and has found life.

- When you see your likeness, you are happy. But when you see your images that came into being before you and that neither die nor become visible, how much you will have to bear!

- How miserable is the body that depends on a body and how miserable is the soul that depends on these two.

The Upanishads

The Katha Upanishad features an exchange between Naciketas, who offers himself as a sacrifice so that his poor father can keep his few worldly possessions, and Death, who tells him the following:

"The all-knowing Self was never born, nor will it die. Beyond cause and effect, this Self is eternal and immutable. When the body dies, the Self does not die. If the slayer believes that he can kill or the slain believes that he can be killed, neither knows the truth. The eternal Self slays not, nor is ever slain. Hidden in the heart of every creature exists the Self, subtler than the subtlest, greater than the greatest. They go beyond all sorrow who extinguish their self-will and behold the glory of the Self through the grace of the Lord of Love. The immature run after sense pleasures and fall into the widespread net of death. But the wise, knowing the Self as deathless, seek not the changeless in the world of change.

"The supreme Self is beyond name and form, beyond the senses, inexhaustible, without beginning, without end, beyond time, space, and causality, eternal, immutable. Those who realize the Self are forever free from the jaws of death. When the ties that bind the spirit to the body are unloosed and the spirit is set free, what remains then? What is here is also there; what is there, also here. Who sees multiplicity but not the one indivisible Self must wander on and on from death to death."

The Isha Upanishad features the following almost literal description of endless consciousness:

"The Self seems to move, but is ever still. He seems far away, but is ever near. He is within all, and he transcends all. Those who see all creatures in themselves and themselves in all creatures know no fear. Those who see all creatures in themselves and themselves in all creatures know no grief. How can the multiplicity of life delude the one who sees its unity? The Self is everywhere. Bright is the Self, indivisible, untouched by sin, wise, immanent and transcendent. He it is who holds the cosmos together."

Tibetan Book of the Dead

The Tibetan Book of the Dead contains the following passage:

"Listen! When the expiration hath ceased, the vital-force will have sunk into the nerve-centre of wisdom, and the knower will be experiencing the clear light of the natural condition. At that time do not fear that bright, dazzling-yellow, transparent light, but know it to be wisdom; in that state, keeping thy mind resigned, trust in it earnestly and humbly. If thou knowest it to be the radiance of thine own intellect—although thou exertest not thy humility and faith and prayer to it—the divine body and light will merge into thee inseparably, and thou wilt obtain Buddhahood. Be not fond of that dull bluish-yellow light from the human world. That is the path of thine accumulated propensities of violent egotism come to receive thee. At that time fear not the glorious and transparent, radiant and dazzling green light, but know it to be Wisdom; and in that state allow thine intellect to rest in resignation. Henceforth the body of the past life will become more and more dim and the body of the future life will become more and more clear. Now the signs and characteristics of the place of birth will come. Enter upon the white light of the devas, or upon the yellow light of human beings."

Socrates

The following quote is attributed to Socrates:

"Like children, you are haunted with a fear that when the soul leaves the body, the wind may really blow her away and scatter her. And is [death] anything but the separation of soul and body? And being dead is the attainment of this separation; when the soul exists in herself, and is parted from the body and the body is parted from the soul—that is death. Then the soul is more like to the unseen, and the body to the seen. The soul is in the very likeness of the divine, and immortal, and intelligible, and uniform, and indissoluble, and unchangeable; and the body is in the very likeness of the human, and mortal, and unintelligible, and multi-form, and dissoluble, and changeable. If the immortal is also imperishable, the soul when attacked by death cannot perish. That soul, I say, herself invisible, departs to the invisible world, to the divine and immortal and rational: thither arriving, she lives in bliss and is released from the error and folly of men. And when the dead arrive at the place to which the genius of each severally conveys them, first of all they have sentence passed upon them, as they have lived well and piously, or not. And these must be the souls, not of the good, but of the evil, who are compelled to wander about such places in payment of the penalty of their former evil way of life."

Appendix B:

The Real Secret: Beyond the "Law of Attraction"

Note: Most of this content also appears as Chapter 23 of *The Enlightened Savage* and in *The Law of Realization*, which is why I relegated it to an Appendix for this book.

Several chapters in this book mention the "Law of Attraction" and present various reasons why this concept is fatally flawed despite its almost overpowering *truthiness*, or seeming correctness based on common sense or intuition. Proponents of the "Law of Attraction" cite quantum mechanics and various Eastern spiritual traditions to support their claims, which do indeed sound perfectly valid based on the Copenhagen interpretation, idealism, and a solid body of philosophy. (See Chapters 29, 31, and 32.) Having mentioned this "law" several times with less-than-favorable reviews, it is only fair that I come clean and explain why it is only true on a technical level—and why the real truth is both far more subtle and far more powerful.

If you believe in the "Law of Attraction," then this Appendix will be tough reading, especially if you are part of the self-help industry and actively involved in helping people master this "law." You would be well within your rights to think that I am wrong and that I am taking undue delight in trashing your beliefs and possibly your profession. I assert, as I did for religion in Chapter 2, that the more you disagree and/or don't want to read this Appendix, the more you should read the whole thing with an open mind. You will be very glad you did. That's a promise.

I believe that most self-help experts who subscribe to the "Law of Attraction" do so in good faith precisely because of the truthiness and the overwhelmingly positive message contained therein that essentially says, "If you can dream it, you can do it." A strict reading of quantum mechanics where a waveform of possibilities collapses into actuality does mean that literally anything is possible, and that the "Law of Attraction" is therefore technically correct. The problem is that this correctness only applies under very esoteric conditions, which effectively renders the "Law of Attraction" null and void.

According to both this "law" and standard quantum mechanics, you could will yourself to fly, manifest tons of money, attract the perfect mate, and more—provided you are willing to endure a wait that would be, on average, far longer than the current age of the universe. If you have trillions of years to wait—or if you think you can be one of the very few of the billions of people who buck these odds in a fashion similar to the handful of people who experienced "miraculous" recoveries after visiting Lourdes—then by all means give it a shot. On the other hand, if you want statistically viable odds of achieving the results you seek sometime soon, then the "Law of Attraction" breaks down and must be replaced with what I call the Law of Collapse (of the quantum waveform). Alternatively, since waveform collapse yields a single reality from infinite possibility, we could call this the Law of Realization for "making real." (Please see Chapter 23 if you need a refresher on the quantum waveform.)

What exactly is the "Law of Attraction" and where did it come from? What claims do its proponents make? Why are these claims all but useless for all practical purposes? What makes the Law of Collapse/Realization different and why should you abandon your belief in the "Law of Attraction" in favor of it? These are all valid questions. Let's answer them.

What is the "Law of Attraction?"

The "Law of Attraction" originated during the New Thought movement that began in the early 19th century and preached that God is truly OOO, that spirit is the ultimate reality, that disease originates in the mind, and that good thoughts cause good results—a philosophy that takes idealism and the Copenhagen interpretation to their logical ends and beyond. If you are thinking that this sounds very similar to the New Age movement that began in the latter part of the 20th century, then you are absolutely correct. One could perhaps say that New Age represents an evolution of New Thought.

The basic idea behind the "Law of Attraction" is that positive and negative thoughts cause positive and negative results, that "like attracts like." For example, if you think that you need more money, then you are attracting the need for money; thinking that you have all the money you need will attract more money. Thomas Troward, a leading New Thought thinker, said that thought precedes physicality and that, "the action of Mind plants that nucleus which, if allowed to grow undisturbed, will eventually attract to itself all the conditions necessary for its manifestation in outward visible form." Or, as mystic Meher Baba said, "Don't worry. Be happy."

A Brief History

The term "Law of Attraction" appeared in the 1906 book *Thought Vibration* by William Walker Atkinson. It appeared again in the 1907 title *Prosperity Through Thought Force* by Bruce MacLelland where he claimed that, "You are what you think, not what you think you are." In 1910, *The Science of Getting Rich* by Wallace D. Wattles used his interpretation of Hindu idealism to claim that believing in and focusing on what you want will realize—or manifest—that goal, while negative thoughts will of course manifest negative results.

Napoleon Hill (1883-1970) was one of the pioneers of the modern self-help industry. His 1928 book, *The Law of Success in 16 Lessons* contained many direct references to the "Law of Attraction" as did *Think and Grow Rich*, which was first published in 1937 and has sold more than 60 million copies to date. Hill explains that one must control one's thoughts and the energy behind those thoughts to attract other thoughts and ultimate success. The beginning of this book features a teaser that talks about a "success secret" along with the promise to reveal portions of this secret at least once per chapter thereafter. Hill never spills all the beans but does explain that "attraction" has something to do with it.

Other authors took this concept and ran with it, but the "Law of Attraction" had to wait until the bestselling 2006 book and movie *The Secret* by Rhonda Byrne to reach the mainstream zeitgeist. Hundreds of thousands of copies sold within hours, with millions more selling over the succeeding years. Dozens—if not hundreds—of self-help experts have also jumped on the bandwagon, and countless seminars, books, movies, etc. have been sprouting up ever since. These days, one is hard pressed to find a system that is not based on "The Law of Attraction" as popularized in *The Secret*. If nothing else, *The Secret* has been a marketing coup; however, just because millions of people believe something does not make it true. For example, millions of people used to believe that the world is flat. The few who cling to that belief today are widely—and rightly—scorned.

Let's take a look under the hood of *The Secret* to see what it actually says, after which I will mention some of the criticism raised about some of its claims as I reveal why the "Law of Attraction" is functionally useless. From there, I will conclude this Appendix by explaining why the Law of Collapse (or Law of Realization, if you prefer) is the *real* secret.

Inside The Secret

The Secret claims that historical luminaries such as, "Plato, Shakespeare, Newton, Hugo, Beethoven, Lincoln, Emerson, Edison, [and] Einstein" guard the ultimate secret to life, love, wealth, anything one could ever want—a statement that has eye-rolling parallels with the Holy Grail conspiracy that powers Dan Brown's very enjoyable *The Da Vinci Code*. The key difference is that *The Secret* says that this secret—the "Law of Attraction"—is quite real. Ms. Byrne describes an astoundingly fortuitous search for the "Law of Attraction" through the pages of history and her success in discovering the "modern day practitioners" of this ancient art. The rest of the book consists of quotes by these "practitioners" with snippets from Ms. Byrne thrown in to explain/add to some of these quotes, which include:

- **Bob Proctor:** The Secret gives you anything you want: happiness, health, and wealth.
- **Joe Vitale:** You can have, do, or be anything you want.
- **John Assaraf:** We can have whatever it is that we choose. I don't care how big it is.

I can almost see Mike Myers as Austin Powers in *The Spy Who Shagged Me* retorting to Mr. Assaraf with, "Well, I want a toilet made of solid gold, but it's just not in the cards now, is it?" Believe it or not, that response goes to the heart of why the "Law of Attraction" is useless for any practical purpose, but I'm getting ahead of myself.

The "Law of Attraction" Explained

We all know that the universe operates by a set of natural laws and forces with extremely precise values. We also know that life as we understand it would be impossible if some of these laws or values were even slightly different (the anthropic principle). Of these, the "Law of Attraction" is by far the most powerful. It takes center stage because we are attracting absolutely everything we have in our lives for good or for ill. As Bob Proctor explains it, "Whatever is going on in your mind you are attracting to you. Every thought of yours is a real thing—a force." All successful people rely on the "Law of Attraction" either consciously or unconsciously by thinking about abundance and wealth while deftly eschewing and avoiding any negative thoughts that could bring the entire house of cards down around their ears. People who go through cycles of rags to/from riches use the "Law of Attraction" sporadically, which explains their ups and downs.

At its root, the "Law of Attraction" is a magnet, and everything in our lives is metal. Thinking about something we really want brings that thing into our lives. Thoughts have frequencies like individual TV stations that radiate out into the universe like messengers carrying a cosmic dinner delivery order that will always be delivered piping hot without any possibility of exception. Change the frequency, and you could end up ordering Chinese, which could really ruin your day if you were hoping for pizza. Thinking, "I can taste the cheesy pepperoni goodness and feel my happy tummy!" is a surefire way to get pizza while thinking, "I really don't want Chinese food!" is the best possible way to guarantee a piping hot plate of chow mein in your imminent future. In other words, focusing on what you want as if you already had it is the secret; thinking about things you don't want brings those undesirables into your life because the universe only hears "Chinese" or "debt" while skipping over the "I really don't want..." or "I need to get out of..." parts, respectively. It's the subject of the thought that counts, not the context. It's that simple.

Idealist Roots

The Secret says that, "Physicists tell us that the entire Universe emerged from thought!" Well, sort of. Some physicists and philosophers do indeed tell us this, and I believe they are correct for the reasons I have detailed throughout this book. Still, it is more than a small stretch to say that such views are mainstream when in fact just the opposite is true; most physicists choose to duck the question and sequester their spiritual/religious views from their scientific views while others are die-hard materialists. That said, in an idealist universe, consciousness and the thoughts that consciousness generates are indeed the stuff of which the universe is made. Does this mean that, "Nothing can come into your experience unless you summon it through persistent thoughts." as *The Secret* claims? Sort of. More on this later.

The Power of Positivism

If *The Secret* is to be believed, it is impossible to have negative thoughts while simultaneously feeling good because feelings are the universe's way of telling you what you are thinking. Bad feelings are warnings to get happy right quick lest bad things be manifested.

- **Bernard Beckwith:** You can begin right now to feel healthy. You can begin to feel prosperous. You can begin to feel the love that's surrounding you, even if it's not there. And what will happen is the universe will correspond to the nature of your song. The universe will correspond to the nature of that inner feeling and manifest, because that's the way you feel.

- **Lisa Nichols:** Your thoughts and your feelings create your life. It will always be that way. Guaranteed!

The Secret asserts that love is the most powerful positive emotion—an assertion I wholeheartedly agree with.

Placing Your Cosmic Pizza Order

According to *The Secret*, the universe is the ultimate waiter on wheels who shares some key traits with Aladdin's genie. All Aladdin has to do is rub his magic lamp to have the genie therein at his beck and call. To continue the pizza example, if Aladdin—or anyone for that matter—wants pizza, all s/he need do is ask, which alerts the universe to your wishes and sets the ethereal cosmic gears into motion. Joe Vitale says this process is, "really fun. It's like having the Universe as your catalogue. You flip through it and say, 'I'd like to have this experience and I'd like to have that product and I'd like to have a person like that.' It is you placing your order with the Universe. It's really that easy."

Placing the order—once will do, no need to keep asking—is step one. Step two is to believe that whatever you want is already in hand with every fiber of your being whether you can see it or not. After all, one need not perseverate after ordering that pizza; one can simply relax, safe in the knowledge that the delivery process is moving towards its inevitable successful conclusion. How these practitioners explain mistaken or misrouted pizzas or other "orders" is not revealed. I for one assume that there is a very high chance that my pizzas will arrive as expected, but this does not always happen; am I "attracting" the wrong pizzas from time to time, because I busy myself with other things and don't give the pizza much thought until the doorbell rings?

Step three is to actually receive that which you have ordered and which you have dreamed/visualized/seen yourself receiving and letting yourself feel gratitude and happiness, which of course is a key ingredient in manifesting another pizza, beach house, yacht, sultry blonde, or whatever you fancy. Did I mention the happiness and gratitude?

Above all, get yourself into the good "frequency" through affirmations such as, "I have the most gorgeous blonde girlfriend who loves me," "I have that pizza in my tummy right now," "I have a million dollars," etc. Facilitate this effort by test driving the car you want, viewing the house you want, anything it takes to make you feel like it's already yours. The more you do this, the more likely you are to receive it (if the car dealership and realtor don't get wise to you first and thwart your efforts by using their own affirmations to ensure that you will actually buy the goods.) Keep it up, and one day there it will be, easy as that!

When the doorbell rings, you must get up to go answer it to get the pizza or welcome the blonde into your love nest. In other words, mere thinking and affirmations are not enough; you will actually need to do something at some point, but don't worry: It won't feel like work.

The local pizza parlor may promise delivery within 30 minutes or a free pie, but the universe has no such guarantees, because time itself is an illusion. *The Secret* says that, "Quantum physicists and Einstein tell us is that everything is happening simultaneously." and indeed they do—if and only if you are a photon traveling at light speed! (See Chapter 24.) This is similar to the concept of Nows we explored in Chapter 28 and is therefore technically correct; however, saying that, "Any time delay you experience is due to your delay in getting to the place of believing, knowing, and feeling that you already have it." ignores the fact that anything slower than a photon must traverse a path of sequential Nows. Your pizza, girlfriend, Swiss bank account, yacht, etc. are out there… but you need to get there first before you can have them. In other words, Veruca Salt from *Willy Wonka & the Chocolate Factory* is absolutely correct to want everything NOW but that does not mean she will get her way every time or even some of the time.

Miraculous Possibilities

What can the "Law of Attraction" do for you? The possibilities are endless, so long as you focus on what you do want and not on what you don't want because the universe does not take no for an answer. Want to lose weight? Visualize yourself at your ideal weight. Buy smaller clothes. Don't think about the weight you need to lose lest it stick with you. After all,

"Food cannot cause you to put on weight, unless you think it can." Pizza and ice cream for everyone! Why not, especially when, "A person cannot think 'thin thoughts' and be fat. It completely defies the 'Law of Attraction.'" More on this later.

Feeling a bit under the weather? Got cancer or any other disease/condition? Blame the bacteria or virus all you want but the real culprit is stress. Disease is a feedback loop for stress, which can be countered by—wait for it—love and gratitude and visualizing yourself perfectly healthy. *The Secret* even includes a testimonial from a woman who used the "Law of Attraction" to cure breast cancer. Bob Proctor says that, "Disease cannot live in a body that's in a healthy emotional state." And to think my partner Jennifer actually wasted ten years of her life on premed, medical school, internship, and residency!

The Secret does point out, as I also do in this book, that our entire bodies are renewed every few years. So how can we possibly degenerate or have illness remain in our bodies? Apparently the Hayflick limit we discussed in Chapter 17 is just as emotionally dependent as one's bank account or love life. Who knew? (I mean, besides Plato, Shakespeare, Newton, and the rest?)

If *The Secret i*s to be believed, the "Law of Attraction" can even be employed to, as *Harry Potter and the Sorcerer's Stone* put it, "put a stopper in death." John Assaraf says that, "Beliefs about aging are all in our minds. Release those thoughts and focus on eternal youth." On his Web site, Mr. Assaraf mentions breaking down in tears when he realized that his son was using the secret and mastering his own life. I can only wonder whether that boy has shown any signs of growing/aging since then, but I suspect that I already know the answer.

I have seen the placebo effect in action on my own son. I am aware of studies proving that patient attitude plays a huge role in medical outcomes. In *The Enlightened Savage*, I explain the prey instinct in humans and how it shapes our species's entire mentality and worldview and how our fear of getting literally killed and eaten drives most of what we do. I know for a fact that our beliefs, emotions, and thoughts greatly affect our health. I know that negative emotions release toxins into the bloodstream. That said, one could argue that Buddhist monks are the most highly trained "Law of Attraction" practitioners on the planet, because of their lifelong focus on finding and connecting with the Godhead. There are many reports of dead monks taking a very long time to begin decomposing, which raises many interesting questions. Still, I have yet to learn of an immortal monk—or anyone, for that matter.

Expect piles of checks in your mailbox instead of piles of bills. When a bill does arrive, see it as a check for ten times the amount due. Whatever you do, don't think about the debt you are in. Expect a check and a check will arrive. Want that Maserati but the bank account won't quite cover it? Forget it because you can afford it! The happier you are, the richer you will be. Don't worry, there is plenty for everybody.

Want a new relationship or to fix up an existing one? Here again, be happy. Focus on having that relationship with that lusty busty blonde and concentrate on her good attributes. The more you love yourself, the more you will find loving people.

The Power of Gratitude

The happier and more grateful you are, the happier the people you will attract to you. You get the idea. (I very much agree with this last point from personal experience, albeit not for the reasons offered by *The Secret*.) Make a list of everything you are grateful for. It is impossible to have more than you have now if you are not grateful for what you do have. All ungrateful thoughts carry that negativity to the universe, which responds with bad feelings to alert you to the imminent danger of additional loss. Don't stop there: Be grateful for the things you are "attracting" as well. Make sure to think happy, grateful thoughts as you drift off to sleep each night and continually find ways to work that into your life and habits.

Always Look on the Bright Side

Never fight against what you don't want. Fight for what you do want. The example in *The Secret* involves anti-war sentiment and the admonition to advocate for peace instead of against war. I'll buy this because working toward a goal does indeed tend to be much easier and more rewarding than struggling against something.

The Universal Mind

According to *The Secret*, the universe consists of the Universal Mind, which "is the attractive force which brings electrons together by the 'Law of Attraction' so they form atoms; the atoms bought together to create molecules, [etc.]" John Hagelin says that, "Quantum mechanics and quantum cosmology confirm that the universe essentially emerges from thought and that all of the matter around us is just precipitated thought." This is idealism at its most fundamental, which as far as I am concerned has been all but conclusively proven. Bernard Beckwith agrees, saying that, "Are there any limits here? No, we are unlimited beings. We have no ceiling, The capabilities and the gifts and the power that is within ever single individual on the planet is unlimited."

At face value, the "Law of Attraction" seems compatible with everything we have discussed in this book, my tongue-in-cheek descriptions notwithstanding. We know we live in a universe built on consciousness, that matter is an illusion, etc. We also know from firsthand experience that attitude is everything. *The Enlightened Savage* stresses getting rid of negative beliefs and emotions to open you up to achieving your goals. Given all of this, how can the "Law of Attraction" be as useless as I say it is? We're about to find out.

The Flaws in the "Law"

If you have come this far and still believe in the "Law of Attraction," fine. It is my hope that the next few pages will replace that belief with a much more powerful truth. If your belief in the "law" has been shattered, don't worry—all will be as well as it can be. If you never believed in the "law" that's also OK; just keep your mind open for what is to come.

The ABC News article *Science Behind 'The Secret'?* says that, "Critics say *The Secret* is not only wrong, it is dangerous, leading people to believe you can get what you want, whether it's getting rich, curing disease, losing weight... simply by thinking positively about it." About the claimed medical benefits of the "Law of Attraction," this article quotes Dr. Richard Wender, president of the American Cancer Society as saying, "I want to be very clear that there is no evidence that people attract cancer by their thoughts." and, "If some person chose to strictly follow the steps in this book, there is a risk that they could die needlessly." Physicist Brian Greene is quoted as saying, "If by 'Law of Attraction,' they have this notion of having a thought and it attracts like thoughts, I can assure you that quantum mechanics has nothing to say about that." As for historical luminaries guarding secret knowledge, Greene says, "Look, I've never met any of those guys, but I have zero evidence that any of them would've held on to any fundamental secret about the world and not shared it."

These criticisms are damning in their correctness for reasons I am about to share. As you read, please keep in mind that any one criticism alone is enough to sink the "Law of Attraction." Taken all together, the only conclusion we can reasonably make is that this "law" is little more than wishful thinking.

Creating our Reality

We know from Chapter 9 that we humans use the ETEAR process to create our own personal realities. Our earliest emotions (that probably begin before birth) create thoughts that form beliefs. These thoughts/beliefs drive our emotions, the addiction to which spurs us to take action to manufacture whatever pattern we are familiar with. I explain in *The Enlightened Savage* and *The Natural Savage* that humans are fundamentally emotional creatures, and that our logic exists to rationalize and validate our emotions to keep us pumping the same addictive chemical cocktails (what Pert calls the "molecules of emotion") through our veins. This process fuels more emotions, and around and around we go.

The assertion that emotions warn us about bad thoughts and act as "manifestation warnings" is wrong, because it assumes that humans are logic-driven when in fact we are emotionally driven. Conscious thought can bypass our emotions and induce us to take different actions, which triggers a withdrawal process that is every bit as real as withdrawing from heroin or other such drug. The real secret is therefore to take action and to persevere despite all obstacles, being sure to reward yourself copiously for taking these steps—the end goal being to replace your current emotional addictions with new ones. As emotional prey animals, we can never escape the fear of being killed and eaten and will always follow our survival instincts to the letter. We can, however, reprogram those instincts.

Inherent Contradictions

The quotes I presented above—all of them lifted verbatim from the pages of *The Secret*—say that we can do, have, and be absolutely anything we want if we follow the three steps of placing the order, visualizing having that order fulfilled, and being happy and grateful no matter what because the "Universal Mind" is OOO and cannot refuse our every wish. The concept of universal mind/Godhead is absolutely correct, as is the assertion that the universe follows extremely precise laws to the last iota. *The Secret* quotes Bernard Beckwith saying that, "We live in a universe in which there are laws..." So far, so good, except that Beckwith goes on to say, "... just as there is a law of gravity. If you fall off a building it doesn't matter if you're a good person or a bad person, you're going to hit the ground." Which is it?

If we can have anything we want to the extent of defying aging and death, then it seems only reasonable to assume that we should be able to wish ourselves a feather-soft landing after stepping off a high balcony. Why can the universe bend its rules to give you anything you want while at the same time being incapable of bending those rules to give you anything you want? This one sentence alone dooms *The Secret* and the "Law of Attraction." I am flabbergasted that the editors let this one in, because most philosophies don't sow the seeds of their own destruction quite so blatantly or obviously.

The universe must obey its own laws, because making one exception risks making another and another until the very fabric of space-time that makes life possible quite literally disintegrates. A universe without the laws of our universe is incompatible with life as we know it. A universe that is self-consistent cannot possibly break its own laws. The extent to which the anthropic principle applies and to which sciences like quantum mechanics are as precise as they are makes this self-evident. The universe and Godhead are rational to their core. You cannot have your cake and eat it too, except under very precise circumstances governed by the quantum waveform in which all things are indeed possible, but some things are far more likely than others.

Don't jump off any tall buildings. If you get sick, seek medical help. Accept the certainty that your life is going to end and (most probably) continue to another form of existence, in which you will be at once not yourself and yet far more "you" than you are now. If you are overweight, eat less and exercise more. If you want more money, take steps to make more money. If you want a new relationship, go out and look for one. If you want to fix an existing relationship, take steps to fix it. In other words, use the laws of nature to your advantage. If any "expert" suggests that you can violate the laws of human nature or physics, the only questions you need to ask pertain to that "expert's" true knowledge and/or motives, period.

Heisenberg Would Not Approve

You already know that the Heisenberg Uncertainty Principle limits the extent of knowledge we can ever have, because there is an element of randomness built right into the foundations of the universe. We can see this as wiggle room for God to act in the universe or not; it doesn't matter. What matters is that we cannot avoid dealing with Heisenberg randomness. the assertion that we can have anything we want simply by "manifesting" it is therefore false on its face. The "Law of Attraction" is again utterly falsified for all practical purposes.

The Universe is Not Stupid

Proponents of the "Law of Attraction" insist that you must focus on what you do want (such as money) and not on what you don't want (such as debt), because the universe--which ultimately consists of the Universal Mind/consciousness/Godhead that is literally the source of everything--only hears the nouns and not the modifiers.

Let me get this straight: The single fractionated consciousness that created the convincing illusions of space, time, energy, matter, you, me, and the incredibly consistent tapestry of natural laws is incapable of hearing and understanding things like, "I want more money." and, "I need to get out of debt." because it can only hear "money" and "debt" and pile it on but never subtract it? The Godhead is that stupid? Really? Somehow, I doubt that!

Shared Reality

I mentioned in Chapter 15 that the Wigner's friend paradox highlights the role of consciousness in creating reality (by collapsing quantum waveforms), which seems to validate the fundamental underpinnings of the "Law of Attraction." There is just one problem: It does no such thing. On the contrary, Wigner's friend blows yet another hole in the "Law of Attraction." Goswami's article *Consciousness and Quantum Physics* says, "Imagine that Wigner is approaching a quantum traffic light with two possibilities, red and green; at the same time his friend is approaching the same light from the perpendicular road. Being busy Americans, they both choose green. Unfortunately, their choices are contradictory; if both choices materialize at the same time, there would be pandemonium. Obviously, only one of their choices counts, but whose?"

The answer that Goswami correctly provides is that the universal consciousness/Godhead gives both Wigner and his friend green lights half the time. This is accurate, in conformance with the quantum waveform, and a direct falsification of the "Law of Attraction." It is simply impossible for everyone to get everything they want all the time, period. We could not have an orderly universe that is fit for life any other way. The single consciousness that forms the universe itself (the Ground/Godhead) cannot and will not contradict itself no matter how hard its illusory subdivisions that we call "you" and "me" try. The only way anyone could get absolutely everything s/he wants would be if s/he were utterly alone in the universe, which kind of puts a crimp in any wish for a relationship!

One can make the argument that different people have different levels of willpower and thus different levels of ability to make things go their way. One can then say that someone with perfect will can indeed have anything s/he wants. This makes perfect sense. Now find me just one person with perfect will. Each of us can indeed change what is possible in our lives and can make dramatic changes in a very short time; however, that is—again—a very far cry from "manifesting" your desires.

Shared reality also presents another problem. The 1999 movie *Mystery Men* includes a character named Invisible Boy who is only invisible when nobody is looking. The lesson here is that you may be convinced you can do something but you must also share aspects of your reality with people who may be convinced otherwise. If you think you can fly and I think you can't, earthbound you will remain.

Disentanglement

Correlated subatomic particles display nonlocality. Correlated people can influence each others' brainwaves. This is true; however, entanglement can exist on many levels. Photons generated under carefully controlled conditions can be extremely tightly entangled/correlated, which is how we get amazing results to experiments such as delayed choice, quantum eraser, and others we discussed in Chapter 23. People who are intimately involved cannot make their partners zig when they zag with any degree of certainty but they can influence thoughts, emotions, and actions—a weaker form of entanglement. I dare say that the entanglement between you and the luxury yacht you found in the pages of the *Robb Report* is orders of magnitude weaker still.

Entanglement never shrinks to zero, but it does become so weak as to be negligible, just like the quantum waveform, which we will examine next. For now, the highly variable nature of entanglement does mean, as *The Secret* asserts, that "we are all one." What *The Secret* will not tell you is that this oneness is less than absolute for all practical purposes, and that the "Law of Attraction" is yet again fatally torpedoed by its own assertions.

The Quantum Tsunami

The dual slit experiment described in Chapter 23 demonstrates both the equivalence of energy and matter and that all things exist in a state of superposition that encompasses all possible outcomes at once. This assortment of possibilities is contained within a quantum waveform that literally extends across the entire universe and that literally never shrinks to zero. This explains both how you may suddenly find yourself orbiting Pluto or how your flat tire may spontaneously reinflate and why you should absolutely not hold your breath waiting for either event—or anything similar—to happen.

Appendix B
The Real Secret: Beyond the Law of Attraction

As I explained in Chapter 23, my son's waveform spreads across the entire universe the moment he leaves my sight every weekday morning to go to school. The moment the door closes behind him, he could quite literally be anywhere in the universe. That said, his waveform—the probability wave that predicts the odds of him being in a given location at any given time—is extremely high along the route to his school before settling over his classroom as he arrives. There is an extremely high probability that chasing after him or visiting his school will find him exactly where he is supposed to be, within negligibly small Heisenberg limits. The GPS tracker in his cell phone allows me to collapse his waveform any time I want. If the tracking feature says he is within a 40-yard radius of his classroom, then I know that his waveform is extremely high above his seat, much lower around the edges of his classroom, lower still in adjoining areas, and 0 at any distance more than 40 yards from his indicated location. My measurement of his location has collapsed his waveform to that 40-yard radius, and my walking into his classroom would collapse his waveform to within a subatomic radius. His waveform instantly spreads back out across the universe the moment the measurement ends but always remains extremely high over his school while the day is in session.

The fact that the waveform is never zero between measurements means that yes, you can indeed manifest anything you want and that the "Law of Attraction" is fundamentally correct in this assertion. You can indeed manifest anything you want from money to health, love, immortality, the ability to leap tall buildings in a single bound, anything at all. I could will my son to Pluto using the methods contained in *The Secret* and it could happen. The only catch is that the odds are extremely low—so low that I could wait countless trillions of years before Logan's waveform randomly fluctuates to make such a thing statistically possible. Even the seeming exceptions to this rule, such as the woman who cured herself of cancer by "attracting" health, only prove the rule. As we saw in Chapter 9, the number of people healed because they visited Lourdes did so at a rate many times lower than the statistical odds of spontaneous remission. Miracle? No. Attraction? Hardly. Quantum waveform and Heisenberg uncertainty (read: raw statics)? Absolutely. The quantum waveform is therefore a tsunami that utterly wipes out the "Law of Attraction."

Paranormality is No Excuse

I fully believe that paranormal phenomena are real for the reasons I presented in both the Introduction and in Chapters 15 through 17; however, these effects are subtle at best. It is one thing to see a chain saw blade or a flat tire or a future event. It is quite another thing to actually cause that event to transpire. The psi research I mention in this book involves very small, almost unnoticeable changes. The odds of these things happening by chance are vanishingly small but so are the effects. Paranormality is therefore not a viable explanation for the "Law of Attraction."

Debunking the "Law of Attraction"

I hope that everything you have read so far proves that the "Law of Attraction" is indeed real, in the sense that anything is possible. Under these very esoteric assumptions and under extremely rarified circumstances, we can truthfully say that the "Law of Attraction" is real. The "Law of Attraction" therefore boils down to a convenient shorthand for everything we have learned about quantum mechanics, consciousness, idealism, and the Godhead throughout this book. It also demonstrates why it is absolutely ludicrous to think that applying this "law" as its proponents would have you do has any viable chance of yielding any practical results.

But what about the many testimonials of spectacular success the self-help experts flaunt at every opportunity? I submit that these success stories have nothing whatsoever to do with attraction and everything to do with collapse.

I fully believe that the experts and "gurus" who sell the "Law of Attraction" do so from a place of wanting to help others, which is the noblest possible pursuit. They are also out to make a living, and many of them do quite well financially. There is nothing wrong with that at all. I freely confess that I would love this book and the entire *Savage* series to become smash hits. That said, it is blindingly obvious that these experts are mostly relying on both each other and some of the outliers of the scientific community for their information, which ultimately boils down to misunderstanding or misapplying what sciences from neurology to psychology and quantum mechanics really have to tell us. This misinformation is gathered and disseminated in good faith, of that I have no doubt; however, the fact remains that this misinformation reveals a lack of true knowledge.

Critics and mainstream science scoff at the self-help industry, and rightly so. The sad part is that a lot of good information with the power to help many people is needlessly tainted and invalidated because of its reliance on the nonexistent "Law of Attraction." The good news is that most if not all of the "experts" I know in the industry can earn removal of the quotation marks around their names and gain mainstream acceptance by updating their material to reflect not attraction... but collapse (or actualization/realization, which has a more positive ring to it.)

The Law of Collapse

We know that the quantum waveform collapses whenever a measurement is taken. Physicists debate the root cause of this collapse—hence the different interpretations we explored in Chapter 29—but the bottom line is that probability does collapse into a single actuality. The Law of Collapse is thus a fundamental law of nature. We can also call it the Law of Realization because reality results from waveform collapse. The difference depends on whether we want to focus on what is happening (collapse of the waveform) or on the results (the actuality created by the collapse). For marketing purposes, I think most people would prefer to call it the Law of Realization because "realization" sounds a lot happier than "collapse." For the sake of brevity and what I feel is slightly greater scientific fidelity, I will refer to this law simply as the Law of Collapse.

The quantum waveform that we can calculate using the Schrödinger equation is both real and central to the Copenhagen, hidden variable, and many worlds interpretations of quantum mechanics. It is also perfectly compatible with materialism, because it involves matter and energy whether or not one allows a role for consciousness. Dualism and idealism put consciousness into the mix, which further validates the reality of the quantum waveform. The waveform and Law of Collapse are thus compatible with all models of reality on a foundational level. One therefore need not adopt any specific belief about how the universe works in order to benefit from the Law of Collapse. Right off the bat, this is a huge advantage over the "Law of Attraction." But does it hold water?

Skewing the Waveform

Let's recap: Objects exist in a superposition of all possible states before measurement and collapse. This superposition is infinite for all practical purposes, but the waveform is much higher for a very select states than virtually all other states. None of us is a "perfect" observer, and it is therefore virtually impossible to "manifest" anything. Harry Potter may be able to wave his wand and say, "Accio <object>!" to have that object immediately fly to his hand. The "Law of Attraction" would have us believe that all of us are Harry Potter incarnate, but that just is not the way of things. I may want my overlarge tummy to vanish while I stuff myself with steak, lobster, and cake, but, as Austin Powers said, "it just ain't in the cards." Sure there is a chance that I may wake up one morning in perfect shape to fit into all of the smaller clothes I purchased as part of my visualization, but that's not the way to bet. There is also the chance that I could keep my excellent health (my total blood cholesterol is 124) without making any changes, but that too is not the way to bet. By contrast, I followed a medically supervised program of diet and weight loss supported by vitamins and FDA-approved appetite suppressing medication and the weight literally melted away at a rate of about five pounds per week.

Before starting this program, I believed in my heart of hearts that a calorie is a calorie and that folks like Dr. Atkins who see carbohydrates as the enemy are quacks. By everything *The Secret* says, what I eat should make no difference. Imagine my surprise when I discovered that I lost more weight during those weeks where I keep my consumption of grains, starches, and sugars to a minimum. A hundred calories of starchy baked potato is not the same thing as 100 calories of things like meat, leafy greens, etc. This is only one example; I am sure that everyone can recount tales of discovering the falsity of some dearly held belief or truth. I am not suggesting that there is an objective reality; quantum mechanics and relativity prevent me from doing so because subjective individual realities is all we ever truly have; however, we share these individual realities with others in the framework of a lawful universe, which is why I really don't recommend defenestrating yourself anytime soon. (In other words, don't jump out a window because you think you can fly!)

My weight loss program had nothing to do with manifestation or belief. It had everything to do with taking action to quite literally skew the waveform to where the possibility of my losing weight is much higher that it could have been before. Sitting on my keister eating bonbons falls under the category of, "Yes, it's theoretically possible, but don't hold your breath." where the waveform is concerned. Getting off my keister and eschewing the bonbons alters the waveform, making the possibility of my losing weight that much greater. The mere fact that I am taking these steps is all it takes. I don't have to believe it will work for the waveform to skew. Yes, I believed the program would work, and I saw fabulous results, which did help tilt both the waveform and my perception of that waveform. Yes, I am happy and grateful to be feeling lighter and for the newfound energy and expanded (contracted?) wardrobe options I am experiencing. That also helped tilt the waveform toward my desired goal because it gave me incentive to stick with it despite my love of most foods sweet and starchy. It is not manifestation; it is a feedback loop where:

1. Initial action skews the waveform toward the desired results.

2. Measurable progress (weekly weigh-ins) prove the results are on their way.

3. This progress increases my determination, happiness, gratitude, etc., which keeps me in the program.

4. I start to believe I can do it, which gives me yet more incentive to stick with the program, and which incrementally reprograms my ETEAR process around food, exercise, and weight.

5. Lather, rinse, repeat.

Over time, my ETEAR process changes to where I will "fear" eating too much and exercising too little in the same way I "feared" eating less and exercising more. The prey instinct I describe in *The Enlightened Savage* is thus reprogrammed to make me achieve my new goal of staying in shape. As an added benefit, my waveforms having to do with diabetes, cardiac problems, kidney stones, and a long litany of other ills skew in a way that greatly reduces my odds of getting these diseases. The fact I began this program from a place of perfect health in all aspects except the excess fat further skews my waveforms in favor of future health.

In short, I have neither wished for nor manifested anything. I took action to alter my statistically probable options in favor of my desired outcome, and the waveform/Godhead/whatever has done the rest. My daily activities alter the waveform toward my desired outcome, and my weekly weigh-ins collapse the waveform into a slimmer, sexier actuality. The Law of Collapse is affirmed.

Want a new relationship? Want money? A nice house? Vacation? Car? Pizza? Chinese food? Act! Take concrete action toward your desired outcome. Your mind and body will resist you because that is what it is designed to do because you are a prey animal who is physically addicted to your survival instincts as directed by your core beliefs, as I discuss in depth in *The Enlightened Savage*. Follow the four-step process of:

1. Making a goal,
2. Breaking that goal into manageable chunks,
3. Taking action on your goal one chunk at a time, and
4. Rewarding yourself for each and every chunk of progress, no matter how small...

...and you will eventually get both results and reset beliefs to boot.

You can only ever experience an outcome in accordance with the statistical odds contained in the waveform, which is why the "Law of Attraction" is worse than useless for practical purposes. By contrast, the Law of Collapse (Realization) is far more powerful because it is part of the very core of the universe's innermost workings. It is far more limited in that it promises only the most statistically probable odds contained in the waveform as opposed to anything you want, any time you want it. It is far more subtle because it achieves the real outcome by dealing with possibilities as affected by altering the initial probabilities. It is also far more powerful because it works. The universe we live could not exist without it.

You may not always get what you want. In fact, Wigner's friend guarantees that nobody will ever get anything they want all the time. That said, you can open doors you never knew existed simply doing something, anything, to make those doors many orders of magnitude more statistically available to you than they would be otherwise.

Be happy and excited about the outcome and you will want to persevere. Reward yourself for every little step on your journey and you will associate your goals with pleasure instead of existential prey-based trepidation. Be grateful for every result you see, and you will be happy for where you are and eager to press on with no "promissory gratitude" for things that have not happened (and that may never happen). All of these things will tip the waveform in your favor and will help you achieve your goal. They will also make you rich beyond imagination right here, right now, no matter what your bank statement says. Loving yourself for where you are will tip the waveform in favor of finding someone to truly love you or for someone to love you more deeply.

The "Law of Attraction" is all about the destination. In this lifetime, the only real destination is death, which may be the end of the line (although I don't think so, hence this book). Life—and the Law of Collapse—are all about the journey. Make your journey a good one!

Beliefs Don't Matter (Much)

In this Appendix, I assert that beliefs don't matter, because action is all that matters. I have also asserted in this book that only reason I am sitting at my computer typing is because I believe I am, that the brain cannot tell dream from reality, and that there may be no difference, etc. It may seem like I am contradicting myself, but that is not the case. Beliefs do matter because they shape our ETEAR cycles and thus our realities. But think about it: Have you ever done something that you truly believed to be absolutely impossible? I submit that the only way that you will jump out a window is if you are convinced to the core of your being that you can fly (again, not recommended) or if you want to end this lifetime and believe that a long fall and resulting sudden stop is the way to do it. I would never have begun my weight loss program if some tiny part of me did not think it possible—however unlikely—for it to work. Belief on a conscious or (usually) unconscious level is always at work. Practically, however, it is the action that matters. The Catch-22 of being unable to change beliefs without results and unable to get results without belief is thus avoided, for the simple reason that we are not perfect.

Accept no Limitations

Many self-help experts know that sitting around wishing it so does not do much to actually make it so, which is why they shield themselves from responsibility (and potential liability) by couching the "Law of Attraction" in terms of taking action. For example, author Alexander Kjerulf says that, "I believe that the 'Law of Attraction' is very real. I've used it on any number of occasions. It works." Kjerulf then goes on to say that, "Changing your thinking changes nothing out there, in the vast universe surrounding you. It changes something inside of you. Changing your perception, your focus, your emotions and your thinking from negative to positive (from what you lack to what you want) has an enormous effect on your motivation, energy and creativity and that's why you will then be more efficient working towards your goals." In other words, if you want something, do something! This sounds a lot less like the "Law of Attraction" found in *The Secret* and its many spin-offs and copycats and a lot more like—you guessed it—the Law of Collapse.

Living the Real Secret

If you want to change something in your life, take action. Find any creative way you can to skew the waveform for whatever it is you want in your favor. For example:

- I love sailing and am actively on the market for a sailboat. My dream boat is a 50-foot Hunter sloop, but I don't have a spare half-million dollars lying around. I can buy a much cheaper used sailboat and then invest significant time, effort, and money into fixing it up and maintaining it... or I can pay a very modest fee for a stake in a Hunter that is very close to my ultimate goal and have all the maintenance, insurance, cleaning, dock fees, registration, etc. taken care of for me at a cost that is very competitive compared to keeping a "cheaper" used boat shipshape. It is also a great way for me to refine my sailing skills and prepare for the day I can afford my own dream boat on which to retire to the South Seas with Jennifer.

- I love flying. Even the worst day goes down as a good one in my memory when I get a chance to fly. I can't afford the cost of buying, maintaining, inspecting, and insuring a plane.... but my local flight clubs have whole fleets of airplanes at my disposal whenever I want, and I get a much greater variety of airplanes to choose from instead of being locked into only one bird.

The list goes on and on. The bottom line is that if you want to do something, there is probably a creative way to do it. It may not meet your exact desires, but a stepping stone is both progress and a great way to make sure that you really do what you think you want. It also a great way to protect your future investment when and if you make it. When I finally buy my own Hunter, I will know all there is to know about Hunter boats, their strengths, weaknesses, how to maintain and operate them, etc. than I ever could by simply buying a new one. That insight will save me time and money down the road and could even save my life.

Be happy about the many blessings you do have in life no matter what changes you want to make. Be grateful for all you do have no matter how little of it you have. Love yourself for all your flaws and foibles. Nothing in life is ever inherently good or bad unless and until you make it so by choice. Learn to consciously decide what is good and what is not good instead of relying on your subconscious processing to do it for you. Choose to see the good.

Do this and you will be healthier, happier, and richer than you are now no matter what happens or doesn't happen. Do this and your waveform will stretch as needed to help you along the way. The Law of Collapse guarantees that something will happen. How you use it has a profound effect on what happens.

Don't manifest or "attract" your dream reality. Collapse it. Make it real!

960 | *The Divine Savage*
Revealing the Miracle of Being

Appendix C:

Additional Resources

This chapter lists some of the many books that have guided me on my journey, and from which I have drawn material for this book.

Books

- *A Brief History of Time* – Stephen Hawking
- *A Brief Introduction to Hinduism* – A.L. Herman
- *A History of God* – Karen Armstrong
- *After We Die, What Then?* – George W. Meek
- *Afterlife Encounters* – Dianne Arcangel
- *Beliefs That Changed the World* – John Bowker
- *Belonging to the Universe* – Fritjof Capra & David Steindl-Rast
- *Beyond Belief: The Secret Gospel of Thomas* – Elaine Pagels
- *Bhagavad Gita* – Swami Prabhupada
- *Biocentrism* – Robert Lanza, MD & Bob Berman
- *Black Holes and Baby Universes and Other Essays* – Stephen Hawking

- *Black Holes & Time Warps: Einstein's Outrageous Legacy* – Kip S. Thorne
- *Body Mind Spirit* – ed. by Charles Tart
- *Breaking the Spell* – Daniel C. Dennett
- *Buddha* – Karen Armstrong
- *Cities of God* – Rodney Stark
- *Climbing Mount Improbable* – Richard Dawkins
- *Consciousness Beyond Life: The Science of the Near-Death Experience* – Pim Van Lommel
- *Creative Evolution* – Amit Goswami
- *Creative Mythology: The Masks of God* – Joseph Campbell
- *Discovering your Past Lives* – Gloria Chadwick
- *Dreams of a Final Theory* – Steven Weinberg
- *Einstein & Buddha: The Parallel Sayings* – Wes Nisker
- *Einstein's Universe* – Nigel Calder
- *Entangled Minds* – Dean Radin
- *Existentialism from Dostoyevsky to Sartre* – Walter Kaufmann
- *Finding Darwin's God* – Kenneth R. Miller
- *Fingerprints of God* – Barbara Bradley Hagerty
- *Flatland: A Romance of Many Dimensions* – Edwin A. Abbott
- *From Science to God* – Peter Russell
- *Genesis and the Big Bang* – Gerald L. Schroeder
- *Ghost Hunters* – Deborah Blum
- *God Against the Gods* – Jonathan Kirsch
- *God Is No Delusion* – Thomas Creane
- *God Is Not Dead* – Amit Goswami
- *God Is Not Great* – Christopher Hitchens
- *God: The Failed Hypothesis* – Victor J. Stenger

- *Holy Blood, Holy Grail* – Michael Baigent, Richard Leigh, Henry Lincoln
- *How We Believe* – Michael Shermer
- *How We Die: Reflections on Life's Final Chapter* – Sherwin B. Nuland
- *Hyperspace* – Michio Kaku
- *I Am a Strange Loop* – Douglas Hofstadter
- *Ideas and Opinions* – Albert Einstein
- *In Search of the Miraculous* – P.D. Ouspensky
- *In Search of Shrödinger's Cat: Quantum Physics and Reality* – John Gribbin
- *In the Beginning* – Virginia Hamilton
- *Is There an Afterlife?* – David Fontana
- *Is There Life after Death?* – Anthony Peake
- *Islam: A Short History* – Karen Armstrong
- *Jesus: A Revolutionary Biography* – John Dominic Crossan
- *Jesus After the Crucifixion* – Graham Simmans
- *Life After Death: Living Proof* – Tom Harrison
- *Life Before Life* – Jim B. Tucker, M.D.
- *Life After Life* – Raymond A. Moody, Jr. MD
- *Lost Christianity* – Jacob Needleman
- *Mary Magdalene: Myth and Metaphor* – Susan Haskins
- *Mere Christianity* – C.S. Lewis
- *Misquoting Jesus* – Bart D. Ehrman
- *On Death and Dying* – Elisabeth Kübler-Ross
- *Once Before Time* – Martin Bojowald
- *Origins of Existence* – Fred Adams
- *Parallel Myths* – J.F. Bierlein
- *Paths from Science Towards God* – Arthur Peacocke
- *Physics Metaphysics and God* – Jack W. Geis
- *Physics of the Soul* – Amit Goswami

- *Pi in the Sky* – John D. Barrow
- *Primal Myths* – Barbara C. Sproul
- *Quantum Enigma: Physics Encounters Consciousness* – Bruce Rosenblum and Fred Kuttner
- *Quantum Evolution* – Johnjoe McFadden
- *Quantum Gods* – Victor J. Stenger
- *Religion Explained* – Pascal Boyer
- *Science and Evidence for Design in the Universe* – Behe, et. al
- *Science vs. Religion: The 500-Year War* – David J. Turell, M.D.
- *Show Me God* – Fred Heeres
- *Soul Search: A Scientist Explores the Afterlife* – David Darling
- *Spook* – Mary Roach
- *Staring at the Sun* – Irvin D. Yalom
- *Stiff: The Curious Lives of Human Cadavers* – Mary Roach
- *Thank God for Evolution* – Michael Dowd
- *The 3-Pound Universe* – Judith Hooper and Dick Teresi
- *The Age of Reason* – Thomas Paine
- *The Afterlife Experiments* – Gary E. Schwarz
- *The Bible*
- *The Biology of Belief* – Bruce Lipton
- *The Blind Watchmaker* – Richard Dawkins
- *The Celestine Insights* – James Redfield
- *The Conscious Mind* – David J. Chalmers
- *The Conscious Universe* – Menas Katafos & Robert Nadeau
- *The Conscious Universe* – Dean Radin
- *The Edge of Evolution* – Michael J. Behe
- *The Elegant Universe* – Brian Greene
- *The End of Faith* – Sam Harris

- *The End of Science* - John Horgan
- *The End of Time: The Next Revolution in Physics* – Julian Barbour
- *The Enlightened Savage* – Anthony Hernandez
- *The Evolution of God* – Robert Wright
- *The Fabric of the Cosmos* – Brian Greene
- *The Fermi Solution* – Hans Christian von Bayer
- *The Field* – Lynne McTaggart
- *The Gnostic Gospels* – Elaine Pagels
- *The God Delusion* - Richard Dawkins
- *The G.O.D. Experiments* – Gary E. Schwarz
- *The God Gene* – Dean Hamer
- *The God Theory* – Bernard Haisch
- *The Grand Design* - Stephen Hawking
- *The Grand Inquisitor's Manual: A History of Terror in the Name of God* – Jonathan Kirsh
- *The Hidden Face of God: Science Reveals the Ultimate Truth* – Gerald L. Schroeder
- *The Holographic Universe* – Michael Talbot
- *The Jesus Dynasty* – James D. Tabor
- *The Jesus Family Tomb* – Simcha Jacobovici & Charles Pellegrino
- *The Jesus Papers* – Michael Biagent
- *The Koran*
- *The Language of God* – Francis S. Collins
- *The Little Book of Atheist Spirituality* – Andre Compte-Sponville
- *The Matter Myth* – Paul Davies and John Gribbin
- *The Message of the Sphinx* – Graham Hancock & Robert Bauval
- *The Mind of God* – Paul Davies

- *The Mind's Sky* – Timothy Ferris
- *The Natural Savage* – Anthony Hernandez
- *The Nature of Consciousness, The Structure of Reality* – Jerry D. Wheatley
- *The New Atheist Crusaders* – Becky Garrison
- *The Origin of Species* – Charles Darwin
- *The Perennial Philosophy* – Aldous Huxley
- *The Physics of Consciousness* – Evan Harris Walker
- *The Physics of Immortality* – Paul A. Tipler
- *The Probability of God* – Steven D. Unwin
- *The Search for Superstrings, Symmetry, and the Theory of Everything* – John Gribbin
- *The Secret* – Rhonda Byrne
- *The Self-Aware Universe* – Amit Goswami
- *The Spiritual Brain* – Mario Beauregard & Denyse O'Leary
- *The Tao of Physics* – Fritjof Capra
- *The Templars* – Piers Paul Read
- *The Templar Revelation* – Lynn Picknett & Clive Prince
- *The Unfinished Universe* – Louise B. Young
- *The Varieties of Religious Experience* – William James
- *The World of Tibetan Buddhism* – The Dalai Lama
- *The World's Religions* – Huston Smith
- *Theories of the World from Antiquity to the Copernican Revolution* – Michael J. Crowe
- *Think and Grow Rich* – Napoleon Hill
- *Vital Dust: Life as a Cosmic Imperative* – Christian de Duve
- *What Happens When We Die* – Sam Parnia, MD
- *What is Life?* – Erwin Shrödinger
- *What the Bible Really Says* – ed. By Morton Smith & Joseph Hoffman

- *When Science Meets Religion* – Ian G. Barbour
- *Where God Was Born* – Bruce Feiler
- *Who Dies?* – Stephen Levine
- *Why We Believe What We Believe* – Andrew Newburg
- *Wrinkles in Time* – George Smoot & Keay Davidson
- *Yearnings: Embracing the Sacred Messages of Life* – Rabbi Irwin Kula
- *Your Eternal Self* – R. Craig Hogan, et. al

Web Sites

I could fill another book with the list of Web sites I visited while researching this book. My general recommendation is to do a search for the keywords you are interested in, such as people, events, concepts, etc. and see what comes up. This will be far more productive than my giving you a list of sites to visit because information and links are always changing.

I do recommend Wikipedia (www.wikipedia.com) as a fantastic jumping-off point. You can enter your search terms directly on the site and review linked articles. If you want to go even deeper, you can read the linked articles and/or refer to the source material used to create that article, which is generally listed at the bottom of the page. This is a great way to get additional information that often goes into much more depth on your selected topic than many books.

968 | *The Divine Savage*
Revealing the Miracle of Being

www.ingramcontent.com/pod-product-compliance
Lightning Source LLC
Chambersburg PA
CBHW080526300426
44111CB00017B/2625